Semiconductors

The authors for the 2nd edition

Chapter 1	Dr. Werner Klingenstein
Chapter 2	Dr. Raimund Peichl Jakob Huber Christian Paul
Chapter 3	Michael Lenz Walter Geffcken Marco Pürschel Dr. Dirk Ahlers Dr. Gerald Deboy Dr. Holger Kapels Joachim Krumrey
Chapter 4	Thomas Stemmer Axel Beier Dr. Klaus Panzer Markus Wicke (OSRAM OptoSemiconductors GmbH
Chapter 5	Dr. Thomas Bever Dr. Udo Ausserlechner
Chapter 6	Gunnar H. Krause
Chapter 7	Hans Sulzer
Chapter 8	Veronica Preysing Jochen Hanebeck Dr. Jochen Schöllmann Dr. Jörg Schepers
Chapter 9	Dr. Alfons Graf Frank Auer Jürgen Englisch Jürgen Hoika Nico Kelling Michael Lenz Patrick Leteinturier Marcus Venmann
Chapter 10	Gerhard Lohninger Jürgen Wondra
Chapter 11	Arno Rabenstein
Chapter 12	Dr. Helmut Gassel
Chapter 13	Dr. Frank Klotz Dieter Födlmeier Reinhold Gärtner Thomas Lindner Thomas Steinecke
Chapter 14	Dr. Klaus Müller Michael Ahr Dr. Klaus Pressel Jürgen Winterer
Chapter 15	Hans Mahler
Glossary	Gunnar H. Krause

Semiconductors

Technical information, technologies and characteristic data

2nd revised and considerably enlarged edition, 2004

Publicis Corporate Publishing

Bibliographic information pubished by Die Deutsche Bibliothek

Die Deutsche Bibliothek lists this publication in the Deutsche Nationalbibliografie; detailed bibliographic data is available in the Internet at http://dnb.ddb.de

http://www.publicis-erlangen.de/books

ISBN 3-89578-071-5

2nd edition, 2004

Editor: Infineon Technologies AG, München
Publisher: Publicis Corporate Publishing, Erlangen

Printed in Germany

Foreword

Dear reader,

I would like to preface our publication with a truism: only the really new is new. Why? The third edition of „Semiconductors“ was urgently needed for two reasons. On the one hand, budding electronic and electrical engineers, just as much as teachers and users, should have ready access to proven knowledge in microelectronics. A purpose for which this book, which in specialist circles is rapidly achieving the status of a standard work, is outstandingly suitable: as a classical teaching medium, reliable compendium, and stimulating reading material.

What is really new about „Semiconductors“ on the other hand is that you will find in it not only the indispensible basic knowledge from „yesterday“. From A as for A/D converters to Z as for Zener effect, about which you can obtain information in depth from the appropriate chapters or simply a rapid overview from the Glossary at the end. No: what's really new is that we have succeeded in setting down for the reader all the latest trends, impulses and developments in semiconductor technology.

That is sometimes easier said than done, if you just think of Moore's Law. Gordon Moore, co-founder of Intel, prophesied back in 1965 that the number of transistors in an IC, an integrated circuit, would double about every 18 months. He was right, as the present situation confirms: today there are 40 million times as many components on an integrated circuit as there were 40 years ago. And by 2085, if we project the calculation out to the bounds of the conceivable? Will a component then consists of a mere half an atom?

That is a speculation which we can leave to the futurologists. What is certain even today, however, is that the semiconductor sector is one of the highest tempo fields of business in the world, whether in communications, in the automobile industry, in relation to broadband and access technology, in mobile communications, or memory products.

For that reason we are, with „Semiconductors“, so much the more dependent on all those who will drive forward the innovations in the information and communications industry in future years and decades. The relevance of these innovations can be seen from the fact that even today there is an Infineon chip looking after safety matters in every third airbag around the world, every second GSM mobile phone is fitted with our components and one in six new PCs and one in five new servers is fitted with a main memory supplied by us.

I shall be delighted if our handbook helps your personal progress. **Never stop thinking!** „Semiconductors“ is a first step in this direction, providing an authoritative accompaniment through to the completion of your studies, and setting the course for your professional future. I wish you every success!

Dr. Ulrich Schumacher
CEO Infineon Technologies AG

Munich, February 2004

Content

1 Semiconductor fundamentals and historical overview

1.1 Introduction

The term semiconductor is applied to solid materials which, due to their structure – their lattice structure – and depending on the temperature have a larger or smaller number of electrons which are free to move, or missing electrons (holes) which are free to move. Due to these mobile charge-carriers, the material has a conductivity of a greater or smaller magnitude or, if we consider the reciprocal of this, the specific electric resistance of semiconductors at room temperature is in a range between 10^{-2} (for example, indium arsenide and gallium arsenide) and 10^{6} Ωcm (for example, selenium). Materials which have no mobile charge carriers, and thus have even higher specific resistances, are referred to as "insulators" (for example, glass (SiO_2), mica, amber, hard porcelain, PVC). Materials which retain their conductivity even at the lowest temperatures, and which in normal circumstances have significantly lower specific resistances, are called conductors (aluminium, copper, silver, gold). In contrast to solid metallic objects, the conductivity of gases or liquids is due to the mobility of their ions, and hence depends on the mobility of the material itself.

The sections which follow are concerned exclusively with semiconductor components, metallic connections used for wiring them, and insulators for separating neighboring components. The ability to combine these elements and the mastering of their cost-effective manufacture, have led to the success of the semiconductor industry and today's high level of integration, of up to 1 billion circuit elements in a single device.

1.2 Historical overview

1.2.1 Semiconducting diodes

As early as 1939, the physicists Walter Schottky and Eberhard Spenke (Siemens & Halske) published a scientific work ("Zur quantitativen Durchführung der Raumladungs- und Randschichttheorie der Kristallgleichrichter" [On the quantitative explanation of the space-charge and surface layer theory of crystal rectifiers]) about how crystal diodes with a metal/semiconductor junction work. Their work was based on intensive fundamental research which showed, both theoretically and in practice, that the junctions described exhibit a rectifying property, i.e. they offer different electrical resistance to an electrical current depending on its direction of flow. Schottky's contribution is honored to this day in that semiconductor diodes with a metallic film are called "Schottky barrier diodes".

The first rectifiers were manufactured using selenium and germanium. For example in microwave technology, particularly in radar engineering, point contact germanium diodes started to come into use from 1942. For simple radio receivers too, the diode was happily used as a detector element. On the other hand, selenium diodes were from an early stage used as rectifier components.

1.2.2 Bipolar transistors

A notable milestone was the development of the transistor, which was started at Bell Laboratories under the leadership of the physicist William Shockley in 1945. In 1947, together with Walter Brattain and

John Bardeen, Shockley constructed a so-called two-point contact diode out of polycrystalline germanium. On the 16th December 1947 the team observed, by chance, that a change in the forward voltage across the first diode resulted in a change in the reverse current through the second diode. They gave this effect the name "transistor effect", derived from "transit" and "resistor". Industrial production initially proved to be difficult, amongst of other because of the inability to control the device parameters: their distribution was extremely wide. An improvement first began to appear with the development of the area type junction transistor. This was based on a revolutionary idea from Shockley, not to make the junctions using contacts, but to implement them into the crystal, by doping the IV-valent germanium crystal selectively with III- (e.g. indium) and V-valent (e.g. arsenic) materials, that is to "impurify" it somewhat, thus changing its conductivity and crystal properties (energy band structure). He described the idea in 1949, and he crafted the first PNP germanium transistor in 1950. This construction principle is referred to as an "alloy-junction transistor", because a drop of indium is, in a sense, "fused into" the germanium and thus (from a metallurgical point of view) an alloy is produced.

The first transistors were still being produced from polycrystalline germanium. This material has the disadvantage that dislocations in the crystal structure and impurities have a detrimental effect on the conductivity. This disadvantage was only eliminated in practice when it became possible to manufacture highly pure and monocrystalline materials, using the invention of floating zone refining.

Another improvement was the development of the diffusion technique by Bell Laboratories and General Electric. This was presented at a symposium in 1955, and had the advantage of reliably reproducible layer thicknesses and doping concentrations. For their ground-breaking invention and the associated research and development work Shockley, Bardeen and Brattain were awarded the Nobel prize for physics in 1956.

Further details of how bipolar transistors work will be found in section 1.3.1, which is about integrated bipolar circuits.

1.2.3 Silicon's victory parade

Germanium has the advantage of a high conductivity so that it is particularly suitable for use at high frequencies. The disadvantages of germanium, on the other hand, are that the crystal structure cannot withstand temperatures above about 75°C without incurring damage. In addition, the reverse current causes interference, even at room temperature. By contrast, silicon crystals will stand temperatures up to 150°C, and its higher band gap together with the higher specific resistance result in an far lower reverse current. Apart from this, silicon occurs naturally to a practically unlimited extent. As early as 1952, G. Teal and E. Buchler published a second process besides, floating zone refinement for the manufacture of monocrystalline silicon from a melt of polycrystalline silicon. Nowadays, this process is known as the Czochralski process. Siemens developed an alternative process between 1953 and 1956. With this process, the so-called CVD process (Chemical Vapor Deposition), bars of silicon, almost pure but polycrystalline, are precipitated out from the gas phase. These bars then pass through the zone melt phase according to W.G. Pfann, in which they are not only purified but also become monocrystalline. This process is still in use today, after undergoing numerous detailed improvements.

In 1954, Gordon Teal succeeded in producing the first silicon bipolar transistors in the Bell Laboratories. He started up industrial production in 1956 at Texas Instruments. For low frequency applications, this component rapidly demonstrated its superiority.

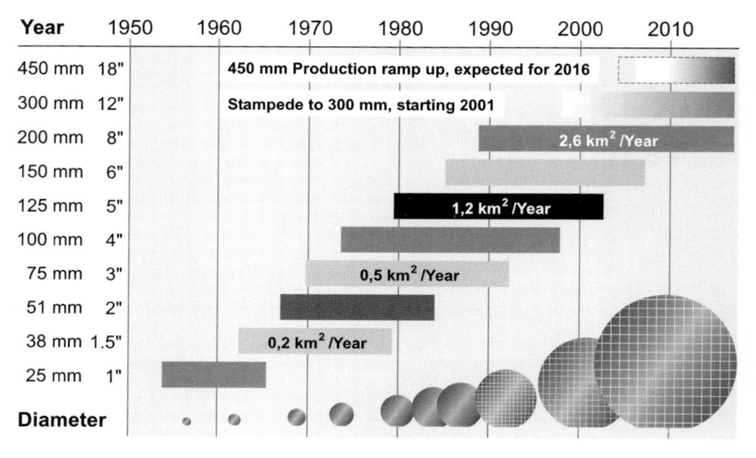

Figure 1.1 Development of the wafer area

In June 1958, M. Atalla, an employee of Bell Laboratories, described the outstanding insulation characteristics of thermally produced silicon dioxide (which occurs naturally in the form of quartz). Using this material, it was possible to construct planar semiconductors, i.e. semiconducting structures which could be built up in layers. The individual conducting layers are separated by an extremely thin silicon dioxide layer. A essential advantage is that the oxide layer can be selectively etched away, permitting specific connections between superimposed layers. In addition, the crystal surface remains flat, which eliminates many of the surface effects which may nterfere. This planar technology enabled, using photolithographic processes and diffusion from the gas phase, to build up on a single silicon wafer numerous individual transistors in the same way in parallel. Fairchild Semiconductors put this process into use in 1959. Apart from this, planar technology opened up the way for the integration of different types of components on one semiconductor die in parallel.

The continuing development of silicon technology and corresponding substrates required for it has led to a doubling of the area per wafer about every 10 years (Figure 1.1). Beginning with 25 mm wafers in the 1960s, through 51 mm wafers in the 1970s, 100 mm and 150 mm wafers in the 80s and 200mm in the 1990s, a changeover to 300 mm "pizza" wafers finally began in 2002. As well as drastic improvements in the defect density, both on the surface and also in terms of doping and crystal imperfections, there are nowadays numerous derivative specializations, in the form of epitaxial layers, silicon on insulator layers (SOI) and strained Si layers on the substrates. The objective of all these developments is to satisfy the performance and cost requirements posed to the semiconductor industry.

1.2.4 Other semiconducting materials and components

In 1953 Heinrich Welker discovered in the research laboratory at Siemens & Halske that certain junctions between three- and five-valent materials, for example gallium and arsenic, possess semiconducting properties. These semiconducting materials have since then become of great importance in opto-electronics. This is above all due to the fact that when electrons and holes recombine in these materials (to a much greater extent than with silicon and germanium) the energy released is emitted as photons. It has been known since as far back as 1956 that a pn junction operated in the forward direction emits radiation. However, it was not until the mid-60s that light-emitting diodes (LEDs) which gave out red light were successfully manufactured. The main problem consisted and consists in getting the light which is emitted to the surface of the crystal before it is absorbed. This is particularly difficult in the case of short-wave blue light. It was not until the 80s that staff in the Siemens laboratory succeeded in constructing LEDs (from silicon carbide) which radiate blue light.

A host of specific and ingenious process technologies have been developed for the production of components and circuits using III-V semiconductors. An important role took molecular beam epitaxy.

The principal applications for III-V semiconductors lie in optical data communication and optical displays. To this end, both the development of semiconductor lasers with numerous layers, some of which are monomolecular, and also the integration of optical switches and filters, have been pushed forward. Current research is putting the GaN in the spotlight, as a base material for a host of different light wavelengths and for components at the highest frequencies.

For the production of high power switches, SiC is on the threshold of commercialization as the basic material. The general requirement which must be met by high power switches are a high thermal stability, a low resistance in the switched-on state, and a high switching frequency.

1.2.5 Field effect transistors

As far back as 1928, J.E. Lilienfeld and O. Heil described the idea of realizing a field effect component not in the form of an electron tube (that is, in a vacuum) but using solids. Their thoughts paved the way for the junction field effect transistor (JFET) and the MOS field effect transistor (MOSFET). However, at that time they could neither be experimentally understood nor industrially implemented, because of the only vague knowledge of solid state physics. Shockley was the first to take up the idea again, in 1952. However, a couple of more years passed before the first junction field effect transistor was actually constructed on the one hand, and Atalla and Kahng (both of Bell Laboratories) presented the first capacitively-controlled field effect transistor made in metal oxide form (the so-called MOSFET) in 1959. For a long time, MOS types with N-channels and P-channels existed alongside each other. With time however, the N-MOS technology proved to be more advantageous, so that the P-MOS technology was a gradually pushed into the background. In 1964 however, RCA introduced a complementary circuit design technology (Complementary MOS = CMOS), in which both types were integrated on a single die. With this circuit variant, one transistor out of a number in series is always blocked, and for this reason the CMOS technology is to this day always used when power saving technology is required. Nevertheless, with increasing clock rates, growing leakage currents with very short transistors and lower supply voltages, which at the same time require lower threshold voltages, even the CMOS technology is now once again pushing against its limits. Especially for portable devices, there is an intensive search for possible improvements in terms of the components and circuitry. Further details of how field effect transistors work will be found in section 1.3.2

1.2.6 Integrated semiconductor circuits

The essential driving forces of the semiconductor industry are

- cost reductions of about 25% p.a.
- packing density doubling approximately every 18 months

- increasing complexity from the integration of peripheral functions on-chip
- speed (cf. clock rates of microprocessors)
- **power consumption!** especially for battery-powered devices
- integration of additional functionality, e.g. in the form of memories, and interfaces in the form of sensors and actuators

Today, the number of transistors and other semiconductor devices which can be integrated on a single chip is up in the billions. The production of integrated circuits is fundamentally based on an ingenious sequence of doping processes, deposition processes for conducting and insulating layers, plus their lateral structuring by means of lithography and etching (Table 1.1). Linked to the semiconductor industry are high levels of capability for the development of suitable materials (homogeneity, selectivity, purity, process compatibility) and specific processing equipment. For photolithographic structuring in particular, over the years processes have been developed which permit structures to be manufactured with sizes which are smaller than the light wavelengths used in their production. For the deposition process, there exists equipment which permit individual atomic layers to be laid down, and the layers used for modern components lie within the range of 3 to 4 atoms thick. Significant difficulties are these of homogeneity across the wafer (e.g. 300 mm), reproducibility and defect free layers. For the small depth of focus associated with the highest resolutions, specific surface smoothing processes (CMP = chemo-mechanical-polishing) have been devised.

The manufacture of an individual finished wafer often requires hundreds of individual steps. Throughout manufacture, attention must be paid to maintaining the highest precision and purity, because the most minute deviations have tremendous im-

Table 1.1 Simplified sequence of processes in the production of integrated circuits

Substrate	
Well	Production of basic doping for n-channel and p-channel transistors in the case of the CMOS technology
Insulation	Electric insulation of neighboring devices basically by SiO_2
Device	Setting of the device characteristics by channel doping, gate oxide thickness and gate material
Device connections	Doping profile for the connection of the channel area to interconnecting wiring levels Low impedance connections to the wiring levels
Local connections	Short wiring runs within a circuit (higher resistances have no detrimental effect on the circuit)
Wiring	General wiring, depending on the complexity of the circuit up to 10 layers recently
Global wiring	Low impedance wiring with relaxed requirements in terms of the structure size, for long connections and power supplies to the blocks
Passivation	Encapsulation of the circuit against environmental influences, and thereby ensuring the long term functionality
Level 0 packaging	For modern technologies, the preparation of the circuits at the wafer level for modern packing technologies: bumping, wafer-level packages, chip-on-board (COB)

plications for the microscopic structures. Going outside the tolerances for an individual manufacturing step can drastically reduce the yield of usable chips.

Because of the clock frequency, the power consumption and the maximum size requirements, imposed on components by the end devices, the classical value chain of Design – Layout – Masks – Wafer – Test – Packaging – Final test is becoming increasingly blurred. Modern solutions call for integrated wafer-package solution, system-in-package concepts and module concepts. To an increasing extent, cost optimization can be achieved in modern special packages by the 3D integration of a number of chips (ASIC + memory + analog interface + MEMS).

Bipolar ICs

The first patent claims for the integration of several components were submitted independently in 1959 by Jack S. Kilby (of Texas Instruments) and Robert N. Noyce (a co-worker of William Shockley at Fairchild Semiconductors). The patent rights dispute which resulted was determined by the US court of appeal to the effect that both were awarded the intellectual property rights in the IC in equal measure. The first integrated circuit (realized by Kilby, using germanium) had a bipolar transistor, three resistors and a capacitor. The greatest interest in the new technology came from the computer industry, which up until then had had to make large numbers of circuits, often identical, individually. In quick succession, there appeared between 1961 and 1963 the resistor-transistor logic, RTL (from Fairchild), the diode-transistor logic, DTL (from Signetics), the transistor-transistor logic, TTL (from Pacific Semiconductors) and the emitter-coupled logic ECL (from Motorola). TTL (particularly the Series 7400 from Texas Instruments) and ECL very rapidly achieved wide distribution in digital circuitry. But there was soon also success – even if only with greater difficulties – in constructing analog circuits. Fairchild and National Semiconductor were particular examples of companies which went over to integrating analog amplifiers.

MOS-ICs

From 1964 onward, the MOS technology was also increasingly integrated, and the trend was towards an ever rising level of integration. One of the steps along this path was the LOCOS technology (Local Oxidation of Silicon), which Philips brought to the market in 1966. This worked with transistor islands, which were separated by silicon dioxide surrounding, and allowed the gaps between the individual transistors to be significantly smaller without this producing parasitic transistors at the same time. Further improvements were brought by ion implantation, using which it is possible to dope selectively exceptionally small areas. Although this process had already been described in 1952 by S. Ohl (Bell Laboratories) and had been patented in 1954 by William Shockley, it was not until 1970 that it became technically possible to implement it.

Applications which combine analog signal processing with analog/digital converters and digital signal processing, demand high speeds, high driving capabilities, high linearity and good signal/noise ratios at the same time as high packing densities for the digital circuit elements. For this purpose, there has been an increasing development of BICMOS processes with CMOS devices and bipolar devices. The leading technologies use SiGe materials to permit the highest transit frequencies/maximum frequencies. Important application areas are transceivers, serializer/deserializers, laser-diode drivers and low noise amplifiers (LNA).

Memory ICs

From 1966 on, even the magnetic core memory used up until then in computers was replaced by semiconductor ICs. The first device of this type was produced by the computer manufacturer International Business Machines (IBM). It had 16 bipolar flip-flops, so that it could store 16 bits. As early as 1968, IBM was manufacturing a 64-bit flip-flop, but this did have 664 components (that is, more than 10 per bit). In the same year, the IC pioneer Robert Noyce together with two Fairchild employees (Gordon Moore and Andrew Grove) founded the company "Integrated Electronics", later renamed as "Intel". By the start of 1969, the company brought to the market a static RAM with 64-bits (in bipolar Schottky technology). Three months later there followed a 256-bit development of this. Both types were a financial flop: by comparison with magnetic core memory they were too expensive. But Intel was on the right path. In this it was critical to reduce costs, by using fewer transistors per bit. IBM had indeed already invented the single transistor memory (DRAM) in 1966, but Intel was the first manufacturer to turn this idea into a commercial success. At the end of 1970 there appeared the 1103, a 1024-bit DRAM device in NMOS technology which, although it could not immediately gain dominance over magnetic core memory in price terms, had by 1972 already become a best seller because of its technical superiority: the 1103 became the most-bought IC in the world. This module was superseded in 1975 by a 4-kbit DRAM, which had no further problems in competing against core memory. After this there appeared, at three to four year intervals, 16-, 64- and 256-kbit DRAMs. Since then, it is now possible to get DRAM devices with more than 1 Gbit. Intel's co-founder Gordon Moore had predicted the trend back in 1964: every 18 months, the number of transistors on an IC doubles. This hypothesis is still essentially true today, and is known as Moore's Law. Further details about DRAM technology will be found in Chapter 6.

Simple transistor based memories had, and have, the disadvantage that they only retain the stored data while they have a voltage supply. By contrast with this volatile storage technology, non-volatile storage has a content which doesn't change upon disruption of supply voltage, and which is programmed either during manufacture (mask ROM) or by the user (PROM, EAROM, EPROM, EEPROM, Flash, PRAM, NROM, Nano Crystal). From 1970 on, in parallel with the RAM modules, ROM and EPROM types came onto the market. The electrically erasable EEPROM also appeared soon after. Although ROM capacities did increase over time, they could not keep up with the development of the DRAMs. In 1987 Intel brought the first flash memory onto the market. In the ten years which followed, this low-cost alternative to the EEPROM conquered a large part of the market for non-volatile storage, and by now many devices (palmsize computers, mobile radio devices, cellular phones, digital cameras etc.) are unimaginable without flash memory. More about this in sections 1.3 and 1.4 of this chapter.

In pursuit of the objective of achieving even higher storage densities combined with reduced supply voltages, a variety new concepts are currently being investigated and developed:

- NROM = FLASH using a dielectric as the data storage medium
- FeRAM = ferroelectric storage
- MRAM = magnetoresistive storage
- PRAM = phase change memory

Here, both the storage requirements for user-friendly image processing and the non-volatility of the data play a critical role.

Microprocessors

In 1969, the Japanese concern Busicom wanted to construct an electronic pocket calculator using five ICs. The company itself was not in a position to manufacture the chips, and made an inquiry to Intel. When the specification was put in front of the developer Marcian E. (Ted) Hoff, he though to himself that it must be possible to realize the specification using only a single IC. After some initial skepticism, Busicom got involved and awarded the order. The chip which Intel then developed was the first microprocessor, with a processing width of 4 bits, which had 2300 transistors and satisfied the specification in all respects. Later, Intel bought the rights from Busicom, and brought the processor unchanged onto the market with the type designation 4004. Only a short time later followed the first 8-bit microprocessor, which with logical consistency was called the 8008. A further development, the 8080, was presented by Intel as early as 1974. By this time, Texas Instruments, Motorola, Fairchild, National Semiconductor, Signetics and Toshiba had also brought their own microprocessors to the market, RCA even offered a power-saving CMOS variant and the co-developer of the 4004, Frederico Faggin, had by then founded the Zilog Corporation and brought out the Z80. Intel remained the market leader, but was coming under increasing competitive pressure, particularly from Motorola, when the latter company struck back with the 68000 processor family, following Intel's presentation in 1978 of the first 16-bit microprocessor in the form of the 8086. Early on, Motorola awarded production and marketing licenses for its processors, and thereby soon secured for itself significant market shares. The company even succeeded in winning over the first microcomputer manufacturer, Apple, to the 68000 concept. As early as 1976 this company, founded by Steven Jobs and Steven Wozniak, introduced the Apple I (with an 8-bit processor from Motorola), and has remained true to Motorola processors to this day. However, Intel successfully offered IBM an 8-bit variant of the 8086, which the computer manufacturer chose for its first microcomputer, the so-called "Personal Computer" (PC). The PC appeared in 1981 and for Intel ensured the long-term market success of the 8086 series and its successors, the 286, 386, 486, Pentium, Itanium, Xscale, Centurino, Opteron etc. Today, Intel dominates with about 85% of the market. AMD supplies an expanding 15%.

Microcontrollers

At the end of the 70s, Intel brought out an 8-bit microprocessor with integrated peripheral functions: the 8048. This provided ROM, RAM, a timer and various input/output ports, and as an option also provided an EPROM instead of the ROM (Type 8748). As the application area for this was mainly in undertaking control functions, the module was soon dubbed a "microcontroller". In 1980 a further development then appeared, the Type 8051, which to this day is the de facto standard for 8-bit microcontrollers. The market conquest by microcontrollers, now also in 16-bit and 32-bit versions, has run in parallel with the development of microprocessors.

Back in 1975, IBM established, as part of a large-scale statistical investigation, that the 80/20 rule also applies to processors: for 80 percent of the time only 20 percent of the available instructions are being used. It soon became clear that it would be more advantageous, particularly for many control functions, to devise a processor with a restricted instruction set. This consideration led to the Reduced Instruction Set Computer (RISC) which was, however, not brought to the market until the middle of the 80s by the company MIPS Computer. Since then, many control and signal processing functions are now handled by RISC processors.

Apart from computing and control functions, fast digital signal processing (DSP) has come to have great importance, in particular, for wireless communication. In this field, the companies Texas Instruments and the IP house ARM had established good positions by the end of the 90s.

Customer-specific ICs (ASIC)

At the start of the 80s, a new dilemma was becoming ever clearer: the standard ICs were an obstacle to the trend towards the integration of entire applications (instead of discrete construction). The functions which circuit developers wanted were only rarely available as standard items on an IC or chipset. But special highly integrated designs were and are expensive, and only cost-effective in large numbers. For this reason, the IC manufacturers developed circuits which could be modified by the user. Here, a distinction must be made between two groups: programmable logic device ICs (PLDs) and application-specific ICs (ASICs). In 1983, Monolithic Memories Inc. (MMI) was the first company to bring a PLD to the market, with its PAL module (Programmable Array Logic). With the PAL technology, a programmable AND matrix controls the inputs to a specified OR matrix. Programming is effected in a similar way as for a PROM. There followed in 1985 the LCA technology (Logic Cell Array) from Xilinx, with which the program is stored in volatile form (in RAM cells). Each time the module is switched on, a copy of the program is loaded into it from a non-volatile storage. The advantage of this is that a relatively cheap standard module can be used for the programming, to control the more expensive LCA matrix. By contrast, an ASIC is an IC with numerous standard blocks, which can be programmed in accordance with certain rules. The wafer in this case is manufactured as far as possible according to a standard process. Only certain steps are customer-specific. This method is cost-saving, but it too is only worthwhile for large production volumes. For small and medium volumes, PLDs/LCAs are the material of choice. More about application-specific circuits will be found in Chapter 12.

Surface mount technology (SMT)

Strictly speaking, surface mount technology is not really semiconductor technology. However, it has made a big contribution to the ability to combine semiconductor structures, which have been getting ever smaller, to form modules which also occupy ever less space. With surface mount technology, simple components such as resistors, capacitors, inductors and transistors no longer have connecting wires or through hole pins, but are mounted directly on the printed circuit board. For this purpose, a new miniaturized form of construction was developed, which had to be space-saving and reliable. Such components are called SMDs: surface mounted devices. With surface mount technology, it also became almost impossible to repair faulty assemblies.

1.2.7 Categorization of semiconductor components

Semiconductor components can be distinguished from various points of view, including by:

- the technology used (bipolar, MOS, CMOS, BICMOS, Schottky, ternary),
- their internal design, i.e. individual semiconductors or integrated circuits (ICs),
- their application area (analog, digital, power semiconductors, signal processing, optoelectronics).

From the user's point of view it is the last distinction which is most meaningful. For that reason, we shall use it in the sections which follow, in that semiconductors will be described according to their application areas.

1.3 The design and functioning of integrated circuits

In the case of an integrated circuit (IC), all the circuit elements and their connecting wires are manufactured together in one production process flow on one monocrystalline semiconductor "chip" or "die" (plural "dice"). The circuit elements, which are arranged in close proximity on the chip, must then be electrically insulated from each other. The chip is enclosed in a package and, in the common packages, the chip connec-

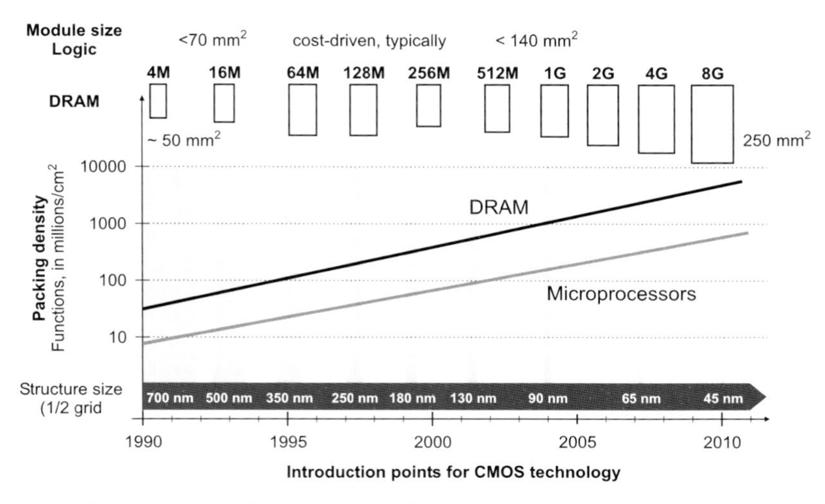

Figure 1.2 Increased packing density makes it possible to manufacture Giga-chips

tions are connected to the package leads by means of very thin wires, which are welded on.

Silicon is the main material used as the semiconductor for integrated circuits and, since other materials (GaAs, GaN, SiC) will for the foreseeable future play a minor role for integrated circuits, the following explanations will be completely restricted to Silicon.

The major advantages of integrated circuits compared to circuits with discrete components are the low manufacturing costs when the volumes are large, high reliability, small space requirement and high speed of working. Because of these advantages, there has been an uninterrupted trend towards ever higher levels of integration, and to an ever higher packing density, speed and complexity of the circuit on a chip. This has been achieved above all by reducing the size of the structures on the chip. On financial grounds, there has also been a constant attempt to make the silicon wafer ever bigger, so that as many chips as possible can be accommodated on the silicon wafer. Figure 1.2 shows how these trends have developed over time.

Corresponding to the basic transistor types, the bipolar transistor and the MOS transistor, a distinction is also made for integrated circuits between bipolar circuits and MOS circuits. The following sections are intended to sketch out the technological construction and the electrical function of the bipolar and MOS circuits.

1.3.1 Bipolar integrated circuits

The PN-junction

Pure silicon is a semiconductor with an electrical conductivity which is very low at room temperature. This is mainly due to the fact that almost no free charge-carriers are present in the material. However, if as few as one silicon atom in a thousand is replaced by a dopant atom (e.g. boron, phosphorus, arsenic), then the conductivity rises by many orders of magnitude, because the dopant atom supplies a charge carrier (which may be an electron or a hole, depending on the valency of the atom), which can be easily released from the atomic bond to move around freely.

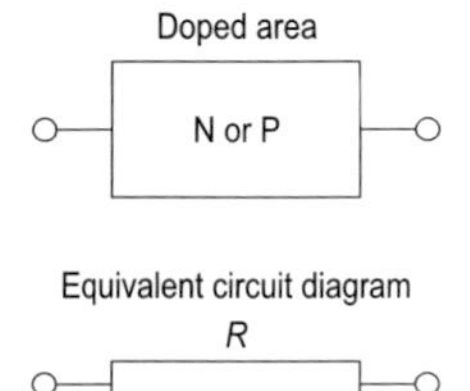

Figure 1.3
Doped area and equivalent circuit

The free electron or hole, as the case may be, leaves behind a hole or electron respectively, which in turn will readily accept other electrons. This process is referred to as recombination. Correspondingly, a doped silicon area will present a resistance of greater or lesser magnitude, depending on the concentration of dopant (Figure 1.3).

The second important semiconductor effect is that the mobile charge carriers, which are responsible for the conductivity, are either negatively charged electrons or positively charged holes, depending on whether the silicon is doped with atoms with a valency of five (phosphorus, arsenic or antimony) or with trivalent atoms (boron). The areas in which the electrons are the charge carriers are called N-type areas (N-areas), and holes are the charge carriers in P-type areas (P-areas).

Both the N- and the P-areas are initially electrically neutral. In the N-areas for example, the free electrons which are responsible for the current flow are compensated for by an equal number of phosphorus, arsenic or antimony ions with a positive charge (donors), which are anchored in fixed positions in the silicon lattice. The positively charged mobile holes in the P-areas are compensated for by fixed-position negative boron ions (acceptors).

At the boundary layer between a P-area and an N-area, (the PN-junction) a potential threshold builds up, because electrons from the N-area diffuse into the P-area, and holes from the P-area into the N-area (Figure 1.4). The electrons which have diffused into the P-area leave behind in the N-area, in a thin layer at the PN junction, a layer depleted of electrons which has a space charge determined by the fixed-position positively charged donors (Figure 1.4a). Correspondingly, in the P-area a negatively charged layer which is depleted of holes arises at the PN junction. The N-area thus acquires a higher potential than the P-area (Figure 1.4c). The electrons diffusing into the P-area must now overcome this potential threshold. When the potential threshold reaches a certain level – for silicon around 0.7 V – the diffusion of the electrons or the holes comes to a halt.

If one applies a negative voltage between the P- and N-areas, then the potential threshold at the PN junction is raised (dashed curve in Figure 1.4c). Only a few electrons or holes, as applicable, can overcome the higher potential threshold; only a very small reverse current flows. Conversely, with a positive voltage between the P- and N-areas, a higher current flows. A PN junction thus has the function of a diode with a forward voltage of 0.7 V. The space charge zone on either side of the PN junction, which is depleted of free charge carriers, functions as a capacitance. Thus the electrical equivalent circuit for a PN junction consists of a diode with a capacitance in parallel (Figure 1.4d).

PN-diodes break down at a certain reverse voltage, which is in the range between 5 V and over 100 V for the PN junctions which are to be found. In some bipolar circuits, this breakdown voltage for the diode (Z or Zener diodes) is used to implement a defined voltage drop which is independent of the current.

Apart from PN diodes, the so-called Schottky diode is used in many integrated circuits. It consists of a contact between aluminium and a lightly doped N-area. The primary difference from the PN diode consists in the fact that Schottky di-

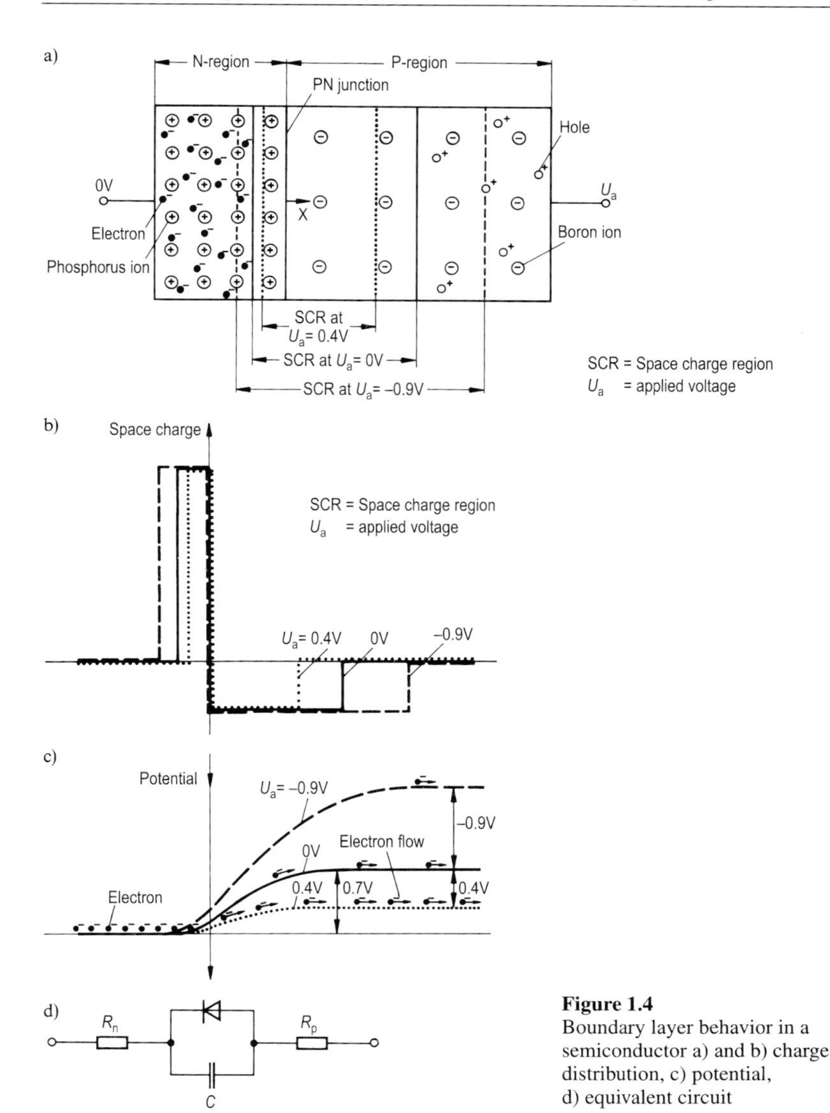

Figure 1.4
Boundary layer behavior in a semiconductor a) and b) charge distribution, c) potential, d) equivalent circuit

odes have a lower breakdown voltage (around 0.4 V) for a less sharply defined threshold voltage.

The bipolar transistor

The NPN bipolar transistor consists of two N-areas, which are referred to as the emitter and the collector, and between them a P-area which is called the base.

Figure 1.5 shows the general structure of such a bipolar transistor, and how it works. When no base voltage is applied (U_{BE} = 0 V), the potential threshold between the emitter and base is 0.7 V, and

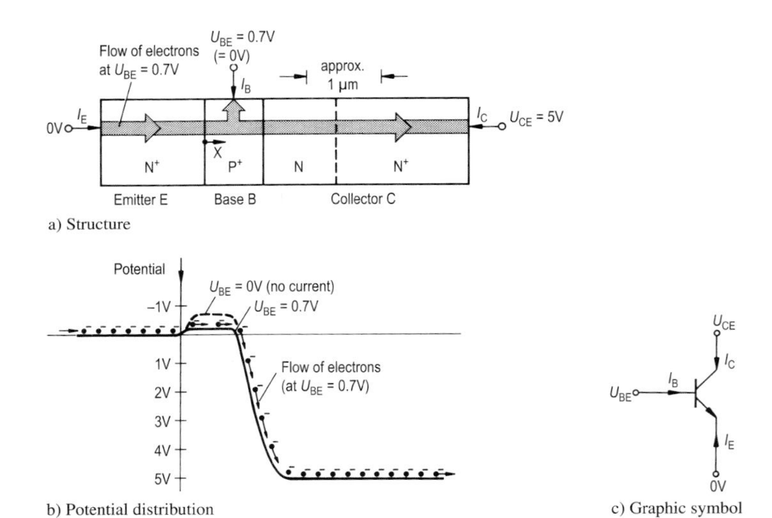

Figure 1.5
Structure and method of working of an NPN bipolar transistor
a) structure, b) potential distribution, c) circuit diagram

only very few electrons can overcome this threshold (cf. Figure 1.4). The transistor is blocked. With an applied voltage U_{BE} of 0.7 V, the potential threshold is reduced so far that an electron current can flow from the emitter area into the base area. If the width of the base – that is the distance between the PN junctions, between the emitter and base, and the base and collector respectively – is only a few tenths of 1 µm, then almost all the electrons will diffuse to the base/collector junction, in that they are accelerated and then, after migrating through the collector area they are conducted away through the collector contact. Only a few electrons 'recombine' with the excess of holes which is present in the base area.

If the emitter/base junction is biased in the direction of flow, then not only will an electron current flow from the emitter area into the base area, but in addition a hole current will flow in the reverse direction from the base area into the emitter area. However, if the dopant concentration in the emitter is very much higher (N-area) than in the base area (P-area), this hole current will be much smaller than

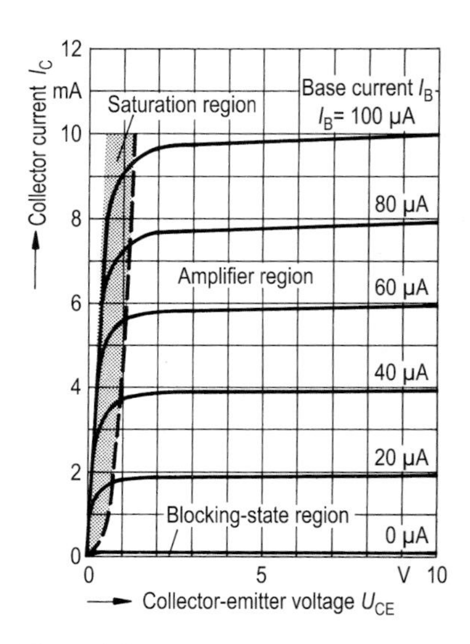

Figure 1.6
Set of characteristic curves for an NPN bipolar transistor

the electron current. The hole current flowing from the base area into the emitter area is subsequently supplied from outside in the base current, Figure 1.5a.

Figure 1.6 shows a typical set of characteristic curves for a bipolar transistor. The important feature is that it is possible to "control" a large collector current using a small base current. The ratio of collector current to base current is called the current amplification. A typical value for this is 100.

The bipolar transistor can also be driven in reverse, that is the emitter and collector can be swapped. However, in this case the current amplification is lower (values of around 1) owing to the lower dopant concentration in the emitter area.

Structure of bipolar circuits

An integrated bipolar circuit consists of numerous bipolar transistors which are arranged beside one another on a silicon chip and are interconnected by conducting tracks in accordance with the required circuit function.

Figure 1.7 shows a typical configuration for an NPN bipolar transistor with its immediately surrounding area in an integrated circuit. In this example, it is assumed that the transistor's collector is connected to the base area of a neighboring transistor by an aluminium interconnect. The P^+ frame around the transistor works electrically like two diodes connected in opposite directions (cf. Figure 1.4) and thus guarantees the mutual insulation of neighboring transistors.

The insulating frame and the P-silicon substrate connected to this frame are held at the lowest potential which arises in the circuit.

In practice, NPN bipolar transistors dominate, because they have better electrical characteristics and are simpler to manufacture than PNP transistors.

The specific layout of the individual doped areas in Figure 1.7 is determined, among other factors, by the external requirements which the integration must meet: the contacts must only be attached to the silicon surface; and to insulate the collector areas from each other each transistor is surrounded by a p-doped area.

For example, under these conditions one requires a highly-doped buried N-island

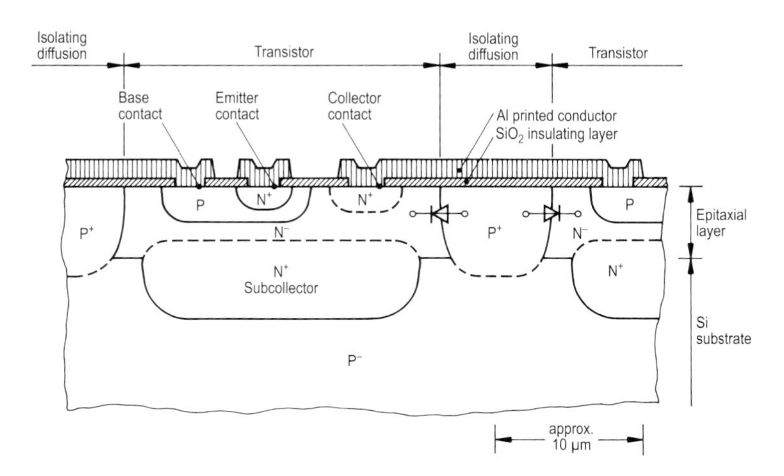

Figure 1.7 Typical integrated NPN bipolar transistor

(subcollector) to keep the collector resistance low. A consequence which follows from such buried islands is that the overlying monocrystalline silicon layer, in which the emitter and base areas are embedded, must be laid down as an epitaxial layer after the buried islands have been created. The P insulation frame which extends through the epitaxial layer occupies a comparatively large space. For this reason, attempts are made in the integrated circuits to lay those transistors whose collectors are connected together in the circuit in a contiguous "N well".

To provide an insulating layer, between the silicon surface and the conducting tracks laid down on top of it, use is made of a SiO_2 layer about 0.5 µm thick, which can be produced by thermal oxidation of the silicon surface. Where contacts are required, photolithographic masking techniques are used to etch holes in the SiO_2 layer. In general, each bipolar transistor in an integrated circuit requires three contacts to the interconnection layer (large space requirement). Only if neighboring transistors have, for example, a shared base area is it possible to save a contact.

The interconnect consist mainly of aluminium. The resistance of a conducting track with a thickness d, breadth b and length l is

$$R = \frac{\varrho \cdot l}{d \cdot b}$$

where ϱ is the specific resistance of aluminium. The ratio l/b can be interpreted as the number of surface squares (□), each having a side of length b, which make up the conducting track. Correspondingly, q/d is the resistance of one such surface square in the conducting track. This resistance is called the surface resistance, measured in Ω/□. For a thickness of 1 µm aluminium tracks have a surface resistance of 30 mΩ/□. This means that, for example, a section of conducting track which is 5 µm wide and 50 µm long (equivalent to 10 □), has a resistance of 0.3 Ω. Advanced technologies refer to Cu or even Au to achieve lower resistivition and better electromigration hardness.

More complex integrated bipolar circuits, such as gate arrays or memory, can no longer manage with a single interconnection layer. By using insulating layers deposited between different interconnection layers, it is possible to realize multi-layer wiring. The contact holes between the interconnection layers are called vias. They are produced in the same way as the contact holes to the silicon.

In contrast to the structure shown in Figure 1.7, modern bipolar circuits exhibit two further developments, namely oxide insulation to provide electrical insulation between neighboring bipolar transistors, and polysilicon connections for the emitter and base contacts (Figure 1.8). Both these developments lead to a significant saving of space, and hence to smaller parasitic capacitances and resistances, which makes faster circuits possible. Furthermore, the use of polysilicon emitter contacts produces better transistor characteristics.

An obvious idea is that integrated bipolar circuits should not only include the transistors but also use e.g. the doped areas individually, or in different layouts, if such arrangements exhibit a required electrical function. An emitter/base junction (possibly including a base/collector short-circuit to increase the forward current), or an Al/N^- contact, could function as a diode. Capacitors can be realized using PN junctions biased in the reverse direction (see Figure 1.4d), or by using a SiO_2 layer () as a dielectric between an N^+ area and an aluminium electrode. Resistances can be created using doped areas with appropriate dimensions, insulated electrically from their surrounds by a PN junction biased in the reverse direction (see Figure 1.3). With heavily doped N^+ areas (see Figure 1.7) it is possible to achieve resistances of between 2 and 40 Ω/□, while typical base doping opens up a resistance range of between 100 and

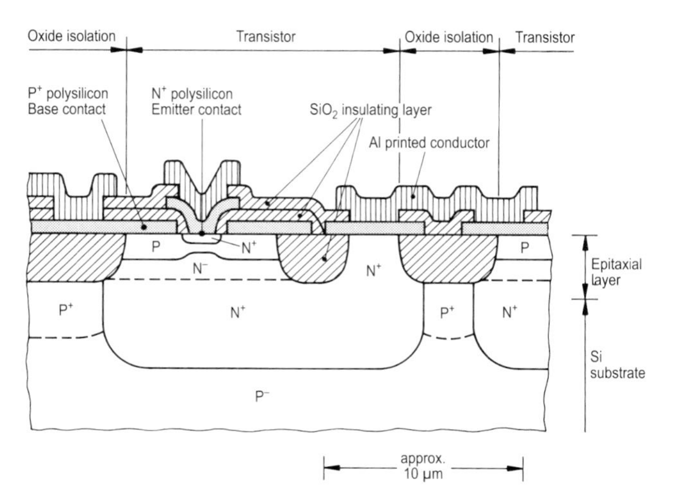

Figure 1.8 Bipolar transistor with polysilicon base and emitter contacts

300 Ω/□. In a bipolar transistor layout (Figure 1.7), if the current is conducted horizontally through the narrow base zone under the emitter, this has a resistance of around 5 Ω/□ (pinch resistance), although this is non-linear and has a low breakdown voltage (5 to 8 V). On the other hand, it offers the possibility of controlling the magnitude of the pinch resistance by means of a positive voltage at the N^+ emitter area (junction field effect transistor). High value resistances in the region of MΩ/□ can be created by low-level doping of polysilicon structures, laid down on a SiO_2 substrate.

In closing, it should be mentioned that PNP bipolar transistors can be realized on the same chip by an arrangement of the doped areas which differs from that in Figure 1.7. If the sub-collector island is omitted, this produces a vertical PNP arrangement (substrate PNP transistor). Another possibility is to arrange two p-doped areas a few µm apart in an n-doped surround (lateral PNP transistor). Both of these types of PNP transistor are significantly inferior to NPN transistors created on the same chip in terms of current amplification and maximum operating frequency, but as the PNP transistors can be created with no additional process steps, it is advantageous in some applications (e.g. for I^2L technology = Integrated Injection Logic) to include them in the integration.

Bipolar circuit technology

Bipolar circuits are used both as analog circuits and as digital circuits. Whereas in analog circuits the bipolar transistors are operated in the amplifying region, in digital circuits they function as switches, with the two states of the switches being respectively in the blocking state region and the amplifier or saturation region of the transistor.

The bipolar circuit techniques make systematic use of the following advantages which the bipolar principle has over the MOS principle:

- the ability to drive large currents,
- the availability of an extremely stable and reproducible threshold voltage in the form of the emitter-base forward voltage of 0.7 V, and
- the short switching time or high maximum operating frequency of bipolar transistors.

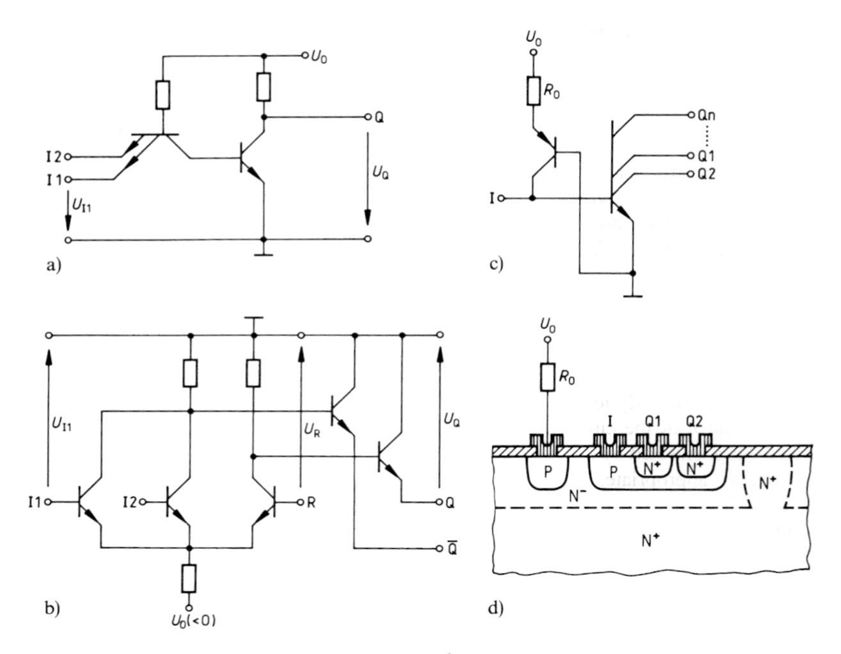

Figure 1.9 Basic gates in TTL (a), ECL (b) and I^2L technology (c and d)

In spite of the large variety of types with the most diverse applications, analog circuits mostly comprise relatively simple basic circuits, such as those already familiar from circuit technology using discrete components. However, particularly frequent use is made of basic circuits which benefit from the fact that the characteristics of the circuit elements and the temperature on a chip are subject to only limited random variation, and that the surface and resistance conditions on a chip can be very precisely maintained. Mention should be made here of current mirror circuits and difference amplifiers, but also multipliers and operational amplifiers. Final amplifier stages, with low impedance outputs, can be particularly effectively integrated using bipolar circuit technology.

In the case of digital circuits (logic and memory circuits) a distinction is made between various families, depending on the form in which the basic gate is designed. The most important of the logic families are TTL (transistor-transistor logic), ECL (emitter coupled logic) and I^2L (integrated injection logic).

Figure 1.9 shows basic gates in TTL, ECL and I^2L technology. Figure 1.9a shows a TTL NAND gate, Figure 1.9b an ECL OR/NOR gate and Figure 1.9c an I^2L inverter. ECL technology is distinguished by its particularly high speed. Gate delays of around 100 picoseconds are possible. Although the I^2L technology is space saving and low-power, it is however slower than the ECL technology.

1.3.2 Integrated MOS circuits

The MOS transistor

The N-channel MOS transistor consists of two N-areas, which are called the source and the drain, and between them a p-doped channel area, above which are attached insulated control electrodes, the gate.

Figure 1.10 shows the general structure of an MOS transistor, and how it works. Figure 1.10a shows the typical voltage applied, Figure 1.10b shows a graph of the potential along the surface of the silicon and the electron current from source to drain, and Figure 1.10c shows the graphical symbol for an MOS transistor. The U_{BS} connection is generally omitted from the symbol. The letter N stands for N-channel. Analogously to the bipolar transistor, control of the current in an MOS transistor is effected by a reduction or increase, as appropriate, in the potential threshold at the edge of the source on the gate side. However, the difference from the bipolar transistor is that the control voltage is applied to an insulated gate electrode above the P-area, and not to the P-area itself. A typical set of characteristic curves is shown in Figure 1.11. These assume the example of a gate oxide thickness of 20 nm, a channel length of 1 µm and a channel width of 10 µm. This different control mechanism is also responsible for the essential differences in the behavior of the MOS transistor in comparison with a bipolar transistor:

- Because the gate is insulated, no static gate current flows. This means that the use of an MOS transistor for control uses no power (a charging current only flows in the switching phase, and charges or discharges the gate-source capacitor).
- Because a positive voltage at the gate attracts the electrons, which are re-

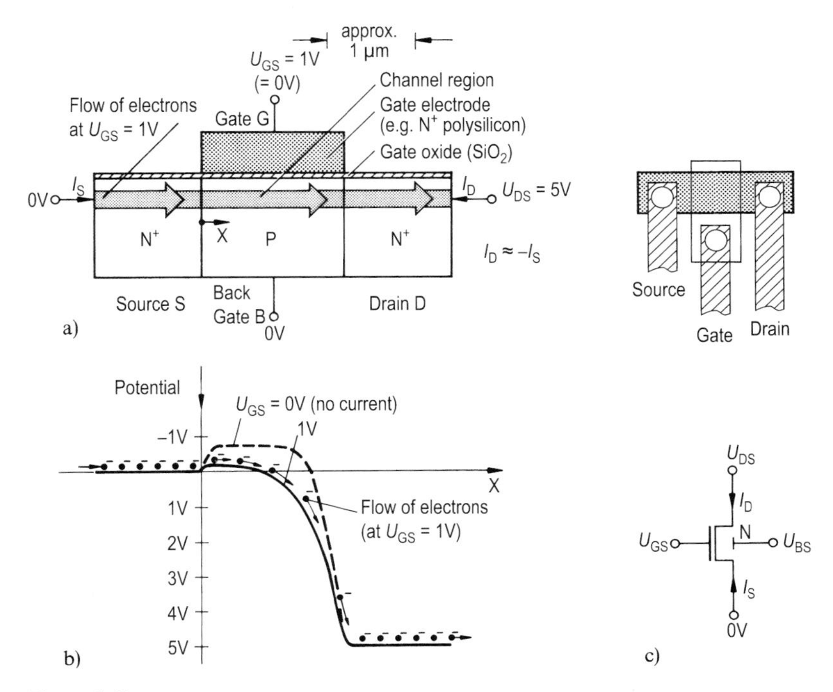

Figure 1.10
Structure (a), method of functioning (b) and graphical symbol (c) for an MOS transistor

sponsible for the current flow from the source to the drain area, to the surface of the silicon, and on the other hand the holes, which are present in excess in the P-area, are repelled from the surface (space charge zone), the current channel together with the source and drain N^+ areas are electrically insulated from the surrounding P-area (underneath and on the sides by the space charge zone, and above by the gate insulation layer). Since the gate is also insulated, the MOS transistor represents a self-insulating – and hence space saving – component, which is outstandingly suitable for circuit integration.

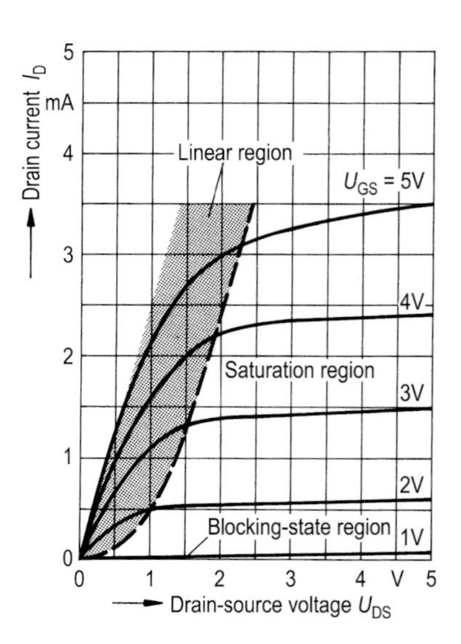

Figure 1.11
Characteristic curves for an N-channel MOS transistor

- Due to the fact that some of the controlling gate voltage drops across the gate oxide and the space charge region beneath the gate is highly dependent on doping concetration. The current startup voltage (threshold voltage), which for a bipolar transistor has an almost technology-independent value of 0.7 V, is technology-dependent in the case of an MOS transistor. Where the dimensions are small the threshold voltage does indeed still depend on the channel length and width (disadvantage). On the other hand, one can selectively vary the threshold voltage for the transistors (advantage). For example, a phosphorus implantation in the channel area of N-channel transistors can be used to move their threshold voltage into negative values.

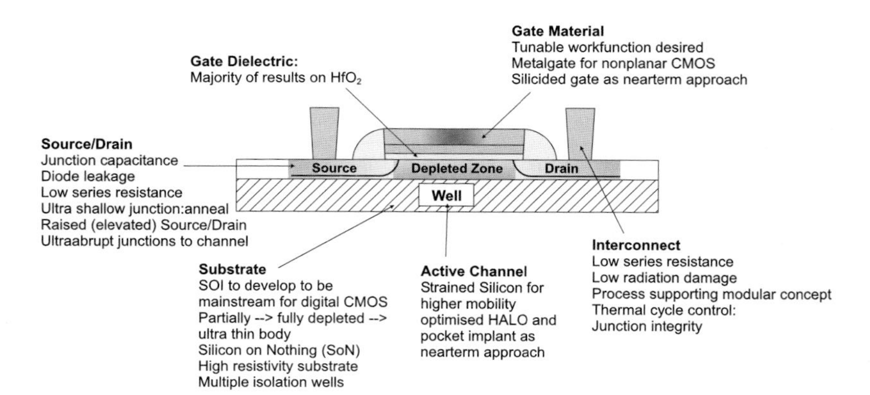

Figure 1.12 Improvements in critical areas with continuous reductions in structure sizes

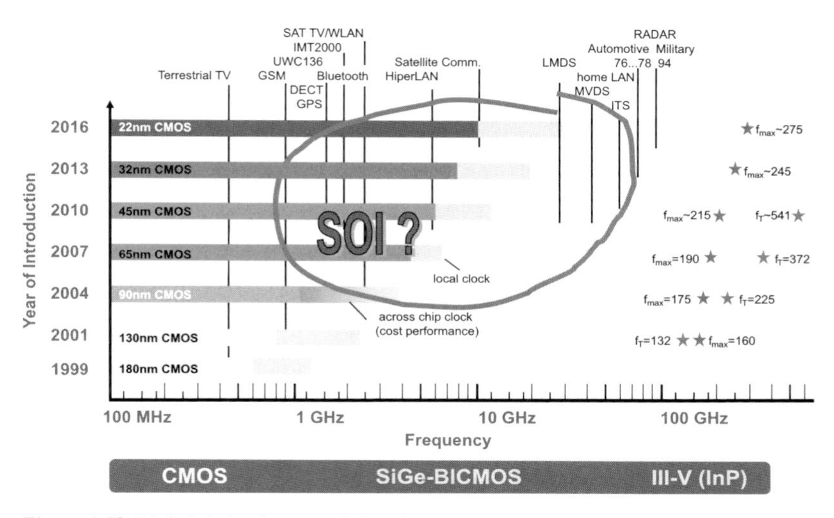

Figure 1.13 Digital design becomes RF analog

- For the same reason as above (voltage drop across the gate oxide), the transconductance $\Delta I_D/\Delta U_{GS}$ under comparable conditions is less than the transconductance $\Delta I_C/\ \Delta U_{BE}$ of a bipolar transistor. The saturation resistance of an open MOS transistor is larger than the saturation resistance of an open bipolar transistor (compare the characteristic curves in Figures 1.6 and 1.11). MOS transistors are therefore less suitable as current drivers.
- In parallel with the continuous reductions in structural size, improvements are being achieved in all the critical areas of MOS transistors (Figure 1.12). By reductions in the supply voltage (5 V … 3.3 V … ~1 V), the power consumed in switching cycles is being reduced. By reductions in the gate oxide thickness and, in future, the use of new materials, the short channel behavior of the transistors is being improved. By the use of low-impedance gate materials, the transistors are becoming more rapidly switchable. By optimization of the doping, the limits of the devices are being pushed out, and the parasitic resistances reduced. A host of such improvements has led to a situation in which CMOS devices, and in future particularly CMOS SOI (silicon on insulator) devices, will also be usable in the highest speed ranges, which were previously reserved for bipolar circuits and devices based on III-V materials (Figure 1.13).

MOS circuits

An integrated MOS circuit consists of numerous MOS transistors, which are arranged alongside each other on a silicon chip and are connected together in accordance with the desired circuit function by metal interconnect.

Figure 1.14 shows a typical design for an N-channel transistor with its immediately surrounding area in an integrated MOS circuit. Electrically, the transistor is completely symmetrical; source and drain are interchangeable. In this example it is assumed that the drain area of the transistor is connected to a polysilicon "conducting track" via an aluminium conducting track.

As already mentioned, the MOS transistor is a self-insulating circuit element. It

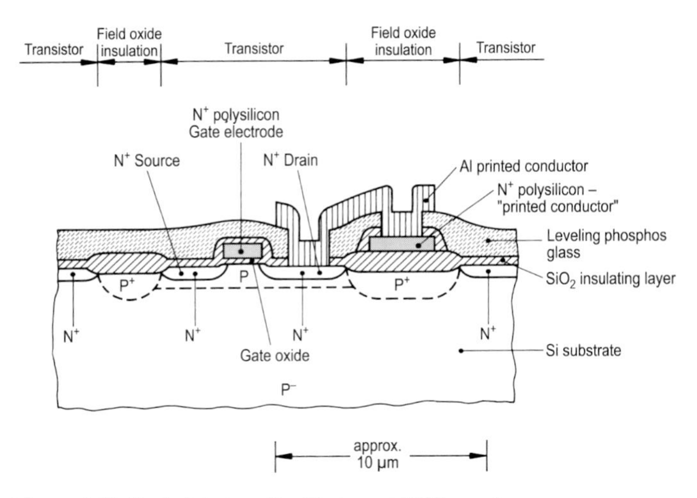

Figure 1.14 Typical design of an N-channel MOS transistor

is only necessary to make the oxide layer between neighboring transistors thicker than the gate oxide, so that the conducting tracks which run over the oxide layer do not act as gate electrodes and induce a conducting channel between the transistors. This results in an arrangement which, by comparison with an integrated bipolar transistor (Figures 1.7 and 1.8), is space saving. A further space-saving effect arises from the fact that it is not necessary to provide a contact to a conducting track for every source, gate or drain connection. Instead, one can realize electrical connections by extending the source or drain area, or the gate electrode, as appropriate, over the actual transistor area. However, the resistance of such "conducting tracks" is around a thousand times higher than that of aluminium conducting tracks (approx. 30 Ω/□). With the help of silicide layers (e.g. $TaSi_2$, $MoSi_2$, $TiSi_2$) it is possible to reduce the resistance of the layer by about one order of magnitude (approx. 3 Ω/□). The possibilities for making connections using N^+ doped silicon tracks or silicide tracks are extensively used for local wiring in MOS circuits on grounds of the space saving.

Depending on a chip contains only N-channel transistors, or only P-channel transistors, or both N-channel and P-channel transistors, we speak of NMOS or PMOS or CMOS circuits, respectively (C = Complementary). Nowadays, CMOS is the dominant technology because it is low power and offers the circuit developer the most options.

If the N-channel transistors of a CMOS circuit are arranged on a p-doped substrate, as in Figure 1.15, then the P-channel transistors must be accommodated in an "N-well", because to be self-insulating the P-channel transistors require an n-doped environment.

In order to keep the PN junctions in the well continuously blocked, any P-substrate or P-wells are held at the lowest potential which occurs in the circuit (in general 0 V), whereas any N-substrate or N-wells have the highest potential (V_{DD}) which occurs applied to them.

In order to avoid interference effects (e.g. latchup effects), a lightly doped epitaxial layer on a heavily doped substrate would be preferable to a uniform lightly doped substrate.

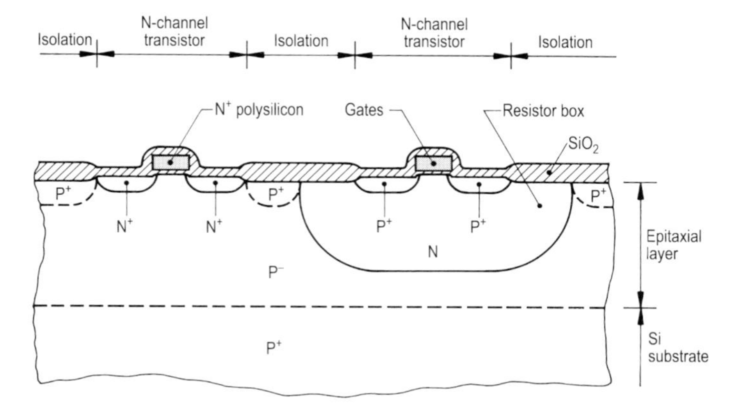

Figure 1.15 N- and P-channel transistors in a CMOS circuit

As in the case of bipolar circuits, with MOS circuits use is again made of the options for realizing diodes, resistances and capacitances by appropriate arrangement of the doped areas which are present, together with the conducting and insulating structures. When doing so, the thin gate oxide permits relatively high capacitances per unit area to be created, such as are required in memory circuits for example. The possibility of realizing EEPROM memory using floating gates will be disscussed somethat later.

With the help of some additional process steps, it is possible to integrate onto one chip both MOS and bipolar transistors (BICMOS technology).

MOS circuit technology

MOS circuits are used above all as digital circuits. However, analog functions are also being realized in MOS to an increasing extent. Because of the small space requirements of MOS transistors, MOS circuits dominate in the field of haighly integrated circuits. Here, because of its low power requirement and designer-friendliness, CMOS technology is dominant.

In circuit design, a distinction is made between standard and application-specific

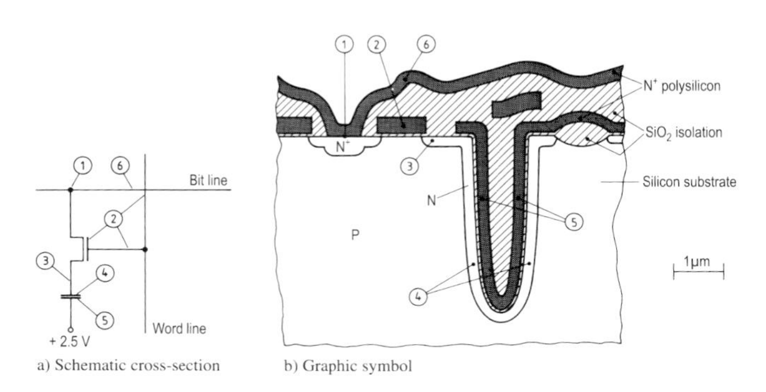

a) Schematic cross-section b) Graphic symbol

Figure 1.16 Dynamic memory cell

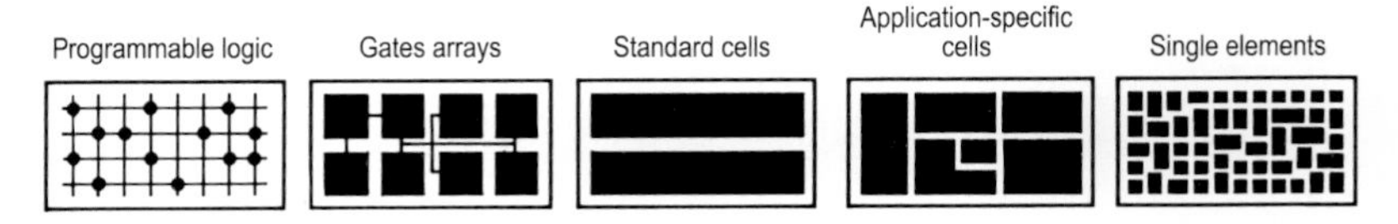

Figure 1.17 Design methods for application specific circuits (ASICs)

integrated circuits (ASICs). Whereas standard ICs such as memory modules (Figure 1.16 Dynamic memory cell) or microcomputer devices (microcontrollers, microprocessors, peripheral modules) are designed using the classical design process, in the case of ASICs use is made of rationalized design processes, which call on previously-developed basic circuits with particular basic functions. Figure 1.17 gives an overview of the various design methods which are in use today for application specific circuits. The larger the basic circuits developed using the standardized basic functions, the greater is the level of rationalization in the design work. Increasingly important are the standard cells, which are collected together in cell libraries. Figure 1.19 shows an example of a standard cell, in this case a NAND gate.

Figure 1.19a shows the circuit diagram for a memory cell, and 1.19b a schematic cross-section through it. In order to reduce the space required for a given memory capacity, the capacitor in the above example of a 4M-DRAM extends approximately 4 µm down into the silicon substrate.

The enlargement of storage capacity for DRAM memory cells has proceeded in a vertical direction, both downwards (trench cells) and also upwards (stack cells) from the cell transistor. In both cases, the reason is to achieve storage capacity which is as large as possible – and hence adequately immune to interference – within the smallest possible area. This use of the third dimension is linked to the accumulation of substantial production experience of extreme width-to-depth ratios, and the mastering of specific materials with layers which are only a few atoms thick.

Apart from dynamic memory mechanisms, such as DRAM and SRAM, nonvolatile memory is increasingly playing an important role. With non-volatile memory, either a charge is stored in an area which is completely insulated under normal conditions (reading), or use is made of changes in the crystal structure for storing data.

Nonvolatile memory cells include, in particular

EPROM Electrically Programmable Read Only Memory

EEPROM Electrically Programmable and Erasable Read Only Memory

FLASH a variant of EEPROM memory cells which is special, in terms of programming and particularly the circuit technology, and space saving

NROM a memory cell in which the charge is stored in a dielectric instead of a conducting layer

FeRAM resistive data storage in a ferro-electric crystal lattice structure

MRAM data storage in a magnetic resistance structure

PRAM (OUM) phase change memory

OUM Ovonyx Unified Memory = data storage in a chalcogenide crystal structure, which is reprogrammed by heating

Polymer data storage in a polymer layer – similar principle to OUM

Quantum dots (usefulness still uncertain)

The main advantage of non-volatile memory is that even if the battery or power supply should fail, the data in the memory is retained. For this reason, important application areas are in mobile telephony, digital cameras and smart cards.

The functioning of the EEPROM / FLASH / NROM-based memory cells is based on the displacement of the threshold voltage for a transistor by the insulated charge in the gate area (Figure 1.18).

This charge is altered during the writing or deletion of the data by hot charge carriers or tunnel currents. In both cases, relatively high voltages or power levels are required. For this reason, an important development objective is to optimize the programming and erasing conditions.

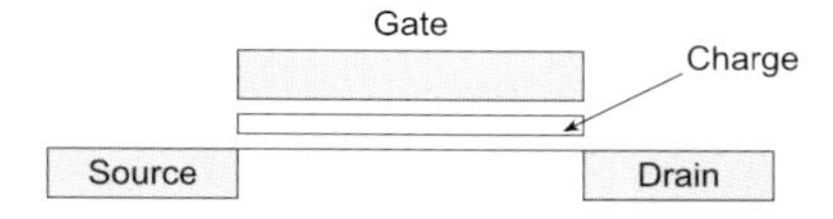

Figure 1.18
Structure of an EEPROM-, Flash- or NROM-based memory cell

The other non-volatile memory cells are based essentially on the change in resistance of an appropriate layer, brought about by electric fields or heat.

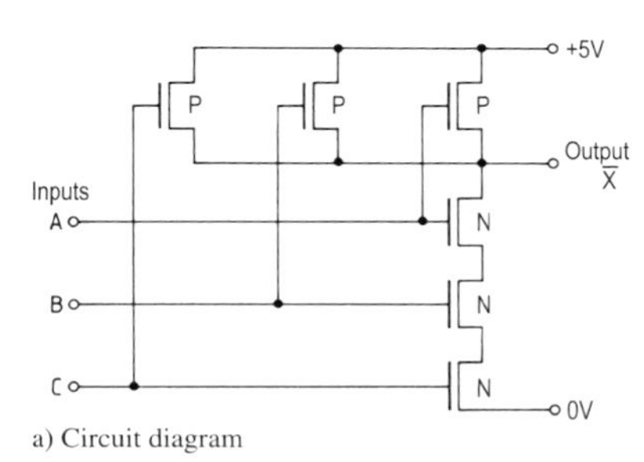

a) Circuit diagram

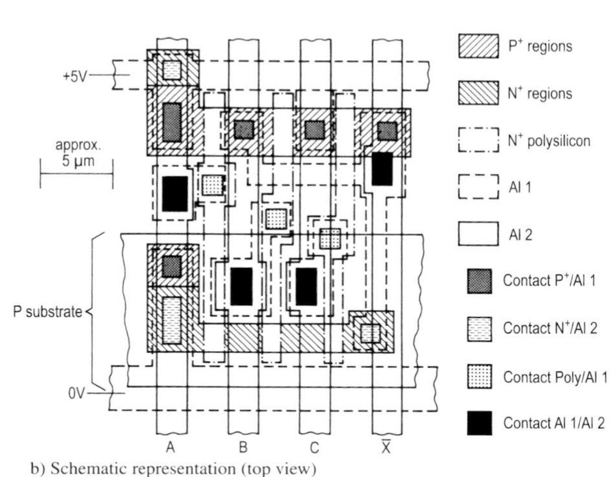

b) Schematic representation (top view)

Figure 1.19
3-NAND gate as an example of a standard cell

1.4 Other semiconductor components

Within the confines of this book, it is not possible to go in detail into all the semiconductor components. However, we wish to include below at least a brief description of some of these other semiconductor devices. In doing so, what has counted has been not so much how commonly the variants described occur, but rather their technical and/or historical significance. Apart from these remarks, the Glossary (Chapter 16) contains further notes.

1.4.1 Semiconductor modules with no special structure

Although most semiconductor modules exhibit some special structure, such as a surface layer, or make use of field effect, on occasions the particular characteristics of semiconducting materials can also be used in other ways.

PTC thermistors

Most conductors are thermistors, i.e. their conductivity decreases with increasing temperature. In general, however, the temperature coefficient of their conductivity is under 1%/°K. Simple temperature data measurement or capture requires components with a much higher coefficient. In the case of one ceramic material (barium titanate), which at the end of the 30s was already being used for the manufacture of ceramic capacitors, it was discovered that it acquired semiconducting characteristics when it was doped with certain materials. The notable feature was that above its Curie temperature it exhibited a steep rise in its temperature coefficient. Components using this material were called PTC thermistors (PTC = **P**ositive **T**emperature **C**oefficient). They are part of the family of thermistors (**thermally** sensitive res**istor**s). Since 1963 it has been possible to vary the Curie temperature, so that it became possible to create PTC thermistors with different characteristic curves. Today's PTC thermistors consist of mixtures of barium carbonate and certain metal oxides. Doped silicon has also been used as a material since the end of the 60s. PTC thermistors are used mainly as temperature sensors.

NTC thermistors

In 1941, in the course of research into magnetically soft ferrite, mixtures were encountered which had a conductivity which increased as their temperature rose. Thermistors made with this material are called NTC thermistors (NTC = Negative Temperature Coefficient). Although semiconductors also exhibit negative temperature coefficients, they cannot be used as ohmic resistors owing to their strong nonlinearity. NTC thermistors are used as temperature sensors, or to avoid excessive switch-on peaks at low temperatures by putting them in series with the PTC thermistor which is to be protected.

Varistors

In practice, most resistances follow Ohms law (for a constant temperature) over many orders of magnitude. Varistors (from **var**iable res**istor**) on the other hand exhibit a voltage dependency (in some cases a strong one) in their ohmic impedance. It is generally true that the resistance decreases with increasing voltage, and that the characteristic curve is symmetric (i.e. there is no preferred polarity). The material used for the first varistors was silicon carbide, and they had a continuously decreasing resistance. Today they are predominantly metal oxides such as titanium oxide and zinc oxide, which have an altogether steeper slope. The material is sintered, so that numerous individual grains are produced, on the bounding surfaces of which the varistor effect occurs. A varistor thus actually consists of a host of micro-varistors connected in parallel and in series. Varistors are mainly used to protect circuits or individual components against overvoltages. Metal oxide varistors can also be used for (simple) voltage stabilization.

Photoresistors

The photoelectric effect discovered by Albert Einstein manifests itself in semiconductors, in that incoming photons which have sufficient energy release valency electrons out of the crystal lattice. These freely movable charge carriers increase the intrinsic conductivity of the material until, after drifting about for a longer or shorter time, they recombine. This is what makes photoresistors possible: they consist of mixed crystals with no junction layer, made up of cadmium selenide or cadmium sulfide, the resistance of which decreases greatly under the incidence of visible light. Structures of lead sulfide or indium antimonide exhibit the same effect under infrared irradiation.

1.4.2 Semiconductor diodes

Semiconductor diodes have a boundary layer which can not only be used as a diode or current block, but also for other applications.

Tunnel diodes (Esaki diodes)

As early as 1954, William Shockley expressed a presumption that in certain materials the speed of migration of the electrons would reduce under the influence of high electric field strengths. Over some regions, a component with this property would then have a negative current/voltage characteristic, that is a negative differential resistance. The Japanese engineer Leo Esaki confirmed Shockley's idea in 1958, when he was experimenting with heavily doped PN junctions. Because the boundary layer in these diodes is very thin due to the heavy doping, even at small voltages (at which the diffusion current is virtually zero) a very high field strength develops. This high field strength initially permits the so-called "tunnel effect", a quantum mechanical effect in which the electrons in a certain sense "tunnel through" the thin boundary layer. As the field strength increases, the tunnel effect actually declines, so that a backward current occurs (see Figure 1.20). The first tunnel diode for microwave applications appeared on the market in 1960. In 1973 Esaki was awarded the Nobel prize for physics, together with the American Ivar Giaver, who had investigated the tunnel effect in superconductors.

Varicap diodes

As the name indicates, a varicap diode is a diode with variable capacitance. This is based on a non-linear effect in the junction layer: as the reverse voltage increas-

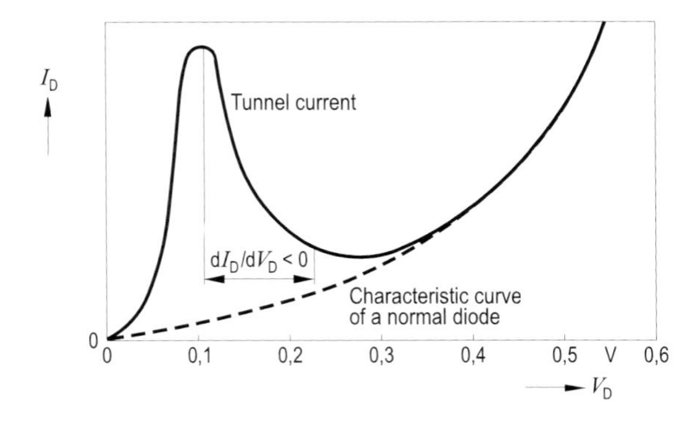

Figure 1.20
The characteristic curve of a tunnel diode

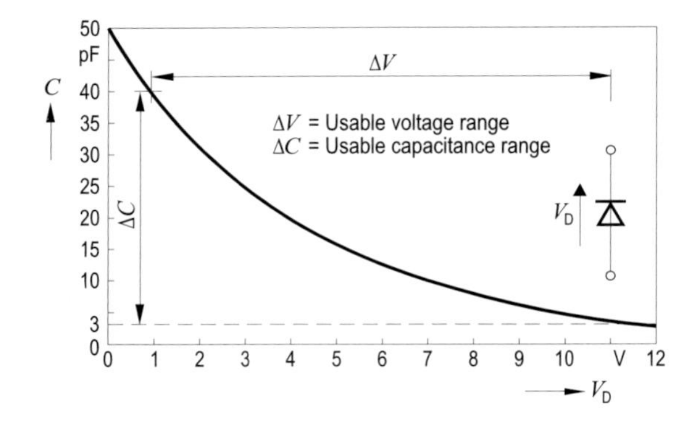

Figure 1.21
Graph of the junction capacitance for a varicap diode

es, the junction becomes wider, which causes its capacitance to drop. If a suitable doping profile is used, the capacitance can be varied over a substantial range (see Figure 1.21). One example of a use for the varicap diode is as a tuning diode in HF circuits.

Zener diodes

During his investigations into the blocking behavior of semiconductor junctions, the scientist C. Zener discovered that at certain doping levels the breakdown voltage drops sharply, and when it is reached the reverse current increases sharply. This effect is because of the fact that free charge carriers, which have moved from the valence band into the conduction band due to a high voltage (the Zener voltage), greatly increase the conductivity. This is called the Zener effect, after its discoverer. Up to the breakdown voltage (also known as the Zener voltage) of 1 – 50 V (depending on the doping), only a low reverse current flows (see Figure 1.22). Above the Zener breakdown, the current increases in an avalanche-like fashion as the reverse voltage increases. For this reason, Zener diodes are suitable for voltage limiting and stabilization in electrical circuits.

Avalanche diodes

With an avalanche diode, an avalanche-like amplification of the Zener effect occurs. If the field strength and migration path lengths are great enough, the free charge carriers released by the Zener ef-

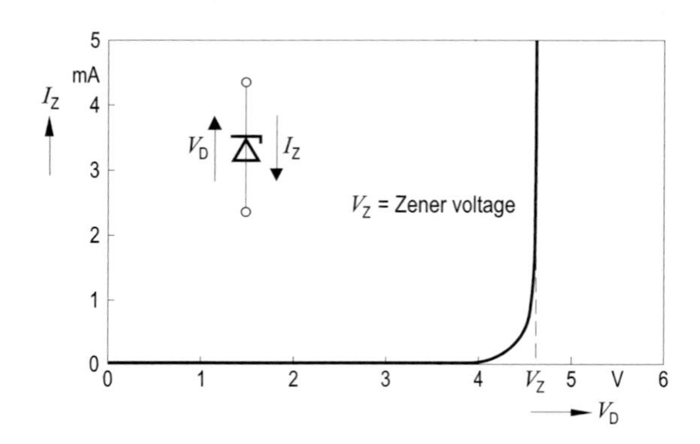

Figure 1.22
The characteristic curve of a Zener diode

fect are accelerated sufficiently so that their kinetic energy is enough to permit secondary electrons to transfer from the valence band into the conduction band.

Schottky diodes

With the development of planar technology, it also became possible to realize diodes with junctions constructed of a semiconductor layer and a metal layer. These so-called hot-carrier diodes are also often referred to as Schottky diodes (after the semiconductor pioneer, Walter Schottky), or more accurately: Schottky barrier diodes. Their advantages are an extremely short switching time, small random variations between individual examples, low noise and high power rating. The area in which they are used is fast logic circuits.

Gunn diodes

In 1963 the British physicist Jan B. Gunn discovered, during his investigations into GaAs and InP resistors in the IBM research laboratory, that above certain voltages their intrinsic noise increased sharply, up to the point of oscillation. This effect was named, after its discoverer, the Gunn effect, although it had earlier been described theoretically by Ridley, Watkins and Hilsum. It results from the fact that above a certain field strength the charge carriers group themselves into domains. Their mobility is thereby reduced. The consequence is that when a current flows an oscillating drift develops, which can be used for HF oscillators with an extremely high frequency (some indeed over 100 GHz). It is actually incorrect to call them "diodes", because Gunn diodes have no pn junction, but consist of material which is n-doped to different extents.

BARITT diodes

In a similar way to the tunnel diode, the Gunn diode and the IMPATT diode, barrier injected transit time (BARITT) diodes have areas where they exhibit a negative differential resistance. In the case of these diodes, charge carriers are virtually injected into the junction. They are used predominantly in microwave circuits.

IMPATT diodes

The impact avalanche transit time (IMPATT) diode is a semiconductor diode operated in the reverse direction, in which impact ionization with an avalanche effect occurs in some areas of the depletion zone, so that the current drops as the forward voltage rises. These diodes thus exhibit – in a similar way to the tunnel diode, the Gunn diode and the BARITT diode – a negative differential resistance. The idea of the IMPATT diode had already been mentioned in 1958 by W. T. Read (a colleague of William Shockley), but it took until 1964 before semiconductor researchers could prevent the diode from being destroyed when the avalanche breakdown occurred (a team around A. S. Tager, in the former Soviet Union, had succeeded in doing this at the start of the 60s, but the development only became known in the West much later). The limiting frequency of the IMPATT diode lies at around 300 GHz. The component can deliver a higher power than the Gunn diode (which is, however, less noisy). It is used predominantly in microwave circuits.

TRAPATT diodes

The trapped plasma avalanche triggered transit (TRAPATT) diode is a semiconductor structure for fast microwave diodes. It uses the avalanche effect in a trapped plasma, one consequence of which is that it has a negative differential resistance in some regions.

Burrus diodes

The Burrus diode is a diode which radiates infrared, with the radiation being, so to speak, extracted in a downward direc-

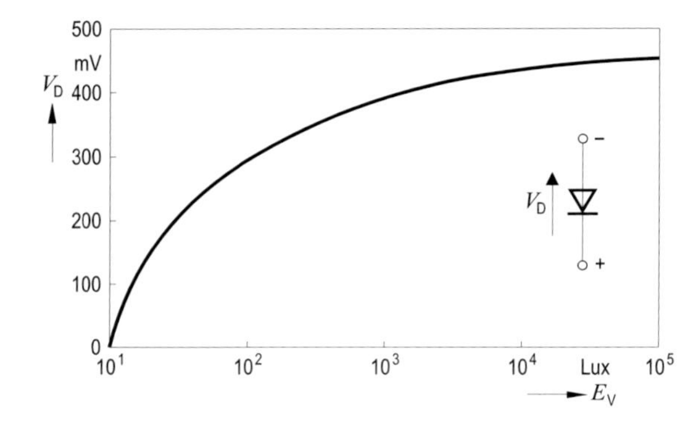

Figure 1.23
Characteristic curve of a photovoltaic cell

tion through a "hole" in the substrate (etched in after production); its PN junction is located on the underside. This produces particularly good conditions for coupling it up to fiber optic cables. As the internal construction of Burrus diodes makes them particularly fast, their use is preferred for fiber optic transmission.

Avalanche photodiodes

The avalanche photodiode (APD) is a photo diode operated in the reverse direction, in which the photocurrent generated releases secondary charge carriers, leading to an avalanche effect. This avalanche effect only occurs at high field strengths, but these can only be built up if the junction layer is very homogeneous. If there are inhomogenities, local breakdowns occur, and these prevent a high field strength building up. This represents a particular technological challenge. Large-area APDs are thus rare and expensive. With APDs, lighting can still be reliably registered, even if it is of very low strength.

Photovoltaic cells

If radiation energy penetrates into the junction area of a pn-boundary, electron-hole pairs are produced, the electrons from which migrate into the p-layer under the influence of the diffusion voltage, while the holes are drawn into the n-layer. A photovoltage arises at the contacts, and this can also be loaded: a photocurrent flows and the photovoltaic cell supplies power. As the intensity of illumination increases, the photovoltage initially increases rapidly, and then (see Figure 1.23) approaches the diffusion voltage asymptotically (logarithmically). Photovoltaic cells are used, for example, in exposure meters.

Solar cells

A photovoltaic cell with a large area is referred to as a solar cell. These structures are making it possible to obtain useful power from sunlight by photovoltaic means.

1.4.3 Transistors

The transistor has revolutionized electronics. Without it, a host of pioneering inventions and developments would not have been possible.

UJT

Even using a single pn-junction, it is possible to realize a transistor-like structure. This so-called uni-junction-transistor (UJT) is similar to the junction FET, but is bistable. If a sufficiently high voltage is applied (the threshold voltage, which is a function proportional to the collector voltage applied) to the control electrode, the base channel becomes conducting. The UJT is occasionally used in trigger circuits.

HEMT

The high electron-mobility transistor (HEMT) is a heterostructure depletion layer FET. The FET channel consists of a quantum well, which produces a two-dimensional cloud of extremely mobile electrons. The high mobility permits a very short switching process. The component is thus suitable for applications with short switching cycles and times.

HJBT

The heterojunction bipolar transistor (HJBT) is the bipolar counterpart of the HEMT.

RHET

The resonance-tunneling hot electron transistor (RHET) is a special transistor structure for microwave applications. It is based on quantum-mechanical tunnel effects of so-called "hot electrons", which are free electrons with a kinetic (thermal) energy lying significantly above kT (k is the Boltzmann constant, T is the temperature in K).

SET

The single electron transfer transistor is a new development, with which a switching operation is initiated by a single charge carrier. This makes the energy consumption per switching operation extremely small. Currently, the principle will only work at very low temperatures (up to around 100 K).

Thyristor (hook transistor)

If an NPN and a PNP transistor are connected together in such a way that the collector of each controls the base of the other transistor, the result is a bistable component which becomes permanently conducting if the base of one of the transistors is given a large enough impulse. A structure of this type can also be manufactured as a single component with three external connections, the anode, the cathode and the gate (control electrode). This is then called a thyristor (from **thyra** = door, and trans**istor**), and is suitable for switching large currents (see Figure 1.24). The first thyristor was constructed in 1958 by the Radio Corporation of America (RCA). The thyristor is often also called an SCR (Silicon Controlled Rectifier). The operation of the component is basically unidirectional, but it is possible to connect (and also to integrate) two components in anti-parallel, and one then has a bidirectional switch. In the normal situation it is only possible to switch this off by the voltage dropping below the operating level, but there are also thyristors with cancellation electrodes, with which a reverse control current switches the component off. These switch-off thyristors are called GTOs (gate-turn-offs).

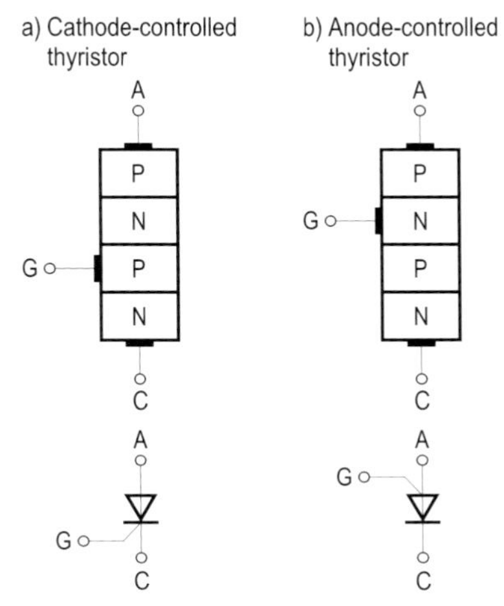

Figure 1.24 Structure of a thyristor

1.4.4 Other integrated semiconductors

CCD

A **c**harge-**c**oupled **d**evice (CCD) is a charge-coupled switching component (also called a charge transfer element or bucket brigade device) with an MOS structure, in which step-wise charges are transported by closely neighboring MOS capacitors and are then evaluated. This makes it possible to construct a shift register for analog signals. The CCD principle was invented in 1970 by two employees of Bell Laboratories, W.S. Boyle and G.E. Smith. CCDs are used mainly as image sensors.

***R**ead **O**nly **M**emory (ROM)*

Read only memories can only be read. In practice they consist of a semiconductor matrix for which a particular pattern of connections between the two layers is implemented during manufacture (for which reason it is often called mask ROM). The content of the memory is thereby fixed indelibly.

***P**rogrammable **R**ead **O**nly **M**emory (PROM)*

Programmable read-only memory consists of a semiconductor matrix in which, when it is delivered, some of the cells have not already been given particular value settings using a mask, as is the case with mask ROM. The individual cells can still be programmed by the user, by destroying irreversibly certain connections, using a programming voltage. This latter is usually applied by an external device (the programmer), so that a memory module cannot be changed after it has been installed.

EAROM

Electrically Alterable ROM (EAROM) is a read only memory which can be altered electrically. Unlike an EEPROM, when it is deleted only part of the data content is deleted (rather like with Flash memory). However, only certain types permit the deletion of individual cells. These are very expensive and are only used when there is no alternative.

2 Diodes and transistors

2.1 High-frequency diodes

In the high-frequency region, the types of diode which are most frequently used are PIN, varactor (variable capacitance) and Schottky diodes. Although all these diodes exhibit rectifying characteristics at low frequencies, very different diode characteristics are exploited in the high frequency region.

Like the varactor diode, the PIN diode is also a bipolar device, whereas the Schottky diode is a unipolar device, which essentially means that only one type of charge carrier (generally electrons) contributes to the current flow.

A PIN diode has a very lightly doped quasi-intrinsic zone between the highly doped p and n connections, with a resistance which can be adjusted over a wide range by means of the voltage or current, as appropriate (Figure 2.1). As the limiting frequency for a PIN diode typically lies far below 100 MHz, at higher frequencies the diode exhibits the characteristics of a linear resistance. As this depends on the DC current, the diode is used as a RF switch as well as an adjustable attenuator. The qualitative features of these diodes are their high frequency resistance together with their blocking capacitance when the polarity is reversed.

In the case of varactor diodes, use is made of the changes in the capacitance of the depletion zone (junction capacitance) as the reverse bias is altered. These diodes are used as frequency-fixing devices, for example in voltage-controlled oscillators (VCOs). A varactor diode is constructed as a pn-diode (Figure 2.2), in which the shape of the doping profile determines the capacitance/voltage characteristics, and hence the voltage-dependence of the resonant frequency in a resonant circuit. By reference to this profile shape, a distinction is made between abrupt and hyperabrupt diodes, with the latter being distinguished by an exceptionally sharp capacitance rise. Figure 2.3 shows a typical graph of capacitance against voltage.

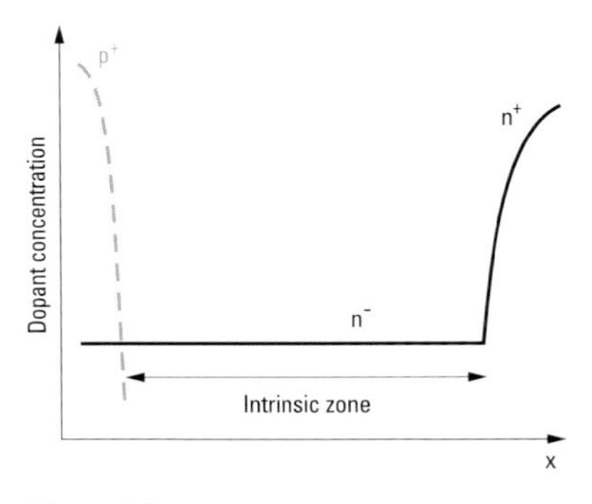

Figure 2.1
Simplified representation of the doping profile in a PIN diode

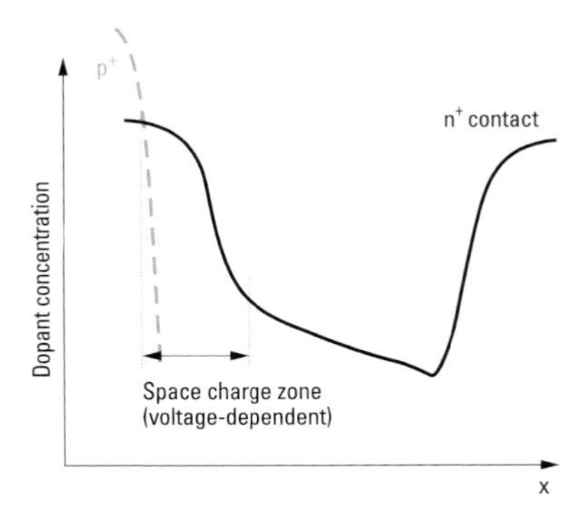

Figure 2.2
Doping profile in a varactor diode

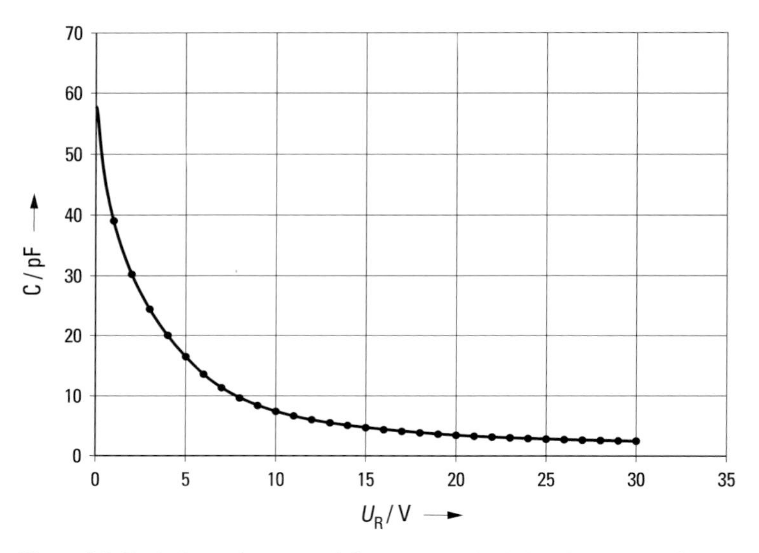

Figure 2.3 Typical capacitance graph for a varactor diode, in this case a BB639C

The device which is used as an actual diode in the high frequency region is the Schottky diode. Its unipolar construction with a metal contact on lightly-doped silicon (Figure 2.4) has the advantage of vanishingly small diffusion capacitances, which gives it limiting frequencies far into the GHz region. A further advantage in many applications is its blocking voltage, which by comparison with pn-diodes is significantly lower and can be modified by the choice of Schottky metal. Schottky diodes are used as detector and rectifier diodes, and also in diode mixers.

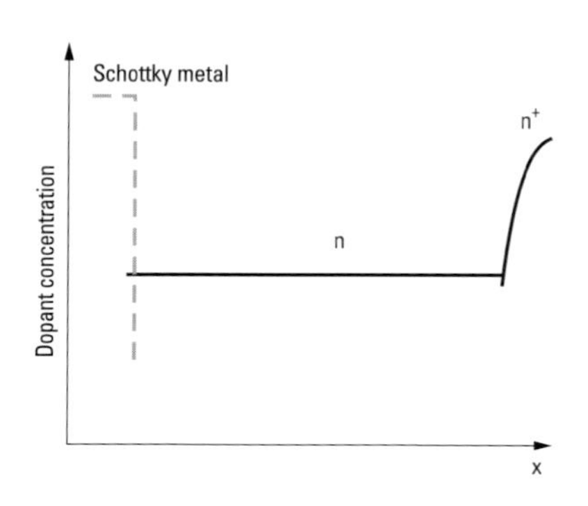

Figure 2.4
Doping profile of a Schottky diode (schematic)

2.2 Charge carrier life and series resistance of high-frequency PIN diodes

In this section, we describe the physical fundamentals of high-frequency PIN diodes, examine the general design considerations and present a method for measuring the life of charge carriers.

PIN diodes are predominantly used as switched or controlled resistances for signals in the middle to upper MHz band and up to the mobile communication frequencies. Figure 2.5 shows the structure of a PIN diode schematically.

A lightly-doped (quasi-)intrinsic zone separates the heavily doped p^+ und n^+ connection areas. Even if no bias voltage is applied, the diode is already largely

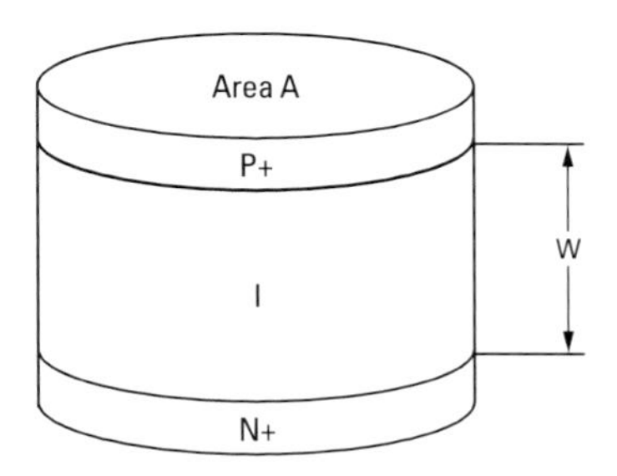

Figure 2.5
Structure of a PIN diode

drained of charge carriers, and for small signals the diode represents a capacitor with a capacitance of

$$C_0 = \frac{\varepsilon \cdot A}{W}$$

where the meaning of the symbols is:

C_0 capacitance for a zero bias voltage

ε dielectric constant of the i-zone material

A area

W thickness (width) of the intrinsic zone (i-zone)

When no bias voltage is applied, or in the blocking direction, the PIN diode blocks high-frequency signals because of the large extent of the space charge zone – and the resulting small capacitance.

In the forward direction, the specific resistance of the i-zone is greatly lowered by the injection of charge carriers from the heavily-doped p^+ and n^+ regions. The concentration of charge carriers which develops is determined both by the forward voltage and by recombination within and on the border of the i-zone. The recombination within the interior of the i-zone is essentially determined by the life of the charge carriers (electrons and holes).

The resulting small signal resistance of the i-zone can be described approximately by the following equation:

$$r_f = \frac{W^2}{(\mu_n + \mu_p) \cdot \tau \cdot I_f}$$

where the meaning of the symbols is:

r_f forward resistance at radio frequencies

I_f forward current

μ_n mobility of the electrons

μ_p mobility of the holes

τ life of the charge carriers

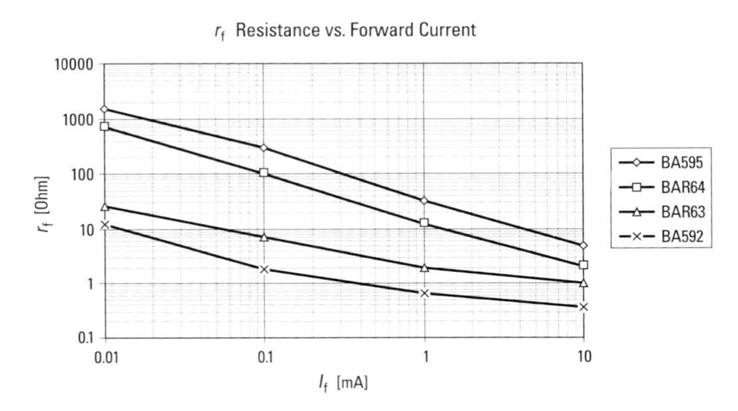

Figure 2.6 Typical values of r_f

Table 2.1 Overview of the diode parameters

	A	W	τ	r_f	C_0
BA595	medium	high	1600 ns	high	230 fF
BAR64	high	medium	1400 ns	medium	300 fF
BAR63	low	low	80 ns	low	300 fF
BA592	medium	very small	120 ns	very small	1200 fF

The above equation assumes that the current is determined primarily by recombination in the i-zone. However, if τ is interpreted as an effective life, then the above equation is also valid if the injection current in the p^+ and n^+ zones, and the recombination at the various boundary surfaces (for example silicon/oxide), are included. This is one of the main reasons why τ is generally a quantity which decreases with the current. Figure 2.6 shows typical values of the resistance r_f.

If a PIN diode is operated at high frequencies then, for a forward bias, it functions primarily as a linear resistance. The rule of thumb here is that the signal frequencies should be substantially greater than the reciprocal of the value of τ:

$$f \geq \frac{10}{\tau}$$

Modulation of the charge carrier concentration is suppressed, thereby largely avoiding inharmonic effects.

If a PIN diode is used as a switch, its qualitative characteristics are a small forward resistance and hence low losses for a forward bias, together with a guaranteed high level of insulation in the blocking direction, due to the low capacitance. Table 2.1 shows a comparison of the qualitative parameters of typical high-frequency PIN diodes.

2.2.1 How can the electrical parameters of a PIN diode be measured?

Under the conditions of a forward bias, r_f and C_t can be determined using an impedance tester, for example the HP4291. To determine the life of the charge carriers requires a different test arrangement (Figure 2.7).

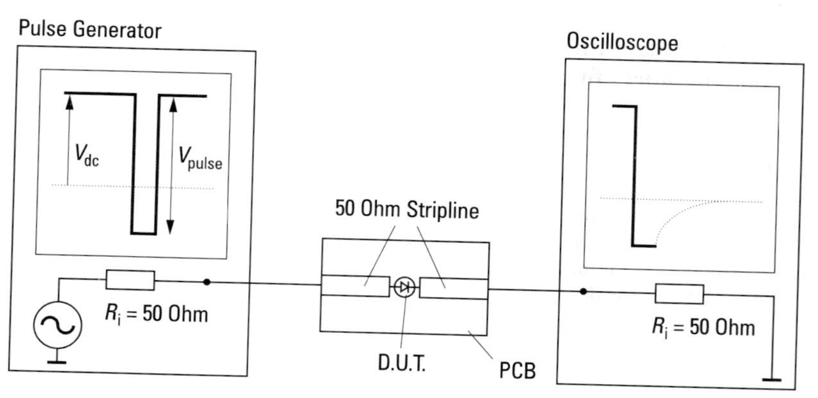

Figure 2.7 Test arrangement for measuring the life of the charge carriers

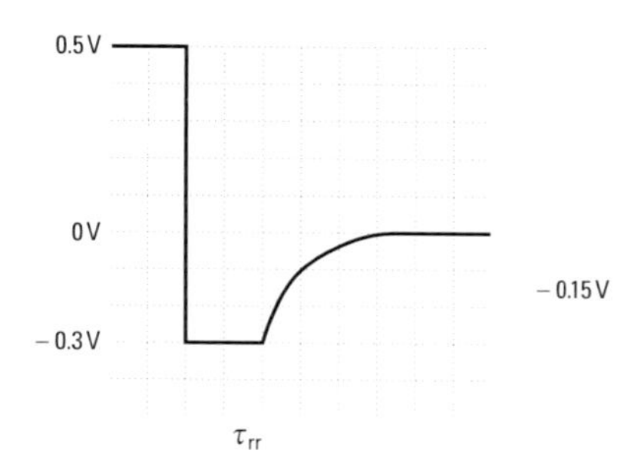

Figure 2.8
Oscilloscope trace, showing a measured charge carrier life

The width of the negative pulse from the generator must exceed the expected life. The DC voltage conditions and pulse amplitude should be set such that the oscilloscope shows a trace like that in Figure 2.8, in this example I_f = 10 mA (0.5 Volt across 50 Ω) in the forward direction, I_r = 6 mA in the reverse direction. The time τ_{rr} (decay time in the blocking direction), measured from the falling edge of the pulse up to the point in time at which a blocking current of 3 mA is reached, gives and estimate of τ, which can be used to infer the relationship between τ_{rr} and τ.

For the chosen current ratio, the logarithm is approximately 1.

2.3 Definition of the capacitances for bipolar transistors

Many definitions are possible for the capacitances in a transistor (Figure 2.9), and from a historical point of view have been used. The simplest is to assume a capacitance between each of the 3 connections. The following section describes the interrelationships.

C_{cb} Capacitance, Collector-Base
C_{ce} Capacitance, Collector-Emitter
C_{eb} Capacitance, Emitter-Base

Figures 2.10 to 2.15 show the various definitions and how they are measured.

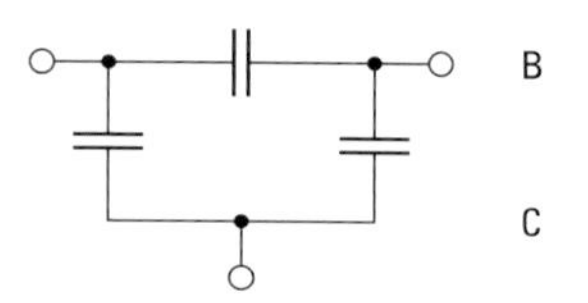

Figure 2.9
General definition: equivalent circuit for a bipolar transistor

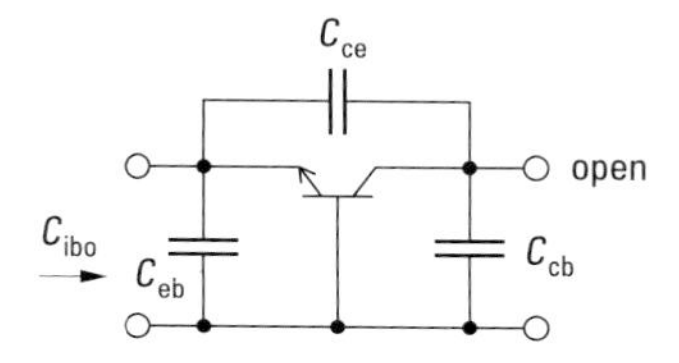

Figure 2.10 Definition of C_{ibo}

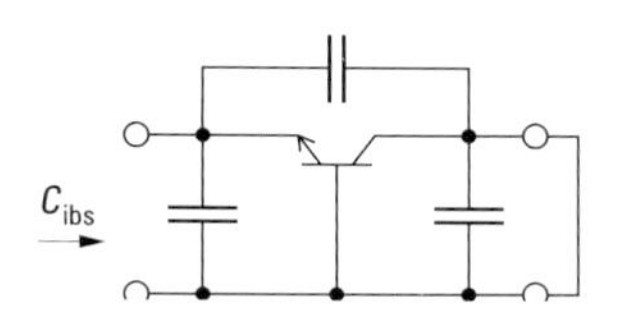

Figure 2.11 Definition of C_{ibs}

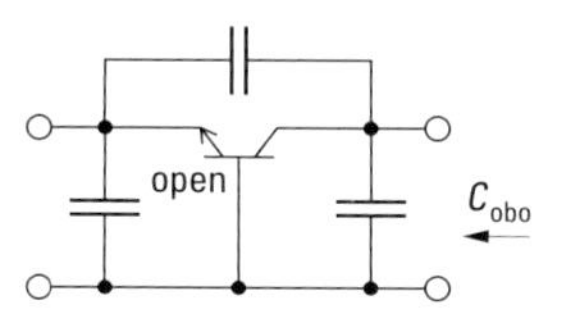

Figure 2.12 Definition of C_{obo}

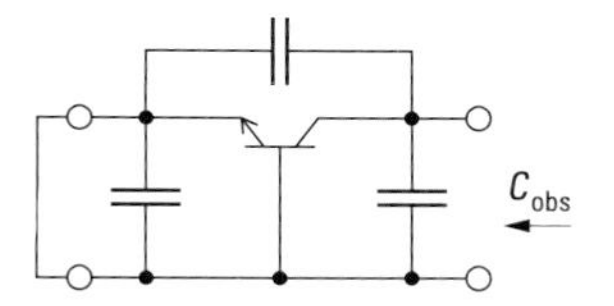

Figure 2.13 Definition of C_{obs}

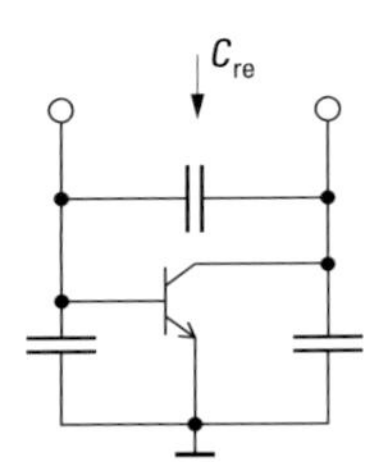

Figure 2.14
Definition of C_{re}

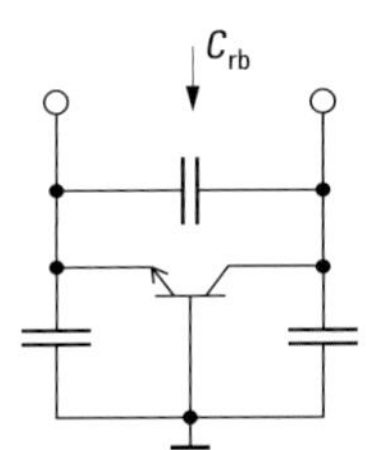

Figure 2.15
Definition of C_{rb}

$\boldsymbol{C_{ibo}}$ Capacitance, Input, Common-Base, Output open
$C_{ibo} = C_{eb} + (C_{ce} \cdot C_{cb})/(C_{ce} + C_{cb})$
also known as: C_{ib}, C_{ebo}, C_e

$\boldsymbol{C_{ibs}}$ Capacitance, Input, Common-Base, Output short
$C_{ibs} = C_{eb} + C_{ce}$
also: C_{ib}, C_{ebs}, C_{11b}, C_{11e}

$\boldsymbol{C_{obo}}$ Capacitance, Output, Common-Base, Output open
$C_{obo} = C_{cb} + (C_{ce} \cdot C_{eb})/(C_{ce} + C_{eb})$
also: C_{ob}, C_{cbo}, C_c

$\boldsymbol{C_{obs}}$ Capacitance, Output, Common-Base, Output short
$C_{obs} = C_{cb} + C_{ce}$
also: C_{ob}, C_{22e}, C_{22b}

$\boldsymbol{C_{re}}$ Capacitance, Reverse, Common-Emitter
$C_{re} = C_{cb}$
also: C_{12e}

C_{eb} and C_{ce} relative to ground have no effect on the measurement of C_{cb} in a capacitance measuring bridge.

$\boldsymbol{C_{rb}}$ Capacitance, Reverse, Common-Base$C_{rb} = C_{ce}$
also: C_{12b}

C_{eb} and C_{cb} relative to ground have no effect on the measurement of C_{ce} in a capacitance measuring bridge.

2.3.1 How are C_{cb}, C_{ce} and C_{eb} measured?

For simple measurements, the options of a capacitance (such as HP4279A) can be used. This bridge can measure the capacitances between the two coaxial outputs, with the capacitances from the co-ax input to ground being of no significance. They are also used for supplying any required DC bias voltages.

- For the measurement of C_{cb}, the emitter is connected to ground. The measurements are made between the collector and the base.
- For the measurement of C_{ce}, the base is connected to ground. The measurements are made between the collector and the emitter.
- For the measurement of C_{eb}, the collector is connected to ground. The measurements are made between the emitter and the base.

The measurement strategy is shown by the above definitions of C_{re} and C_{rb}. If the third connection is grounded, only the capacitance between the other two connections is measured.

2.4 Definition of a small signal RF transistor by the measurement of three parameters

The increasing pressure to reduce development times means that developers fall back more and more on simulation aids. These can only provide satisfactory results if sufficiently precise and definite test values and device data are available.

The aim of this section is to contribute to an understanding of the source of the required values, and is intended to assist readers in their own analyses or in making measurements.

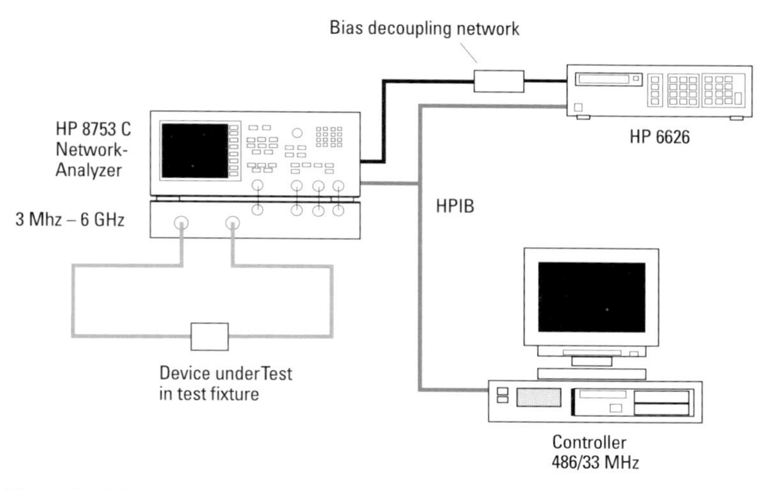

Figure 2.16 Test arrangement for measuring the S parameters

2.4.1 Measuring the S parameters

a) Using a test fixture

The S parameters of FETs or bipolar transistors are generally measured in a precisely-manufactured 50 Ω test fixture in a common-emitter or common-source arrangement. When doing so, a complete 2-port calibration must be performed on both ends of the RF cable. In order to obtain the S parameter at the reference plane of a transistor, the network analyzer incorporates the previously determined S parameters of the test fixture into the calculation for the 12-term calibration (Figure 2.16).

The attenuation at Port 1 must be set sufficiently high to permit small signal operation of the device under test (DUT) even at the lowest collector currents and frequencies. Interference can be eliminated by a 20 dB lower attenuation at Port 2 together with a very low setting of the IF bandwidth.

Figures 2.17 and 2.18 show the position of the reference planes for two different types of package.

The HP6626 current sources are set up in accordance with HP Application Note 376-1. This describes the advantage of the

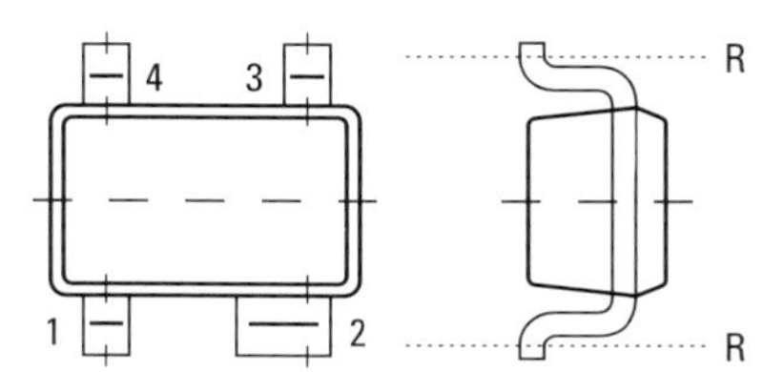

Figure 2.17
The S parameter reference plane for an SOT343 package

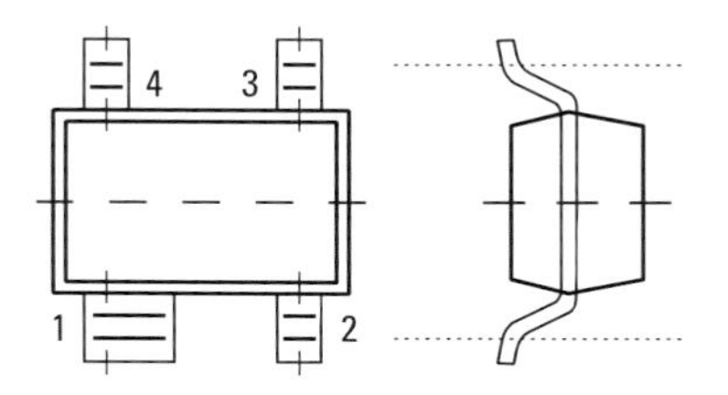

Figure 2.18
The S parameter reference plane for an SOT143 package

collector-base voltage configuration and of setting the emitter current directly.

b) Using microstrip circuit boards

Many users prefer to measure the S parameter of the transistor using low-cost SMA connectors on a microstrip circuit board. After a 3.5 mm calibration, a correction is made for the length between the connector and the transistor by using the "electrical delay" function of the tester. Inaccuracies due to the plug connector, impedance errors and attenuation are neglected.

A more accurate method is to carry out a TRL calibration using the appropriate standards, a precise 50 Ω line and special plug connectors.

Another problem with this method is that the measured values include the inductances of the via holes for connecting the emitter to ground. These are not taken into account in the datasheet values for the S parameters.

2.4.2 Measuring arrangement for determining the noise figure for a transistor

The arrangement described here can be used at up to 2000 MHz, and requires no mixing stage between the preamplifier and the noise figure meter (Figure 2.19). The transistors are measured in a precision test fixture.

The source impedance can be tuned to F_{min}, the impedance of the load for maximum amplification. The use of circulators can be avoided by reducing the impedance changes for the noise source using a 10 dB attenuator and using an amplifier with good matching.

One method for avoiding a special calibration path is to interconnect the tuner directly without the device under test in a 50 Ω tuner setting. In this case, the tabulated values for the ENR before the input must be reduced by the attenuation between the noise diode and the device under test. In the case of BJT measurements, the additional attenuation can in most cases be rendered negligible by the use of a tuner with a high Q-value.

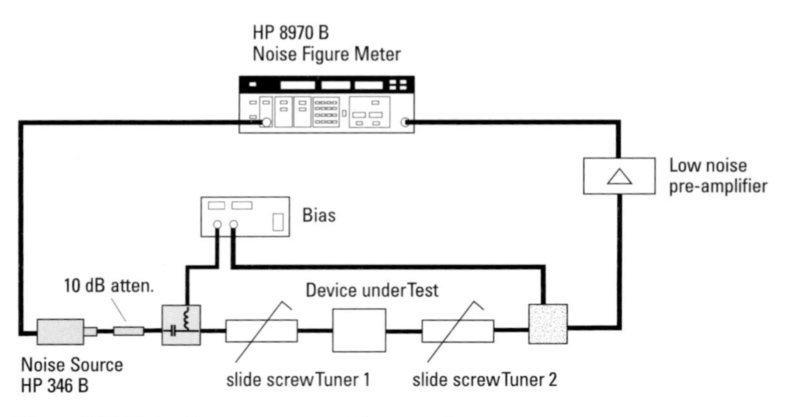

Figure 2.19 Noise figure measurement for a transistor

2.4.3 Measuring arrangement for determining the noise figure for a mixer

The frequency offset of the ENR values for the calibration and the indication of the measured value are effected by the noise figure setting SP 1.3 on the meter, the input frequency f_{IF} is selected using SP 3.0. One combination of f_{LO} and f_{IF} is the desired measurement frequency – the other is an interfering image frequency at a distance of $2f_{IF}$. This must be suppressed using a filter, for example an adjustable bandpass filter.

As the noise effect of the local oscillator cannot be eliminated by calibration, it is very important that the chosen oscillator has a very clean spectrum, with low phase and AM noise. Examples of such oscillators are tube or quartz oscillators (see Figure 2.20).

2.4.4 Measuring the IP3 value (3rd order intercept point)

Although the rule of thumb that $IP3_{out} = P_{-1dB} + 10$ dB suggests the use of a compression point P_{-1dB} as a direct replacement, it is highly recommended that a characterization of the intermodulation (IM) is performed directly (Figure 2.21). In practice, the operating conditions are scarcely comparable, because $IP3_{out}$ is a class A parameter while P_{-1dB} is a parameter which is affected by limiting effects at high power.

The 3rd order distortions of the device under test generate IM products at $2f_1 - f_2$ and $2f_2 - f_1$. The intercept point at the output, $IP3_{out}$, is defined as the point where the extrapolation of the 3rd order IM product (3:1) intersects with extrapolated fundamental mode (1:1), as a function of the input power.

The intercept point measured at the output can also be defined for the input.

$$IP3_{out} = P_{out} + dIM/2$$

3rd order intercept point at the output [dBm]

$$IP3_{in} = P_{in} + dIM/2$$

3rd order intercept point at the input [dBm]

$$IP3_{out} - IP3_{in} = \text{Amplification} = P_{out} - P_{in}$$

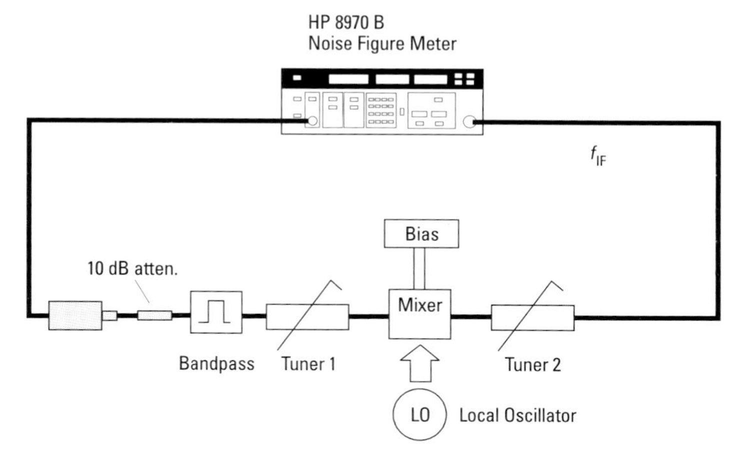

Figure 2.20 Measuring arrangement for determining the noise figure for a mixer

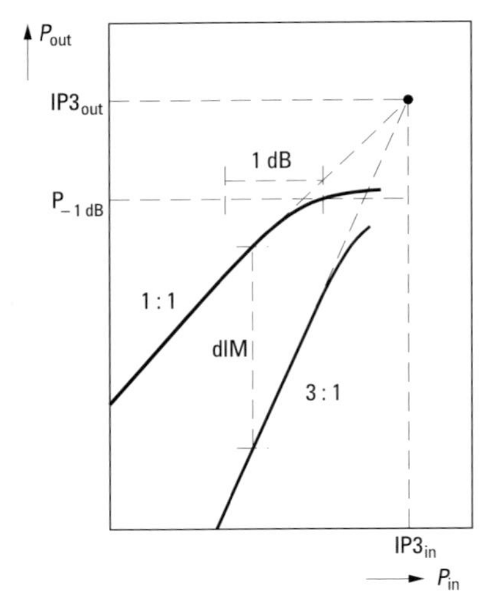

Figure 2.21
Definition of the intercept point

P_{out} and P_{in} are signal levels for <u>one</u> frequency. The dIM value is the difference between the fundamental mode and the IM product in dB.

In order to give dIM values between 50 and 60 dB, the measurement is carried out with a low signal level. The attenuation between the device under test and the power meter must be precisely taken into account. Figure 2.22 shows the measuring arrangement.

Although the typical use of the system described above is to characterize transistors or amplifiers, it may also be appropriate to use it for mixers. In this case, the spectrum analyzer must be tuned to $f_1 - f_{LO}$, but the level definitions are not affected. The power meter shows the value for the two carrier waves. In order to calculate IP3, the power of one carrier only is required (–3 dB).

2.5 Bipolar RF transistors

Since the development of RF transistors began, small signal bipolar transistors have been mounted with the rear of the chip on the collector connection. Using the SIEGET® technology (Siemens Grounded Emitter Transistor), the 4th generation of RF transistors achieve frequencies up to a limit of 25 GHz. This

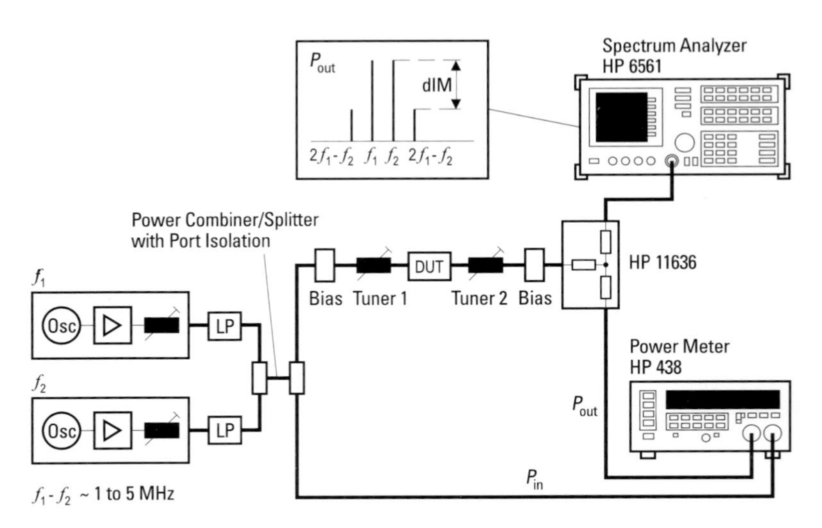

Figure 2.22 Measuring arrangement for determining the intercept point

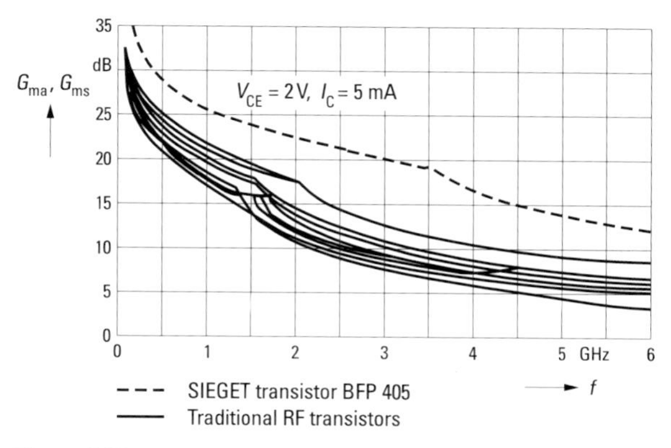

Figure 2.23
SIEGET transistors achieve a significantly higher amplification than previous RF transistors

makes them particularly suitable for use in mobile communications (Figure 2.23). Today, Infineon AG can already supply transistors of the 5th (B6Hfe) and 6th generations (silicon-germanium), with transition frequencies of up to 70 GHz.

The manufacture of RF transistors started at Siemens in 1964. The immediate impulse for the change over, from the germanium which was then dominant to silicon, was the highly demanding requirements in terms of thermal stability and linearity, which are necessary for broadband transmission links. This 1st generation had a limiting frequency (f_T) of 2 GHz. As early as 1975, the arrival of ion implantation produced an opportunity to use arsenic instead of phosphorus as the dopant for the emitter, which enabled f_T to be raised to 5 GHz for the 2nd generation. Eventually, the use of the optical wafer stepper made possible the development of the 3rd generation. By using structures down as small as 0.8 µm, these transistors achieve a transition frequency of 8 GHz and permit the construction of amplifiers up to around 2 GHz. In order to meet the market requirements for better reproducibility of the devices, and massive cost pressures, the production process for these older generations was redeveloped on the basis of the 4th generation, and changed over to a modern IC line. The wide spectrum of types of 6 generations, which cover circuits from 100 MHz up to 12 GHz, supports a host of applications, not only in mobile communications but also in the consumer field and the automotive sector.

2.5.1 SIEGET: standing things on their heads

Until now, bipolar npn RF transistors have been built up on a n-doped substrate (Figure 2.24). As a result, the rear side of the chip is unavoidably the collector connection, and as a consequence the transistor chip is always mounted on the collector pins of the package. The base and emitter are located on the front side, and have a finger-shaped structure: they must be bonded to the adjacent package contacts using fine gold bondwires. These bondwires have an inductance which, at high frequencies, cannot be neglected. The emitter bond in particular represents a negative feedback in the ground path, which can cause a loss of gain of up to 10 dB. With its SIEGET principle, Siemens solved this problem by a revolutionary approach.

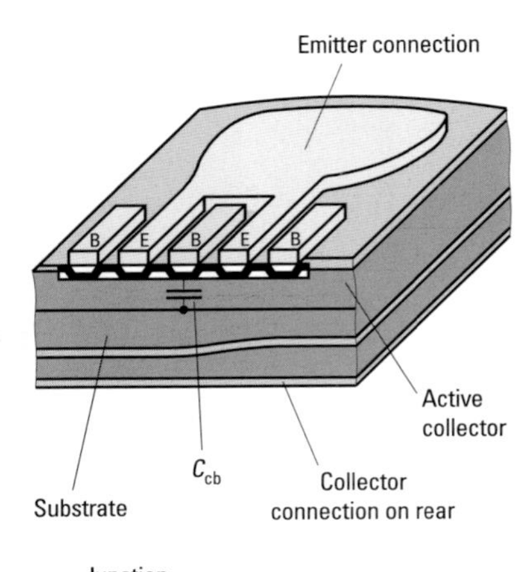

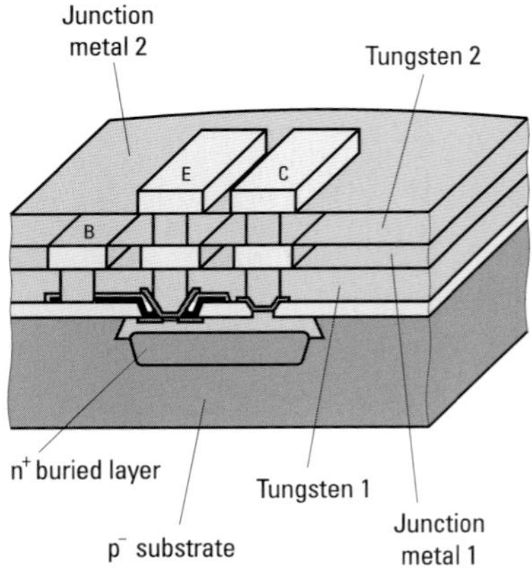

Figure 2.24
The parasitic base-collector capacitance (top) of conventional RF bipolar transistors is reduced by the SIEGET technology (bottom)

SIEGET transistors are manufactured in B6HF technology, a process line for integrated RF ICs, with a transition frequency of 25 GHz. The actual transistor is realized on a lightly doped, p-conductive substrate. The active transistor thus remains insulated from the substrate, and hence it is possible to mount the transistor chip on a ground connection of the package. In contrast to earlier generations however, the collector must now be led out towards the top and bonded there. In exchange, the emitter bonding can be effected downward by the shortest path, using several bondwires. In this way, the emitter inductance is reduced from about 1 nH with the old mounting technology to about 0.25 nH. The emitter is thus led directly against the RF ground. The traditional arrangement may be likened to having a person stand on his or her head, whereas now things are put "right side up". The production process is based on doped polysilicon layers, a self-aligning process using spacer techniques and a radical shrinkage of all the lateral and vertical dimensions. As a result, the base thickness is a mere 80 nm and the emitter width 400 nm. The collector-emitter blocking voltage has been reduced to 4.5 V, which permits higher current densities and this leads to a relative reduction in the internal reverse transfer capacitance, which is one of the factors that substantially determine RF gain. For example, with BFP405 the remaining capacitance is only about 50 fF. Another factor making a significant contribution to reducing the capacitance is the fact that the new transistor technology no longer uses a base well, but instead provides the base connection by means of a p-doped polylayer on the oxide base, immediately beside the emitter. The emitter and base contacts are separated only by a fine spacer.

The SIEGET concept also solves a problem which is as old as the hills, which has troubled circuit designers from the very beginning: for bipolar transistors, the heat generated by the transistor always arises at the collector. In this situation, the user is confronted by the awkward task of attempting to handle not only the voltage supply but also the access to the RF signal and the heat dissipation, all independently of each other. With a SIEGET transistor this difficulty quite clearly disappears of its own accord.

In order to make full use of the exceptionally high performance of the new transistor chip (Figure 2.25), it was necessary to develop a special package compatible with RF. From preference, the SIEGET

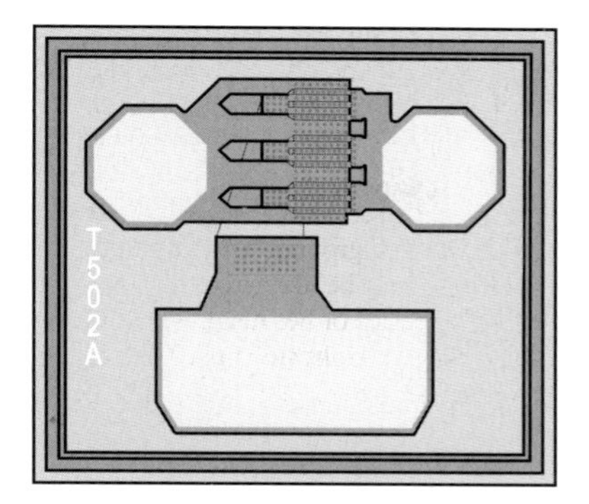

Figure 2.25 Chip layout of the BFP 420

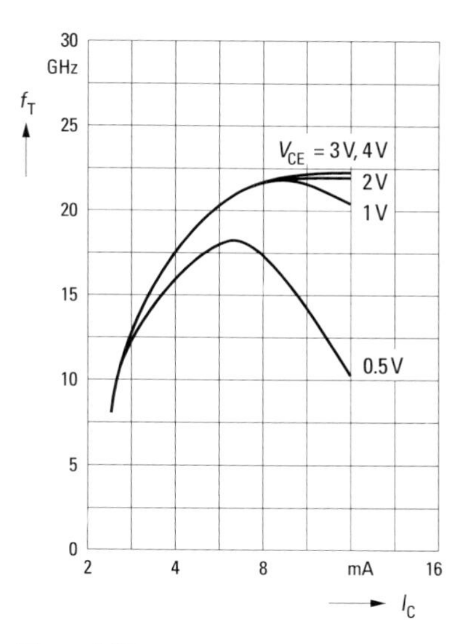

Figure 2.27
Transition frequencies for SIEGET transistors

line is offered in a specially modified super-mini package, SOT 343. The 2 mm × 1.3 mm sized package body allows it to be used even in extremely small mobile telephones. Most recently, these transistors are also available in the thin-small-flat package, TSFP-4, with even smaller dimensions of 1.4 mm × 0.8mm. The installed height now amounts to < 0.59 mm, so that it can also be used for module applications.

In order to reduce the emitter inductance, the SIEGET chips are mounted on a wide continuous copper bar in the SOT 343 super-mini SMD package; for comparison Figure 2.26 shows, on the left, the internal arrangement in the package for conventional RF transistors with the SOT 143 package.

The technical data for the transistors gives the RF developer significantly more latitude. For a typical graph of f_T against current, (Figure 2.27), down to a VCE of 1 Volt the transition frequency drops off hardly at all by comparison with higher voltages. Even at V_{ce} = 0.5 V, due to a still high f_T of 17 GHz there is an ample scope for the circuit designer. This means that at V_{ce} = 0.7 V and I_c = 1 mA, the BFP 405 SIEGET transistor provides significantly greater amplification at f > 2 GHz than a transistor of the 3rd generation at V_{ce} = 7 V and I_c = 10 mA. While conventional RF transistors hardly continue to function at supply voltages below 2 V, SIEGET transistors exhibit their optimal performance at 2 V.

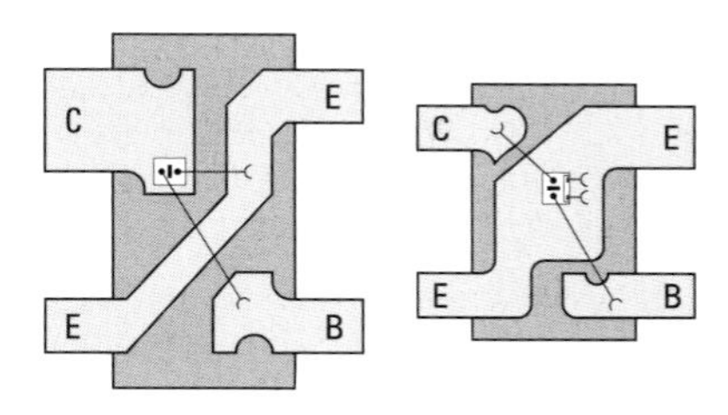

Figure 2.26 Arrangement of the SOT 343 package (right) and the SOT 143

As a result, new possibilities open up even for pager applications with V_{ce} < 1 V. Normally, a problem with amplifiers which are optimized for minimal RF noise figures is the loss of amplification associated with mismatching. SIEGET transistors provide very low noise figures at the same time as extremely high amplification (Figure 2.28). As a result of the optimal impedance characteristics of bi-

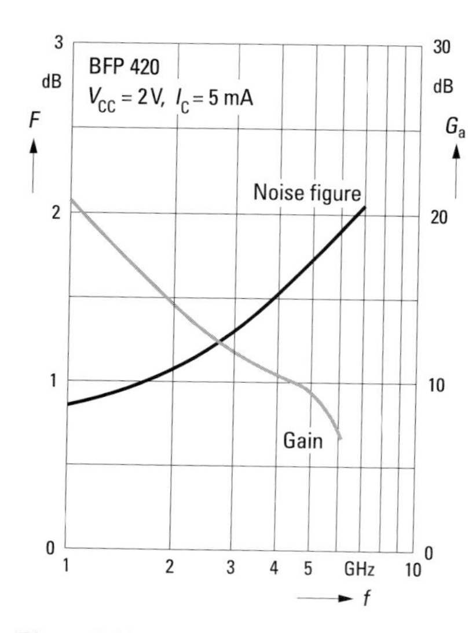

Figure 2.28
Noise figures and gain of SIEGET transistors

polar transistors it is possible, for example, to operate the BFP 420 for many applications on a broadband basis, completely without matching. Their easy matching to 50 Ω is one of their advantages compared to GaAs FETs. This gives the developer yet more latitude, so that if necessary it is even possible to optimize an amplifier for perfect matching and/or stability at low amplification. At frequencies of around 6 GHz, the data achieved is the same as that exhibited until now by the transistors at about a mere 2 GHz.

2.5.2 Applications

The three SIEGET types BFP 405, BFP 420 and BFP 450 cover a current range from below 10 mA up to 100 mA (Table 2.2).

Here, the great advantages compared to earlier RF transistors are in applications with lower supply voltages and low current consumption. But even for a collector-emitter voltage of about 2 to 3 V, the transistor operation is problem-free, with the remaining voltage dropping across the collector resistance. The data achieved under these circumstances is not matched by earlier transistors, even at far higher operating voltages. One application, which until now has been impossible using bipolar transistors in a plastic package, can now be constructed with the SIEGET BFP 405: a volume production DRO (dielectrically stabilized oscillator) for 12 GHz LNCs. It provides an output power of > +3 dBm at 10 GHz, and by comparison with previous oscillators using GaAs FETs it generates more than 5 dB less sideband noise. Figure 2.29 shows a trial construction of a dielectrically stabilized oscillator at 10 GHz.

2.5.3 SiGe transistors

With the SIEGET transistor it appeared briefly that once again the ultimate performance had been achieved, and no further improvement in the transistor technology would be possible.

Table 2.2 Data for the SIEGET transistors BFP 405, BFP 420 and BFP 450

BFP 405 ($I_{c,\,max}$ = 12 mA)	**BFP 420** ($I_{c,\,max}$ = 35 mA)	**BFP 450** ($I_{c,\,max}$ = 100 mA)
• Oscillators up to 12 GHz • Low noise amplifiers at low currents • High amplification at low currents • “Low current” type	• Oscillators up to 9 GHz • Low noise amplifiers • High amplification • Universal type	• Low noise stages with high modulation capability • Driver stages with high amplification, with P_{out} = +19 dBm at V_{ce} = 3 V and f = 1.9 GHz • “Medium power” type, driver

Figure 2.29
Laboratory version of a dielectrically stabilized 10 GHz oscillator using the BFP 405

It was indeed possible again to derive a 5th generation (BFP520 / BFP540) from the 4th by the consistent application of the known theoretical optimization strategies – mainly by shrinking all the vertical and lateral structures. But with the use of present production methods – disregarding extremely expensive possibilities such as those of photolithography – a purely silicon technology has reached its limits. The fact that it has been possible to jump even this hurdle is all thanks to an ingenious idea from the first days of semiconductor history. In 1954 Herbert Kroemer published a suggestion for modifying the bandgap in the base of the transistor in such a way that the charge carriers emerging from the emitter "see" an internal built-in field. This field accelerates the electrons so that the propagation time through the base is greatly reduced, which corresponds to a raising of the transition frequency. For his idea, Kroemer was honored with the Nobel prize, in the year 2000 (!).

The modification of the bandgap is best achieved by the selective incorporation of germanium atoms in the Si lattice, so that from the emitter towards the collector there is a steadily increasing concentration. Because Ge atoms are larger than Si atoms, massive tensions arise in the crystal, which tend to form cracks. Hence it was many years before the first laboratory trials resulted in a functioning production process. For a complete and generally comprehensible presentation, refer to the journal "Elektronik" 18/2002 (Lohninger: "Diskret, aber nicht trivial – Teil3" [Discrete but not trivial – Part3]). Since 2000, Infineon has offered the BFP620 with a transition frequency of 70 GHz. It has been possible to reduce the noise figure to 0.7 dB at 2 GHz. For such a high limiting frequency there is, unfortunately, a price to be paid, in the form of a lower breakdown voltage. Because the limiting frequency is not determined solely by the propagation time in the base, but also in the collector. As the speed of the electrons is fixed, the only remaining possibility is to make not only the base but also the collector thinner. But a thinner collector also has a lower breakdown voltage. Consequently, the BFP620 can only be operated up to a maximum of about 2.5 V. As this often implies increased circuit costs, and for circuits up to 3 GHz the extremely high limiting frequency is not required, the process has been modified towards a higher operating voltage. This has produced the two latest types of the 6th generation, the BFP640 and the BFP650. These have a 4.5 V blocking voltage and correspondingly the limiting frequency is reduced to 40 GHz. This does not alter the noise figure.

2.6 Silicon MMICs simplify RF development

Radio frequency (RF) circuits with complex requirements in respect of operating point control, thermal response, ESD protection or switching functions, fail when discrete components are used if too many active components must be used where some of them are spatially closely coupled together or insufficient area is available on a circuit board or in a module. These requirements can be significantly better satisfied using Si-MMICs (silicon monolithic microwave ICs), because with these it is possible to integrate both active components (e.g. bipolar transistors, diodes) and also passive components (e.g. resistors, capacitors, inductors) on very small areas on a chip. The use of Si-MMICs thus enables the development times, problems and risks of discrete RF applications to be significantly reduced. Examples of possible applications are low-noise amplifiers (LNAs), variable gain amplifiers (VGAs), voltage controlled oscillators (VCOs) or mixers.

In general, a more or less complex DC bias circuit regulates the operating point of the RF stages so that the desired characteristics, for example constant amplification or constant current consumption, are achieved regardless of the temperature, supply voltage, process variations or RF input power. At the same time, the RF stages are optimized for the special requirements. With LNAs, for example, these are a minimum noise figure combined with power matching, i.e. matching to 50 Ω at the RF input and output.

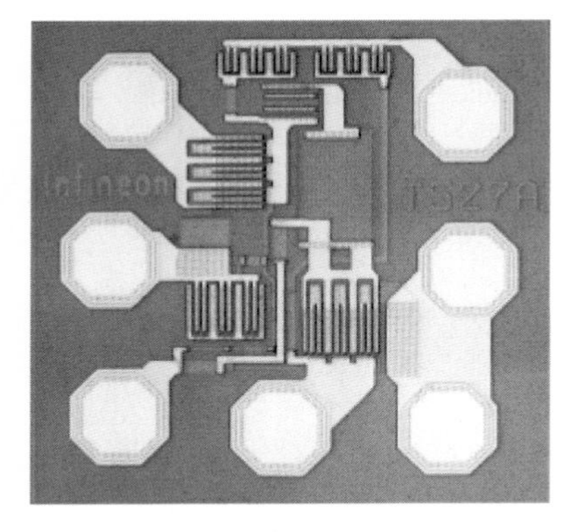

Figure 2.30
The T527 chip measures 330 μm × 360 μm

Various MMIC amplifiers are available, based on the diverse bipolar technologies developed by Infineon, such as SIEGET®-25, SIEGET®-45 or the 70 GHz silicon germanium (SiGe) technology, with unconditional stability in the frequency range from 0 to 10 GHz and at an operating point matched to 50 Ω. With excellent noise and gain parameters even at voltages below 3 V, they are above all suitable for use in mobile communication. The Si-MMICs exhibit RF characteristics (noise figure, amplification, etc.) similar to those of discrete transistors from the same line of technology, i.e. they can benefit from the technological advances in the discrete transistors. With the advances in technology, for example, the collector-base capacitance can be significantly reduced by making the base area extremely small. The accompanying reduction in the feedback from collector to base results in the high gain values up into frequency ranges of several Gigahertz. At the same time, optimized process steps make it possible to achieve lower base resistances, and hence significantly lower noise figures.

Figure 2.30 shows the chip layout for a two-stage MMIC. The active transistor structures can be identified by finding the emitter fingers. The p-doped substrate means that the active structures in the substrate are electrically insulated, and the ground potential can be applied directly to the rear of the chip. As a result, the bondwire to ground can be made extremely short, because it can be bonded downward to the leadframe directly alongside the chip. When combined with a suitable leadframe, this results in a smaller inductance to ground, and thus to a higher gain at high frequencies. Metallization with gold increases the reliability of the MMIC. The packages used for MMICs are SMD packages from the SOT range (small outline transistor) and the significantly flatter and smaller TSLP range (thin size leadless package) (Figure 2.31).

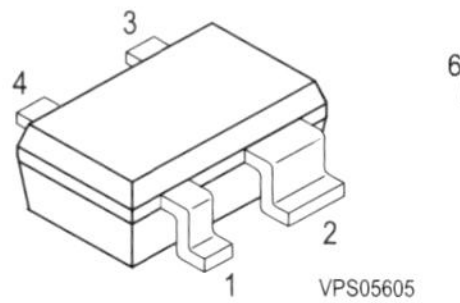

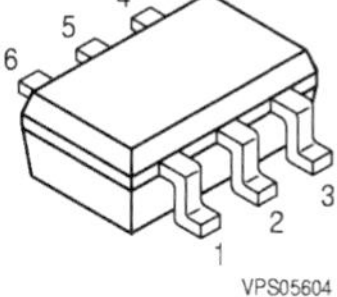

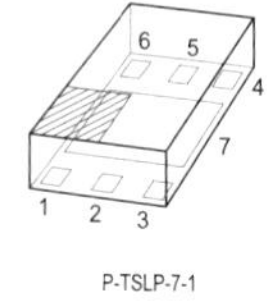

Figure 2.31 SOT and TSLP packages for MMICs (4-7 pins)

Variants

The BGA420 single-stage amplifier (Figure 2.32a) consists of active structures designed for optimal matching and a resistive bias network. Here, the RC feedback from the collector to the base guarantees input matching over the entire bandwidth from 0 to 3 GHz. The collector resistance reduces the production-dependent distribution of the current gain (voltage-current feedback). On the other hand, it ensures broadband output matching of the MMIC to 50 Ω.

With the two-stage design of the BGA427 (Figure 2.32b), the input stage is identical with that of the single-stage design, BGA420. The collector resistance of the first stage simultaneously sets the operating point of the second stage, which is in the form of an emitter follower. For various reasons, including to make the MMIC more flexible to use, there is no collector resistance for the second stage. One function of the emitter resistance of the second stage is to provide negative feedback and stabilize the operating point, independently of the spread of the parameters for the active part of the second stage. Another is to ensure the broadband output matching of "Out A" to 50 Ω. By comparison with the BGA420 single-stage device, the two-stage BGA427 amplifier clearly exhibits higher gain values (Table 2.3).

Unlike the BGA420 and the BGA427, the BGA428 is a relatively narrowband amplifier. Thanks to the SIEGET®-45 technology, it is able to achieve higher gain and a significantly lower noise figure at high frequencies than is the case for BGA427. The BGA428 is used as an LNA in various mobile communication systems, such as DCS 1800 and PCS 1900. It is a two-stage LNA in a cascaded

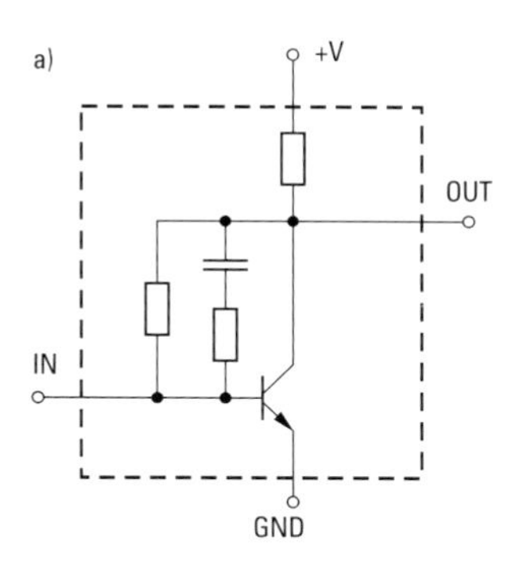

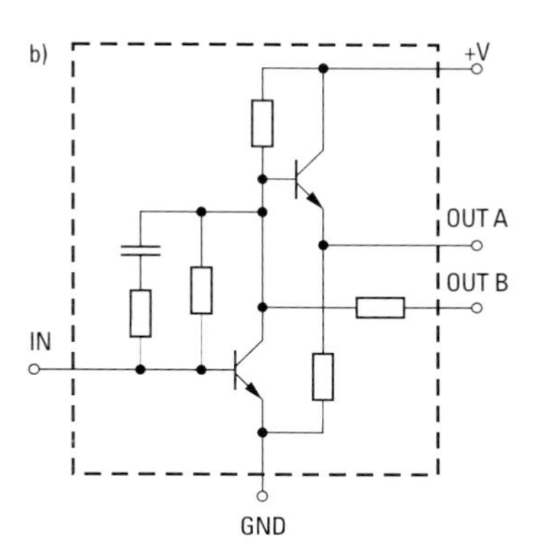

Figure 2.32 MMICs as single-stage (a, BGA 420) and two-stage amplifiers (b, BGA 427)

Table 2.3 Characteristic data for selected MMICs in various technologies at a glance

		BGA420	BGA427	BGA428	BGA622
No. of amplifier stages		Single-stage	Two-stage	Two-stage	Two-stage
Technology		SIEGET®-25, 25 GHz f_T	SIEGET®-25, 25 GHz f_T	SIEGET®-45, 45 GHz f_T	SiGe, 70 GHz f_T
Package		SOT343	SOT343	SOT363	SOT343
$\|S21\|^2$ in dB at	0.1 GHz	19	27	-	-
	1.0 GHz	17	22	-	-
	1.8 GHz	13	18.5	20	14
	2.1 GHz	-	-	18	13.5
	2.4 GHz	-	-	17	13
Noise figure in dB at	0.1 GHz	1.9	1.9	-	-
	1.0 GHz	2.0	2.0	-	-
	1.8 GHz	2.2	2.2	1.4	1.1
	2.1 GHz	-	-	1.4	1.1
	2.4 GHz	-	-	1.5	1.15
IP3 at output in dBm at	0.1 GHz	11.0	8.5	-	-
	1.0 GHz	10.0	8.0	-	-
	1.8 GHz	9.5	7.0	11	-
	2.1 GHz	-	-	12	16
	2.4 GHz	-	-	13	-
Supply voltage, in Volts		3.0	3.0	2.7	2.75
Current in mA, typical		6.4	9.5	8.2	5.8

emitter circuit. The first transistor stage achieves its noise matching as a result of the sophisticated matching to the second stage which follows, with simultaneous power matching at the RF input. Due to this, the input matching is better than 10 dB for a noise figure of a mere 1.4 dB at 1.8 GHz. In addition, the BGA428 offers a so-called gain-step mode, in which the LNA acts as an attenuator with an insertion loss of about 14 dB. The insertion loss is lower than when the amplifier is completely switched off, because selective couplings via the base-collector capacitances of the transistor are exploited. This mode is activated in a mobile phone when the RF input signal level is very high, to avoid overdriving the following stages.

The BGA622 single-stage LNA is manufactured in the B7HF 70 GHz SiGe technology. Due to its optimized design it has a noise figure which at a mere 1.1 dB is exceptionally low compared to other silicon LNAs, combined at the same time with power matching. Because the single-stage LNA is already well-matched at both the input and the output, due to an on-chip inductor, it is easy to use and is deployed in many modern mobile communication systems (GSM, GPS, DCS, PCS, UMTS etc.). In addition, it can be switched off by a DC voltage applied at the output.

2.6.1 Three application circuits

Depending on the MMIC and requirements, various applications are possible; examples of three of these are presented below:

Application 1: an LNA using the BGA420 or BGA427 with minimal external circuitry

An example of a circuit diagram for an LNA, which can be realized using the BGA420 or BGA427, is shown in Figure 2.33 in this case using the two-stage BGA427 device in the SOT343 package. This particularly simple design requires only three capacitors for the external circuitry. For optimum functioning, it is important to have the best possible RF short-circuit parallel to the supply voltage, so that the output signal can be tapped off fully at the "Out" or "Out-A" pin. For this reason, the capacitor C3 should be located as close as possible to the "+V" pin. The capacitors C1 and C2 serve to block the DC voltage applied to pins 1 and 4.

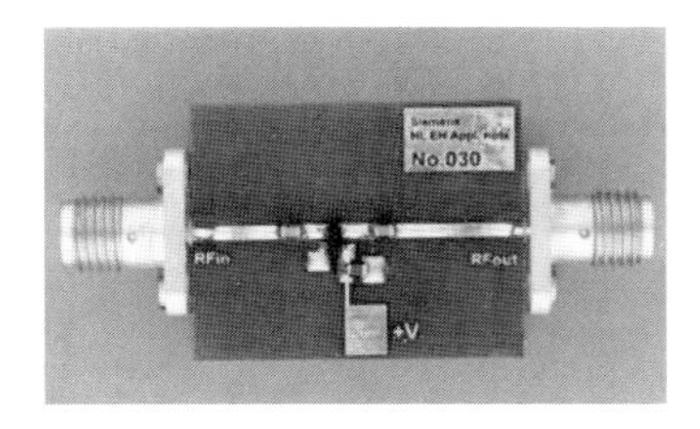

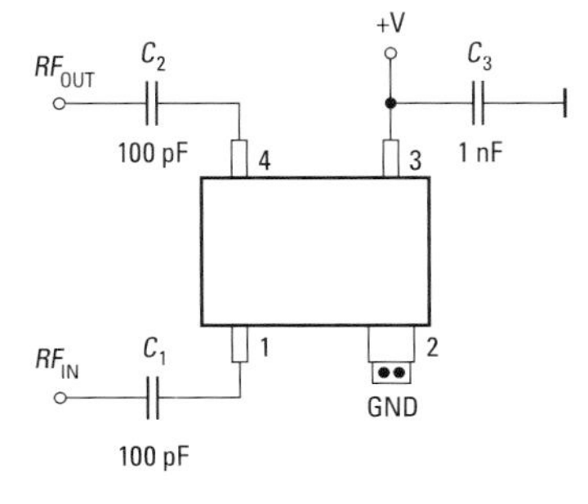

Figure 2.33
An LNA using the BGA420 or BGA427 with minimal external circuitry

Application 2: an LNA using the BGA428

The BGA428 requires five external passive components to ensure it performs optimally (Figure 2.34). The capacitor C1 at the input only blocks the base voltage from the first amplifier stage, for RF matching at 1.8 GHz it is not necessary. The inductor L1 and the capacitor C2 act as matching elements at the output. In combination with the parasitic components of the SOT363 package, they transform the external 50 Ω load impedance in the reference plane of the collector to a higher impedance with a real part of about 100 Ω, which enables the compression point of the second stage to be raised. The capacitors C3 and C4 serve to block off the RF signal from the DC voltage source, and ensure that the RF signal sees a defined impedance. For this reason, they should be located close to the BGA428 and beside L1. Finally, the resistance R1 at the gain-step pin defines the current and hence the insertion loss in gain-step mode.

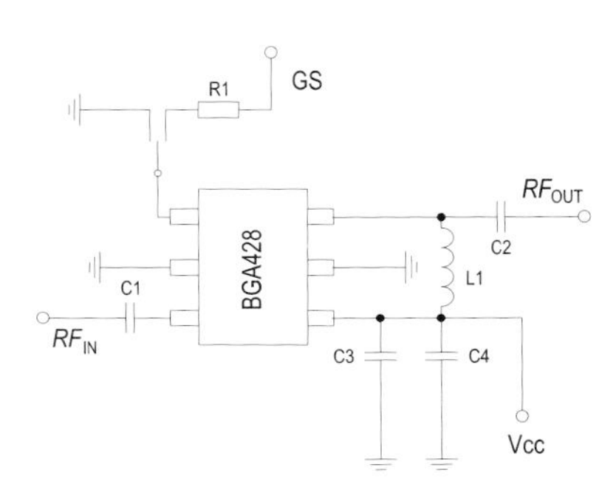

Figure 2.34 An LNA using the BGA428

Application 3: an LNA using the BGA622 in a circuit for improved IP3

Due to its optimized circuit design, in a minimal application at 2.1 GHz the BGA622 requires only two external elements (C1 and C4, Figure 2.35). The inductance L1 is necessary if it is required to improve the input matching below 2.1 GHz (e.g. for GPS, 1575 MHz). For numerous applications, an additional requirement is for a high third order intermodulation point (IP3, 3rd order intercept point), because generally there are several RF input signals applied to the LNA at the same time. The problem here is the low frequency mixed products which, due to the non-linear characteristic curve of the transistor, are produced from the frequency differences in the input signals. These cause an unwanted amplitude modulation of the base, which is superimposed on the RF signals. At the LNA output it is no longer possible to distinguish the amplitude modulation from the third order distortions, the interference effect being the same. Hence, the objective is to reduce these mixed products without affecting the DC operating point and the RF matching. This is achieved by an LC element to ground from the input to the LNA (L2 and C2). To the mix products in the region of several MHz, the capacitor C2 represents a very low impedance, and the inductance L2 effects the decoupling of the RF signals. This increases the linearity of the application, and improves IP3 by several dB, while the remaining RF characteristics are scarcely affected.

2.6.2 Mobile phones are not the only use

Apart from the major field of mobile communication, other MMICs are available for other application areas, such as for example as an LNA for CATV, as IF amplifiers in LNBs for satellite receivers, or as oscillators in a frequency range of several hundred Megahertz. The BGA420 module is outstandingly usable, for example with a supply voltage of 2 to 5 V, as a low-cost fully-matched buffer amplifier, with very good insulation characteristics. Due to the flexible circuit concept of the BGA427, the modules can also be used for IF amplifiers or as transimpedance amplifiers in transceivers for optical communications. With the BGA416 cascode, it is very easy to design oscillators with integral output buffer amplifiers. Because of the high backward isolation of the output stage, the oscillator circuit is then optimally separated from the load impedance, which significantly improves load pulling.

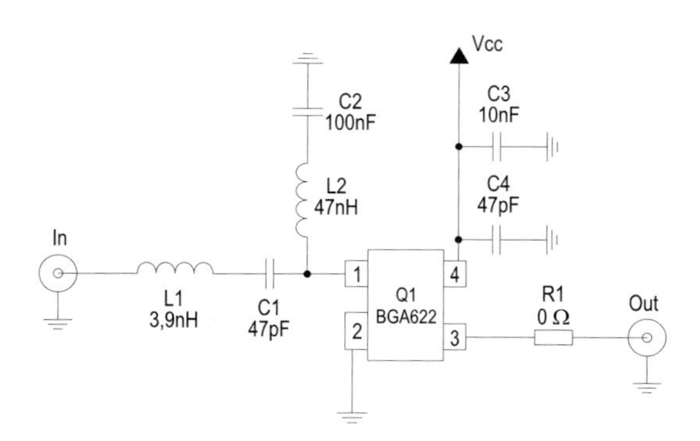

Figure 2.35 An LNA using the BGA622 in an application, with improved IP3

2.7 Stabilizing current with the BCR 400 operating point stabilizer

Devices and systems for mobile uses require a stable operating point over a wide temperature range. At the same time they work at low operating voltages with low current consumption. For these applications, Infineon Technologies offers the BCR 400W active operating point stabilizer.

Mobile telephones, vehicle electronic systems, and portables in consumer electronics, are classical examples of devices with mobile use. They have in common a requirement for low supply voltages, low current consumption and constant operating points, even under large temperature fluctuations. In addition, however, all this must be accommodated on the smallest possible area. "Classical circuits" for stabilizing the operating current (parallel or series connected) sometimes exhibit fluctuations in the stabilized operating current which are no longer tolerable when there are voltage fluctuations, and also due to the current amplification range of the transistor used. Optimizations of these circuits require a larger number of components, and as a result therefore also more space in the module. Moreover, additional power is again required in this case. Hence, as a result of the experience gained from the development of transistors with integral resistors ("digital transistors"), the BCR 400 was developed as an active operating point stabilizer in the form of a single chip solution. The main examples of the uses for this circuit are in the RF stages of mobile telephones (Figure 2.36).

2.7.1 Method of working

The way that the BCR 400 works can be illustrated by the example of collector current stabilization (Figure 2.36a) together with the internal circuitry (Figure 2.37): the collector current I_C of the npn RF transistor which is to be regulated, together with $I_{E(pnp)}$ across the resistance $R_{Tot} = R_{Int} \parallel R_{Ext}$, produces a voltage drop $U(I)$. This determines the emitter potential of the pnp transistor in the BCR 400W, the collector of which supplies the base of the transistor which is to be regulated. A reference voltage U_{Ref} is obtained across the two diodes, and this defines the base potential of the pnp transistor. When current flows, $U_{EB(pnp)}$ amounts to about 0.65 V, U_{Ref} around 1.3

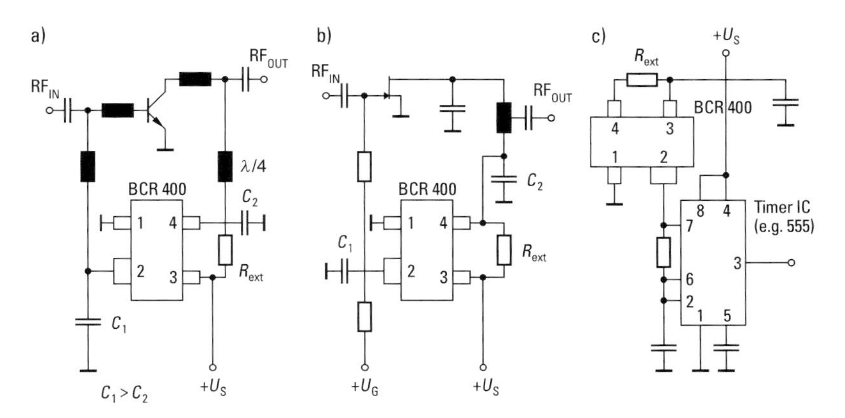

Figure 2.36 Circuits using the BCR 400 operating point stabilizer

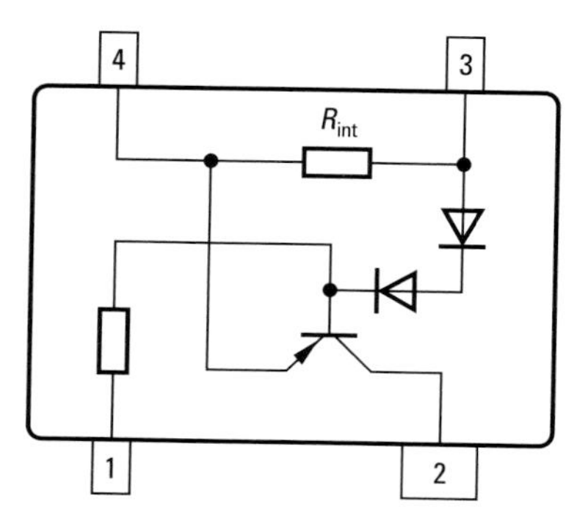

Figure 2.37
Internal circuitry of the BCR 400

V. As a result, within the operating region $U(I)$ is also determined as about 0.65 V. $I_{E(pnp)}$ is about the same as the base current, $I_{B(npn)}$, of the transistor to be regulated, which can be neglected by comparison with $I_{C(pnp)}$ for $B_{npn} > 40$. Hence, to an first approximation $I_{C(npn)} = 0.65\ V/R_{GES}$ and $U_{CE(npn)} = U - 0.65$ V.

2.7.2 Control response

Fluctuations in the operating voltage

When the operating voltage U increases, the current through the reference diodes rises, but U_{Ref} remains approximately the same because of the exponential characteristic curve of the diodes. Hence, $U(I)$ also scarcely changes, I_c remains approximately constant, but the U_{ce} values for both transistors increase directly with U.

Distributions of the component characteristics for the transistor to be regulated

The distributions of the component characteristics for the transistor which is to be regulated mainly affect its current amplification B. However, this has hardly any practical effect on the values of the mesh which determines the current. Smaller or larger values of B only change the magnitude of $I_{E(pnp)}$, which is in any case small

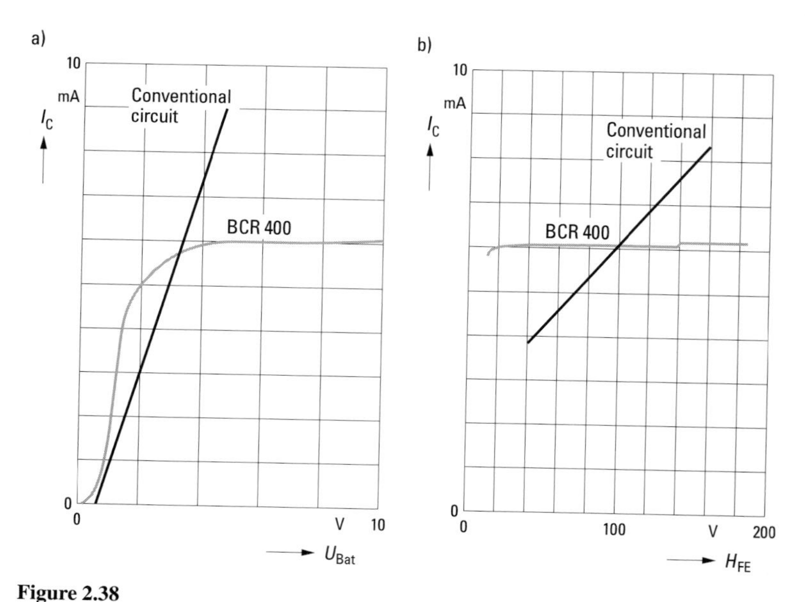

Figure 2.38
Correction of (a) operating voltage fluctuations and (b) the current amplification of the transistor which is to be stabilized, using the BCR 400 or a conventional circuit

Table 2.4 Limiting data and characteristic values for the BCR 400

Limiting values		
Operating voltage (final version)	U_S	15 V
Control current	I_{contr}	10 mA
Control voltage	U_{contr}	8 V
Characteristic values		
Additional current consumption at $U_S = 3$ V	I_0	max. 40 mA
Minimum stabilized current at $U_S = 3$ V	I_{min}	0.1 mA
Change in the collector current for – temperature fluctuations – fluctuations in operating voltage ($U_S > 3$ V)	$\Delta I_c/I_c$	 0.2%/K 0.15 $\Delta U_S/U_S$

relative to $I_{C(pnp)}$. Even in the extreme case of $B_{npn} = 10$, for example, this would only reduce the value of I_C by about 10%.

Temperature fluctuations

On the one hand, temperature fluctuations cause massive changes in the value of B; however – as described above – this is largely without consequence. On the other hand, the forward voltages of the pn junctions change by a few millivolts, with the effect of one of the diodes being compensated by that of $U_{EB(pnp)}$ and only the temperature coefficient of one diode remains. The value of this is around −2 mV/K and consequently it leads to an error in *U(I)* (that is, also an error in I_C) of at most ±15% for a temperature variation of ±50 K. The fluctuation in $U_{EB(npn)}$ has no effect, because it does not influence the mesh which determines the current.

This regulatory characteristic of the BCR 400 is shown in Figure 2.38. By comparison with conventional circuits, the BCR 400 is significantly better at correcting fluctuations in either the operating voltage or even the current amplification of the transistor which is to be regulated.

Characteristic data

The limiting values and characteristic values of the BCR 400W are listed in Table 2.4. Here we should highlight the low voltage drop of a mere 0.7 V over an application range from < 200 mA to > 200 mA.

The BCR 400 can be supplied in two versions: BCR 400R in the SOT143 package (dimensions as for the SOT 23) and as the BCR 400W in the SOT 343 mini-SMD package (dimensions as for the SOT 323). The two versions are pin-compatible.

3 Power semiconductors

In order to define a power semiconductor and its uses, we will start by considering control units in devices, so-called ECUs (electronic control units). In an ECU, an input variable from a sensor is converted into an electronic signal. This signal is then further manipulated by signal processing facilities to produce suitable control signals. Finally, the signal processing unit controls an adjusting unit, the actuator, to achieve the desired result.

An example will clarify this: the heating system in a motor vehicle registers the temperature inside the passenger compartment by means of a temperature sensor. This sensor – usually a PTC – supplies a signal voltage which is compared against threshold values (over- and undertemperature). As an output variable, the signal processing facilities supply a switching signal which, for example, if the temperature is too high closes the switch to operate the air-conditioning compressor.

In short: measurements, calculations, settings are made; or in other words: There is always a need of sensors, microcontrollers and power semiconductors.

3.1 Classification

Figure 3.1 shows a generally-applicable block diagram of an ECU.

If an ECU is subdivided into electronic function blocks, it can be split into a power supply unit, a communication interface, the microcontroller with its periph-

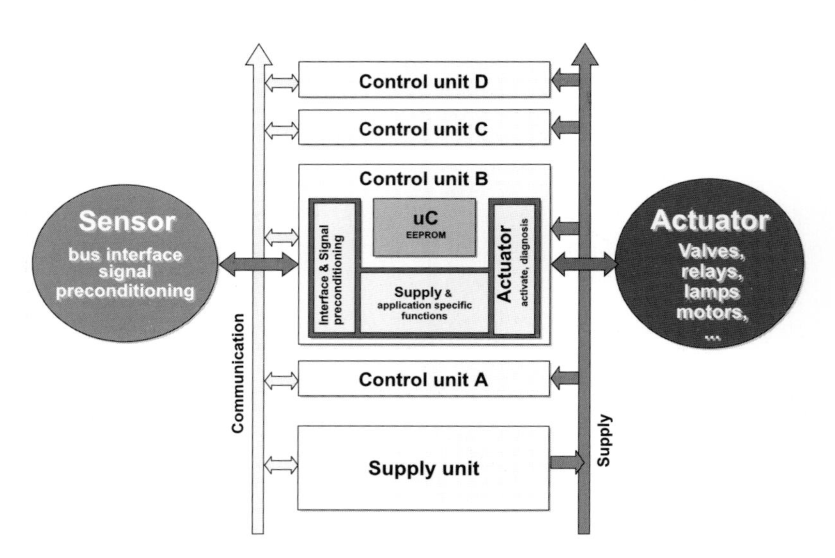

Figure 3.1 Block diagram of an electronic control unit (ECU)

erals (memory etc.), any further signal processing functions (AD converters) and the actuators. Power semiconductors are used mainly in the actuation systems and in the supply.

Such electronic controllers can be found in countless applications; apart from laptops, mobile 'phones and other consumer goods, also in automobiles and in many industrial applications. Ever more frequently, electronic solutions are replacing electromechanical solutions, such as relays. As they do so in the more complex devices, the reliability demands (protection against short-circuit and overload) on an individual component are ever increasing. Reparability in the event of a malfunction must be supported by diagnostic functions. For this reason, the control and regulation semiconductor chips in ECUs are often linked in distributed networks. These may take a very simple form or, in extreme cases, may act as highly-complex controllers. In this context, the term power semiconductor is greatly dependent on the application which is being considered, because the switching capacity can differ by several orders of magnitude, depending on the field of application.

In every case, however, one of the functions of these components is to set into operation the actuators, such as motors, lamps, heating resistors and other electromagnetic drives. Their other function is to provide the complete control device with energy (conversion, power supply). In this context, efficiency has recently become much more important as a parameter. With the increasing functionality, the energy consumption must continually be further reduced. Looking at it from the point of view of miniaturization, one finds that the energy density in W per m^3 in a modern microprocessor recently passed that of a fuel rod in a nuclear reactor. These trends are continuing apace, and present new challenges for power semiconductors.

To clarify the aspects specific to each field of application for power semiconductors, we will consider in more detail the three main fields of use, industrial, automobile and consumer electronics.

The most important parameters in this connection are shown in Figure 3.2 as rays out from an origin, with their weightings for each application. For each appli-

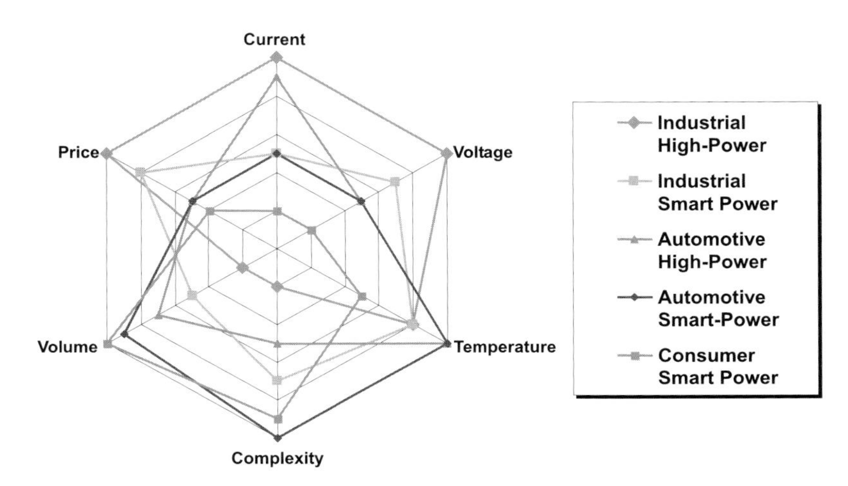

Figure 3.2 Application profiles for power semiconductors

cation field, this gives a loop that represents the application profile.

The relevant technical parameters are the current, voltage, temperature range and electromagnetic compatibility requirements (EMC). Market specific aspects are incorporated into the assessment in the form of the volume (number of pieces) and achievable price parameters.

In addition, the complexity of the products has been introduced as a parameter. This parameter allows the industrial and automobile electronics market to be split into two sub-groups: high-power and smart-power.

3.1.1 Categorization of power semiconductors by their parameters

From Figure 3.2 it is possible to draw the following conclusions:

- Current: Here, the industrial field is dominant, together with the newly-emerging field of automotive high-power (integrated starter-alternator and battery management systems; see also Chapter 9), with current values in the kiloampere range. The smart applications in automobiles and in industry can handle currents from 5 A to 100 A. For consumer applications in computers, mobile phones, TV or radio, the currents required are generally in single figures. An exception is the high-current power supplies for modern processors, with currents up to 100 A.
- Voltage: Here industrial applications are clearly out in front, with their higher (20 - 200 V) and highest values (kV). However, automobiles will in future require higher voltages for lighting (HID), ignition and for so-called piezo-actuators (injection systems). The same applies to applications which lie further in the future, in automobiles with electric drives or which use fuel cells. Today's standard automotive semiconductors require 8 V - 18 V in automobiles and up to 36 V in the field of commercial vehicles. Short term peak values of over 100 V can be reached as a result of malfunctions, e.g. battery disconnection (load-dump). In the consumer field, the voltage values lie mostly at lower values. An exception here, and a relatively large market segment, is represented by switched mode power supplies (SMPS) and so-called power-factor controllers (PFC), which are connected to the mains power network (220 V AC). Here, voltage values of a few 100 V are required.
- Temperature: A distinction is made between the ambient temperature in operation, T_a, and the junction temperature T_j. It is necessary to make this distinction in considering power components because as a general rule active power losses arises in a power semiconductor. This leads to a significant temperature differential across the heat transfer resistance between the "junction" and "ambient" (R_{thj-a}). The absolute values of the ambient temperature range T_a are highest in the automotive field, from -40°C to +135°C. In industrial electronics, the range is usually limited to -25°C to +85°C. And for consumer electronics a range of 0 to 85°C usually covers everything. Users and customers want the greatest possible temperature compatibility in power semiconductors (T_j values), because a reduction in the (j-a) temperature differential just mentioned can rarely be achieved without additional costs (heat sinks, etc.). For this reason, all Infineon's power semiconductors are specified for junction temperatures of up to 150°C with peaks to over 200°C.
- Complexity: Power semiconductors with intelligence (smart-power) for automotive, consumer and industrial applications are significantly ahead of high-power components in respect of this parameter. The rule is: the more complex, and powerful in the sense of switching-technology options, a wafer technology is, the more expensive it

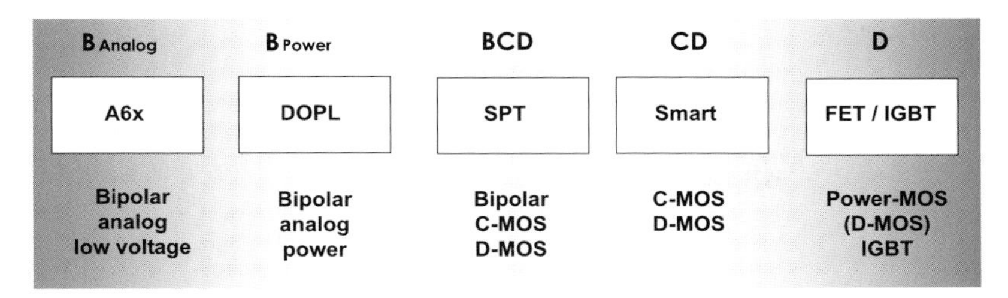

Figure 3.3 Wafer technologies for the automotive field of applications

is. With the chip-on-chip approach developed by Infineon, subfunctions of the product are implemented in whichever technology is optimal for them. This makes it possible to manufacture intelligent high-power semiconductors with optimized price/performance ratios.

- Volume and price (market): The highest volumes are to be found in the consumer field, followed by automotive and industrial smart-power devices. Rather lower volumes are found in the case of high-power applications. Nevertheless, the value, the so-called "silicon content" or the average prices of these applications, is very high. In the smart-power segments the price pressures are relatively high. The lowest prices achieved are in the consumer segment.

The different profiles can only be provided for in an optimal way by the use of specifically adapted front-end (wafer) and back-end (package) technologies. As an example, the research and development technologies required for the automotive field are shown in Figure 3.3. For this reason, Infineon is in the position of being able to offer optimized products for all the sub-segments. Infineon's presence is thus that of a so-called broadliner, a supplier of system solutions. The Automotive Power, Industrial Power, Microcontrollers and Advanced Sensors business units each ensure coverage of the system sub-segments.

3.2 Product development

Product development starts with the product idea. At this point in time, a target datasheet is agreed with one of more potential customers. After this, the feasibility of the product is tested by simulating it, and possibly also the complete system, i.e. the customer brings to the process his application know-how, and the specialists at Infineon design the appropriate product (map the application requirements onto the optimal front-end and back-end processes).

Following a successful "feasibility study", realization of the product begins with a plan of the commercial technicalities. After this plan has been released by the management, the product concept is developed. The front-end and back-end implementations are then defined and specified.

When all the relevant parameters have been fixed, a first wafer lot production takes place. If no redesign work is required, this completes the definition and realization phase. Following on from this, the evaluation and marketing phase starts.

Following completion of the wafer throughput and assembly in the back-end, the first silicon is measured, tried out in the application and, if the result is satisfactory, is released for sample production. Product freeze only occurs when all the necessary amendments have been implemented, the datasheet has been achieved in every respect, and robust production has been demonstrated. At this point in time, the Preliminary Datasheet is ready. Customers are supplied with samples for the so-called design-in.

In addition to this, some customer receive an evaluation package. This consists of the application board, application reports and notes, any software which may be required to operate the product, and details of the responsible specialist(s) who is/are the contact for all technical queries. This 'field application engineer' supports the customer throughout the complete duration of the project. If problems arise, he can consult further product specialists at headquarters. For example, together with the customer, complete electronic control units (ECUs) can be simulated thermally in this way. Or, if interference radiation problems arise in the application, the corresponding ECU-block can be optimized in collaboration with the EMC specialists from Infineon.

After the samples have been manufactured for qualification under the supervision of Quality Assurance, the qualification of the product begins. With its release for delivery, the ramp-up for series production can start. After a sufficiently large number of production lots, the final datasheet is released by the product marketing, and large-scale deliveries start up.

3.2.1 Differences during product development

There are major differences during product development, depending on whether the product to be developed is a standard product (commodity), a so-called application-specific standard product (ASSP) or an application specific integrated circuit (ASIC).

In the case of standard products, there is very little differentiation. The objective is that these products should be used by as many customers as possible, in as many applications as possible. As such products are easy to copy, a competitive situation develops rapidly, with relatively strong price pressures.

ASSPs on the other hand are customized for an application, and are in general used by several customers. ASSPs have clear differentiating attributes. They can only sensibly be used in particular applications for which they have been optimized. Examples of the ASSPs which have been realized today would be such items as the electronic accelerator pedal (Electronic Throttle Control, ETC) optimized motor bridge TLE7209, a special airbag supply IC such as TLE6711, or a chip set for a vehicle front door, consisting of the communication/supply IC TLE6263 and the multi-function driver IC TLE7201.

The trend towards the ASSP is exemplified by a door IC of the future, shown in

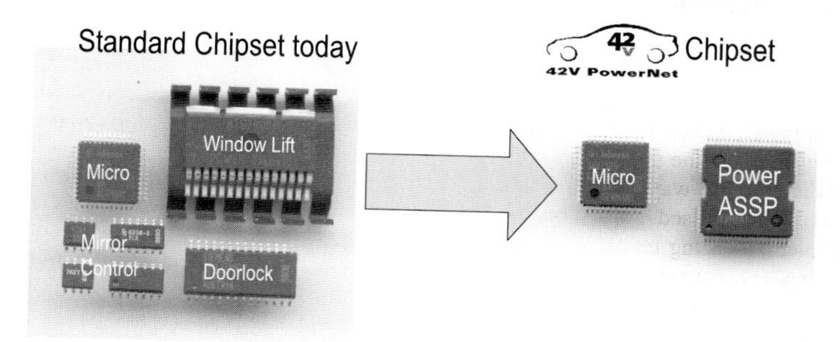

Figure 3.4 Door ASSP for a 42V vehicle

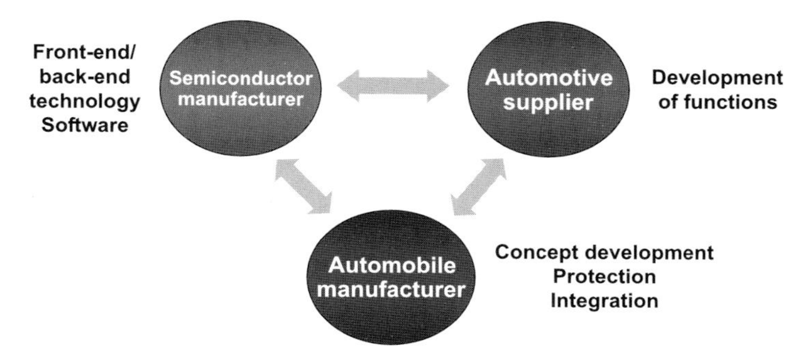

Figure 3.5 Triangular relationship in future system developments

Figure 3.4. Using the 42V vehicle power supply, the complete power functionality can be incorporated in just a single package. And in principle the microcontroller could also be included in the package.

ASICs on the other hand are completely tailored to the customer's requirements. The focus is on implementing what the customer wants. Only rarely is it possible to work with the customer to further optimize the system. In this case, it would be important for the so-called original equipment manufacturer (OEM), e.g. the automobile manufacturer, to participate in the discussions (Figure 3.5).

Because this is the only way to eliminate any 'frictional losses' which there may be, and which inevitably arise with a "pipelineorganization" OEM-tier1-2nd tier (tier1: automotive supplier; 2nd tier: semiconductor supplier).

On the other hand, in the case of ASSPs, and even more so for commodities, it is necessary to survey a large number of customers in order to achieve the greatest possible coverage of the application volume. This is the only way that complete chipsets can be discussed. The organization at Infineon (sensing technology, microcontrollers and power- ICs under one roof) makes it possible to achieve board dominance (i.e. if possible, to be able to offer all the chips on the PCB as a system chipset) in many applications.

3.3 The product groups

In order to categorize power semiconductors into product groups, we consider the product performance as a function of the level of integration (complexity). Figure 3.6 shows the simplest components, the MOS transistors and IGBTs, at the lower end of the performance (complexity) ray.

If one adds snubber circuits to the MOS transistor or the IGBT, in the simplest case one gets the so-called Temp-FET (temperature-protected MOSFET). Power semiconductors with additional smart status-monitoring are generally called smart FETs. A further distinction which is made is between the PROFETs (protected FETs) or high-side switches (the switch is connected to the plus terminal of the power supply and one side of the load is connected to ground) on the one hand, and the HITFETs (highly integrated FETs) and the low-side switches (see section 3.4.3) on the other.

If a number of these smart FETs are combined in the array with additional functions, this gives us the PICs (smart power integrated circuits). This product group covers the so-called multi-channel switches, the half- and full-bridges (a half-bridge consists of two switches connected in series between the plus and minus rail of the power supply), the supply ICs and the line-driver ICs.

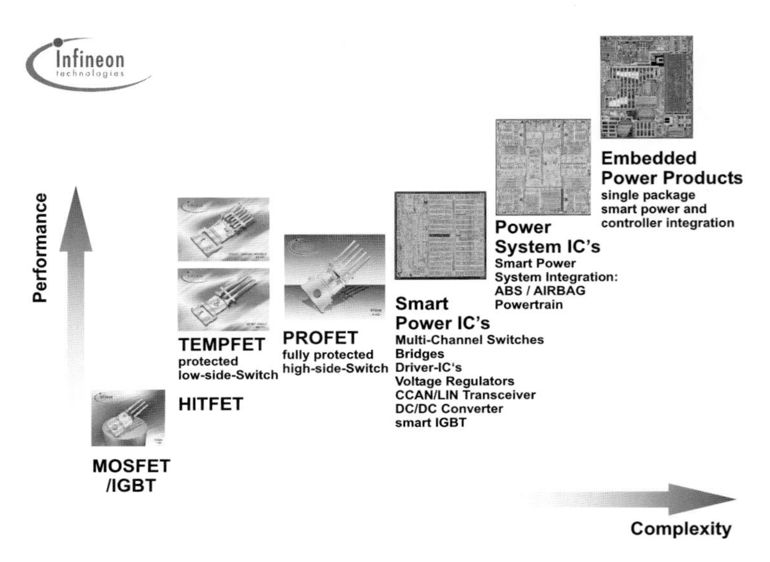

Figure 3.6 The power semiconductor product groups

An even higher level of integration leads finally to the so-called power systems ICs. These are manufactured for particular applications (for example for an airbag or an ABS control device) as cost-optimized products.

The highest level of integration is achieved by the embedded-power products. These, at least theoretically, implement a monolithic integration of a system on one piece of silicon or in a package as a multi-chip product. However, it is nowadays only in rare cases that this is a commercially sensible route to take.

3.4 Wafer technologies (front-end)

Each individual product group can be realized in an optimal way using a particular technology. Power electronics calls for a very wide range of technologies. These comprise the basic bipolar MOS, complementary MOS (C-MOS) and power MOS (D-MOS) technologies, and combinations of them. Figure 3.7 shows their association with the product groups.

3.4.1 Basic processes

CMOS processes

CMOS (= complimentary MOS) is the name given to a process which covers only p-channel and n-channel MOS transistors, and does not include any bipolar or other components. The transistor is constructed from a p-well, an n-well and a polysilicon gate. It is possible to use the polysilicon layer as a resistor, and to construct capacitors by using the polysilicon and a doped substrate as the plates of the capacitor, with the gate oxide as the dielectric medium. CMOS processes are optimized for the construction of logical functions. Essentially, the process can produce a group of devices with very low voltages (5V, 3V, 1.8V), and hence permits the use of very small components with a high integration density. The transistors can also be used to a limited extent for analog functions. Infineon offers numerous logical processes (C5, C6, … ,

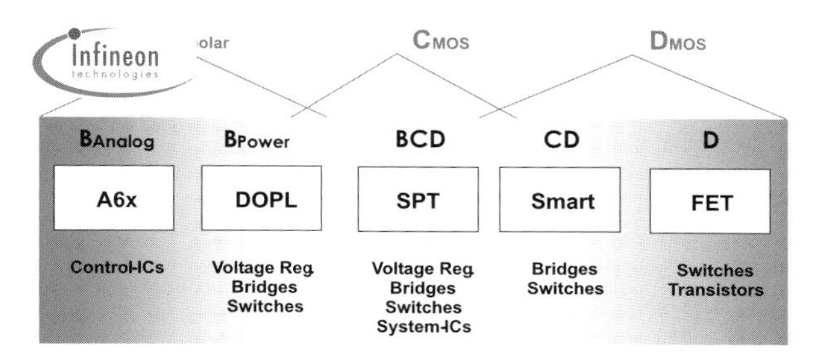

Figure 3.7 Overview of technologies and their associated product groups

C11), including some with integrated E2PROM or flash memory.

Bipolar processes

Bipolar processes use npn and pnp bipolar transistors as the active circuit components. For a pure bipolar process, polysilicon gates are not necessary. In this case, therefore, the process can be very cost-effective with few process steps. The integration density depends on the voltage class of the process. By varying the transistor sizes, various voltage classes are possible for the transistors. DOPL is a bipolar process from Infineon.

DMOS processes

DMOS (double diffused MOS) transistors are transistors which are optimized for switching high currents, and are designed for high voltages. The component has a long channel so that high breakdown voltages can be achieved, and is constructed so that several parallel cells (parallel devices) permit high currents (low on resistance) and high energy densities. DMOS processes have a thicker gate oxide layer than logical processes, and this makes it possible to produce more robust components. Infineon offers various DMOS processes, which are optimized for particular applications. Examples of these are the PFET and SFET processes.

If these basic processes are combined in a logical way, one gets the following interesting process variants, which are suitable for particular applications as a result of their specific characteristics:

BC processes

BC processes combine bipolar and CMOS components in one process. This combination permits various analog functions, such as very precise reference voltage circuits. Infineon offers various BC processes which are, for example, optimized for RF applications.

CD processes

CD processes combine CMOS and DMOS components in one process. This enables high currents and high power to be combined with logical functionality in one IC. The 'smart' process is an example of a CD process from Infineon.

BCD processes

BCD processes combine bipolar, CMOS and DMOS components in one process. It is possible to produce components for various voltage classes. CMOS permits a high logic density. It is thus possible, for example, to integrate a microcontroller. The combination of bipolar and CMOS makes possible very precise reference voltage circuits. DMOS transistors make it possible to switch high currents and high voltages (up to 20A and up to 80V).

In some cases, more than one gate oxide layer is used, to make possible high integration densities for low voltage logic circuits (e.g. submicron logic). It is possible to have several resistive layers of polysilicon. With advanced BCD technology, it is possible to have more than 25 photo-sensitive layers (masks). However, this does make the process more cost-intensive than simpler processing methods, such as for example CMOS.

3.4.2 Power MOSFET

Power MOS transistors are used in many cases as switches. Only rarely does their analog operation play any part. The following operating states can be distinguished:

• Transistor switched off:
The least possible current should flow at the highest possible voltage. The associated parameters are the blocking (breakdown) voltage and the leakage (reverse) current.

• Transistor switched on:
In this state, the on resistance should be as low as possible for the highest possible current levels. The associated parameters are the drain- source ON-resistance and the maximum current.

• Transistor switching on or off:
The switching time should be as short as possible and the charge quantities to be moved as small as possible. The associated parameters are the switching times, their transconductance and the magnitude of the gate charge.

Today's technologies are designed for temperatures of over +200°C. The highest possible operating temperatures make for savings in terms of the silicon chip area and cooling costs in the application. For this reason, the trend is towards higher junction temperatures combined with greater lifetime and active and passive cycle robustness.

Additional important "robustness parameters" which must be mentioned are the resistance to irradiated RF interference (EMI) and resistance to electric discharges (ESD).

Infineon has developed the OPTIMOS technology specially for automotive applications, to be able to offer optimized products for all application situations.

Figure 3.8 shows the possibilities which have most recently been opened up by the development of the most important parameter, the "on-state resistance R_{on}".

This shows the last four generations of a product, an 18 mOhm switch with its as-

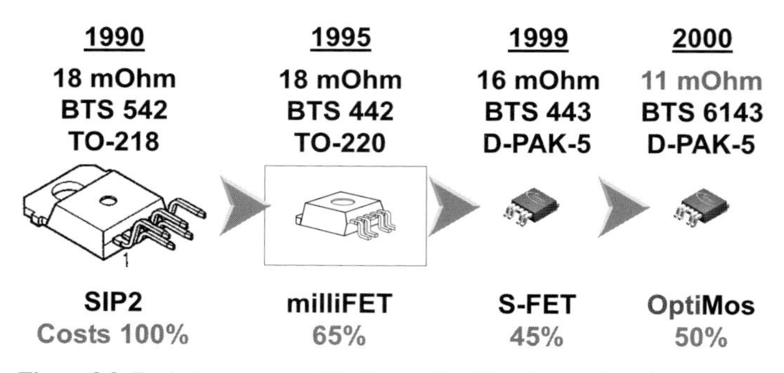

Figure 3.8 Evolution as exemplified by an 18 mOhm high-side switch

sociated power MOS wafer technology, and the corresponding total costs. The price/performance ratio has improved dramatically; ever smaller, cheaper chips can be utilized to switch ever more power. The associated packages can also be manufactured at lower prices, because the power loss which arises is lower.

3.4.3 Smart FETs

If one enhances the power-MOS technology with additional p- and n-wells, it is possible to integrate additional functions monolithically. The MOS transistors become "smart". For example, protection and monitoring functions can be integrated. The range of components which results can be seen in Figure 3.9.

As one can see, there are no bipolar structures. We therefore also speak of CD technologies (C-MOS and D-MOS). The key parameters of these technologies are determined by the specification of electronic switches.

In the case of the CD technology, it is of interest how many and what types of additional components it is possible to integrate. This parameter is called the "feature size", roughly speaking the size of the components, alternatively called the packing density, and is connected with the gate oxide thickness and additional process steps. In every case, the rule is: processes with a smaller feature size are more difficult to manufacture, and hence more expensive.

As in the case of the power MOS transistors, the current flow through the power element is vertical. The rear of the chip is therefore identical with the drain of the power structure. The on-state resistance is lower than for technologies with an insulated drain. It can be further reduced by grinding the wafer down to a thickness of a few 10 µm. Several power structures, for which the drains are inevitable joined together, can be produced in this way. With these common-drain products, the heat slug on the package must be insulated from the heat sink, because the drain potential is generally not the same as the ground potential of the system. Figure 3.10 shows the current flow and Figure 3.11 examples of these relationships.

In terms of applications, it is possible to construct so-called low-side and high-side switches using the CD technologies. The terms low-side (LS) and high-side (HS) derive from the circuit diagram (Figure 3.11), in which a series circuit of the switch and the load is shown. In general, the negative voltage is connected to the system ground in this circuit, as is the

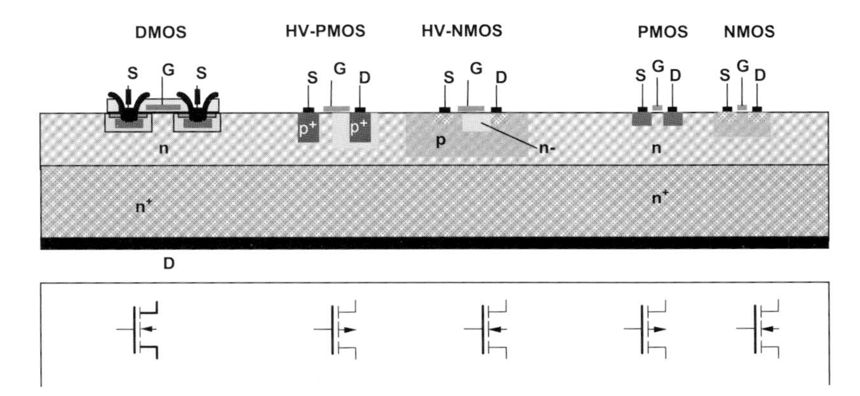

Figure 3.9 Components in Infineon's smart-power technologies

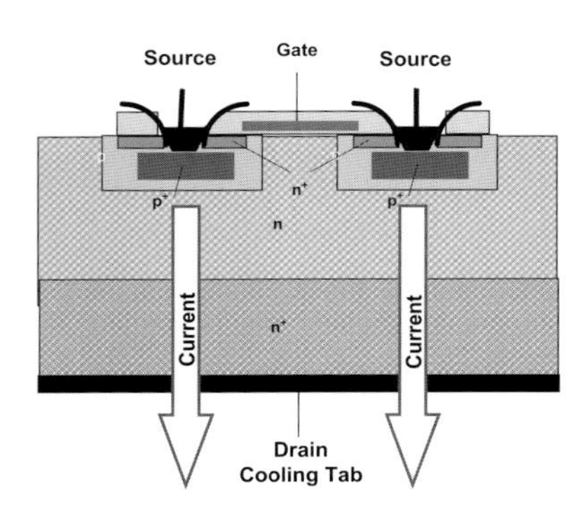

Figure 3.10
Vertical current flow in the common-drain technology

Highside Switch:
• Load to GND
• Drain to +Vs

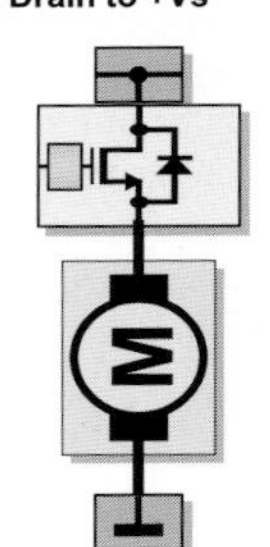

Figure 3.11
HS and LS switches; block circuit diagram

case for example with an automobile power supply.

If the load is connected to the system ground (e.g. in an automobile), this saves one wire because the current flows back to the battery via the chassis. If the load is connected to ground, the switch must be inserted between the load and the plus terminal of the battery; in the above circuit diagram this is on the "high-side". A low-side switch, on the other hand, always has its source connected to ground. A typical representative of each of Infineon's families, the low-side and the high-side, is shown in Figure 3.12.

It is not only switches which are required in electronic form but also, because so-called reversing drives are with increasing frequency being realized in electronic form, so-called bridge circuits have been implemented. Figure 3.13 shows the principle of the so-called H-bridge. This can be used to connect the load, generally an DC motor, to the supply voltage with either polarity.

Apart from the monolithically integrated bridge circuits for low currents, the so-called TRILITHICs have also been developed. In these devices, MOS transistors and/or CD chips are combined in one package to form full-bridges. "TRI" because the optimal way to realize a full bridge circuit is with two low-side chips and one double high-side chip. Figure 3.14 shows a typical product. The leadframe can be seen as the chip carrier. The molding compound has been omitted, so that the internal construction can be seen.

Infineon has been mass producing CD products for over 20 years. The latest

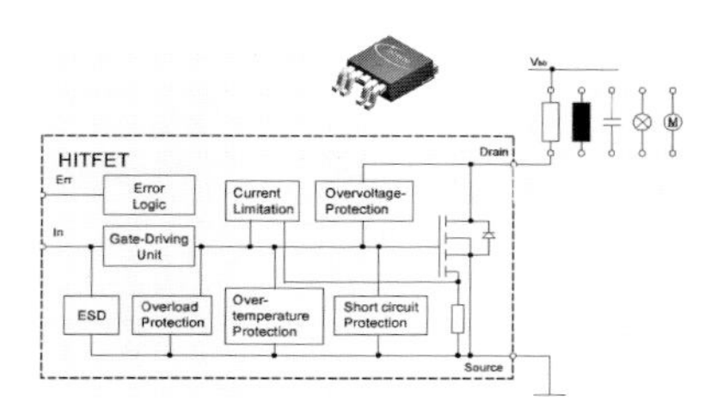

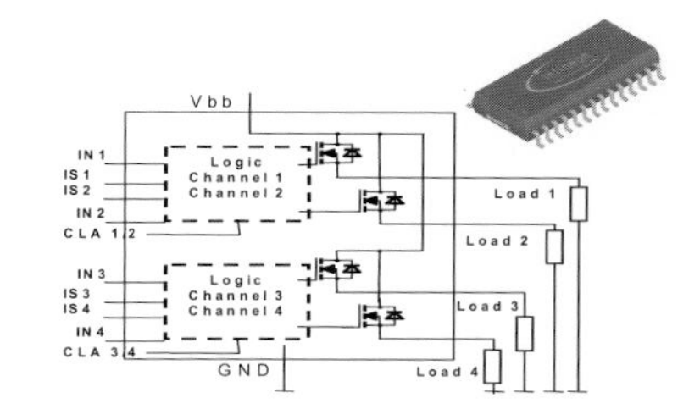

Figure 3.12 A low-side switch BTS3150 (left) and a high-side switch BTS5140

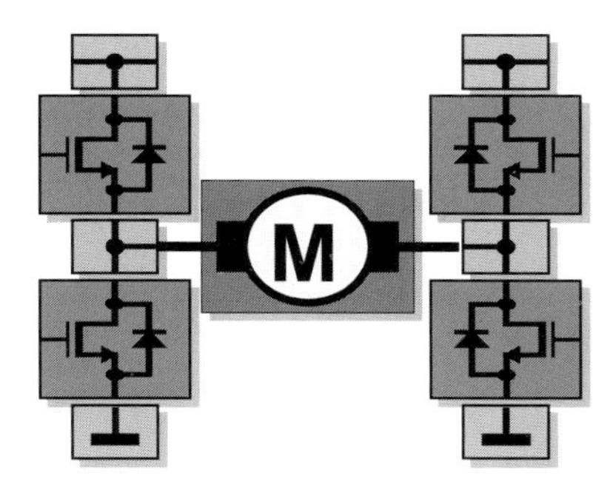

Figure 3.13
H-bridge circuit diagram for reversing drives

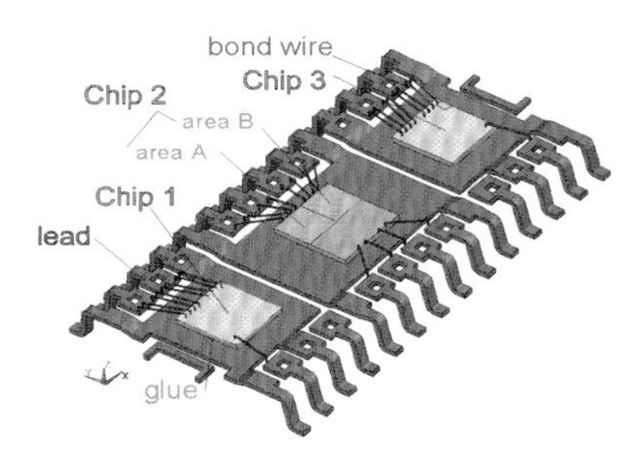

Figure 3.14 Typical TrilithICs

Smart-5 generation has an extremely low on-state resistance combined with a significantly reduced feature size. For the first time, this makes possible high power switches with complex logic. There are now no remaining obstacles to prevent the production of smart switches with comprehensive diagnostics.

In pursuit of the slogan "Silicon instead of relay", there will from now on be many applications with more demanding diagnostic requirements where the only acceptable implementation will be using smart switches. An example from practice can be seen in Figure 3.15. To make the shrinkage effect clear, this figure shows true to scale chips with the same functional range. On the left is the old generation, and the chip alongside on the right is in the latest smart-5 technology.

Of course, whenever the CD technology is to be used one should check that a monolithic integration is worthwhile. Infineon has taken this fact into account in developing its core competences, "chip on chip" and "chip by chip". In doing so, a standard MOS transistor has been used as the so-called base chip. Only the relatively limited area of the intelligent top-chip needs the complicated (expensive) process. The two chips are assembled one on top of the other or one beside the other in a special process.

The result: from a defined current or on-state resistance (R_{on}) threshold, which is determined by the production costs, so-called high current switches are manufactured as chip-on-chip (CoC) or chip-by-chip (CbC) devices.

Today, from a R_{on} value of approximately 20-50 mOhm the chip-on-chip approach

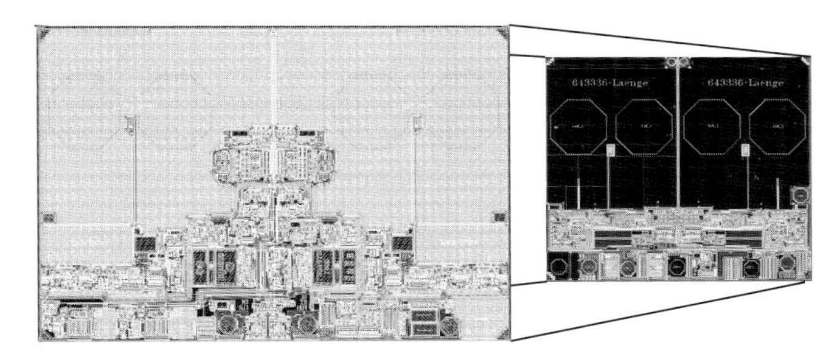

Figure 3.15 Smart 5 chip shrinkage

is cheaper to manufacture. At lower currents, it is possible to accommodate several switches in one package. In this range, the call is for technologies with freely-connectable transistors, the so-called smart-power IC technology.

3.4.4 Smart-power ICs

With the CD technologies, there is no analog capability, and the feature size is also not particularly great. This gap is filled by the so-called smart-power technologies (SPT). With the help of these, it is possible to develop integrated circuits with power output stages, so-called power ICs (PICs).

Depending on the technological variant, the set of components is significantly more extensive than the minimum shown in Figure 3.16.

With this technology, freely-connectable power transistors are possible. The current flow is, as shown in Figure 3.17, initially vertical, down to the so-called "buried layer". This buried layer now conducts the current laterally over to the drain connection. The vertical portion of the resistance, to the surface, is kept as low as possible by means of a so-called N^+ sinker. This enables an insulated tap to be made to the drain of the power connection on the chip surface. However, this is at the cost of a significantly higher on-state resistance.

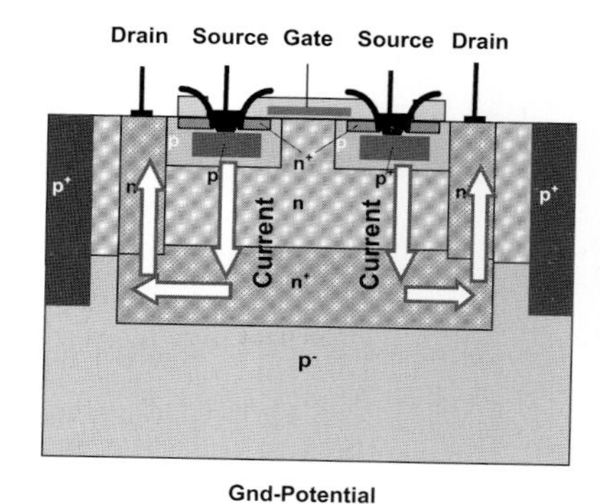

Figure 3.17
Current flow and insulation from the substrate in an SPT power device

Consequently, the SPT devices are suitable for medium level currents. To insulate the drain from the rest of the circuit, the choice is a p-substrate which, when connected to ground potential, holds the pn drain-substrate junction blocked. It is easy to see that the pn junction between this layer and the component wells will only remain blocked if the p-substrate is always at the lowest electrical potential in the circuit.

For cooling purposes, the rear side of the chip in this device can have a conducting

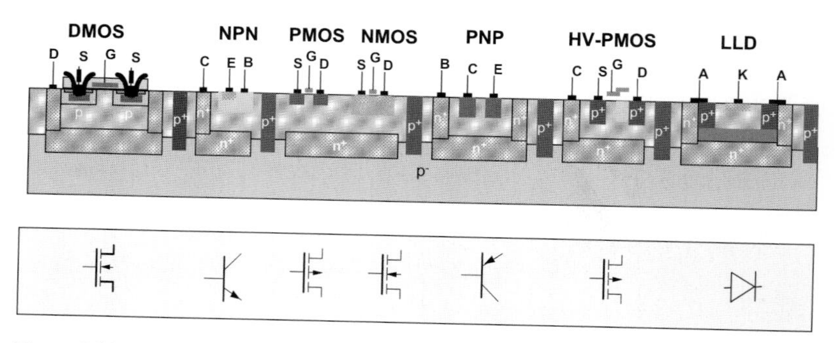

Figure 3.16 Set of devices in the smart-power technology, SPT

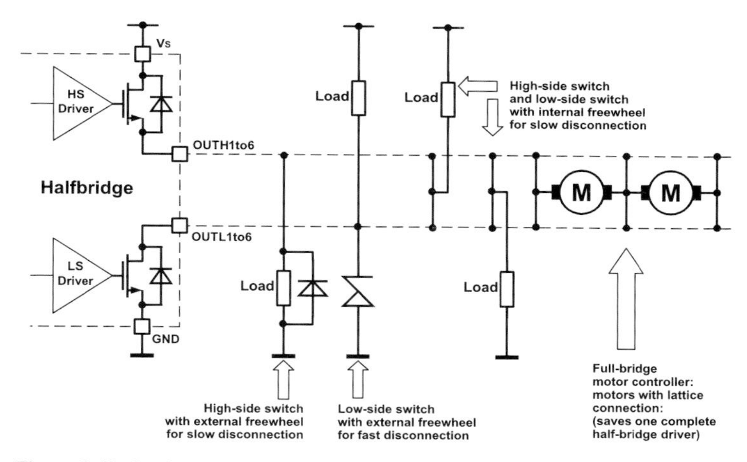

Figure 3.18 Configuration options for switches, half- and full-bridges

attachment to ground. There is no need for additional insulating sheets. The wide variety of options is clearly shown in Figure 3.18.

Apart from the high-side and low-side switches already discussed, the exceptionally important half-bridge circuit can also be realized in this way. It consists of two fast-switching cascaded MOS transistors.

In addition, it is possible to make virtually any required combinations of analog and digital circuits. Typical representatives of these ICs are the TLE6263 supply CAN transceiver, with numerous precision analog circuits and a digital control

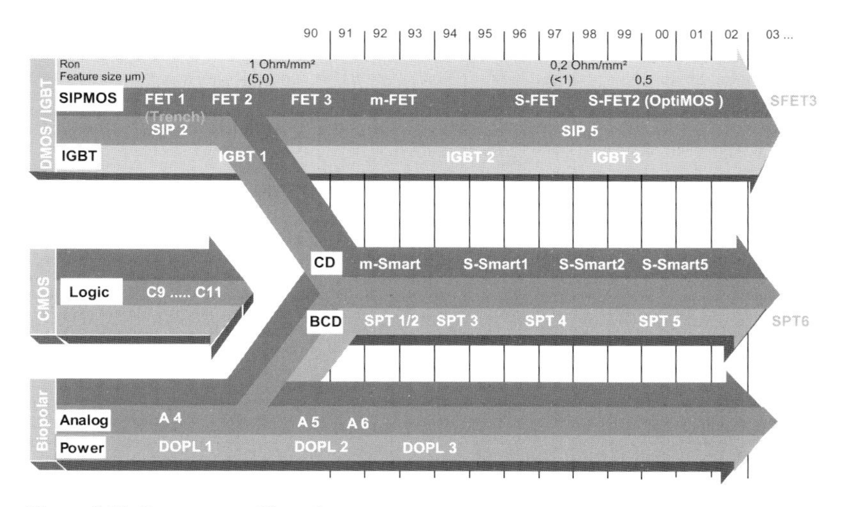

Figure 3.19 Smart-power IC roadmap

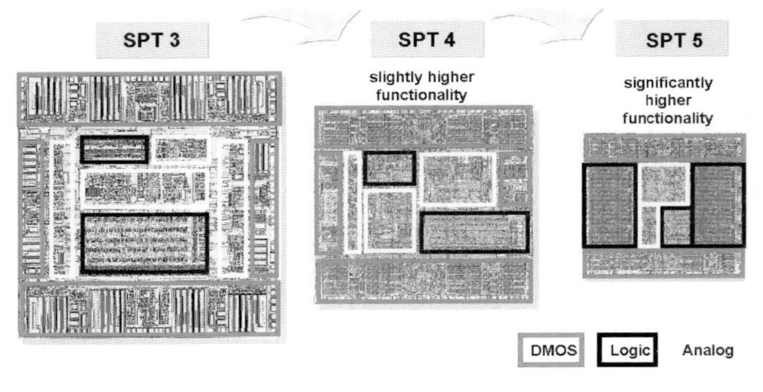
We have been able to offer our customers chips with constantly higher functionality and lower size
SPT 3
SPT 4
slightly higher functionality
SPT 5
significantly higher functionality
DMOS
Logic
Analog

Figure 3.20
The advantages of large-scale integration from the development of integration capabilities

interface (SPI), and the TLE6288, a multi-channel low-side switch with many additional functions.

Infineon has been producing SPT devices for about 20 years; currently the technology used is the fifth generation, SPT5. The roadmap in Figure 3.19 shows the development of the most important characteristics, R_{on} and the feature size, over this period of time.

One can see how the power-MOS technologies (above) and the bipolar lines (below) have come together with the CMOS technologies to give the SPT.

The advances in the SPT technology are shown in summary in Figure 3.20. This shows a system IC with comparable functionality, in the third, fourth and fifth generations of SPT. It can be seen that significantly more functionality can now be integrated onto the same chip area.

Requirements for increased complexity are being posed for the chip area in today's generation of technology. So, on a standard SPT IC today 30% of the chip area is taken up by power elements, about 40% by analog functions and the remaining 30% by digital functions. If the complexity of each of these area increases further, the chip will increase in size threefold. The most significant share of this will go to the digital part, which would become six times larger. The shrinkage effect of the next technology generation would more than compensate for this effect. The corresponding split on the chip would then be 46% for power elements, 34% analog and 20% digital functions. There is thus further potential for the integration of significantly more logic and system functions. This is realized in the smart power-system ICs.

Smart power system ICs

The large-scale integration of the smart power ICs produces the smart power system ICs. In order to understand the optimal extent of integration for an ECU, the characteristics of the functional units must be considered. These are shown in Figure 3.21 for a typical ECU. A distinction is made here between logic (digital), analog and power.

The current, voltage and temperature requirements determine which subfunctions can be realized in which technology, and in which package the chips are then housed. Figure 3.22 shows that, in partic-

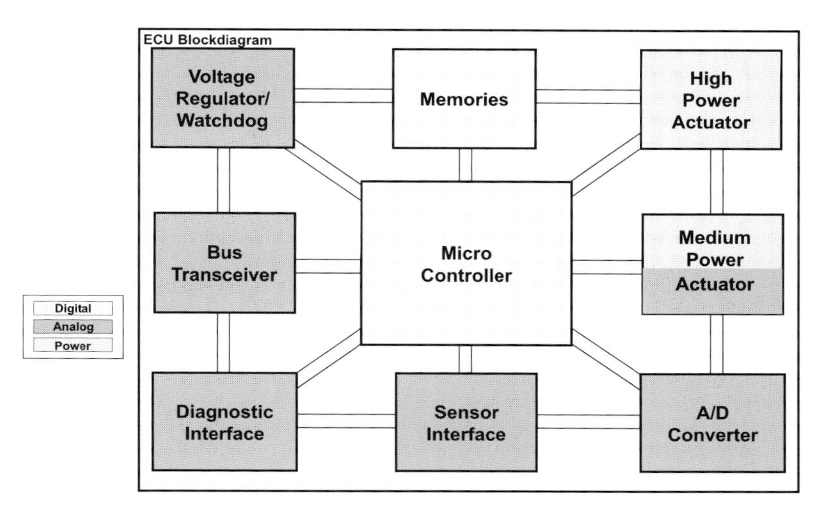

Figure 3.21 Functional units in an electronic control unit (ECU)

ular, the interface between the controller and the shell of high-voltage resistant power semiconductors which surrounds it is especially interesting and difficult.

Here, finding the technical/commercial optimum requires very close cooperation, for example between the automobile manufacturer, his suppliers and the semiconductor manufacturer. As a broadliner, Infineon has the complete range of know-how about this system optimization process under one roof. Not for nothing therefore is Infineon the first port of call when it comes to system solutions and their "smart partitioning".

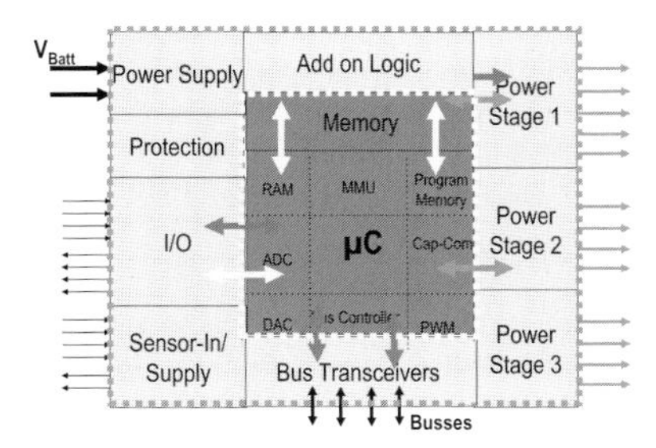

Figure 3.22 Smart partitioning of an ECU

3.4.5 Outlook and trends

The requirements which future technologies must meet are essentially determined by the need for performance which is higher, in some cases significantly, combined with higher efficiency. This implies, on the one hand, greater computing power on the controller side and, on the other hand, a need for more optimal (lower resistance) switches in the case of the power semiconductors. Figure 3.23 shows the individual parameters involved – temperature, EMC, system capacity (current and voltage) – for power devices and microcontrollers.

What is evident here is that the requirements are developing in different directions.

An analytical view of this qualitative situation is shown in Figure 3.24. It is evident that, depending on the combination of performance and current, there are different technological variants, in some cases equally promising.

This makes it necessary to have a deep knowledge in the technological field, and to be able to offer a complete portfolio.

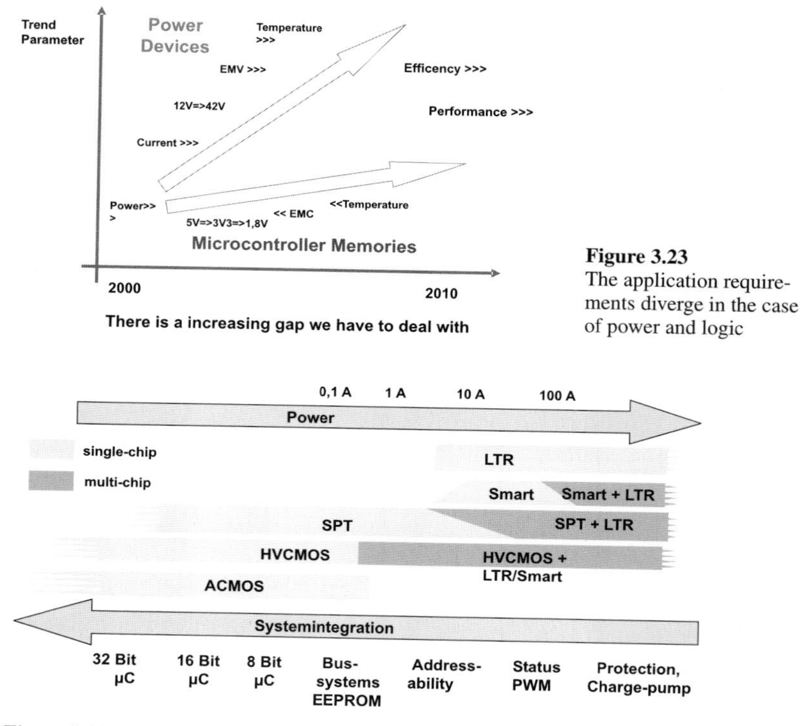

Figure 3.23 The application requirements diverge in the case of power and logic

Figure 3.24 Comparison of the technologies in terms of current and logic capabilities

Because only then is it possible to determine and realize the most appropriate (smart) partitioning in each particular case.

The need to equip complex systems with comprehensive diagnostic capabilities is clear if one considers the probability of failure of the system as a whole. Thus, for example, when a modern top-class automobile is started up, more computing power is required than was necessary for the complete flight to the moon. Because of this fact, significantly higher logic capabilities will be demanded for future technologies.

Another important influencing factor is intelligent networking, and the associated decentralization of controllers. This is the driving force behind so-called mechatronization, i.e. the complete integration of the electronics into the mechanical system. A particularly appropriate example seems to be the co-called electronic muscle, which applies a mechanical force. This can be a solenoid or an electric motor. For example one can, as indicated in Figure 3.25, build into an electric motor a complete controller, and thus turn it into an intelligent actuator. Here, the transfer of control commands in one direction and of diagnostic data in the other takes place via a serial single wire interface. All that is required in addition is two further wires for the energy feed (plus and minus). Particularly in the electrification of an automobile, this is of great importance. The stumbling block here is often the additional costs of the local intelligence.

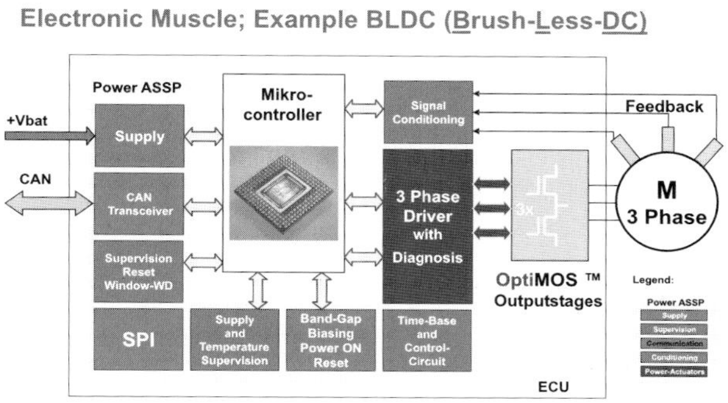

Figure 3.25 Application of the future – an "electronic muscle"

The next step would be what has already been implemented in industrial electronics, the feeding of control data over the power supply wires.

With the decline in the importance of analog functions, pure power switches combined with a high-voltage CMOS line will be one of the technologies of the future. This applies particularly in the case of systems with lower complexity; for the mechatronic systems mentioned above. However, the implementation of large and medium computing powers will remain the preserve of the pure CMOS lines for some considerable time yet.

3.5 Packaging technologies (back-end)

Packaging technology is of particular importance in the case of power semiconductors. One reason for this is the active production of power losses and the high dynamics when the devices malfunction. Thus, the power loss of an MOS transistor, which normally lies around one to two Watts, can in the event of a short circuit rise by over three orders of magnitude. It is easy to understand that in this situation the material stack – chip, solder, heat slug – is stressed to the limit by the effects of thermal expansion.

A second reason is that the connecting leads carry high currents, or are subject to high voltages. For example, if one were to calculate the current in a normal commercially available connecting cable with a cross-section of 0.75 mm^2 at the same current density as in the bondwires, it would need to be able to carry currents of 1000 A. This means that the materials are being operated at the limit of the physically feasible. As a result of the advances in the MOS technologies it has become possible to produce extremely low resistance devices. For this reason, the internal resistances of the package itself (the bondwire, chip/heat slug connection and the heat slug itself) will be responsible for an ever higher proportion of the power loss which occurs.

3.5.1 Categorization of semiconductor packages

In general, there are two different groups of packages. The first group has a cooling surface on the chip carrier (leadframe), which can be directly soldered on. The thermal resistance of this package, between the chip and the cooling surface, is called Rthj-c (junction-case), and has very low values.

The second group has a "thermally enhanced leadframe". With these packages, metal bridges lead out from the chip carrier to the pins of the package. Purely from the outside, one cannot distinguish this package from the standard devices,

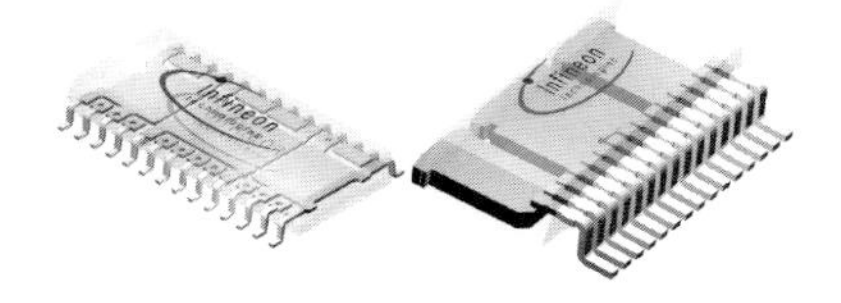

Figure 3.26
Difference between a thermally-enhanced SO package and a heatslug package

because the molding compound hides these details. Figure 3.26 shows the two types of package, exemplified by a P-DSO-28 (3 separate leadframes with the corner pins as the cooling path) and a P-TO263-15.

Many users are now posing the question as to the size of the cooling surface with increasing frequency. In particular, when modern SMD devices are to be used. The trend away from through-hole packages to the low-cost SMD applications is dictated by the improvements in chip technology. In many cases this makes "silicon instead of heatsink" possible, with the printed circuit board (PCB) itself forming the heat sink.

In calculating the PCB as the cooling surface, many factors must now be taken into consideration. Earlier concepts involved attaching (by screws or clamps) a solid heat sink to the power package. The thermal resistance could then be determined with relative ease, simply from the geometry of the heat sink. With SMD concepts on the other hand, sizing is much more difficult, because it is necessary to analyze the thermal path: chip (junction) – leadframe – package contact (case or pin, as applicable) – footprint – PCB materials (base material, thickness of the copper layer) – PCB volume – ambient. The possible ways of cooling SMD devices is shown schematically in Figure 3.27.

As the layout of the PCB makes an essential contribution to the result in this situation a new method must be used. Below, we explain for users the steps required for this method, using in each case a representative example of the package groups mentioned (tab package and thermal enhanced package).

3.5.2 Static characteristics of power packages

To explain the static characteristics of a power integrated circuit (PIC), the internal structure of a PIC is shown in Figure 3.28. together with the associated method of mounting on a printed circuit board (PCB) or a heat sink, as applicable. The PIC consists of one chip, which is mount-

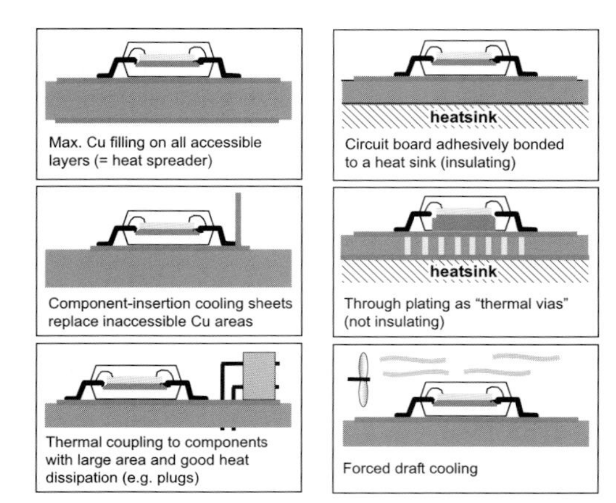

Figure 3.27
Cooling SMD IC packages

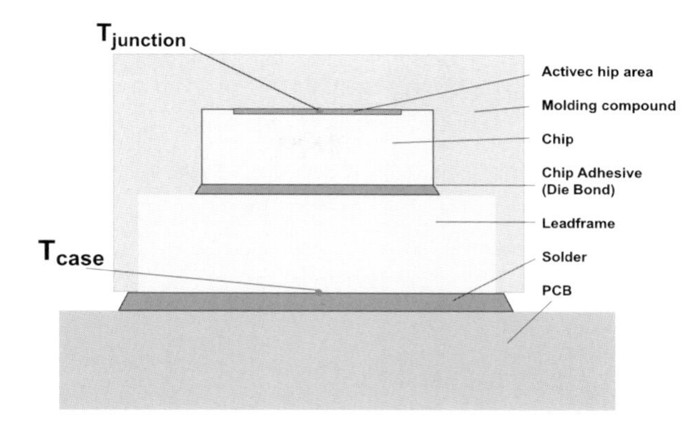

Figure 3.28
Internal structure of a PIC

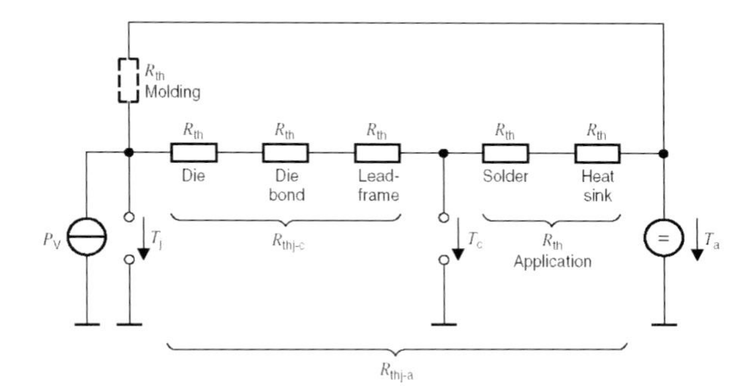

Figure 3.29
Summary equivalent circuit for a power IC

ed on a chip carrier using a metal solder or contact adhesive. The circuit board consists of a material with a good conductivity, such as copper, and can be several millimeters thick.

The associated equivalent static circuit is shown in Figure 3.29. In this, the following analogs of electrical variables are used:

The power loss P_V which arises close to the surface of the chip is represented as a current source. The thermal resistances are represented as ohmic resistances. The electrical equivalent circuit is, in principle, a series circuit of thermal partial resistances. As a first approximation, the resistance for the molding compound (dashed outline) in the parallel circuit can be neglected. The ambient temperature is represented by a voltage source.

Corresponding to this analogy, the heat flow $P_V = Q/t$ can now be calculated with

Table 3.1 Thermal-electric correspondences

Thermal	Electric
Temperature T in [K]	Voltage V in [V]
Heat current P in [W]	Electrical current I in [A]
Thermal resistor R_{th} in [K/W]	Ohmic resistor R in [V/A]
Thermal capacity C_{th} in [Ws/K]	Capacitance in [As/V]

the help of the heating equivalent of Ohm's law:

$V = I \cdot R$ corresponds to
$T_j - T_a = P_V \cdot R_{thj-a}$.

In summary, Table 3.1 provides an overview of this process.

In explaining the application as a whole, the function $P_V = f(T_a)$ is of critical importance. From the above we get: $P_V = -T_a / R_{thj-a} + T_j / R_{thj-a}$. This is a straight line which falls with a slope of $-1/R_{thj-a}$, and with a value of zero at T_j.

$I \sim P_V$, $R \sim R_{th}$, $V \sim T$

Specification of R_{th}, P_V, T_j for thermally enhanced power packages

Figure 3.30 shows this function for a P-DSO-14-4 package (thermally enhanced power package) mounted on a standard application circuit board. From the above function it is possible to deduce directly the permissible power loss for any ambient temperature.

For example, at $T_a = 85°C$ the permissible power loss amounts to approximately 0.7 W. The exact value can be calculated using the following equation:

$$P_V = (T_j - T_{max})/R_{thj-a} =$$
$$= 65\ K/(92\ K/W) = 0{,}7\ W$$

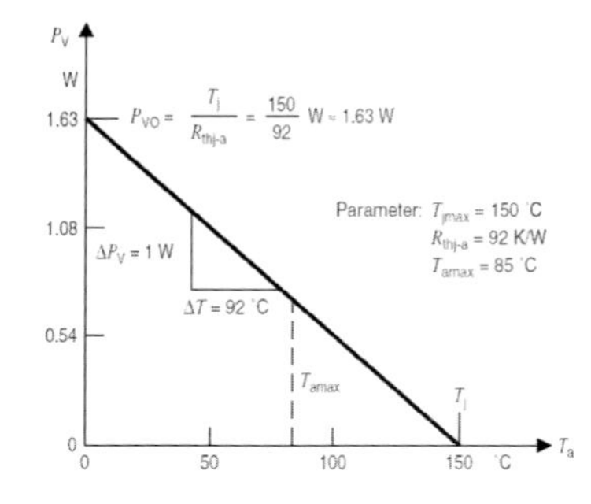

Figure 3.30
Permissible power loss for a P-DSO-14-4 package mounted on a PCB with a cooling area of 400 mm^2 (as a function of the ambient temperature)

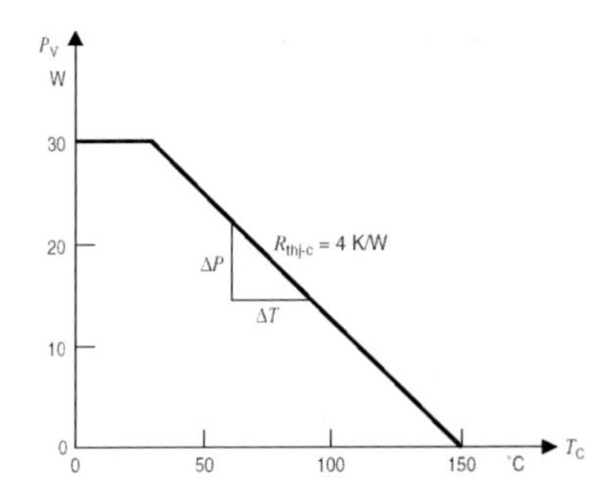

Figure 3.31
Permissible power loss for the P-TO252-3-1 (as a function of the package temperature)

Thermal tab packages

In the datasheets for the PICs, the power loss is specified as a function of the package temperature T_c, because the manufacturer does not know the specific thermal resistances. As before, this function is again a falling straight line. In this case, the slope has the value $1/R_{thj-c}$. The zero point remains at T_j. As an example, this function is shown in Figure 3.31 for the P-TO252-3- 1 package.

It is interesting to ask why $P_{Vmax} = 30$ W remains true. A higher power loss is prevented by the intervention of the internal current limiters within the chip. For this reason, the value for the power loss remains constant at lower temperatures.

3.5.3 Dynamic properties of power packages

The thermal behavior of PICs changes if dynamic phenomena (pulsed power mode) are taken into account. This behavior can also be represented by a thermal capacity C_{th}, which is directly proportional to the volume V (in cm^3) of the material concerned, its density ρ (in g/cm^3), and a proportionality factor for the specific heat, c in Ws/(g × K):

$$C_{th} = c \cdot \rho \cdot V = m \cdot c$$

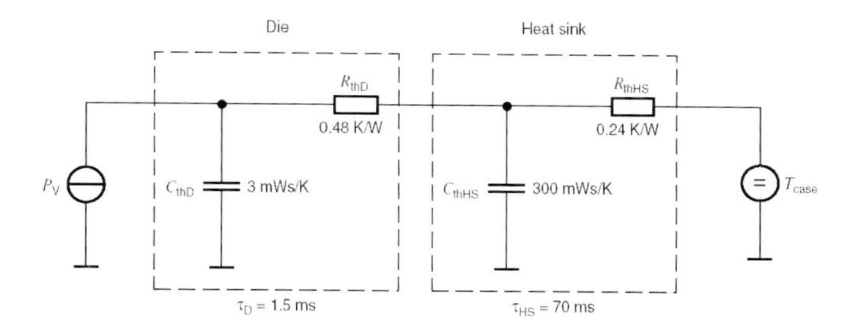

Figure 3.32 Thermally equivalent circuit for the P-TO263-7-3 package (simplified)

The meaning of this is: the thermal capacity of a body with a mass $m = \rho \times V$ corresponds to the quantity of heat required to raise the temperature of the body by 1°K.

In order to calculate the quantity of heat from ΔT, we must use the equation for the electric charge in a capacitance C. This equation is:

$$V \cdot C = I \cdot t = Q$$

The equation for the quantity of heat is analogous to this:

$$\Delta T \cdot C_{th} = P \cdot t = Q$$

The meaning of this is: just as the current strength $I = Q / t$ represents the transport of charge per unit time, the power loss P represents the transport of thermal energy per unit time. It follows from this that:

$$\Delta T = \frac{P \cdot t}{C_{th}}$$

The equivalent circuit for the P-TO 263-7-3 power package, with the heat capacities added in, is shown in Figure 3.32. The thermal capacities, calculated from the material and its volume, are inserted in parallel with the thermal resistances.

In making the calculations for the components of a circuit, it is necessary to know the thickness d, the cross-sectional area A and the thermal conductivity L in W/(m · K), in order to determine the corresponding thermal resistance R_{th}. The formula for doing so is:

$$R_{th} = \frac{d}{L \cdot A} \quad \left[\frac{K}{W}\right]$$

In order to calculate the thermal capacity C_{th}, it is necessary to know the volume $V = d \cdot A$, the specific density ρ in g/cm^3 and the specific thermal capacity c in Ws/(g × K).

Then:

$$C_{th} = m \cdot c \quad \left[\frac{W}{K}\right]$$

Dynamic thermal behavior

Corresponding to the analogy with electrical systems, the thermal response of the chip can be regarded as the voltage rise across an R-C network which is powered by a current pulse generator.

$$V(t) = R \cdot I \cdot \left(1 - e^{\frac{t}{R \cdot C}}\right)$$

And for the temperature rise:

$$T(t) = R_{th} \cdot P \cdot \left(1 - e^{\frac{t}{R_{th} \cdot C_{th}}}\right)$$

The heating and cooling process is shown qualitatively in Figure 3.33 (only valid for $t_p >> 2$ ms).

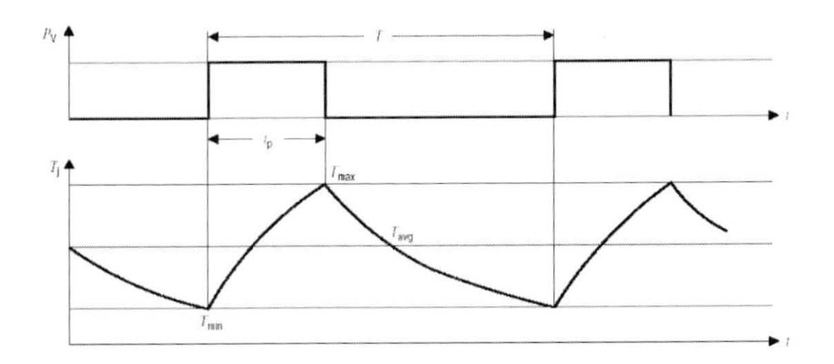

Figure 3.33
Chip temperature T_j as a function of time for periodically pulsed operation

The chip temperature oscillates between T_{min} and T_{max}. The level of the variation depends on the magnitude of the current impulses and their duration.

Specification of Z_{th}, t_p, DC

These temperature transients can be represented by a function, where the dynamic thermal impedance

$$Z_{th} = \frac{(T_{max} - T_{min})}{P_V}$$

is shown as a function of the pulse width t_p. The duty cycle (DC) is a parameter in these graphs (see Figure 3.34).

A special case of this representation is the dynamic thermal impedance for single pulse operation (DC = 0). Figure 3.35 shows the thermal impedance for single pulse operation of the medium power package P-DSO-14-4 with three different cooling surfaces on the PCB. From this function, the ranges over which the different time constants of the chip, the circuit board and the PCB each dominate can be clearly seen.

The time constant of the chip, τ_D, lies in the millisecond range, whereas the circuit board dominates in the region of several 100 ms, and the time constant of the PCB covers the 100-second range.

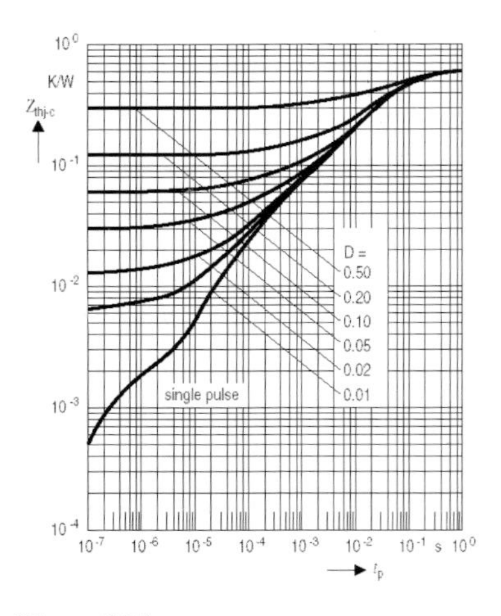

Figure 3.34
Dynamic thermal impedance Z_{thj-c} of a P-TO263-7-3 package

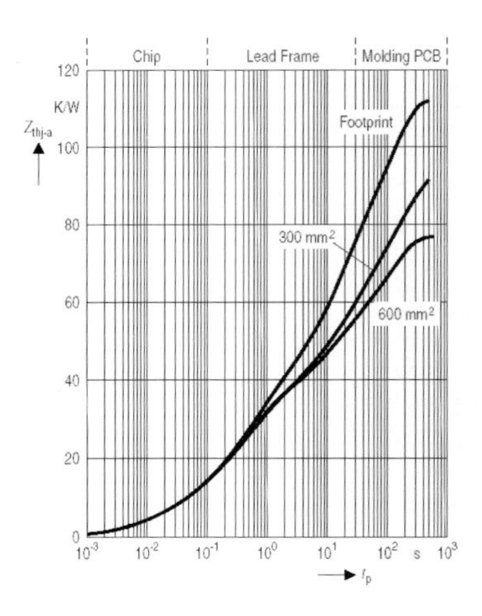

Figure 3.35
Thermal impedance of the P- DSO-14-4 package for single pulse operation

Computational example

The results of the calculations for the P-TO263-7 package will serve as an example (Figure 3.36 and Table 3.2).

The chip attachment (die bond) and the molding compound are neglected, because they do not materially affect the calculation of R_{thj-c}. For the sake of completeness however, the details are listed here:

R_{thDB} = 0.01 to 0.1 K/W

C_{thDB} = 0.1 to 0.5 mWs/K

τ_{DB} = 1 to 50 ms

R_{thM} = 100 K/W

C_{thM} = 0.64 Ws/K

τ_{M} = 64 s

(Die-bond subscript = DB; Molding compound subscript = M)

With the help of the measurement described below, the real thermal resistance can be determined. In order to determine R_{thj-a}, we must know the difference between the chip temperature T_j and the ambient temperature T_a. The following equation applies:

$$R_{thj-a} = \frac{T_j - T_a}{P_V}$$

The power loss P_V and the ambient temperature T_a can be simply calculated or determined in an oven.

The measurement of the chip temperature (T_j) calls for a slight trick: there must be a

Table 3.2 Data on the parameters of the P-TO263-7- 3

Parameter	Symbol	Value	Unit
Chip			
Area	A_D	5	mm^2
Thickness	d_D	360	µm
Thermal conductivity of silicon	L_{Si}	150	W/(m × K)
Thermal resistance of chip/device	R_{thD}	0.48	K/W
Density of silicon	ρ_{Si}	2.33	g/cm^3
Mass of chip	m_D	4.2	mg
Specific thermal capacity of Si	c_{Si}	approx. 7.0	Ws/(g × K)
Thermal capacity of chip	C_{thD}	approx. 3	mWs/K
Thermal time constant of chip	τ_D	approx. 1.5	ms
Heat Slug			
Area (effective area: 64 mm^2)	A_{HS}	14	mm^2
Thickness	d_{HS}	1.27	mm
Thermal conductivity of copper	L_{Cu}	384	W/(m × K)
Thermal resistance of heat slug	R_{thHS}	0.24	K/W
Density of copper	P_{Cu}	8.93	g/cm^3
Mass of heat slug	m_{HS}	0.8	g
Specific thermal capacity of Cu	C_{Cu}	0.385	Ws/(g × K)
Thermal capacity of heat slug	C_{thHS}	310	mWs/K
Thermal time constant of heat slug	τ_{HS}	70	ms

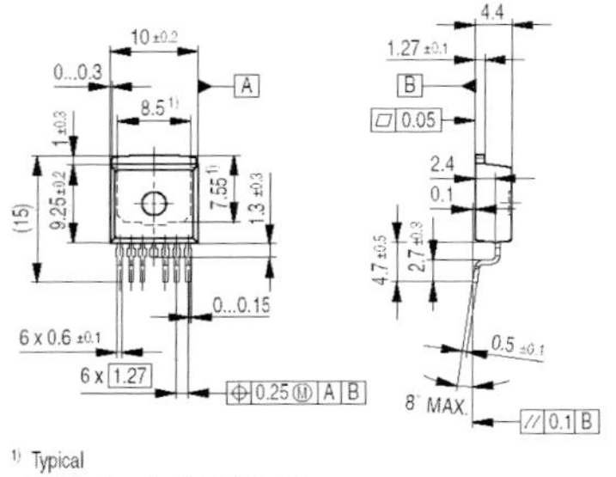

1) Typical
Metal surface min. X = 7.25, Y = 6.9
All metal surfaces tin plated, except area of cut.

Figure 3.36
Dimensioned drawing of the P-TO263-7-3 power package

temperature sensor on the chip which can also be read out while it is in operation. With many products, it is possible to use a substrate diode on the outputs (status, reset etc.) to measure the chip temperature. To do this, the forward voltage V_F for the diode is measured at constant current, as a calibration curve. Because of the thermal characteristics of the forward voltage, which has a negative temperature coefficient of about – 2 mV/ K, the relevant chip temperature can then be determined.

Process steps

As an example, the process will be explained for the voltage regulator TLE 4269 in a virtual application.

1. Measuring the calibration curve

The calibration curve is measured in a circulating air oven. The power loss should be kept as low as possible, to ensure that the chip temperature is the same as the ambient temperature. Figure 3.37 shows the calibration curve for a TLE 4269 GM (P-DSO-14-4 package) voltage regulator (measured at the diode on the reset output (RO), Pin 7). Figure 3.38 shows the corresponding measurement circuit.

2. Inject a defined power loss into the device at $T_a = 25°C$

If switch S1 is closed and the output voltage V_Q = 5 V, then an output current of 5/35 A flows.

The power loss $P_V = (V_I - V_Q) \cdot I_Q$ in the chip of the voltage regulator now amounts to 1 W.

Measurement of $V_F = V_{F25}$ at $T_a = 25°C$; result: V_{F25} = 600 mV.

3. Change T_a (for example, to 85°C)

S1 remains closed, hence: P_V = 1 W.

Measurement of V_{F85}; result: V_{F85} = 400 mV (e.g.)

4. Determine T_j for T_a = 85°C, P_V = 1 W

The calibration curve allows one to determine: T_j = 125°C @ V_F = 400 mV

5. Calculation of $R_{thj\text{-}a}$

The exact thermal resistance for the real application is calculated from these values by applying the following formula:

$$R_{thj\text{-}a} = \frac{T_j - T_a}{P_V}$$

Parameters such as the airflow, for example, can be changed without reducing the accuracy of the measurements. For example $R_{thj\text{-}a}$ = (125 K – 85 K) /1 W = 40 K/W.

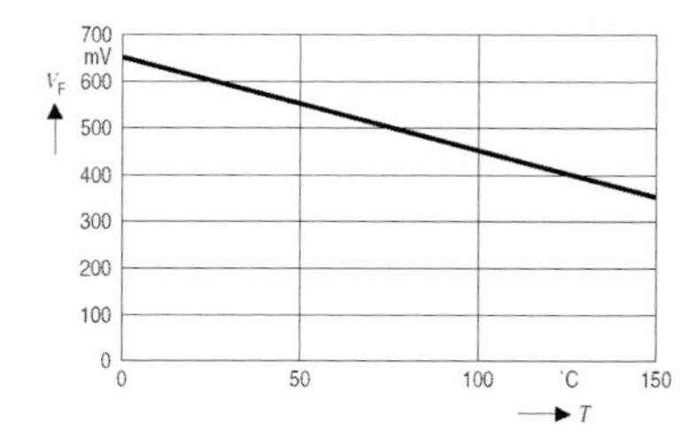

Figure 3.37
TLE 4269 GM calibration curve for IRO = –500 mA (current consumption at Pin 7; RO)

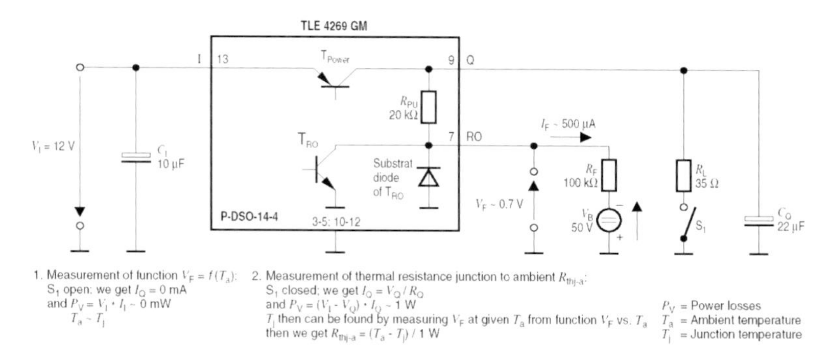

Figure 3.38 Measurement circuit with TLE 4269 GM

3.5.4 Analyzing power semiconductor packages with the finite-element method

With the help of finite element analysis, extensive thermal measurements can be eliminated. In order to determine the data for the calculation of the thermal resistance, the geometry of the package under investigation and the chip it incorporates is input into a simulation. In cases where the bondwires make a significant contribution to the cooling, they are also modeled. Figure 3.40 shows, upper left, the model of a tab-package with chip; complete and without molding compound. Below on the left is the tab-package in the application (on the PCB) and beside that it can be seen in operation, but without the molding compound. On the right one can see the half-symmetric model and the use of an enhanced SO package, both with and without the molding compound.

Further possibilities of FEM simulation are shown in Figure 3.39. The temperatures of the individual components, chip, leadframe (die-pad), plastic envelope (molding compound) and the connecting contacts can be considered individually or in combination. This enables valuable information to be gained during the design of the package itself and for the product in practical use. Over and above this, it is possible to optimize the complete application without having to construct an expensive prototype.

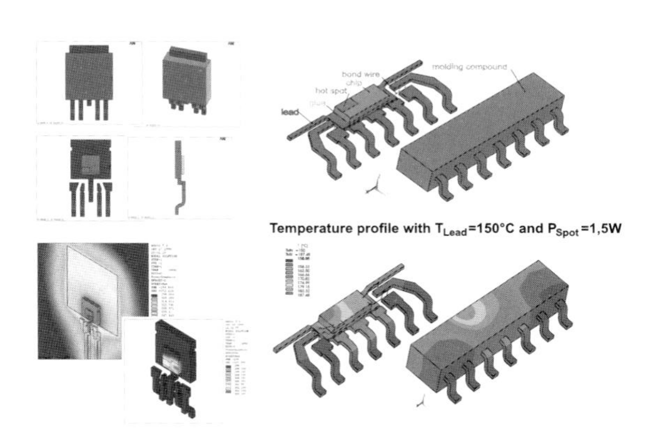

Figure 3.39 Example of FEM modeling; left a tab-package; for comparison on the right, a thermally enhanced SO package

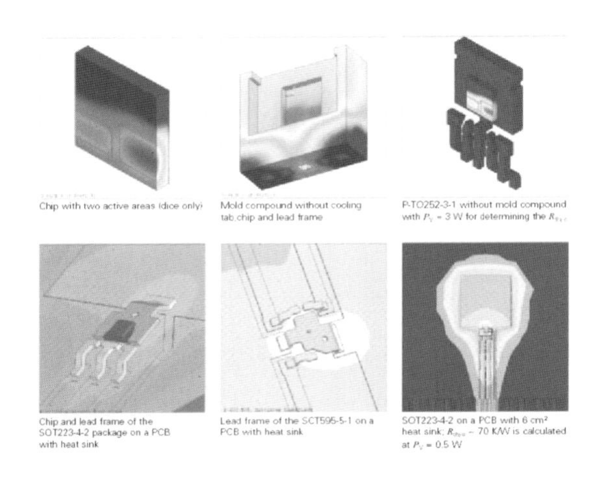

Figure 3.40
FEM analysis opens up a host of fascinating possibilities

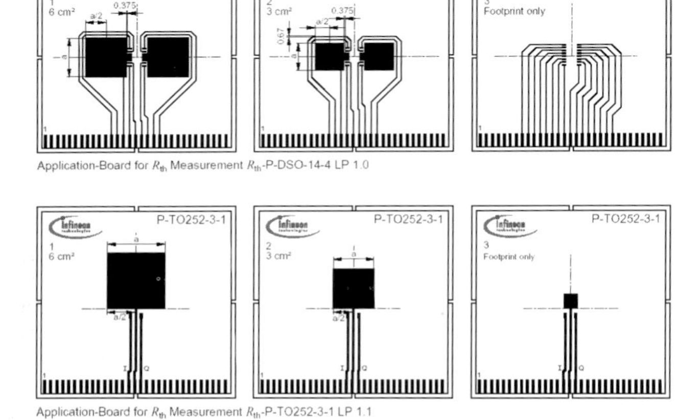

Figure 3.41
Layout of the thermal resistance measurement boards, for determining the thermal resistance of the system to environment as a function of the copper area on the PCB

In order to obtain data which can be exploited in daily practice, three different PCBs were created for each of the package models. These differ in the size of the cooling areas with a length of 'a' (heat sink), which are bonded to the heat dissipating parts of the package (die pad in the P-TO-252 or center pins for the P-DSO-14) (see Figure 3.41).

The FEM simulation provided the following results for the static thermal resistances:

- R_{thj-a} (junction-ambient = system-environment) following resistances and also included in this the
- R_{thj-c} (junction-case) for packages with cooling-pad, or

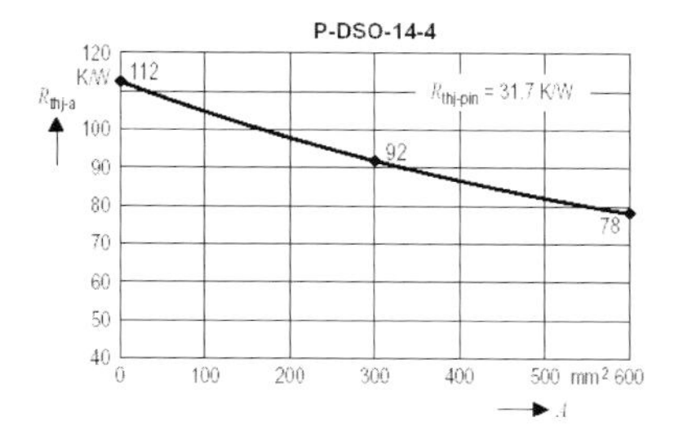

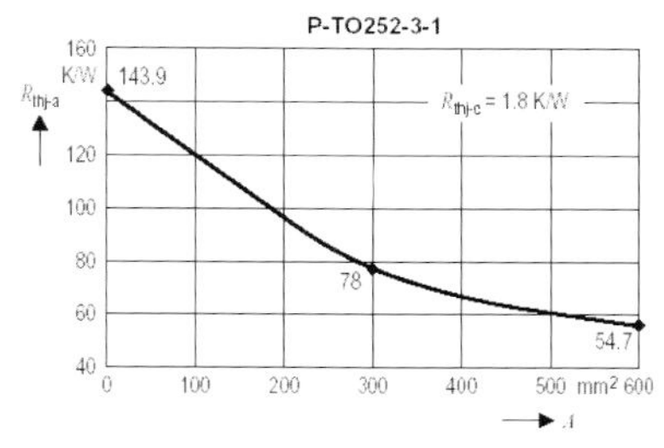

Figure 3.42
Showing how the thermal resistance between the junction and ambient, R_{thj-a}, depends on the available cooling area A

- $R_{thj-pin}$ (junction to a defined pin) for P-DSO packages with no cooling-pad.

As these values depend on the active chip area, in each case a chip of medium size (> 2 mm^2) was simulated. As the chip resistances are small compared to the PCB resistances, the changes can be neglected for the application when the power loss is in the medium range (< 5 W).

If one plots the static values of the thermal resistance R_{thj-a} (junction- ambient = system-environment) against the PCB cooling area, one gets a function which is exceptionally important for the application of the device (see Figure 3.42).

By estimating the cooling area in the real application, the user can thus easily determine the value to be expected for R_{thj-a}, particularly as the values shown were calculated for still air and ignoring any thermal radiation.

They thus represent the "worst case". With real applications, the values for the thermal resistance will sometimes be significantly lower. For example, with an airflow of 500 lin ff/min (linear feet per minute), the value of R_{thj-a} for the P-DSO-14-4 is about 30% lower (Figure 3.43).

FEM analysis can also be used in analyzing the dynamic behavior. The dynamic thermal resistance is defined as the ratio

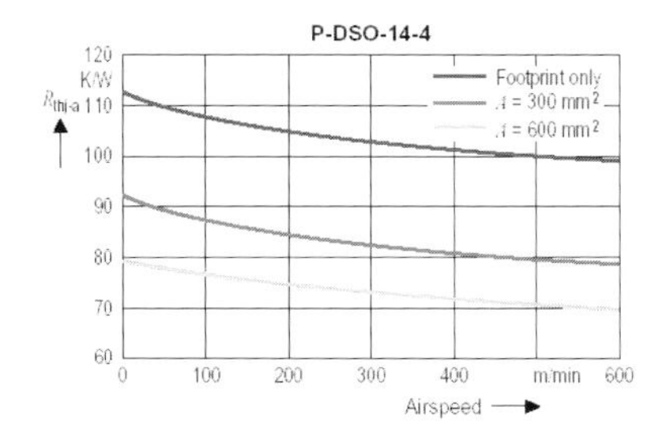

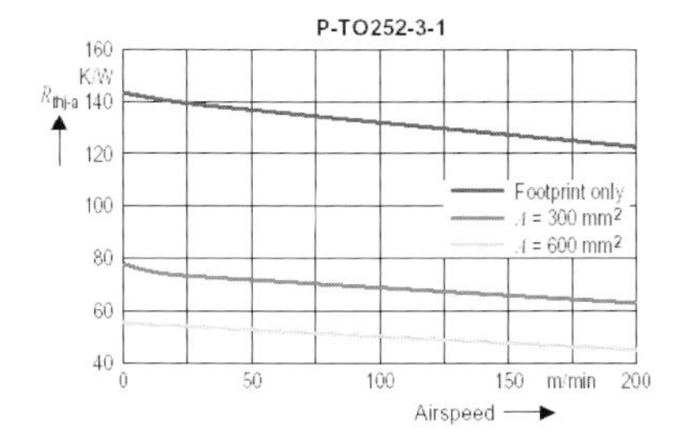

Figure 3.43
Showing how the thermal resistance between the junction and ambient, R_{thj-a}, depends on the speed of the airflow

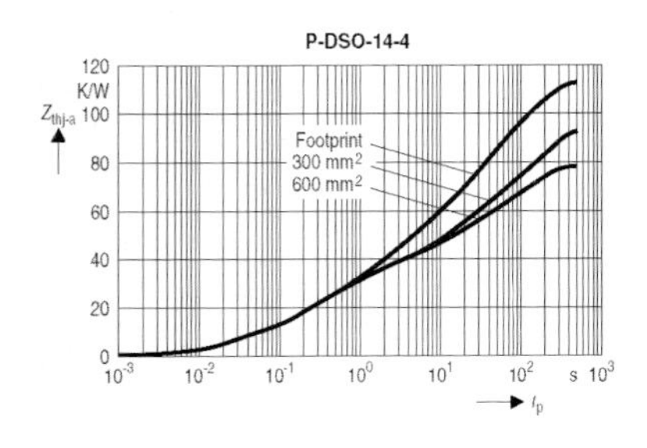

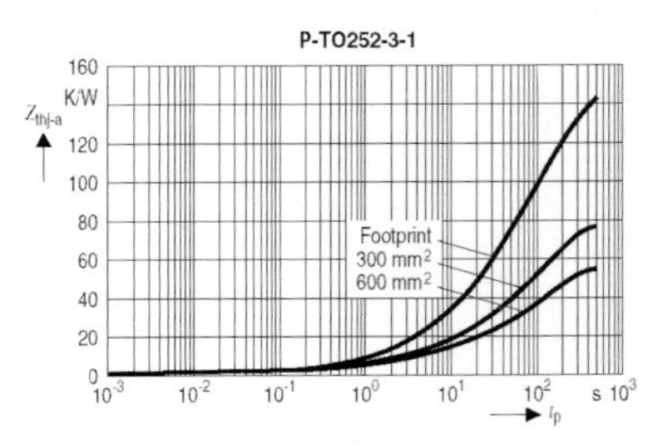

Figure 3.44
Thermal impedances for the P-DSO-14-4 and for the P-TO-252 (D-pack), each for various PCB configurations

of the temperature difference reached at time t_p:

$\Delta T = T_j - T_a$

If one carries out a transient FEM simulation, one can easily obtain the graph $Z_{thj-a} = f(t_p)$ (dynamic thermal resistance as a function of the pulse width t_p). From this curve, it is possible to calculate the peak temperatures.

An example, using the following data, should clarify this:

- P-TO-252 (D-pack); 3 cm^2 cooling area;
- Power loss $P_V = 1$ W;
- Pulse width $t_p = 100$ s;
- Ambient temperature $T_a = 85°C$.

From the middle curve in Figure 3.44 one can read off the value of R_{thj-a} at $t_p = 100$ s as being about 50 K/W. From this one gets the temperature rise $\Delta T = P_V \cdot R_{thj-a} =$ 50 K.

The peak temperature T_{jmax} is thus given as $T_{jmax} = 85°C + 50°C = 135°C$.

3.5.5 The "Thermal and Package Information" datasheet

In order to give the user an overview of the parameters relevant for an application, Infineon makes available "Thermal and Package Information" datasheets for its power semiconductor packages. This summary contains all the important data for the package concerned. For example, the datasheet for the P-DSO-14-4 gives the user numerous items of data. These start with the footprint and package dimensions. The various variants of the PCB for which simulations have been performed are then shown. This is followed by the FEM thermal distribution images, and finally the results of the FEM simulation in diagrammatic form.

The diagram of the static thermal resistance R_{thj-a} as a function of the PCB cooling area A also shows the thermal resistance R_{thj-c} (junction-case) associated with the package, or $R_{thj-pin}$ for the SO packages (tabless), as applicable.

In addition, the user will find a diagram of the dynamic thermal impedance Z_{thj-a}, with three curves for the various PCB cooling areas, as a function of the duration of the single pulse, t_p.

At this point, it should be mentioned again that these values really represent "worst cases", because convection and heat radiation are not taken into account. So, in using this data, the user is always on the safe side. If the layout is very tight, a detailed simulation should clarify matters.

3.5.6 The product-specific characteristics of power semiconductor packages for automotive applications

FET/IGBT

Mosfets and smart-FETs are offered in the common heatslug packages for high current applications. Figure 3.45 shows the important packages. Some of the molding compound has been shown as transparent, to reveal the internal structure.

As the channel resistance falls, due to improved wafer technologies, the portion due to the package becomes relevant. For this reason the bondwire diameter and the number of parallel bondwires has been increased, on the one hand. On the other hand, in order to reduce the vertical channel resistance, the thickness of the power-MOS wafer has been reduced below 100 μm.

In most applications nowadays, packages capable of surface mounting are required. Infineon's offering includes all the common types, and on top of this has developed its own special packages. For example, the TO252-15 has three heatslugs, insulated from one another, and is outstandingly suitable for the integration of a so-called H-bridge circuit (see also section 3.6.4).

Very recently, in order to switch very high powers and currents several MOS transistors have been combined (generally connected six at a time as a 3-phase half-bridge), as is common in industrial electronics. If a 42 V power supply is implemented in automobiles these new applications will be required and produced in high volumes.

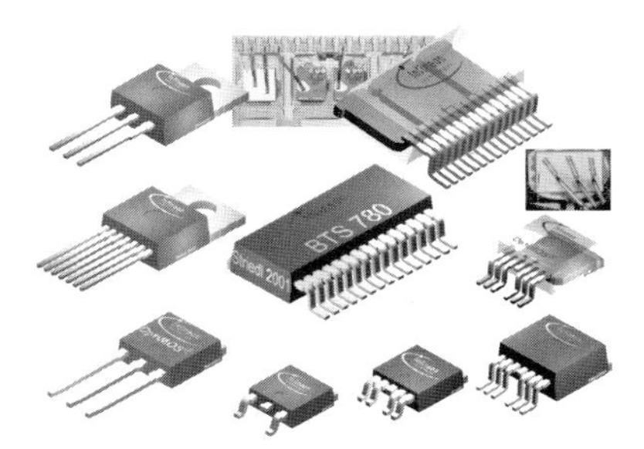

Figure 3.45 Packages with heatslugs

Figure 3.46
Infineon's most important Power SO packages

Smart FETs

If it is required to implement multi-channel switches, more than seven connections will be required. For this reason, the range of packages with heat slugs has been extended with a series of so-called Power SO packages. Infineon can call on a range of packages, providing anything up to 100 connections. Figure 3.46 shows some important representatives of this type.

It is not necessary in every case to dissipate high thermal power loss, and in many cases may even be undesirable, because the heat dissipation to the environment is prevented by design characteristics (plastic packages). Nor is the very high thermal capacity of the heatslug packages required in all cases, because they result in very high power loss in the event of a malfunction (short circuit or overload), which must then be dissipated through the PCB.

If the aim is to optimize the use of materials, one obvious idea is to reduce the weight of the heat sink. In addition, it is in many cases sufficient to work with a rela-

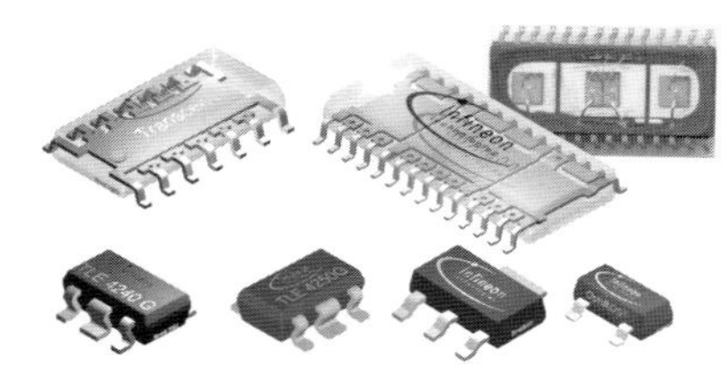

Figure 3.47
Examples of Infineon's "Thermally Enhanced Packages"

tively small thermal capacity. This is the case, for example, when switching on incandescent lamps. These are heated up by high peak currents in a few 10 ms to an operating current which is an order of magnitude lower. In such applications, if one reduces the ON-resistance by using more silicon, one can work with significantly larger thermal resistances. Infineon has encapsulated this pointedly in the slogan: "Silicon instead of heatsink".

The consequence has been the development of a wide range of so-called "thermally-enhanced" SO packages. This type of package has exactly the dimensions of the standard SO packages. However, the insides have been critically altered. From the leadframe, the chip carrier, thermally conducting links have been created, in that individual pins take the form of bars, feeding directly onto the leadframe. Figure 3.47 shows some of the SO Enhanced Packages.

An optimally priced package is obtained by combining the aluminium thick-wire assembly with the enhanced SO technology. This world-leading technology has been applied with great market success by Infineon in the SO TRILITHIC product family, the BTS77xx series (this can be seen nicely in the partially exposed device at the top right of Figure 3.47).

Smart power ICs and smart power system ICs

As in the case of multi-channel switches, smart power ICs also require packages with many connections. Here, a heatslug or a thermally enhanced package is selected depending on the application.

As large scale integration advances, so ever more heat sources are put together. For the manufacturer of complex ECUs, it is helpful to be able to create a virtual prototype, using a thermal simulation. This makes it possible to make preliminary estimates for the many interactions between the integrated subsystems.

Nevertheless, the complete ECU must be modeled with its application environment. Infineon offers this service, which sometimes involves extremely high costs, for its target customers.

Using the application data, such as the current versus time curves, PCB material, position of the connector (can be used for cooling), package construction (e.g. as a worst case: enclosed plastic shelter with no contact to the chassis) Infineon's product specialists work out the partitioning. In doing so, the following questions are answered, or are discussed with the customer:

- how many power transistors, switches, motor bridges are required?
- in which technology will the devices be realized?
- how high/low have the ON-resistances been specified?
- which packages best satisfy these requirements?

The positioning of the devices on the PCB is then optimized. For the application situations which have been determined jointly with the customer, he is given a temperature profile of the ECU, cross-sectional views and peak temperatures for the individual devices. If required, transient simulations can also be carried out.

A typical simulation for an ECU will now be shown. To carry this out, the individual device ICs, the PCBs, the package and the connector must be appropriately modeled. As an example, Figure 3.48 shows

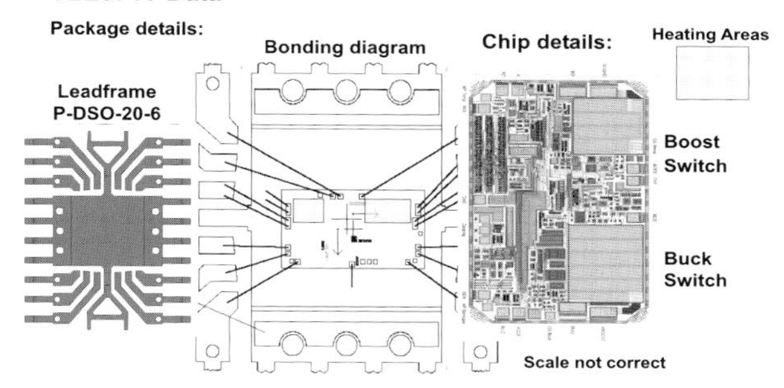

Figure 3.48
Modeling data for the TLE6711

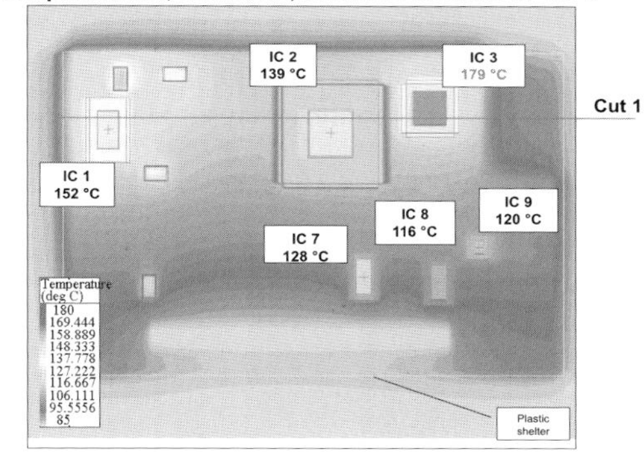

Figure 3.49
Thermal simulation of an automotive ECU (top view)

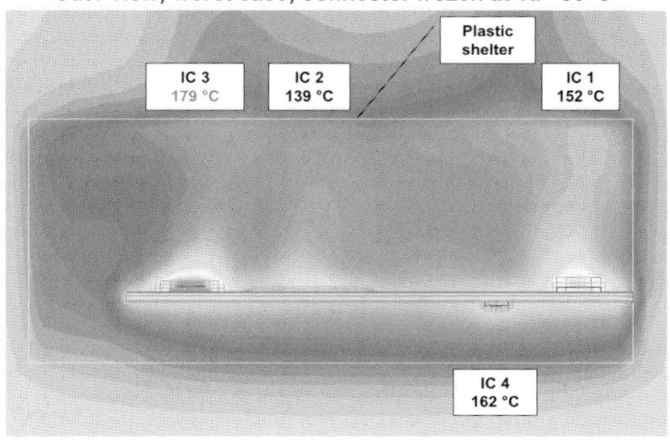

Figure 3.50
Side view through the "Cut1" plane of the same ECU

the TLE6711GL, a buck/boost switched mode power supply IC (IC 1 in the thermal simulation), with the required data.

The leadframe, the molding compound and the chip structure with each of the chip areas which produce power loss are represented in the model. The results of such a simulation can be seen in Figures 3.49 and 3.50.

The thermal enhanced rectangles are the devices which generate the power loss (in this example, ICs, diodes and switched

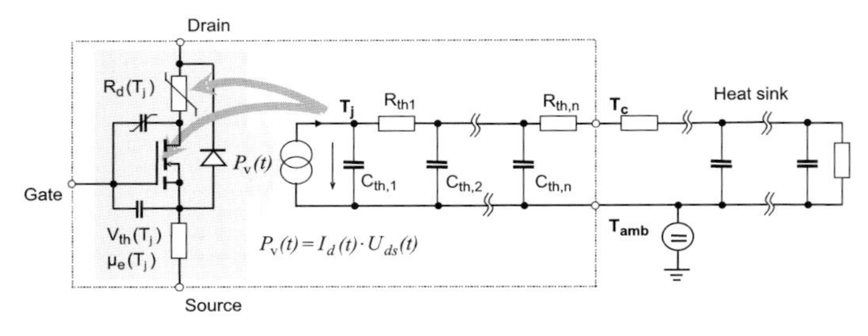

Figure 3.51
Electro-thermally linked simulation model for P-Spice/Saber; Example: a power MOS transistor

mode power supply inductances). The connector to the wiring harness can be seen as the elongated "cool" rectangle along the lower edge of the PCB. In this example IC 3, with a junction temperature of almost 180°C is overheated. However, the power density on the PCB is greatest in the neighborhood of IC 1.

The side view adds valuable details about the heat flow due to convection within the plastic shelter (Figure 3.50). One can thus see, for example, that the convection dome above IC1 extends out to the package cover. An alleviation of this situation can easily be realized by changing the arrangement of the devices.

As the electrical characteristics change with temperature (specifically, for example, R_{on} for a D-MOS transistor in automotive applications, which doubles over the very wide temperature range from –40°C to +175°C) it is desirable to link the electrical simulation with the thermal one. This is achieved by a dynamic model such as that shown in Figure 3.51. Infineon supplies such models for all the common types.

The direct feedback of the junction temperature (T_j) in the model to the electrical device parameter R_{on} enables the external cooling provisions to be simulated and, ultimately, dimensioned.

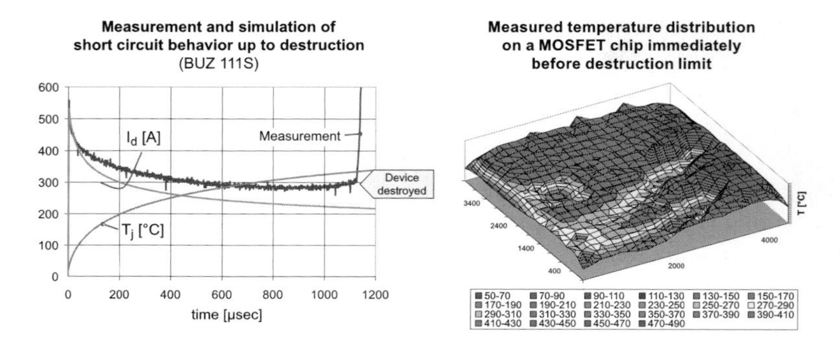

Figure 3.52
Measurement of the temperature distribution in a high power MOS switch, operating at high currents

Another very difficult area is the simulation or measurement, as applicable, of the ultimate limits for a power device. Figure 3.52 shows a measurement of the temperature or current distribution in an MOS transistor at high currents. Inhomogeneities, due to the gate contact and the heat dissipation by the bondwires, can be clearly recognized. With results like these, one can create a model which permits the cell densities to be optimized.

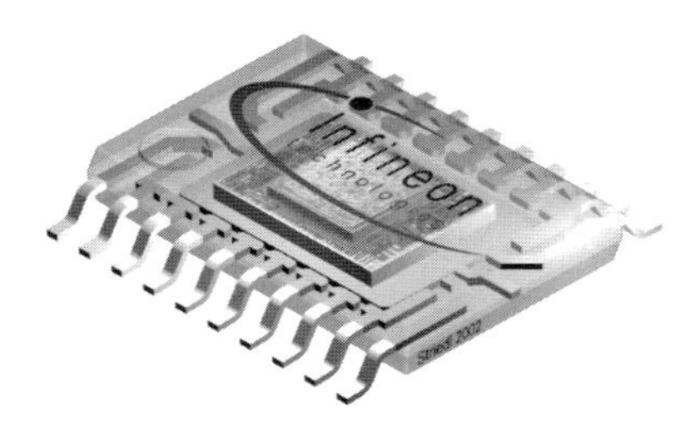

Figure 3.54
System in a package; a T-SSOP- 20 package with chip-on-chip mounting

3.5.7 Multichip packages and trends

Infineon has developed its chip-on-chip technology as a core competence for high current switches and motor bridges. Figure 3.53 shows one of the most modern high current PROFETs in the world, for peak currents up to about 100 A

The base chip can be made in the power D-MOS technology using only a few masking steps. On the left, one can see the source contacts to four external pins, effected by the four thick aluminium bondwires. As a result, each individual wire can still be checked. The complex top-chip has been affixed to the base chip in a suitable thermally-conducting way. The chip is connected to a few external connections via gold bondwires, which are so thin they can hardly be seen in the figure (and thus can be neglected in the thermal model). In addition, chip-to-chip bonds are used to effect a connection to the base chip, to control the gate and to enable the current to be measured.

This technology is, of course, equally usable for the coming generation of highly-integrated system chips. For example, a power/high-voltage chip can carry a highly-complex microcontroller chip. Figure 3.54 shows such a "system in a package".

In the near future, the large-scale integration of systems will demand new packages which can be integrated into the application as mechatronic elements. This means that lower thermal resistances must be realized for an increasing number of connections. "Chip-scale" is the magic word. With this, the dimensions of the

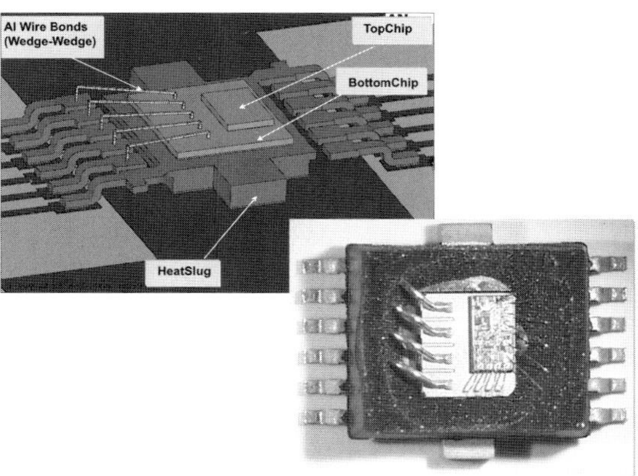

Figure 3.53
High current switch – model and the reality of the chip-on-chip technology

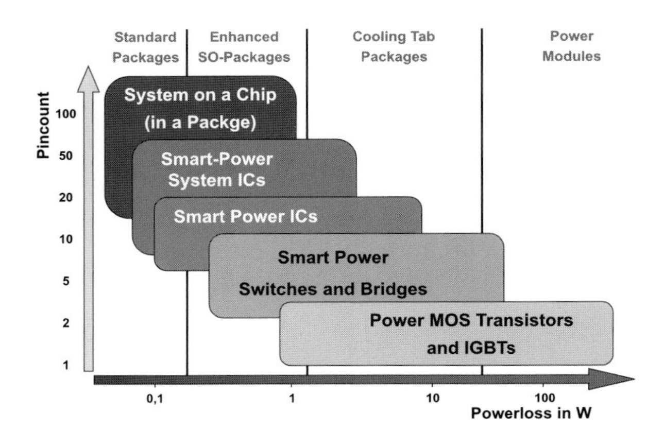

Figure 3.55
Pin-count against power loss for various product groups

package are only slightly larger than those of the chip. The so-called VQFN packages with an "exposed leadframe" (the leadframe is accessible for soldering on the underside of the package) are good examples of this new technology.

By way of summary, Figure 3.55 shows the relationship between pin-count and power loss. In the smaller power ranges, standard packages and the SO enhanced versions are used. They are principally used for more complex products. Following on from the range in which tab packages are used, for switches and bridge applications, are the power modules (generally transistor or IGBT sextets) as high-power packages.

The mounting of "naked" chips (bare-die) is an approach which is being more and more used. However, doing so requires expensive clean-room assembly. The same applies for so-called flip-chip assembly, in which a supplementary process step for the wafer applies 'bumps' to the bond spots on a chip. The chip can then be turned over (flipped) and soldered directly onto the carrier (generally an expensive ceramic substrate). A major advantage of the flip-chip technology is that the bond pads can be located almost anywhere on the chip surface. With the present commonly-used assembly methods on the other hand, the pads must be located on a pad rim around the circumference of the chip. On the other hand, arranging the cooling for a flip-chip is difficult, because only the bumps would conduct the heat. This is therefore alleviated by underfilling the chip with thermally conductive "underfill materials". Another problem which may not be neglected is that of ensuring a "Known Good Die", the guarantee from the semiconductor manufacturer that a good chip has been supplied.

3.6 Automotive power devices

3.6.1 MOSFETs and IGBTs

In the case of power semiconductors, it is power MOS transistors which represent the basis for the wide range of possible products. Users want to make use of a device which comes as near as possible to the ideal switch. In the 'on' state, the loss voltage (the voltage drop) should be as small as possible. This feature is characterized by the on-state resistance between the drain and the source, know as the $R_{DS(on)}$.

When switched into the off state, on the other hand, the transistor must be able to apply a secure block against the highest possible voltages, with no current flow-

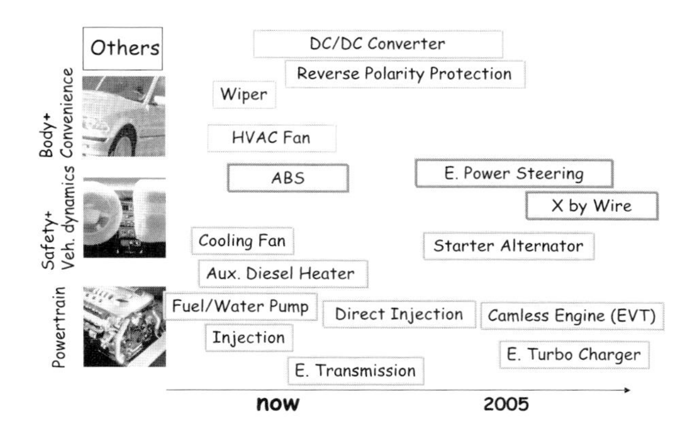

Figure 3.56 Applications for power MOS transistors

ing. These characteristics are quantified by the breakdown voltage between the drain and the source, V_{BRDS}, and the leakage current characteristics. The ideal switch can be switched on and off with no power consumption. A real MOS transistor requires a final charge which must be moved each time there is a switching operation.

In automotive engineering, the efficiency has a direct effect on the fuel consumption, and for this reason an MOS transistor with a low R_{on}, a negligible leakage current across the supply voltage range and a favorable switching characteristic, will find possible uses in countless applications.

Figure 3.56 shows a summary of the most important applications.

Whereas DC-DC converters tend to embody smaller, higher-volume applications, the high current applications such as a starter/alternator (ISAD) integrated into the power train require many transistors in parallel for a single controller, in order to achieve an acceptable R_{on}.

Many applications in an automobile replace conventional mechanical or hydraulic solutions. The so-called "front-drive accessory", the belt drive for driving the water pump, the alternator, the power steering pump and various other units, can be completely replaced by electrical drives. Linear drives, such as valve positioners, also require MOS switches to actuate them. In short: wherever force is required, one can nowadays find an "electronic muscle" – a device which controls the electric current and thereby drives an electromagnetic positioner.

In the near future, controllers in an automobile will operate using distributed intelligence. An approach in which the electronics are located directly in the actuator; the mechanical components together with the electronics become mechatronics. Ever more low-resistance switches make possible new applications with ever higher current consumption; the 12 V vehicle supply is coming up against its physical limits. For this reason, from 2005 on the voltage will be increased by a factor of three to 42 V; the power levels which can be handled thus rise by almost an order of magnitude. As the MOS technology is scarcely capable of yielding yet lower resistance switches, closer consideration must be given to the packaging technology.

The first step is the optimization of the bond technology. At Infineon, the Power Bond has been developed, a special high-current triple bonding. Figure 3.57 shows

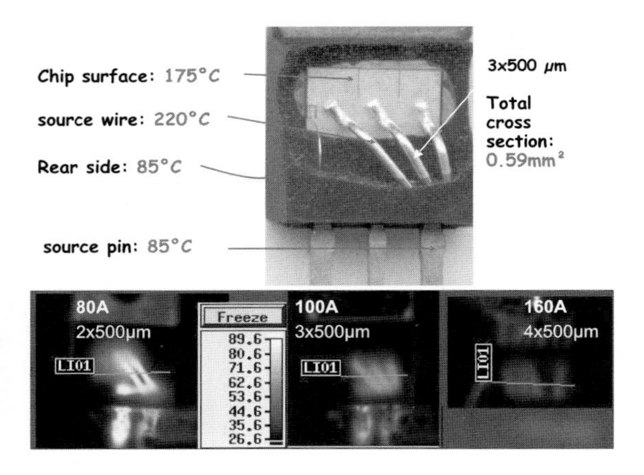

Figure 3.57
The power bond permits very high currents in the conventional TO220 and D-Pack packages

the temperatures of the individual parts of the package during high current operation. As can be seen, the bondwire is the "weak link" in the system.

For very high currents, one nowadays constructs several MOS transistors in one module. If even this is not enough, the bondwire must be replaced by flat contact areas. A "bondless" package design study for an MOS half-bridge (two MOS transistors in series) can be seen in Figure 3.58.

Why is this arrangement of two MOS transistors so important? If one considers the efficiency of the application in the case of an automobile, it is a fact that the fuel consumption is becoming one of the most important financial-ecological factors. For this reason it is necessary to realize almost all of the energy conversion processes in switched mode. With the free running mode being implemented by means of synchronous rectifiers. In almost every application, regardless of whether a motor is being driven, or a buck or boost converter is being operated, MOS half-bridges are used in every case.

In terms of both the temperature range and also the voltage and short-circuit requirements, the requirements to be met in

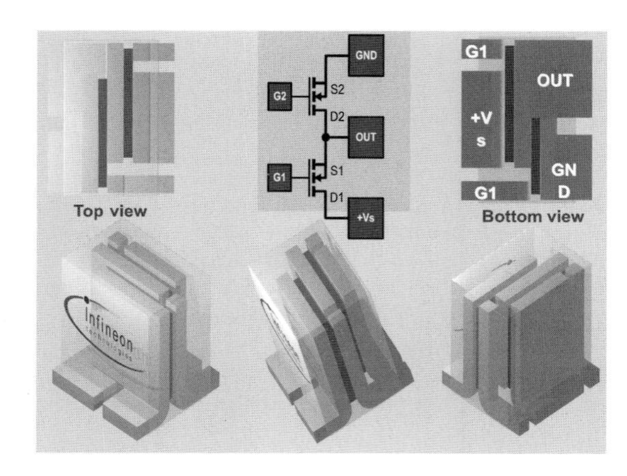

Figure 3.58
Bondless package design study for an MOS half-bridge

an automobile are more demanding. Consequently, the safe operating area (SOA) of modern transistors is especially important. For this reason, Infineon's OPTIMOS transistors have been specially tailored and developed to the needs in an automobile. Thus they can withstand junction temperatures of up to 200°C without difficulty. In another special technology, IGBTs have been developed for ignition applications. These devices will withstand high voltages, and can absorb high energies.

If one wishes to increase the functionality of the devices, additional functions must be realized within one package. This leads on to the smart switches.

3.6.2 Smart FETs and smart IGBTs

The use of power MOS transistors calls for additional circuit elements to drive the gate and for status evaluation. When low-side switches are driven with low switching frequencies, the gate can be switched directly from a standard controller port. At higher frequencies, or when a high-side switch is being driven, the current efficiency or withstand voltage respectively of the commercially available controller outputs are insufficient. Status evaluation, such as for overcurrent, sudden load disconnection, overtemperature and short circuit, involves costs which may be substantial. The problem can be solved with an additional driver circuit with diagnostic capabilities. In many cases, the use of smart FETs or smart IGBTs is the cost-effective solution. These advanced devices have switching functions integrated with them, either by chip-on-chip assembly or, indeed, monolithically.

Infineon offers a very wide range of smart switches.

The TEMPFETs were the first intelligent switches in the world to have an integral protective function. A controller chip bonded adhesively to the power chip clamps the gate voltage when the device becomes overheated. From underneath, the device cannot be distinguished from a normal MOS transistor, and is just as easy to use. Figure 3.59 shows the most important fast TEMPFETs from Infineon, the Speed TEMPFETs. To enable the constructional details to be better seen, the chosen view omits the molding compound.

HIT-FET

There is an increase in functionality when, in addition to pure temperature monitoring, current measurement and overvoltage measurement are required. With these, the device can have defined protection within a safe operating area (SOA). In this case, Infineon's HIT-FETs (High-

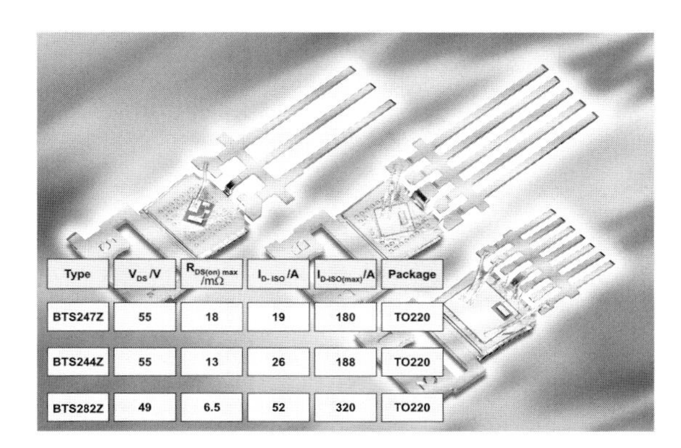

Type	V_{DS} /V	$R_{DS(on)\,max}$ /mΩ	$I_{D\text{-}ISO}$ /A	$I_{D\text{-}ISO(max)}$/A	Package
BTS247Z	55	18	19	180	TO220
BTS244Z	55	13	26	188	TO220
BTS282Z	49	6.5	52	320	TO220

Figure 3.59
The Speed-TEMPFETs: fast, protected low-side switches, in packages with 3, 5 and 7 connections

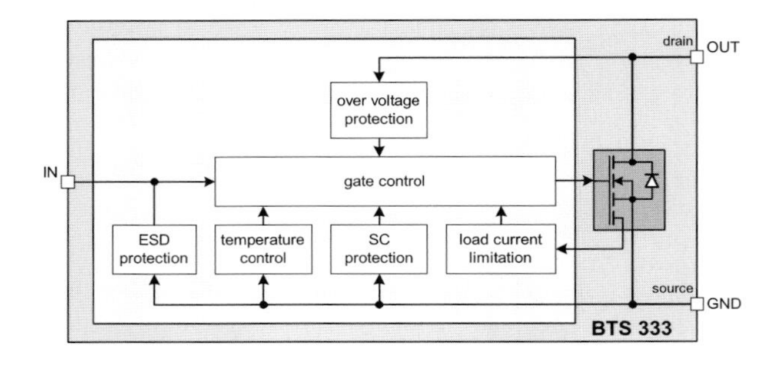

Figure 3.60
Block diagram of a HI-FET with its driver and diagnostic functions

R_{ON} (m Ω)	1 Channel		2 Channel	
	Latch	Restart	Latch	Restart
2000		BTS3205		
500		BSP75		BTS3408
200	BTS3110	BSP76		BTS3410
100	BTS3118	BSP77 BTS118		
50	BTS3134	BSP78 BTS134		
28	BTS3142	BTS142		
18		BTS333		

Figure 3.61
The HIT-FET family for use in automobiles

ly-Integrated FETs) will be used. A block diagram of a typical representative of this group of devices is shown in Figure 3.60.

Apart from the driver circuit, which operates the power MOS gate, the drain current of the power transistor is sampled by means of special sensing cells. In addition, there is an integrated overvoltage protection. All the protective circuits except for the ESD input protection circuit exert an appropriate influence on the gate controller.

The range of these switches, which are optimized for automotive applications, can be seen in Figure 3.61.

The family contains one- and two-channel switches, some with latched protection circuits and alternatively a series with automatic sampling protection circuits (restart). Whereas the higher resistance switches, in the range from 2 Ohms down to about 100 mOhm, are made using smart power technology (SPT), for the very low resistance switches the smart technology with vertical current flow makes its appearance.

Smart IGBTs

In a certain sense, the smart IGBT represents a special case of a low-side switch. In an automobile, it is required for driving the ignition coil. In today's automobiles, the high voltage is generally generated centrally using standard IGBTs, and then distributed. This requires very expensive high-voltage wiring. To save this expense,

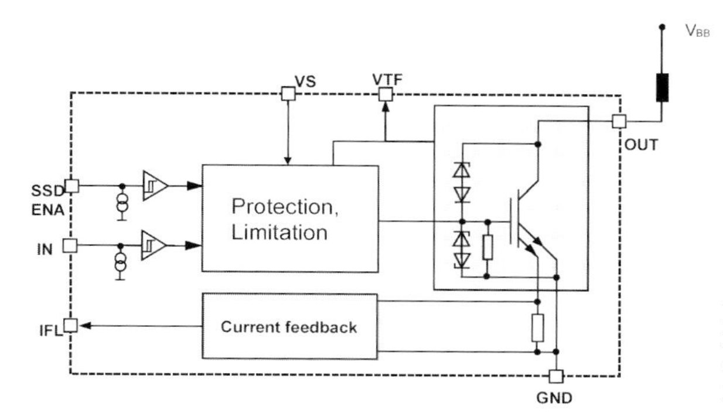

Figure 3.62
Block diagram of the chip-on-chip smart IGBT BTS2145

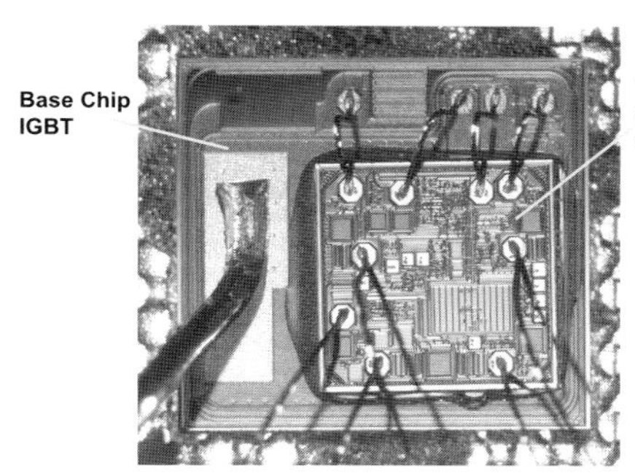

Figure 3.63
Internal construction of the BTS2145 smart IGBT

ignition coils have been developed which are mounted on the spark plugs (pencil coils) and which are driven by smart drivers. Because smart drivers are more reliable, and help to save on size and therefore weight, they are superior to conventional discrete solutions. Infineon has developed the BTS2145, a smart IGBT which is constructed as a chip-on-chip. Figure 3.62 shows a block diagram of this product.

In addition to a comprehensive diagnostic interface, the device contains all the necessary circuits to protect the power IGBT from overvoltage, overcurrent and overtemperature. The insides of this chip-on-chip device can be seen in Figure 3.63.

PROFET

When high-side switches are required, smart technology can be used to effect the advantageous and cost-optimal monolithic integration of one or more switches, including the driver circuits, level conversion, protection circuits and a diagnostic facility. As complexity increases, diagnosis is assuming a key role. If this functionality were omitted, even today it would scarcely be possible to effect rapid and cost-effective repairs. Infineon has developed a wide range of switches with diagnostic functions for high-side applications, the so-called protected FETs (PROFETs). These devices contain chips which are manufactured in the smart technology, with a vertical current flow.

Because very many of the functions in an automobile are decentralized, a vehicle requires countless numbers of these switches with diagnostic capabilities, for almost every field of application. The principle of their application and the advantages accruing to the user from this

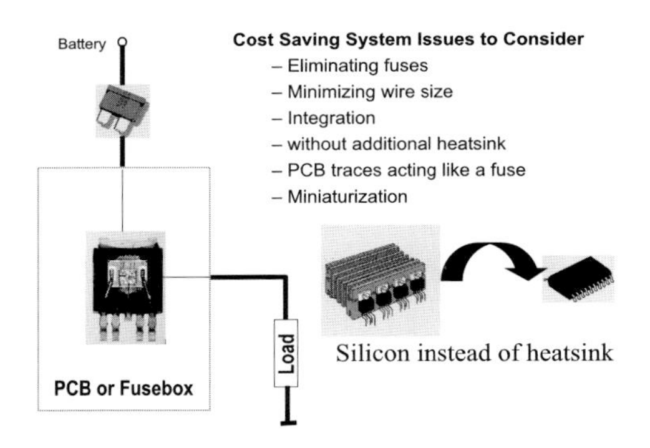

Figure 3.64
A PROFET as a high-side switch replaces a relay, with significantly greater reliability

family of switches are shown in Figure 3.64.

However, the main field of application is in 'body electronics', because PROFETs are outstandingly suited to the switching of incandescent lamps, motors and resistive consumers.

They can be used to switch nominal currents of up to 50 A. In many applications, in particular in the switching of lamps, no heat sinks are required because high pulse powers only occur at the instant of switching on. By limiting the peak currents, the cross-sections of wires can be reduced and hence, ultimately, the weight and fuel consumption can be reduced. PWM operation is also possible with no difficulties. Even the expensive fuses can in many cases be cost-effectively replaced. However, the high pulse powers do lead to high thermal loading on the devices. These thermal cycles call for very robust packages and a great deal of know-how in the layout of the power chips. A typical block diagram of a PROFET with four switched outputs is shown in Figure 3.65.

As in the case of the HITFETs, the current, voltage and temperature are parameters which are determined by appropriate circuits. However, the gate controller is

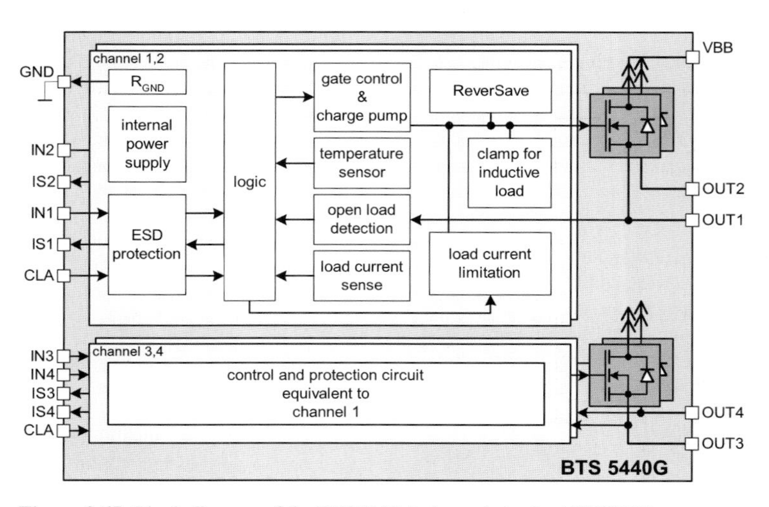

Figure 3.65 Block diagram of the BTS5440 4-channel standard PROFET

significantly more costly, because the gate must be "pumped" from the supply voltage by a charge pump, in order to use the on-state resistance of the MOS transistor. In order to keep the application costs low, all the required devices have been monolithically integrated. The safe commutation of inductive loads is achieved by a special Zener circuit. A further patented function is the Reversafe function. This ensures that in the event of polarity reversal the module does not suffer thermal damage.

Signaling to the control stage (the microcontroller) is effected by analog sampling of the output current, or via one or more digital status flags. The output stage for these flags is realized in open drain form, so that, if required, several flags can be read out using one IO port ("wired or"). The latest Smart5 technology even permits a serial peripheral interface (SPI) to be monolithically integrated. Appropriate multi-function switches which, for example, undertake the complete control of the lighting for vehicle headlights, are under development.

If even higher currents are required, the chip-on-chip technology is used, in the form of high-current PROFETs. With these products, an intelligent top chip is adhesively bonded onto a power MOS chip. This enables the user of Infineon's BTS555, which is unique throughout the world, to switch currents of up to 1000 A. Particularly when such high currents are switched off, the inductance of the supply conductors represents an additional challenge; because the energy stored in them must, for example in the event of a short circuit, be absorbed by the PROFET.

The high-current PROFETs are a nice example of how the flexible use of front- and back-end technologies can produce cost-optimized products. For example, the source of the entire voltage in an automobile, the battery terminals, can now be switched. This permits power management for the many previously independent loads.

The block diagram of a high-current PROFET can be seen in Figure 3.66. This device is one of the latest generation, and is supplied in the cost- and space-saving D-Pack package. And it is, furthermore, ready for use with the future 42 V vehicle power supply.

The device is switched on by pulling the input pin IN down to ground potential.

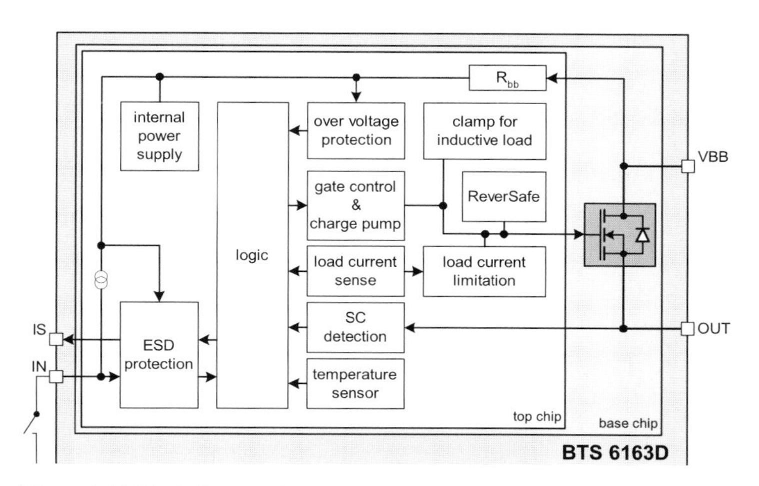

Figure 3.66 Block diagram of the BTS6163D high-current PROFET

Table 3.3 Data for the BTS61xx family

Type	$V_{DS(AZ)}$ [V]	$R_{on(max)}$ [mOhm]	$I_{L\text{-}SC(typ)}$ [A]	Current Sense Ratio	Shutdown overtemp./short circuit	Package
BTS6133D	39	10.0	75	10 000	restart / restart	DPAK5
BTS 6143D	39	10.0	75	10 000	restart / latch	DPAK5
BTS 6144P	39	9.0	90	12 500	restart / latch	DPAK5
BTS 6163D	63	20.0	70	9 000	restart / latch	TO220/7

The IS connection provides an image of the output current. By using a resistance to ground, it is then easy to make a real time current measurement. Table 3.3 shows a summary of the data for the BTS61xx product family, which will have a great future.

For the countless PROFET applications, Infineon has a very wide product range in mass production. Figure 3.67 shows the PROFET products which are offered in packages with heatslugs, in the form of a selection tree. This makes it easier for the user to find the desired product.

The products which do not have heatslugs (thermally enhanced SO packages) are shown as a selection tree in Figure 3.68.

It is, of course, possible to combine HITFETs and PROFETs to form half- or full-bridges. Infineon launched devices of this type on the marked in 1998, as TRILITH-ICs.

3.6.3 Multi-channel switches

In contrast to the smart switches, multi-channel switches are manufactured in the smart power technology, SPT. Using this it is possible to integrate power stages up to continuous currents of 5 A DC (peak current 10 A) monolithically with analog functions and complex logic. It is possible to combine both high-side and low-side switches. The main automotive applications are in the area of the power-

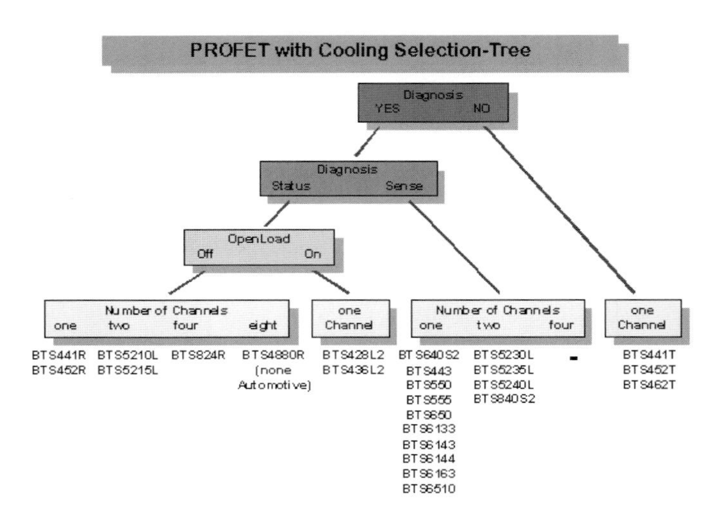

Figure 3.67 PROFETs in packages with heatslugs

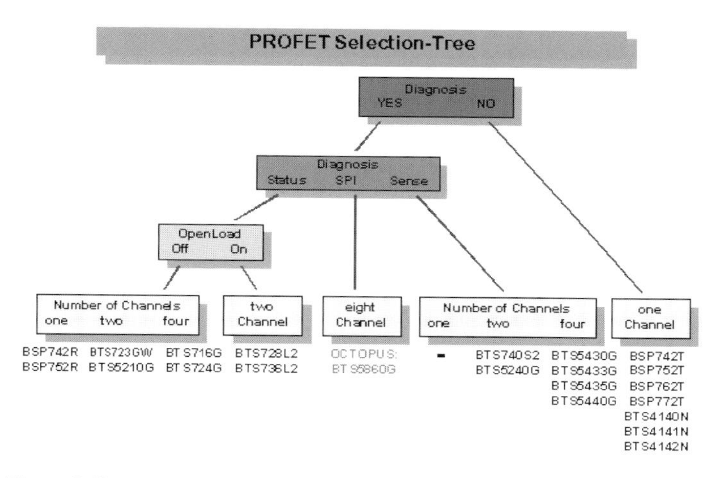

Figure 3.68 PROFETs with "thermally enhanced SO packages"

train, engine management and gearbox control. In the safety field, multi-channel low-side switches are used as valve drivers in ABS and chassis applications. In these applications, the current must generally be regulated in high-frequency switched mode. The universally high inductances present a particular challenge in terms of long-term reliability, because with every commutation operation a considerable power loss must be absorbed in the power MOS switch. The requirements for low radiation switching processes (soft clipping for PWM and commutation), for the temperature range (gearbox controls) and for the diagnostic capabilities are therefore particularly great. The main characteristics of Infineon's multi-channel low-side switches are:

Basic characteristics

- Comprehensive protection
  - Short circuit + overload
  - Overtemperature
  - Overvoltage
  - Electrostatic discharge
- Current limitation
- Open load detection
- Direct parallel control of 4/6/8 … channels for PWM applications
- 5 V supply voltage
- 3.3V µC-compatible

Product-specific characteristics

- SPI interface
- P&H current regulation
- Detection of short circuits to ground
- General FAULT connection
- Standby mode
- Enhanced supply voltage range (4.5V- 32V)
- Open load programming pin
- Current monitoring
- Overload shutdown
- µs-bus
- Configurable for high-side/low-side

In power train applications these ICs are typically used to operate all the low-side loads in the range from a few mA up to several Amps. From pure input/output through to PWM applications. In the body field, multi-function switches find uses as relay drivers and in diverse low power applications. What is required here is generally pure input/output operations.

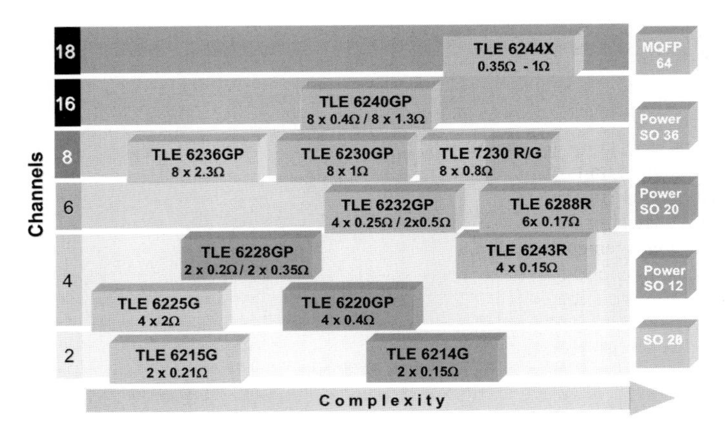

Figure 3.69 The range of low-side switches from Infineon

In order to achieve the highest possible market coverage, Infineon has developed a wide range of multi-channel low-side switches, which are offered in the widest variety of packages. Figure 3.69 shows the products in a complexity/channel-count matrix.

It will be seen that everything is available, from simple 2-channel switches up to highly complex system switches for engine management in modern automobiles. Because of the severe automotive specifications (wide temperature and voltage ranges, etc.), these modern component devices can, of course, also be used with no difficulties in industrial and consumer areas

The latest generation of these products is equipped with precise current regulation; such as is required, for example, in new gearbox controllers.

In order to increase their possible uses, some of these new products can be configured as high-side or low-side switches. This enables bridge applications or high-efficiency synchronous rectifiers to be constructed.

As early as 1996, Infineon introduced to the market the first HS-LS product in the world: the "quarter-bridge switch", TLE6208-6. This incorporates six HS and six LS switches, which can be actuated and diagnosed via a serial interface. The

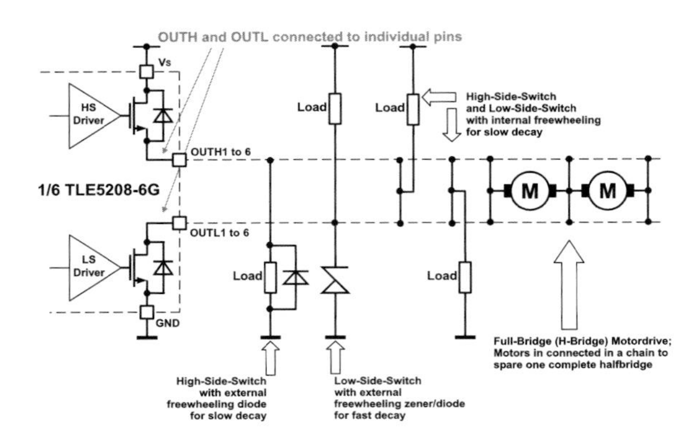

Figure 3.70 Possible uses of the TLE6308-6 quarter-bridge switch

diverse possibilities for switching the outputs are shown in Figure 3.70.

The target applications for this device are in the body field. For example, the TLE6208-6 can control up to five motors in an air conditioning unit quasi-synchronously. Further, it is suitable for driving relays, or for switching ECU functions on/off (battery management) such as sensors etc. As an automobile contains a host of reversing (right/left rotating) drives, controlling them electronically requires a wide range of motor bridge control ICs.

3.6.4 Bridge circuits

In automobiles there is a sharply rising number of electric motor drives. Depending on whether the motor is operated unidirectionally or bidirectionally, different driver circuits are required. Figure 3.71 shows various concepts. The unidirectional drive types, A, B and C, can of course also be constructed using high-side switches.

Unidirectional applications can be provided for by either low-side or high-side switches. In the higher-frequency PWM mode, half-bridges are used because they permit a greater efficiency. For bidirectional drives, a bridge circuit is required. One high-side and one low-side switch, in a series circuit between the supply voltage and ground, form a half-bridge. Between two of these half-bridges therefore, a bipolar load (a motor) can be supplied

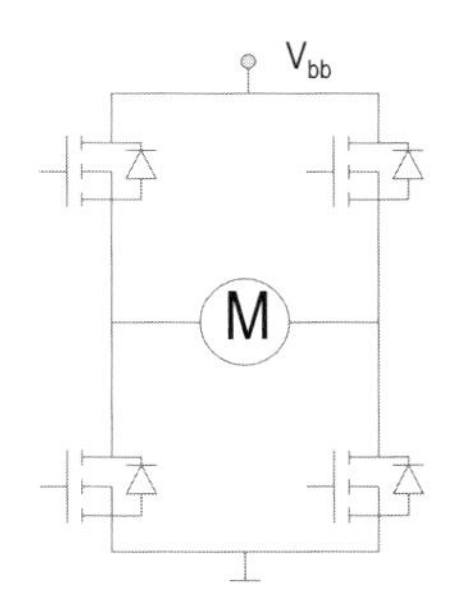

Figure 3.71 Driver concepts for DC motors

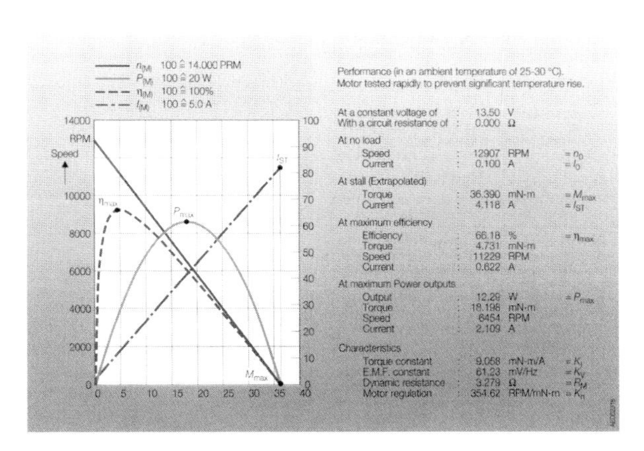

Figure 3.72
Set of characteristic curves for a DC motor

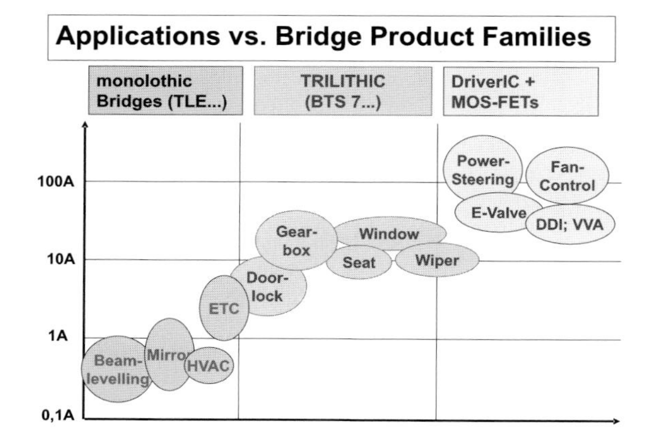

Figure 3.73
Motor applications clustered by working current

with forward and backward current.

In dimensioning a drive, the parameter which is regarded as most important is the torque. For an electric motor, the torque and the current in the armature winding are directly proportional, and within wide limits are linearly related. This can be seen from the set of characteristic curves for a DC motor, in Figure 3.72.

The values specified are those for a typical small door lock motor. Further details will be found on Infineon's homepage, as a 'special subject book': "DC motor drives".

Figure 3.73 shows the automotive applications, arranged by nominal current.

The three groups result from the different technologies which are used to manufacture the driver stages.

The smaller currents up to about 10 A can be serviced by monolithically integrated motor bridges, manufactured in smart power technologies. This group is shown in Figure 3.74 as a selection tree.

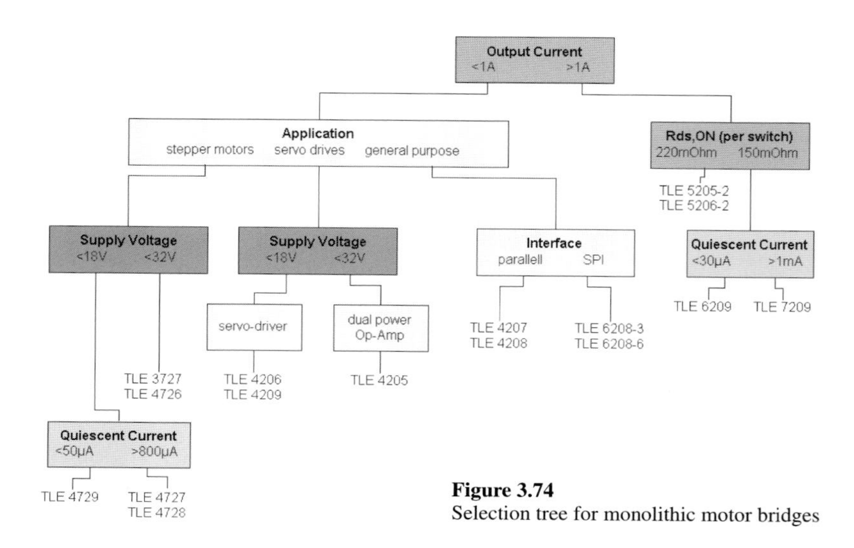

Figure 3.74
Selection tree for monolithic motor bridges

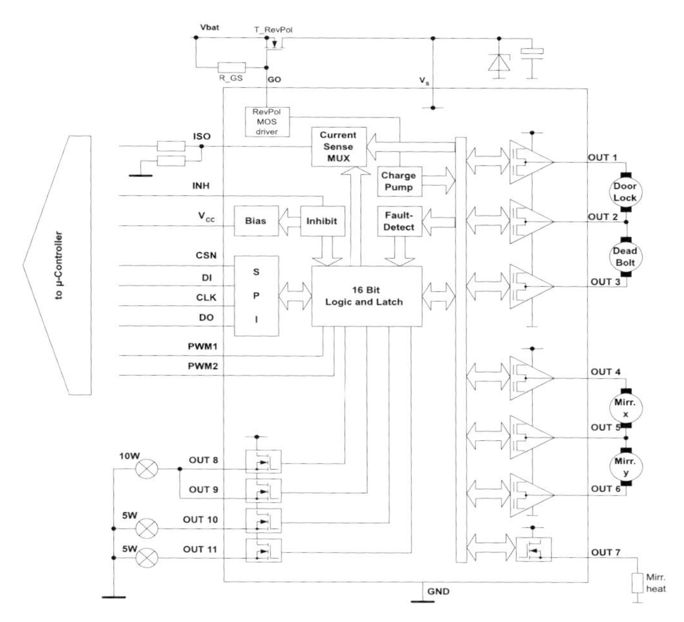

Figure 3.75
The multi-function power IC TLE 7201R, with all the actuation functions for an automobile front door monolithically integrated

In this group, the motor bridges with nominal currents of more than one Ampere will be found on the right hand side. In a vehicle these bridges are used, for example, in an electronic throttle (electronic gas pedal, E-GAS) application. The significantly larger group of motor drives under 1 A can be subdivided in turn into servomotor drives (drives with analog feedback, e.g., the headlamp beam adjustment (leveling) which was introduced as a legal requirement in the EU area in the 1980s), standard DC motor bridges, and the specific group of stepper motor drivers. Stepper motor drivers are used to avoid the need for feedback, generally a potentiometer, and the complete analysis circuit. Their main areas of application are so-called linear actuators (stepper motors with worm drives as valve positioners) and diverse servo-drives.

With the large-scale integration of these products into electronic controllers in an automobile, the motor bridges are incorporated into system ICs, such as the TLE7201R multi-function bridge circuit. The block diagram of this module is shown in Figure 3.75.

For currents of over 10 A up to about 100 A, low-resistance switches are required. For these, smart switches are used in combination to form motor bridges, the so-called TRILITHICs.

New bonding techniques are used in the so-called thermally enhanced SO packages to permit larger currents compared with gold nailhead multi-bonds, with significantly increased reliability. Figure 3.76 shows the bonding of a TRILITHIC in an SO28 package.

For the high-volume applications of window lifts, sun roof and seat adjusters, Infineon has developed the BTS78xx series of TRILITHICs.

The new generation has 250 µm Aluminum bond wires:

Figure 3.76
High-current bonding of TrilithICs in an SO28 package

This second generation is accommodated in a package specially developed for the TRILITHICs, with three separate cooling areas, the power pack.

In many new applications such as power steering, variable valve train (VVT), fan control, diesel direct injection (DDI) etc., the currents required are sometimes well in excess of 100 A. These currents must in addition be controlled using fast PWM regulators. High gate currents are required for this purpose, which necessitates MOS driver ICs which can supply these high pulsed currents and drive power MOS transistors in a half-bridge configuration, so that a cross-current (both transistors conducting, a so-called shoot-through) is reliably avoided. As many of the applications in an automobile must conform to strict security standards, further protection and diagnostic functions are required.

Infineon has completed its wide range of high-current switches at the top end with a series of driver ICs in the TLE628x series, the most important characteristics of which are shown in Table 3.4.

Table 3.4 Data for Infineon's TLE628x series of power MOS drivers

	TLE 6280/7 GP 3 Phase driver	**TLE 6283 Q** 3 Phase driver	**TLE 6281 G** H-Bridge driver	**TLE 6282 G** H Bridge driver/ half bridge driver
Package	Power SO36	MQFP 44	SO 20	SO 20
Vehicle power supply	12 V	12 V	12 & 42V	12 & 42V
Diagnostic	•	•	•	•
SC protection	•	•	•	•
100% DC		•	•	•
Current sense*		•		
Switching time	300 ns	300 ns	600 ns	600 ns
Main application	EPS, EHPS, Fan control	EPS, EHPS, Fan control	Wiper, Gearbox Window lift	EVT, VVA, Common rail

* Calculated for SPB 80N04S2-H4 (TO263, 40 V, 4 mOhm, Normal Level Device)

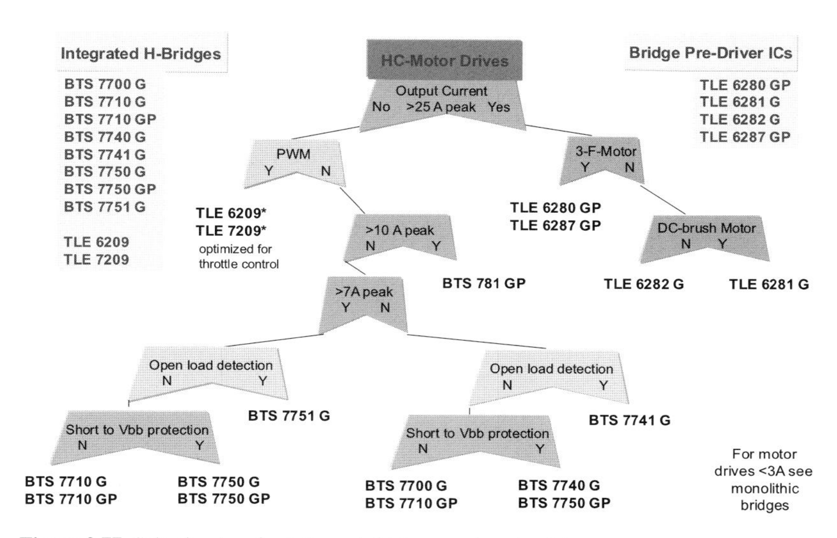

Figure 3.77 Selection tree for Infineon's high-current motor drivers

To summarize, Figure 3.77 now shows the selection tree for the high-current motor drivers.

On the left of this are shown the TRILITH-ICs in the BTS77xx and BTS78xx series. Down the left-hand branch are added the E-GAS bridges, TLE6209 and TLE7209, as fast PWM drivers.

Down the high-current line is the TLE628x series, the MOS driver ICs.

These drivers will also be integrated into system ICs in the near future, provided that their flexibility is not too greatly restricted by doing so.

The wide range confirms strikingly: motor drivers are to be found in almost every application segment.

3.6.5 Supply-ICs

This product group brings together the analog voltage regulators, DC-DC converters and current regulators.

Because a power supply module is required for each controller, products from the analog voltage regulator group are used in most applications. The most important of these are:

- ABS/ASR systems
- Airbag systems
- Air conditioning/heating
- Seat control
- Car radios
- Dashboards
- Door electronics
- Motor management
- Immobilizers
- Keyless entry systems
- Chassis/body
- Power train

In some areas, "stand-alone regulators" are integrated into system ICs. This is so, for example, with ABS and airbag systems, the "mature" applications. This term refers to the applications which have now been in automobiles for many years, which are subject to constant optimization activities and are therefore under heavy cost pressures.

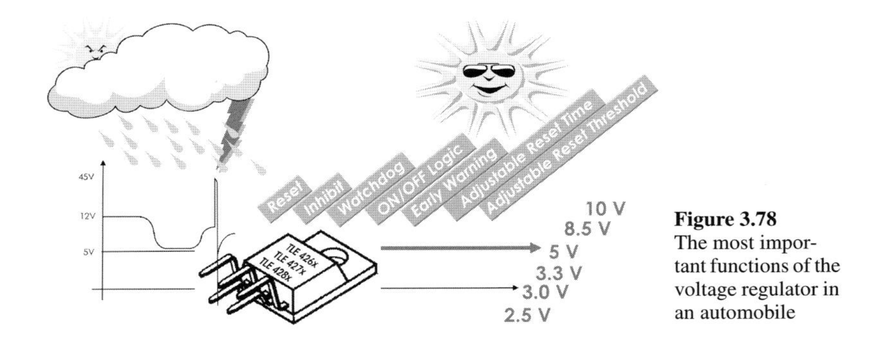

Figure 3.78 The most important functions of the voltage regulator in an automobile

At higher currents, DC-DC converters are increasingly being used on grounds of the power loss. Most recently, constant current regulators have become necessary to control LEDs.

The voltage regulators in an automobile must be able to withstand the harsh environment of automobile engineering, and be able largely to suppress the interference signals on the vehicle power supply.

Figure 3.78 shows that the voltage regulator must both protect from damage the sensitive electronics, the microprocessor or smart sensor, throughout high temperature and voltage fluctuations, and must at the same time monitor them with various diagnostic systems.

The most important monitoring/diagnostic functions of Infineon's voltage regulators are the early-warning function,

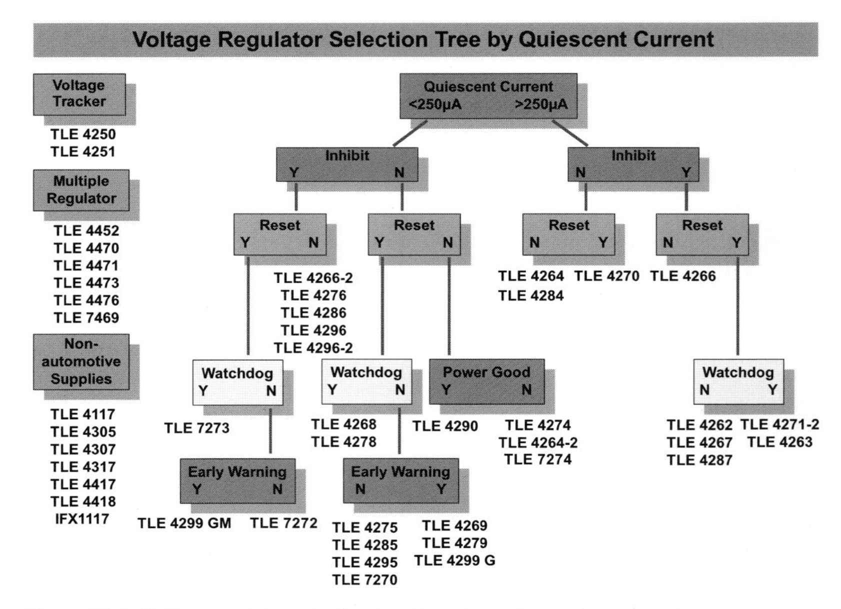

Figure 3.79 Voltage regulator selection tree, by quiescent current consumption

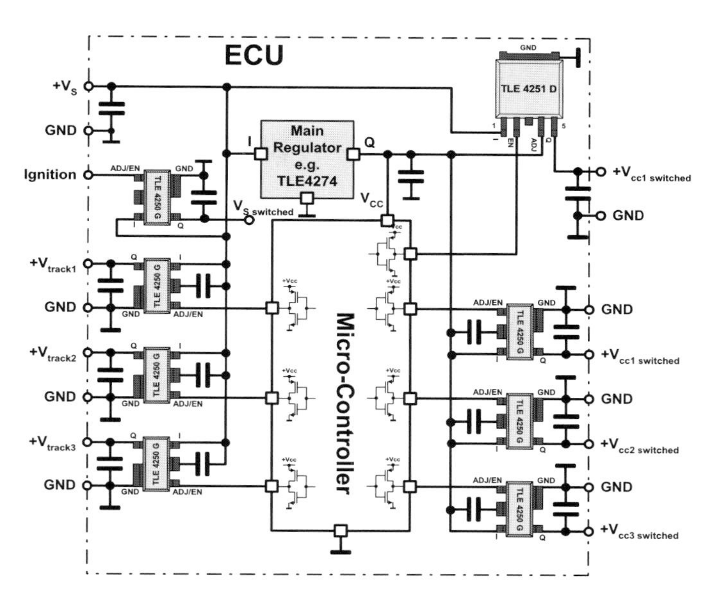

Figure 3.80
Application examples for the Voltage Trackers, TLE4250 and TLE4251, for uniform distribution of the power loss on a PCB

the watchdog function and the reset function. For example, with the early-warning function one can monitor the battery for an undervoltage condition, and thus initiate a data backup before the system is reset by the undervoltage reset. The watchdog function monitors the processor for "deadloops", and if necessary frees it from an endless loop in the software by a reset.

In an automobile, the quiescent current for the voltage regulator plays an essential role in many applications. Particularly those systems which remain active when the vehicle is parked are developed for the lowest possible current consumption. The selection tree in Figure 3.79 therefore distinguishes right at the top between low quiescent current regulators and those which have a higher current consumption.

This highly diverse tree shows that the range of voltage regulators from Infineon can satisfy virtually every requirement of an automotive customer. At the top right will be found the Voltage Tracker, which can be used to make existing system voltages available in buffered form to additional applications. Figure 3.80 shows the various application possibilities for these universal products. Apart from the single-channel versions there are two further versions, the TLE 4252 for one 200 mA supply and the TLE 4452 for two 200 mA supplies, are in preparation.

One major advantage here is that, in the event of a short-circuit of the "tracked" voltage, the master voltage is not affected. In addition, it is advantageous to distribute the power loss on the PCB across several packages. The main fields of application for Infineon's Trackers are, for example, sensor controllers. Because the output voltage from the Tracker can be switched off using an 'inhibit' input, the current can be simply reduced, and with it the fuel consumption. It is, of course, also possible to use the Tracker ICs as high-side switches for small loads.

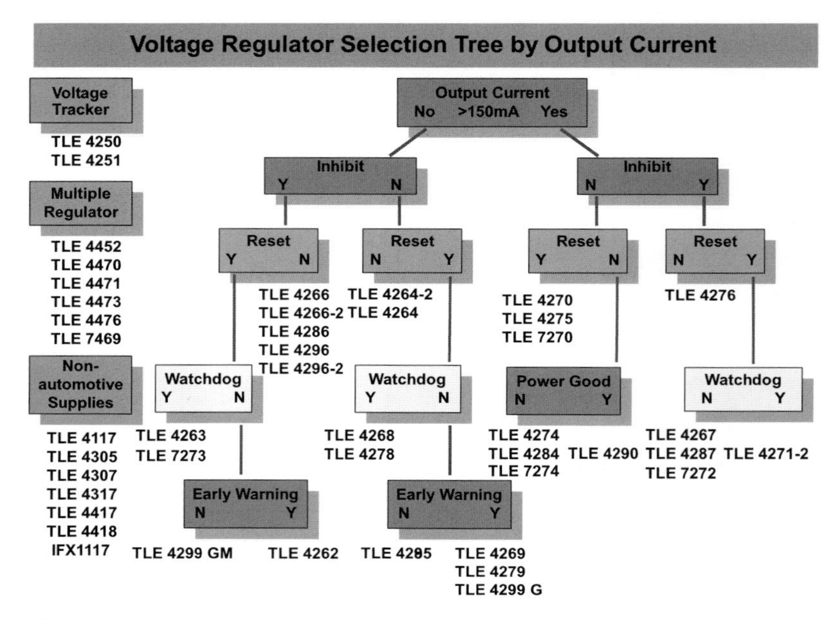

Figure 3.81 Voltage regulator selection tree, with the output current as the search criterion

The selection tree below in Figure 3.81 enables a user to look for the appropriate product for a known output current. Both of these selection trees are kept up-to-date on the Internet.

DC-DC converters in automobiles

In applications which require powerful computers (power train, safety and infotainment), current consumption is rising, and with it the power loss in the analog regulators. For this reason, increasing use is being made in this area of switching regulators, the so-called DC-DC converters. A distinction is made between the buck converters, which reduce the voltage, and boost converters which create a voltage which is higher than the supply voltage. Figure 3.82 shows a comparison

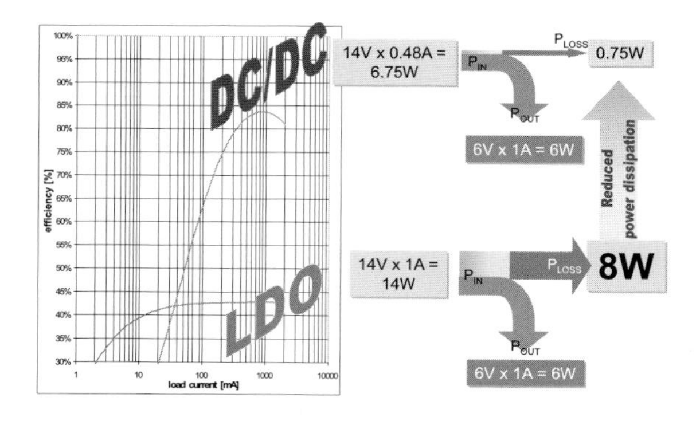

Figure 3.82 Efficiency of an analog voltage regulator compared to that of a DC-DC converter

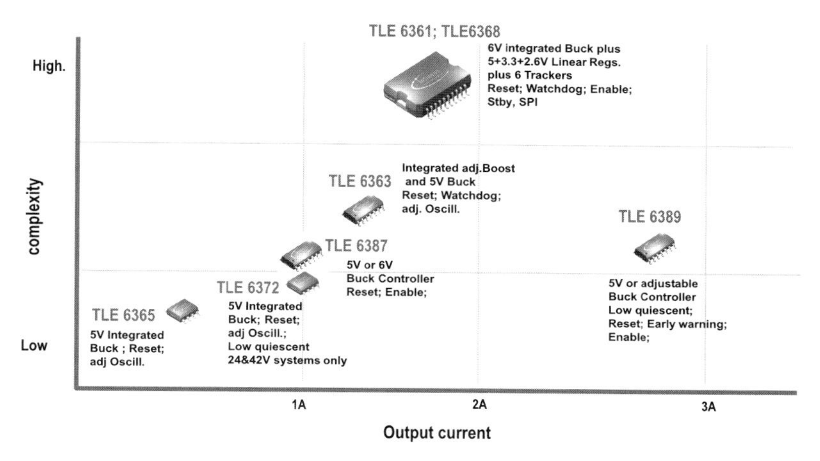

Figure 3.83 The main DC-DC converters for automotive applications

of the efficiencies (analog regulated / switched) for a typical engine management application.

The substantial gain in efficiency cancels out the additional costs required for switched mode at an output current as low as about 1 A. With the introduction of the 42 V vehicle power supply into automobiles, DC-DC converters are going to put analog regulators under pressure across a broad front. For this reason, Infineon has today already developed a range of DC-DC converters with 42 V compatibility. The overview in Figure 3.83 shows the main products.

It will be seen that there are simple monolithically integrated buck regulators (TLE6365 and the TLE6372) and combinations of buck and boost regulators (TLE6363). In addition, drivers for external P-channel MOS transistors (TLE6389) and more complex system voltage regulators with analog post-regulators will be found in this overview.

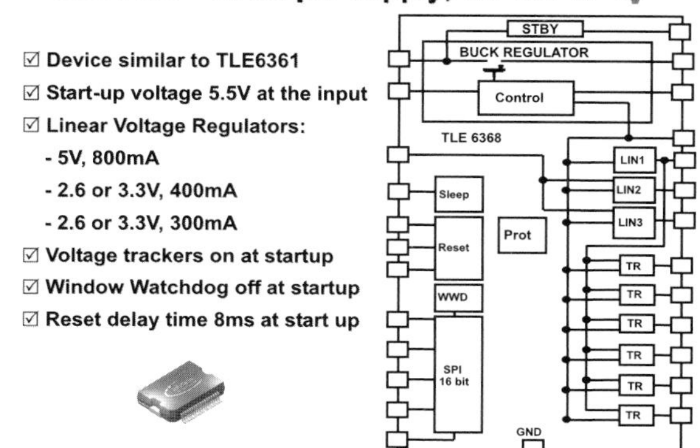

Figure 3.84
The TLE6368 system-voltage regulator for supplying power to an automotive application, with its powerful processor

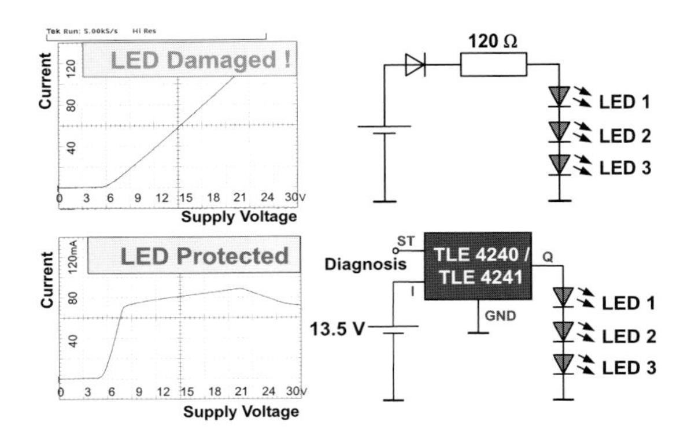

Figure 3.85
LED driver in an application, a comparison against simple current limitation by a series resistor

The block diagram and the main features of the latest product, the TLE6368, are shown in Figure 3.84.

A range of further products are under development; understandably, these cannot be shown here.

Current regulators for controlling automobile LED-lights

Another wide field of application is covered by the current regulators in the TLE424x series: LED lamps in an automobile. With the latest generation of powerful LEDs, many applications of incandescent lamps can be provided for. Because LEDs reach their operating light intensity much more quickly than incandescent lamps, they are particularly well suited as brake lights. In addition, there are constructional characteristics of LED lights which are advantageous (design freedom and installed depth).

Infineon's TLE4240, the first LED regulator in the world, is already in high volume in several applications. Its main advantages are:

- constant current regulation
- constant brightness
- extended service life
- diagnostics

Figure 3.85 illustrates that, apart from the advantage of constant brightness, when compared with simple current limiters using a series resistor there is another important improvement in reliability: by regulating the current, the LED can be effectively protected against overload. In addition, any break in the LED chain is reported to the controller by a diagnostic flag, ST.

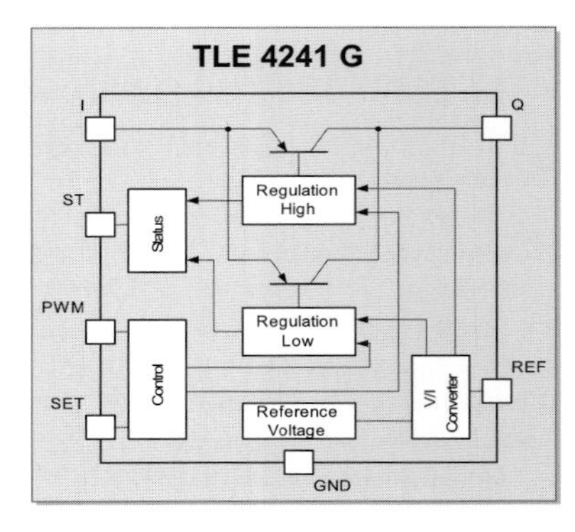

Figure 3.86
The TLE4241, one of the latest generation of analog LED drivers

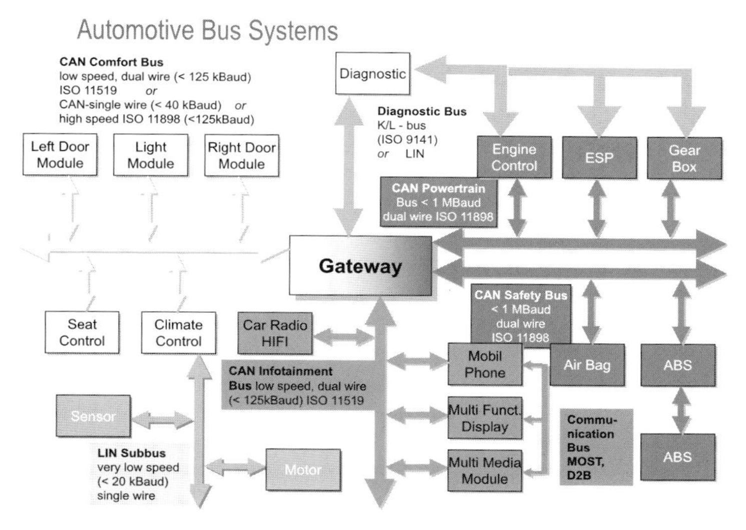

Figure 3.87 Overview of the bus systems in an automobile

Figure 3.86 shows the newest product, TLE4241, an LED driver with switchable current level and PWM input. Because the set current for this regulator can be programmed using an external resistance, it can be used for a standard tail- or brake light and in various auxiliary light applications in many vehicles.

In the near future, switched concepts will be developed here as well, because particularly in the lighting area power loss is a serious problem.

Another group of products with diverse application possibilities are the transceivers, which provide for the communications between the electronic controllers.

3.6.6 Transceivers

By networking the controllers in an automobile, using various bus systems, it is possible to save a great deal of wiring and weight. Because the requirements to be met by the applications are very different, various cost-optimized bus systems are used. For example, engine management calls for real time characteristics, requiring a fast bus to be selected. For the rear-view mirror or seat adjusters on the other hand, a slower bus is adequate. The various bus systems in a modern automobile are shown in Figure 3.87, with their main parameters.

A distinction is made between the high-speed, the low-speed and the LIN bus. In addition to these there is a very fast databus in the infotainment area.

To connect a microcontroller to the so-called "physical layer", the wiring, use is made of transceivers (from transmit and receive). These ICs must screen off the sensitive microcontroller from the harsh external world, and must have sufficient driver power to be able to drive the bus load adequately. The wires used for communication generally take the form of unshielded twisted pair wires. By using reverse polarized signals on the two wires of the twisted pair, the emitted radiation can then be kept low. The requirements to be met by the transceiver driver stages in terms of symmetry are correspondingly high. As high field strengths arise in an

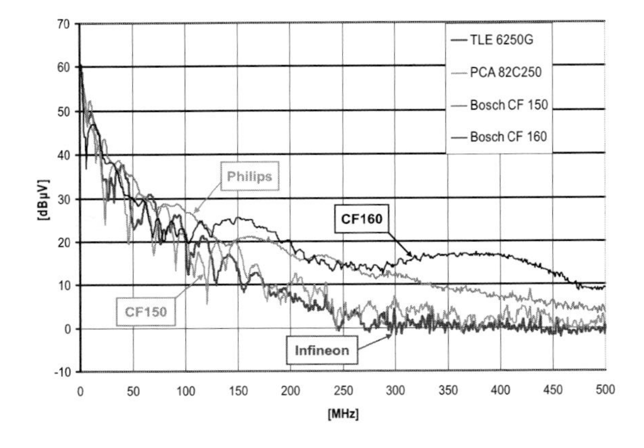

Figure 3.88
Comparative measurements of the radiation characteristics of various high-speed transceivers

automobile, the immunity of the transceiver inputs and outputs from radiated interference is also an important parameter. It is correspondingly difficult to operate reliably in one network transceivers from different manufacturers. In order to ensure compatibility therefore, comprehensive conformance tests have been developed. The test results from such comparative measurements are shown in Figure 3.88 for the TLE6250 high-speed transceiver.

The TLE6250 clearly shows the best results. This is due to a special output stage with especially good symmetry of the switching edges, and of the low and high levels.

The main functional blocks of a transceiver will now be explained, referring to the block diagram of the TLE6250 in Figure 3.89.

The bus wires of the CAN network are connected to the left hand side, at the CANH and CANL connections. With the help of the output stage, a microcontroller can drive the bus wires by applying logic signals at the transmit input, TxD. The CAN bus data items are read from the bus using the receiver (bottom of block diagram), the signal levels are converted to asymmetric ones (signal against GND) and the logic levels, and are then available at the Receive output, RxD.

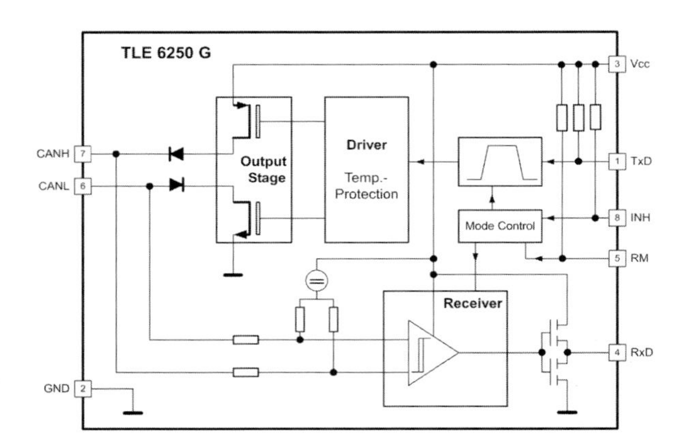

Figure 3.89
Block diagram of a TLE 6250

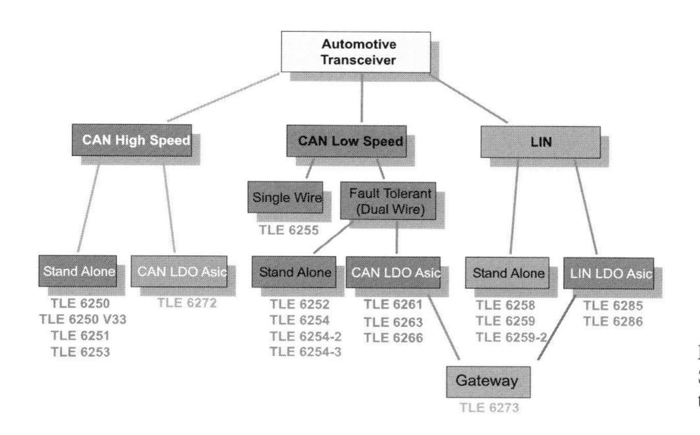

Figure 3.90
Selection tree for transceivers

Infineon has developed and released appropriate transceivers for all areas of CAN networks. The selection tree in Figure 3.90 shows a clear overview of the various products.

Like voltage regulators, CAN transceivers are to be found in almost all applications, so an obvious idea is to build the two products into one package. At Infineon, this step has already been taken consistently for all three transceiver areas. At the second level in the search tree, there is in each case a branch offering the choice of stand-alone or CAN-LDO.

As its name reveals, a CAN-LDO contains not only the standard CAN transceiver but also a low dropout voltage regulator which can be used for many applications.

For example, the TLE6266 was developed for the low-speed fault-tolerant bus. Its block diagram and main characteristics can be seen in Figure 3.91.

This modern device from the second generation of CAN-LDOs will be used, for example, in the coming generation of door modules. It is distinguished by having an extremely low current consumption in sleep mode.

The first generation of the product (TLE6263) has for years now been in

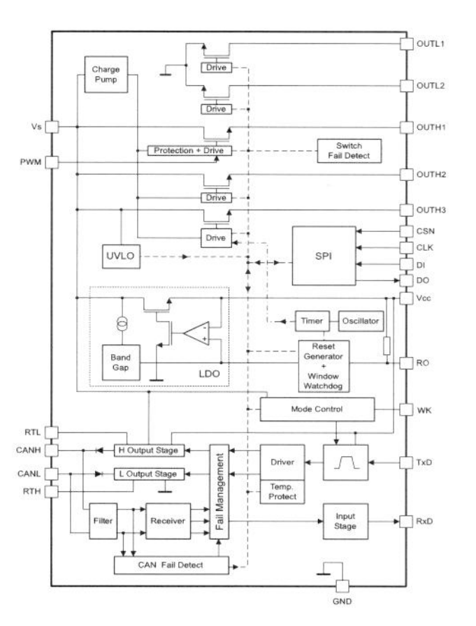

TLE 6266

Main Features:

- **LS CAN transceiver**
- **RxD-only mode**
- **LDO (50 mA external)**
- **low quiescent current**
- **16 bit SPI**
- **Reset circuit**
- **Window watchdog**
- **Early warning**
- **3 HS- & 2 LS-switches**
- **Wake-up input**
- **Fail safe functions**
- **P-DSO-28 package**

Figure 3.91
The TLE6266 low-speed CAN-LDO for applications in the body area

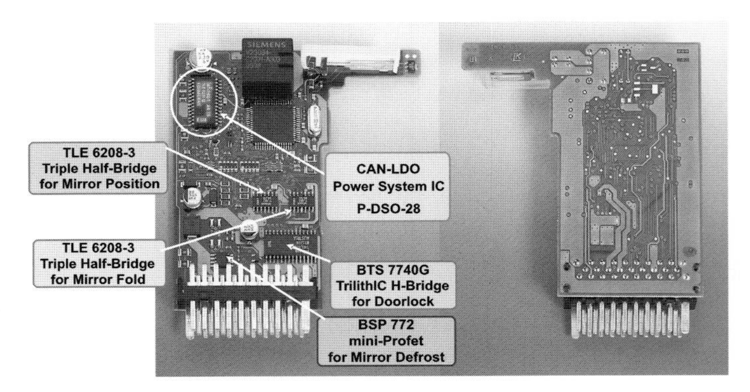

Figure 3.92
The first generation CAN-LDO as an important part of Infineon's chip set for door modules

door modules produced in high volumes (Figure 3.92).

The TLE6263 set a standard. The rear of the PCB shows that components only have to be inserted from one side, which helps to save on costs. The next generation of door modules will include, apart from the new TLE6266, the TLE 7201R multi-function bridge IC, which has already been mentioned.

For the High-Speed CAN bus, the TLE6272 realizes a chip-by-chip approach. A standard voltage regulator chip in the proven bipolar technology has here been put together with a standard transceiver chip in the latest SPT technology, in a standard SO14 package. As the chips are only connected via the GND contact to the leadframe acting as the chip carrier, each of these modules has a very high interference immunity, and conformity with the stand-alone CAN chip is ensured by the fact that the same chip is used. The block diagram and the most important characteristics of this product, unique throughout the world, can be seen in Figure 3.93.

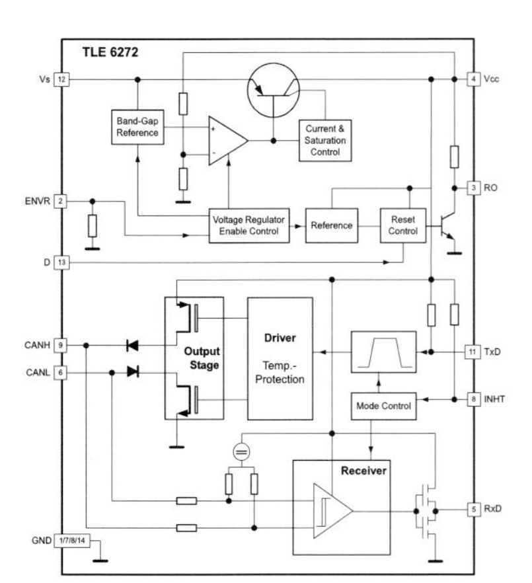

TLE 6272 G

Main Features:

- **HS CAN**
- **RxD-only mode**
- **LDO (150 mA external)**
- **low current consumption**
- **Inhibit Output/ Enable LDO**
- **Adjustable reset delay**
- **Reset Output**
- **P-DSO14- package**

Figure 3.93 The TLE6272 – a chip-by-chip high-speed CAN product

For the field of simple applications using the single-wire LIN bus, the TLE6285 LIN-LDO has been developed. Like the high-speed CAN module, this has also been conceived as a chip-by-chip product, and is distinguished by having a particularly low quiescent current, because it implements the modern standard voltage regulator TLE4299.

3.6.7 Smart power system ICs

In the near future, system ICs in the form of application-specific smart power ICs will come to dominate the market. For example, the door module of an automobile now already realizes a monolithic integration of the supply and communication functions of the TLE6266 CAN-LDO and those of the TLE7201 multi-function bridge IC. Looking at the functions of an ECU from this point of view, we can distinguish functional blocks for supply, communication, computing & memory, signal conditioning, and actuators. In Figure 3.21, these blocks have been color-coded by category: digital, analog, and power. As described there, it is only possible to optimize the breakdown of the product – "smart partitioning" – by making use of the know-how of both the user and the semiconductor manufacturer. The various options (the level of large scale integration in each case) are shown in Figure 3.94.

By its nature, the domain of the power system ICs lies in those applications which have been present in automobiles for a long time now. One of the most important of this type is the airbag controller. Today, almost every automobile has four or more airbag firing devices installed in series production. For this application Infineon has a complete chipset, shown in Figure 3.95, which is a market success.

This consists of sensors, the microcontroller, and a complete set of supply, communications and squib driver ICs. The next generation of this chipset, which shows Infineon's vehicle system dominance, is under development.

Another application in which system ICs are used in very large volumes is in ABS-ESP systems. Here again, Infineon can show high system coverage. The supply, communication and signal conditioning functions have in this case already been highly integrated, using an L-chip approach. The next step would be the integration of the safety controller and/or the driver (output) stages. However, because of the penalties in terms of system flexi-

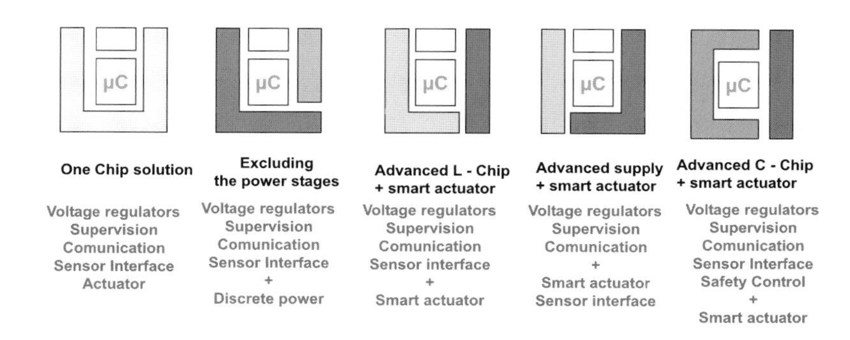

Figure 3.94
Partitioning, carried out professionally by the user and the semiconductor manufacturer, is a prerequisite for developing optimal chipsets

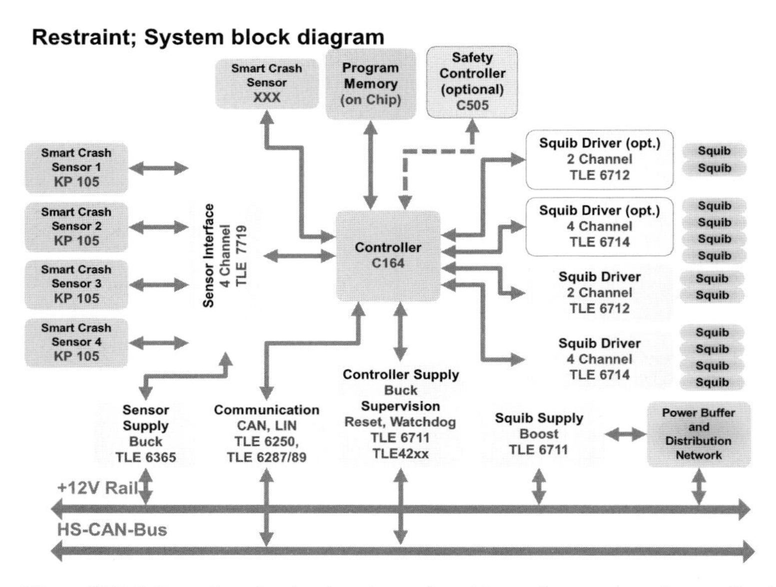

Figure 3.95 Infineon has also developed complete chipsets for restraint-safety applications

bility, and because of the concentration of the large power losses, this step is a matter of great discussion, above all in respect of commercial considerations.

In engine management too, Infineon is able to offer a service, with a complete chipset (Figure 3.96). Around the TC 1765 or TC 1775 microcontroller there is a complete ring of power ICs for actuation, voltage supply and communication.

Up until now, the applications have ignored large-scale integration of the microcontroller, because the computing power required, that is the complexity, clearly excludes it. However, in the case of very small simple applications, defining the objectives requires closer consideration. These embedded-power applications are exemplified by the motor drives (see "electronic muscle" in section 3.4.5). A system block diagram of the most important representative of this group, the BLDC motor drive, can be seen in Figure 3.97. This 3-phase brushless DC motor drive will replace the conventional DC motor and many hydraulic and mechanical solutions in mechatronic applications.

For a drive using an electronically commutated motor there are openings in countless applications. As an alternative to belt drives or hydraulically powered units, it is often the lower priced and almost always a fuel-saving solution. To combine increased reliability with improved interference characteristics (the brush sparking from a DC motor represents a strong noise source) the maintenance-free BLDC drives will many times displace the heavier and less reliable DC motors. The reason is that every 100 kilograms of weight reduction saves 0.3 liters of fuel consumption per 100 km in a conventional automobile.

The partitioning can be – depending on the commercial cost – either as a monolith, as two chips or even into several chips. Here, the power MOS transistors are never mounted in the same package but, as already described in section 3.6.1, are accommodated in appropriate special packages.

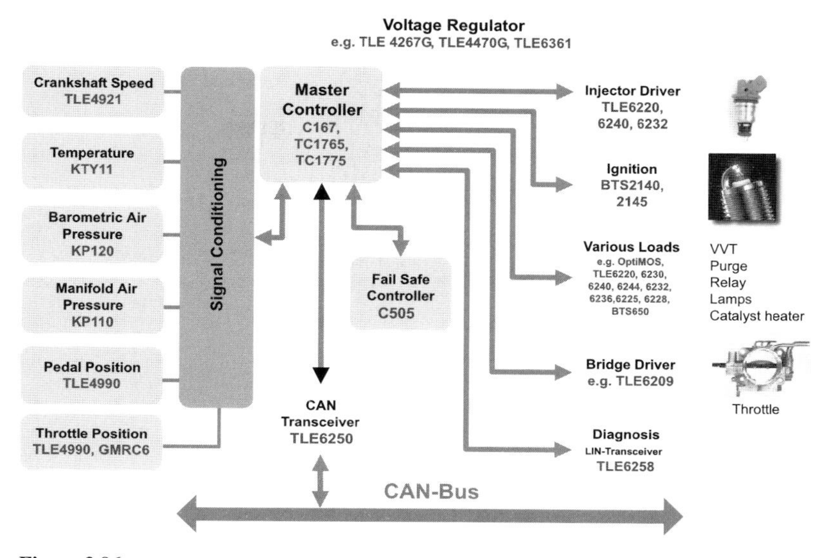

Figure 3.96
In the most important application in the power train area, engine management, Infineon's ICs again provide comprehensive coverage of the functions

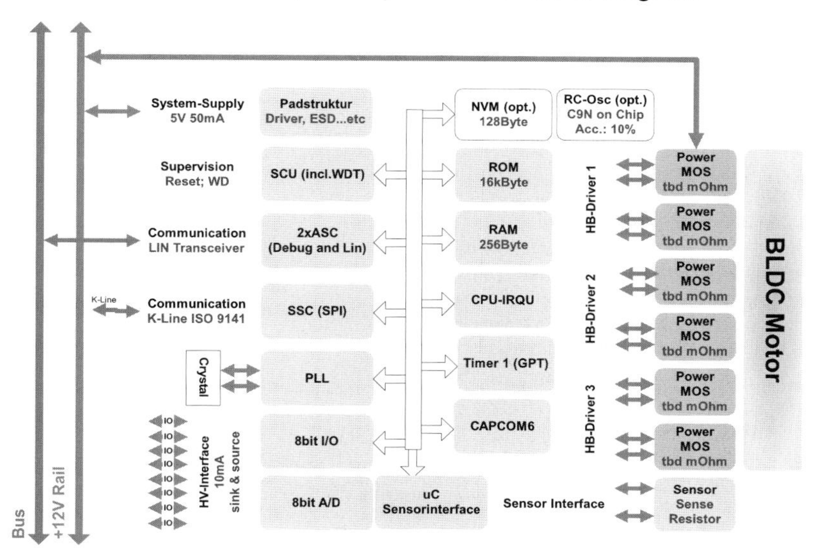

Figure 3.97 Block diagram of the brushless DC motor drive

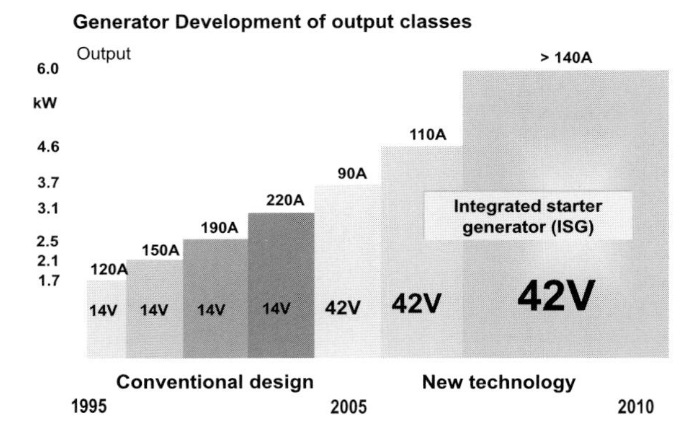

Figure 3.98
The rising energy consumption in automobiles calls for a higher supply voltage of 42V

An interesting variant on the integration is the combination of a high-voltage chip in smart power technology with a highly-integrated C-MOS chip. The two are mounted together in one package. For this purpose, Infineon has developed the SOLID process, by which the front sides of the two chips are joined together conductively without using wire bonds.

Due to the preference for using electrical consumers in an automobile, the energy requirement will rise sharply over the next few years. Figure 3.98 shows the generator power and the peak current for a medium-class vehicle over this time.

However, for each kilowatt of electrical energy the fuel consumption per 100 km increases by 1.5 liter with a conventional 14 V vehicle power supply. If the generator voltage is increased to 42 V, this increase can be reduced by 0.3 liters, down to an increase of 1.2 liter per 100 km. For

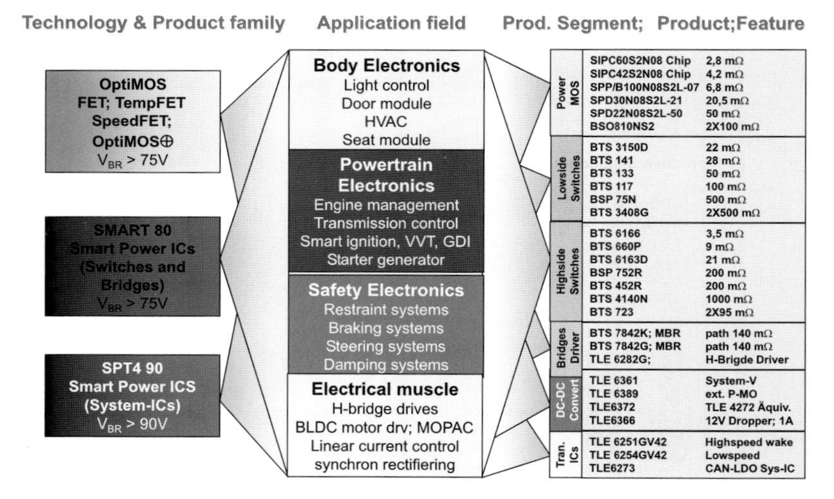

Figure 3.99
Overview of the smart power technologies and products which have been developed for a 42V vehicle power supply

this reason, the use of electronic devices will undergo a significant increase. Because, with a more powerful generator in the vehicle, many new applications become possible, for example electronic valve train, electro-mechanical steering, electro-magnetic brakes, or even electrical screen or interior heating. For all these applications, and for those which are common today, it will be necessary to arrange all the smart power ICs, which have been optimized for 14 V, for 42 V. Infineon is the first semiconductor manufacturer in the world to do so, and since the end of 2002 has fully developed and released for use the necessary smart power technologies and an initial complete set of devices (see Figure 3.99).

The classical application fields of body, power train and safety are fully covered. In addition, the new field of "electronic muscles" can be handled. On the right are shown the product groups required for this purpose, with the products concerned and their most important characteristics. More extensive details of these and all the applications will be found in Chapter 9.

3.6.8 Trends in automotive applications

In an automobile, power semiconductors are indispensable in virtually any area when the requirement is for innovations. These should, on the one hand, help in saving costs, and on the other hand should increase the performance of the vehicles. One of the most important parameters by far in the future is likely to be the fuel consumption efficiency. Power semiconductors can make important contributions to this.

If one looks at the development of the world market for semiconductors installed in automobiles, see Figure 3.100, there is a clear growth of up to 9% per annum and more (CAGR; cumulated annual growth rate).

The safety area is and remains the largest, and its annual growth of 9.1% is the greatest. Looking at semiconductors in isolation from microcontrollers, signal ICs and sensor ICs, then some of the growth rates are even significantly higher. In addition to this, the introduction from 2006 of the 42 V vehicle power supply will bring a further growth in the semiconductor sales volumes. Because many applications only become possible with 42 V.

As discussed above under smart power system ICs, the automobile is more and more becoming a network with distributed intelligence. Some of today's controllers are linked together, but only using CAN or similar networks. In the near fu-

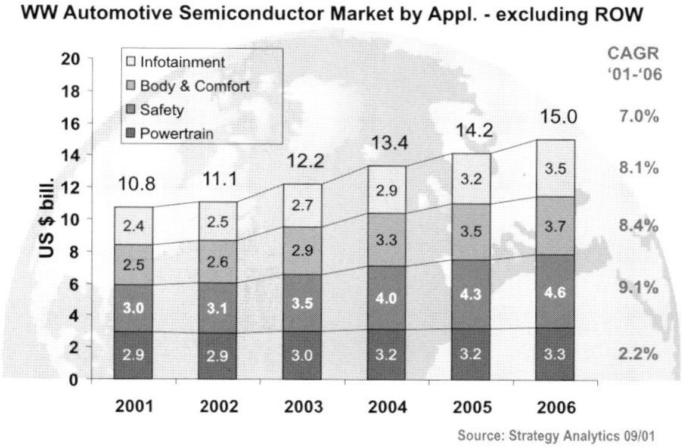

Figure 3.100 Development of the automotive semiconductor market over the next few years

ture the automobile will become a "neural network" on wheels. Controllers will exchange data of every sort at almost any required speed. The computing power can then also be optimally deployed.

This has not, of course, failed to make its mark on the semiconductor manufacturers' development processes. In the near future they will no longer be developing only ASSPs, but will turn increasingly to the development of systems. Many examples of system chipsets have been shown in section 3.6.7. The trend indicates a second dimension: hardware integration vs. product integration, for example TMPS (Tire Pressure Monitoring System).

Looking a little further into the future, one can see a third dimension, software integration. This gives us a space in which to develop a new, important differentiating feature, such as TCG (Telematics Communication Gateway).

3.7 Power supply and drive applications

Today's lifestyle can no longer be imagined without the provision and processing activities which use electrical energy, and its supply. A host of devices in everyday use are operated either directly from the mains power grid, from the vehicle power supply in an automobile or using accumulators.

The electronic circuits in modern devices in entertainment, data processing or industrial electronics are mostly supplied with direct voltages from 12 V down to below 1 V. To be able to operate these devices from the common alternating voltage mains network, or to charge up the internal accumulators, a power supply is required.

3.7.1 Switched mode power supplies – topologies and products

Conventional power supplies consist of a mains transformer for voltage reduction and galvanic isolation from the mains, a rectifier for producing a direct voltage and a bulk capacitor for voltage smoothing (Figure 3.101).

A stabilized current or voltage supply, which is independent of fluctuations in the input voltage and load, is achieved by means of a stabilization section. Stabilization circuits generally have considerable losses, and call for a correspondingly large mains transformer (Figure 3.102). This increases the total losses, and causes heating of the device. In total, this results in an unsatisfactory efficiency, because when a mains overvoltage occurs the lin-

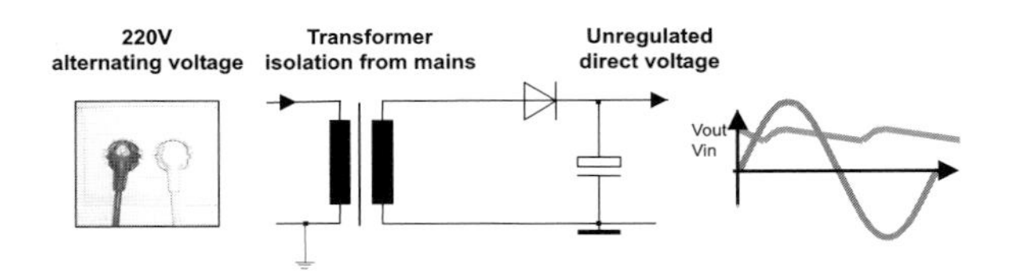

Figure 3.101 Conventional power supply without stabilization

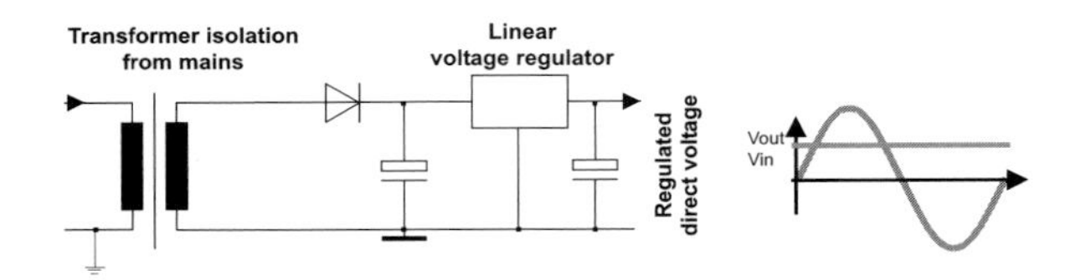

Figure 3.102 Conventional power supply with stabilization of the output voltage

Figure 3.103
Smaller and lighter components in a switched mode power supply compared to a conventional power supply of the same power

ear voltage regulator must absorb the entire power difference.

To an increasing extent, these conventional power packs are being superseded by so-called switched-mode power supplies (Figure 3.103). These have the following advantages:

- lower volume and weight, due to a smaller transformer and smaller capacitors
- better efficiencies, and hence less heating
- lower losses in standby mode
- the possibility of operating the power supply over a wide input voltage range (85- 245 V)
- electronic safety functions for fault situations, such as short circuits

In the age of globally marketed electronic devices, the wide possible input voltage range of these power supplies is especially necessary. As an example, a device developed in Germany may be sold and operated in countries with a 110 V mains power supply (including the USA, Japan) or in Europe or Asia (220 V mains power supply).

Ever more electronic devices are no longer disconnected from the mains by a mains switch when they are not in use (TVs, video recorders, PCs, chargers etc.). Instead, the devices are put into a standby mode, and thus consume unnecessary energy. There is thus an increasing demand for the power consumption in standby mode to be reduced, which has now become very efficiently feasible, using switched mode power supplies.

This development is promoted by new and even more efficient passive and active components. Examples of these are innovative high and low voltage MOSFETs (CoolMOS™ and OptiMOS™), new types of diodes and transistors based on silicon carbide (ThinQ™) and special, highly integrated open and closed loop control circuits (CoolSET™).

Basic principle of the switched mode power supplies

Switched mode power supplies (SMPSs) are power supplies which chop the rectified and filtered mains voltage at a frequency which is significantly higher than the 50 Hz of the mains alternating current (Figure 3.104).

By using the semiconductors exclusively as switches, only switching and forward losses arise. This is responsible for the characteristically high efficiency of a pulsed power supply by comparison with analog methods. Regulation is effected either by altering the pulse duty factor at a constant frequency, or by changing the frequency for a fixed or variable pulse duty factor. The voltage chopped in this way can be transformed to any other required voltage, and rectified.

The frequency of this alternating voltage, which may be rectangular, trapezoidal or sometimes even sinusoidal, lies in a range from a kHz up to several 100 kHz. As a consequence of this high working frequency, smaller transformers with ferrite cores can be used. The ferrite core transformer serves not only to make the required voltage conversion and provide galvanic isolation from the mains supply

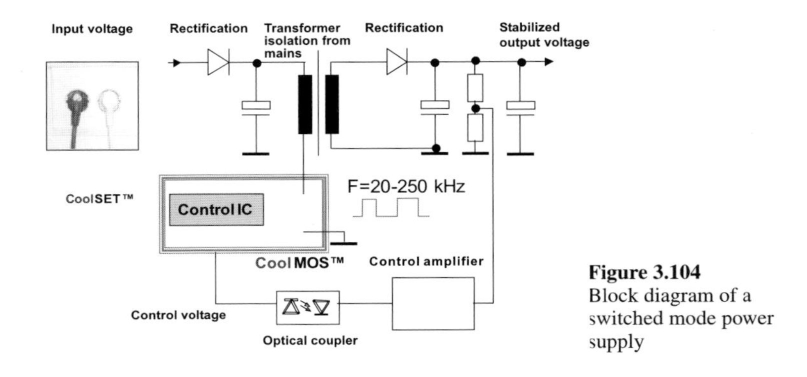

Figure 3.104
Block diagram of a switched mode power supply

but also, depending on its working principle, to store the magnetic energy.

With SMPSs, the pulsing results in harmonic waves, which can cause radiated interference to radio and TV reception, as well as to communication transmissions. Legislation demands a limit on the interference signal levels for all electrical devices and systems which produce high frequency energy.

An SMPS which is powered from the alternating voltage mains and which supplies a DC voltage at its output is also called an AC-DC converter. If a direct current source (e.g. an AC-DC converter, or a battery) is connected to the input, then we speak of a DC-DC converter, or a switching regulator. If the voltage is not rectified at the output, the device is a DC-AC converter or inverter. If the SMPS is supplied from the mains, and there is no rectification on the output side, we have an AC-AC converter, i.e. an alternating current converter.

The main application fields for SMPSs are:

- Consumer electronics: TVs, DVD players, video recorders, set top boxes, satellite receivers, chargers and external power supply units.
- Electronic DP: PCs, servers, monitors, notebooks
- Telecommunications: mobile communication base stations, switching stations, mobile phone chargers
- Industrial electronics: open and closed loop control engineering, measuring instruments, auxiliary power supplies, battery chargers etc.

3.7.2 Switched mode power supply topologies

The working principle of an SMPS essentially determines its characteristics and production cost. The basic circuits of the most frequently used AC/DC and DC/DC converters will now be described.

A basic distinction is made between two different conversion principles: the feed forward converter and the flyback converter (Figure 3.105).

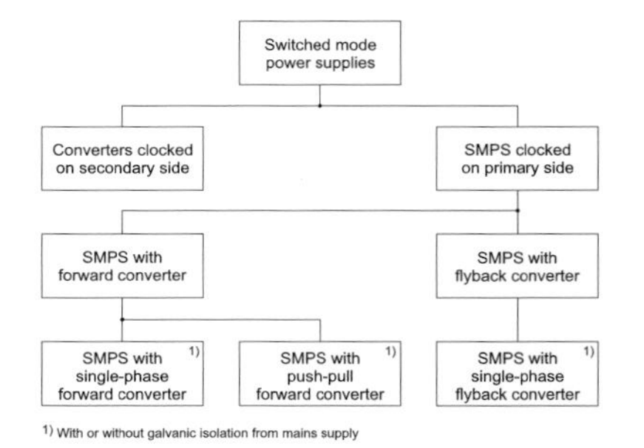

Figure 3.105
Overview: switched mode power supplies

The name feed forward converter is explained by the response characteristics of the arrangement, with which an energy flow arises between the primary circuit and the secondary circuit during the semiconductor's blocking phase. On the primary side, the load current is superimposed on the magnetizing current. For this reason, conditions must be created to allow the transformer to be demagnetized again. With a single-phase feed forward converter this occurs during the semiconductor switch's conducting phase. With push-pull and bridge circuits, the conducting phase of one semiconductor switch is followed, after a short blocking phase for both semiconductor switches, by the conducting phase for the second.

With a flyback converter, energy is accumulated in the converter during the conducting phase of the semiconductor switch, and is then given up on the secondary side during the blocking phase.

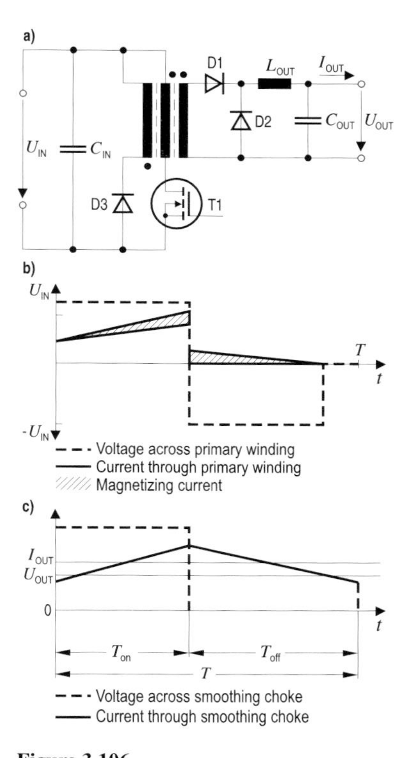

Figure 3.106
Single-phase feed forward converter:
a) circuit diagram, b) and c) response characteristics in steady-state operation

The feed forward converter

Figure 3.106 shows the circuit diagram of a single-phase feed forward converter.

On its input side, this converter has a smoothing capacitor C_{IN}. This performs the functions of smoothing the rectified mains voltage, providing a low-inductance supply of the pulsed currents required by the converter, and absorbing the magnetizing current which is fed back from the transformer. The transformer of the feed forward converter has a ferrite core with no airgap in order to achieve a high magnetic coupling of the windings. The primary winding is connected to the input voltage by a switching transistor (CoolMOS™). When the transistor is conducting, an induced voltage arises in the secondary winding with a rectangular waveform and magnitude corresponding to the transformation ratio, which produces a current flow in the secondary winding through the rectifier diode D_1 and the smoothing choke L_{OUT} on the output side (Figure 3.106b and c: time period T_{on}).

The current in the secondary winding induces a current in the primary winding corresponding to the transformation ratio. In addition to this load current, superimposed on it in the primary winding is the so-called magnetizing current.

The magnetizing current must be allowed to decay during the transistor's blocking phase. This is effected by a demagnetization winding, which serves as a third transformer winding. It has the same number of turns as the primary winding, but with a smaller conductor cross-section because only the magnetizing current flows through this winding, during the transistor's blocking phase. Compared to the primary and secondary windings, the demagnetization winding has reversed polarity, indicated in the circuit diagram by the dots at the start of each winding. The demagnetization winding is connected directly to the input voltage, via a diode D_3. During the transistor's conducting phase, the same voltage is induced in the demagnetization winding as there is in the primary winding, so that twice the input voltage arises at the diode $D_{3,}$ in its blocked direction.

During the transistor's blocking phase, the energy which has been built up in the transformer core as a result of the magnetizing current must be released again, so that the magnetizing current does not grow arbitrarily large and the ferrite core does not reach magnetic saturation. For this reason, a freewheeling diode D_2 is provided in the secondary circuit, through which the current flows on through the smoothing choke L_{OUT} when the voltage in the secondary winding drops to zero or goes negative. During this phase of operation, the diode D_1 decouples the current loop on the secondary side from the transformer. This allows the polarity in the windings to reverse.

The magnetizing current now flows back through the diode D_3 and the demagnetization winding into the smoothing capacitor on the input side. As a result, twice the input voltage is now applied to the transistor as the blocking voltage. The demagnetization of the transformer is assured if the voltage/time area, that is the area enclosed under the voltage in the primary winding along the time axis for the duration of the demagnetization, is at least equal to the corresponding area for the switched-on state. For this reason, the duration of the on-state for a single-phase feed forward converter may not amount to more than 50% of the cycle duration.

The purpose of the smoothing choke L_{OUT} is to generate a steady energy flow from the current or voltage signal, as appropriate, which arises across the secondary winding of the transformer during the switched-on period T_{on}, and to limit the current rise in the transformer. The smoothing capacitor C_{OUT} at the output smoothes the current ripple from the choke, and acts as an energy store during load changes. The response characteristics of the feed forward converter are described by the following formula:

$$U_{OUT} = U_{IN} \cdot \frac{n_1}{n_2} \cdot \frac{T_{on}}{T}$$

n_1 number of turns in primary winding,
n_2 number of turns in secondary winding.

The flyback converter

These converters also have a smoothing capacitor C_{IN} on the input side, with the functions of smoothing the rectified mains voltage, and providing a low-inductance supply of the pulsed currents required by the converter. Unlike the single-phase feed forward converter, here the magnetizing current is not fed back to the input capacitor, but to the smoothing capacitor C_{OUT} on the output side (Figure 3.107).

In its basic form, the flyback converter has two windings, with opposite polarization. When the switching transistor T_1 (CoolMOS™) is switched on, the anode-cathode voltage for the rectifier diode D_1 is negative, i.e. no current flows in the secondary winding of the transformer. The magnetizing current flows in the primary winding. In a flyback converter, a ferrite core is used which has an air gap, as a result of which a considerably larger inductive current flows and this establishes a magnetic field in the air gap. In addition, when the transistor is switched on magnetic energy is stored in the transformer of the flyback converter (predominantly in the air gap).

When the transistor is blocked, the voltage across the windings reverses. The voltage across the secondary winding rises until the rectifier diode D_1 becomes conducting, i.e. at the value of the output voltage U_{OUT}. Because the magnetic flux

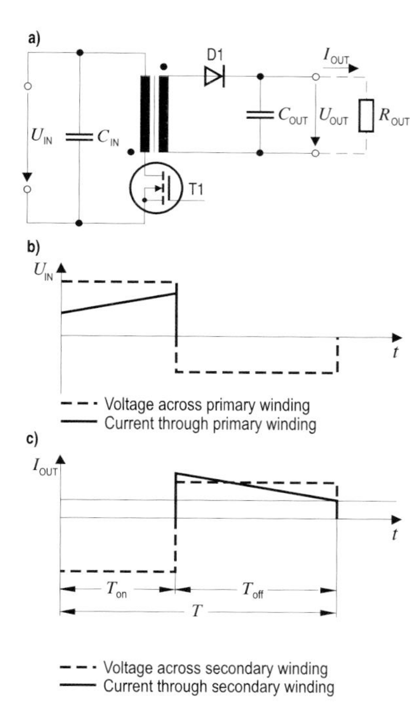

Figure 3.107
Flyback converter: a) circuit diagram, b) and c) response characteristics in steady-state operation with trapezoidal current waveform in the transformer windings

in the transformer is steadily changing, at the point in time when the transistor is blocked a current, corresponding to the primary current transformed by the transformation ratio, will flow in the secondary winding. For this reason, the rectifier diode D_1 must feed directly to a capacitor C_{OUT} which is capable of accepting the high current.

During the time the transistor is switched off, the blocking voltage applied to it will be the input voltage, plus the output voltage transformed in accordance with the transformation ratio. With the standard dimensioning, this represents somewhat more than twice the input voltage. We now distinguish between forward converters with a trapezoidal current waveform in the transformer, and those with a triangular current waveform. With the trapezoidal waveform (Figure 3.107b and c), the transistor is switched on again at a point in time which is before the current in the secondary winding reaches zero. An important characteristic of this form of operation is that the maximum values which occur for the current are, relative to the output current, significantly lower than when operated with a triangular current waveform.

The response characteristic for a trapezoidal current waveform is represented by the formula:

$$U_{OUT} = U_{IN} \cdot \frac{n_2}{n_1} \cdot \frac{T_{on}}{T} \cdot \frac{1}{1 - \frac{T_{on}}{T}}$$

From this it will be seen that the output voltage U_{OUT} changes if the pulse duty ratio T_{on} / T is altered. However, the relationship between the output voltage and the pulse duty ratio is not linear, but hyperbolic. This means that the output voltage will become infinitely large if the pulse duty ratio approaches close to 1. For this reason, flyback converters must not be operated without a load resistance or a closed control loop, because the output voltage, and with it also the blocking voltage for the transistor, can assume high values.

Switched mode power supplies with several output voltages

With a switched mode power supply which has several output voltages, one of these voltages must be chosen to provide the controlled variable, as a consequence of which this will be the best regulated voltage (Figure 3.108, voltage U_{OUT1}).

The regulation of the other voltages will be less exact. For this reason, it may be necessary to stabilize them by means of longitudinal regulators (voltage U_{OUT3}) or pulsed secondary regulators (voltage U_{OUT4}). The last named switching regula-

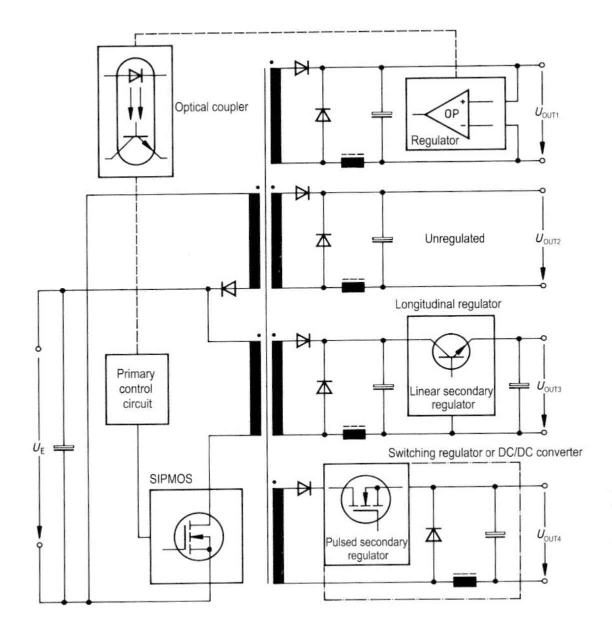

Figure 3.108
Single-phase feed forward converter with several different regulated output voltages

tors represent converters with a secondary clock pulse control.

3.7.3 Selection criteria for switched mode power supplies

Table 3.5 provides a comparison of feed forward and flyback converters, Table 3.6 shows the powers which can be achieved by the different types of converter. Tables 3.7 and 3.8 contain the circuit diagrams, voltage and current waveforms and the advantages and disadvantages of the different variants of flyback and feed forward converter.

3.7.4 Integrated circuits for switched mode power supplies

For these applications, Infineon offers a wide range of semiconductors, e.g. EPCOS, from passive components onward. In addition and apart from the datasheets, a host of application reports and dimensioning software are available to the developer via the Internet (http://www.infineon.com). PSPICE models are provided for discrete power components and some of the ICs.

In terms of reliability, quality, characteristics and price, integrated circuits with their sophisticated switching technology, high complexity and compact construction are far superior to controllers of discrete construction. In addition, they permit simplifications in the switching technology of switched mode power supplies. The development costs are reduced, so that even small production quantities are profitable.

Depending on the application and the switching topology used, there is a choice of various driver ICs for SMPSs.

Free-running driver ICs

Types TDA 4605-3, TDA 16846 and ICE1QS01 are ideally suited for quasi-resonant flyback converter switched mode power supplies.

Table 3.5 Comparison of feed forward and flyback converters

	Feed forward converter	Flyback converter
Number of components required	Greater	Fewer
Smoothing choke and freewheeling diode	Required	Not applicable
Transformer core	Without air gap	With air gap
Magnetic coupling during switching processes	Better	Worse
Voltage excess during switching processes	Smaller	Larger
Susceptibility to influence of magnetic fields	Smaller	Larger
Current amplitudes relative to load current	Significantly smaller	Substantially larger
Pulsed current loading of components	Smaller	Larger
Interference suppression and smoothing of input and output quantities depending on interference modes	Simpler	More expensive
Energy flow controlled by changing the pulse duty ratio by	Varying the voltage / time integral	Storing variable proportions of energy
Creation of several strictly regulated secondary DC voltages at the same time, by adding further secondary circuits	Possible to a limited extent (choke current must be continuous)	Easily possible
Regulation dynamics of output quantities	Slower (due to choke smoothing)	Faster

Table 3.6 Power ranges for the various converter types

Power (W)	≤ 100	100…300	300…1000	1000…3000	≥ 3000
Single-phase flyback converter	×	×			
Single-phase feed forward converter	×	×			
Half-bridge converter		×	×		
Full-bridge converter		×	×	×	
Push-pull converter		×	×	×	×

This topology ensures that the switching transistor (CoolMOS™) can be switched on at the minimum voltage of the resonant oscillation, which arises at the drain connection of the blocked transistor (Figure 3.109). This reduces the switching losses and at the same time decreases high-frequency interference.

The working frequency of the circuit is load-dependent, which also leads to a spreading of the interference spectrum.

The ICs have numerous integral safety functions, and can be regulated using either opto-coupler feedback (from the secondary side) or, if the voltage stability requirements are lower, also from the primary side. The ICE1QS01 in particular has an extremely efficient circuit (burst mode) for minimizing the current consumption in the so-called stand-by mode.

These devices are ideally suited for consumer electronic devices, but also for

Table 3.7 Flyback converter variants

No.	Circuit diagram	Description	Advantages
1	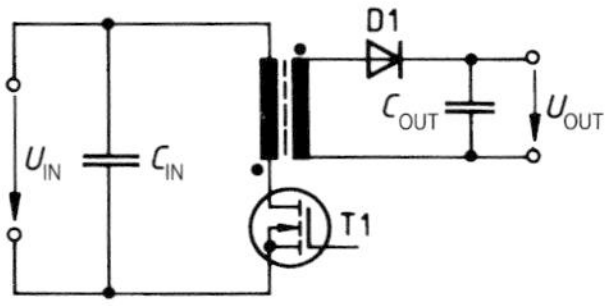	Flyback converter	• Several output voltages can be regulated simultaneously • Large regulation range for operatin voltage changes ("stepless transfor-mation ratio")
2	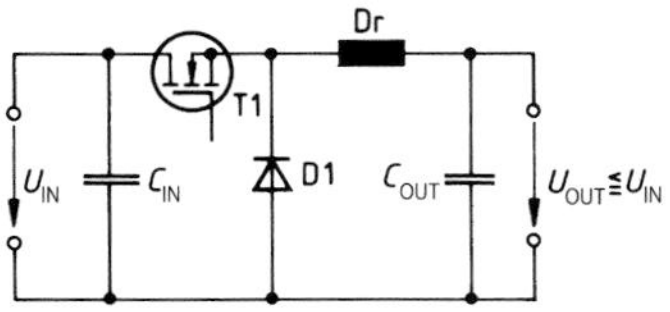	Buck converter	• SIPMOS blocking voltage $U_{DS} \approx U$ • Simple choke • No problems with magnetic coupli • Low stresses on the output capacitc • A pulse duty ratio $T_{on} / T = 1$ is possible
3	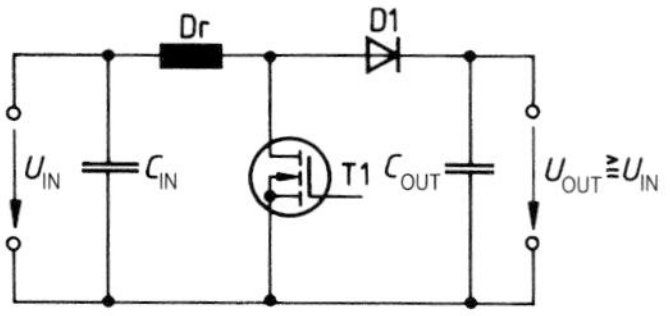	Boost converter	• Simple choke • No problems with magnetic coupli
4	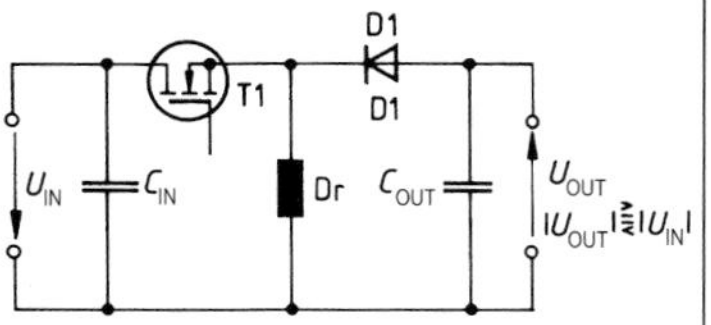	Buck-boost converter	• Simple choke • No problems with magnetic couplir
5	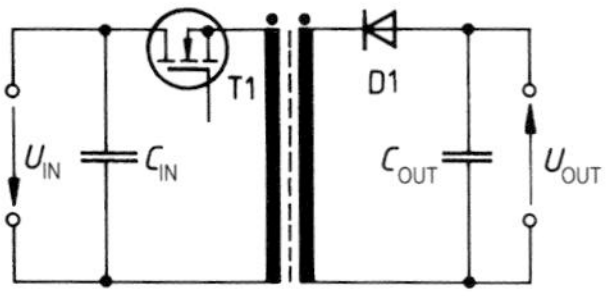	Flyback converter	Buck-boost converter with galvanic isolation between input and output voltage is identical to the flyback converter in the above diagram

isadvantages	Pulse duty ratio	a: Voltage waveform at transistor b: Current in secondary winding or choke	c: Current waveform in input capacitor d: Current waveform in output capacitor
Commonly, power transistor blocking voltage $U_{DS} > 2\ U_{IN}$ High stress on the capacitor and diode at output Good magnetic coupling required Requires large cross-section with air gap Problems from magnetic radiation and eddy currents	$\frac{T_{on}}{T} = 0,5$	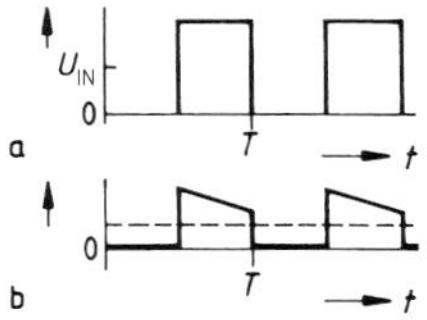 a, b With a trapezoidal current waveform in the transistor or primary winding, as applicable	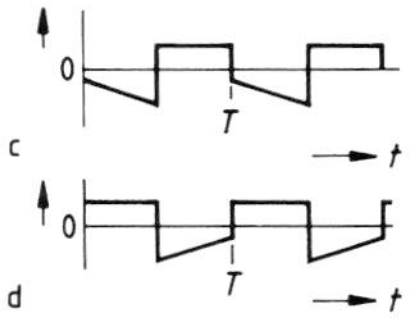 c, d
No galvanic separation between input and output voltages Drive must "float"	$\frac{T_{on}}{T} = 0,5$	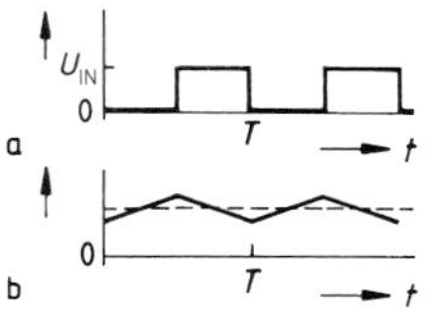 a, b	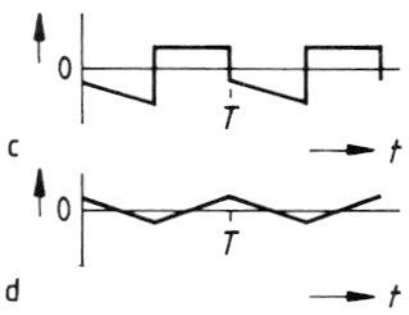 c, d
Power transistor blocking voltage $U_{DS} \approx U_{OUT} > U_{IN}$ No galvanic separation between input and output voltages Moderate stress on the output capacitor	$\frac{T_{on}}{T} = 0,67$	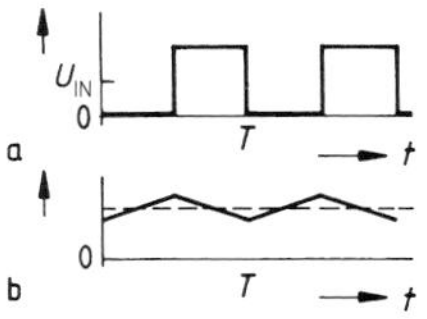 a, b	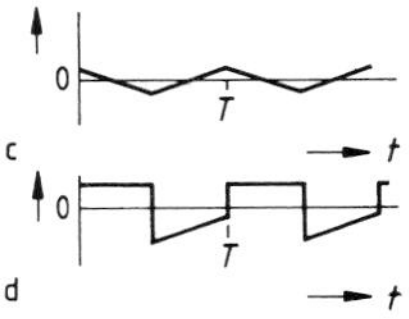 c, d
Power transistor blocking voltage $U_{DS} \approx U_{IN} + U_{AOUT}$ No galvanic separation between input and output voltages High stress on the output capacitor Drive must "float" Output voltage negative to input voltage	$\frac{T_{on}}{T} = 0,5$	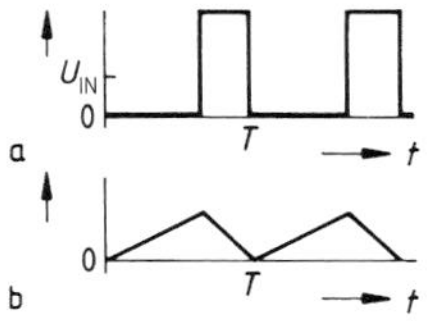 a, b	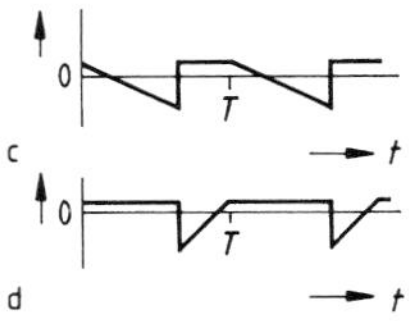 c, d
	$\frac{T_{on}}{T} = 0,5$	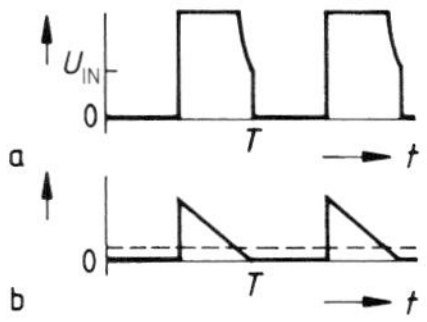 a, b With a triangular current waveform in the transistor or primary winding, as applicable	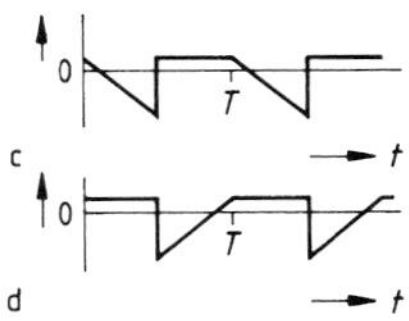 c, d

Table 3.8 Feed forward converter variants

No.	Circuit diagram	Description	Advantages
1	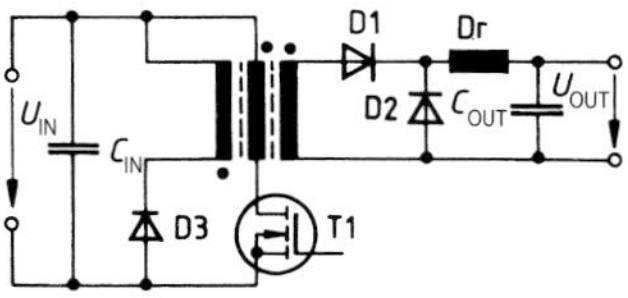 D1 Dr D2 C_{OUT} U_{OUT} U_{IN} C_{IN} D3 T1	Single-phase feed forward converter Single transistor forward converter	• No problem in demagnetizing core • Low cost
2	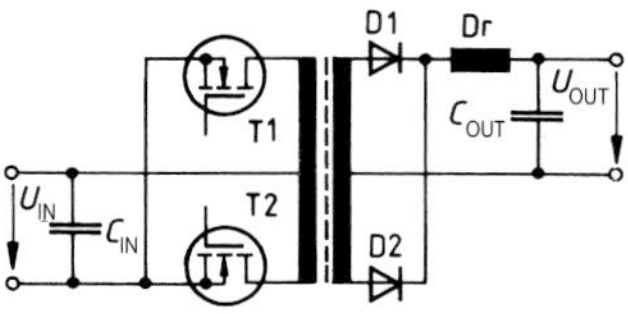 D1 Dr C_{OUT} U_{OUT} T1 T2 U_{IN} C_{IN} D2	Push-pull feed forward converter with two-way rectifier Push-pull converter	• Both transistors are driven at the same potential
3	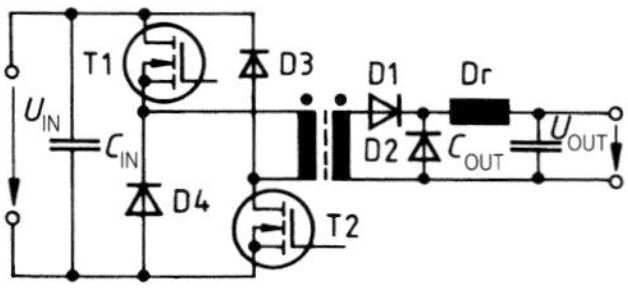 T1 D3 D1 Dr U_{IN} C_{IN} D2 C_{OUT} U_{OUT} D4 T2	Asymmetric half-bridge feed forward converter Two-transistor forward converter	• SIPMOS blocking voltage $U_{DS} \approx U$ • No problem in demagnetizing core • The transformer may have a high stray inductance
4	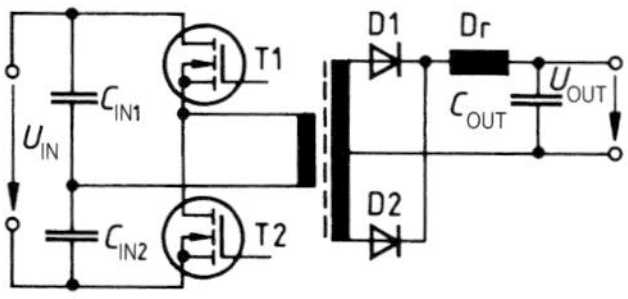 D1 Dr T1 C_{IN1} C_{OUT} U_{OUT} U_{IN} D2 C_{IN2} T2	Symmetric half-bridge feed forwarvd converter Single-ended push-pull converter	• SIPMOS blocking voltage $U_{DS} \approx U$ • The transformer may have a high stray inductance
5	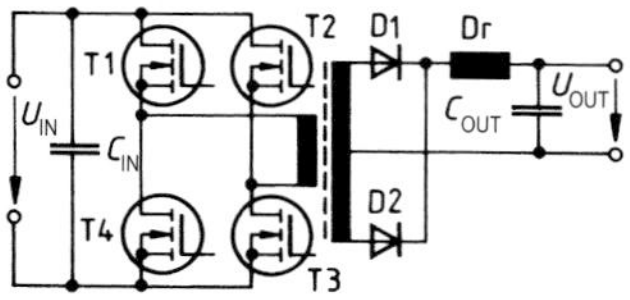 T2 D1 Dr T1 C_{OUT} U_{OUT} U_{IN} C_{IN} D2 T4 T3	Single-phase feed forward converter Single transistor forward converter	• No problem in demagnetizing core • Low cost

Disadvantages	Pulse duty ratio	a: Voltage waveform at transistor b: Output or choke current	c: Current waveform in input capacitor d: Current waveform in output capacitor
• Power transistor blocking voltage $U_{DS} > 2\ U_E$ • Requires a demagnetizing winding • Good magnetic coupling required between primary winding and demagnetizing winding	$\frac{T_{on}}{T} = 0{,}5$	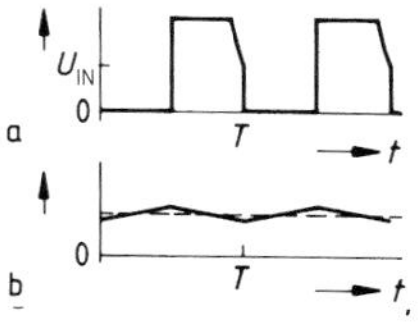	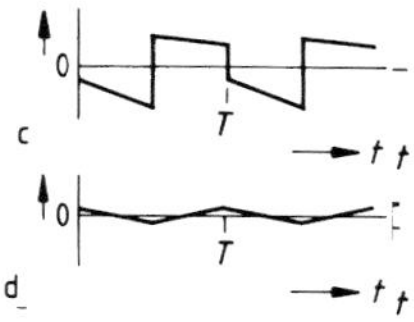
• Power transistor blocking voltage $U_{DS} > 2\ U_E$ • Problems achieving symmetry • Good magnetic coupling required between two primary windings • Danger of both transistors conducting simultaneously	$\frac{T_{on}}{T} = 0{,}42\,(2\text{x})$	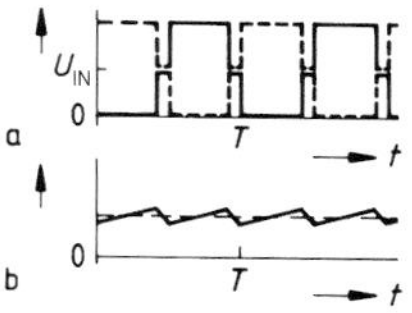	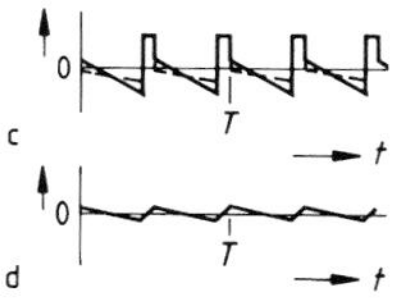
• Requires galvanically isolated drive	$\frac{T_{on}}{T} = 0{,}5$	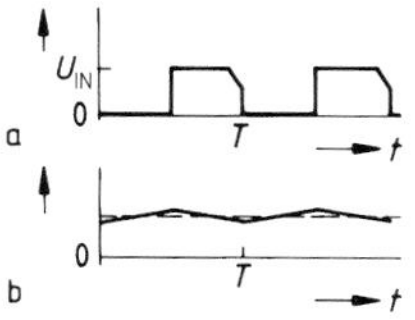	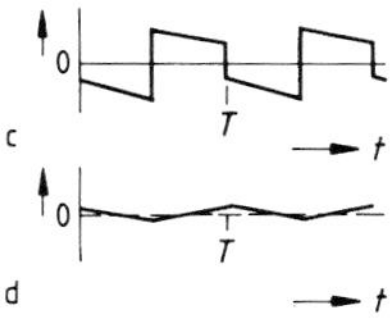
• Symmetry problems • Danger of both transistors conducting simultaneously • Requires galvanically isolated drive	$\frac{T_{on}}{T} = 0{,}5$	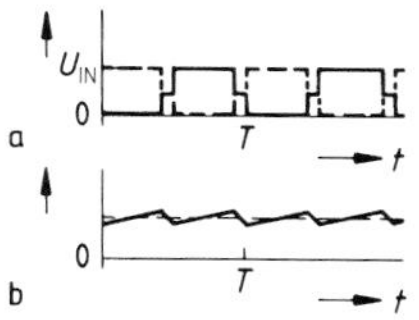	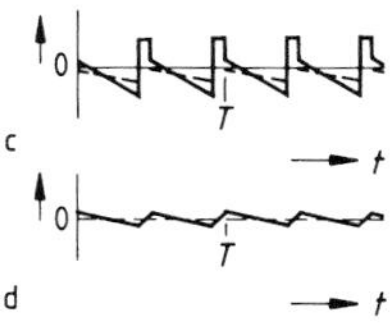
Symmetry problems Danger of both transistors in a half-bridge conducting simultaneously Requires galvanically isolated drive	$\frac{T_{on}}{T} = 0{,}42\,(2\text{x})$	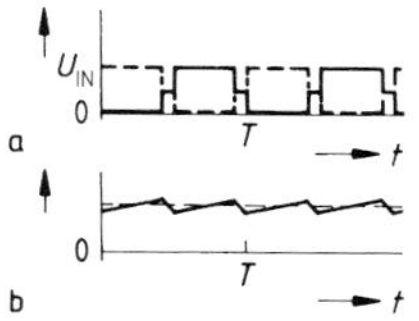	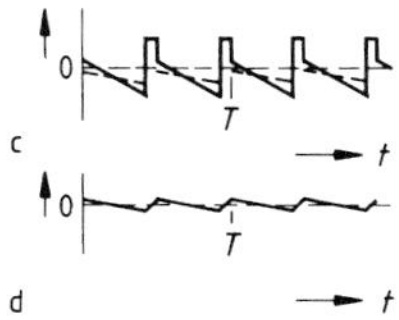

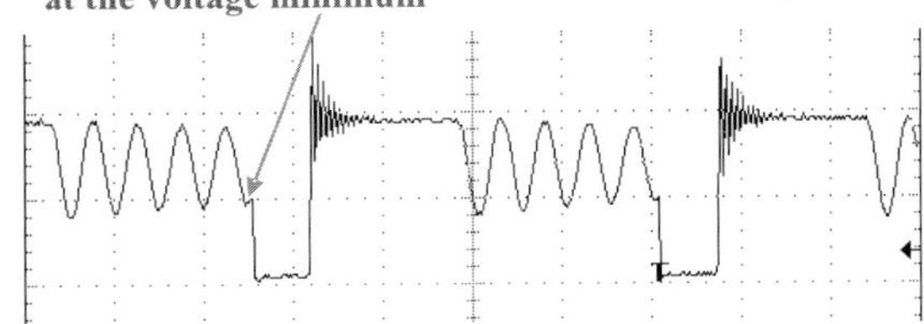

Figure 3.109
Switching on the CoolMOS™ at the voltage minimum

standard power supply units up to about 300 W (Figure 3.110).

Fixed frequency driver ICs

Fixed frequency driver ICs (ICE2A(B) S01, ICE3DS01, TDA16850 and TDA 16888) are suitable for chargers, adapters (LCD monitors, printers etc.), monitors, lower powered consumer electronic devices, but also for PC power supply units and industrial applications. They are best suited for flyback converter topologies, but also for forward converters in the case of the TDA 16888. Here too, the protective functions have been optimized. Even the ICE2A(B)S01 has outstanding values for its low stand-by power consumption. With the ICE3DS01 a matchless 100 mW power consumption has been achieved at 0 Watt output power. Thereby satisfying all the international energy-saving regulations.

The TDA 16888 has a special position: in addition to the driver control circuit for a switched mode power supply, it also has an integral PFC stage driver (see also section 3.7.5).

Fixed frequency driver ICs with integral CoolMOS™

To reduce the numbers of components and circuit board size, and to simplify the design of switched mode power supplies, an innovative family of devices is offered, combining the functionality of the ICE2A(B)S01 driver IC and of the CoolMOS™ transistor in one DIP-8 or TO-220 package, as appropriate.

The advantages of the two-chip construction are optimized technologies for the driver IC and the power transistor. Unlike monolithic solutions, no cross-talk effects arise here. The uniquely low area-dependent forward resistance of the CoolMOS™ switching transistor (see section 3.7.8) can be used to realize power supply units for up to 67 W in a DIP-8 without a

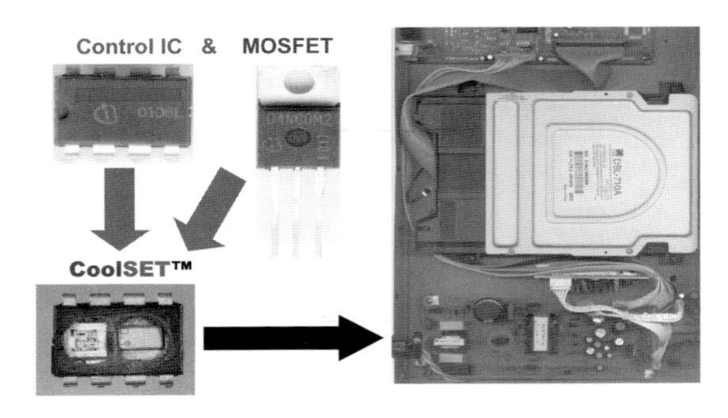

Figure 3.110
CoolSET™ in a DVD application

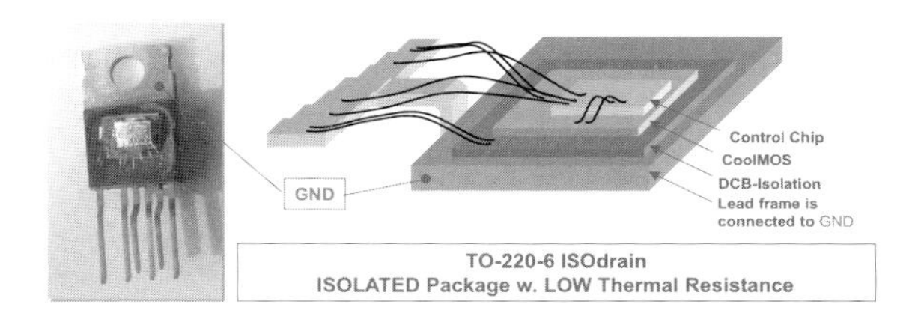

Figure 3.111 Structure of the CoolSET™ in its insulated TO-220 package

heat sink. The TO-220 variant is fully insulated, i.e. the device can be directly attached to a heat sink (Figure 3.111). The chip-on-chip technology has been used for its construction, to enable the lowest possible forward resistance to be realized for the CoolMOS switching transistor

Table 3.9 shows an overview of the available CoolSET types.

3.7.5 Power factor

The power factor is the technical term for the ratio of the real power to the complex power, and takes a value between 0 and 1. The power factor is therefore defined as the cosine of the phase difference between a sinusoidal voltage and the associated current waveform, i.e. PF = cos (φ_{UI}) if the system only contains linear loads. With a purely ohmic load, e.g. an incandescent lamp or a radiator element, the voltage and current are in phase, and thus the power factor is 1. With an inductive load (e.g. an asynchronous motor) or a capacitive load, there is a phase shift and the power factor is < 1 (Figure 3.112).

However, a power supply unit with an input rectifier and a downstream smoothing capacitor (Figure 3.113) represents a non-

Table 3.9 Overview of the available CoolSET™ types

V_{DS} [V]	f_{OPER} [kHz]	Type	$R_{DS(on)}$ [Ω]	$P_{OUT(max.)}$ [1] [W]	Package
650	100	ICE2A0565	6.0	13	P-DIP-8-6
		ICE2A05656Z	6.0	13	P-DIP-7-1
		ICE2A165	3.0	18	P-DIP-8-6
		ICE2A265	1.0	32	P-DIP-8-6
		ICE2A365	0.5	45	P-DIP-8-6
		ICE2A765P	0.5	130	P-TO220-6 ISO
650	67	ICE2B0565	6.0	13	P-DIP-8-6
		ICE2B165	3.0	18	P-DIP-8-6
		ICE2B265	1.0	32	P-DIP-8-6
		ICE2B365	0.5	45	P-DIP-8-6
		ICE2B765P	0.5	130	P-TO220-6 ISO
800	100	ICE2A180Z	3.0	17	P-DIP-7-1
		ICE2A280Z	0.8	31	P-DIP-7-1

[1] R_{th} = 56 k/W (~ 6 cm² copper area), T_a = 75°C, T_j = 125°C, V_{in} = 85 V … 270 V
[2] R_{th} = 2,7k/W, T_a = 75°C, T_j = 125°C, V_{in} = 85 V … 270 V

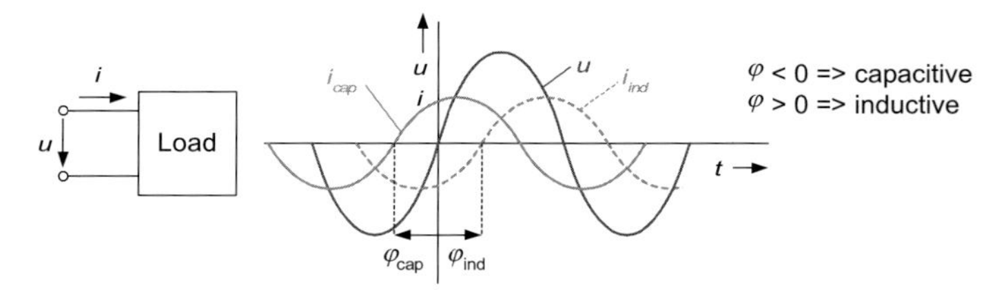

Figure 3.112
Sinusoidal alternating current with examples of inductive and capacitive currents

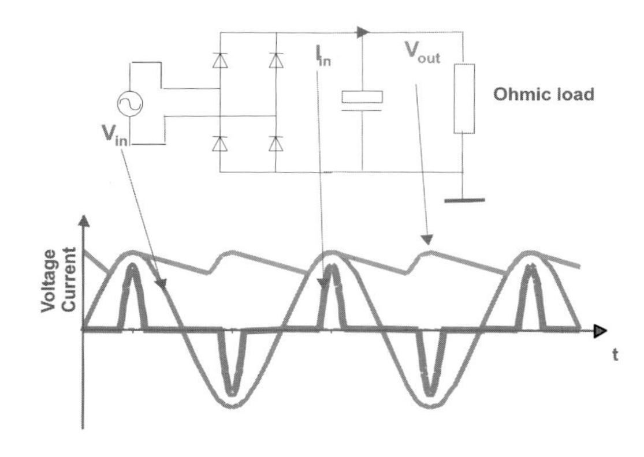

Figure 3.113
Current and voltage waveforms for a typical power supply input circuit

linear load. The current which flows is pulsed, and is in phase with the input voltage provided that this input voltage is greater than the voltage across the smoothing capacitor.

The current flow is indeed broadly in phase with the input voltage, but the spectral distribution shows a strong harmonic content. Apart from the fundamental wave, therefore, the current also has components with frequencies which are an integral multiple of the fundamental wave. These deviations from a sinusoidal current are referred to as the THD (total harmonic distortion):

$$\mathrm{THD} = \frac{\sqrt{(I_{\mathrm{RMS}}^2 - I_1^2)}}{I_{1\mathrm{RMS}}^2}$$

I_1 = Effective value of the fundamental oscillation

I_{RMS} = Effective value of the total current

For a purely sinusoidal current, THD = 0. The greater the deviation is from the sinusoidal shape, the larger is the THD.

The power factor (PF) is then determined, taking account of the non-linear components, as:

$$\mathrm{THD} = \frac{\cos(\varphi_1)}{\sqrt{1 + \mathrm{THD}^2}} = \frac{I_{1\mathrm{RMS}} \cdot \cos(\varphi_1)}{I_{\mathrm{RMS}}}$$

where φ_1 = phase shift between the input voltage and the fundamental current wave.

From this it can be seen that if there is strong distortion of the current the power factor will be < 1. Typical values for switched mode power supplies lie around PF = 0.6 … 0.7.

An example of a practical implication of the high THD value is that considerable transient currents flow in the house wir-

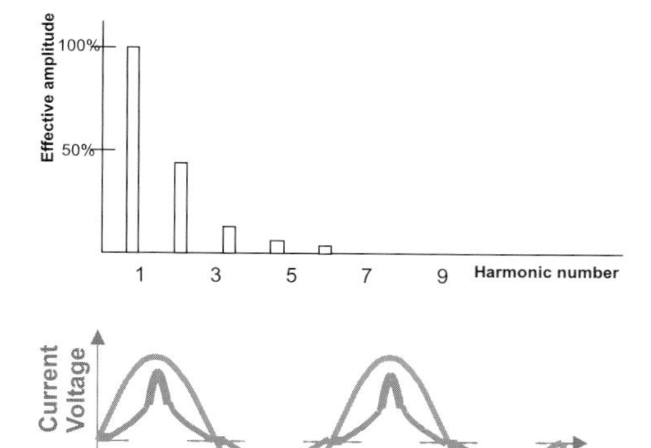

Figure 3.114
Amplitudes of the harmonics (top) and current or voltage waveform for a choke PFC solution

ing, requiring wires with a larger cross-section.

In general, the effect of a lower power factor is that power stations must make available a considerable reactive power, which is practically unused in the load. As a result, power stations cannot be operated optimally, imposing additional loads on the environment.

PFC (Power Factor Correction) circuits represent a solution to this problem, making a power factor of virtually 1 possible in the ideal situation.

In 2001 the norm EN-61000-3-2 (IEC 1000-3-2), on limiting values for harmonic currents in the mains power supply, was introduced. According to this, devices such as TVs, monitors and PCs with a power consumption of > 75 W must adhere to certain limiting values for harmonics. All fluorescent lamps are covered by the norm. A similar standard applies in Japan, so far in the USA only fluorescent lamps are subject to a PFC condition.

Circuit principles for the PFC stages

Passive PFC stages: the simplest solution, and generally the least expensive, consists in inserting an appropriately sized iron-cored choke before the input rectifier of the power supply unit. In this way, power factor values of < 0.9 can be achieved, and harmonic amplitudes which lie just beneath the limiting values (Figure 3.114).

The disadvantages of this solution are the high weight and volume of the PFC choke, while the sensible power limit is at < 200 W.

A similar quality is achieved by using a so-called charge pump, which can be realized using Infineon's TDA 16846 or ICE1QS01 pulsed mode power supply IC (Figure 3.115).

The choke sits behind the rectifier, and is exposed to the pulse frequency of the switched mode power supply, so that a ferrite-cored choke can be used which has significantly smaller dimensions than for an iron-cored choke. The power limit lies

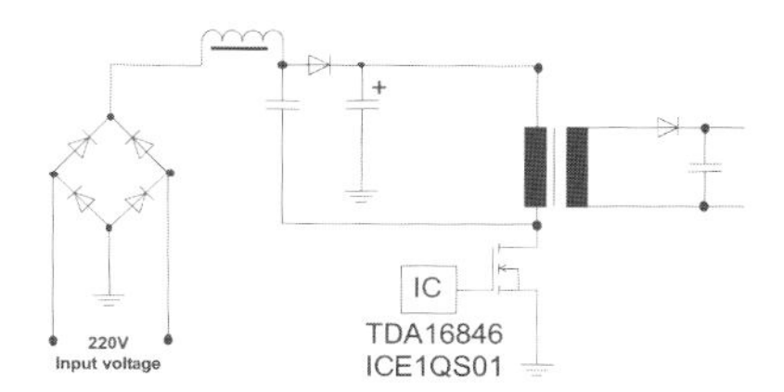

Figure 3.115 Charge pump PFC circuit

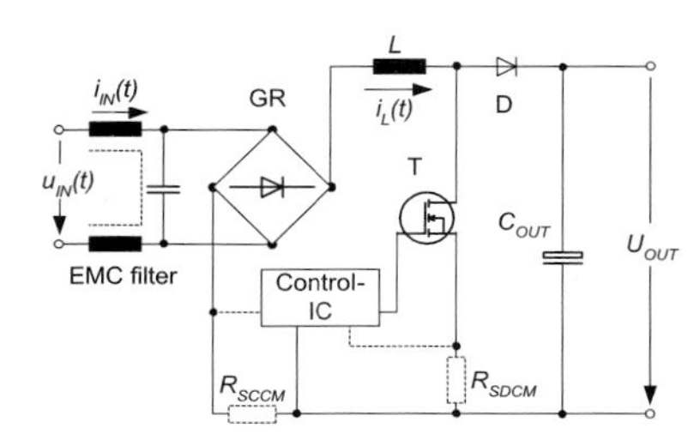

Figure 3.116
Schematic design of a boost converter for PFC operation

at about 250 W. Consequently, this solution is very well suited to applications which must satisfy the standard but for which an optimal power factor correction is not being sought. The additional weight and volume are small. Typical applications are power supplies in TVs or adapters.

Active PFC stages: a power factor of almost 1 can be achieved using active PFC circuits. By comparison with the passive solutions, optimal power factor correction and a reduction in the harmonic content are achieve over a significantly wider range of input voltages and loads, combined with a higher efficiency. To the mains supply, the circuit ought to appear as an ohmic load. Weight and volume are similar to that of a charge pump circuit, that is significantly more favorable than the iron choke solution. The active PFC circuits are connected upstream of the switched mode power supply. Basically, three topologies are used: boost converter, flyback converter and buck/boost converter. These topologies have already been introduced in connection with the switched mode power supply circuits (Tables 3.7 and 3.8). However, the most-used principle is the boost converter (Figure 3.116), which converts a sinusoidal input voltage into a direct voltage output, with a value greater than the peak value of the sinewave voltage.

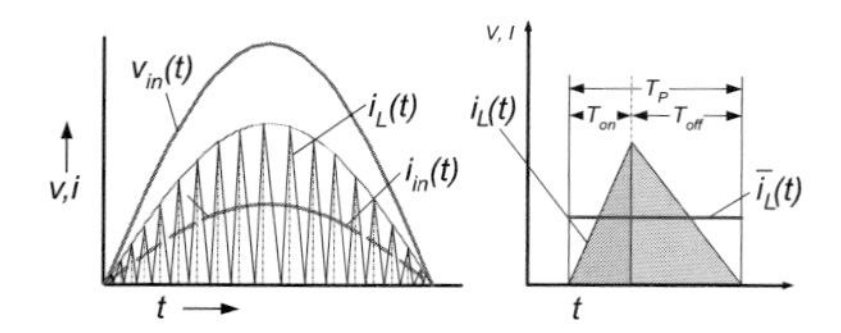

Figure 3.117
Current waveform for free-running operation

This constant high input voltage to the following switched mode power supply has the advantage that the switching transistor (CoolMOS™) and the input capacitor can be made smaller than in the case of a power supply for a wide range of input voltages which does not have a PFC boost converter. In general, the input capacitor for the converter is completely omitted, because the output capacitor C_{OUT} can, as shown in Figure 3.116, completely replace it. The drive is either free-running, i.e. with a variable pulse frequency, or at a fixed pulse frequency. With a free-running PFC converter, within each pulse period the boost converter choke L is first magnetized and then is always fully demagnetized. This is shown by the fact that the choke current $i_L(t)$ at the end of a pulse period has returned to 0 again. Immediately after this, a new pulse period begins, with another magnetization of the choke. The result is that a triangular current waveform $i_L(t)$ arises. The converter is thus operated right up to the point of discontinuity in the choke current, so that this mode of operation is also called discontinuous conduction mode (Figure 3.117).

The time trace of the choke current over one pulse period can be described in accordance with the law of induction as:

$$i_L(t) = \frac{1}{L} \cdot \int_0^t u_{in}(t)dt = \frac{u_{in}(t)}{L} \cdot t \qquad (1)$$

$$0 \leq t \leq T_{on}$$

$$i_L(t) = I_0 - \frac{U_{out} - u_{in}(t)}{L} \cdot (t - T_{on}) \qquad (2)$$

$T_{on} \leq t \leq T_p$

Here, I_0 is the final value of the choke current which is reached during the magnetization period up to the time T_{on}, i.e. $I_0 = u_{in}(t) \cdot T_{on}/L$. As the duration of a pulse period, T_P, is much smaller than that of a mains supply cycle, i.e. $T_P << T_{Mains}$, one may assume as a first approximation that within any one pulse period the values of the voltages and currents are constant. Taken with (1) this gives the mean value of the choke current as:

$$\bar{i}_L(t) = \frac{1}{T_p} \cdot \int_0^{T_p} i_L(t)dt = \tag{3}$$

$$= \frac{1}{T_p} \cdot \frac{I_0 \cdot T_p}{2} = \frac{T_{on}}{2L} \cdot u_{in}(t) =$$

$$= konst \cdot u_{in}(t)$$

Hence, within a pulse period T_p, the mean current $i_L(t)$ through the choke, which is also the input current, depends linearly on the instantaneous value of the input voltage $u_{in}(t)$. A prerequisite for this is, however, that the time T_{on} for which the MOSFET T is switched on is correspondingly constant. If one now puts an infinite number of pulse periods one after another, one gets a continuous current flow which is proportional to the input voltage.

If a PFC converter is driven using a fixed frequency, the boost converter choke is never fully demagnetized or discharged, i.e. the transistor switches into the next pulse period while there is still a choke current $i_L(t)$ flowing. The choke current is therefore trapezoidal in form and has a significant ripple. Operation in this manner with no breaks in the current is also referred to as continuous conduction mode (Figure 3.118).

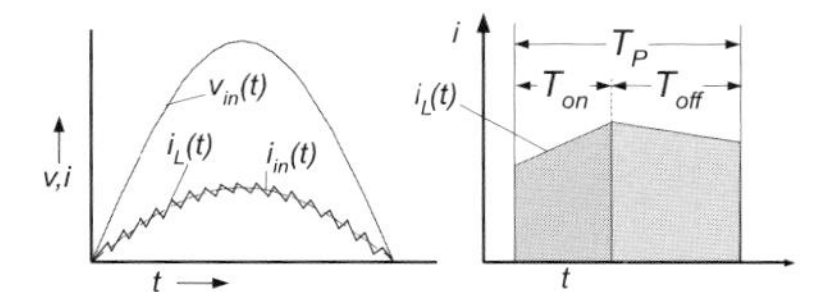

Figure 3.118
Current waveform for fixed-frequency operation

In contrast to the free-running PFC concept, the current between ground and the negative connection on the bridge rectifier is measured via the resistance R_{SCCM}, and is shown in Figure 3.118. This measures the entire choke current. This signal is smoothed and multiplied by a signal which corresponds to the instantaneous value of the input voltage. The set value for the pulse width modulator is thus

$$i_{set}(t) = I_{AVG} \cdot k_U \cdot u_{in}(t) = konst \cdot u_{in}(t) \tag{4}$$

The value I_{AVG} here represents the smoothed value (mean/average) of the choke current, which is assumed to be constant over a pulse period. For this reason, this type of control is also called average current control. The factor k_U is a scaling factor with the units 1/V. From equation (4) it is easy to see the linear relationship between the current set value and the input voltage. The current set value is then converted by a pulse width modulator into a drive signal for the switching transistor.

The decision as to which method is most suitable for which application depends ultimately on the system costs. Very broadly, the discontinuous conduction mode is today preferred for powers up to about 200 W, and above this the continuous conduction mode. Because of the high peak currents, the cost of interference suppression is higher for the discontinuous conduction mode, but on the other hand the requirements to be met by the diodes, in terms of blocking characteristics, are much more critical for the continuous conduction mode. For this reason, silicon carbide diodes (SiC) are especially suitable for this mode of operation.

Integrated circuits for PFC applications

To supplement the product range of switched mode power supply ICs, there is

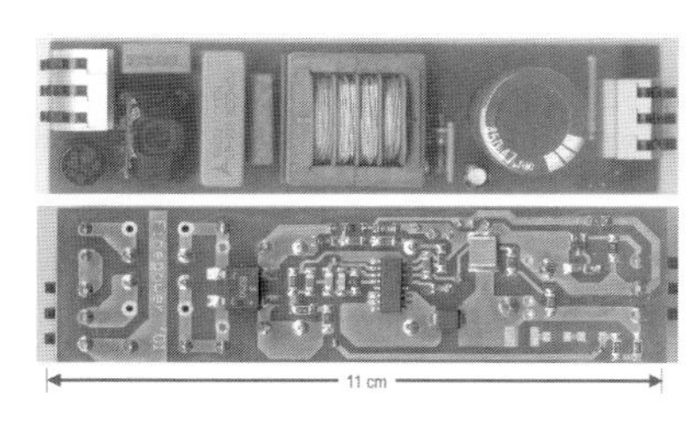

Figure 3.119
Compact PFC circuit using a CoolSET™ ICE1PD265G (50W, wide range)

a range of ICs which have been designed for driving active PFC circuits.

Discontinuous conduction mode PFC: types TDA 4862 and TDA 4863 are suitable for active PFC circuits in lighting ballasts, power supply units for notebooks, LCD monitors and other applications in the power range up to about 150 W. They make possible a power factor of almost 1 with a very precisely regulated output voltage, even over the wide input voltage range of 85-265 V. These devices work in free-running mode with triangular currents, incorporate a quadrant multiplier for precise power factor correction, and safety functions such as overvoltage monitoring, supply voltage limitation, output voltage limitation during load discontinuities, and fast current limitation (cycle-by-cycle).

The device ICE1PD265G is a CoolSET version (see section 3.7.4) which combines the TDA 4863 driver IC and a CoolMOS transistor with a forward resistance of 1.1 Ω in a P-DSO-16 package (Figure 3.119). Depending on the input voltage range, PFC circuits for up to 100 W can be implemented. The advantages of this form of construction are their small space requirement, the elimination of the heat sink and the reduction in the number of components.

Continuous mode PFC: for higher powers, preference is given to the use of continuous conduction mode PFC circuits. A typical driver module for these is the TDA 16888 (see section 3.7.4), the so-called "Powercombi", which represents a combination of a PFC driver and a switched mode power supply driver (PWM) in one IC. It works at a fixed clock rate which can be set between 15 and 200 kHz. This PFC clock rate is strictly linked to the clock rate of the switched mode power supply driver. In order to minimize electromagnetic interference, the PFC stage switches at the rising edge of the clock system, the PWM section with the falling edge. Apart from the outstanding safety functions, mention should also be made of the "modulated gate-drive". By using a non-linear time trace of the gate drive when the transistor is switched on, the steepness of the switch-on edge of the CoolMOS™ is somewhat reduced, which simplifies interference suppression for the system. Applications for this device are any pow-

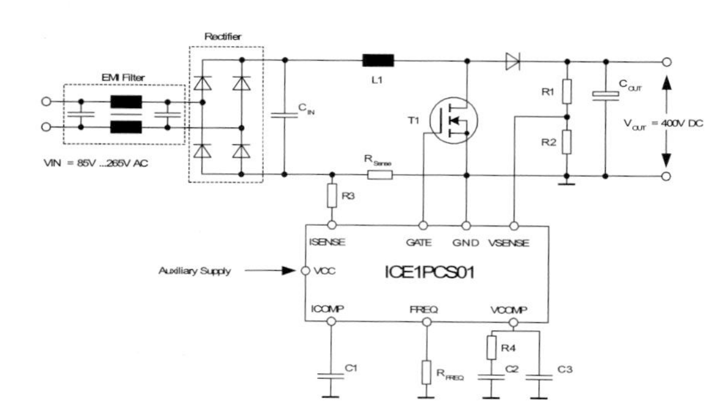

Figure 3.120
Continuous conduction mode PFC stage using an ICE1PCS01

er supply units with PFC, in the range 150 W up to > 1000 W.

The ICE1PCS01 (Figure 3.120) represents a new class of PFC driver devices. It has been specially designed for very cost-sensitive applications. It is offered in a DIP-8 package, but in spite of this has an external frequency adjustment and all the necessary safety functions. The number of external components is minimal, and the development of the PFC circuit is therefore greatly simplified. Typical applications are PCs, servers, adapters, smaller drives and universal power supply units.

3.7.6 Drives – rotation speed control and power electronics

Power electronics is the link between energy generation and energy use. With its objective of being able to supply DC, a.c. and three-phase drives from any required mains supply, drive technology represents an important area. Here, converter connections with power semiconductors such as diodes and IGBTs are used to continually adjust the voltages and currents. From a fixed mains power supply they generate a variable voltage system optimized for the drive.

From a historical point of view, converter circuits began with mercury vapor rectifiers. Selenium rectifiers opened up the lower power ranges of drive technology. As a result of the continuous development of semiconductor devices, it has been possible to offer ever more compact, lower cost and more efficient solutions. Today, modern drive concepts combine digital processing with power semiconductors in modular solutions, but also discrete constructions with packaged devices.

Inverter applications

With the development of the (DC) inverter it became possible to operate any motor from any required mains supply. From drive motors for washing machines through to traction drives for rail locomotives, inverter uses encompass a wide power range. Typically nowadays, IGBTs are used with antiparallel diodes as switches and valves for such applications, either as discrete constructions for lower powers or as modules for higher powers. However, the method of functioning of these motor drives is in principle always the same, and will be briefly described below.

Unlike a direct current machine, for which the rotational speed can be changed by a change of voltage, in order to change the rotational speed of an alternating current machine it is necessary to vary the frequency of the current.

The synchronous rotational speed is directly proportional to the frequency ω and inversely proportional to the number of poles p, and can be calculated from the following formula:

$$n_{\mathrm{d}} = \frac{\omega}{2 \cdot \pi \cdot p}$$

When $p = 1$ we speak of a 2-pole machine. However, the pole count can be higher, which reduces the rotational speed.

In order to control the rotational speed of the motor, the frequency of the current must be varied. Frequency converters of inverters are used for this purpose.

Figure 3.121 shows one phase of a frequency converter, omitting the input rectifier, which from the mains power supply value provides a constant intermediate

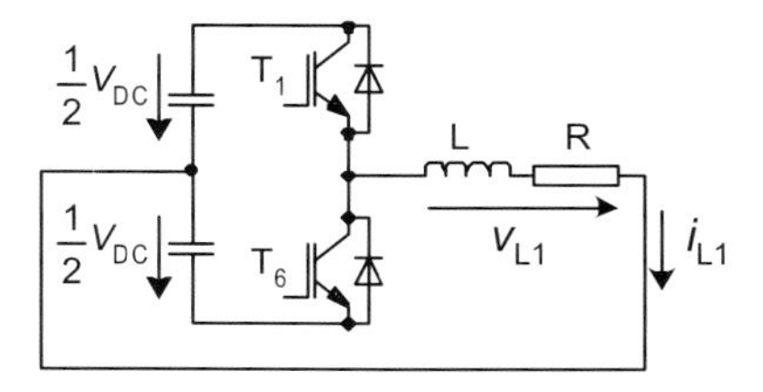

Figure 3.121 One phase of a converter

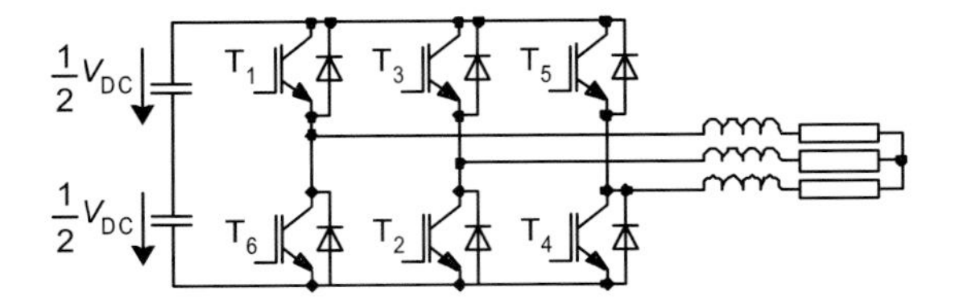

Figure 3.122
Complete 3-phase converter

circuit (link) voltage. Three such phases form the typical converter for a three-phase motor load, as shown in Figure 3.122. The switches T1 to T6 represent the IGBTs with anti-parallel free-wheeling diodes.
The way in which such a frequency converter functions will now be clarified, referring to Figures 3.123 and 3.124.

The switch in the converter uses the PWM (pulse width modulation) method to chop a rectified link voltage V_{DC}. Integration of this chopped signal produces a sinusoidal voltage, the frequency of which can easily be set using the PWM waveform. The ohmic-inductive load due to a motor acts as an integrator for the current, and thus produces sinusoidal

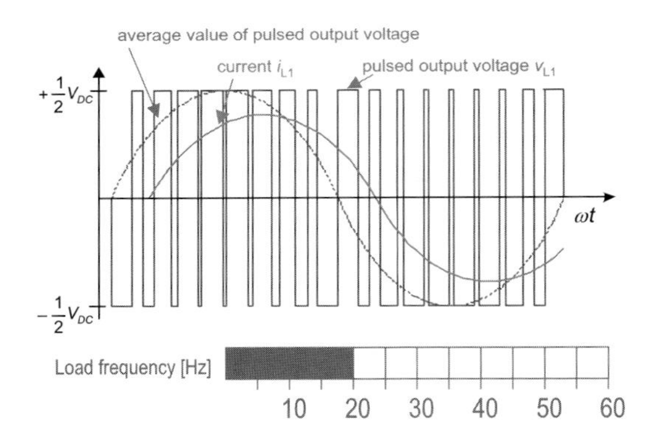

Figure 3.123
PWM waveform for 20 Hz

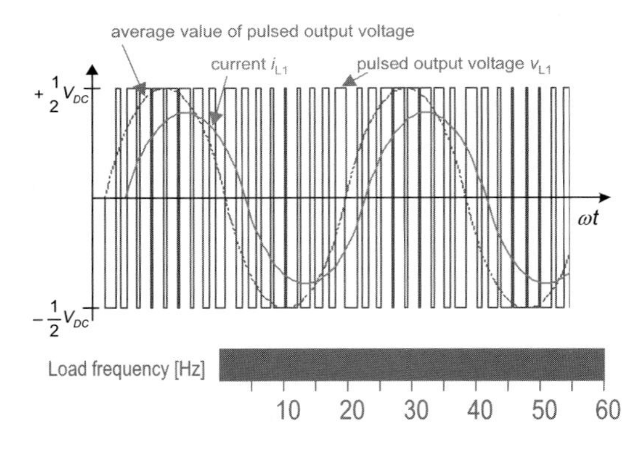

Figure 3.124
PWM waveform for 60 Hz

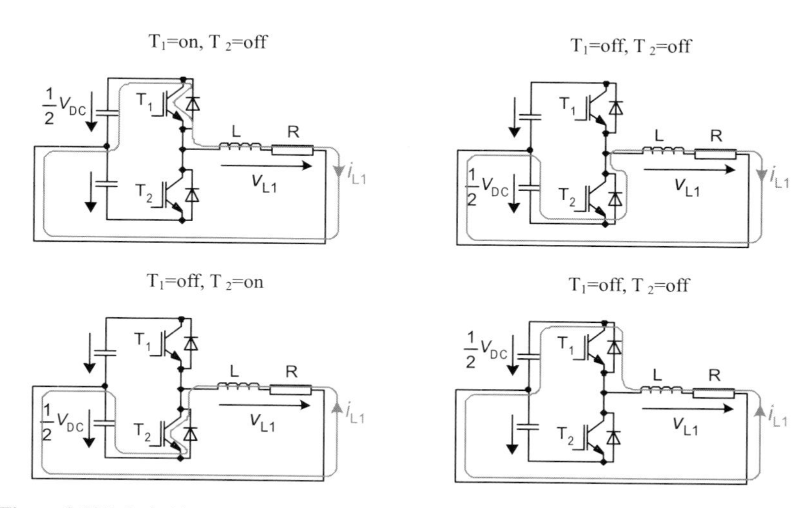

Figure 3.125 Switching sequence and current path in one phase of an inverter

waveforms, phase-shifted relative to the voltage.

Figure 3.125 shows the switching sequence and the current path for one phase of a converter. Because of the phase-shift between current and voltage, a free-wheeling path is always required via the IGBTs. In addition to the switch, the converter application requires a valve which is anti-parallel with the active component, a free-wheeling diode.

In order to increase the efficiencies and power densities, new technologies are constantly being developed. One approach is to improve the topologies or to optimize the switching technology, another is to develop the semiconductors further. As a semiconductor manufacturer, Infineon Technologies focuses on the second of these, and as a technological leader in the field of IGBTs introduces new concepts for use in inverters.

3.7.7 Low-voltage power transistors: OptiMOS™

As the diversity of electronic applications has grown over recent years, so too the need for specially customized power supply concepts has increased greatly. Modern generations of processors impose ever higher requirements in terms of the constancy of the supply voltage, while the current demand has risen steadily. Portable devices such as mobile phones, PDAs and laptops must use the available battery charge with as little loss as possible, to achieve long standby and operating times. And in automobiles, the electronic revolution has only just begun. Mechanical and hydraulic functions are more and more being replaced by electro-mechanical solutions, to increase safety and convenience and to reduce fuel consumption. All these applications call for a power supply drive system which cannot be realized without low-voltage switches.

Unlike bipolar transistors, for which – as the name indicates – both types of charge carrier are involved in the current flow, in the case of MOSFETs only the majority charge carriers contribute to the current flow: with n-channel MOSFETs only the electrons, with p-channel MOSFETs only the holes. This characteristic has two important advantages:

1. The drain-source voltage drop for a low-voltage MOSFET is smaller than can be achieved at the saturation curve of a bipolar transistor. Particularly at low supply voltages and high currents, this implies a substantial reduction in the losses.
2. As a unipolar device, the MOSFET has no storage charge in the conducting state. The tail current which is usual with bipolar components during switch-off processes is thus absent. Extremely fast switching times and significantly lower switching losses therefore make the MOSFET the ideal power switch for pulsed applications such as power supplies and motor control units.

MOSFET – conductivity

The forward resistance is the most important parameter of the component. It is determined as the sum of all the resistive elements in the MOSFET structure. The individual contributions to the resistance are indicated in Figure 3.126 for a planar MOSFET. The insulated planar gate and source connections are located on the top of the chip, the underside forms the drain connection. The largest contributions in the case of low-voltage transistors arise from the channel resistance, $R_{channel}$ (approx. 30%), from the resistance of the intrinsic junction FET which is formed between the two p-wells, R_{JFET} (approx. 25%), and from the resistance of the drift zone, R_{epi} (approx. 30%). This last is determined by the blocking capability required of the transistor. The sheet resistance rises more than quadratically as the breakdown voltage rises, being $\sim V_{br}^{2.5...2.6}$. These three resistance elements together produce over 80% of the total resistance. New technological developments are therefore based on cell concepts, which are directed at these three component elements.

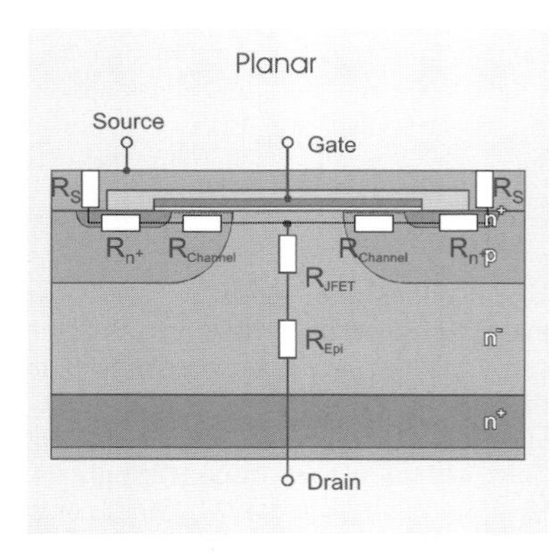

Figure 3.126 Cell of a vertical MOSFET

The channel resistance is proportional to the channel length and inversely proportional to the channel width.

However, because of the penetration of the drain potential into the p-well, the channel cannot be made arbitrarily short without having a massive effect on the blocking characteristics and threshold voltage of the transistor. The channel width, on the other hand, can be significantly increased by a higher density layout. The rapid developments in photolithography make it possible today to realize ever smaller structures in silicon. Power electronics may be limping along a few years behind the very large scale integration in microelectronics, but it too moved into the sub-µm region some years ago. By using ever narrower cell geometries, it has been possible in the last few years to substantially increase the channel width per unit area, and thereby significantly lower the channel resistance as a proportion. Modern cell concepts nowadays achieve channel widths of about $1 m/mm^2$. However, with the smaller cell structures the p-wells also get closer together, so that the JFET proportion increases, and the planar concept comes up against its limits.

It soon became necessary to look for new cell concepts. In an analogous way to the introduction in power electronics of the vertical MOSFET instead of the lateral one, there was a switch over to a vertical

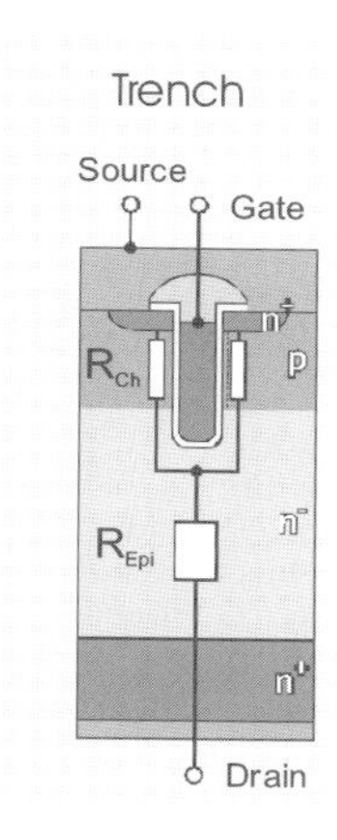

Figure 3.127
Cell of a trench MOSFET

gate structure, so-called trench geometry. This concept brings at the same time three quite critical advantages:

1. Depending on the structure, the JFET component is eliminated. The p-zones are no longer in the form of wells, lying in a regular arrangement along the surface. The p-zones are now separated by trenches, on the surface of which the channel is constructed in the vertical direction. The voltage drop in the forward direction thus no longer causes a narrowing of the current path between neighboring p-zones, and the flow of charge carriers to the drain is unimpeded.
2. As a result of the vertical arrangement of the channel, the channel length no longer imposes any restrictions on the spacing of the cells. This makes it possible to make further significant increases in the channel width. Today, the lower limit for the cell spacing is only determined by the available lithography and the precision of the trench process.
3. With the planar concept, bends in the p-wells leads to field peaks, which means that the doping in the drift zone must not be too high. With the adjacent p-zones in the trench concept, this problem is also reduced, and by increasing the doping a further improvement in the forward characteristics is achieved.

For the trench cell construction shown schematically in Figure 3.127, the elimination of the JFET is clear and the enormous potential for increasing the channel width is easy to imagine. The gate is located in the trench, the channel runs perpendicularly to the chip surface. By this means, it has been possible to realize an enormous reduction in the area-specific on-state resistance of power MOSFETs in the last few years.

MOSFET – switching characteristics:

The second fundamental difference between MOSFETs and bipolar transistors is their quasi-zero-power control via the insulated gate electrode. Whereas with bipolar transistors a constant base current is required for them to conduct, which for power transistors is correspondingly larger and can reach a substantial proportion of the total current, with a MOSFET it is sufficient to apply a voltage above the threshold voltage in order to switch the transistor on. Losses occur, during switching on and off, due solely to the changes in charge in the MOS capacitances.

Regardless of whether the geometry is planar or trench, a MOSFET can be described to a good approximation by three capacitances. The drain-source capacitance is determined by the capacitance of the pn-junction between the source and the drain. The gate-source capacitance is formed by the overlap of the gate electrode with the n^+ zone of the source connection, and with the p-well which is short-circuited to the source on the one hand and, to a much smaller extent, by the overlap with the source metallization over the gate. The gate-drain capacitance, also known as the Miller capacitance, consists ultimately of two parts, in series: the so-called oxide capacitance between the gate and the n^- zone, with the gate oxide as the dielectric; and the capacitance of the

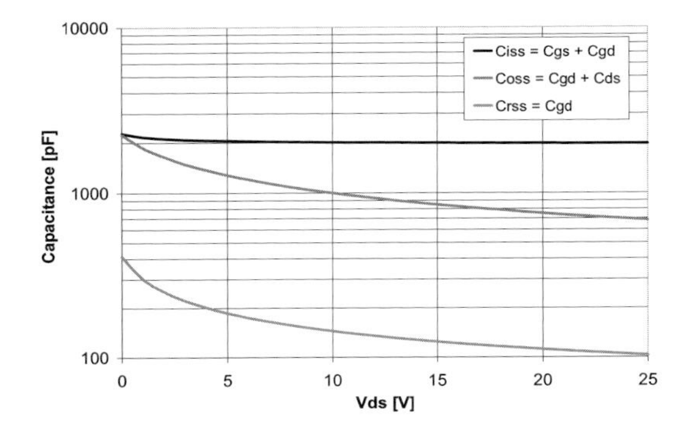

Figure 3.128
Capacitances of a MOSFET as a function of the applied drain-source voltage

space charge zone, which forms when there is a negative voltage between the gate and the drain. Over the transistor's voltage range, the gate-drain capacitance is therefore very voltage-dependent, and can vary over several orders of magnitude. In the datasheets, these capacitances are often specified as the input (C_{iss}), output (C_{oss}) and reverse transfer (C_{rss}) capacitances, but these can easily be calculated from the transistor capacitances. A typical graph of the capacitances is shown in Figure 3.128. The gate-drain capacitance shows a strongly non-linear behavior.

In order to switch on a MOSFET, for an n-channel transistor a positive voltage must be applied to the gate. The progress over time of the switching process with an inductive load with a free-wheeling circuit can be subdivided into various sections (see Figure 3.129a). First, the gate-source capacitance is charged up until the threshold voltage is reached. During this time, the blocking characteristics do not yet change in any way. The time this lasts is called the delay time, and it characterizes the response time of the transistor. When the threshold voltage is exceeded, the channel is formed, and the drain current starts to flow. Further increase in the gate current is now opposed by the Miller capacitance C_{gd}. The gate charge required to charge up the capacitances must be supplied by the gate driver, and determines the losses in the control circuit. Only when the Miller capacitance has been fully charged by the reduction in the space charge zone will the transistor switch on, and the forward resistance and

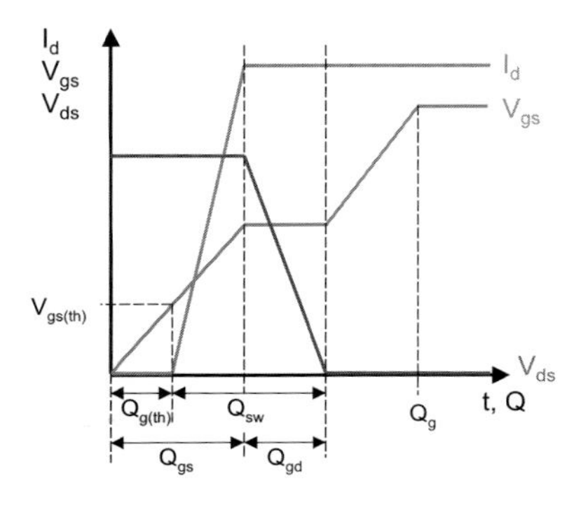

Figure 3.129a
The switch-on process in a MOSFET, for an inductive load with a free-wheeling circuit

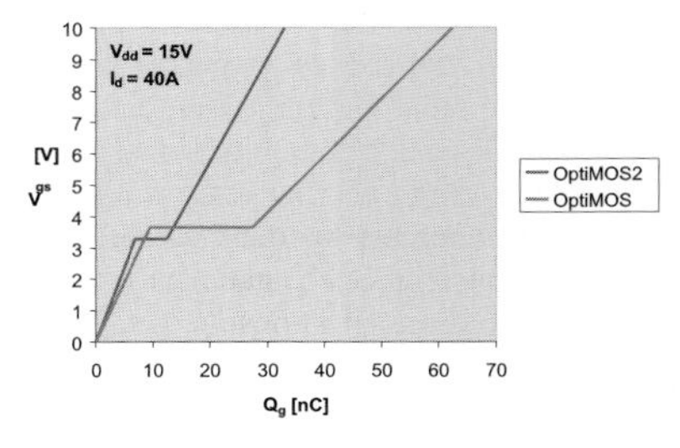

Figure 3.129b
Typical development of the gate charge for components with the same R_{on} in OptiMOS and OptiMOS-2 technologies

load current determine the voltage drop across the transistor.

Figure 3.129b shows the gate voltage against the gate charge for two transistors, in the OptiMOS and OptiMOS-2 technologies. It as been possible to reduce the length of the Miller plateau, and with it the gate-drain capacitance, by more than 50%. It is easy to read off from this diagram how large the individual proportions are of the capacitance or charge. The gate-source capacitance is the main determinant of the slope at $Q_g = 0$. The size of the Miller charge can be determined directly from the length of the plateau phase. And finally, the gate-source and oxide capacitances affect the last branch of the diagram.

Modern DC/DC converters for power supplies to μ-processors today operate at frequencies of several 100 kHz. The smallest possible Miller capacitance is therefore a prerequisite for keeping the switching losses in the transistor, and also the driver losses, within acceptable limits, and hence also the dimensions of the control loop. The objective in the development of the OptiMOS-2 technology was therefore to reduce the contribution of the Miller capacitance as far as possible, and hence to create an ideal component for these applications.

By the use of the trench concept and an enlargement of the channel width per unit area, the area-specific forward resistance can thus be reduced. However, the price to be paid for doing so lies initially in a significant increase in the capacitances per unit area. A consequence of the greater channel width is a larger overlap of the gate with the n^+ region of the source, which produces an increase in the gate-source capacitance. The gate-drain capacitance also rises, as a consequence of the greater channel width brought about by the higher cell density. However, this effect can be compensated if one restricts the overlap of the gate and drain to the trench floor. The source-drain capacitance is the only one which becomes smaller for the same transistor area as a result of the planar pn-junction.

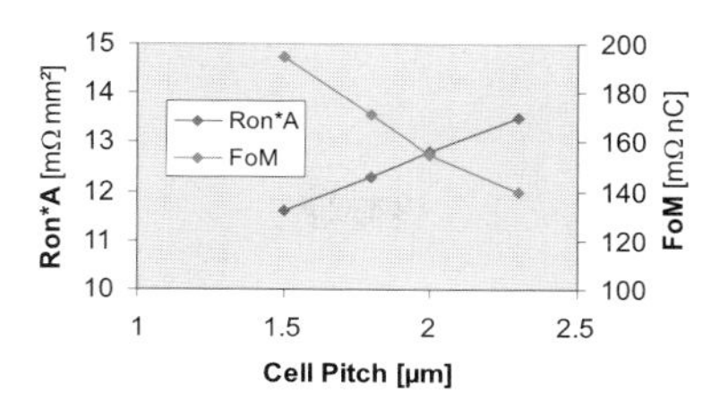

Figure 3.130
Area-specific resistance $R_{on} \cdot A$ and the FOM as a function of the cell size

In developing low-voltage power MOSFETs, a compromise must therefore be found between an optimal area-specific resistance $R_{on} \cdot A$ and an area-specific gate charge Q_g/A which is as low as possible. If one defines a so-called figure-of-merit (FOM) as

$$\text{FOM} = (R_{on} \cdot A)\,(Q_g/A)$$

then one has an area-independent characteristic quantity for the technology concerned. Figure 3.130 shows a model calculation for $R_{on} \cdot A$ and the FOM for various cell sizes.

Depending on the requirements of the application, the technology can be optimized for $R_{on} \cdot A$ (conduction losses) or for the FOM (conduction and switching losses).

MOSFET storage charge (reverse recovery charge)

In the most widespread topology for DC/DC converters, the buck converter, the backward diode of the MOSFET is also used to conduct the current in the free-wheeling circuit. In this operating state, the transistor behaves like as bipolar component and the current is carried by the minority and the majority charge carriers. If the polarity is now reversed to the blocking direction, then the charge stored in the transistor, the so-called reverse re-

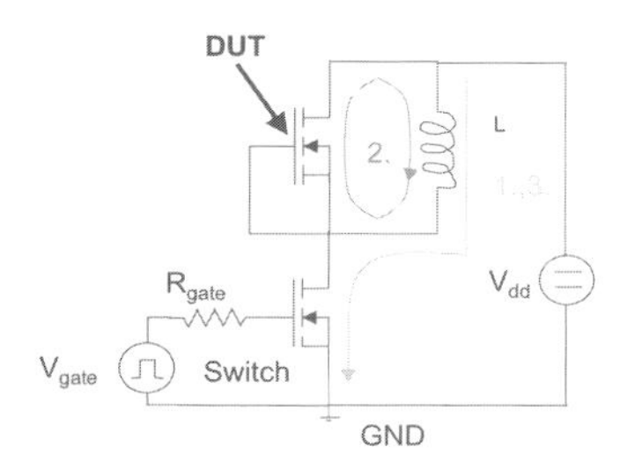

Figure 3.131a
Test circuit for measuring the reverse recovery charge

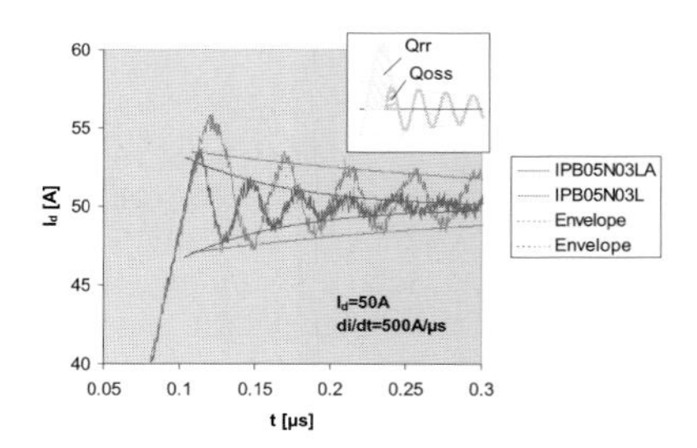

Figure 3.131b
Typical reverse recovery charge for the OptiMOS and OptiMOS-2 technologies

covery charge, must first be dissipated. In an application, this leads to additional losses due to current peaks, which represent a danger for the circuit. A low reverse recovery charge is thus a third important characteristic of modern low-voltage transistors.

Figure 3.131a shows a test arrangement for measuring the reverse recovery charge. In Phase 1 of the test, with the lower transistor (Switch) switched on, the current in the coil increases linearly. In Phase 2, after it is switched off, the current flows in the upper free-wheeling circuit, through the diode of the transistor device under test (DUT). In Phase 3, the lower transistor is then switched on again. The current commutates from the free-wheeling circuit to the load circuit. As this happens, the polarity of the upper transistor is reversed from diode mode to the blocked state, and the reverse recovery charge must be dissipated via the lower transistor in addition to the load current.

Figure 3.131b shows a typical measurement for Infineon's OptiMOS and OptiMOS-2 technologies at a switching speed of 500A/µs. The reverse recovery charge for these technologies is so low that only exceptionally small current peaks arise. The oscillation arises from the resonance of the output capacitance (C_{oss}) with the stray inductance of the test rig. In order to determine the reverse recovery charge, the oscillation, shown in the small inset diagram, must be subtracted. The remaining area corresponds to the reverse recovery charge Q_{rr}. For the OptiMOS-2 technologies, typical values lie under 10 nC.

MOSFET – avalanche resistance (ruggedness)

Under extreme operating conditions, it can happen that the transistor is driven to an avalanche breakdown. Apart from the circuit being switched off by a fuse or a protective circuit, this should not cause any damage to the transistor. For this reason, the specification of the avalanche energy specifies how much energy the transistor can withstand in the event of a breakdown – the transistor then behaves rather like a Zener diode.

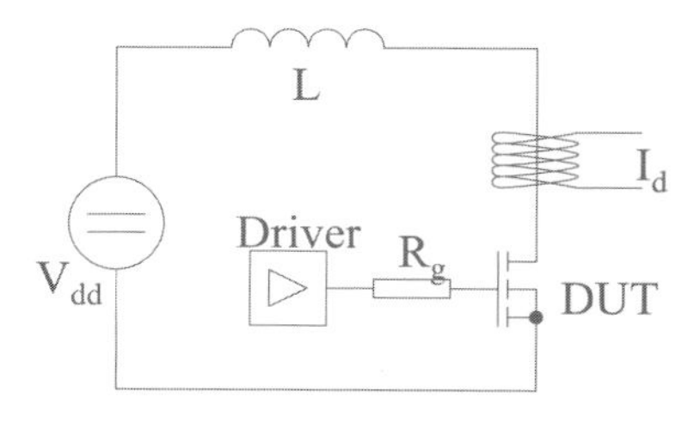

Figure 3.132a
Test arrangement for measuring the avalanche capacity of MOSFETs

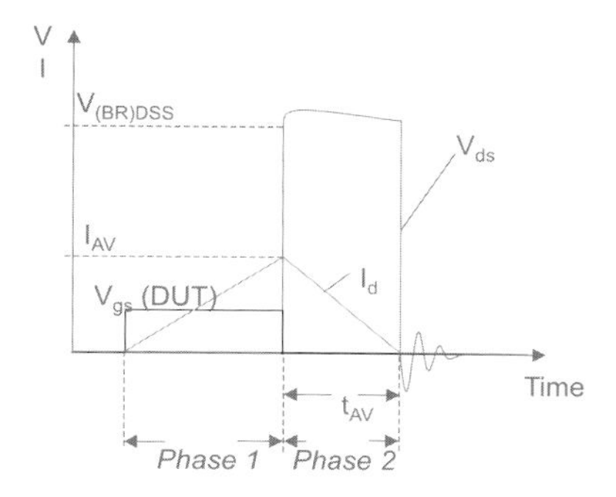

Figure 3.132b
Current / voltage graph for the avalanche test

A test circuit for determining the avalanche resistance is sketched in Figure 3.132a. Initially, the transistor is switched on. The current in the circuit increases in a way determined by the inductance L and the supply voltage V_{dd}. If the transistor is now switched off, the inductance forces a continuing current in the circuit. The voltage across the transistor rises, until eventually the breakdown voltage is reached, and the current can flow away through the transistor's intrinsic body diode. Figure 3.132b shows the two phases of the avalanche test schematically: in phase 1, energy is stored in the coil, and this must be absorbed by the transistor in the breakdown of phase 2.

For the simple test circuit, it is easy to calculate the duration of the avalanche event from the breakdown voltage of the transistor, its inductance and the current:

$$V_{br} = V_{dd} - L\frac{di}{dt} \Rightarrow t_{av} = \frac{L \cdot I_{as}}{V_{br} - V_{dd}}$$

The duration of the avalanche event is proportional to the inductance and to the current, and inversely proportional to the voltage drop across the coil. In the low-voltage range in particular, the small voltage difference between the breakdown voltage and the supply voltage leads to relatively long avalanche phases. The energy stored into the device is given by:

$$E_{as} = \frac{1}{2}V_{br} \cdot I_{as} \cdot t_{av} = \\ = \frac{1}{2}L \cdot I_{as}^2 \cdot \frac{V_{br}}{V_{br} - V_{dd}}$$

With modern power MOSFETs, suitable technological measures can be used to avoid the parasitic npn transistor which is formed at switch-on by the source connection, channel and drift zone. The avalanche resistance is limited solely by an increase in the temperature of the device. Under this assumption, it is possible to derive some fundamental dependencies from the thermal conduction equation. To do so, we begin by considering the average power dissipated in the transistor, which is given by:

$$\overline{P}_{av} = \frac{1}{2}V_{br} \cdot I_{as}$$

The thermal resistance for a particular pulse length t_{av} is given by the material properties for silicon and the active chip area A_{active}:

$$Z_{th}(D = 0; t_{av}) = 2K\sqrt{t_{av}}$$

where

$$K = \frac{F(\text{silicon material parameters})}{A_{active}}$$

From these two quantities, the temperature rise compared to the initial temperature T_0 is given by:

$$\Delta T_j = T_j - T_0 = Z_{th} \cdot \overline{P}_{av} = \\ = K \cdot V_{br} \cdot I_{as}\sqrt{t_{av}}$$

The intrinsic conductivity of silicon increases exponentially with increasing temperature. When the number of thermally generated charge carriers eventually reaches the order of magnitude of the background charge, the device loses its blocking ability and is destroyed. The destruction thus always occurs at a fixed junction temperature

$$T_{j,destr} = \text{const}$$

This gives the following relationships for the avalanche current and the avalanche energy:

$$I_{as} \propto A_{active}(T_j - T_0) \cdot t_{av}^{-1/2}$$

$$E_{as} \propto A_{active}(T_j - T_0) \cdot t_{av}^{1/2}$$

$$E_{as} \propto A_{active}^2(T_j - T_0)^2 \cdot I_{as}^{-1}$$

If measured destruction currents are plotted as a function of the initial temperature T_0 for different inductances, or pulse durations, as appropriate, the assumptions of the model can be confirmed to a good approximation (see Figure 3.133a). This gives a destruction temperature of about 380°C, very close to the intrinsic temperature of silicon. The avalanche resistance it thus only limited by the ability of the device to withstand thermal loads.

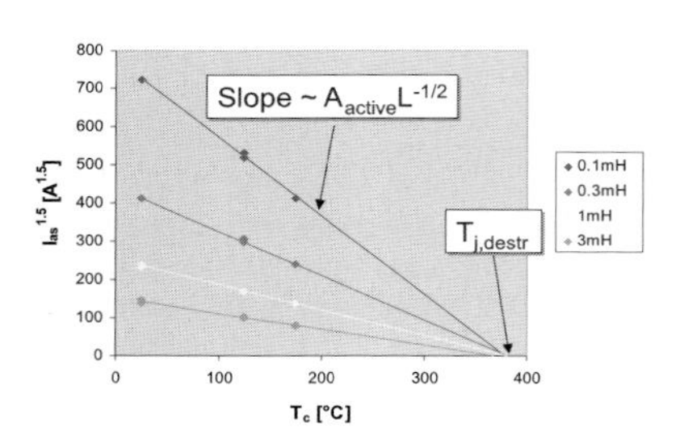

Figure 3.133a
Maximum avalanche current as a function of the inductance and initial temperature T_0. All the lines intersect at the destruction temperature $T_{j,destr}$

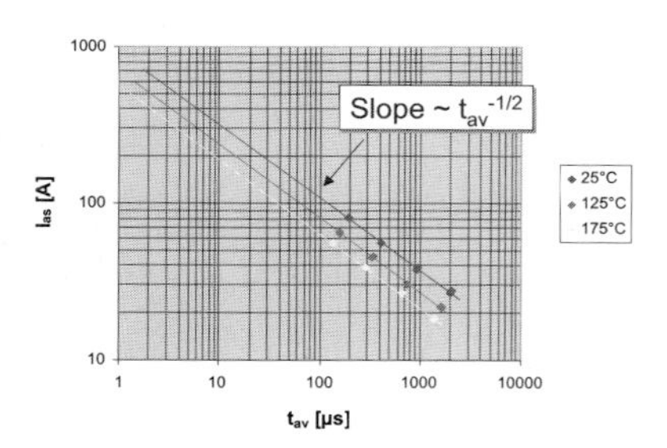

Figure 3.133b
Avalanche current as a function of the duration of the avalanche event, at various temperatures

Summary and outlook

Apart from the rapid development of microprocessors and memory in recent years, the power semiconductor field has also undergone a revolution. This has been carried forward by the introduction of innovative cell concepts and the availability of high-resolution exposure systems for smaller structural sizes. Modern low-voltage MOSFETs are now distinguished by an extremely low on-state resistance, low switching losses and gate charge, and ruggedness in extreme operating states. The optimal balance between the individual parameters is already, and to an ever increasing extent, being determined by the application concerned. The development of application-specific technologies, with quite specific characteristics, will therefore become increasingly more important in future.

3.7.8 High-voltage transistors: CoolMOS™

Concepts for high-voltage switches and the route to the compensation principle

The challenge of the high-voltage switch is that it should combine a high blocking capability with very good characteristics in the on-state. Further, it should be simple to control, fast-switching, overload resistant and, of course, cheap. If one excludes mechanical relays (service life of a few seconds at 75 kHz), all that remains is the world of semiconductor solutions. Here, the implication of the requirement for "high breakdown voltage" is low doping and a relatively large thickness of the layer to which the voltage is applied. But the desire for good forward characteristics demands high doping and small thickness of the active layer. This contradiction was resolved for the first time by high-voltage bipolar transistors, in which a lightly-doped active layer is flooded in the conducting state by the injection of a highly conducting electron-hole plasma. In the on-state, these switches require a base current which is only one or two orders of magnitude less than the load current, which makes their control very complex. In addition, there are relatively large fluctuations in the parameters, which today limit the fields of application for this concept to a few applications in the area of line scanning in TVs and in energy-saving lights.

The combination of the bipolar transistor principle with zero-power control leads to the insulated gate bipolar transistor, or IGBT for short. Here, instead of the p-well being controlled (by the base), the electron current is switched by a lateral or vertical MOS channel. This electron current leads to the injection of holes from a pn-junction with forward polarity, located on the rear of the device. This in turn enables the low conductivity of the basic doping to be raised in the on-state by several orders of magnitude, by the injection of an electron-hole plasma. This conducting plasma must then be removed from the active zone again when the device is switched off. This inevitably results in thermal switch-off losses, because while the voltage is rising across the transistor a drain or tail current continues to flow.

IGBTs have now come to dominate a wide field of application, extending from the control of all types of electric motors through to their use in pulsed mode power supplies. At medium and high frequencies, because of the reverse recovery charge, the switch-off losses of an IGBT dominate the losses in the conducting state, when it is switched on.

This range is therefore the classical domain of unipolar switches, i.e. power semiconductor devices in which the current is carried by only one type of charge carrier. The main representative of this group of devices is the conventional power MOSFET, as developed in the mid-1970s. As in the IGBT, the electron current is controlled by a lateral or vertical MOS channel, and in power transistors usually flows vertically through the layer to which the voltage is applied, to the drain contact on the rear. In order to achieve a high blocking capability, the doping of this layer is appropriately low. As no additional charge carriers are injected in the on-state, this concept exhibits a very high area-specific resistance which rises non-linearly as a function of the voltage, proportional to a power of between 2.4 and 2.6. This serious disadvantage leads to attempts, in the application, to manage with the lowest possible breakdown voltage capacity, for example 450 V for the American mains supply or 500 V for a 230 V mains supply. For MOSFETs, the established FOM is the so-called area-specific resistance, $R_{DS(on)} \cdot A$, which specifies the on-state resistance which can be achieved by an area of 1 mm^2 for a given breakdown voltage.

The further developments of the concept in the 1980s and 1990s have so far not enabled this fundamental disadvantage to be overcome; suggestions have included a doping profile which becomes higher with greater depth, a close cell pitch, with increased doping between the cells, flatter p-wells, or combinations of these options. In the most favorable situation, one can just achieve the so-called silicon limit, given by the solution of the two-dimensional optimization problem, by attempting to simultaneously maximize the conductivity and blocking ability by means of a doping profile. Figure 3.134 shows the corresponding approaches.

In switched mode power supplies, the application areas of the IGBT and the MOSFET overlap, depending on the topology and requirements, so that many manufacturers include both device concepts in their portfolios.

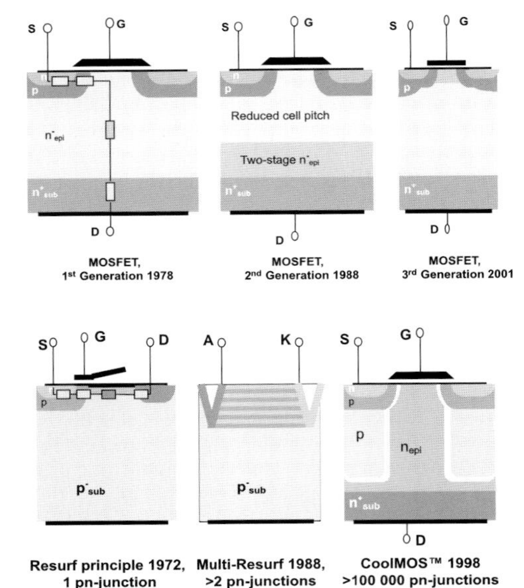

Figure 3.134
Development of the conventional MOSFET

Figure 3.135
Development steps on the way to the compensation device

Both device concepts are thus coming up against the physical limits, which stand in the way of a basic optimization of high-voltage switches to become ideal switches: the reverse recovery charge of an IGBT prevents rapid switching, the high on-state resistance of a MOSFET – determined by the silicon limit – is the main obstacle to low-loss current flow.

It follows that improvements in high-voltage switches will not be achieved by evolution, but only by the adaptation of a new principle. The injection of bipolar charge-carriers does not seem to be the ideal solution here. On the other hand, in the switched-on state one requires many charge carriers, but in the blocked state few. The revolutionary solution to this problem consists rather in separating the two types of charge carrier from each other spatially in the device, so that in the blocked state their net charge is mutually balancing to almost zero, while in the on-state an unreduced doping with one of the two charge-carrier types is available for the current flow. This provides a simple way of breaking through the silicon limit, because the limitation of the conventional MOSFET, that one doping profile must optimize the resistance and breakdown voltage, no longer exists. This idea, referred to as the compensation principle because of the balancing effect of the p- and n-zones, has for a long time been known for lateral transistors as resurf, but its adaptation for vertical transistors has long been considered technologically infeasible. Its first successful application in a commercial product came from Infineon Technologies, under the registered trademark of the CoolMOS™ transistor. Figure 3.135 shows how the above device concept has developed over time.

The CoolMOS transistor's new type of structure permits the on-state resistance to be drastically reduced, because the doping level of the conducting n-zones is now decoupled from the breakdown voltage requirement for the device: thus, when increasing the blocking voltage of the device only the zone to which the voltage is applied does need to have a greater thickness together with correspondingly deeper reaching p-columns.

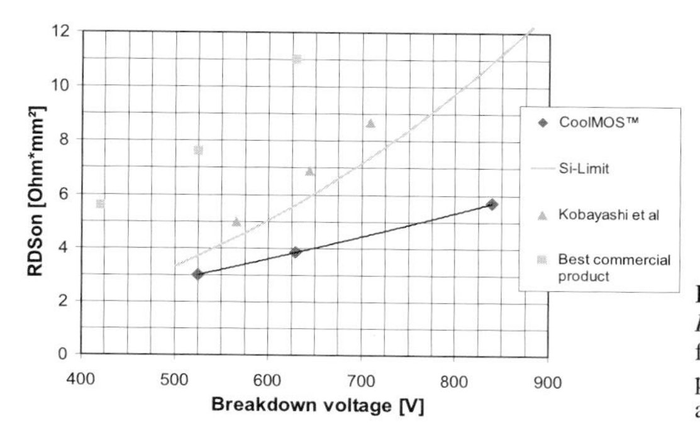

Figure 3.136
$R_{on} \cdot A$ for CoolMOS™ as a function of the voltage, compared against the competition and the silicon limit

Decrease in doping level is no longer needed. This has the effect that the rise in the on-state resistance as the breakdown voltage rises, which in MOSFETs is non-linear, is reduced to a simple linear increase. Figure 3.136 shows the area-specific resistances achieved in the CoolMOS™ product family as a function of the breakdown voltage, compared to that in conventional solutions and compared to the competition. "Kobayashi et al" here refers to the result from a Japanese group, published in 2001 on the ISPSD, while "best commercial product" refers to the products from a competitor, freely obtainable from the trade.

At 600 V, CoolMOS™ achieves a resistance which is lower by a factor of three than the best commercially obtainable conventional MOSFETs. This means that for a given area or in a given transistor package a correspondingly lower on-state resistance is possible, e.g. in a TO220 190 mOhm instead of 600 mOhm, or for a given resistance a correspondingly smaller chip or package, as applicable. This potential shrinkage is a great advantage particularly for applications with high power densities.

The challenge of the CoolMOS concept lies in the manufacture of the densely packed p-column structures, which penetrate deep into the active zone, and the exact control of the total charge in the p- and n-columns respectively. This charge control has a major effect on the breakdown voltage of the device, and today is a given factor of the process technology which represents a limit on compensation devices. For this reason, the development of these devices towards yet lower on-state resistances will demand accompanying improvements to the individual processes involved in their manufacture.

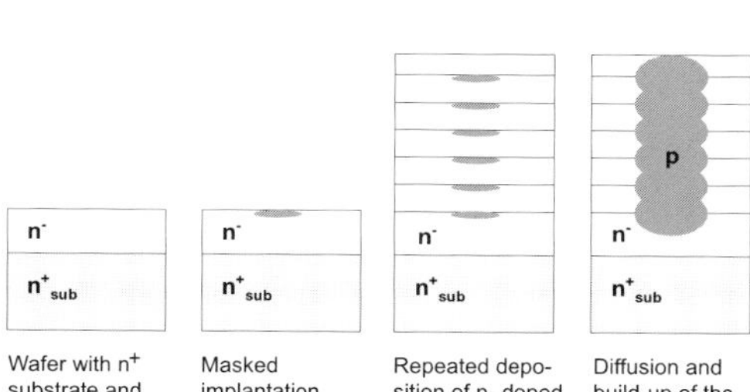

Figure 3.137
Individual steps for building up the columnar structure of a compensation device, using the multi-epitaxy method

The manufacturing process which has now taken over for all the commercially obtainable products is the multi-epitaxy method. Figure 3.137 is a diagram of the sequence of individual process steps required.

Characteristics of compensation devices

The outstanding characteristic of compensation devices is their negation of the relationship between breakdown voltage and on-state resistance which applies to classical MOSFETs. This is achieved by the spatial separation of the p- and n-zones within a transistor cell. The doping levels of the two zones are set in such a way that, within the limits of manufacturing tolerance, their charges virtually cancel each other out – they are mutually compensating.

If a blocking voltage is applied to the device, the charge carriers are initially cleared out along the boundary of the pn junction in the two doped zones. A space charge zone develops along the column structure, creating a field component which is primarily lateral. This forces the charge carriers in the p- and n-columns out of the drift space from two sides. Even at relatively low voltages (typically < 50 V) the column structure is completely cleared, and the space charge zone acts as a quasi-intrinsic layer (Figure 3.138). As soon as the entire column structure has been cleared, the application of the reverse voltage continues with a rise in the vertical electric field in the space charge zone and by a vertical expansion of the space charge zone into the remaining drift space. A field distribution develops in which lateral and vertical components are superimposed. For the maximum breakdown voltage combined with the best possible forward characteristics, the vertical and lateral field components should ideally be of the same magnitude.

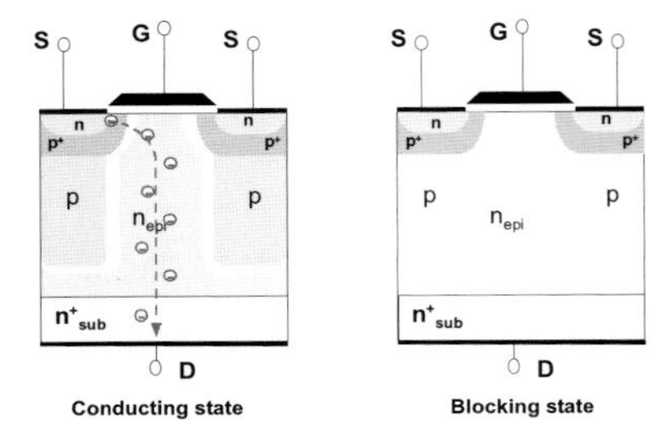

Figure 3.138
Conducting state (left) and blocking state (right) for a compensation device

The space charge zone, which builds up as described over the column structure of the p- and n-zones when a reverse voltage is applied, offers important advantages not only for the blocking state, but also for the process of switching on and off. Because the p- and n-zones are completely cleared even at relatively low reverse voltages, that is to say at the start of the switching process, the space charge zone is already established after a very short time. This leads to a strongly non-linear development of the drain-source capacitance C_{ds} or output capacitance C_{oss} because unlike the classical MOSFET not only the width of the space change zone changes with rising voltage, the surface of each of the p- and n-zones is linked to the drain-source voltage too. As soon as the p- and n-columns have been cleared out, at about 50 V, the output capacitance drops to a value which, because of the smaller chip area, is very small and is now virtually independent of the voltage. Because of the rapid expansion of the space charge zone into the n-columns, the gate drain capacitance (C_{gd}) or reverse transfer capacitance also develops in a highly non-linear way, with very small values above 50 V. The gate-source capacitance C_{gs} benefits mainly from the reduction in area compared with conventional power MOSFETs, made possible by the compensation principle. Figure 3.139 shows a comparison of the transistor capacitances of a conventional MOSFET and CoolMOS™, in each case for 600 V blocking devices.

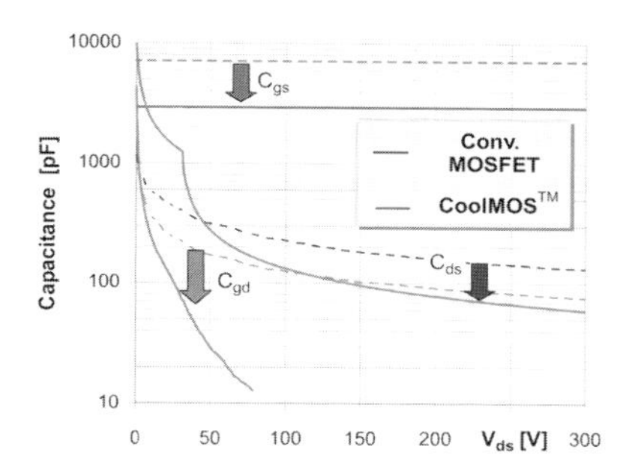

Figure 3.139
Comparison of device capacitances for CoolMOS™ and standard transistors

In heavy-duty switching operation, the energy stored in the output capacitance C_{oss} is essentially converted into loss heat, so that in order to optimize the switching losses it is imperative to take into account this stored energy. Figure 3.140 therefore shows the energy E_{oss}, which is the result of integrating the product of C_{oss} against the voltage. Because of the relatively strong voltage weighting of the capacitances (lower for CoolMOS™), has voltages of 350 to 420 V are typical in applications, the energy stored in the output capacitance is reduced by around 50%.

In applications with resonant switching, here the most important is the phase shift zero voltage switching (phase shift ZVS) topology, the energy stored in the output capacitance comes from an inductive element: here, a small value of E_{oss} helps to keep the resonance condition for turn-on up to very low currents. Over and above this, the strongly non-linear output capacitance helps with the commutation of the current from one arm of the bridge to the other.

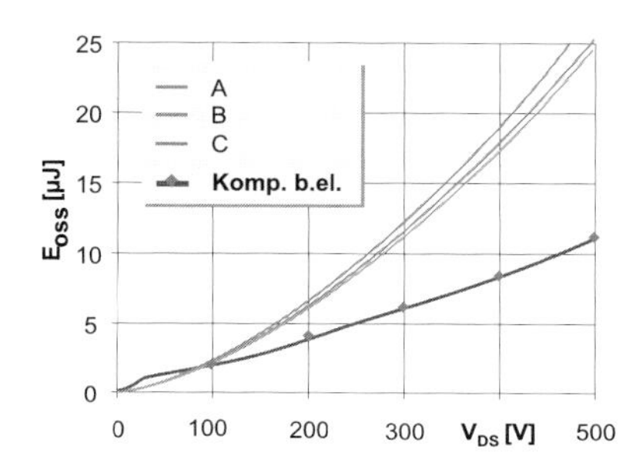

Figure 3.140
The energy E_{oss} stored in the output capacitance of a CoolMOS™, compared to various standard transistors

Apart from the capacitances, the switching speed also plays a role of course, and the resulting time overlap of the current and voltage. When the device is switched off, the electron flow via the MOS channel is cut off. When this occurs, the mobile charge carriers in the p- and n-columns flow as majority carrier drift currents to their respective contact zones, the p-well linked to the source and the n^+ substrate on the rear linked to the drain. In doing so, the charge carriers do not cross the space charge zone in which the voltage is rising. In other words, the current passes nearly no voltage drop in the device, and hence produces no thermal losses. The switching speed therefore depends essentially on the charge changes at the gate electrode, and hence on the gate-source and gate-drain capacitances discussed above, and on the performance of the gate driver stage. Typical switching times are around 5-7 ns.

During device turn-on, the space charge zone must be dissipated again by the neutralization of the charged acceptors and donors in the p- and n-columns. For this process, the electrons are supplied by the channel current, the holes on the other hand flow down as a drift current from the p-well into the p-column. Consequently, a high switching speed demands a low-resistance connection, and thus in the manufacturing process a correspondingly good adjustment and linkage of the individual p-implantation islands. Typical switching times again lie in the region of a few nanoseconds.

As a result, compensation devices like the CoolMOS™ are among the fastest high-

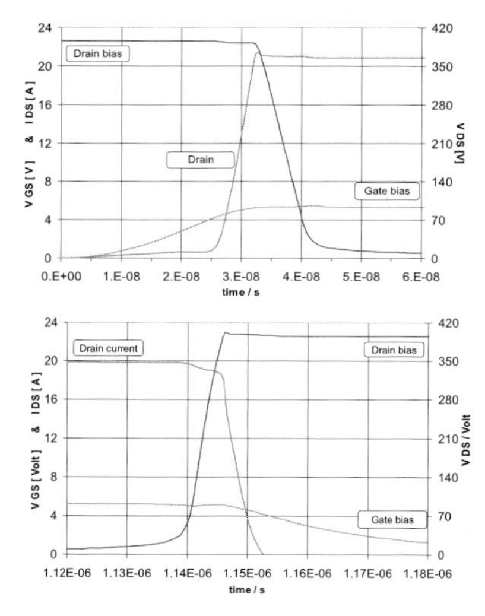

Figure 3.141
Idealized turn-on and turn-off characteristics of a 500 V CoolMOS™

voltage switches on the market. Figure 3.141 shows the turn-on and turn-off characteristics.

One consequence of the area reduction with the CoolMOS technology is that the current density in the device is much larger than in conventional transistors. This makes it more difficult to meet the ruggedness requirements, for example the short circuit handling. The graph of the short circuit behaviour of the standard MOSFET is characterized by an insufficient current saturation, where the short circuit current increases with rising drain-source voltage. Even for moderate voltages or voltages around 400 V, which are typical in applications, the short circuit current reaches seven times the nominal current. Conventional power MOSFETs therefore provide only insufficient short circuit behaviour, because low current values are desirable for safe operation at the device limits.

The reason why compensation devices have significantly better characteristics is the columnar structure of the p-zones, which represent a vertical junction FET (JFET) with a stronger cut-off behaviour with increasing voltage. The short circuit current is thereby very effectively limited, and has an almost constant value over the entire "safe operating area" (SOA). CoolMOS™ here specifies the full breakdown voltage at up to three times the nominal current.

By reducing the gaps between the p- and n-zones, and at the same time raising the doping in them, it is theoretically possible to continuously reduce the device resistance for a constant breakdown voltage. In reality however, one limiting factor for maximum improvement in $R_{DS(on)}$ is the photo-lithography and the process control for the p- and n-doping. Another one is that at very small structure widths the JFET described above already puts restrictions on the column conductivity in the on-state even when very low voltages are applied.

Selected applications

The continuous trend towards more compact switched mode power supplies, together with a strong increase in output power density, are resulting in two main requirements for the power switches: high current handling capability, i.e. small packages which can handle the device losses without additional cooling; and low switching losses, i.e. higher switching frequencies that the passive components can be reduced.

A compensation device satisfies both requirements in an almost ideal way. As a result it opens up completely new options for compact constructions. This affects switched mode power supplies in the field of servers and telecommunications, with continuously increasing power densities, just as much as new applications for battery-powered devices where, for example, the charger can be integrated into the plug.

Furthermore, it becomes possible to combine entire systems on package level. For

Figure 3.142
Comparison of two system solutions

example, a controller and a power switch can be combined in one package to form a smart solution, as a "system in a package"; Figure 3.142 shows an example. The conventional solution (left) has 3 W power losses, and requires cooling facilities. The solution using CoolMOS™ (right) has much lower power losses of only 1 W, and operates without additional cooling effort.

3.7.9 Silicon carbide – the basis for high power densities

The majority of the semiconductor devices manufactured today are based on silicon. Due to their physical properties it is suitable for the realization of both unipolar and bipolar devices. However, with the steadily increasing demand for power semiconductors combining high blocking voltages together with high switching speed, devices based on silicon are reaching their physical limits.

For this reason power semiconductor devices based on silicon carbide (SiC) have been the subject of research since the beginning of the 1990s. The first Schottky diodes are commercially available since 2001. The significantly lower static and dynamic losses are enabling much larger power densities than possible with silicon devices.

Physical characteristics

The crystal structure of silicon carbide is built up of parallel layers of silicon an carbon atoms. Due the strong electronic bond between both atoms, the material has a very high mechanical hardness. In this structure, each silicon atom has bonds to four carbon atoms in an almost tetrahedral arrangement, and vice versa. As a result of slight displacement of the Si atom from the center of this, double layers of silicon and carbon atoms are produced. Depending on the mutual orientation and stacking sequence of these double layers, different stable crystal structures are produced, the polytypes. Depending on the alignment and sequence, up to 200 different polytypes are known, with different material parameters (for a selection, see Table 3.10).

Apart from cubic and hexagonal SiC, which are the most used forms, there are diverse intermediate forms. The best know are the 3C-form with a cubic lattice arrangement, and the 4H- and 6H-forms, with hexagonal lattices. Here, the numeric prefix specifies the number of atom layers within each of the regularly repeating vertical periods. For each of these forms there are thus also different application areas. 6H-material is predominantly used in opto-electronics, and 4H-SiC in power electronics.

The material parameters of silicon carbide

Due to its very tight crystal structure, silicon carbide has a higher band gap than silicon, and a greater breakdown field strength, which makes it possible to real-

Table 3.10 Material parameters of silicon carbide

	Si	GaAs	6H-SiC	4H-SiC
E_G [eV]	1.12	1.42	3.02	3.26
E_{crit} [V/cm]	$3.0 \cdot 10^5$	$4.2 \cdot 10^5$	$25 \cdot 10^5$	$22 \cdot 10^5$
n_i [cm^{-3}]	$1.4 \cdot 10^{10}$	$1.8 \cdot 10^6$	$1.6 \cdot 10^{-6}$	$5.0 \cdot 10^{-9}$
$v_{sat,n}$ [cm/s]	$1 \cdot 10^7$	$1 \cdot 10^7$	$2 \cdot 10^7$	$2 \cdot 10^7$
μ_n [cm^2/Vs]	1500	8500	400	1000
μ_p [cm^2/Vs]	450	400	101	115
ε_r [ ]	11.9	13.1	9.7	9.7

ize the same blocking ability with thinner semiconductor layers.

$$W_{Drift} = \frac{2U_{BR}}{E_{crit}} \qquad (1)$$

The reduced layer thickness makes it possible at the same time to increase the doping concentration, so that silicon carbide is particularly suitable for unipolar semiconductor devices.

$$N_{Drift} = \frac{\varepsilon_0 \cdot \varepsilon_r \cdot E_{crit}^2}{2q \cdot U_{BR}} \qquad (2)$$

The ohmic resistance of the drift zone is then determined by the thickness and doping respectively the breakdown voltage and the maximum electric field strength.

$$R_{Drift} = \frac{w_{Drift}}{A \cdot \mu_n \cdot q \cdot N_D^+} = \frac{4U_{BR}^2}{A \cdot \mu_n \cdot \varepsilon_s \cdot E_{crit}^3} \qquad (3)$$

For the same blocking voltage, the 10-times larger breakdown field strength of silicon carbide enables a doping concentration two orders of magnitude lower than for silicon and an active larger thickness which can be reduced by one order of magnitude. (Figure 3.143). Therefore, about 300 times smaller drift zone resistances can be realized.

As a result of the low intrinsic charge carrier concentration, even at very high temperatures the reverse leakage currents are very low. Therefore, silicon carbide is well suited for applications in high-temperature environments. Additionally, the high thermal conductivity ensures good dissipation of the losses. However, due to limitations in the assembly technology, this potential cannot yet be fully used commercially.

Devices made of silicon carbide

The much smaller on-state resistances enable Schottky diodes, JFETs and MOS-

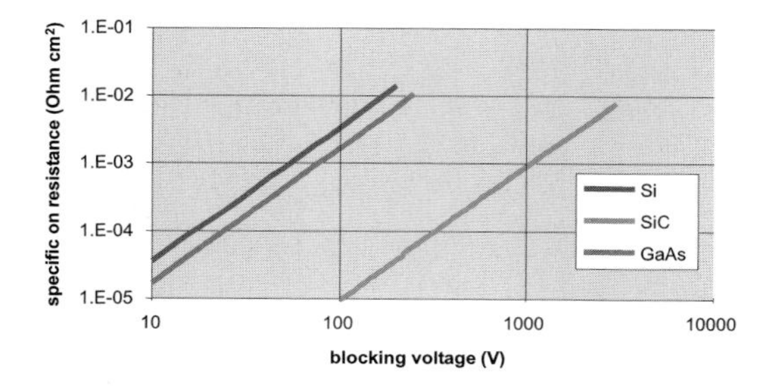

Figure 3.143
Relationship between the achievable on-state (forward) resistance and blocking voltage for various materials

FETs of higher blocking voltages than before. They can be used in applications with larger bus voltages, in which until now only bipolar Si devices could be used. With SiC devices, the change-over to bipolar Si devices, such as pin-diodes and IGBTs, can be increased up to blocking voltages > 2000 V. At blocking voltages larger than 2 kV, the drift zone resistance of unipolar devices made of silicon carbide then also rises so sharply that bipolar injection becomes necessary. For bipolar pn-junctions in silicon carbide there is a diffusion voltage of approximately 2.7 V, due to the higher band gap.

Almost all device structures using silicon carbide which have been announced up to now have been vertical ones. In the bulk of the device the electron mobility is only about 20% lower than in silicon. However, immediately beneath the surface it is greatly reduced, mainly caused by the larger surface roughness, so that SiC-MOSFETs with planar gates have a very high channel resistance. At the same time, the higher breakdown field strengths is leading to much larger gate oxide stress and reduces the SiC-MOSFETs reliability so far.

For realizing transistors in silicon carbide, two different concepts are used today. On the one hand, there is a concentration on UMOS and VDMOS structures with the gate arranged vertically or obliquely in order to be able to utilize the higher charge carrier mobility in the bulk material. On the other hand, by using JFET or SIT structures it is possible to avoid gate oxide entirely. The gate is buried in the device and the channel is controlled by the width of the space charge zone. Because of their channel structure, the device is conducting the current without applied gate voltage. They are therefore called 'normally on' devices. However, normally-on devices nowadays play no role in power electronics. The reason is the limited availability of control devices, the insecure normally-off state and the associated greater risk, for example, of bridge short-circuits in bridge designs.

The speed of development of SiC will also be determined by the quality of the substrates. The expensive sublimation method for manufacturing the substrates, and the associated density of defects (micropipes or dislocations) makes it difficult to increase the wafer diameter. In the mid-1990s 1″ wafers could be economically manufactured.

As a result of the steadily rising demand for basic materials for power electronics and opto-electronics, the wafer diameter could be increased up to 3″ in 2000. However, the continuing high density of imperfections in the substrate also limits the economically useable device area. Therefore, the commercially available rated currents are only a few ampere.

SiC Schottky diodes

Schottky diodes were the first commercially obtainable devices made of silicon carbide. They consist of a SiC substrate wafer, on which a field stop layer and the base zone, which is determing the device characteristics, are deposited epitaxially. The Schottky junction is realized on the surface by means of a suitable metal. The edge structure is formed by implantation (Figure 3.144).

The Schottky junction: if two materials with different work functions, for example a metal and a semiconductor, are combined, a so-called heterojunction is build up. The work function is the energy (e.g. $q \cdot \Phi_u$) which must be supplied to an electron in order to remove it from its atomic bond. At the contact between the two materials, the vacuum potential remains constant, band bending occurs in the semiconductor, due to the different work functions and a space charge zone is build up. A Schottky junction is formed on an n-semiconducting material if the work function of the metal is greater than that of the semiconductor.

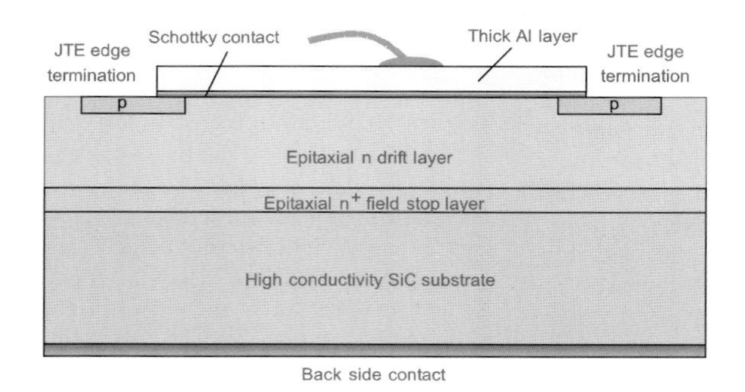

Figure 3.144
Structure of a SiC Schottky diode

When a positive voltage is applied, the barrier height is reduced and current flow becomes possible. In this case, the current transport is purely carried by majority charge carriers (Figure 3.145).

Because of the step-shaped rise of the conduction band at the junction, the diffusion current model no longer applies for the charge transport. The current transport is predominantly determined by thermoionic emission. The current flow starts up as soon as the kinetic energy of individual charge carriers is sufficient to overcome the barrier. The current density is determined by the height of the barrier, the externally applied voltage, temperature and the effective Richardson constant. For an n-semiconductor, the following then applies:

$$j_{\mathrm{n}} = \left(A^{*} \cdot T^{2} \cdot e^{-\frac{q \cdot \phi_{\mathrm{S}}}{k \cdot T}}\right) \cdot \left(e^{\frac{q \cdot U_{\mathrm{a}}}{k \cdot T}} - 1\right) \quad (4)$$

A: Richardson constant

An important conclusion from (3) and (4). is that high blocking voltages and low on state resistances are only possible in the same technology by larger device areas. But larger device areas are simultaneously resulting in increasing reverse leakage currents, too.

Static behavior: Schottky diodes made of silicon can today only be realized for blocking voltages up 200 V. Above this level, the reverse current rises so steeply that large thermal losses arise. A blocking capability of up to 250 V can be realized by means of so-called merged-pin Schottky diodes. These consist of a combination of a Schottky junction and a network of islands of locally opposing doping, to reduce the maximum electrical field at the surface. At the same time, the diode's resistance to overcurrents is reduced by bipolar injection at high current densities. For higher blocking voltages, bipolar pin-diodes are typically used. In the on-state, the base zone is flooded with both types of charge carrier reducing the device re-

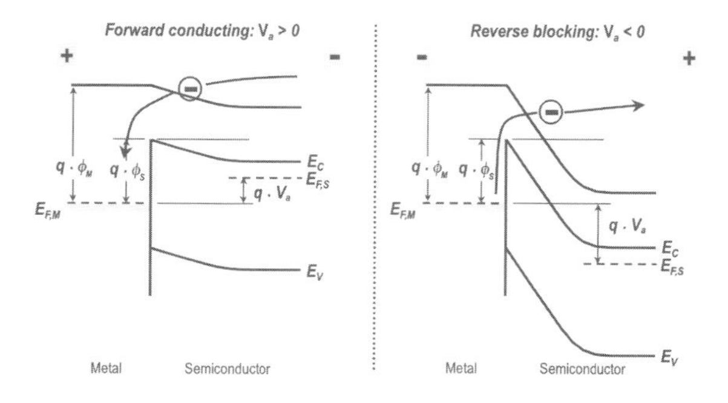

Figure 3.145
A Schottky junction under positive and negative voltage

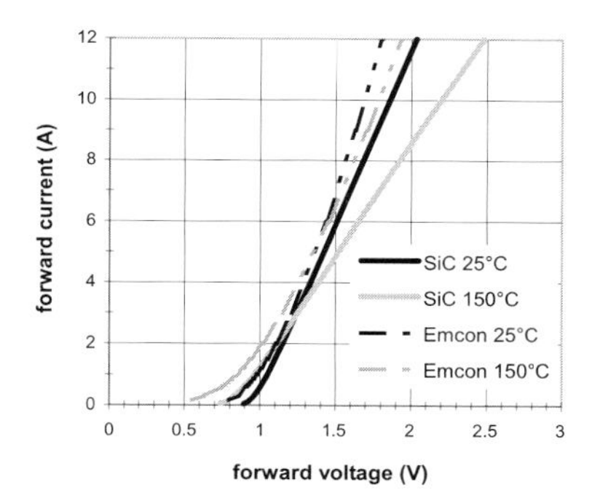

Figure 3.146
Comparison of the forward characteristics of a 600V Si pin- diode (IDP06E60) and a SiC Schottky diode (SDP06S60)

sistance. But during device turn-off this large amount of carriers has to be removed from the device resulting in larger delay times and switching losses.

In silicon carbide, the utilizable blocking voltage for unipolar Schottky diodes is much higher. In particular, as a result of the greater breakdown field strengths and the smaller reverse currents, it is possible to extend the usable voltage range for Schottky diodes up to around 2000 V. In doing so, the diodes can be so dimensioned that the static characteristics come close to those of Si diodes in comparable voltage classes (Figure 3.146).

Because of their unipolarity, Schottky diodes have a positive temperature coefficient, which simplifies parallel connection. For the diode shown here, the resistance of the drift zone increases in proportion to $T(\mathrm{K})^{2.5}$, thus doubling between 25°C and 125°C. As a proportion of the total resistance, the drift zone resistance is about 40% at 25°C, and rises up to 70% at 125°C. At the same time, the voltage drop across the Schottky barrier falls by about 1.6 mV/ K.

The reverse current has an exponential relationship with voltage predicted by the thermoionic emission theory.

If a voltage of 400 V, the typical operating voltage in applications, is applied, thermal values comparable to those of pin-diodes made of silicon are achieved. If the blocking voltage is icreased up to 600 V then, although the reverse current is about 20 times higher than for a Si diode, but because of its very low value far below the forward losses.

Dynamic characteristics: Schottky diodes are particularly suitable for fast switching. Because of its unipolar behavior, the current is carried only by electrons, no additional flooding charges need to be dissipated when it is switched off, before the diode is able to accommodate the blocking voltage. A SiC Schottky diode behaves like a small voltage-dependent

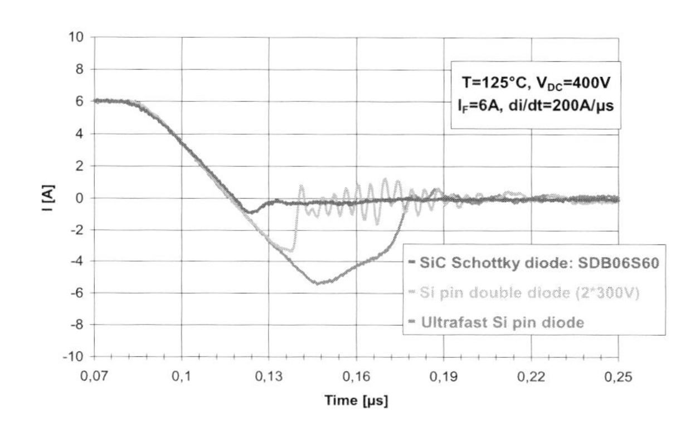

Figure 3.147
Comparison of the switching characteristics of a SiC Schottky diode and bipolar Si diodes

capacitance, so that the only losses occurring during the switching process are those due to changes in charge in the capacitance. There is no reverse current peak in the conventional sense, as the voltage builds up only a displacement current can be seen (Figure 3.147).

By contrast, with bipolar Si diodes a substantial reverse current peak arises during switch-off, due to the excessive charge carriers which must be dissipated, this being strongly dependent on the forward current, the speed of switch-off and the temperature of the device. The switch-off characteristics of a SiC Schottky diode is, on the other hand, largely independent of the forward current and temperature.

Applications of SiC Schottky diodes

On the one hand, the material characteristics of silicon carbide enables devices with high blocking voltages combined with fast switching. On the other hand, the expensive manufacturing methods and the number of defects, still high by comparison with silicon, limits the economic reasonable device area to only a few amperes of rated current. This makes application areas with rated voltages greater than 300 V, and a power range from a few 100 W up to a few kW, attractive. These will be briefly presented by reference to three examples.

In the voltage range of 250V - 300 V, Schottky diodes are used mainly for rectification on the secondary side in power supply units for 48V networks in telecommunications systems (Figure 3.148). Here, Schottky diodes made of either GaAs or SiC can be used.

Both types of diode have low switching losses. However, GaAs has a lower thermal conductivity than silicon carbide and a greater temperature coefficient, so that overall the SiC system generates the lowest static losses (Figure 3.149).

SiC Schottky diodes in the 600V class are used predominantly in PFC (Power Factor Correction) stages of modern switched mode power supply units. Power supply units with an output power of over 75 W are legally required to have power factor correction. Basically this is a boost converter, which distributes the power taken from the mains supply in packets over the mains supply waveform (Figure 3.150).

The interference pulses are reduced, harmonic content and reactive power are decreasing. A distinction is made between discontinuous (DCM) and continuous (CCM) operation. Although the largest dynamic loads on the components arise during the continuous current mode, but

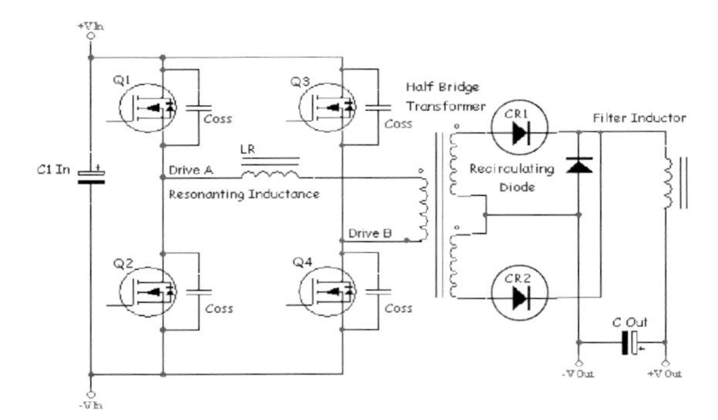

Figure 3.148
Use of SiC Schottky diodes for secondary-side rectification

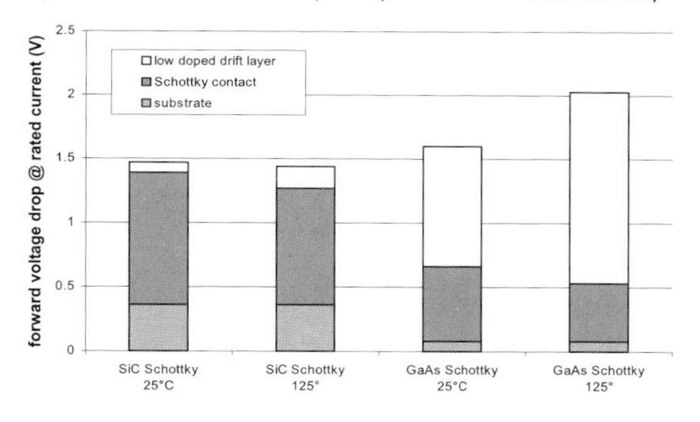

Figure 3.149
Split of the forward voltage drop in SiC and GaAs Schottky diodes

the efficiency and reliability is increased. In addition, the passive components can be reduced with increasing switching frequency (Figure 3.151).

SiC Schottky diodes are particularly suited to this application, because their low static and dynamic losses allow smaller switching transistors, an increase in the switching frequency and hence a further miniaturization.

In the 1200V class, fast diodes are found primarily in motor controllers and higher power switched mode power supply units, such as for example uninterruptible power supplies (UPS) for servers (Figure 3.152). On their mains side, most uninterruptible power supplies have a voltage doubling circuit, so that 1200V diodes are required.

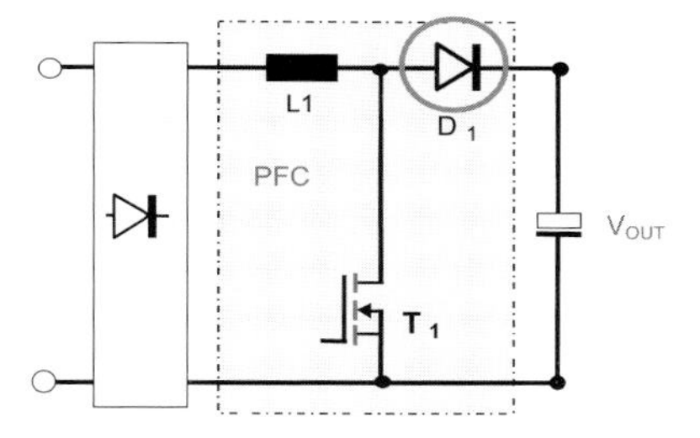

Figure 3.150
Example of a SiC diode in a PFC circuit

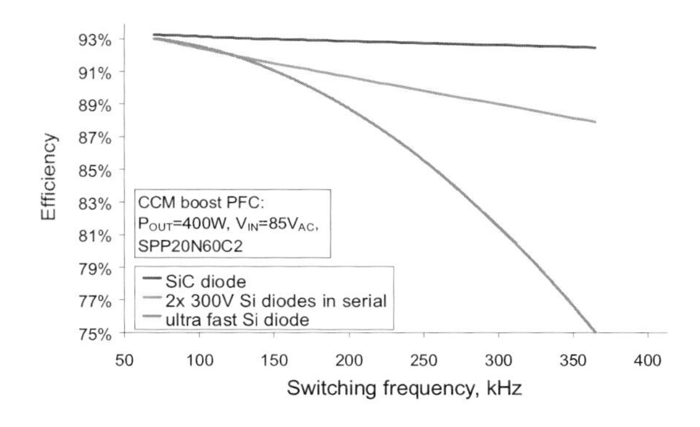

Figure 3.151
Efficiency of various diodes in a PFC stage

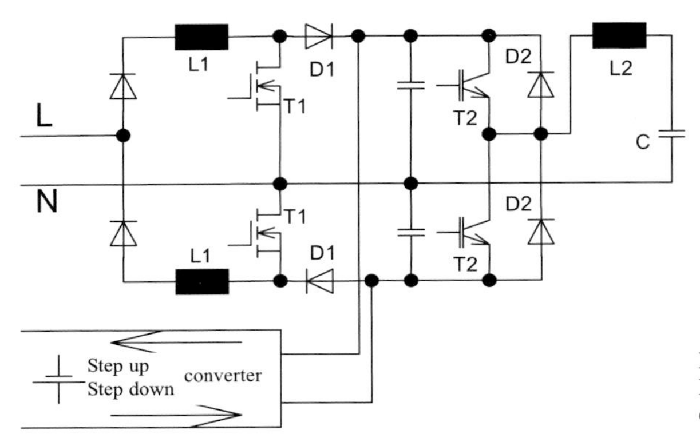

Figure 3.152
Example of a UPS circuit

Using bipolar Si diodes, switching frequencies of 20 kHz are nowadays achieved in these applications. By using 1200 V SiC Schottky diodes, the dynamic losses can be reduced and the switching frequency can be increased. At the same time, size of the passive components can be reduced too.

Transistors made of silicon carbide

For transistors made of silicon carbide, two device concepts are essentially used: the SiC MOSFET and the SiC JFET. The fact that until now the SiC MOSFET has been unable to establish itself by comparison with silicon, even though it has a normally-off characteristic, is due to the low channel mobility of planar gate structures in silicon carbide, and uncertainties to the reliability of the gate oxide when high electric field strengths are applied. In contrast, the SiC JFET has a normally-on characteristic, and no gate oxide. By using a cascode circuit comprising a SiC JFET and a Si MOSFET with low blocking voltage, it is possible to realize a normally-off characteristic (Figure 3.153). The hybrid technology construction here requires slightly more effort than the single chip assembly.

Static characteristics: the basic structure of the SiC JFET, shown in Figure 3.153, has a lateral channel, which is controlled via the top gate and a buried gate. Here. the channel resistance is reduced on the source side by an additional n-implantation. The primary current path runs vertically down to the drain contact on the rear of the chip. The direct gate contact on the upper side of the chip avoids large gate lead resistances. For a blocking capacity of 1500 V, this configuration allows on-state resistances < 15 mΩcm^2.

Dynamic characteristics: due to the favorable properties of silicon carbide as base material, the dynamic characteristics of a

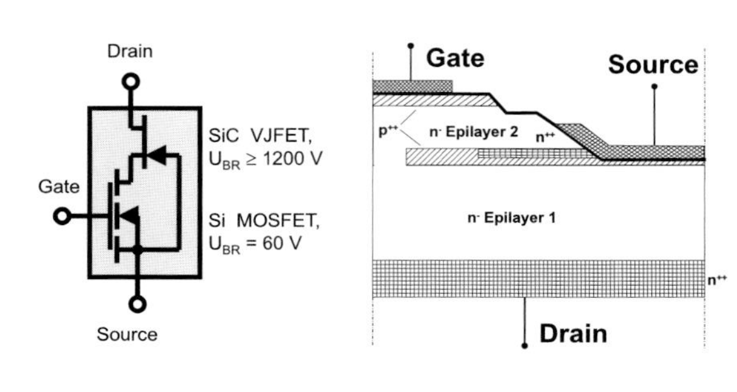

Figure 3.153
a) Hybrid cascode circuit comprising a SiC JFET and a Si MOSFET
b) Structure of a SiC JFET

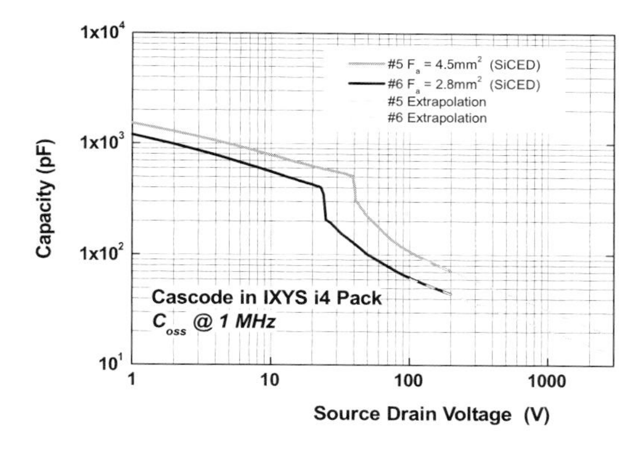

Figure 3.154
Effect of different JFET chip areas on the output capacitance of a 1500V SiC JFET cascode

SiC JFET cascode are essentially determined by the Si MOSFET. To ensure reliably turn-off behaviour of the JFET cascodes the blocking voltage of the Si MOSFET must be larger than the channel pinch-off voltage of the SiC JFET. The output capacitance waveform of the SiC JFET cascode thus shows a steep decrease at the point of the channel pinch-off of the SiC JFET. Beneath this pinch voltage, the graph of the capacitance is determined by the source-drain capacitance for the Si MOSFET (Figure 3.154).

This characteristic can also be observed in the turn-off waveform of the SiC JFET cascade. By using MOSFET devices of different generations, it is possible to show visually the effect of different delay times (Figure 3.155).

The SiC JFET cascode thus enables fast switching times at high blocking voltages comparable to those of Si MOSFETs with much smaller blocking voltage.

Applications for SiC JFET cascodes

The classical application for unipolar high voltage devices are auxiliary power supplies for motor controllers, where today 1500V Si MOSFETs are used. These MOSFETs can be replaced by SiC JFET cascades without modifications in the circuit topology.

Furthermore, as a result of the historically limited availability of cheap high voltage transistors, bridge circuits have been invented using several transistors with half or less the required blocking capacity. For

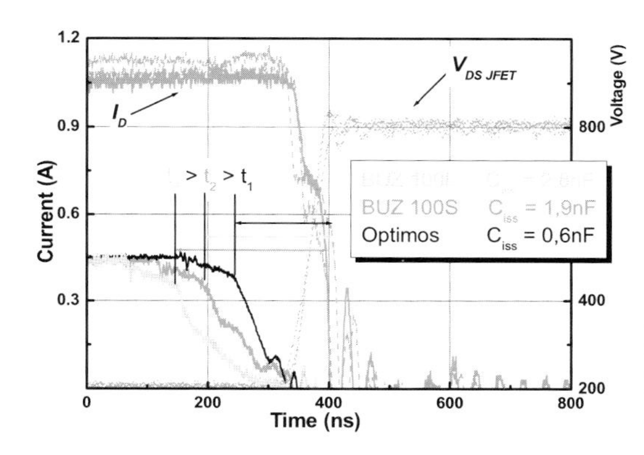

Figure 3.155
Effect of different Si MOSFETs on the switching characteristics of a 1500V SiC JFET cascode

example, the voltage stress across the transistor for single switch topologies typically is 2-3 times the input voltage and therefore requires a power switch with significantly higher voltage rating. The on-state resistance of a conventional Si Power MOSFET increases more than quadratically with higher rated voltage leading to low efficiency.

$$R_{\text{Drift}} = \frac{4U_{\text{BR}}^2}{A \cdot \mu_{\text{n}} \cdot \varepsilon_{\text{s}} \cdot E_{\text{crit}}^3} = \qquad (5)$$

$$R_{\text{Drift}} = \frac{(5{,}93 \cdot 10^{-9})}{A} \cdot U_{\text{BR}}^{2{,}5}(\text{Si})$$

Most attractive among all single switch solutions seems to be the Resonant Reset Forward topology, which is used only for power levels up to 40 W so far. In contrast, the favorable material characteristics of silicon carbide enables power transistors with on-state resistances of 0.5 Ω and a blocking capacity of 1500 V, which can extend the application area for these topologies up to 1500 - 2000 W.

By comparison with conventional solutions, this means that it is possible to replace two to four 600 V MOSFETs with just a single SiC switch, so that the complexity of the circuit can be reduced and the power density increased (Figure 3.156).

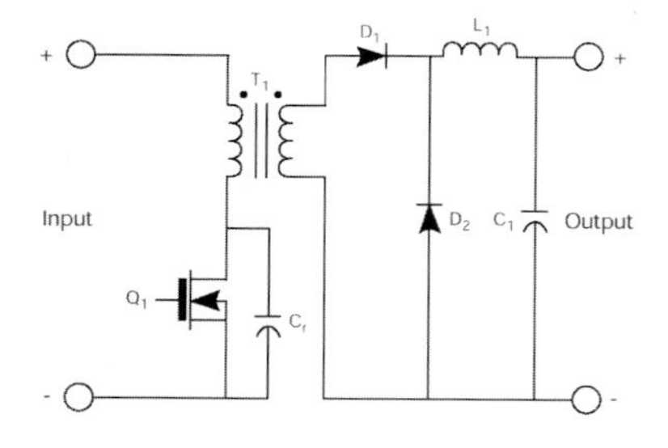

Figure 3.156
Equivalent circuit of a single switch resonant reset forward converter with just one 1500V transistor

For this reason, silicon carbide devices will continuously increase their market share in power electronics. Compared to silicon devices smaller device areas are required and less power losses can be realized. Together with the low intrinsic carrier concentration SiC devices would also be well suited for high temperature electronics. But due to assembly related restrictions this potential can't be completely realized today. The examples quoted show that, although the nominal currents which can be realized by using silicon carbide are still inhibiting its use in the high-power electronics, but they do nevertheless fulfill the requirements for diodes and transistors in the large application field of switched mode power supplies.

3.7.10 High-voltage power IGBTs

In the field of modern power electronics, the bipolar counterpart to the MOSFET is the IGBT (insulated gate bipolar transistor), which is used mainly in drive applications. The typical application range of discrete IGBTs contains devices for speed control of washing machines or air-conditioning systems, larger IGBT modules are used for traction drives, such as rail locomotive motors. The following section gives an insight in the different IGBT technologies starting with the fundamental differences to MOSFET devices up to Infineon's technological leadership in this field, with its new Field-Stop concept.

IGBT versus MOSFET

Just like the MOSFET, the IGBT is also a field effect transistor with an insulated gate connection. Thus, it is turned on by a positive gate-emitter voltage as soon as the threshold voltage for the MOS channel is reached and the pn-junction on the anode side is polarized in forward direction. As in the case of a capacitor, the gate electrode becomes charged up and, via the electric field which build up, this switches on the channel between the gate and the p-body of the IGBT. To switch it off, one applies zero volts to the gate so that it can discharge, and the channel then closes.

Looking at the cell structure of the two devices, no significant difference can be detected. In both cases, the gate is insulated by polysilicon. In vertical power devices, the source contact (MOSFET) or emitter contact (IGBT) is located on top of the chip, the drain (MOSFET) or collector (IGBT) contact on the backside of the device.

The essential difference in device behaviour of both concepts is caused by the additional p-layer on the backside of the IGBT structure. In the on-state, the electron current is travelling through the base zone towards the backside pn-junction causing there an injection of minority charge carriers into the drift zone. An electron-hole plasma is built in the base zone leading to a strong reduction in device resistance.

In contrast to the unipolar MOSFET the IGBT is therefore called a bipolar device. Both majority charge carriers (electrons) and also minority charge carriers (holes) contribute to the current flow. A MOSFET is a unipolar device with exclusively majority charge carriers conduction in the on-state. Figures 3.157 and 3.158 show cross-sections of MOSFETs and vertical power IGBTs respectively, and the difference in their technologies.

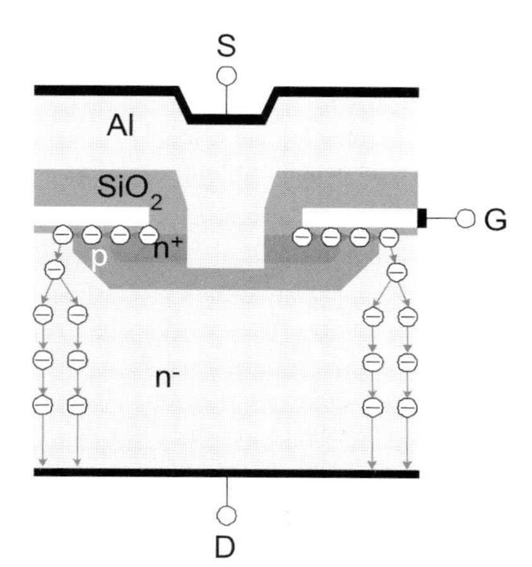

Figure 3.157
Cross section through a MOSFET

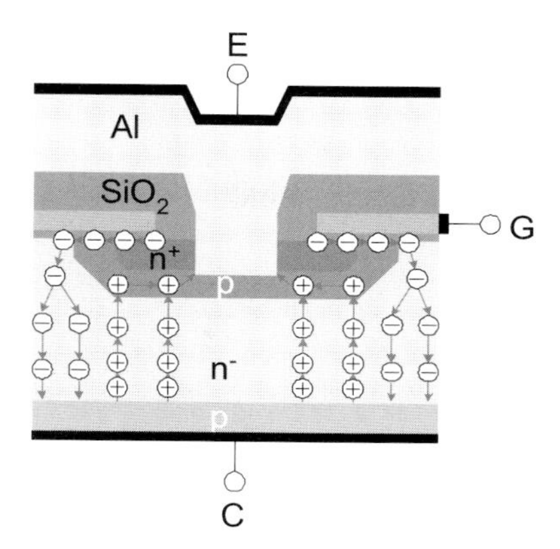

Figure 3.158
Cross section through an IGBT

This bipolar behavior has an enormous effect on the characteristics of the IGBTs. While a MOSFET is in the on-state essentially determined by its $R_{DS(on)}$, an IGBT exhibits a diode build-in-voltage, knee-point voltage and a differential re-

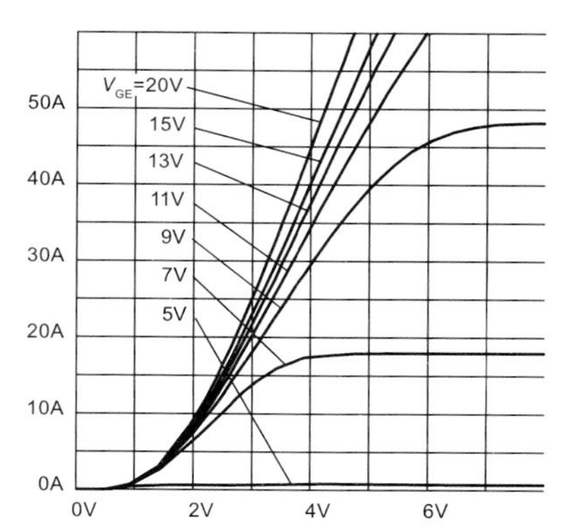

Figure 3.159
Output characteristic of an IGBT

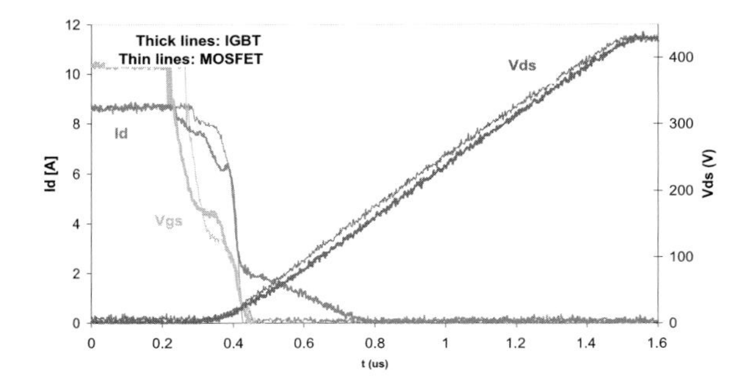

Figure 3.160 Switching characteristics of IGBT versus MOSFET

sistance. These characteristics are shown clearly in Figure 3.159.

In addition, an IGBT shows a different switching characteristic compared to a MOSFET, which is shown in the waveforms in Figure 3.160.

During turn-off, the IGBT is characterized by a slowly decreasing current tail caused by the extraction and recombination of the stored electron-hole plasma which must be completely removed out of the device. In contrast, the MOSFET doesn't show this behaviour due to it's unipolar device structure, resulting in lower turn-off losses. The characteristics of the switch-on process are similar for the two devices.

PT and NPT technologies

The objective of the semiconductor manufacturer is to optimize the losses of an IGBT – either to shorten the tail current during turn-off or to minimize the forward voltage drop.

Today, there are essentially two different concepts available on the market, the PT (**P**unch **T**hrough) IGBT and the NPT (**N**on **P**unch **T**hrough) IGBT. Cross-sectional views of these are shown in Figures 3.161 and 3.162 respectively. In principle, the cell structures are the same, the difference is in the substructure of the device.

On its rear side, the PT IGBT has a very strong emitter, which leads to the device being showered with minority charge carriers in the on-state. In the blocking state the electric field has a trapezoidal distri-

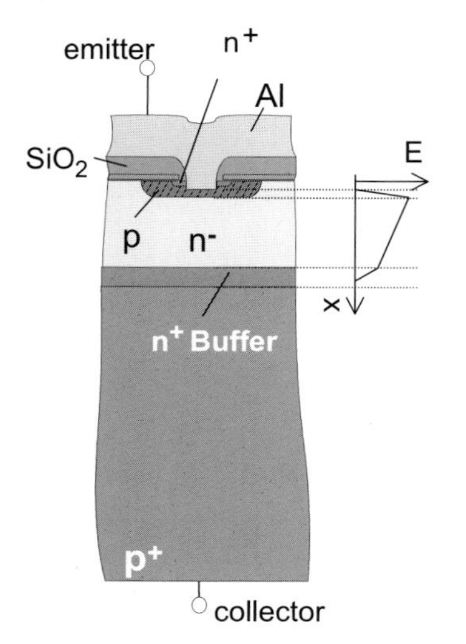

Figure 3.161 PT technology

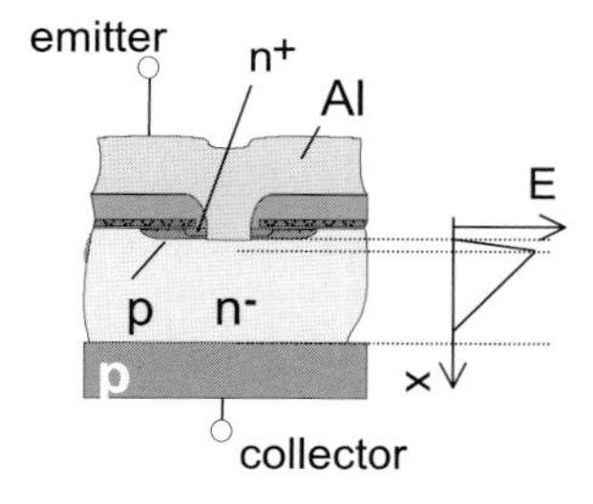

Figure 3.162 NPT technology

bution, and drops to zero in the highly-doped n^+ buffer zone. As a consequence of this, the n^- layer can be kept very thin. This results in a low forward voltage drop, so that the PT IGBT has minimal forward losses. However, a high concentration of holes and electrons when the device is in the on-state inevitably leads to a high and long tail current when it is switched off. The electron-hole plasma must be completely removed before the device completely blocks the current flow. As a countermeasure, a high lifetime-killing doping can be implanted, with the disadvantage of higher forward losses.

In contrast to this concept, the NPT technology has a very weak emitter efficiency on the rear. While it is in the on-state, the device is not so heavily showered with minority charge carriers, which enormously reduces the turn-off losses. However, with this concept the electric field in the off-state has a triangular distribution, and must be completely dissipated in the n^- zone, correspondingly thick. The disadvantage which has to be of this concept are larger forward losses.

Insofar as the losses affect the choice of device for any particular application, selection criteria can be defined by reference to them:

- maximum voltage which may occur plus a safety margin (generally 20%)
- current load must lie within the safe operating area (SOA) for the device
- package must meet with the usual provisions about tracking and spark gaps
- the junction temperature of the chip must be within the specification
- While the first three points can easily be determined without computational effort, in determining the chip temperature it is necessary first to consider the thermal conditions applying for the application and secondly to calculate the power loss of the device.

Calculation of an IGBT's losses

As the first step in evaluating whether a selected IGBT lies within the thermal specification for its chip temperature, the external conditions of the application must be considered.

The maximum allowable power losses of the IGBT is determined by the ambient temperature and cooling environment, together with the maximum permissible chip temperature for the device:

$$P_{max} = \frac{T_{J(max)} - T_A}{R_{thJA}}$$

However, the more difficult part of this calculation is what losses are to be expected from the IGBT in the application concerned.

The losses from an IGBT can be made up from three individual components. The static losses, based on leakage currents, can in general be neglected, because their contribution is unimportant:

$$P_{Leakage} = I_{CES}(V_{CE}, T_J) \cdot V_{CE} \cdot D$$

where D is the duty cycle for the device.

The loss parameters which turn out to be critical are the forward (conducting) losses while the IGBT is in the on-state and the switching losses, during both the turn-on and also the turn-off processes:

$$P_{tot} = P_{cond} + P_{switch}$$

Forward losses are easy to calculate, because they are determined only by the set of characteristic curves for the IGBT output (datasheet) and the current which occurs in the application.

The output characteristic curve can be approximated by two straight line sections, with hardly any effect on the accuracy of the result (Figure 3.163).

For a current with a trapezoidal waveform, the forward losses can be calculated as follows:

$$P_{cond} = D \cdot \left[\frac{1}{2} \cdot VTO \cdot (K_{min} \cdot i_{peak} + i_{peak}) + \frac{1}{3} \cdot RCE \cdot [(K_{min} \cdot i_{peak})^2 + K_{min} \cdot i_{peak}^2 + i_{peak}^2]\right]$$

On the other hand, the calculation of the switching losses is very difficult, because they are depending on the collector-emitter voltage, the collector current and also on the external gate resistance. These relationships can be taken from the datasheet, as the basis for a calculation. In spite of this, it is not possible to calculate the switching losses exactly, because parasitic effects can only be taken into account to a limited extent and for the test arrangement concerned.

From the details in the datasheet, the switching losses can be calculated from:

$$P_{switch} = \left[E_{on}(I_c) \cdot \frac{E_{on}(R_{G,user})}{E_{on}(R_{G,datasheet})} \cdot \frac{V_{DC,user}}{V_{DC,datasheet}} + E_{off}(I_c) \cdot \frac{E_{off}(R_{G,user})}{E_{off}(R_{G,datasheet})} \cdot \frac{V_{DC,user}}{V_{DC,datasheet}}\right] \cdot f$$

The relationship between the switching losses and the collector-emitter voltage can be assumed to be linear, and can easily be scaled.

In order now to check if the selected IGBT meets the requirements of the application, its total losses must be less that the maximum permissible losses determined by the external thermal conditions:

$$P_{max} \geq P_{tot}$$

In making the optimal choice, the objective is to find the IGBT which comes as close as possible to this condition, and which can be driven as hard as possible.

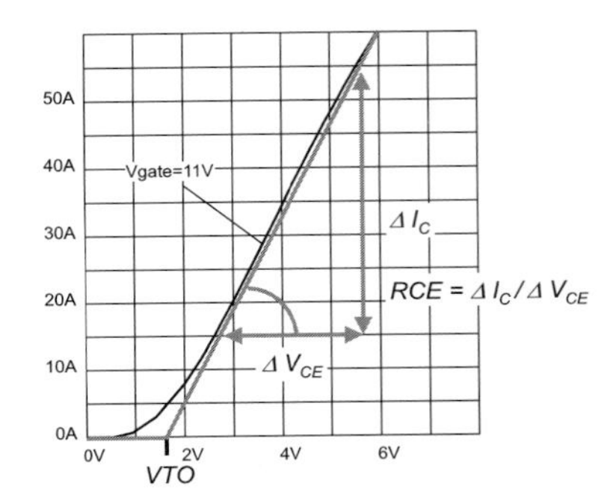

Figure 3.163
Determining straight line approximations to the characteristic curves for an IGBT output

Trade-offs for an IGBT

Choosing the right IGBT for a particular application is made more difficult by the wide range of different IGBT technologies and optimizations.

Whereas the objective with a MOSFET is to achieve the lowest possible area-specific $R_{DS(on)}$, in the case of IGBTs there is the possibility of changing the balance between switching losses (tail current) and on-state losses (V_{CEsat}) using the same technology. Depending on the switching frequency required, it is appropriate to use an IGBT with the smallest possible tail losses, or the smallest possible on-state losses.

Figure 3.164 shows the different technological approaches for such a process, here including the additional parameter of the short circuit time. In practice, the ratio of the switching and on-state losses will be set by the doping of the emitter on the rear (p-zone).

Figure 3.165 shows the technologies now available from Infineon Technologies, for the case of a 600 V blocking voltage. The continuous line shows the NPT technology optimized in three different ways: a high speed IGBT, for higher frequency applications, a fast IGBT, and a low V_{CEsat} device for low frequency applications. The dotted curve represents a new tech-

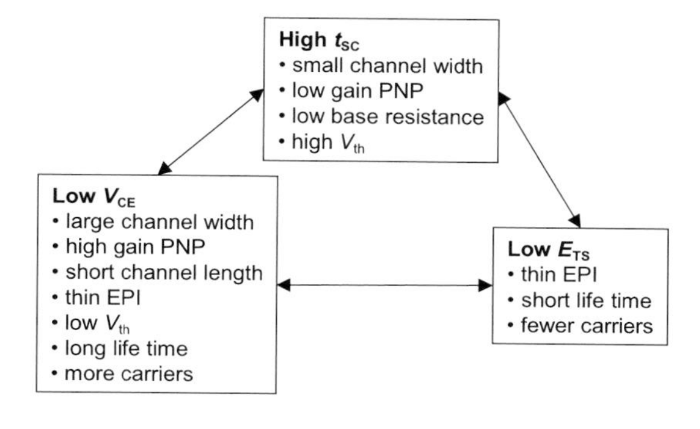

Figure 3.164
Trade-offs in optimizing an IGBT

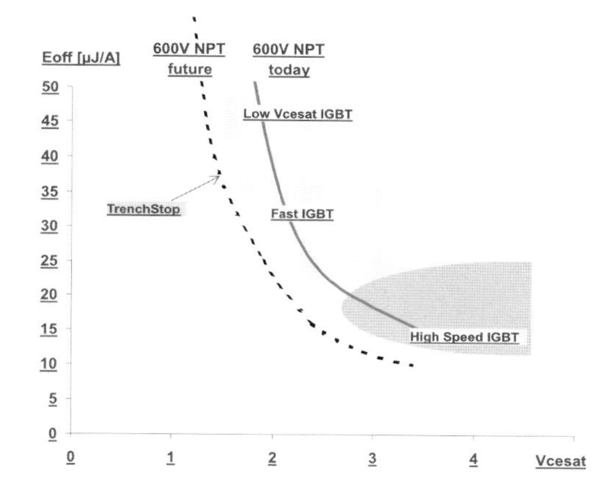

Figure 3.165
Trade-off curve for the IGBT technologies from Infineon Technologies

nology the Field Stop IGBT technology. This trade-off has a clear displacement parallel to the original trade-off curve towards lower values, both for the switch-off losses and for the on-state losses.

The field stop (FS) or trench stop (TS) IGBT

The 600V field stop principle represents an evolution of the NPT technology. Figure 3.166 shows a cross-section of this new concept. One significant difference is the cell structure, where instead of a planar structure a trench cell concept is used. The advantage of this concept is a stronger charge carrier accumulation at the top of the chip, and hence a more homogeneous distribution of the charge carriers across the device. For the user, this results in a lower saturation voltage and reduced conduction losses.

The second difference is in the substructure, in which an n^- field stop layer has been implemented. At first sight, this concept looks very similar to the PT concept, but there are significant differences. The advantages of the two familiar technologies are combined: a trapezoidal electric field profile in the off-state, which results in a thin n^- layer and hence low conduction losses, and a weak emitter on the rear side which results in low turn-off losses and hence a short tail current.

Because the field stop layer is lightly doped, the sole function which it undertakes is to reduce the electric field to zero

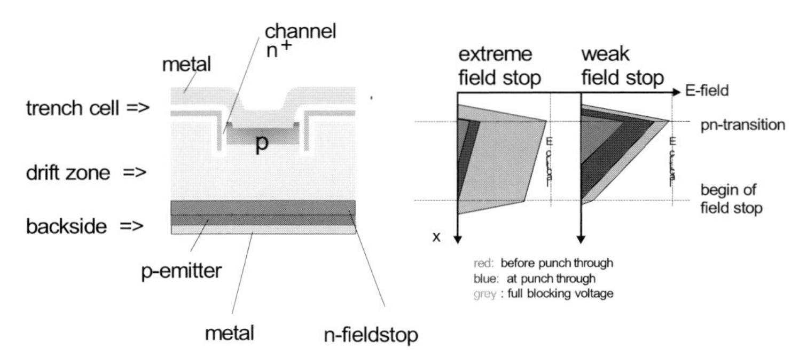

Figure 3.166 FS technology, with its trench cell

in the off-state. When the device is in its on-state, it does not affect the emitter on the rear side, so that this latter may be weakly structured.

For this purpose, a high-ohmic substrate with a processed front side is ground down to a thickness of 70 µm. An additional n-doped layer is implanted in the rear side of the wafer. The resulting electric field distribution in the cross-section of the IGBT is almost rectangular. As with the PT technology, this permits the thickness of the drift zone to be reduced.

As a result of the high-ohmic substrate, the drift zone is cleared out at voltages (approx. 100 V) significantly below the full breakdown voltage. Resulting in a cutoff of the tail current, the integrated losses being given by:

$$P = \int U(t) \cdot I(t) dt$$

These are well below those of a conventional NPT device with tail currents at higher voltages.

As a further result of this technology, it has been possible to reduce the thickness of a 1200 V device by 1/3 compared to that of a thin-wafer NPT technology. Numerically, this means the thickness is reduced from 175 µm to 120 µm.

Table 3.11 summarizes the differences between the IGBT technologies once again.

Table 3.11 Comparison of the different IGBT concepts

	PT IGBT	**NPT IGBT**	**FS IGBT**
P-Emitter	Very highly efficient	Less efficient	Less efficient
n^- layer	Thin	Average	Thin
Additional n-layer	Buffer layer = highly doped Efficient reduction of the high emitter Stops the electric field	No	Field stop layer = lightly doped Stops only the electric field
Charge carrier life	Short (lifetime killing)	Long	Long

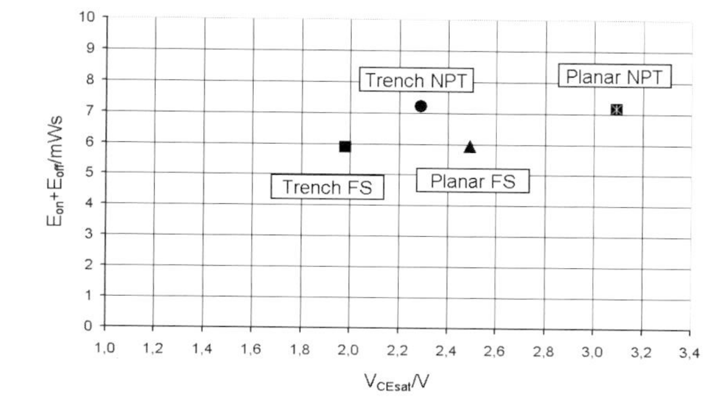

Figure 3.167
Total switching losses of NPT technology compared to the new FS technology

Figure 3.167 makes clear what a strong effect the advantage of the field stop technology combined with a trench cell structure has on the device losses. It has been possible to significantly reduce both the conduction and switching losses.

Outlook and summary

It is supposed, that, the loss comparison between IGBTs and MOSFETs shows fundamental differences. While the IGBT has lower on-state losses due to the bipolar carrier modulation, it exhibits its higher switching losses because of its characteristic tail current. New concepts are intended to reduce this tail current, to get as close as possible to the switching characteristics of a MOSFET. From all of these concepts, the field stop technology is today the most advanced. Using it, new applications markets can be opened up for IGBTs away from their classic application in inverters.

Switched mode power supplies, such as those used for supplying power to PCs or servers, are more and more becoming the focus.

4 Opto-semiconductors

4.1 The physics of optical radiation

4.1.1 Fundamentals and terminology

In optoelectronics, the term 'light' is used to cover the visible region and the adjacent region of the spectrum of electromagnetic radiation (360 to 830 nm).

The optical properties of detectors and emitters are well specified by the receiving and radiation characteristic in the far-field region. Here, there are two diagrams which are important for an LED. The angular intensity diagram (Figure 4.1) specifies how much signal energy is received by a detector, located on the optical axis, when the LED is rotated away from the optical axis by ϕ degrees. The irradiance diagram shows how much signal energy is received by a detector, located on the optical axis, when the LED is moved in a plane perpendicular to the optical axis.

The 'color temperature' of a light source is the temperature which a black-body radiator would need to have in order to produce the impression of having the same color as the light source.

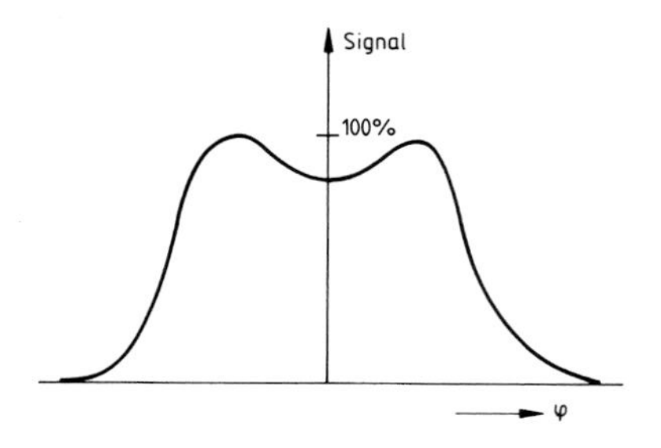

Figure 4.1 Radiation intensity diagram

Radiators

Thermal radiators (e.g. incandescent lamps) are characterized by the fact that the energy supplied to them is initially converted into heat, and is then emitted as radiation. Thermal radiators are continuous-spectrum radiators, i.e. the radiated energy is distributed continuously over a large range of wavelengths.

Luminescent radiators (e.g. LEDs) store the energy supplied as potential energy (electrons are raised from the valence band to the conduction band) and then emit it as radiation. They are capable of high radiation emissions at low temperatures. Luminescent radiators are generally line radiators, with a narrow emission spectrum.

The Planck radiator is an ideal continuous-spectrum radiator (thermal radiator). It is defined as a heated cavity with a small hole through which the radiation is emitted. Its total radiation is defined by the Stefan-Boltzmann law of radiation, and the maximum of its spectral emission by Wien's displacement law.

DIN A standard light, defined in IEC 306 part I and DIN 5033, is produced using a Tungsten filament lamp specially built for the purpose, at a filament temperature of 2856 K. In the visible range, it approximates to the light from a black body (Figure 4.2).

Spectral sensitivity of the eye

Within the optical radiation wavelengths, from 100 to approx. 10^5 nm, the visible

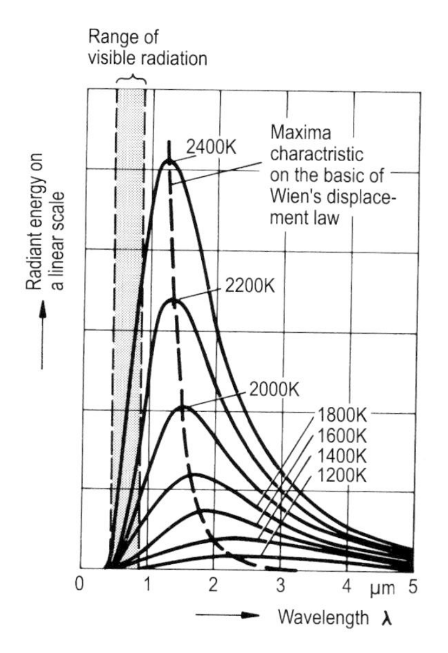

Figure 4.2
Distribution of spectral energy for a black body at different temperatures

region extends from about 380 to 780 nm. The spectral sensitivity of a light-adapted human eye is at its highest in the central, green region, and falls off quickly towards the blue and red ends. It is tabulated for a standard observer in DIN 5031.

For a dark-adapted eye (typically at: $< 3 \cdot 10^{-3}$ cd/m^2) the sensitivity curve is shifted by up to 50 nm towards the violet end.

In the visible range, the units used for quantifying the luminous intensity and the luminous flux are lumens (lm) and candelas (cd). These are based on the perception of radiation by the human eye. For the remaining wavelength ranges, radiometric units from radiation physics are used (e.g. Watt/sterad etc.).

Table 4.1 summarizes all the radiometric and photometric quantities.

Radiant flux

The radiant flux is the total power (in units of Watts) emitted in the form of radiation. The luminous flux (in units of lumens, lm) is the radiant power perceived by an eye with a sensitivity of V_{lambda}.

Typical values are:

1 m^2 surface of the sun:
Φ_e = 60 Megawatt

40 W fluorescent lamp:
Φ_v = 750 to 3200 lm

100 W incandescent lamp:
Φ_v = 1600 lm

red LED:
Φ_v = 5 lm at 50 mA

At the point of maximum sensitivity of the eye, at 555 nm, 1 Watt radiant flux corresponds to a luminous flux of 683 lm.

Radiation intensity

The radiation intensity I_e (units: W/sr) is the radiant flux per steradian. The associ-

Table 4.1 Radiometric and photometric quantities and units

Radiometry			Photometry		
Quantity		Unit	Quantity		Unit
Radiant power / flux	Φ_e	W	Luminous flux	Φ_v	lm
Radiant intensity	I_e	W/sr	Luminous intensity	I_v	lm/sr = cd
Irradiance	E_e	W/m^2	Illuminance	E_v	lm/m^2 = lx
Radiance	L_e	W/m^2 sr	Luminance	L_v	cd/m^2

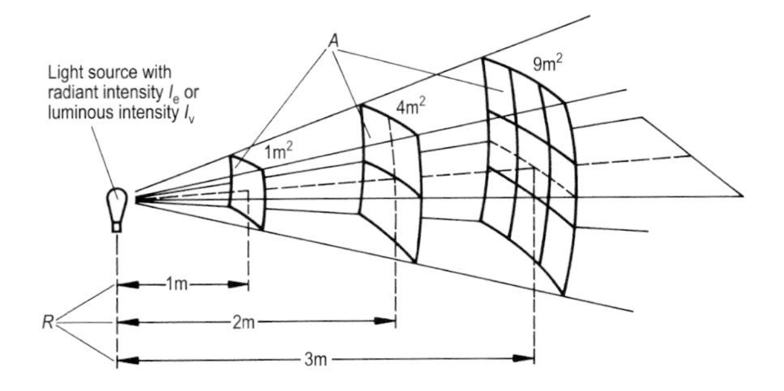

Figure 4.3
Definition of the solid angle Ω = 1 sr (steradian)

ated quantity, weighted by the eye sensitivity V_{lambda}, is called the luminous intensity I_v, (units: candela cd, 1 cd = 1 lm/sr).

Typical values are:

100W incandescent lamp:
I_v = 110 cd

100W headlight:
I_v = up to 10^6 cd in the direction of radiation

Semiconductor laser 880 nm, 2 mW (with no additional optics):
I_e = 2 to 5 mW/sr

LED fur signaling purposes (10 mA):
I_v = 1 to 1000 mcd

IRED for a remote control (100 mA):
I_e = 10 to 300 mW/sr

Solid angles

The solid angle Ω (units: sr) encloses a part of space which is bounded by the rays emanating in a cone-shape from a point P (e.g. the source of the rays), and terminating on a closed curve in space. If this curve lies on a unit sphere (radius 1 m) with P as its center, and if it encloses an area of 1 m^2, then the corresponding solid angle Ω = 1 steradian (1 sr), $\Omega = A/R^2$ (Figure 4.3).

Radiance

The radiance L_e (units: W/m^{2}sr) is the radiant flux per unit area and solid angle. The associated photometric quantity is called the luminance L_v (units: cd/m^2). The human eye perceives differences in luminance as differences in brightness. Luminance is measured in units of the Nit (nt), the Lambert (L) and the footlambert (fL).

Typical values are:

Surface of the sun:
1.5×10^9 cd/m^2

Filament of an incandescent lamp:
5 to 35×10^6 cd/m^2

Modern fluorescent lamp:
0.9 to 2.5×10^4 cd/cm^2

Night sky:
around 10^{-11} cd/m^2

Irradiance

The irradiance E_e (units W/m^2) is the incident radiant flux per unit receiving area. The associated photometric quantity is called the illuminance E_v (units lux: 1 lx = 1 lm/m^2) or the footcandle fc (1 fc = 10.76 lx).

Typical values are:

Sunshine at midday, outdoors:
max. 100 mW/cm^2
100000 lx

Office workplace:
500 lx

Clear full-moon night:
0.2 lx

An illuminance of 1000 lx corresponds to an irradiance of 4.76 mW/cm^2 for DIN A standard light.

4.1.2 Photodiodes

If photons penetrate a photodiode with sufficient energy, then electron-hole pairs are created within the semiconductor (internal photoeffect). The electric field in the space charge zone splits up the charge-carrying pairs. The result is that a current flows in the external circuit (Figure 4.4).

From the point of view of their electrical operation, a distinction is made between the photodiode mode (with a bias voltage of U_R, quadrant B) and photovoltaic cell mode (quadrant A). The photovoltaic cell acts as a current source, which converts radiation energy into electrical energy (Figure 4.5).

The open-circuit voltage U_L is independent of the diode area and increases logarithmically with the illuminance, and for silicon diodes rises to a value of approx. 0.5 V at 1000 lx. The short-circuit current I_K is proportional to the illuminance and the area. The permissible reverse voltage is low (approx. 1 V), so that operation as a photodiode in the reverse direction is only possible to a limited extent.

For optimal power extraction from a photodiode, the load resistance must be of the order of magnitude of U_L/I_K.

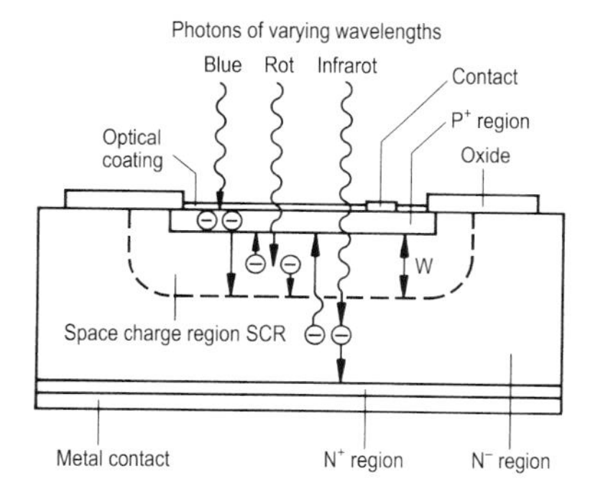

Figure 4.4
Schematic diagram of a planar silicon photodiode

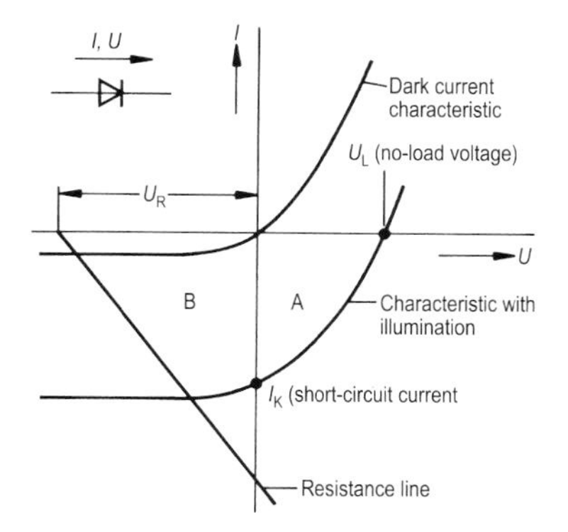

Figure 4.5
The characteristic curve of a photodiode

4.1.3 Silicon photodiodes

If appropriately dimensioned, silicon photodiodes have very low dark currents ($<10^{-11}$ A/mm^2). Consequently they are suitable for detecting very low levels of illuminance. However, they are slow, because the charge carriers must first migrate by diffusion into the narrow space charge zone produced by the doping. Their photoelectric current is proportional to the illuminance over many powers of ten.

Where the speed is critical, Si-PIN photodiodes are used. With these, most of the light is converted in the relatively wide space charge zone, so that with an appropriate bias voltage the charge carriers drift at the saturation speed (approx. 50 to 100 km/s). Because of their wide space charge zone, the junction capacitance of PIN photodiodes is low (a few pF) so that, when used with a low-impedance external load resistance, very low time constants are possible.

The sensitivity S of photodetectors is the ratio of the photocurrent to the incident radiation power or luminous flux, as applicable, and exhibits a distinct wavelength dependency which varies from one semiconductor to another. Figure 4.6 shows the graph of this spectral sensitivi-

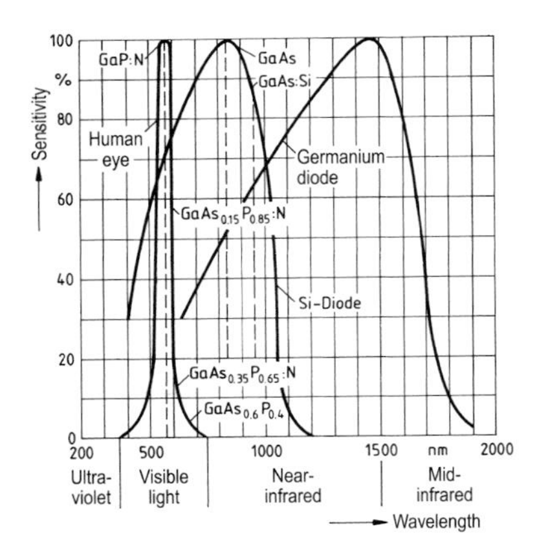

Figure 4.6
Relative sensitivity of Si and Ge photodiodes

ty S (units: A/W) for germanium and silicon photodetectors.

It is apparent that silicon, with its maximum sensitivity at about 850 nm, is well matched to the emission from GaAs, whereas germanium with its broad maximum around 1.5 µm even covers the mid-infrared region up to about 2 µm. In addition, there are semiconductor compounds which permit optimized detectors extending far into the mid-infrared region: (e.g. CdS, PbS, InSb, GaInAsP, HgCdTe etc.) in some cases as photodiodes, in other cases as photoresistors.

4.1.4 Phototransistors

In a phototransistor, the collector-base link is constructed as a photodiode. Its photocurrent, multiplied by the current amplification factor of the transistor (approx. 100 to 1000), appears as the collector current. Hence phototransistors supply strong signals and require less post-amplification than photodiodes. However, their operation is less linear (owing to the non-linearity of the current amplification) and is slower because of the Miller effect combined with the relatively large area of the collector-base diode (Figure 4.7).

The average number of charge-carrying pairs freed by one photon is called the quantum efficiency η. The maximum achievable sensitivity S_{max} of a non-amplifying photoreceiver would be achieved at $\eta = 1$, i.e. when each incident photon creates a charge-carrying pair. All the charge carriers are accumulated and thus contribute to the photocurrent. As the energy of a photon

$$E = \frac{h \cdot c}{\lambda}$$

(e_0 = electronic charge, λ = wavelength, c = speed of light, h = Planck's constant), it follows that

$$S_{\lambda max} = \frac{e_0}{E} = \frac{\lambda}{1.24} \text{ (A/W) } (\lambda \text{ in µm})$$

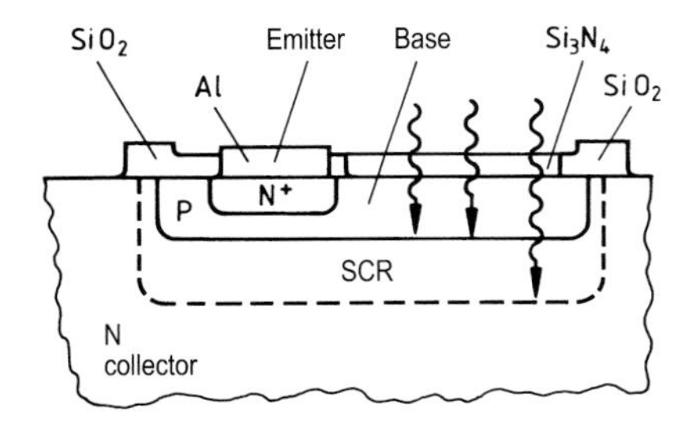

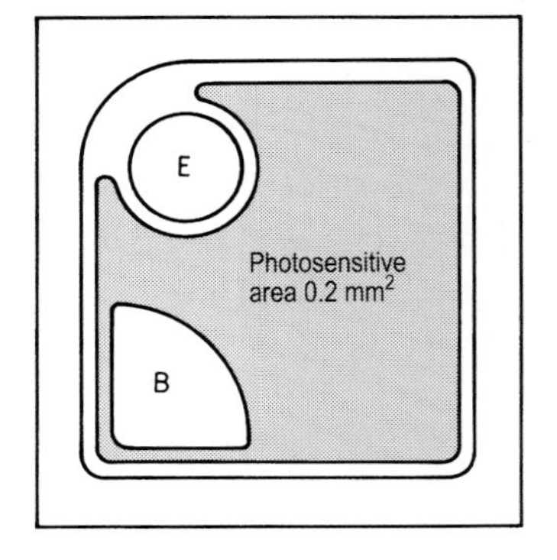

Figure 4.7 Bipolar phototransistor

thus is, the maximum limit on the sensitivity increases linearly with the wavelength. Practical silicon photodetectors achieve up to 90% of this value at $\lambda = 850$ nm.

In the case of detectors for daylight or artificial light, it is the photosensitivity in nA/lx which is of interest. This is relative to a particular illuminance, generally DIN A standard light. On the other hand, the photosensitivity of detectors for infrared applications is generally specified in μA. This is measured relative to a particular irradiance (e.g. 0.5 mW/cm^2) at a particular wavelength (e.g. 950 nm).

In the case of detectors for very small radiation signals, the quantity of interest is the noise equivalent power (NEP), dimensions W/$\sqrt{\text{Hz}}$ (Hz). This is the radiation power which generates a signal, at the output of the detector, which is of the same magnitude as the noise. Here, the measurement conditions (wavelength range of the light, modulation frequency or bandwidth as applicable) have an effect. This definition stems from the fact that, on the one hand, the signal current has a proportional correspondence the optical signal, while on the other hand the noise mechanisms involved cause an effective noise current proportional to $\sqrt{\Delta f}$.

The reciprocal of the NEP for a given radiation-sensitive area A is referred to as the detectivity D^* (also called the detectability limit).

$$D^* = \frac{\sqrt{A}}{\text{NEP}} \left[\frac{\text{cm/Hz}}{\text{W}}\right]$$

The term dark current is used for the current which flows when there is no incident radiation. In the case of photodetectors, this is the reverse current of the diode, for phototransistors and photo-Darlington transistors it is the amplified leakage current for the collector-base link which acts as the photodiode. The dark current increases with the temperature (roughly doubling for every 10 K in the case of Si) and can interfere in the case of phototransistors, and particularly in the case of photo-Darlington transistors.

4.1.5 Light emitting diodes

Light emitting diodes are semiconductor diodes which emit radiation when a current flows in the forward direction. The process involves the injection of an excess of charge carriers into the neutral N and P region, where some of them recombine with the emission of a photon (injection luminescence).

In contrast to an incandescent lamp, their emission spectrum is confined to a narrow range of wavelengths, which is essentially determined by the band gap E_g ($\lambda = (h \cdot c)/E_g$) of the semiconductor. The materials used are III-V semiconductors, because they have appropriate band gaps and because, unlike the indirect semiconductors Si and Ge, their energy band structure effectively permits radiative recombination.

Infrared emitting diodes (IREDs, Figure 4.8) are mostly manufactured from GaAs (E_g = 1.43 eV) or GaAlAs, and they emit radiation in the near infrared, between 800 and 950 nm, close to the maximum of the spectral sensitivity of Si photodetectors. They are used in light barriers, remote controls, optocouplers and also in

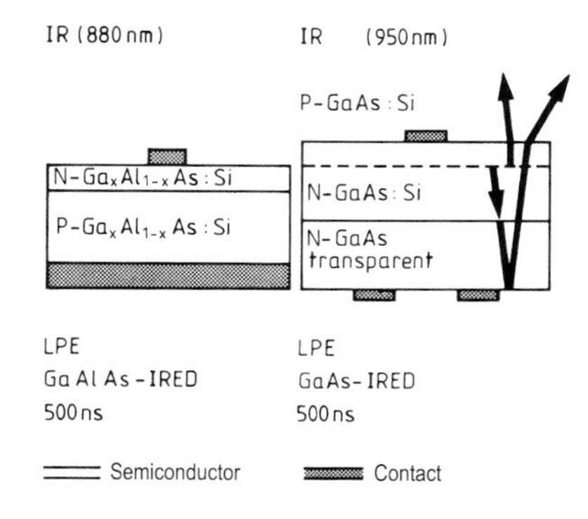

Figure 4.8
Schematic structure of an IRED chip

Table 4.2
Common composition of LEDs (TSN stands for Transparent Substrate Nitrogen, doped)

	λ (nm)	Substrate	E_g (eV)	Active layer
Infrared	950	GaAs	1.3	GaAs:Si
	800-900	GaAs	1.4	GaAlAs
Red	700	GaP	1.8	GaP:Zn,O
	660	GaAs	1.9	$GaAs_{0.6}P_{0.4}$
	635	GaP	2.0	$GaAs_{0.35}P_{0.65}$:N TSN
Yellow	590	GaP	2.1	$GaAs_{0.15}P_{0.85}$:N TSN
Green	565	GaP	2.2	GaP:N
Blue	465	SiC	2.7	InGaN
		Sapphire		
Green-red	560-640	GaAs	2.0	InGaAlP
Blue-green	450-540	SiC or Sapphire	2.7	InGaN

sensing technology. Their advantages are: mechanical compactness and robustness, low operating temperatures, ease of modulation up into the MHz region, plus TTL compatibility and high efficiencies of several percent.

Some of the radiation exits from the diode directly "upwards" or to the side, but even the radiation emitted towards the substrate is usable if the substrate itself is transparent and the radiation is (partially) reflected from the underside.

Again in the case of diodes which emit (visible) light (LEDs), the radiation is generated by the recombination of charge-carrying pairs in a semiconductor with an appropriate energy band gap E_g.

The substrates used are GaAs (absorptive for visible light), GaP, SiC and sapphire (transparent). Depending on the wavelength, the active zones consist of GaAsP of varying composition or GaP.

More recent material systems are InGaN and InGaAlP. Using these two materials it is possible, depending on the composition, to cover the entire visible wavelength range. Table 4.2 shows an overview of the various material systems. Depending on the application, the size of the chips varies from a side length of 150 µm to 1 mm.

To enable as much as possible of the light generated in the semiconductor to be fed out of the chip, the geometric arrangement of the chip is particularly important. A large proportion of the light generated in the semiconductor is reflected back into the semiconductor, because of the critical angle for total reflection at the chip surface. Figure 4.9 shows the surface of an InGaAlP chip.

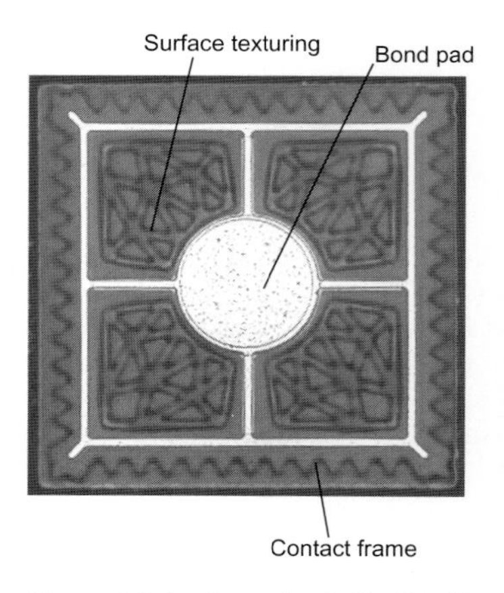

Figure 4.9 Surface of an InGaAlP chip

Figure 4.10
Side view of an InGaN chip with pyramid-shaped side surfaces

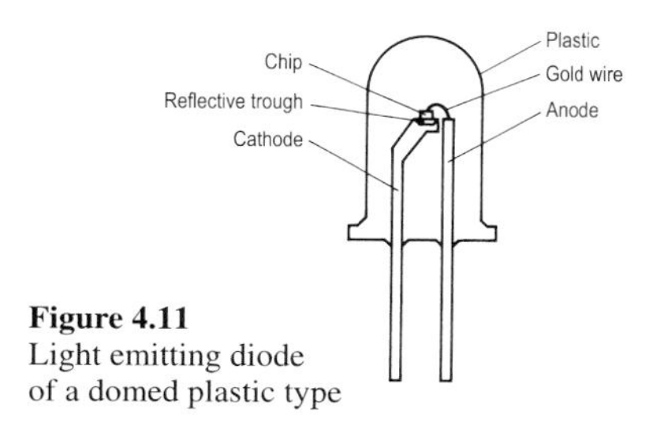

Figure 4.11
Light emitting diode of a domed plastic type

The surface texturing which is applied reduces the reflection back into the chip, so that more light is fed out at the surface. Furthermore, the contact frame distributes the electrical current over the surface of the chip in such a way that the light is not generated where it would be obscured, underneath the bond pad. Similarly, the geometric form of the SiC substrate of an InGaN chip can be so arranged (Figure 4.10) that a large part of the light generated in the semiconductor can be fed downwards through the side surfaces of the substrate. This light can then be reflected upwards by an external reflector.

To protect the semiconductor chip from environmental influences, and to simplify handling and electrical contact-making in a circuit, the chips are primarily used in plastic packages. In addition, the plastic reduces the critical angle for total reflection, and thus increases the radiation power emerging from the semiconductor. The chip is mounted in a reflective trough, so that even the side radiation is redirected upwards. In the case of domed types (Figure 4.11), the curved surface acts as a lens and collimates the radiation along the direction of the axis. Nowadays however, types for flat soldering, for so-called surface mount technology, have become dominant, because these types are not inserted into the circuit board but instead are placed on its surface. Among these, there is a host of different types, depending on the application area (Figure 4.12). In particular, there is a trend here toward very small form factors, as well as toward larger form factors for larger chips (1 mm^2) with a high radiation power.

Visible light emitting diodes are mainly used for backlighting instruments, such as dashboards, navigation systems etc. in the automobile field, or for illuminating LCD displays, for example in mobile telephones. In recent years, the brightness of LEDs has risen to such an extent that more and more application areas are opening up in exterior lighting. For a few years now, the third brake lights

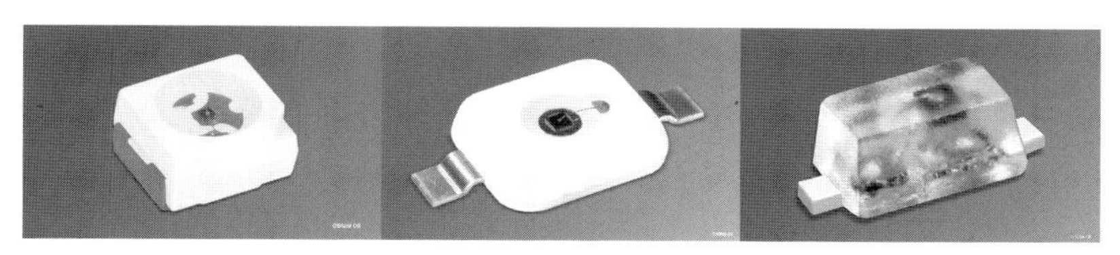

TOPLED®: 3.2*2.8*1.9mm　DRAGON: 7.0*6.0*1.8mm, for chip sizes up to 1mm^2　SmartLED™: 2.0*1.4*1.3mm

Figure 4.12 Various SMT types with details of their dimensions

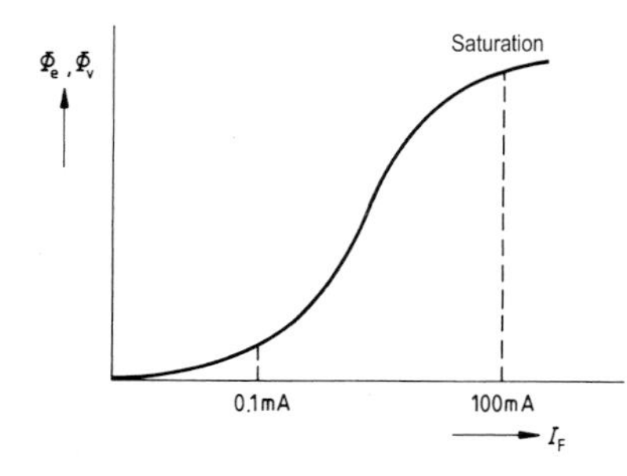

Figure 4.13
Typical photoelectric current characteristic curve for a light emitting diode

(CHMSL) on vehicles have been made almost exclusively with LEDs. Apart from their longer service life, there is another safety aspect here, in that LED brake lights light up significantly quicker than conventional filament lamps. In indicator flashers and rear lights too, some LEDs are now already being used. Further application areas are full-color displays, comprising from several thousand up to a million LEDs, and the backlighting of advertisements.

The light/current characteristic curves for IREDs and LEDs exhibit a section of the curve over which non-radiative recombination predominates, an approximately linear section, and a section over which saturation of the emission (heating, saturation of the radiative transitions) starts to appear (see Figure 4.13). As the temperature of the crystal rises, the wavelength of the emission increases and the efficiency deteriorates.

All radiation-emitting semiconductors have in common a performance degradation, i.e. a declining output power over their operating lifetime. A quality characteristic used to measure ageing is the elapsed time (lifetime) until the radiation, at a constant current, falls to half of its initial value. The phenomenon has not yet been completely explained. However, it is reasonably certain that the degradation is connected with the migration or spreading of imperfections in the crystal.

The relative spectral emission specifies how the distribution of the radiation from an LED depends on the wavelength. For the user, this curve is generally not very helpful, because its measurement is technically difficult and time consuming. LED spectra are therefore generally specified by the following user-friendly parameters:

λ_{peak} = wavelength at which the maximum of the spectral emission occurs.

$\Delta\lambda$ = width of the emission spectrum, measured at the 50%-points.

λ_{Sp} = CoG [German: Schwerpunkt] wavelength: specifies the center of gravity of the emissions and is an important concept for applications in spectroscopy.

λ_{Dom} = dominant wavelength: specifies the color tone of an LED, as perceived by the human eye.

4.2 Semiconductor lasers

The importance of semiconductor lasers has increased rapidly over the last 20 years. Even if gas lasers continue to be utilized, for high powers and coherence together with short wavelengths, there are an increasing number of areas in which these are outweighed by the advantages of the laser diode:

- low costs
- small dimensions
- high efficiency
- long service life

In telecommunications technology laser diodes have become a fundamental element, and even in applications which demand high optical performance semiconductor lasers are making ever deeper inroads. The use of frequency-doubling crystals means that even visible wavelengths, right down to the UV region, are no longer a technical problem.

4.2.1 Fundamentals of the semiconductor laser

The basic physical mechanism involved in the generation of radiation by semiconductors is the interaction of photons with charge carriers in the valence and conduction bands of the crystal. For the occurrence of radiative recombination to be possible, the semiconductor must have a band structure as shown in Figure 4.14. The electron can then jump directly from the conduction band into the valence band, with the release of a photon. The energy of this emitted photon corresponds to the band gap for the semiconductor, and is what determines the wavelength of the radiation emitted.

In the case of a semiconductor such as germanium or silicon, radiative recombination is not possible, because the intrinsic momentum of the photon is not sufficient to satisfy the law of conservation of momentum. When recombination occurs, the momentum balance can only be effected by means of a third partner, such as phonons (the lattice vibrations which occur in the crystal) or defects. This process is referred to as spontaneous emission, and is utilized in light emitting diodes (LEDs and IREDs).

The laser is based on the effect of stimulated emission. This occurs when a photon with a wavelength corresponding to ΔE enters into an interaction with a conducting band electron with energy E_2. This photon then stimulates the electron transition (Figure 4.14b). The feature of this which should be particularly noted is that the photon generated corresponds, in terms of its wavelength, phase, polarization and direction of propagation, to the photon which stimulated it. The stimulated emission corresponds to an amplification of the incident photon. Hence the name "Laser": **l**ight **a**mplification by **s**timulated **e**mission of **r**adiation.

In a thermal equilibrium, the number of electrons in the valence band with energy E_1 is significantly greater that the number of electrons in the conduction band E_2. In this state, therefore, the probability that an incident photon is absorbed is far greater than that it produces the desired stimulated emission. To increase the probability of the process of stimulated emission taking place, it is necessary to achieve an inversion of the charge carriers in terms of the valence and conduction bands, i.e. the number of electrons in the conduction band with energy E_2 must be

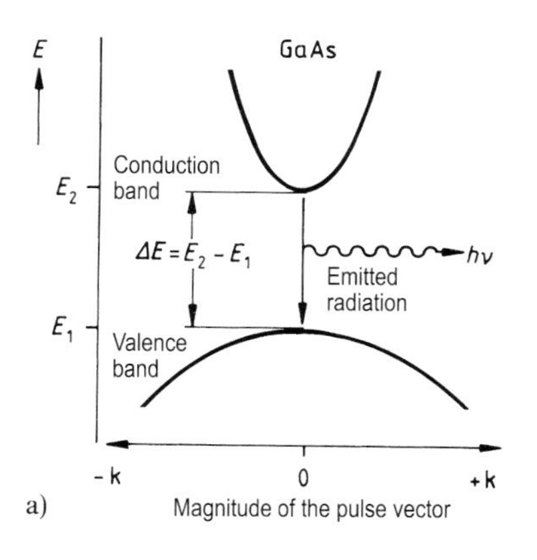

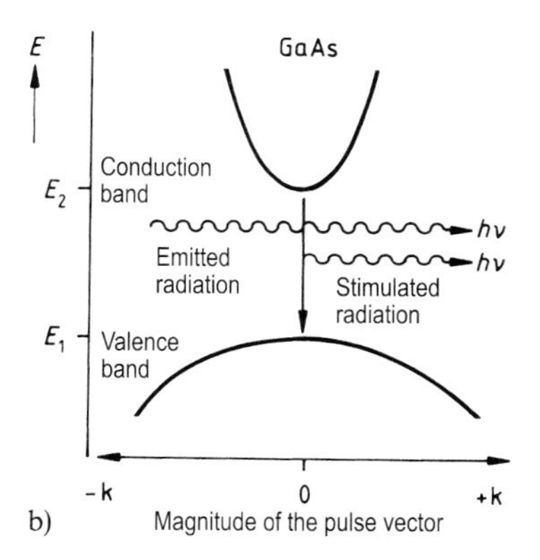

Figure 4.14
Energy band model to illustrate the recombination processes and generation of radiation in the case of gallium arsenide, a) spontaneous emission, b) stimulated emission

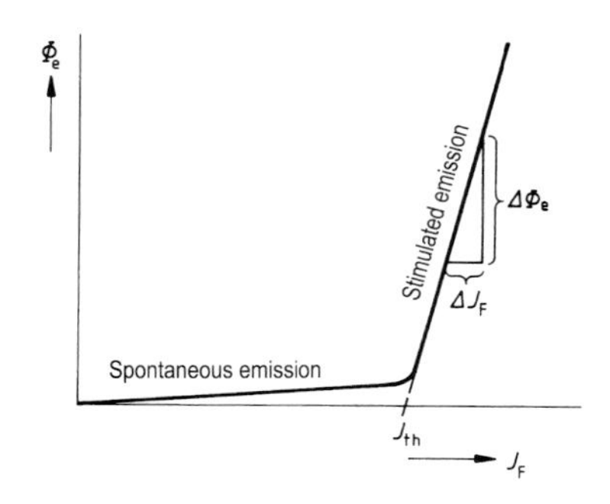

Figure 4.15
Characteristic curve for a laser

greatly increased. For a semiconductor laser, this population inversion is achieved by the injection of charge carriers into an active crystal space which is structured as a PN junction. To make it possible to achieve a sufficiently high occupancy in the conduction band, a certain minimum injection current is required. This is referred to as the threshold current. Figure 4.15 shows the characteristic curve for a laser. When the threshold current I_{th} is exceeded, the emitted radiation Φ_e changes from an initially spontaneous occurrence to become stimulated emission. I_F is the forward current through the laser diode.

Below this threshold current, the laser diode behaves like a light emitting diode. Only spontaneous emission occurs. After the threshold current has been exceeded, the stimulated emission starts up, and the emitted radiation power rises linearly. Another important characteristic quantity for a semiconductor laser can be read off from Figure 4.15, namely the so-called differential efficiency $\Delta\Phi_e/\Delta I_F$. This denotes the steepness of the characteristic curve for the laser.

Depending on the wavelength region required of the semiconductor laser, two material systems are used nowadays. In laser diodes for the wavelength range 1300-1550 nm (mainly for fiber optic transmission, section 4.4) the active zone is InGaAsP on an InP substrate. For the wavelength range 600-880 nm (for plastic optical waveguides and high power laser diodes, section 4.2.3) use is made of AlGaAs for the active zone, and for wavelengths greater than 880 nm InAlGaAs, on a GaAs substrate.

4.2.2 Structure of an oxide strip laser

In order to guarantee a reliable service life, even at high temperatures, the following conditions are particularly important:

- So that a sufficiently high charge carrier density can be produced to reach inversion, it is advantageous to put vertical limits on the injection of charge carriers.
- The active layer must be constructed as a dielectric waveguide, so that a high photon density can develop.
- The active crystal solid must be constructed as an optical resonator, within which the propagating wave is partly reflected and thereby further amplified.
- To permit a high power density (MW/cm^2) at the laser mirrors combined with high reliability and service life, the mirrors must be optimally coated.
- Efficient generation of radiation, and low ageing rates, require a high level of perfection in the crystals.

Vertical limitation is achieved by a so-called double heterostructure: The active layer, which is only 0.1 to 0.2 µm thick, is located between two sheathing layers, each with a higher band gap, as shown in Figure 4.16. As a result, potential barriers are formed, which limit the injected electrons and holes to the active zone. The larger band gap reduces the refractive index, so that a dielectric sheet waveguide results, similar to an optical fiber (with no initial restriction on the lateral width). There are many conceivable ways of providing the lateral restriction. In the case of the oxide strip laser, the current path is limited to a strip with a width of about 3 μm by the application of an insulating oxide layer, as shown in Figure 4.16.

The concentration of the charge carriers into a narrow zone results in a lateral profile of the imaginary refractive index, which maintains a stable lateral fundamental mode (wave propagation in an anisotropic medium). Waves can only propagate and be amplified within this zone. Any wave which does not run parallel to the boundaries is absorbed and cannot propagate. This form of waveguide is called a gain guide. Laser diodes of this type are therefore also called "gain guided" lasers.

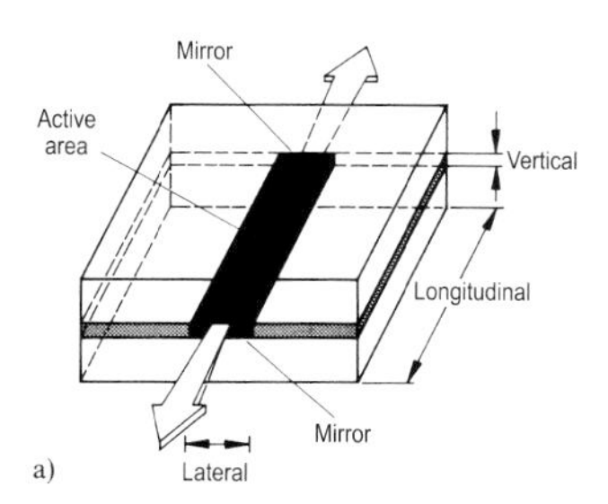

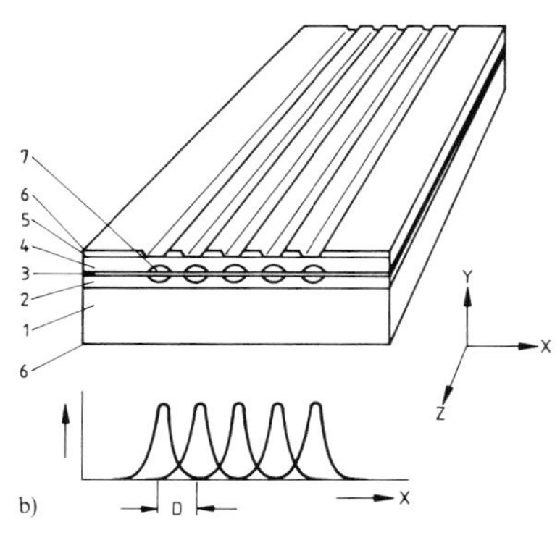

1 Substrate
2 N GaAlAs heterolayer
3 Active zone
4 P GaAlAs heterolayer
5 Oxide insulation
6 Metallization
7 Active mirror surface of the strip

Figure 4.16
Structure of a GaAlAs/GaAs oxide strip laser

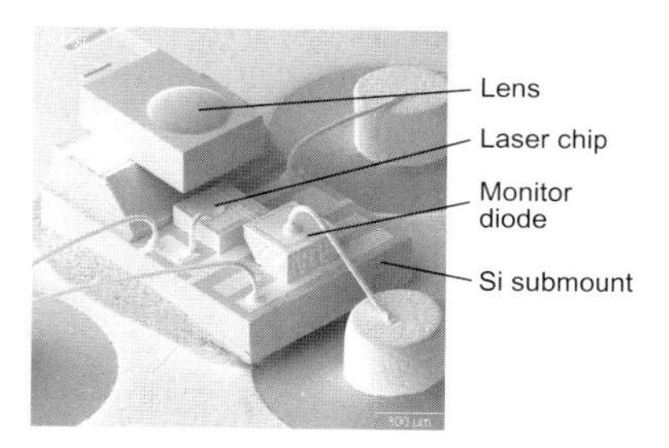

Figure 4.17
Si submount with laser chip, lens and monitor diode

To dissipate the heat losses, semiconductor diodes are used in various metal housings. Figure 4.17 shows a possible construction for applications in telecommunications and data communications. Here, the InGaAsP laser chip is soldered onto a Si submount, so that the heat losses are distributed over a large cross-sectional area. A glass prism redirects the sideways radiation from the laser diode upwards. A lens, soldered onto the glass prism, can focus the radiation onto a single-mode optical waveguide. On the other side of the laser diode is mounted a monitor diode, which captures the very low radiation from the backward mirror. The radiation power of the laser diode is regulated by means of this monitor diode. The complete unit can be glued into a metal housing and an optical waveguide connection (receptacle) can be adjusted into the laser diode's radiation path (Figure 4.18).

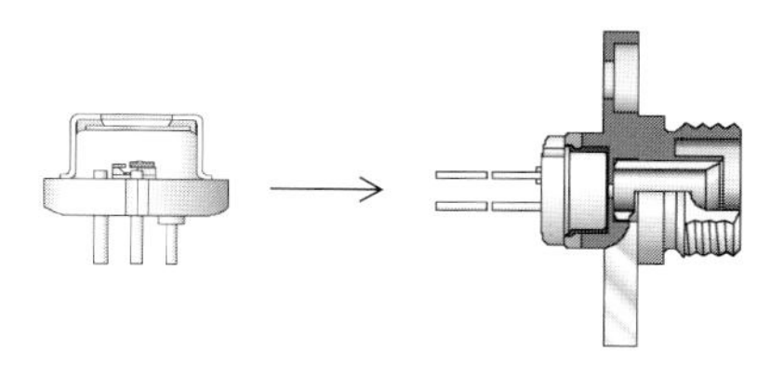

Figure 4.18
Si submount in a TO46 metal housing and with an optical waveguide connection (receptacle)

4.2.3 Laser arrays

Laser arrays are arrangements of several laser diodes of the same type in a single crystal, and are used mainly to increase the power output. The following description relates to a 12-strip arrangement (GRINGSCHSQW structure).

Figure 4.19 shows a photograph of a GaAIAs laser module. The laser chip (circled) is accommodated together with the monitor diode, an NTC thermistor and Peltier cooler, in a TO-3 package. The monitor diode captures the radiation from the backward mirror (about 10% of the usable radiation) so that it can be used for regulating and monitoring the radiation power. The array is mounted on a Peltier cooler which, with an NTC thermistor used as a temperature sensor, enables constant temperature operation to be achieved by means of an external regulation circuit. This is particularly important if the application is such that a very constant wavelength is required (e.g. for pumping neodymium-YAG lasers), when the temperature can be used for precise adjustment of the wavelength and to keep it constant. Section 4.2.2 describes a regulation circuit for this purpose.

In continuous wave (CW) mode, the QW-12 strip laser array offers a continuous output power of 250 mW at a typical threshold current of only 280 mA. In CW mode, the differential efficiency is around 0.7 W/A. This, and the low series resistance (approx. 0.5 Ω) result in an efficiency of over 20%. If the output power is increased up to the point where the characteristic curve for the laser saturates, at around 600 to 800 mW, an overall efficiency of up to 40% can be achieved.

Figure 4.19 GaAlAs high-power laser

Laser diodes are also available as modules in hermetically sealed metal housings. The connections are arranged on the DIL pattern.

Table 4.3 shows the main characteristics of the laser array.

Laser array diodes for high power outputs are suitable for CW operation, and thus open up new application areas. This includes pumping neodymium-YAG lasers. Until now, such lasers have been pumped using high energy flashlamps. However, the emission spectrum of a flashlamp is very wide, and on the other hand a neodymium-YAG crystal has a very narrow absorption spectrum, so that the pumping efficiency of the system is low. On the other hand, the wavelength of a GaAIAs laser can, by means of the arrangement and structure of the active laser layer, be matched exactly to the maximum of the absorption spectrum of a neodymium-YAG laser. This produces several of advantages:

- As a result of the efficiency of the semiconductor laser, which is already high (>20%), and the advantageous pumping efficiency which results from optimal wavelength matching, the system efficiency is high.
- The low thermal loading on the neodymium-YAG crystal permits a better radiation quality, a reduced line width and significantly smaller and lighter cooling equipment.
- Semiconductor lasers manage with a simple power supply, whereas flashlamps require a high-voltage mains power device.

Table 4.3 Characteristics of the 12-strip laser array

		Type	SFH 4801	SFH 48E1	SFH 48R1
Maximum ratings	ϕ_{eCW}	mW	200	200	1000
	ϕ_{epuls}	mW	300	300	1200
	V_R	V	3	3	3
	T_{sub}	°C	10 … 65	−10 … +60	10 … 65
Characteristic values for $T_{sub} = 25°C$ $t_p \leq 10$ µs $D \leq 0.01$	λ_{peak}	nm	805	805	805
	$\Delta\lambda$	nm	2	2	4
	η	W/A	0.35	0.35	0.35
	I_{th}	mA	400	400	2000
	ϕ_{eCW}	mW	150	150	800
	ϕ_{epuls}	mW	250	250	1000
Features			12-strip	12-strip	5 × 12 strip lasers

- Because of the high efficiency, the use of GaAIAs high-power lasers as the pumping source enables long service lives to be achieved for the entire system. Again, a sudden total failure, such as typically occurs with flashlamps, is very unlikely. In the case of laser diodes what happens is rather that a continuous decline takes place in the optical performance.

Regulation circuit for the SFH48EI laser array

Integrated with the laser diode in the SFH48EI high-power laser array are a monitor diode for controlling and regulating the power, a Peltier element for cooling purposes and an NTC thermistor as the temperature sensor. Together with an external regulating circuit, these components enable the laser array to be operated under optimal conditions.

The TCA 2465 power operational amplifier (2 OPs in one housing) supplies up to 2.5 A output current. It is therefore very well suited to this application. The circuit described here maintains the laser array in a stable operational state, protects the laser against polarity inversion, overcurrent and overvoltage.

The wavelength of the emitted light depends on the operating temperature. Within certain limits, therefore, it can be adjusted and held constant as required for the particular application. At the same time, the light power which is output is controlled by the current flowing through the laser diode.

The laser array, together with the monitor diode, a precision NTC thermistor and the Peltier cooler, is accommodated in a TO-3 package. The array is mounted on the Peltier cooler. The Peltier element transports heat from one side to the other, depending on the magnitude and direction of the current flowing through it. Together with the integral NTC thermistor, it enables the external regulating circuit to operate the system under load at a constant temperature.

The operation of the laser requires two current control circuits operating independently of one another: one to regulate the temperature and keep it constant, and one for the laser radiation power. In doing so, the maximum ratings for the current

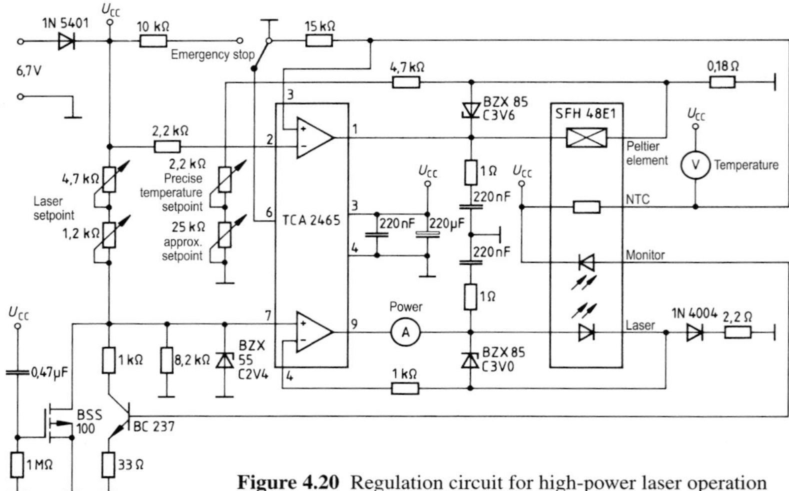

Figure 4.20 Regulation circuit for high-power laser operation

and voltage to the module must under no circumstances, even for short periods, be exceeded.

As shown in Figure 4.20, the two OPs act directly on the laser diode and the Peltier element. Their output currents are fed back, via current shunts, as a voltage applied to each corresponding negative input. This means that the circuit acts as a proportional (P) controller. Potentiometers permit target values to be set for the temperature and light power.

The NTC thermistor which is integrated into the laser package measures the temperature, which can be displayed using a voltmeter. The built-in delay only releases the current to the laser about 1 second after switch-on. This ensures that the cooling is provided in good time.

To effect an emergency shutdown, the power supply to the laser and the Peltier element is cut off by means of the inhibit switch on the OP. The circuit is designed for continuous operation of the laser. If a higher current is required for pulsed operation, the TCA 1365B OP (4-A output) is suitable for the purpose.

4.2.4 Further applications of semiconductor lasers

- Because of their coherence and the high modulation frequency which is possible (direct modulation is possible), the high-power lasers described here are also suitable for optical transmission in free space.
- Because of the high optical power, which can be fed into thick multimode fibers, GaAlAs semiconductor lasers are also suitable for potential-free energy transmission. Using modern photodiodes, high efficiencies can be achieved in converting this back to electrical energy.
- Another possibility is the exactly simultaneous triggering of photosensitive high-voltage thyristors, using a fiber optic network.
- Because of the high output power, it is possible to achieve frequency doubling in non-linear crystals. By generating the second harmonic, it is possible to produce coherent blue radiation from infrared radiation.
- As already described, semiconductor lasers are used as pump lasers for

neodymium-YAG lasers for material processing, such as welding panels in the manufacture of automobiles.

- Because of the advantages which a semiconductor component has over gas or solid-state lasers (geometry, service life, modulability, cost), GaAlAs high-power lasers are well-suited for laser printers.
- Another familiar application is the laser pointer. Until 10 years ago, the preference was for an easily visible red emission, mainly from helium-neon lasers with their typical wavelength of 633 nm. Since then, it has become possible to obtain laser diodes with wavelengths around 600 nm, which can be used to produce a very bright, almost point-shaped, light source with an optical power of 1 mW (which is still low enough not to harm eyes).
- For many years now, pulsed laser diodes have been used for the measurement of distances and speeds, e.g. in an automobile to measure the distance from the vehicle in front. Pulsed laser diodes react to a very short but powerful current pulse (around 50-200 ns with a peak current of up to 100 A) with an equally short and powerful (up to 100 W) infrared light pulse (about 800-1000 nm). The pulse duty ratio is very small (at most a few thousandths), but on the other hand repetition frequencies in the KHz range are possible.
- There are pulsed laser diodes which, although they cannot generate such a high power, can supply pulses in the region of up to 100 μs. Pulses of this type can contain so much energy that they will detonate explosives. If it is possible to transmit the optical power over an optical waveguide, remote firing is possible without the danger of misfiring due to electromagnetic influences.
- Section 4.4 goes into more detail about the possible uses of telecommunications and data communications laser diodes in fiber optic engineering.

4.3 Optocouplers and solid state relays

Structure

Optocouplers can transmit direct and alternating current signals up to a few MHz while providing complete electrical isolation. To do this, a photoemitter converts an electrical input signal into optical radiation (visible or infrared). Within the housing, this is incident on a photodetector (phototransistor, photo-IC, photo-triac), which converts the optical signal back into an electrical one. Solid state relays (SSRs) are used as replacements for mechanical miniature relays, and as the detector they are given a photodiode array with a control circuit, plus two MOSFETs.

The form of housing which has become most common is the DIL plastic package with 4 / 6 / 8 / 16 connections. Here, up to 4 channels are integrated into one package. Insofar as the application permits it, the trend nowadays is towards smaller types of package, such as SOIC-8 or SOT223.

If a high voltage is applied between the input and output sides of an optocoupler, then internally field strengths of up to 10^4 V/cm can arise. As a result of this field strength, it is possible for ions to precipitate out onto the phototransistor chip, which then leads to a change in the transistor's characteristic curve being caused by field effects. In individual cases this can lead to total inoperability of the optocoupler. With Infineon's optocouplers, it has been possible to completely eliminate this effect by the application of a weakly conducting **TR**ansparent **IO**n **S**creen (TRIOS®) onto the endangered surface. To improve the properties of the surface, an additional silicon nitride layer is applied under the TRIOS layer, which is an integral part of the phototransistor (Figure 4.21). This additional layer acts as a passivating agent against foreign atoms and as an optical surface coating.

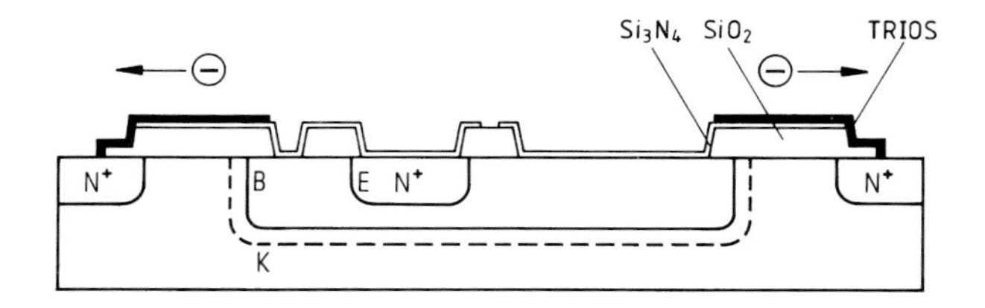

Figure 4.21
A phototransistor with TRIOS

Applications

The field of uses for optocouplers is very diverse, with the following being widespread:

- switched mode power supplies in industrial and consumer areas,
- programmable logic controllers (PLCs),
- telecommunications (modems, DAA, datacomms),
- PC peripherals,
- power electronics (IGBT controllers),
- medical technology.

Parameters

In the case of optocouplers with phototransistor outputs, the current transfer ratio (CTR) specifies the ratio of the output current to the input current. In the data sheets, a value is often specified at I_F = 10 mA (emitter) and V_{CE} = 5 V (phototransistor). In the case of optocouplers with digital outputs or photo-triacs and SSRs, instead of the current transfer ratio, a threshold I_{FTH} (threshold) is specified for the photoemitter, being the value at which the optocoupler must have switched on or off.

Safety

Optocouplers and SSRs are so arranged that a potential difference of several hundred volts between the input and output sides, maintained for years, has no negative effect on the insulation capability. Nor may transient overvoltages in the kV range, which are entirely possible occurrences for brief periods in 250 V supplies, result in a breakdown of the insulating section.

During its manufacture, each optocoupler is tested with the insulation test voltage specified in the datasheet. Optocouplers for mains isolation are qualified in accordance with DIN/VDE 0884 at the VDE, and in the testing bay undergo a test for freedom from partial discharges. Additional checks at other international testing institutes, including for example UL, CSA, BSI, FIMKO, ensure the highest degree of safety for the user.

The main characteristics of optocouplers

The predominant area for the application of optocouplers is in the transmission of signals with electrical isolation. In this respect, the following characteristics determine the possible uses of optocouplers:

- high switching speed,
- transmission of CW (continious wave) and alternating current signals,
- large temperature range,
- can be subjected to high temperature cycle loading,
- small dimensions,
- high reliability,
- resistant to high voltages,
- limited ageing of the light emitting diodes,
- stability of the transistor characteristics.

Stability of the transistor

In service use, there may be a large potential difference between the gallium arsenide light emitting diode and the silicon phototransistor. If there is in addition a high operating temperature (e.g. 90°C), a field effect may become noticeable in the phototransistor.

Optocouplers which must meet highly demanding requirements may optionally be subject to a 100-percent burn-in. That is to say, they will only be dispatched when their electrical and optical parameters have stabilized.

Ageing of the light emitting diodes

As a quality characteristic, the ageing of the light emitting diodes used is measured by the elapsed time until the radiation, at a constant current, falls to half of its initial value. Figure 4.22 shows typical graphs based on the investigation of a large number of components. This shows that the half-life is over 200000 operating hours. The graphs show the relative decline in the current transfer ratio when the coupler diode is loaded with a current flow of I_F = 60 mA (measured at T_A = –25°C, I_F = 10 mA and U_{OF} = 5 V).

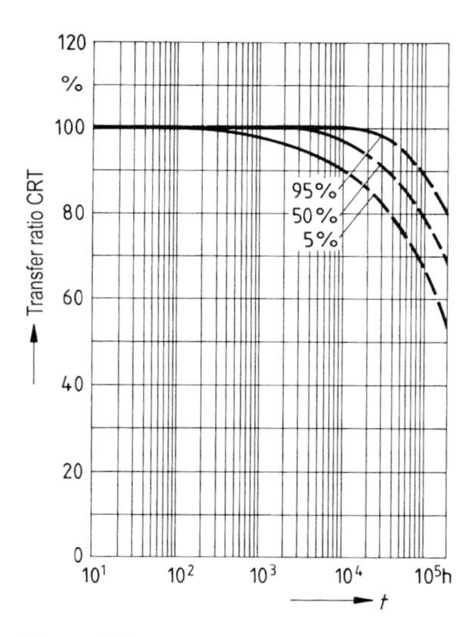

Figure 4.22
Decline in the current transfer ratio

The half-life of the current transfer ratio is also referred to as the service life of the component.

4.4 Optical waveguides

New application areas are continually opening up for optical transmission using optical waveguides made of glass or plastic fibers. Apart from the classical applications in long-range transmission engineering, it is being used in LANs (Local Area Networks) e.g. for data transmission from mainframe computers to the peripherals in onboard systems (aircraft, ships, vehicles), to an increasing extent in entertainment electronics, in metrology, in open and closed loop control engineering and for cable TV and community antenna installations.

The advantages of this transmission technology are:

- high bandwidth (high transmission capacity),
- electrical isolation between the transmitter and receiver,
- not sensitiv to electromagnetic interference,
- security against tapping,
- no stray signal radiation, so no crosstalk,
- no earth loops,
- no possibility of sparking,
- low weight and small space requirement,
- unlimited availability of the materials.

The media used for optical transmission are glass or plastic. For both types of transmission there are specific applications, semiconductors and coupling elements. The use of glass fiber is preferred for transmissions at high transmission rates, and over long distances or in local areas. Plastic fibers are suitable for low transmission rates in local areas, and for the solution of numerous applications in control engineering.

Because the plastic fibers have a larger fiber diameter (1 mm), the coupling elements for them are not as critical as for glass fibers, and so they are significantly cheaper. The last section of this chapter describes a few applications for light transmission over plastic fibers.

4.4.1 Optical fibers as a transmission medium

Glass fibers for use as dielectric waveguides are manufactured either from pure silica glass or from glasses doped with foreign atoms to alter the refractive index. The basic principle can be described as follows, by reference to step-index or single-mode fibers (see Figure 4.23 in this connection): if a beam is incident on the end of the fiber within a particular limiting angle γ (the numerical aperture, NA), it enters into the fiber and is propagated onwards within the system as a result of total reflection at the core/sheath boundary surface, provided that the angle α is less than the critical angle for total reflection.

A distinction is made between three types of glass fiber (see Figure 4.24):

- Multi-mode step-index fibers
  These large-core fibers guide a range of modes, which results in a pulse lengthening of 20-50 ns/km. The attenuation/loss is relatively high (a few dB/km). An advantage is the ease (and hence low cost) of signal injection (launch) as a result of the large core diameter. Their disadvantage is the bandwidth limits arising from the significant pulse lengthening.
- Multi-mode graded-index fibers
  The parabolic graph of the refractive index across the core causes the light to be guided by refraction at the individual core boundary surfaces (instead of total reflection at the core/mantel boundary), and thus significantly reduces the modal dispersion. The two standard types have core diameters of 50 and 62.5 µm. The attenuation is 2.5-3 dB/km at 850 nm and 0.6-0.8 dB/km at 1300 nm. Again, the ease of signal launch as a result of the large core diameter is an advantage. The fiber is somewhat more expensive than a single-mode fiber. The bandwidth for standard fibers is 500 MHz*km at a transmission wavelength of 1300 nm. Fibers which have been optimized for transmissions using lasers have bandwidths which permit transmission at 1 GBit/s over distances from 300 m to 1000 m.
- (Step-index) single-mode fibers
  In the core of these, which is typically

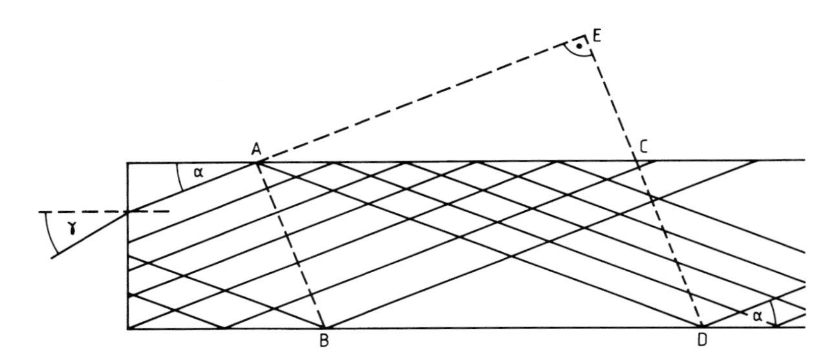

Figure 4.23 Light guidance within an optical fiber

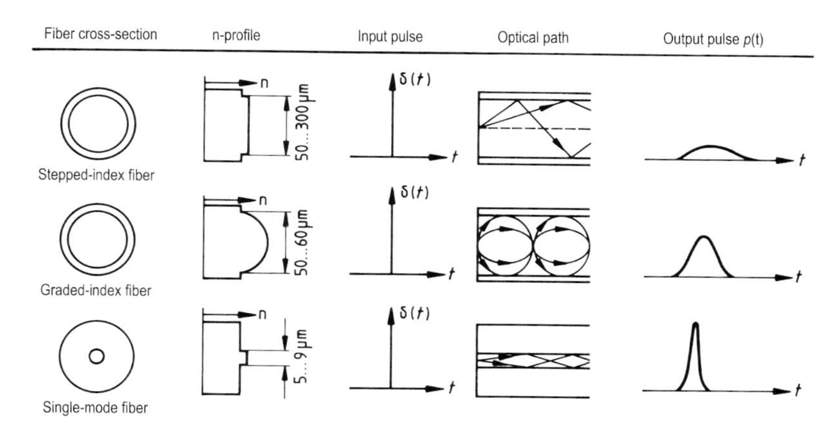

Figure 4.24 Common fiber types and their characteristics

only 9 µm thick, it is only possible to propagate one mode. The attenuation is 0.15 to 0.25 dB/km. The limiting frequency lies at over 10 GBit/s, which makes it possible to transmit over distances of over 100 km at a correspondingly high data rate. The fiber is very inexpensive, but does require appropriately precise transmission and receiving components and connections (plugs, cable splices).

The glass fibers have been internationally standardized in the standard IEC 60793.

Figure 4.24 shows the common fiber types and their characteristics.

4.4.2 Transmission and receiving modules for optical waveguide applications

Because of the spectral attenuation curve for today's glass fibers (Figure 4.25), transmitters and receivers are manufactured for wavelengths around 850 nm, based on Si, GaAs or GaAlAs, and for wavelengths around 1300 and 1550 nm based on Ge, and quaternary compounds such as InGaAsP/InP.

The optomechanical joint imposes highly demanding requirements in terms of the alignment between the semiconductor components and the optical waveguide (in the case of single-mode fibers a mere ±0.1 µm). For this reason, complete modules, which are often customer-specific, are offered with the transmitter and receiver integrated into a single module.

In the development of optical systems, there is an increasing tendency to use so-called transceivers (TRXs), which on the transmission side are equipped with an LED or a laser. The wavelengths which are most often used in this situation, 850 nm and 1300 nm, are subdivided into the following distance categories: 850 nm for local area networks (LANs) up to about 2 km; and 1300 nm for metropolitan area networks (MANs) up to about 15 km and for wide area networks (WANs) up to about 100 km.

On the receiver side normaly PIN diodes are located which are then connected to a receiver circuit. When the signal strength is very low, as happens when large distances are being covered, it may be necessary to use an APD receiver circuits.

Whereas LED transceivers are typically only used for data rates of up to 300 MBaud, the uses for laser transceivers go far beyond this. With current data rates of

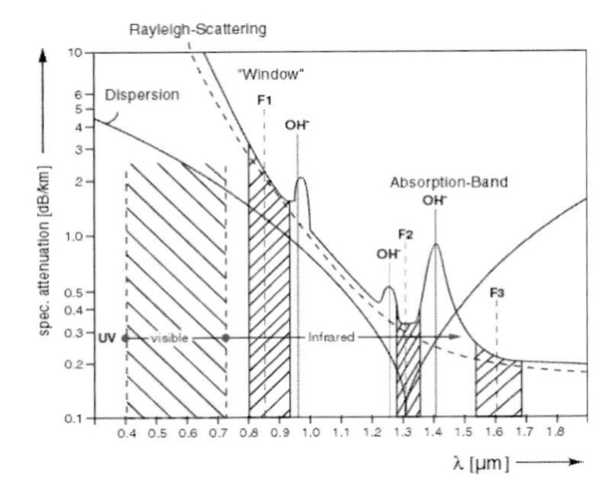

Figure 4.25
Attenuation of a silica glass fiber

up to 10 GBit/s, they are at currently the most frequently used components for opto-electrical conversion in transceivers. With the increasing use of laser modules, not only has the data rate of transceivers increased, but also the transceivers have become smaller, by a factor of 2 (see Figure 4.26).

Apart from the previously common supply voltage of 5 V, the modules in use today are now almost exclusively 3.3 V modules, and the trend is towards even lower supply voltages. As well as this, the TTL interface, which is absolutely standard with LED products, has been superseded by faster interface standards, such as PECL, CML or LVDS.

Figure 4.27 shows the data sheet for a typical 1.25 GBaud transceiver with a laser as the optical transmission element.

Another difference compared to LED transceivers is the requirement for laser safety, and also for increasing monitoring of the transceiver characteristics.

Because of its large radiation angle of almost 180 degrees, an LED is almost incapable of damaging the human eye, unlike a laser. The lasers used in the TRX have a radiation angle of only a few degrees. A consequence of this is that the entire radiated light from the laser can pass through the pupil into a human eye where, depending on its intensity, it can also produce damage effects.

Laser TRXs are categorized into laser classes, depending on the power and wavelength of the light they radiate. So that the light power does not exceed the limiting value for its class, each TRX is equipped with a monitoring circuit which prevents the limiting value being exceeded.

Apart from the safety of the laser, which the TRX must also maintain in the event of a fault, the customer also has other parameters for monitoring purposes available. These are: the input power of the light at the receiver, a signal when data is being received, the bias current of the laser and the output power of the transmitter (laser).

Figure 4.26
A 1 × 9 multistandard and an SFF module compared

Recommended Operating Conditions

Parameter	Symbol	Min.	Typ.	Max.	Units
Ambient Temperature	T_{AMB}	0		70	°C
Power Supply Voltage	V_{CC} - V_{EE}	3.1	3.3	3.5	V
Transmitter					
Data Input Differential Voltage	V_{DIFF}	250		1600	mV
Receiver					
Input Center Wavelength	λ_C	770		860	nm

Transmitter Electro-Optical Characteristics

Transmitter	Symbol	Min.	Typ.	Max.	Units
Launched Power (Average)	P_O	–9.5	–6	–4	dBm
Optical Modulation Amplitude[3]	OMA	156	450		µW
Center Wavelength	λ_C	830	850	860	nm
Spectral Width (RMS)	σ_I			0.85	
Relative Intensity Noise	RIN			-117	dB/Hz
Extinction Ratio (Dynamic)	ER	9	13		dB
Reset Threshold[2]	V_{TH}	2.5	2.75	2.99	V
Reset Time Out[2]	t_{RES}	140	240	560	ms
Reset Time, 20%-80%	t_R			260	ps
Supply Current			45	65	mA

Receiver Electro-Optical Characteristics

Receiver	Symbol	Min.	Typ.	Max.	Units
Sensivity (Average Power)[1]	P_{IN}		–20	–17	dBm
Saturation (Average Power)	P_{SAT}	0			
Min. Optical Modulation Amplitude[8]	OMA		19	31	µV
Stressed Receiver Sensivity 50 µm Fiber	SPIN 50 µm		24	55	µW[6]
			–17	–13.5	dB[7]
Stressed Receiver Sensivity 62.5 µm Fiber	SPIN 62.5 µm		32	67	µW[6]
			–16	–12.5	dB[7]
LOS of Signal Assert Level[2]	P_{LOSA}		–21	–18	dBm
LOS of Signal Deassert Level[3]	P_{LOSD}	–30	–22		
LOS of Signal Hysteresis	P_{LOSA} - P_{LOSD}	0.5	1		dB
LOS of Signal Assert Time	t_{ASS}			100	µs
LOS of Signal Deassert Time	t_{DAS}			350	
Receiver 3 dB cut off Frequency[8]			1.25	1.5	GHz
Receiver 10 dB cut off Frequency[8]			1.5	3	
Data Output Differential Voltage[4]	V_{DIFF}	0.5	0.7	1.23	V
Return Loss of Receiver	A_{RL}	12			dB
Supply current[5]			60	90	mA

Figure 4.27 Data sheet for a typical 1.25 GBaud transceiver

However, to increase the signal level on long-distance lines, increasing use is also being made of fiber amplifiers which eliminate the procedure previously required, of conversion into an electrical signal followed by amplification and then conversion back to an optical signal. With these, a fiber doped with erbium is brought into an excited state by "pumping" it with light. The wavelength of the pump must correspond to one of the absorption wavelengths of erbium (e.g. 1480 nm). When a signal photon arrives with a wavelength of around 1555 nm, several electrons leave the high-energy state. As they do so, a 1555 nm photon is emitted. The result is an amplification effect. For the 1300 nm window, the dopant used is praseodymium.

4.4.3 Transponders for optical waveguide applications

Apart from the transceivers mentioned under 4.4.2, which convert the serial data signal into light and vice versa, use is also made of so-called transponders and parallel-optical modules. These offer the system developer a further simplification, because in addition to the purely serial opto-electrical conversion they also include additional combinations of functions.

A transponder uses a multiplexer to combine data channels, which are connected in parallel to it, to form a serial signal, and then feed this into the optical waveguide as an optical signal.

Using a demultiplexer, the receiver section converts the serial signals received back into the original parallel data signals, and makes them available to the user for further processing.

By doing this, the electrical data signals which are input in parallel are read into the multiplexer at a clock rate which is generated in the module. After this the data is read out of the multiplexer as a serial signal, using a clock rate which is higher by a factor equal to the number of electrical input channels. Then, in exactly the same way as with a transceiver, this signal is then fed to the driver circuit for the LED or laser, where it is converted into light.

The serial optical signal undergoes an opto-electrical conversion. Afterwards it is fed to the demultiplexer which retrieves the clock rate from the incoming data.

This clock rate is used to convert the incoming data, in the reverse sequence to that of the multiplexer, back into parallel data signals, which again are then made available to the user.

Another possibility for transmitting parallel data signals is offered by parallel optical modules. These have the same number of optical output channels as they have electrical input channels, so that they are basically only transceivers connected in parallel, but about 6-12 times more compact than these. As these modules are used mainly for the distribution of bus signals, synchronous transmission is particularly important. To achieve this it is necessary that no differences arise between the propagation times of the signals on the glass fibers which run in parallel. With pulse widths of 400 ps (2.5 GBit/s) and line lengths of up to 100 m this implies a permissible difference in the lengths between all 12 fibers of 0.01%, which means a difference of only about 10 mm.

On the optical side of the modules mentioned above, use is made of plug connectors conforming to different standards, which are described in section 4.4.4.

4.4.4 Connections for glass fibers

For connecting glass fibers, two basic formats have become common: inseparable connection (splicing) and separable connection (plug, semi-permanent splices).

Splices

When glass fibers are spliced, the ends of two fibers are fused together by heating them in a voltaic arc. The high mechanical precision and control of the thermal process is achieved by using fully-automatic splicing devices, Figure 4.28, which permit splice losses of down to 0.05 dB to be realized with single-mode fibers. The splice point must be protected. The main applications are for connecting multi-fiber trunk cables to each other along their route, and to appropriate terminating connectors (plugs, transmitters, receivers, in each case with pigtails) in buildings.

Plug connectors

Plug connectors are used to connect fibers together in a rapid and repeatable manner, and so that the connection can be quickly separated at any time.

As technical development has proceeded, several systems have established themselves on the market, some of them having superseded others. The common plug connectors are internationally standardized by the standard IEC 60874.

The basic principle of the common plug connectors is as follows: the ends of the fibers which are to be connected are fed

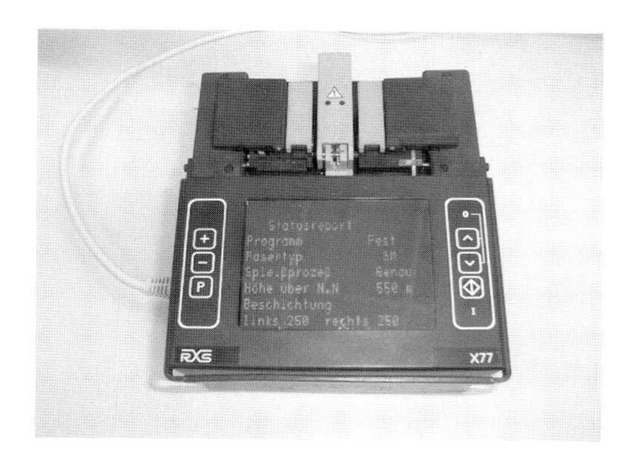

Figure 4.28
Fully-automatic splicing device

into a guide element. This ensures (possibly using additional guide elements) the exact mechanical alignment of the two fibers (axial, radial, angular). The ends of both fibers are highly polished, to ensure the best possible light interface and the lowest possible back reflections (return losses). So that there is no residual air gap between the two fibers, which would result in a Fresnel attenuation of around 0.3 dB and return losses of up to 14 dB, they are polished convex and pressed together by spring loading (so-called physical contact).

The widespread second generation of individual connectors is based on the principle that the fiber is fed into a precise ceramic ferrule with an external diameter of 2.5 mm. The fiber is glued into an axial hole in the ceramic ferrule. The end face is processed. The two ferrules are centered relative to each other by means of a high-precision ceramic sleeve. The outer parts of the connector then permit two fiber faces to be pressed together, and also the connector to be latched into appropriate adapters.

This generation includes the following connectors: ST (bayonet fastening), FC, DIN-LSA (both screw fastening), SC see Figure 4.29 (so-called push-pull interlock, with no thread), E2000 and as the biplugs ESCON and FDDI-PMD (all with locking hooks, with no thread).

The third generation is based on the same principle, but the diameter of the central components, the ferrule and the sleeve, has been reduced to 1.25 mm. This correspondingly makes smaller connector housings possible (so-called small-form-factor). This generation includes the LC and MU connectors (see Figure 4.30).

The insertion loss for these ceramic ferrule connectors ranges from under 0.5 dB down to 0.05 dB, depending on the precision of the fiber type. Return losses of less than 50 dB can be achieved.

The fourth generation is a group of multi-fiber connectors. The most widespread

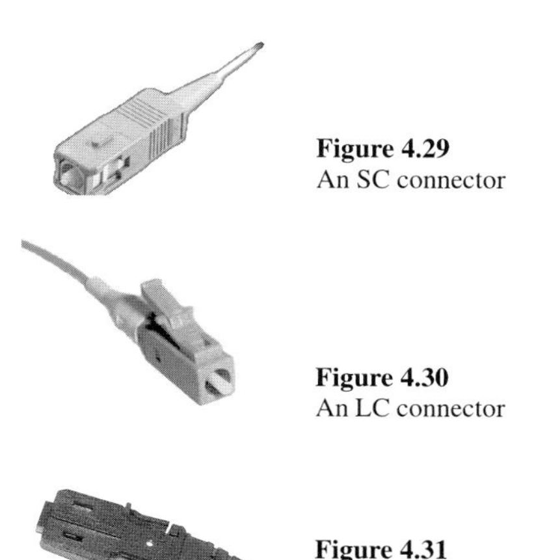

Figure 4.29
An SC connector

Figure 4.30
An LC connector

Figure 4.31
An SMC connector (12-fiber)

principle, of the so-called MT plug connector, is based on an arrangement of the fibers in a rectangular ferrule with a grid-spacing of 250 µm. The two ends of the plug connectors are adjusted relative to each other by means of guide pins or holes, located in the sides of the ferrules. Using this basic principle, numerous housing versions have evolved: MPO (push-pull, up to 12 fibers), SMC (see Figure 4.31 / grid spacing, up to 12 fibers), MT-RJ (so-called small-form-factor connector, up to 4 fibers locked together like the electrical RJ-45 connectors).

The insertion loss for these multi-fiber connectors is also down to less than 0.5 dB per fiber interface.

4.4.5 Coupling elements for plastic fibers

These have a guide bore which is matched to the cross-section of the optical waveguide fiber. The fiber must merely be inserted into the guide bore and fixed with a drop of adhesive.

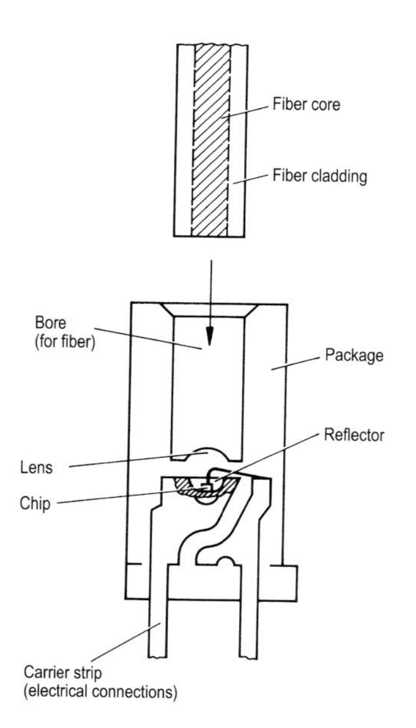

Figure 4.32 Transmission and receiving diodes for plastic optical waveguides

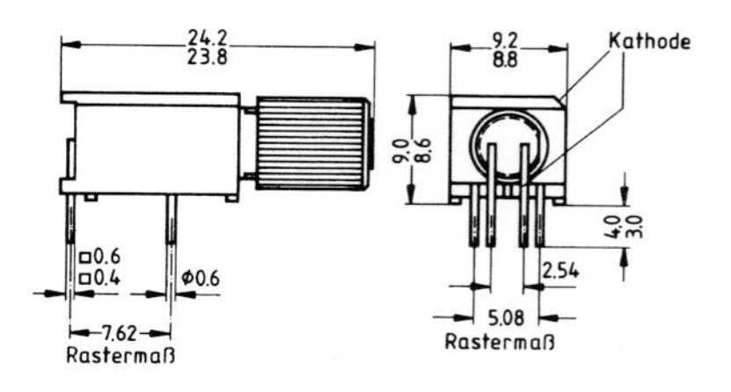

Figure 4.33
Optical waveguides with screw fastenings

The large diameter of the light-conducting core of a plastic fiber (1 mm) imposes significantly less critical requirements on the mechanical characteristics of the optical coupling elements. The transmission and receiving diodes in the SFH range are inexpensive and the screw connectors permit optical fiber to be coupled and then easily decoupled. To make the connection, the fiber must merely be pushed into the opening as far as the stop, and the nut then screwed up. In neither case is it necessary to remove the protective sheath on the plastic fiber.

Figure 4.32 shows the construction of the diodes in the SFH range, Figure 4.33 the construction of the components with a screw fastening.

4.4.6 Typical applications of plastic fibers

Because of their advantages, increasing use is being made of plastic optical waveguides/fiber (POF) in industrial open and closed loop control engineering.

A few examples of this are:

Where extremely high voltages are involved, it is no longer possible to effect electrical isolation by an optocoupler integrated into a housing. A plastic optical

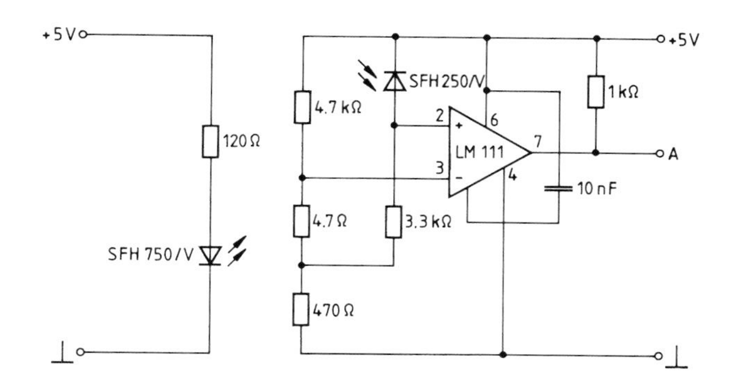

Figure 4.34
Optical 1 MBd transmission up to 20 m

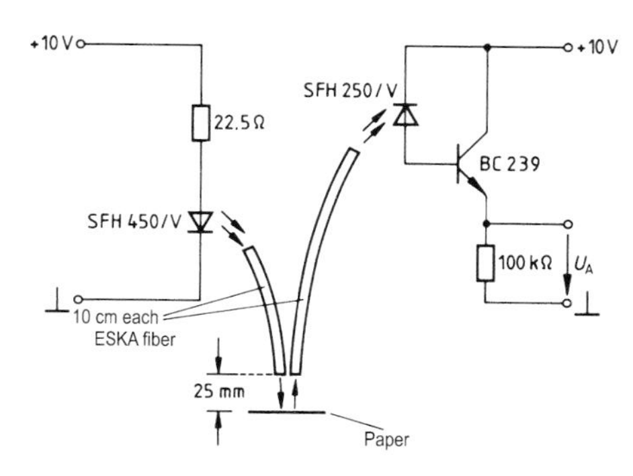

Figure 4.35
Optical reflected-light barrier

fiber solves this problem. Figure 4.34 shows the circuit for a transmission link. This permits a transmission rate of 1 MBaud over 20 m.

Apart from their use as optical display elements, plastic optical fibers can also be used as sensors. Its temperature, and bends in the optical fiber material, lead to changes in the form of the light transmitted. This can be used for measurement purposes. By separating out the fibers, it is possible to build up various fiber-optic light barriers (hybrid, reflected-light). This makes it possible, for example, to determine the direction of movement of objects at the measurement site while maintaining electrical potential isolation between the measurement and analysis locations.

The open ends of the fibers can be up to 5 mm apart.

In a printer, several light barriers are normally required. By using plastic fibers, it is possible to install the electronics in a single place. It is not necessary to run electrical connections to all the light barriers. This is operationally more secure, is more cost-effective and less sensitive to interference; the radiated interference characteristics are better, there is more flexibility in the mechanical design.

Figure 4.35 shows a reflected-light barrier as an example.

Typical attenuation levels for a commercially-available plastic fiber for different distances and transmission wavelengths are:

660 nm ±15 nm: at 1 m – 0.3 dB/m;
at 20 m – 0.25 dB/m;
at 50 m – 0.2 dB/m.

950 nm ±15 nm: at 1 m – 4 dB/m.

4.4.7 The use of optical transmission technologies using plastic fibers in vehicles

The first attempts to use optical communications with plastic fibers in vehicles were already being made nearly 10 years ago.

However, because of the low data rates of the bus systems for the various control systems in vehicles – such as CAN – the use of light for transmissions was not cost-effective.

As a result of the increasing introduction of multimedia applications and the introduction of new communication services, the volume of data to be processed has grown so strongly that transmission using light has become absolutely essential.

In the meantime too, it has become technically feasible to integrate a plastic fiber into the conventional cable harness. Initially, various topologies were studied (ring, double ring and star, active or passive). For multimedia (MOST), the decision came down in favor of a ring, with a safety system (*byteflight*) being realized as a star.

To establish the optical transmission, it needs a transmitter diode with a driver which supplies the necessary current. For use in vehicles, the conventional large-area LED at 650 nm has proved itself. It is robust, fast enough, and has the required reliability and thermal stability. Greater link lengths will be possible in future, using green-light transmission elements. As it always has been, a particular challenge is posed by the large temperature range which is demanded for applications in a vehicle.

In constructing the receiver, a Si receiving diode has been combined with a low-noise transimpedance amplifier (preamp). Downstream from this preamplifier another amplifier (postamp) is connected, which brings the signal to the desired logic level. For low data rates the preamp and main amplifier have already been successfully accommodated in one IC. Work is currently going on to make the same possible for higher data rates (up to around 500 MBit/s). The next step is the integration of the photodiode on the amplifier chip. This will result in a further reduction in the size of the components, and above all, a lower sensitivity to interference.

Construction of MOST transceivers

The individual components, such as the photo and transmission diodes, together with the ICs and capacitors, are mounted on a refined metal frame (leadframe). The electrical contacts between these components are established by gold bond-wires.

The housing for these optical components must have two important characteristics: apart from the mechanical requirements, it must also have best optical characteristics over the temperature range from −40°C to 85°C.

To achieve this, the packaging uses casting technology which has proven itself for millions of LEDs, and which enables components suitable for automotive uses to be produced. The component is cast in a small mold, with the optical window (cavity) initially being closed (see Figure 4.36).

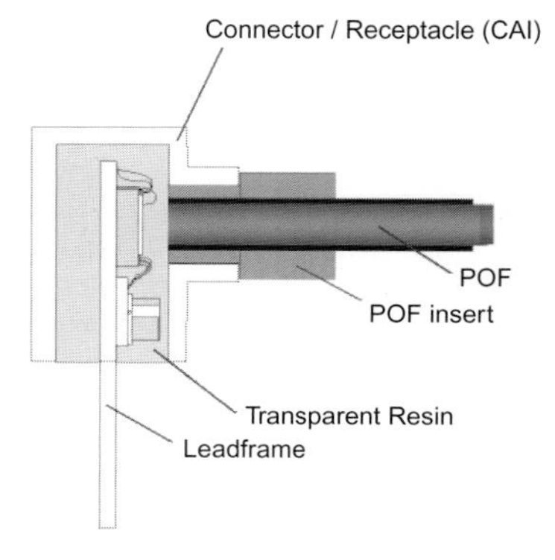

Figure 4.36 Cavity As Interface (**CAI**)

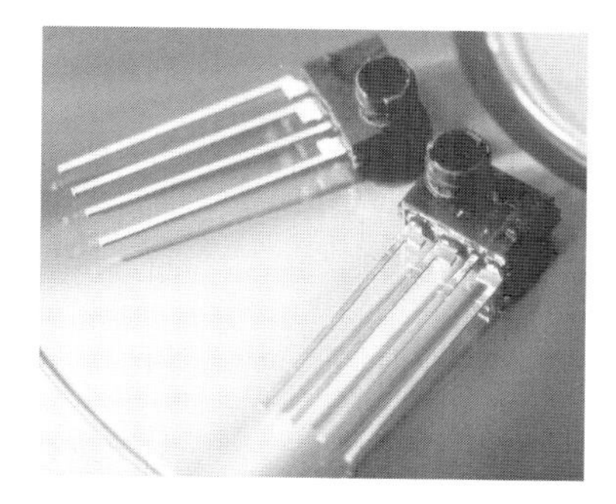

Figure 4.37
MOST transmitter (left), receiver (right)

After casting, this window is opened and forms the interface to the plug connector. For this reason, the construction is referred to as CAI (**c**avity **as** **i**nterface).

Figure 4.37 shows finished cast components, on the left a MOST transmitter and on the right a MOST receiver.

In Figures 4.38 and 4.39 the front of the CAIs has been opened after manufacture, to allow the leadframes to be seen. They show a MOST transmitter with its LED and driver IC, and the receiver with its photodiode, receiver amplifier IC and capacitor module.

These components will later be incorporated into a connector housing which ensures the optical and electrical shielding.

The Byteflight transceiver

For Byteflight too, the CAI technique has been used as the basis for the development of a component, which permits bidirectional operation without the high stray losses of a fiber coupler.

In principle, this takes advantage of the large diameter of the core of a plastic fiber, which is approximately 1 mm. A large-area photodiode at the end of the

Figure 4.38 A MOST transmitter and LED

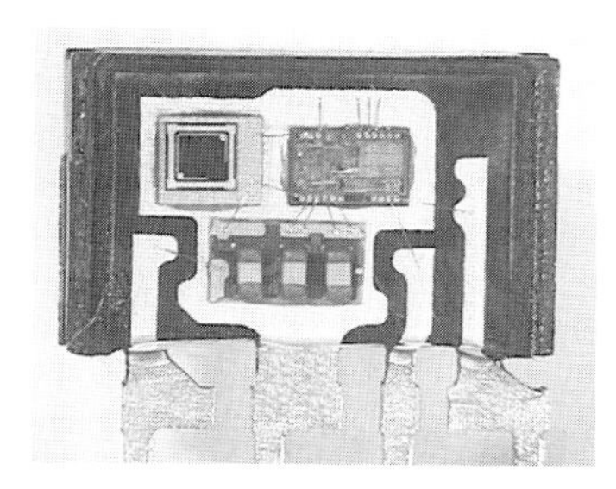

Figure 4.39 Receiver and photodiode

Figure 4.40 The complete Byteflight

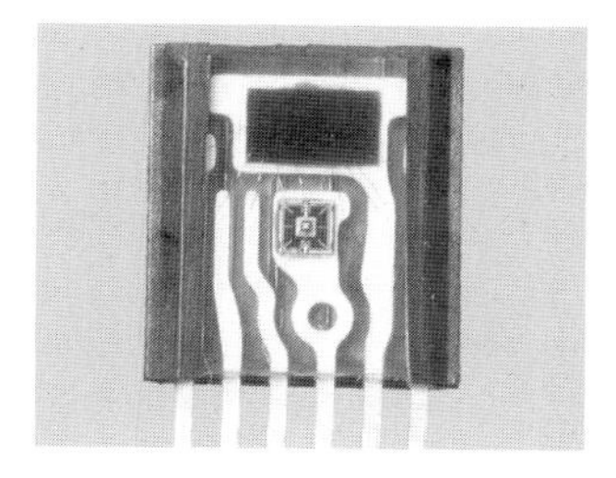

Figure 4.41
Interior, with the transmit/receive IC

link converts all the incoming light into current.

Transmission diodes are significantly smaller. This makes it possible to launch a closely-bundled light beam into the fiber without significant interfering with the receiving direction.

With the Byteflight transceiver, advanced chip-on-chip technology enables the relatively small LED to be located directly on the photodiode. This allows the transmitter to be connected up without loss. The area of the photodiode, which is obscured represents a loss of around 1 dB compared to a photodiode with all its light-sensitive area exposed.

Figure 4.40 shows the finished Byteflight transceiver in CAI form. Figure 4.41 shows a view of its interior, with the transmit/receive IC, which will have a light-tight cover to exclude interference from stray light.

When the transmitter is working, light also falls on the receiving diode beneath it. As the power of this light is significantly above the minimum receivable power, the component in this form cannot transmit and receive at the same time, which would be necessary for full-duplex operation. This was taken into consideration with Byteflight: a system designed for half-duplex operation.

Power budget

To achieve stable and reliable operation of the system, it is very important to plan the power budget. This takes into account the amount of power available for transmission and the minimum power which is required at the end of the link. The following factors must be taken into account by the planning:

- The transmission power is specified for an optimally connected fiber, e.g. after 30 cm and at 10 mA.
- In doing so, possible additional losses over the first meter of POF, compared to the normal fiber attenuation, are taken into consideration.
- The driver current has a certain tolerance, which affects the light power.
- Due to the temperature coefficient of the power, the temperature has an effect on the effective outgoing transmission power, and so the minimum and maximum values should be taken into consideration.
- The losses at the transmitter diode interface.
- The length of the fibers results in a degree of attenuation, which in turn depends on the wavelength of the light.
- A splitter plug connection or any repair connector which may be fitted will produce attenuation.
- Cable-laying losses and aging of the fiber.

Together with the minimum and maximum values for the transmission power, the minimum and maximum values for the cable attenuation give the minimum and maximum values of the received power.

The photocurrent at the input to the amplifier can only be determined after taking the losses at the interface to the receiving diode and the efficiency of the photodiode into consideration. In general, system reserve is set up, so that a variable element of up to 20 dB (factor of 100) may be required.

Without going into the details, the result is the following power data: typical values at room temperature, minimum values over the entire temperature range and service life (see Table 4.4).

IEEE1394

IEEE 1394 is a flexible, simple and low-cost digital interface, designed to link the world of consumer electronics with personal computers. In HAVi (Home Audio Video Interoperability), practically all the leading manufacturers of infotainment devices have come together to define a

Table 4.4 Typical performance data

	Value	Dimension
Data rate	Maximum 45	MBit/s
Transmission power	Typically –6 (min. –10)	dBm
Receive power requirement	Typically –26 (min. –24)	dBm
Tolerable attenuation	Typically 20 (min. 14)	dB

unified instruction set, so that devices from different manufacturers will work together.

The copper cable version of this can already work today with S400 (500 MBit/s).

The IEEE 1394 standard defines the medium, the topology and the protocol. The advantages are:

- hot plugging – connection and disconnection while in operation;
- scalability – speeds of 100, 200 and 400 MBit/s are currently possible, the optimal speed is automatically selected;
- flexibility – devices can be organized in any required way, a ring is not necessary;
- fast, guaranteed bandwidth – IEEE 1394 supports guaranteed assignment of data for critical data.

The relatively high data rates necessitate the use of optical transmission technology for long links.

Appropriate transceivers are now available to interface between 45 MBaud MOST and S100 (125 MBaud).

With the help of high-power RCLEDs and a new highly-integrated receiver, 100 m links can easily be established using standard POF.

Development work is in progress around the world on the realization of S200 (250 MBaud) and S400 transceivers. In the case of S400, conventional plastic fiber transmission technology is up against its limits, especially when the demand is for longer links.

Here the new fibers, which have already been presented at numerous conferences, must be developed to the stage of mass production.

4.5 IrDA – data transmission using infrared radiation

The use of infrared is nowadays an almost everyday matter, for example when we pick up the remote control for a TV, video recorder, HiFi system or other electronic devices. By now, indeed, it is almost the rule in many households that there is more than one IR remote control including, for example, those for automatic garage doors, or the central locking on newer vehicles.

A large number of these remote controls contain IR transmission components from Infineon. However, the use of infrared for data transmission will in future not be confined to the consumer field, but will also make increasing inroads in the office and in mobile communications: infrared data transmission systems for PCs, laptops, CD players, printers, mobile phones, and many other devices, are not only problem-free, safe and reliable, but also cheap. Another point in favor of infrared radiation is the fact that – unlike the use of RF as a transmission medium – its use is in no way restricted by post and telecommunications conditions. Thanks to a standard, which is the same throughout the world, the user does not need to concern himself with special adapter cables nor with interface problems.

For these new markets of the future, the Optoelectronics Division of Infineon has developed a range of infrared transceivers, which have been specially adapted to the requirements of data transfer (IrDT). These IrDT products transmit at data rates between 2.4 KBit/s and 4 MBit/s, and are compatible with the "physical layer" specification of the *Infrared Data Association* (IrDA).

Figure 4.42 The IrDA's logo

4.5.1 IrDA – one world standard for all devices

The IrDA organization was established in 1993, to develop standards for the universal exchange of data using infrared (Figure 4.42). Since then, more than 140 international companies have registered as official members. They include Infineon, which supports the organization. In the meantime, various specifications exist, for both the physical and the software layers, which permit unrestricted exchange of data between any devices. They are equipped with the appropriate interfaces. Today, IrDA is the largest and most effective consortium in the world for the development of IR systems and technologies. As a member of IrDA, Infineon has the ability to participate actively in the formulation of future strategy.

Laptops communicate with mobiles, PCs or printers

The possible applications for IR transceivers are very broad. One very useful application is without doubt the transfer of data between laptops and peripheral devices. It can already be anticipated that all new laptops will be equipped with IrDA interfaces. Laptop owners will then be able to transfer data from their laptop to a desktop or printer, without there being any wired connection between the devices. Many users of small PDAs (Personal Digital Assistants) already have this option nowadays, as long as the peripheral device is equipped with an IR interface. Other products which could benefit greatly from this technology are mobile phones and pagers. IR data transmission between mobiles and laptops will, for example, make it possible for users who are traveling about to check their own office e-mailbox, or to communicate with business partners by e-mail. A large manufacturer of digital cameras has announced the intention to equip its products with IrDA compatible interfaces.

IrDA has advantages even in toolmaking and engineering, and in the service field

IrDA systems have great advantages not only in the consumer field but also in industrial electronics. Increasingly, high-performance PDAs for industrial applications are now coming onto the market. Just like the screwdriver or pliers, they are becoming an ever-more indispensable standard aid in the toolbox of a service technician. IrDA systems are also being used for other demanding applications, such as diagnostic devices for vehicle engines, and data input devices for dentists, nurses, teachers, warehouse staff etc.

In the case of applications with high security requirements, such as the "electronic purse", the ownership of money changes invisibly by means of infrared. Computer games, already great favoritestoday, will in future manage with no connecting wires to the controlling devices. Basically, any device which today uses wires for the transmission of control data is a potential "candidate" for infrared data transfer. The IrDA module, which can transmit and receive data, is a new key component, ready for such advanced IR data transmission systems.

Remote controls for radios and TVs which use infrared light contain the familiar IR emitter as the transmission diode, this generally being modulated by an IC. In this case, the data is only transmitted in one direction.

Figure 4.43
The modules IRM 6000 and IRM 3105

4.5.2 Full IrDA standard

With the appropriate software, the signals will conform to the full standard defined by the IrDA (Infrared Data Association). With dimensions of only 13 mm × 6 mm × 6 mm, this module can be accommodated in almost any housing, which is not always the case for alternative solutions using discrete components.

For applications where the space for installation is extremely small, the IRM 6000 module (Figure 4.43, left) is available, with dimensions of only 9.1 mm x 4.1 mm x 4.3 mm, so that it is ideally suitable, for PDAs (Personal Digital Assistants), pagers or mobile phones. Its technical data is broadly comparable with the modules IRM 3001/3105 (Figure 4.43, right), and for the first time, two-way communication is possible with a single component. This permits half-duplex transmit and receive operation, and is intended for transmissions over short distances (up to 1 m). When used, even over very short distances, no overmodulation occurs. And because some of the modulation electronics is incorporated, the user can save the external components, which would otherwise be required.

All the types can be obtained with single-in-line connections or also in DIL form, which is more suitable for SMT assembly. If required, the type series IRM 300X is also available with lateral guide pins for automatic insertion.

Under development are the modules IRM 3401 and 3405, for even higher data transmission rates of up to 4 MBit/s. The modules are also fully IrDA-compatible.

For customers who are looking for a complete solution, we offer the IRM 7000 encoder/decoder-IC, which greatly simplifies connection to a UART interface.

The Optoelectronic Division of the company is orientated toward the satisfaction of special customer-specific wishes, in cases where an optimal solution cannot be found using the available standard components.

More on the Internet

Data sheets and further application notes for these products will also be found on the Internet, under: http://www.infineon.com; notes about the IrDA organization under http:// www.irda.org.

5 Sensors

5.1 Overview

Sensors convert physical quantities such as pressure, temperature, magnetic field strength and others, into electrical signals which can be processed. Depending on the requirement, the complexity of the output signal ranges from a resistance change to a calibrated output voltage in digital form.

Sensors in the KTY series have resistances which change in a highly reproducible way with the temperature. Pressure sensors from the KP200 series consist of a simple resistance bridge which is supplied with an input voltage and provides an output voltage proportional to the pressure.

The incorporation of the sensor element into an integrated circuit, on the other hand, opens up possibilities for signal processing directly in the sensor module. It is thus possible to calibrate sensor ICs to a specified output characteristic curve, or to switch them into different operating states for diagnostic purposes. Sensor ICs can analyze complex input signals and switch between digital output states as a result.

The technical integration of sensors into silicon technology has already been implemented for a few sensor types. Among the first integrated sensors are magnetic field sensors. Here, the Hall effect in silicon is utilized. With the development of surface micromechanics, pressure sensors have also been integrated into modern CMOS processes, and can process measured values digitally.

Compared with sensor solutions constructed of discrete components, intelligent sensors have become established as an economically competitive approach above all in automotive electronics. In addition, they offer wide-ranging possibilities for optimizing the overall system in terms of communication, precision and fault monitoring.

There are semiconductor sensors for many physical quantities and applications. Mention should be made here of acceleration and rotation rate sensors, for which there are also versions using monolithic integrated circuits. The silicon microphone also belongs to the family of semiconductor sensors.

A special form of technical sensor is represented by the so-called "fingerprint sensors", about which more details will be found in Chapter 8.

5.2 Magnetic field sensors

5.2.1 Discrete Hall effect sensors

Fundamentals

The Hall effect, named in 1879 Edwin Hall, is the result of the Lorentz force on moving electrons which are exposed to a transverse magnetic field. Figure 5.1a shows a representation of the current flow in a material with no magnetic field. It can be seen that points of equal electrical field strength are located along a straight line perpendicular to the direction of current flow. Figure 5.1b shows the current flow in a material which is exposed to a magnetic field at right angles to the Hall

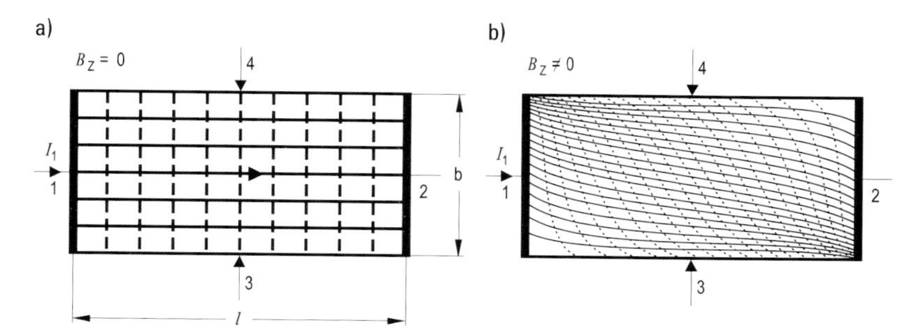

Figure 5.1
Equipotential lines for a current flow
a) with no transverse magnetic field b) exposed to transverse magnetic field

plate. The Lorentz force on the moving electrons is given by:

$$\vec{F} = -q \cdot (\vec{v} \times \vec{B})$$

where:

$-q$ electronic charge
$\vec{v}$ velocity of the electrons
$\vec{B}$ magnetic induction

The Lorentz force is perpendicular both to the direction of current flow and the magnetic field. Its consequence is an electrical field across the conductor, which corresponds to a voltage, the so-called Hall voltage.

When a magnetic field is applied the angle between the equipotential lines (dashed lines in Figure 5.1) and the current (full lines in Figure 5.1) changes. This angle is a function of the mobility μ of the electrons, and thus to the drift velocity of the electrons. For most conducting materials, μ is so small that the Hall effect has no practical significance. However, there are a few exceptions in the case of semiconductors such as silicon and germanium, but especially in the case of substances with III-V compounds, such as gallium arsenide and indium antimonide.

The structure of Hall sensors

Figure 5.2 shows a schematic representation of the principle. A strip-shaped plate of suitable material, with a depth d, length l and width b conducts a current I_1 along its lengthwise direction, and is exposed to a magnetic field B_z, which is perpendicular to the plate. An electric voltage develops between the points 3 and 4, this being referred to as the no-load Hall voltage V_{20} and having a magnitude which is given by:

$$V_{20} = \frac{R_H}{d} \cdot I_1 \cdot B_z \cdot G$$

where R_H is the Hall constant of the material.

The effects of the current contacts and the Hall voltage taps is taken into account by the geometry factor G. The ideal case of point-shaped contacts would give $G = 1$. However, as real contacts have a finite size, in practice G is between zero and one.

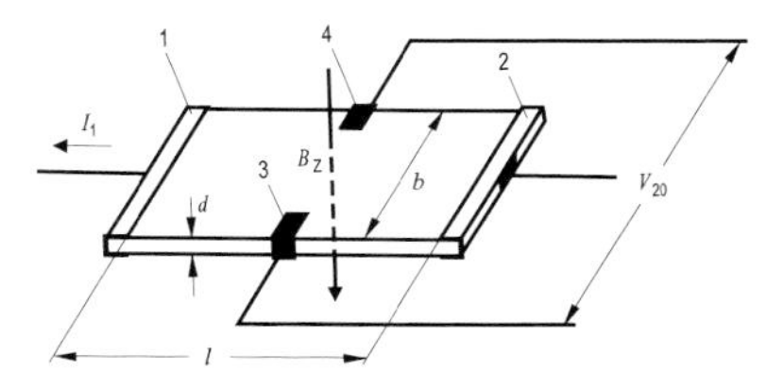

Figure 5.2
Illustrating the principle of a Hall sensor

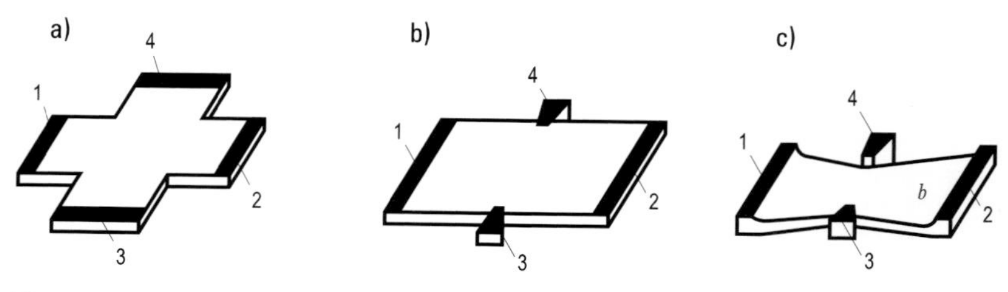

Figure 5.3 Most common forms of Hall sensors
a) Cross-shaped, symmetrical Hall sensor with high magnetic field sensitivity
b) Rectangular Hall sensor, which exhibits a high Hall voltage
c) Butterfly-shaped Hall sensor which, due to the current pinching, exhibits a high sensitivity to magnetic fields

The current is conducted in and out via contacts 1 and 2, and the Hall voltage is measured between points 3 and 4. The effective area of the Hall sensor is set by the limits of the electrode tips. By changing the dimensions and geometry of the Hall sensor, this can be optimized for various functions. Figures 5.3a to 5.3c show the three most common forms.

Concepts and general product data

Nominal current I_{1N}

The nominal current is set at a level such that when operating in still air the semiconductor Hall sensor reaches equilibrium at an overtemperature of 10°C to 15°C. The effects of this temperature rise on the Hall coefficient, and hence on the no-load Hall voltage, is shown in Figure 5.4 and is specified in the relevant data sheets as the TC_{V20} value.

The Hall coefficient R_H is a material constant which, as shown in Figure 5.4 is temperature-dependent. On the other hand, within certain limits it is independent of the magnetic field strength. In the case of gallium arsenide this limit lies at 1 Tesla.

No-load Hall voltage V_{20} (open-circuit voltage)

The data sheets contain values for the no-load Hall voltage, that is the output voltage V_{20} which is generated between the electrodes of an unloaded Hall element if, as described, the nominal current of I_{1N} is applied in the presence of a transverse magnetic field B_z.

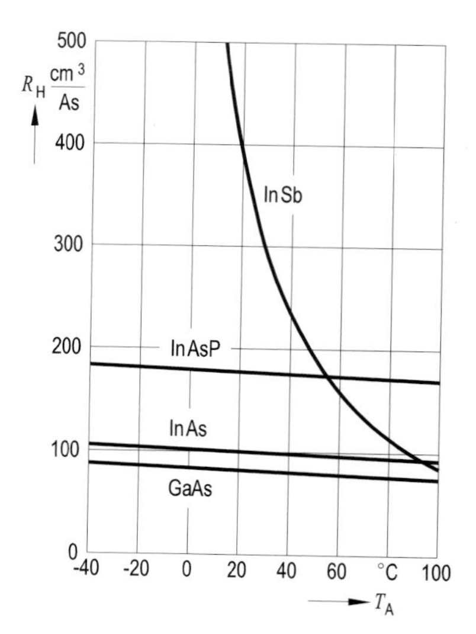

Figure 5.4
Temperature-dependence of the Hall coefficient

Open-circuit induction sensitivity, K_{BO}

The no-load sensitivity to magnetic induction is defined by the following equation:

$$K_{BO} = \frac{V_{20}}{(I_{1N} \cdot B_z)} \quad [V/AT]$$

Ohmic d.c. component

As a result of the production tolerances for the chip geometry, and due to inhomogeneity of the material of the Hall plate, a voltage develops which is ohmic in nature and for $B_z = 0$ results in a d.c. component, which overlays the Hall voltage. Its magnitude is determined by the relationship $V_{2RO} = I_1 \cdot R_{10}$. The maximum value of R_{10} is given in the data sheet, and can be compensated by a high-resistance potentiometer, as in Figure 5.5.

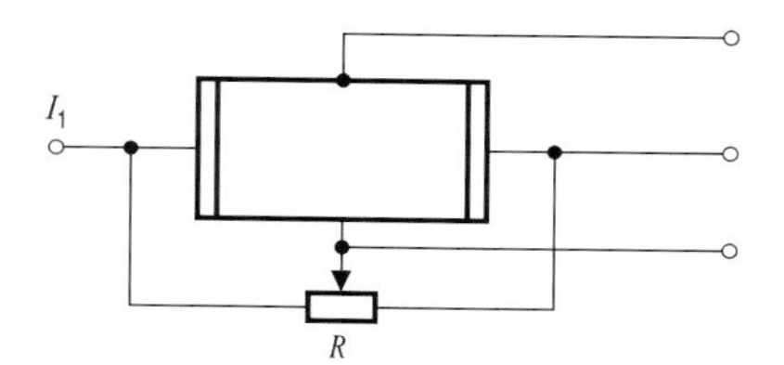

Figure 5.5
Compensation circuit for the ohmic d.c. component

Inductive d.c. component A_2

Inevitably, the connecting wires to the Hall electrodes form a loop which covers an area A_2 which, even with the most careful conductor routing, cannot be reduced to zero. The consequence is that, even when the through current $I_1 = 0$, a fluctuating magnetic flux density induces in this loop a voltage, which can be measured between the Hall electrodes and is determined by the following expression:

$$V_{10} = A_2 \cdot \frac{dB_z}{dt}$$

This is referred to as the inductive d.c. component, and is expressed as the area of the loop in cm^2. It depends on the temporary magnetic flux, its amplitude and frequency.

Temperature dependency

The temperature dependency of the Hall sensor is caused by two basic effects: the temperature dependency of the Hall coefficient and consequently the no-load Hall voltage with its temperature coefficient TC_{V20}, and the temperature dependency of the specific resistance. that is to say the internal resistance of the Hall device, with its temperature coefficient TC_{R10}. The mean values for these coefficients are specified in the data sheets.

In the open-circuit state, only TC_{V20} has any effect, but under load both TC_{R10} and also TC_{V20} will apply.

Permissible maximum value of the control current I_{1M}

The maximum permissible control current for a Hall sensor is strongly dependent on the housing used and the mode of operation, that is to say on the cooling method used and the ambient temperature. The data sheets specify the highest value in still air. If this value is exceeded without adequate cooling measures overheating can result, and consequent damage to the sensors.

The thermal conductance G_{thc} between sensor material and housing

To enable the maximum permissible control current to be calculated for specific cooling methods, the data sheets contain details of the thermal conductance between the sensor and the housing surface. The values quoted relate to heat which is dissipated on both sides of the housing.

Practical applications

Signal sensors

Hall elements are used as signal sensors in a host of applications. In contrast to inductive sensors, their output signal is independent of the operating frequency, so that it is possible to use them at very low frequencies, right down to zero. The main field of application is in brushless d.c. motors, in which the Hall sensor is driven directly by the magnetic field of the stator. In other applications, Hall sensors are driven by bar magnets or magnetic strips containing data, or by magnetic fields produced by currents through nearby conductors.

Position detection

This can comprise the simple recognition of a position which has been reached (limit switch) or the fact that an object has occupied a particular position (i.e. detection of a presence, or counters), and it can also be the continuous reporting of movement or offset, for example to measure a force, pressure, torque or acceleration.

In general, the measurement of this parameter does not require linear sensor characteristics, provided that the non-linear characteristics of the transducer (e.g. a diaphragm for pressure measurement) can be compensated for or linearized electronically.

Frontal mode

Figure 5.6 shows a KSY-14 sensor, which is driven in frontal mode by a small SmCo magnet with a diameter of 4 mm and height of 2 mm. That is to say, the sensor is positioned in front of the pole of the magnet. The graph shows the Hall voltage as a function of the size of the air gap.

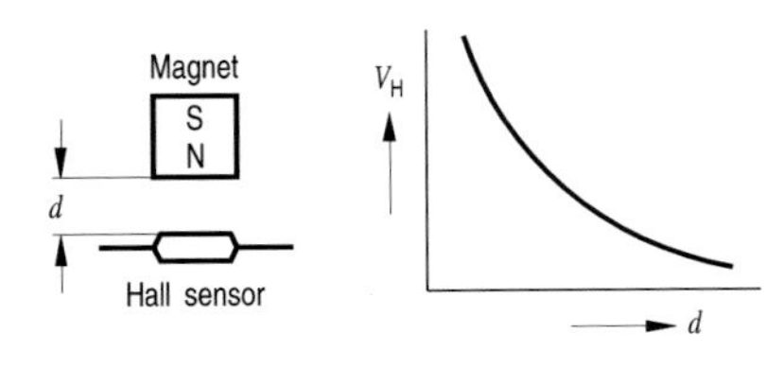

Figure 5.6
Frontal mode for a KSY 14 Hall sensor

5.2.2 Integrated Hall sensor ASICs

The Hall sensors described in 5.2.1, in the form of discrete components, involve relatively high costs to the user in the subsequent signal processing. For this reason, industry has demanded mainly customized sensor ICs. These ASICs (Application Specific Integrated Circuits) offer the following advantages for the module manufacturer:

- Simple offset elimination: using the method of the *Spinning Current Hall Probe*, the ohmic d.c. component (offset) can be separated from the useful signal which is proportional to the magnetic field, and for practical purposes can be adequately eliminated. This method, shown in Figure 5.7, functions for symmetric Hall probes, for which the input and output side electrodes can be interchanged. To do this, in the first clock phase the current is injected into the Hall probe through contacts 1 and 2, and the output voltage $V_{20} + V_{2R0}$ is sampled at contacts 3 and 4. In a second clock phase, the current is injected into the Hall probe through contacts 4 and 3, and the output voltage $V_{20} - V_{2R0}$ is sampled at contacts 1 and 2. It can be seen that the sign of the offset V_{2R0} changes, while the sign of the signal V_{20} which is proportional to the magnetic field does not change. Hence, by forming a simple mean the disruptive offset is eliminated.

  Although this principle has been known for a long time, it has only been possible to make use of it commercially since the mid-90s, as a result of advances in the field of analog signal processing, using standard CMOS technologies. Using this technology, it is possible to keep the equivalent magnet-

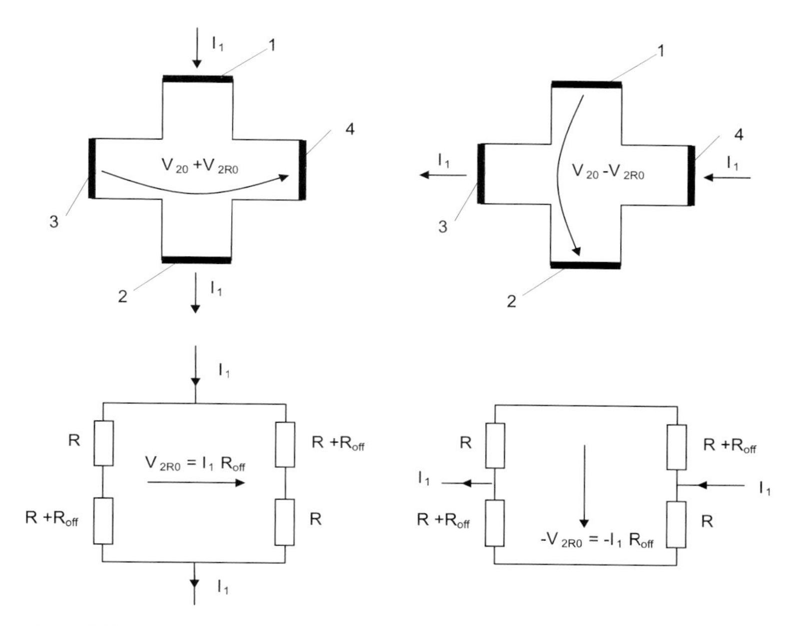

Figure 5.7
Eliminating the ohmic d.c. component by the Spinning Current Hall Probe method

ic offset of the entire system, comprising the Hall probe and signal processing, below 50 μT.

- Offset-free signal amplification: the output signals from a Hall probe are of the order of magnitude of millivolts, which means that they must be amplified by a factor of at least 1000. Conventional d.c. amplifiers with such high amplification factors also suffer from offset. However, in Hall probe ASICs which use the Spinning Current Hall Probe method, the Hall probe is already being operated in switched mode, so one possibility which this offers is discrete time signal processing using offset-free chopper amplifiers. As the implementation of such concepts demands an expertise in analog circuit engineering, it is advantageous for the user if the semiconductor manufacturer offers this complex signal processing together with the integrated Hall probe, as a monolithic solution.
- High temperature tolerance: integrated Hall sensors are often used under harsh operating conditions. It is standard for the operating temperatures to range from −40°C up to +150°C, but for special applications they can extend up to +210°C. In order to minimize thermal leakage currents and parasitic conductor loops, through which noise impulses can be introduced into the signal path, all the parts of the electronic circuit which are susceptible to interference – particularly the Hall probe and signal processing parts – should be as small as possible. Integrated construction enables the inductive d.c. component to be reduced to a negligibly small value. The antenna effect of the chip wiring is also significantly less than that of the connecting pins on discrete components, which are long by comparison. For special EMC requirements, such as micro-breaks, it is also possible to integrate discrete filter ca-

pacitors into the housing. This is shown in Figure 5.8 for the ABS sensor TLE4942C in the P-SSO-2-2 housing. Furthermore, conducted interference pulses are greatly attenuated on-chip by protective diodes and stabilized supply voltages. As a result of integral reverse current protection diodes, short-circuit proofing and built-in protection against overtemperature, Hall sensor ASICs are generally exceptionally simple to handle, and will not be destroyed even if incorrectly installed (e.g. interchange of the supply pins).

- Calibrated digital output signal: during assembly of a magnetic field sensor into a module, assembly tolerances arise. In addition, the sources used for the magnetic field are generally low-cost permanent magnets, some of which have remanence which is subject to substantial variance. These effects make it necessary to calibrate the sensor ASIC after it has been assembled into the module. To permit this, the sensor IC is equipped with a memory for the calibration data and with an interface for the exchange of data during the calibration procedure; in this context, we speak of a *smart sensor*. The memory used is generally in the form of cavity fuses, such as those described in section 5.3.2.
- Whereas in the past the signal from a discrete Hall probe would undergo analog amplification and would be evaluated and calibrated by a microprocessor, modern Hall sensor ASICs provide a ready-calibrated, digital output signal which can also contain information about any faults (such as overtemperature of the sensor IC). This relieves the load on the microprocessor, and distributes the intelligence in the entire system, which results in an increased data throughput combined with greater reliability.

As a representative of modern integrated Hall sensors, the TLE4990 linear Hall sensor ASIC will be described. This module converts the component of the magnetic field which is perpendicular to the surface of the chip into an electric output voltage V_{out}.

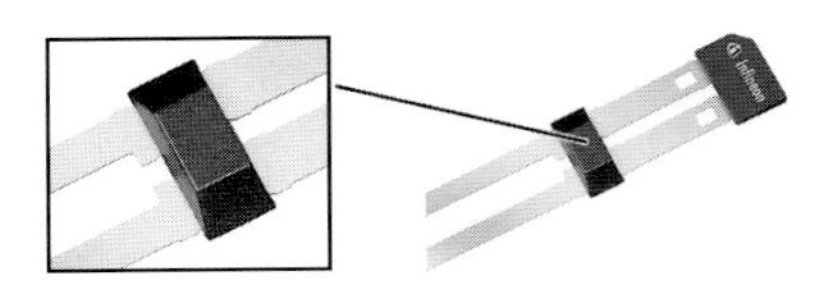

Figure 5.8
The P-SSO-2-2 housing for the TLE4942C ABS sensor integrates the filter capacitor (detail inset) on the leadframe of the IC

$$V_{out} = S \cdot B_z + V_{zero}$$

To achieve this, the Spinning Current Hall Probe method explained above is used to eliminate the offset of the Hall probe (see Figure 5.9). The signal is then amplified by a factor of about 5000. The module has 30 bits for configuring it: 13 bits are used to set the magnetic sensitivity S between 15 mV/mT and 180 mV/mT. 11 further bits allow the output voltage at 0 mT (= V_{zero}) to be set. In addition, the temperature coefficient of the magnetic sensitivity can be set so that, for example, it compensates for the thermal response of the permanent magnets used in the module. The special feature of this module is its outstanding resolution, ratiometry, linearity and the low drift of the output voltage versus temperature, and lifetime. Here, the term ratiometry means that the analog output signal is directly proportional to the operating voltage. This is particularly advantageous if the output voltage V_{out} is being digitized by an ADC (analog-digital converter), with the operating voltage of the Hall sensor ASIC as the reference voltage for the ADC. If the operating voltage changes by 10% from the nominal 5 V, the TLE4990 supplies an output voltage which also changes by 10%. However, the digital output code from the ADC remains unchanged, because it is calculated as the ratio of the sensor output voltage to the reference voltage.

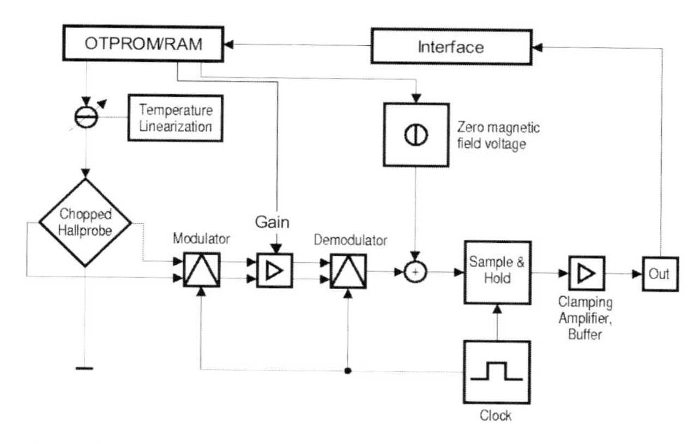

Figure 5.9 Block circuit diagram of the TLE4990 linear Hall sensor IC

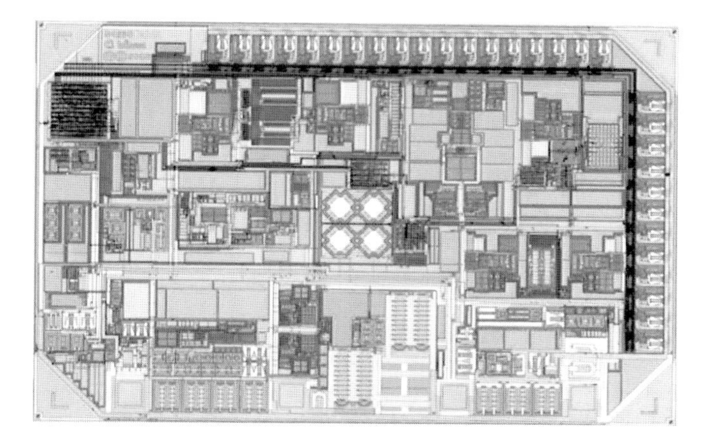

Figure 5.10 Photo of the chip for the TLE4990 linear Hall sensor IC

Figure 5.10 shows a photograph of the chip for the TLE4990. In the middle of the chip can be seen 4 Hall probes; the 30 cavity fuse cells are arranged along the right-hand and upper edges. On the lower edge are 4 pads for contacting the 4 connecting pins: TST, OUT, GND, VDD (from left to right), where TST is only used for test purposes. The TLE4990 is supplied in the P-SSO-4-1 housing. The P-SSO series of housings was specifically developed by Infineon for magnetic field sensors, so as to allow assembly in magnetic circuits with air gaps of only 1.1 mm.

5.2.3 GMRs

Magnetosensors are outstandingly suitable for all types of contactless detection of position data, for gaps, speeds, speed and direction of rotation, and for the non-contact measurement of electrical currents and powers. In performing this function, they guarantee operability even under harsh environmental conditions due to dirt, abrasion and high temperatures. These characteristics have resulted in the widespread use of magnetic field sensors, particularly in automotive and industrial applications, and the diversity of sensor types on the market is increasing steadily.

GMR sensors overcome a weakness shown by conventional magnetoresistors and Hall sensors in many applications, with their high sensitivity to fluctuations in the airgap. Because all the conventional magnetosensors react to the strength of the magnetic field, even the smallest variations in the gap between the magnet and sensor will often result in substantial changes in the signal, which can only be avoided by substantial assembly costs or complex signal processing.

By contrast GMR sensors, which Infineon has developed specially for position sensing applications, measure only the direction of an external magnetic field, largely independently of its intensity, which permits both very large measurement gaps and adjustment tolerances. For the user, assembly is thus substantially simpler and cheaper. When suitably driven, air gaps of up to 25 mm are possible, which opens up completely new applications to the advantages of magnetosensing.

The following sections describe first the fundamentals of the GMR effect and the construction and operation of a GMR sensor. This is followed by application examples, which show the diversity of functions which can be simply and effectively handled by GMR sensors.

Fundamentals

At the end of the 80s, a change of resistance in magnetic fields of over 50% was discovered in stacks of very thin layers of iron and chromium at low temperatures. Because stacked layers of iron and other magnetic metals react particularly sensitively to magnetic fields (Figure 5.11), in specialist circles they were given the name Giant Magneto Resistor (GMR).

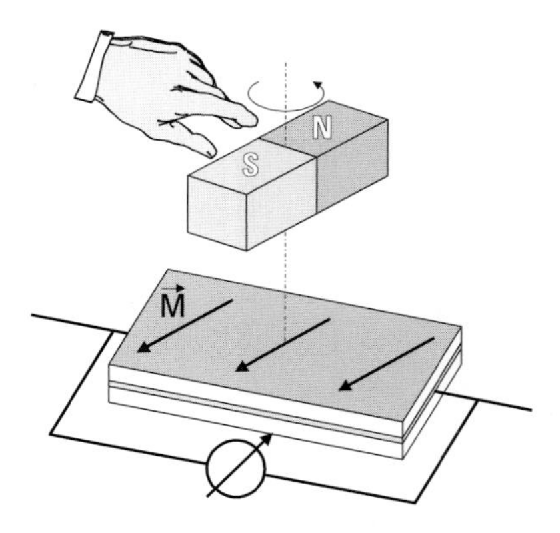

Figure 5.11
Orientation of the outer layers in an external magnetic field

The thicknesses of the individual layers are of the order of magnitude of a few nanometers (millionths of a millimeter). Non-magnetic copper separates the layers of iron and the permanently-magnetic ferromagnetic cobalt (see Figure 5.12). This separation is so thin that the cobalt layers couple to form an artificial antiferromagnet (AAF). The outer layers are of magnetically soft iron, and align themselves to an external magnetic field, whereas the magnetically hard cobalt layers retain their permanent magnetization.

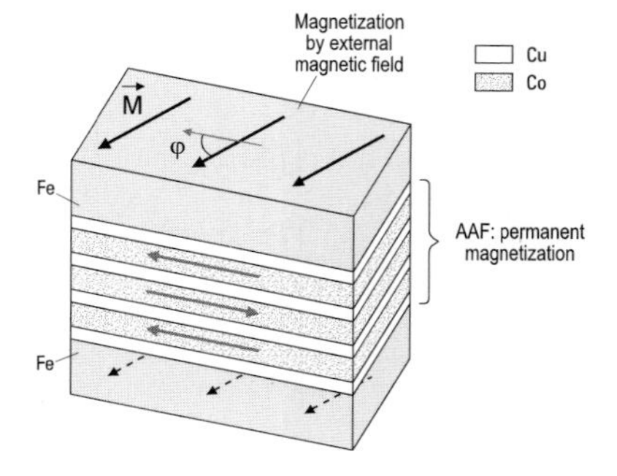

Figure 5.12
Layer sequence: outer layers of iron, artificial antiferromagnet (AAF) of copper and cobalt layers

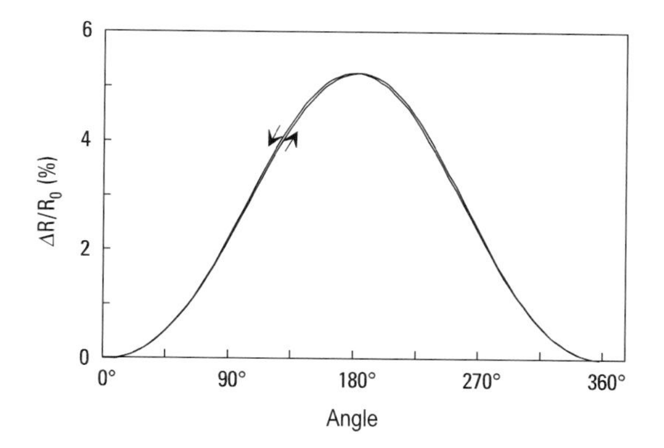

Figure 5.13
Relative resistance change as a function of the angle between the magnetically hard and soft layers

The spins of the electrons in the outer layers also align themselves with the magnetic field. The mean free path of electrons with spins parallel to the magnetization in the artificial antiferromagnet exceeds the thickness of the series of layers, so that only low scattering losses occur. The electrons with antiparallel spin, on the other hand, contribute to a raised resistance, because they are scattered more often within the layer structure. If the magnetic orientation of the magnetically soft and hard layers is the same, the electrons suffer from less scattering and the resistance is at its minimum; if the orientations are exactly opposite the resistance is at its maximum (Figure 5.13). The GMR effect is independent of the direction of the current; the angle between the orientations of the magnetization in the magnetically hard and soft layers is the only factor which determines the total resistance of the system.

Within a wide magnetic window, in which the soft magnetic layers rotate with the external field but the hard magnetic layers remain unaffected by it, the resistance depends only on the direction of the external magnetic field (saturation mode).

With increasing temperature, the GMR effectiveness $\Delta R/R_0$ drops due to the thermal excitation of lattice vibrations and spin waves, on the one hand because the basic resistance R_0 of the sensor increases and on the other hand because the general spin alignment decreases. Experimentally, the temperature coefficient is found to be constant.

Construction

The ultra thin layers, which enable the antiferromagnetic coupling to develop, call for the latest sputtering technology. The system used by Infineon, consisting of eleven layers, has a total thickness of only 25 nm. Magnetic cobalt layers and nonmagnetic copper layers as spacers form an artificial antiferromagnet. Soft magnetic layers of iron cover the top and bottom of the antiferromagnet.

In order to guarantee a basic resistance of the order of magnitude of over 700 Ω, meandrous current paths are etched out on the planar system of layers (see Figure 5.14). The change in resistance $\Delta R/R_0$ due to the GMR effect is more than 4%.

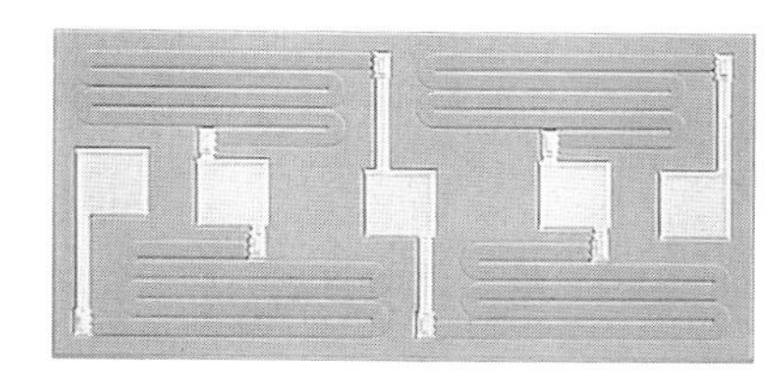

Figure 5.14 The GMR B6 full-bridge chip

Table 5.1 Variants of GMR sensors

Type		Meanders	Alignment of magnetization		Package
S4, S6	Individual sensor	1	0°		SOH, SMT(MW-6)
B6	1 full-bridge / 2 anti-parallel half-bridges	2 + 2	0° 180°	180° 0°	SMT(MW-6)
C6	2 crossed half-bridges	2 + 2	0° 180°	90° 270°	SMT(MW-6)

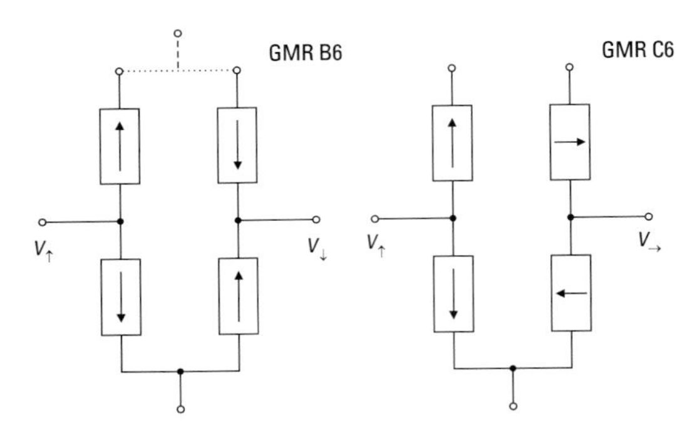

Figure 5.15
Arrangement of the sensor elements on the bridge chips. GMR B6: 2 antiparallel half-bridges or 1 full-bridge (- - -). GMR C6: 2 crossed half-bridges

A strong external magnetic field applied during the manufacturing process orients the artificial antiferromagnet, and establishes the hard magnetization of the individual current paths. GMR resistors can be used as individual sensors and integrated bridges (Table 5.1).

Half-bridges consist of two resistors with antiparallel magnetization, connected in series. The sensor GMR B6 (see Figure 5.15a) contains two half-bridges connected in parallel and magnetized in opposite directions, and these can also be connected as a full-bridge. The crossed half-bridges of the sensor GMR C6 (Figure 5.15b) are hard-magnetized at 90° to each other. The arrows indicate the direction of the internal magnetization.

Operation

As a measure of the applied magnetic field, the resistance of the GMR sensor changes with the angle between the fixed internal magnetization and the soft magnetic layer (see Figure 5.13), which tracks the magnetic field with a hysteresis of less than 2°. This change is detected as a change in voltage via an individual sensor or a bridge circuit. The form of the signal corresponds to a cosine function with an extended linear region.

Bridge circuits

Whereas the resistance of individual sensors in a magnetic field only varies by a few percent of the total value, a voltage can be tapped off from a bridge circuit which corresponds only to the change in resistance, with no offset. The bridge tapping values at Sens 1 and Sens 2 (see Figure 5.16b/c) are evaluated immediately or after amplification.

The bridge sensors GMR B6 and GMR C6 each consist of four individual sensors, integrated on the chip, which form two half-bridges with different pre-magnetization. A half-bridge consists of two individual sensors with an antiparallel

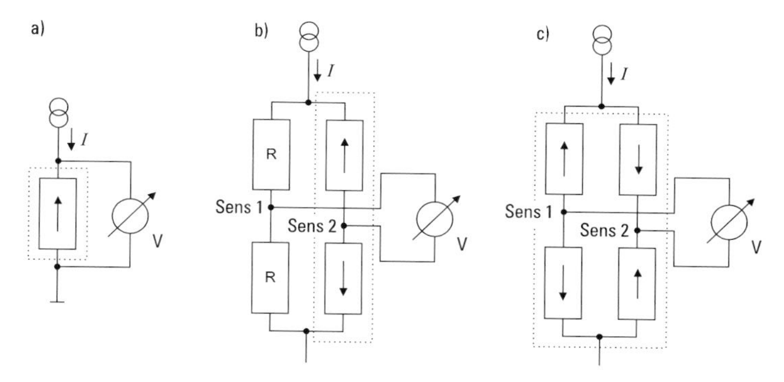

Figure 5.16
Measurement of the bridge voltage a) Measurement of the voltage drop across an individual sensor. Measurement of the bridge voltage b) across a half-bridge with a voltage divider and c) across a full-bridge

orientation (see Figure 5.16). The signal from a half-bridge is produced as the difference from two fixed resistances, and varies symmetrically about the zero point. The different preset orientations of the fixed magnetization is indicated by the phase relationship of the signal:

The two signals from the half-bridges (GMR C6) at 90° to one another (crossed) permit the direction of a magnetic field to be unambiguously determined round a complete circle (see Figure 5.17 above).

The full-bridge, made up of two antiparallel half-bridges (GMR B6), requires no external comparison resistances. The two half-bridges work in opposite senses, so that their difference produces twice the signal strength (see Figure 5.17, below).

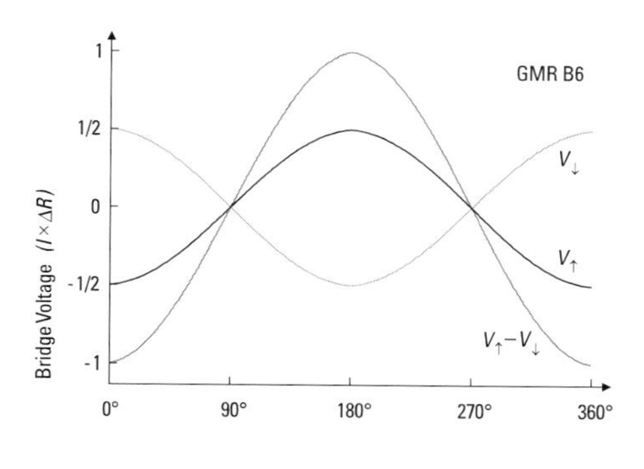

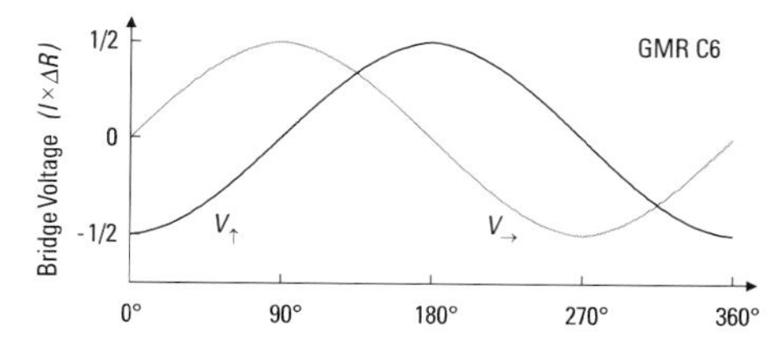

Figure 5.17
Bridge voltage from half-bridges ($V_\uparrow$, $V_\downarrow$, $V_\rightarrow$) and a full-bridge ($V_\uparrow - V_\downarrow$) as the external magnetic field is rotated

Spatial arrangement

The full signal level variation sets in as soon as the external magnetic field reaches a level sufficient to keep the soft magnetic measurement layer of the GMR sensor rotating with it (Figure 5.18). The working region ends at the point where the external magnetic field is so strong that it affects the hard magnetic layer, and irreversibly reduces the sensitivity.

Within the "magnetic window" from 5 - 15 kA/m the variation in signal level is independent of the field strength: the GMR sensor measures only the direction of the applied field.

The distance between the indicating magnet and sensor is totally irrelevant, as long as the field strength lies within this window. For example, Figure 5.19 shows the axial and lateral distances from a permanent magnet of samarium-cobalt at which 100%, 75% and 50% of the maximum variation in signal level is achieved. The tolerable air gap which this implies is unusually large for a magnetic sensor, and enormous spatial tolerance on the construction.

Magnetic primary elements

As the sensor or magnet moves, the magnetic field which the GMR sensor detects will change its direction. Depending on the nature of the movement, the size of the air gap and the required resolution, the magnetic source used may be a per-

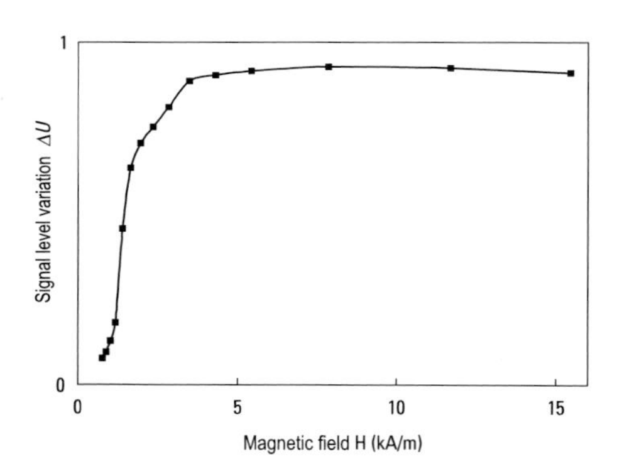

Figure 5.18
Maximum variation in signal level ΔU as a function of the magnetic field strength

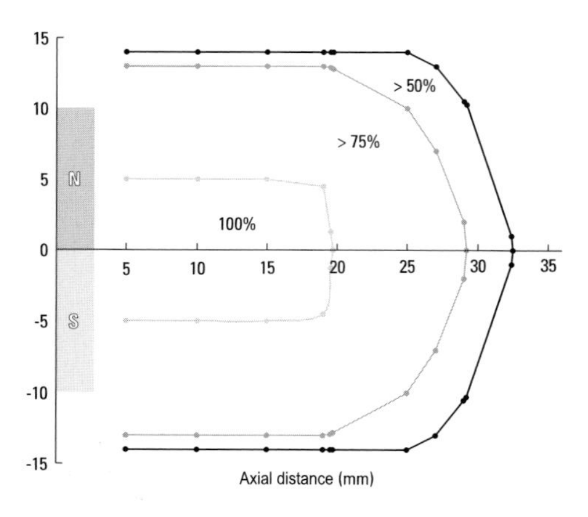

Figure 5.19
Relative strength of the GMR effect in the neighborhood of a $20 \times 10 \times 5$ mm^3 permanent magnet of samarium-cobalt

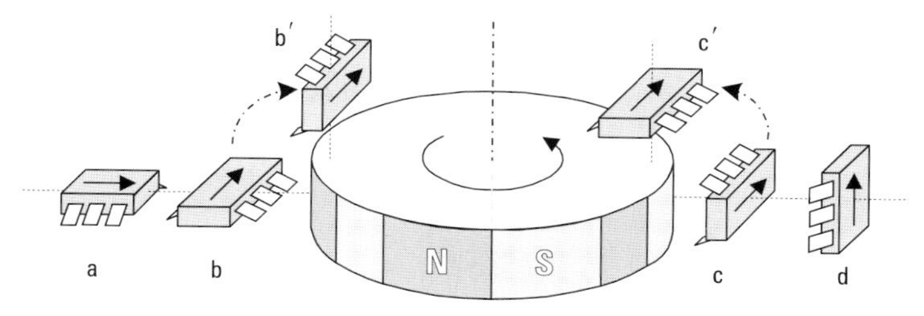

Figure 5.20
Variants on the orientation and magnetization (arrow) of a sensor, in the plane of the magnetic rotor (a-d), and to one side and within the circumference of the magnetic rotor (b′, c′)

manent magnet in the form of a simple bar magnet (dipole, see Figure 5.11) or a magnetic rotor.

Magnetic rotors consist typically of a sequence of magnetic north and south poles around the rotor circumference. One complete rotation of a magnetic rotor generates a complete signal cycle for each pair of poles. The conversion of this permits a higher resolution.

The arrangement of the sensor relative to the magnetic rotor determines the form of the output signal (see Figure 5.20, Table 5.2). Numerous signal forms are possible, because the waveform of the magnetic field in space varies and the sensor only registers that component of the field which rotates in the plane of the chip.

Depending on the requirements, the form of the signal can be varied from a sinewave through triangular to a series of peaks. The rectangular form (see Table 5.2 c) arises when the field rotates perpendicularly to the sensor (instead of in the sensor) and only the component parallel to the magnetization is detected. Rotation of the sensor about the perimeter of the magnetic rotor and out of the plane of the magnetic rotor (b → b′, c → c) gives this signal form.

Magnetic rotors with other variants on the magnetization can lead to a particularly effective arrangement in individual cases.

Applications

Infineon's Giant Magneto Resistor is suitable for a host of different position sens-

Table 5.2
Signal forms for different arrangements of a sensor (see Figure 5.20) magnetized in the direction of the arrow

	Normal to the surface	Magnetization	Approximate form of signal
a	Parallel to axis of rotation	Radial	Sinewave
b		Tangential	Triangular
c	Radial	Tangential	Rectangular
d		Parallel to axis of rotation	None
b′	Perpendicular to axis of rotation	Tangential	As for b
c′	Parallel to axis of rotation	Tangential	As for c

ing applications for linear and rotary movements. Within a wide window of magnetic field strengths, it measures solely the direction of the applied field, while permitting very large gaps and enormous adjustment tolerances. Depending on the application, use is made of permanent magnets in the form of bar magnets or magnetic rotors (see section on *Magnetic primary elements*).

Please note: strong permanent magnets must be kept a minimum distance away from GMR sensors, to avoid damaging the hard magnetic layer. The magnetic field inside the GMR chip must not exceed 15 kA/m!

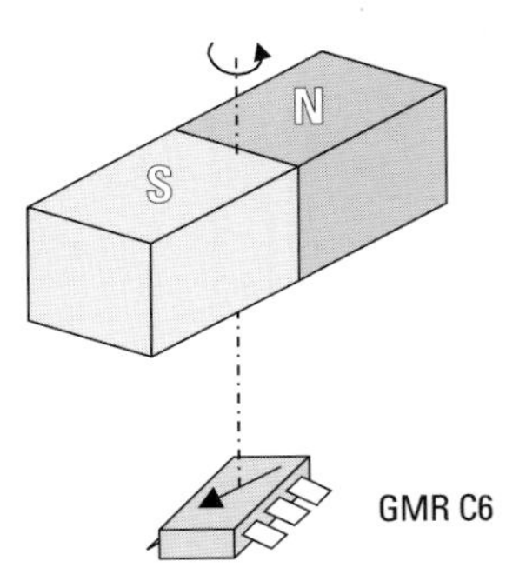

Figure 5.21
Rotation of the magnetic primary element over the GMR sensor

Absolute angle sensor

The resistance of the GMR sensor changes as a function of the external magnetic field, and thus provides an absolute angle sensor. The angular range detected and the resolution depend on the magnetic primary element (number of pairs of poles), on the type of the GMR sensor (individual sensor, full-bridge, crossed half-bridge) and on the processing. The crossed half-bridges of the GMR C6 sensor detect the rotation of a simple bar magnet unambiguously through 360° (see Figure 5.21).

The two half-bridge signals (see Figure 5.17) are offset by 90° and can be assigned by a simple comparison of the 4 quadrants (see Figure 5.22). Within a quadrant, the signal with the greater slope ($V_{\uparrow}$ in A and C, V in B and D) defines the angle by linear interpolation, or better from tabulated values (Table 5.3).

With a supply voltage of $V_{IN} = 5$ V and a GMR effect $\Delta R/R_0 > 4\%$, the fluctuation in the signal level – without amplification – amounts to at least 200 mV. The point of intersection of the two measurement curves determines the threshold value

$$V_{thr} = \frac{200 \text{ mV}}{\sqrt{2}}.$$

An angular accuracy of 2° corresponds to 180 points around a full circle or 45

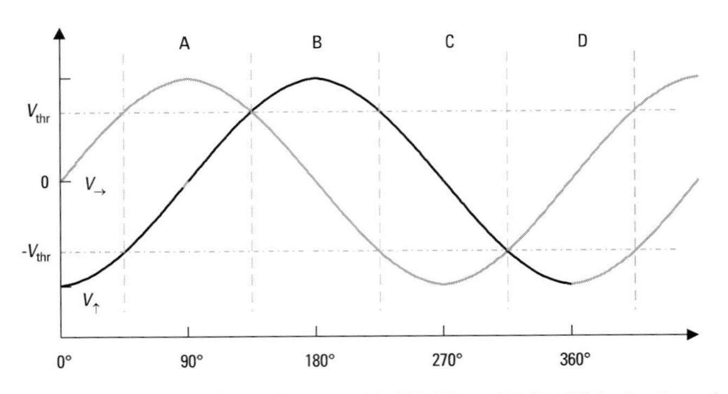

Figure 5.22 Signal from the crossed half-bridges (GMR C6) in the 4 quadrants

Table 5.3 Signal processing for the GMR C6 crossed half-bridges

Quadrant	Condition	Angle (linear approximation)
A	$V_{\rightarrow} > V_{thr}$	$90° + V_{\uparrow} \cdot (45°/V_{thr})$
B	$V_{\uparrow} > V_{thr}$	$180° - V_{\rightarrow} \cdot (45°/V_{thr})$
C	$V_{\rightarrow} < V_{thr}$	$270° - V_{\uparrow} \cdot (45°/V_{thr})$
D	$V_{\uparrow} < V_{thr}$	$360° + V_{\rightarrow} \cdot (45°/V_{thr})$

points per quadrant, and calls for an electrical signal resolution of

$$\frac{200 \text{ mV}}{45\sqrt{2}} = 3 \text{ mV}$$

This resolution of 1.5% is limited by the temperature coefficient of the GMR effect, which is about 0.1%/K. For temperature fluctuations greater then 7°C a compensation circuit is necessary to achieve this target resolution.

Variants

The use of a magnetic rotor with N pairs of poles raises the resolution by a factor of N and restricts the angular range to $360°/N$.

The GMR B6 full-bridge can only unambiguously resolve signals over a range of 180°, but does output twice the signal strength.

Typical uses

Gas pedal, steering wheel, seat rake, flip-phone, potentiometer.

5.3 Pressure sensors

5.3.1 Surface micromechanics, pressure sensors with a digital output (KP100)

With the growing demand for intelligent electronics, so too there is a growing demand for ever more complex sensors, not only in industry but also in household and automotive electronics. In these fields, silicon sensors have come to dominate in pressure measurement applications, thanks to their small dimensions and the outstanding properties of silicon as a material compared to conventional sensors. Modern sensors offer significant potential for improvements, with their possibilities for the integration of sensor elements and processing electronics (temperature compensation, A/D conversion etc ...) on one chip.

A critical step was taken in this direction with the mastery of surface micromechanics. In this technology, the complete sensor elements are created on the surface of the Si wafers, which enables processing to be performed on a standard BiCMOS line. This makes it possible to implement not only the sensor cells but also the complete signal processing and digitization on a single Si chip. In applications this leads to a clear reduction in the components required, which is reflected positively in the circuit's complexity and security against failure.

With the KP100 pressure sensor (Figure 5.23), the first product using this technology went into series production in 1998. This pressure sensor is used in satellite systems for side airbags, in the doors of automobiles. Because the output signal is digital, a microprocessor can communicate directly with the sensor. In the event of an accident, the sensor detects the pressure wave in the side door, and thus starts the triggering routine. Using an appropriate analysis algorithm, the microprocessor ensures that incorrect triggering of the airbag, for example by vigorous closing of the door or banging against it, is avoid-

Figure 5.23
View looking down onto the KP100 pressure sensor in a P-DSOF-8 SMD package

ed. The decision to fire the airbag can be made significantly faster using a pressure signal than with an acceleration signal. This is a critical advantage in view of the limited crumple zones for side impacts.

The KP100 sensor is based on the capacitance principle, i.e. a pressure change in the surrounding medium causes a change in the capacitance on the chip. To make this conversion of physical magnitudes possible, a pressure-sensitive membrane is produced over a hermetically sealed cavity. This membrane forms the upper electrode of a capacitor, with the associated opposite electrode being located in the substrate.

The manufacture of this membrane (Figure 5.24), with a size of 70 µm × 70 µm and thickness of about 400 nm, is the fundamental innovation of this sensor because, unlike conventional silicon pressure sensors, it takes place entirely within a normal 6 inch BiCMOS line. To achieve this, a field oxide is produced over the lower electrode of the capacitor (substrate), which in the next processing step is covered by a doped polysilicon layer. This polysilicon forms the opposite electrode of the capacitor, and is initially "perforated" by a dry etching step. Through these holes, the underlying field oxide (sacrificial layer) is etched out by a liquid chemical process using hydrofluoric acid. This produces a freestanding polysilicon membrane. An oxide boss in the center of the membrane increase the rigidity of the membrane. This seals the cavity underneath it hermetically, at a defined pressure (Figure 5.25). The sensitiv-

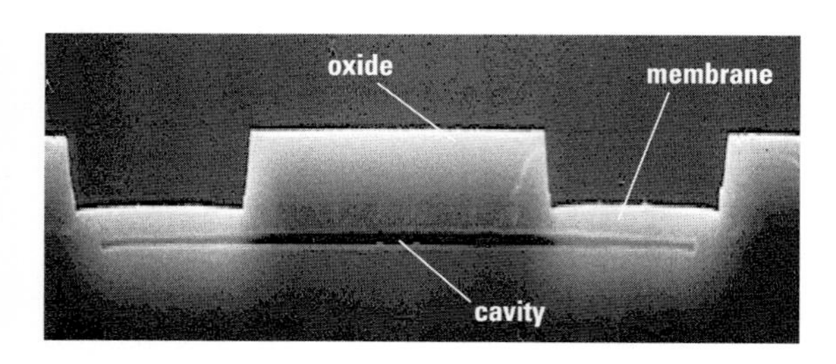

Figure 5.24
Scanning electron microscope image of the cross-section through a sensor cell

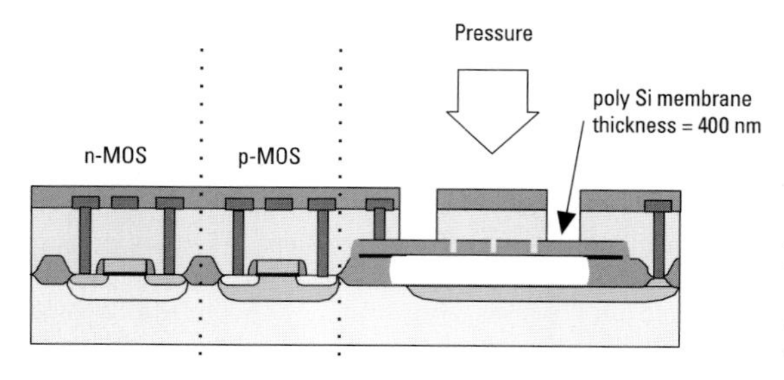

Figure 5.25
Schematic structure of a sensor cell, seen in cross-section

ity of the sensor is largely determined by the geometric parameters (area and thickness) of the membrane. As the processing steps are very well controlled, the sensors can be manufactured with very high reproducibility.

When a pressure pulse occurs, the membrane is displaced, which leads to a change in the capacitance. To increase the magnitude of the signal, an array of 4 panels is created, each with 14 membranes, which are connected in parallel; two of the panels are pressure sensitive and two serve as a reference. The signal is formed as the difference between the pressure sensitive panels and the reference panels.

This signal undergoes an analog-digital conversion which is completely integrated on the chip. This guarantees both a very good signal-to-noise ratio and also high accuracy. The module is externally clocked, at an optional speed of 4 MHz or 8 MHz. The analog sensor capacitance is first converted into a digital bitstream by a sigma-delta modulator. A decimation filter then supplies the bitstream in the form of a series of 16-bit words. High frequency noise is eliminated from the pressure signal which is to be evaluated by means of a low-pass filter with an upper limiting frequency of 360 Hz. The data is then written into a serial shift register (SPI = serial peripheral interface) at a rate of 7.8 kHz, from which it can finally be read by the microprocessor at a rate of up to 500 kHz. A further shift register can be used by the microprocessor to determine the functional mode (normal or diagnostic mode).

The sensor is designed for a pressure range from 60 kPa to 130 kPa, which is resolved with a precision of 12 bits.

It works over a temperature range of −40°C up to +90°C. At a nominal operating voltage of 5 V, the supply current is a maximum of 2.5 mA.

In order to match up to the high safety requirements of automotive electronics, three different diagnostic modes are implemented on the sensor. This makes it possible to carry out different variations of a self-test for the sensor, and to check continuously on its operability. The diagnostic modes allow checks to be made on the membrane arrays, the complete signal path or the digital section alone.

Checking the complete signal section: instead of reading the sensor capacitance, a fixed capacitance which is integral on the chip is read into the signal path. As this capacitance is pressure-independent, provided that the electronics are working correctly the sensor must supply a defined digital value at its output.

Checking the digital section: defined codes are generated and read into the digital decimation filter. The 16-bit words which are then available at the output must be identical with the prescribed words in the specification.

Checking the membrane arrays: after complete assembly of the sensor into the application, an offset is established between the pressure-sensitive sensor arrays and the reference arrays, this being characteristic for the particular sensor. This offset can be read out and saved. Any change in the offset indicates possible mechanical damage to the sensor membrane.

Using these diagnostic modes, the sensor can be monitored for correct functioning while it is in operation, for example in the case of a side airbag module when the ignition key is turned. In addition, the data transmission can be checked by evaluation of a parity bit.

So that the sensor can be assembled as cheaply as possible, a special SMD package has been developed. This P-DSOF8-1 plastic housing has 8 connecting pins and is open on the upper side. After the chip has been bonded in and the contacts made using gold wire, the chip is covered with a silicon gel. The ambient pressure is transmitted to the sensor surface through this gel. In addition it protects the chip from

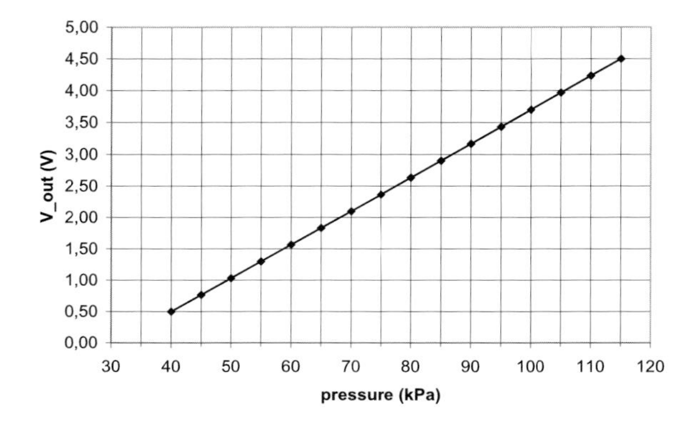

Figure 5.26 Characteristic curve of the KP120

environmental influences. The package is outstandingly suitable for automated insertion into circuit boards, and thus represents a critical (cost) advantage over conventional housings.

5.3.2 Pressure sensor with analog output (KP120)

It continues to be the case that most applications use pressure sensors with analog outputs, i.e. at its output the sensor supplies a voltage proportional to the pressure. The KP120 is such a sensor, and is used for example in the engine management system of automobiles. In this case, on the one hand the ambient pressure (BAP, Barometric Air Pressure) and on the other hand the air intake pressure (MAP, Manifold Air Pressure) are detected. A precise knowledge of the pressure ratios is required when driving, in order to calculate the optimal amount of fuel.

These applications call for a precisely calibrate output signal, which is proportional to the prevailing pressure (see Figure 5.26). The offset, sensitivity and linearity of the sensor cells vary due to production differences, so that each sensor must be calibrated (for an illustration, see Figure 5.27). To effect the calibration, the sensor is measured by the manufacturer at different pressures and temperatures. From the measured values, parameters are determined for the individual sensor, which the sensor then uses when it is operating to "calculate" the output voltage associated with the measured pressure.

This is done either by connecting capacitances into the circuit (for example to compensate for the offset and the temperature dependence) or in the case of the linearization of the characteristic curve by referring to a "look-up table". The parameters required for this process, as determined by the calibration, are read into the chip via a digital interface and are stored in the sensor's PROM (Program-

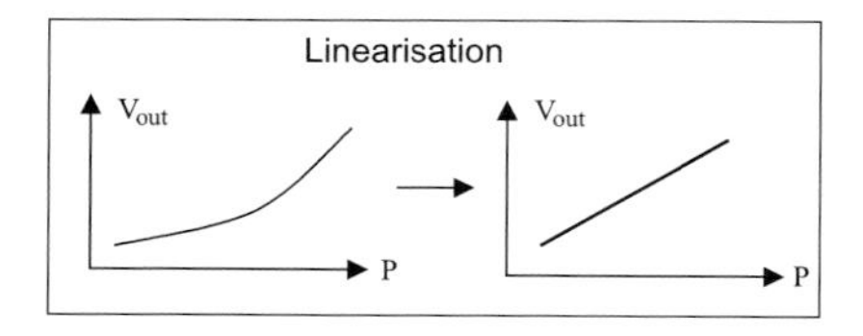

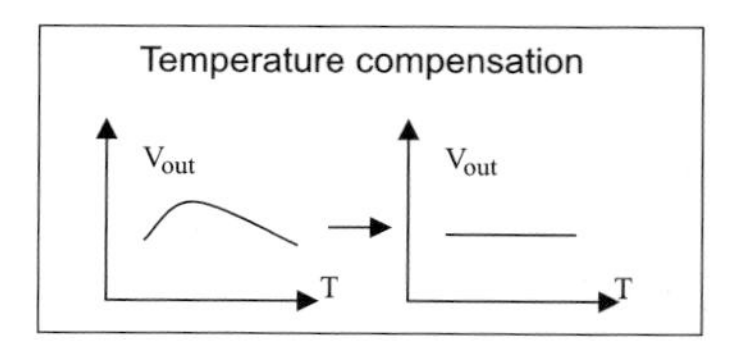

Figure 5.27 Linearization and temperature compensation

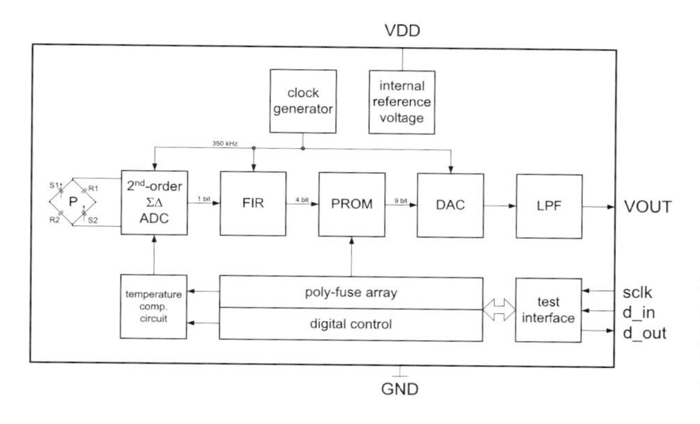

Figure 5.28 Signal path for a pressure sensor with analog output characteristic curve

mable Read Only Memory). After the data has been written into the PROM, test parameters for analyzing the sensor can also be input via this interface.

The architecture of the KP120 chip is similar to that of the KP100 side airbag sensor (see Figure 5.28). Here again, the sensor cells, which work on the basis of capacitance (capacitance C_S), are connected in a bridge circuit with the pressure-insensitive reference cells (C_R). With an applied voltage of V_{ref} a charge $(C_S - C_R) \cdot V_{ref}$ develops at the input to the sigma-delta converter.

To compensate for the offset, and the thermal responses of the offset and sensitivity, capacitors are inserted downstream in the circuit (not shown in the figure).

The changes in the charge (as the sum of the charges which come together at the node) are converted to a 4-bit wide digital signal by the sigma-delta modulator and the downstream decimation filter. This signal is corrected for the offset and is already temperature-compensated. Using the parameters of the linearization table stored in the PROM, the programmed characteristic curve is now generated and finally the bitstream is transformed into an analog output signal.

Technically, the sensor's PROM consists of so-called "cavity fuses" (see Figure 5.29). These are poly-silicon conducting tracks, which run through an evacuated cavity. These cavity fuses have a short narrowed section, and are largely thermally insulated in the cavity. A short cur-

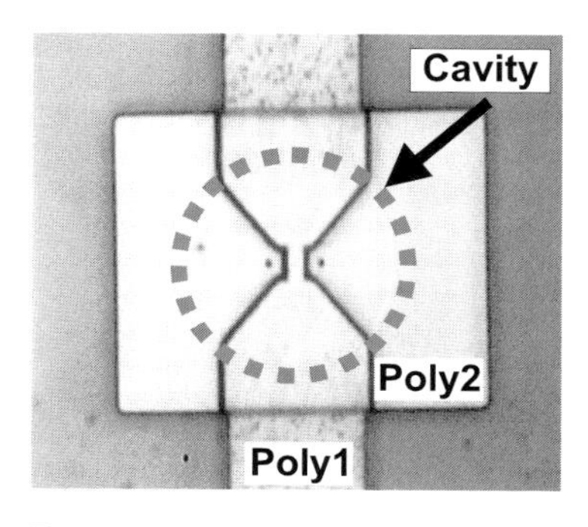

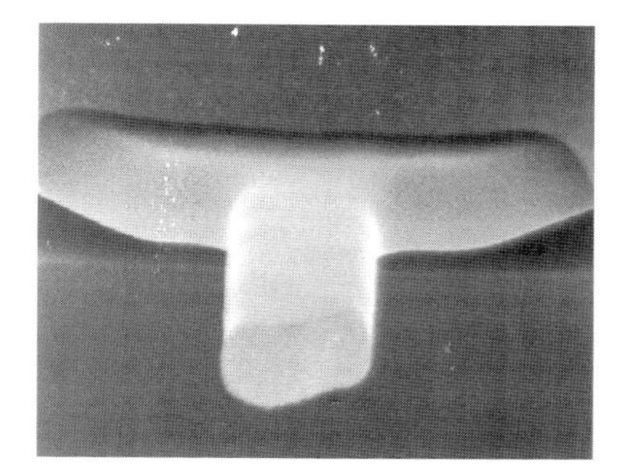

Figure 5.29 Photograph of a cavity fuse

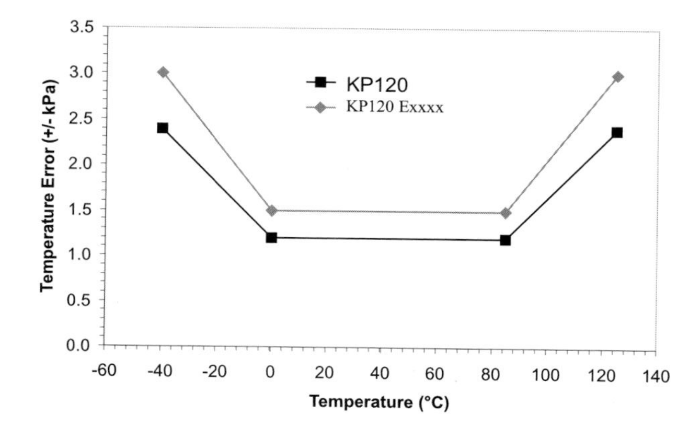

Figure 5.30
Accuracy of the KP120 curve

rent pulse (typically 50 mA, 10 msec) is sufficient to melt such a fuse, and in this way to change a data bit. This low power makes it possible for the sensors – unlike when the otherwise customary "laser trimming" is used – to be programmed via their leads even after they have been assembled in their housing. This allows the accuracy of the sensor to be increased, because assembly generally leads to small but measurable changes in the characteristic curve. Another advantage of this type of data storage is its high reliability and thermal endurance compared to a standard EPROM.

The KP120 series of sensors achieve a particularly high accuracy (see Figure 5.30). Their characteristic curves can be programmed over the range 40 to 115 kPa (0.4 to 1.15 bar). They are mounted in the same housing as the KP100 and can be operated up to 125°C.

5.3.3 Piezo-resistive pressure sensor in an SMD package (KP200)

Piezo-resistive pressure sensors, which have proven themselves in the most demanding of industrial uses, are also offered in a low-cost SMD package.

Unlike the capacitive sensors, piezo-resistive pressure sensors supply a high primary signal. The piezo-resistive effect – the change in electrical resistance caused by mechanical strain – is very marked in silicon. For this reason, it is used in most of the semiconductor pressure sensors in current applications. The material properties of crystalline silicon offer a significant advantage, in that it is free of fatigue almost up to its point of breakage. Because of their accuracy and high long-term stability, pressure sensors of this type (Infineon KPY series) are used mainly in industrial metrology.

In contrast to surface micro-mechanics, in bulk-micromechanics the Si wafer is etched away from the rear (Figure 5.31). With this technique, thin membranes of crystalline silicon are exposed. Depending on the pressure range, the sensitivity and resistance to breakage are optimized by the choice of membrane thickness and diameter. Typically, the membranes are 20 µm thick with a diameter of 1.5 mm. Under pressure, the membrane bends slightly. Piezo-resistors, implanted in the membrane, indicate the mechanical strain which results by a change in resistance. The resistors are in standard bipolar technology, protected against external interference factors (e.g. electric charges) by passivation layers. A Wheatstone bridge of four piezo-resistors supplies a linear pressure-dependent output signal. At the nominal pressure with a 5 V supply volt-

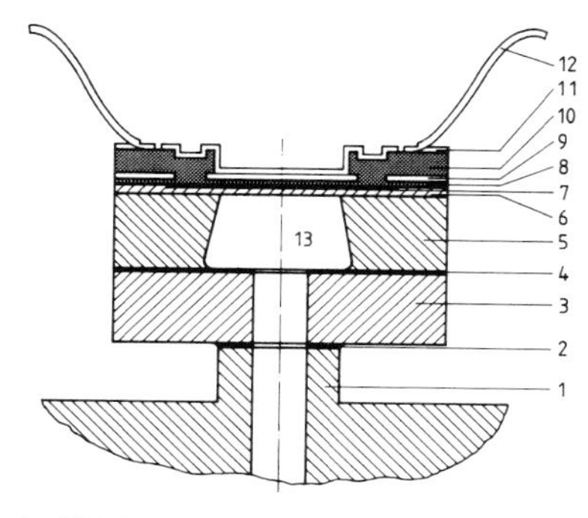

1 Metal mounting base
2 Au – Sn solder
3 Silicon substrate
4 Metallic connecting layer
5 Silicon substrate
6 Silicon epitaxial layer (corresponds to a pressure sensitive membrane)
7 Implanted resistors
8 Silicon oxide
9 Silicon nitride
10 Metallization
11 CVD nitride
12 Aluminium bond wires
13 Cavity, pressure equalization with rear of package

Figure 5.31
Structure of a piezo-resistive pressure sensor

age the typical output voltage is 100 - 200 mV.

To avoid any assembly effects, the sensor chip is attached to a substrate chip with a helium-tight seal using AuSi wafer bonding. This keeps the temperature and pressure hysteresis errors low. To manufacture relative pressure sensors, an opening is etched in the substrate chip, so that pressure differences between the front and rear of the membrane can be detected. For absolute pressure sensors, the ambient pressure is measured against an enclosed reference vacuum.

A new feature in the KP200 is its assembly in a P-DSOF-8 package. This optimized assembly, using special silicon adhesives, is cost-effective and avoids any loss of quality. To protect against mechanical loads on the bond wires, the sensor is covered with gel.

Furthermore, a temperature-sensitive resistor, integrated into the package, enables very easy temperature compensation, accurate to a few percent. For this purpose, the sensor is in circuit with two resistors. If the resistances used have fixed values, as is necessary for automatic insertion, accuracies of around 2% are achieved in a temperature range from −20°C to 60°C. The value of the compensation resistances can be matched to the individual thermal response of the sensor, which further significantly improves the temperature compensation.

The KP200 is the most inexpensive version of very accurate pressure sensors with long-term stability for high volume low-cost applications. Their target market includes the household goods industry with applications in vacuum cleaners (monitoring the suction power) and washing machines (water level).

5.4 Temperature sensors

As an alternative to metallic resistors of nickel or platinum, silicon temperature sensors are in widespread use. They are lower cost, offer a higher sensitivity and come close to reaching the close tolerances and reproducible characteristic curves of metal temperature sensors.

The chip of a so-called semiconductor thermistor has a very simple construction, and essentially consists of neutron-doped silicon with two contacts. This basic material is produced by irradiation with neutrons in a reactor, which converts a small proportion of the silicon atoms to phosphorus. This process establishes a defined level of donor phosphorus. The resistance of neutron-doped silicon has a very reproducible thermal response, and is used as the output magnitude in the KTY series sensors.

The following data characterize the silicon sensor:

Temperature range
  –50°C to 150°C

Resistance at 25°C
  2 kΩ ± 2%

Long-term stability
  ±0.2%

Resistance ratio

$$\frac{R_{(105°C)}}{R_{(25°C)}} = 1,67 \pm 1,2\%.$$

Within the resistance tolerance of ±2%, subgroups check-measured to ±0.5% can be supplied, so that measurement accuracies of 1°C are possible with no additional check measurements on the components.

The dependence of the resistance value on the direction of the current, which often causes problems with semiconductor temperature sensors, has been reduced by technological measures to values < 2 Ω.

Hence, if temperature is to be measured over a wide temperature range, and if measurement with an accuracy of a few tenths of one °C is adequate, the KTY series of silicon temperature sensors offer a low-cost solution. The sensor elements are manufactured in various forms of packaging (KTY10 to KTY16) and resistance groups (KTY10-3 to KTY10-9). Due to the yield distribution, the nominal values of the resistance groups lie in each case at distances of 30 Ω around mean values of 2000 Ω. Sensor resistances can be supplied with nominal values in tolerance classes of 0.5, 1, 2, 5 and 10% at the nominal temperature of 25°C.

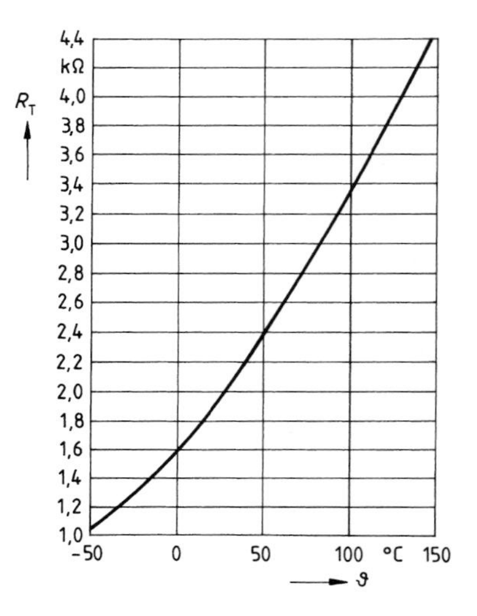

Figure 5.32
Typical characteristic curve for the KTY 10-6

Figure 5.32 shows the typical resistance-temperature characteristic of the KTY10-6.

Examples of applications using temperature sensors will be found in Chapter 9 *Automotive Silicon Solutions*.

6 Memory

The function of memory is to hold data for a shorter or longer period of time. Technological improvements, and above all dramatic cost reductions and the concomitant price reductions, have enabled data storage to proceed on a triumphal march. By now, there are scarcely any areas in everyday life in which data is not stored digitally in one form or another.

There are four main areas in which the various types of data storage are used:

- Processors (in some cases with integral caches)
- External caches (SRAM, common unit is kBytes)
- Main memory (DRAM, ommon unit is MBytes)
- Mass storage (hard disk, ommon unit is GBytes)

6.1 Types of data storage

Apart from their fields of use, another common way of categorizing data storage types is by their technologies (see Figure 6.1).

The first distinction which is made is between volatile and non-volatile storage: volatile storage media lose the stored data when the supply voltage is removed, non-volatile ones retain the data even after its removal.

Within these two categories, subgroups are distinguished by the different technologies used.

6.1.1 Mechanical storage

The first data stores were mechanical, and were used mainly for storing operational statuses in production plants, or to control activity sequences. The audio disk is an example of mechanical analog storage. Mechanical storage ceased to play any part in digital technology a long time ago. It has been superseded by storage based on the other technologies.

6.1.2 Magnetic storage

A non-volatile storage technology which is used predominantly for long-term data storage. In general, such a store consists of a transport mechanism (drive) and a moving magnetic medium which is either permanently installed (for example: a hard disk) or is exchangeable (for example: a diskette, magnetic tape). The data is transferred via a read/write head. With disk stores this head is moveable, so that such stores have relatively short access times. In the case of tape drives, the tape is fed past a fixed-location head. For this reason, tape drives have extremely long access times. The disadvantages of magnetic forms of storage are the access times, which are relatively long compared with processor speeds, and the need for fault-prone moving mechanical parts. Their critical advantages are the low price per unit of storage and their non-volatility.

6.1.3 Optical storage

This non-volatile storage technology is used almost exclusively for long-term storage. The medium almost always consists of a disk, which is scanned using a laser beam. It can be written either repeatedly (CD-RW) or once only (CD-R), or is "pressed" in a one-off factory process (CD-ROM). Currently, the DVD (Digital Versatile Disk) is becoming increasingly

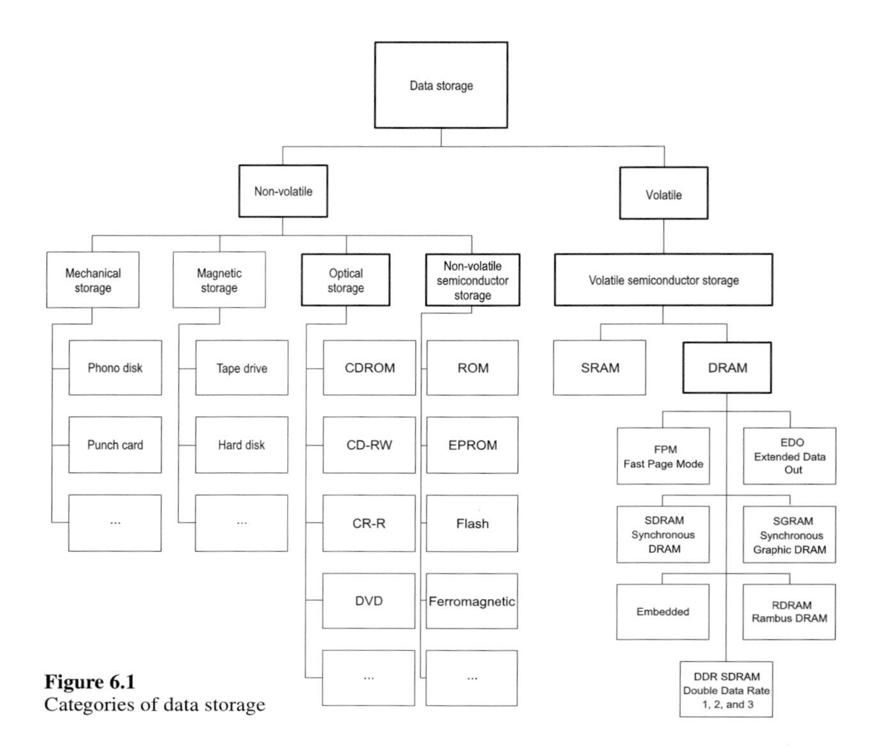

Figure 6.1
Categories of data storage

more widespread and it is superseding magnetic tapes in the area of gigabyte storage. There are also hybrid forms of magnetic and optical storage. The disadvantages are the relatively long access times and the need for fault-prone moving mechanical parts. The advantages are comparable with those of magnetic forms of storage.

6.1.4 Semiconductor storage (memory)

Under this heading, a distinction is made between the storage of fixed values (read-only memory, ROM) and the various forms of programmable/writable semiconductor storage, both non-volatile and volatile, which come in different variants (DRAM, SRAM, EPROM, Flash etc.). Further details of these will be found in the Glossary (Chapter 16 of this book). In general, semiconductor memories have the advantages of fast or very fast data access depending on the type concerned, small space requirement, a low current consumption and great ruggedness: they are practically wear-free and work without moving parts, so that they can also be subject to strong acceleration and vibrations during operation without this causing problems. However, their disadvantage is the price per unit of storage, which is still relatively high. This chapter is concerned mainly with more detail of the DRAM technology.

6.2 Fundamentals and areas of use for DRAMs

In computer technology, the DRAM is the most frequently used integrated circuit. The reason for this is its low price per bit by comparison with other semiconductor memories, its small dimensions and the high speed of access to the data content.

The direct competition, in terms of price per bit and installed size, comes from hard disk drives which, because of their mechanical nature, are far from being able to offer such high access speeds. For this reason such drives are used mainly for mass storage, for the permanent storage of programs and data.

In terms of speed, SRAMs have the advantage over DRAMs. However, the more expensive construction of an SRAM means a higher price per bit, a smaller amount of storage in relation to its unit volume, and larger package dimensions. The storage capacity of SRAMs always trails at least one generation behind that of DRAMs. If the comparison is made against very fast SRAMs, then the available capacity per chip is still substantially smaller. SRAMs are therefore used primarily as cache memory (fast buffer storage) between the processor and DRAM main memory. Figure 6.1 shows an overview of the categories of storage.

6.2.1 What are SRAMs and DRAMs?

Both types are RAMs i.e. random access memories. This means that it is possible to access directly any cell of the memory, chosen at random. This is in contrast to serial access methods, with which it is only ever possible to obtain the data in a particular sequence.

An SRAM is a static RAM. This means that the data in the cells continues to be stored until it is replaced by new data, or until the supply voltage is turned off. If one considers only the storage, the current consumption is very low because the maintenance of data in fully-switched flip-flop circuits constructed using CMOS technology is almost loss-free. Larger currents are only required for accesses to the cells. No precautions have to be taken for the retrieval of data. When doing this, the memory cells are addressed by passing over all the required addresses at the same time.

A DRAM is a dynamic RAM. The term dynamic here refers to the way in which the data is stored in a cell. By contrast with the stable data in an SRAM cell, in the case of a DRAM the data is stored in a capacitor which, over time, loses its charge. The length of time over which a cell can reliably retain an item of data is generally less than one second. In order for it to be constantly available, therefore, the data must be repeatedly refreshed. In spite of this costly refresh mechanism, the construction of a DRAM can be smaller than an SRAM cell, because of its much smaller single-transistor cell. This implies space saving and relatively low costs, so that these chips can be used in large numbers.

Apart from their smaller cell structure compared to an SRAM, to enable the package to be made smaller use is made of the address multiplexing technique to address the individual cells. With a DRAM, access to the memory cells takes place in two steps; first a memory row is sensed and then the column is selected. This permits the address to be split into row and column addresses and supplied one after the other (multiplexed). With an SRAM, all the required addresses are passed to the chip at once, which means that by comparison with an SRAM of the same size a DRAM saves half of the address lines.

Because of their simple cell structure and their larger number per chip, DRAMs have the advantage that they are very suitable for trying out and introducing new semiconductor process technologies. The test results in terms of the numerous memory cells permit good conclusions to be drawn about the quality of the processing. However, because of the different target parameters for the CMOS transistors in DRAM and logic circuits, the DRAM technology is no longer regarded so strongly as the leading technology.

6.2.2 DRAM types

For DRAMs there are several distinguishing criteria.

1. Storage capacity

Depending on the currently available technology, DRAMs are manufactured and categorized in particular sizes, e.g. 256 Mbit, 512 Mbit, or 1 Gbit.

2. Input/output organization

Here we make a distinction between ×1, ×4, ×8, ×9, ×16, ×18 and ×32. For example in the case of a 512 Mbit memory, "×4" (read as: times four) means that one quarter of the 536,870,912 memory cells is assigned to each of the four inputs/outputs (128M ×4). For the corresponding "×16" variant (32M ×16), sixteen 32M blocks are supplied with data in parallel via 16 inputs or outputs.

3. Access time

The time which elapses, from activation of the chip for an access to an arbitrary cell until the desired data is available, is the so-called random access time. With each generation it has been possible, mainly by making the structures smaller, to open up new speed classes. Whereas for the 256K generation it was 120 ns, and for the 16M generation 60 ns, so with the 256M generation a reduction down to around 30 ns has been achieved. Nevertheless, within each generation there have also been improvements in this value, due to developments in the technology and design. However, because of process variability, there is always a spread of access times between fast, standard and slow.

4. Power supply and I/O voltage

Because the semiconductor structures are becoming ever smaller, the trend is towards ever smaller supply voltages. From an original 5 Volt TTL interface through a 3.3 Volt LVTTL interface, we have gone over to the 2.5 and 1.8 Volt SSTL interface. The lower voltage and the new interface definitions have been accompanied by a reduction in the power loss in the I/O area.

5. Interface class

For many years, the asynchronous memory interface was the standard. Developments of this were the Fast Page Mode (FPM) and the Extended Data Out (EDO) operating mode. Since the middle of the 90s, the synchronous interface has become established. We now speak of SDRAMs, that is synchronous dynamic RAM. A doubling of the data transfer rate for the same clock speed has led to the Double Data Rate (DDR) SDRAM, as a result of which the earlier SDRAMs are now retrospectively designated as Single Data Rate (SDR). The standardization within JEDEC now makes a distinction between three numbered generations of DDR SDRAM, called DDR1, DDR2 and DDR3. The lower entry-level data transfer rate of each generation is double that of the predecessors (200, 400 and 800 Megabit per pin and second). Alongside these there are also several generations of the Rambus Interface Definition, which only temporarily had a measure of importance in the memory market.

6. Derivates

From the standard DRAMs named above have been derived special chips, which build on the available memory capacity and contain special functions. So there is, for example, graphic memory optimized for data throughput and special low-power variants for mobile applications.

6.2.3 The specification

In order to ensure that the component a customer gets is the one required for his application, a data sheet (specification) is available for each chip. If the customer uses such a data sheet in buying a component he can, on the one hand, match his circuit to the component and, on the other hand, can rely on the chip having the characteristics specified in the data sheet. The specification sets out the characteristics of the chip and how the user must handle it for it to function properly.

To persist in international competition, DRAM data sheets are largely standardized. This gives the user an opportunity to procure the product from various manufacturers. This offers him a major safeguard in terms of the availability of the chips. For the manufacturer, the main objective in developing a chip is adherence to the specification. In general, a manufacturer who goes it alone with a deviant specification, even if this represents a product improvement, can usually only with difficulty establish it on the mass market. For this reason, developments in the memory market proceed through internationally staffed committees, such as JEDEC, or by bilateral and trilateral cooperative agreements between memory manufacturers.

6.2.4 Mechanical construction of a DRAM

A DRAM is produced in various housings (packages). The DIP package (Dual-in-Line Package) and the ZIP (Zigzag-in-Line Package) are now obsolete, having been superseded by the SMD package (Surface Mounted Device) for surface mounting. For the 1M and 4M generations there was the SOJ package (Small Outline J), where the "J" stands for the pin shape. In common use today are the compact TSOP (Thin Small Outline) Package and the FBGA (Fine-Pitch Ball Grid Array). The FBGA is the first family of SMD packages for memory on which the connections are not arranged outside on the perimeter of the package. In this case, we speak of a chip-size package (CSP), because it is now the size of the silicon chip which determines the package size and not the standardized spacing of the connecting pins.

The voltage supplies and the digital signals are fed through connections which

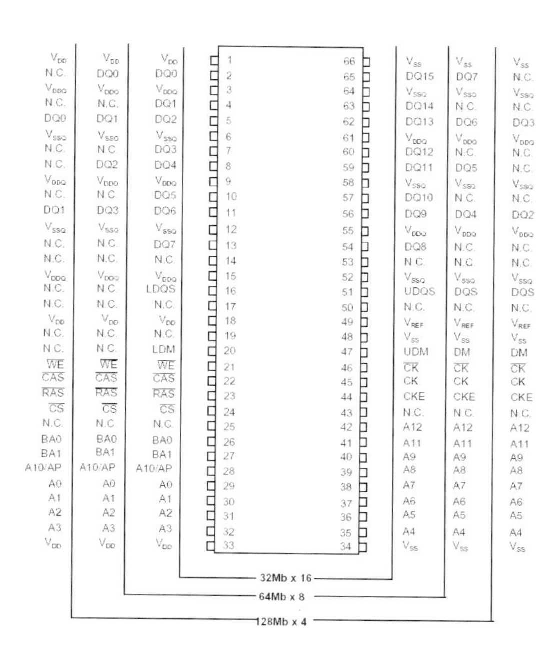

Figure 6.2
TSOPII-66 of a 512 Mbit DDR SDRAM

Table 6.1 Pin assignment of a TSOP package on a DDR SDRAM

V_{DD}	Positive supply voltage	RAS, CAS, WE	Command bus for encoded commands to the memory
V_{SS}	Negative supply voltage, generally referred to as "Ground" (GND)	BAn	Bank address
VDDQ, VSSQ	Supply voltage for the output drivers (Q = Query, the read data requested)	An	Address signal
CK	Clock	DQn	Data input and output (Data & Query)
CKE	Clock Enable signal	DMn	Data masking (Data Mask)
CS	Chip Select signal	DQSn	Data clock (DQ Strobe)

take the form of pins or balls (Figure 6.2, Table 6.1).

6.2.5 Functions of a DRAM as exemplified by an SDR SDRAM

A DRAM has three main functions, namely the writing of an item of data into a cell (write), the retrieval of this data (read) and the refreshing of the cell contents (refresh). In addition to these there are various power-saving modes and an initialization phase.

The functions consist of two parts:

- Cell addressing
- Working mode (Read, Write, Refresh)

Since the introduction of SDRAMs, now only one clock pulse is used to control the memory interface. Whereas previously, for asynchronous memory interface, it was the edges of the control signals which determined the timing, because they defined the time point and change of function, it is now the (rising) edges of the clock pulse which specify the time points, at which the then prevailing logical levels of the control signals determine the command to the memory.

READ cycle

The *Activate* command causes the row address to be read in, and the internal processes are started for sensing the cells in a row (see Figure 6.3).

The *Read* command causes the column addresses to be read in. This determines which cell in the row is selected. The earliest that this can happen is after the time specified in the data sheet, the RAS-CAS-Delay. After two or three clock cycles, the so-called CAS-latency, the data outputs DQ are switched to the values contained in the cells. The number of data words output sequentially can be programmed in the SDRAM as the burst length, and in this example is set to 4.

The *Precharge* command closes the open row of memory cells again.

WRITE cycle

A write access is very similar to the read access (see Figure 6.4). Unlike a read operation, the WE (write enable) signal is active for a write (that is 0, because WE is low-active). Following the write command, the data is read in in time with the clock pulses.

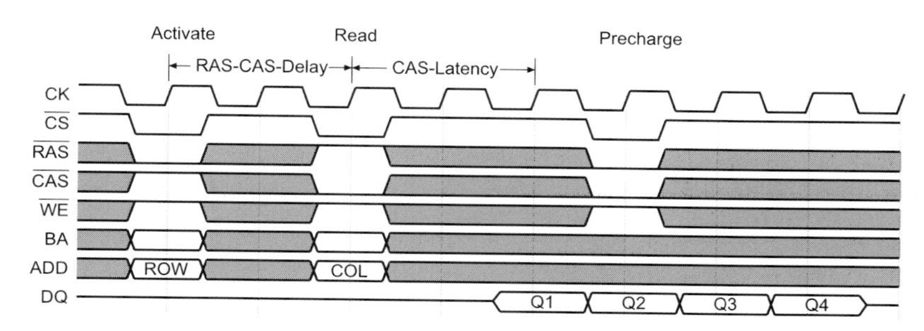

Figure 6.3 The read cycle for a SDRAM

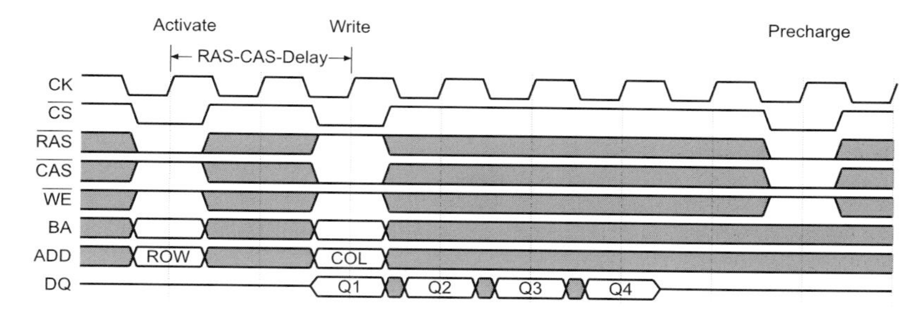

Figure 6.4 The write cycle for a SDRAM

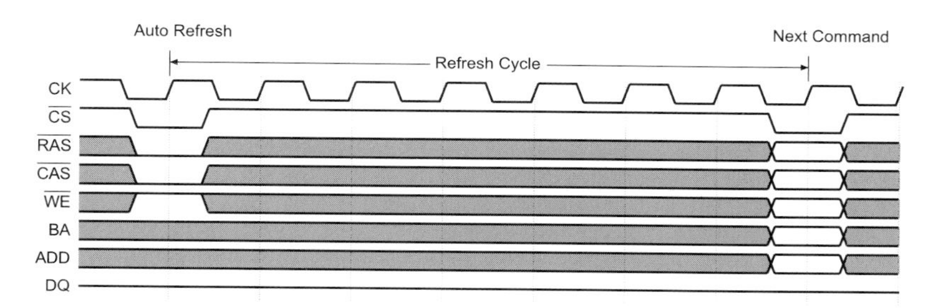

Figure 6.5 The refresh cycle for a SDRAM

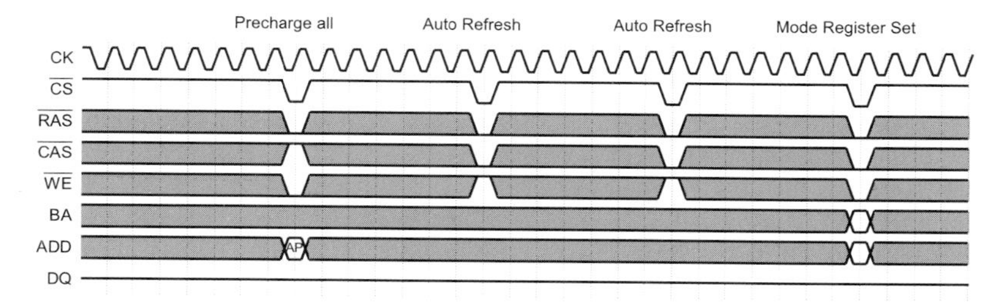

Figure 6.6 The initialization cycle for a SDRAM

In the case of both the read and also the write cycle, data can be masked clock by clock using the DQM signal, not shown here. Furthermore it is possible, between an Activate command and the associated Precharge, to perform several read and write commands. However, when the data transfer direction changes care must be taken that the memory controller and the SDRAM never drive data simultaneously.

Refresh cycle

Any dynamic memory guarantees the storage of an item of data in a cell only for a short time (32 or 64 ms). Refreshing it restarts this time interval from the beginning for the cells addressed.

The *Auto Refresh* command starts the refreshing of a row of cells in each bank of the SDRAM. The row address is specified internally by a counter, so that the external address bus ADD is ignored during this command. As all the banks are refreshed simultaneously, it is unavoidable that all the banks must be in the quiescent state at the beginning of the command. The earliest that the next command can be input is after the specified refresh cycle time, the timing of all the processes in the memory required for the refresh being controlled internally (see Figure 6.5).

Alternatively, all the rows can be refreshed with Activate Precharge commands, because each time a row is activated the contents of all its cells are sensed and by the time of the Precharge command have been fully restored. However, two commands are required for each bank, as opposed to one Auto-Refresh command for all the banks.

Initialization cycle (Power Up sequence)

Before the SDRAM is used for data storage, there is a prescribed initialization sequence (see Figure 6.6). First, all the banks are closed with the command *Precharge all*, so that the Auto Refresh which follows will find all the banks in the idle state. The number of prescribed Auto Refresh commands in the sequence is at least two. It now only remains to set the burst length, CAS latency and burst mode, using the Mode Register Set command. The burst mode determines the way in which the internal address counter addresses the cells during a data burst. This is only significant if the address used for the burst start address is not the lowest address for the cells associated with the burst.

6.2.6 Technology

Today's generations of DRAMs are made exclusively in CMOS technology. By contrast with the NMOS technology of earlier generations of DRAMs, for which the processing is simpler, the CMOS technology offers substantial advantages in the reduced power consumption of the memory chips.

Unlike bipolar transistors, which are current-controlled, field effect transistors or MOS transistors are voltage controlled. A distinction is made between p-channel and n-channel transistors.

The gate is insulated by a thin oxide layer on the surface of the semiconductor (Metal Oxide Semiconductor). However, polysilicon is nowadays used for the gate material, instead of metal. If an appropriate gate voltage is applied, a conducting channel can form on the surface of the semiconductor, connecting the source and drain regions and hence permitting current flow.

The processes taking place within an n-channel transistor

The substrate is p-doped, i.e. an excess of holes is created by donors. If a positive voltage relative to the substrate is applied to the gate, the reaction to the positive charge on the gate electrode is the development of a space-charge zone in the immediate vicinity of the oxide layer, in which the holes are depleted. Because

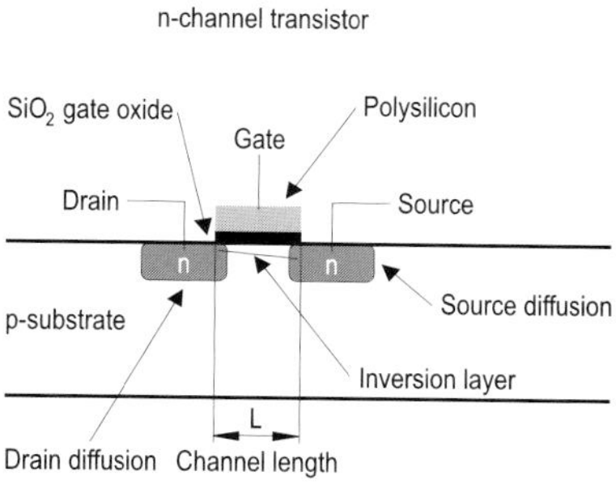

n-channel transistor
SiO_2 gate oxide
Gate
Polysilicon
Drain
Source
n
n
p-substrate
Source diffusion
Inversion layer
L
Drain diffusion
Channel length

Figure 6.7
Schematic diagram of an n-channel transistor

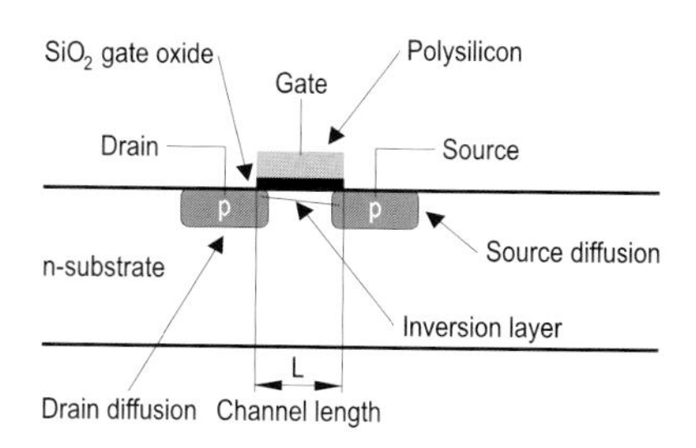

Figure 6.8
Schematic diagram of a p-channel transistor

charge carriers are formed and disappear again throughout the semiconductor, as a result of generation and recombination, this also occurs in the space-charge zone. So at the boundary surface, electrons migrate to the insulator and holes to the substrate connection. The electrons at the oxide layer form an electron layer with polarity inverted relative to the charge carriers in the substrate material, the so-called inversion layer (see Figure 6.7).

Because of the closeness of the PN junctions of the source and the drain (heavily doped regions, which permit a conducting link to the contact connection), additional electrons are potentially available for the extremely rapid formation of an inversion layer. Because the source-bulk voltage U_{SB} (bulk = substrate) and the drain-bulk voltage are not negative, the PN junctions are polarized relative to the substrate, and hence are insulated.

Once the inversion layer has been established, it forms a conducting link between the source and the drain. In order to create an inversion layer, a gate voltage is required which must be $> U_t$, the threshold voltage.

The threshold voltage also depends on U_{SB}, because of the controlling influence of the substrate. This influence arises from the fact that the gate charge, which is constant when the inversion is strong, distributes itself differently between the opposing charge of the inversion layer and that of the space-charge zone. If the voltage U_{SB} is increased, the breadth of the space-charge zone increases, and with it the charge, and that of the inversion layer decreases. This raises the threshold voltage, which is an unwanted effect.

In principle, the same applies for p-channel transistors (see Figure 6.8) as for the n-channel transistor just described, only with inverted signs. However, account must be taken of the fact that in a p-channel transistor the charge carriers are holes rather than electrons, and they have a lower mobility. That is to say, all other things being equal n-channel transistors produce approximately 2-3 times as much current amplification.

In a CMOS circuit (see Figure 6.9), n- and p-channel transistors are combined on one substrate. In order to be able to realize a p-channel transistor in a p-substrate, a so-called n-well is created locally for the p-channel, by redoping.

For transistors with short channel lengths, less than 1 µm, the threshold voltage reduces because the charges in the charge zones of the source and drain regions reduce the charge in the space-charge zone. This raises the charge on the inversion layer, and the threshold voltage drops. An additional factor is that the threshold voltage is also affected by the voltage U_{DS}. The consequence of this is that the shorter

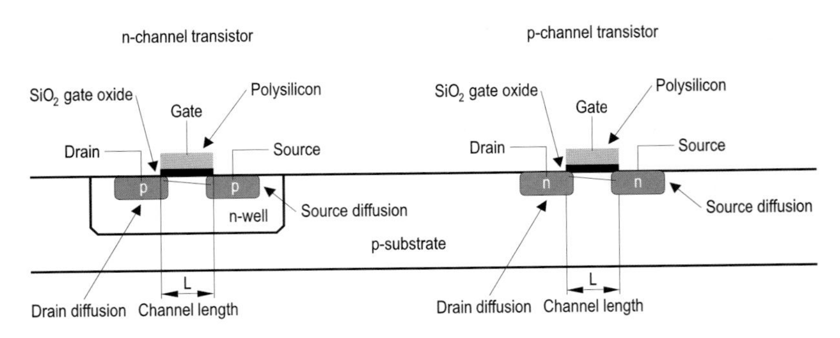

Figure 6.9 Schematic diagram of a CMOS transistor pair

the channel lengths and the greater the U_{DS} is, the lower is the threshold voltage. Ultimately this can lead to the space charge zones of the drain and source touching each other. This produces a current flow which can hardly be controlled any longer by the gate voltage. This effect is called "punch through", or simply "punching".

ESD sensitivity (dielectric breakdown)

The small thickness of the dielectric is responsible for an additional source of faults, dielectric breakdown. This happens if the gate-substrate voltage is so high that conducting channels are created in the oxide, which means permanent damage to the transistor. These voltage spikes are a cause for concern. They can arise from an electrostatic charge build-up on personnel and machines which are discharged across the CMOS device. Damage of this type is referred to as ESD (ElectroStatic Discharge) damage. The problem is confined mainly to those transistors which come into contact with the external world, i.e. the input and output circuits on a DRAM or other MOS device.

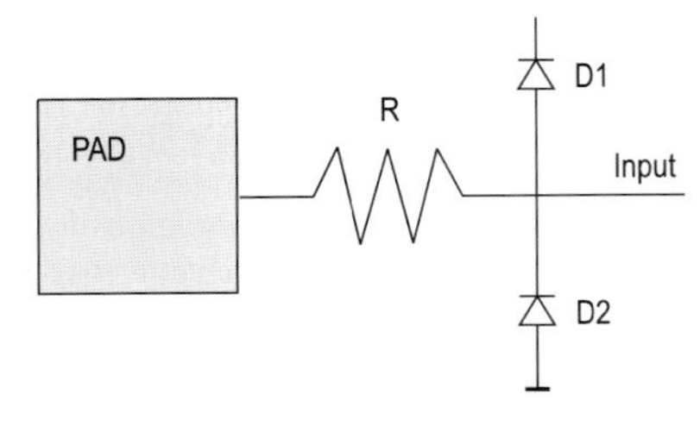

Figure 6.10
ESD protection circuit for DRAM contacts

In order to counteract damage due to ESD, precautionary measures are taken for DRAMs, in the form of ESD protection circuits (see Figure 6.10).

The resistance R limits the peak current through the diodes. The diodes D_1 and D_2 switch to conducting if a voltage above V_{DD} or below V_{SS} is present at the input.

Latch-up effect

In CMOS circuits there are parasitic bipolar transistors, which are normally no problem because the PN junctions are operated in the high-resistance (backward) direction. However, it is possible for the voltage U_{SB} to go negative as a result of various factors, producing a PN junction which is switched in the forward (conducting) direction. If there is a succession of npn- and pnp-junctions, (which is frequently found in CMOS circuits), the configuration of a thyristor (a four-layer transistor) can arise and this, once triggered, will no longer close without external intervention (latch-up). This generally implies the destruction of the device due to the high short-circuit currents between V_{DD} and V_{SS}.

Countermeasures are taken, where possible, by spatial separation or by guard rings of wells or substrate contacts. These rings reduce the resistance between the base and the emitter, which makes self-triggering more difficult.

Latch-up can also be very effectively reduced by the use of a heavily-doped substrate (low resistance) with an EPI layer (high resistance). A deep, heavily-doped N-well (retrograde N-well) can render the use of guard rings superfluous.

Switching characteristics of a CMOS circuit

Because switching a MOS transistor requires at least a threshold voltage U_t, it is clear that, for example, when a "1" condition is set the maximum voltage that an n-channel transistor will experience at the drain is lower by the threshold voltage. On the other hand, with a p-channel transistor a "1" can be set with no losses, because the gate voltage is at the level "0" and hence the voltage U_{GS} is far larger than the threshold voltage U_{tp}. When switched to the value "0" on the other hand, the n-channel transistor has the advantage that the "0" can be held without loss, because U_{GS} is again much larger than the threshold voltage U_{tn}. In this case, for a p-channel transistor, the drain voltage is reduced by U_{tp}. If one wants to produce a transfer element with negligible losses, which allows "0" and "1" through unhindered, then one n-MOS transistor and one p-MOS transistor are connected in parallel and are fed complementary clock pulse signals. This will cause a "1" to be switched through completely by the p-MOS transistor and a "0" by the n-MOS transistor.

It is also possible to take advantage of the switching characteristics of n-MOS and p-MOS in inverters. Thus when a "1" is applied at the input, the n-channel transistor will be switched fully through, and the p-channel transistor remains completely blocked. This produces a clean "0" at the output. If a "0" is applied at the input, then the p-channel transistor will be opened and the n-channel transistor closed, and a clean "1" is produced at the output. This complete through-switching on the one side and total blocking on the other prevents cross-currents from flowing in the switched state. However, it also means that no unnecessary current is required and a CMOS circuit works with exceptionally low losses.

Of course, even CMOS circuits do consume current. In order to switch a transistor, charges must be displaced at the gate. This requires that a current flows through the circuit which drives the transistor until the charges have accumulated in the right place. However, there are also circuits which, intentionally (e.g. in a regulator) or unintentionally (due to a "slight" design error) maintain a cross-current path when operating. As a result, even in static mode their CMOS circuits will never work totally without loss.

That the majority of the current consumption nevertheless takes place in the active mode can be seen from the fact that a DRAM consumes between 50 and 500 mA in the active state, and less than 5 mA in the inactive stand-by mode.

Unlike the "clean" switched states in a CMOS circuit, with n-MOS circuits resistances are required in inverter circuits. Although these are high-ohmic, they are far from being as loss-free as a CMOS circuit.

6.2.7 Internal structure and functional principles of a DRAM

The storage is organized as a matrix. This matrix consists of rows. In the storage array, the rows correspond to the so-called wordlines and the columns to the bitlines. The actual assignment of the addresses fed in externally to the internal structure is determined by the physical arrangement of the cells on the chip, and the space and performance requirements.

Addressing the DRAM

Roughly speaking, the addressing of storage is by binary coding of the cells on the memory chip. This means that if there are 1M (1,048,576) cells, 20 address lines are required (with one line it is possible to address two cells, with 2 it is 4 cells, with 3 it is 8 cells and with 20 it is possible to address 2^{20} cells). Unlike a static RAM, with a DRAM the address multiplex method is used. This means that a 1M DRAM does not have 20 address lines, but only 10. To do this, use is made of the fact that the storage is addressed by its two-dimensional matrix. A distinction is also made between the X and Y addresses, or as they are usually described for a DRAM, the row and column addresses. So with 10 row addresses it is possible to address 1024 rows and correspondingly with 10 column addresses 1024 columns. The matrix with 1024 rows × 1024 columns thus contains the 1,048,576 cells of a 1M DRAM.

The time multiplexing of the addresses is controlled by means of the commands to the DRAM. The row address is passed to it with the Activate command, and the column address with the Read or Write command. In the time between the two, the sensing of the selected line is carried out, as explained in more detail below.

With the introduction of the SDRAM Interface, 'banks' were established as a third dimension of the storage matrix. The density of integration for memory modules resulted in it being possible to achieve the desired amount of memory with fewer memory modules than before. In exchange however, there was a loss of flexibility, because for each memory module it was only possible to hold open one line, so the number of lines open in parallel fell with the number of memory modules. The bank concept now produced an imaginary integration of two, four or more memory modules in one package. Shared use is made of all the functions and blocks which in operation are only used once, such as for example the input stages and the refresh logic. Apart from this, each bank is independent. Instead of an individual selection signal for the discrete memory modules, there is now a common select signal, and bank address signals by which the banks are addressed in binary form.

Organization

DRAMs were developed, among other reasons, because their packages are small relative to their storage content. As a result, initially not only were the addresses multiplexed but also only one input and one output were provided. As from the 64K generation, the idea emerged of subdividing the storage into four areas each with its own input/output. Here again, integration was the driving force: for the same address range, one chip of the appropriate newer generation in the form of a ×4 version would replace four modules in the form of the ×1 version of the preceding generation. This simplifies a direct changeover to a new generation of memories with smaller areas and current demands.

Apart from the addressing, the size of the resulting memory is an important matter. For example, if 128 I/Os are used in the graphics area, then one requires 32 of the ×4 modules which, for the 256M generation corresponded to a total of 8 Gbits, that is 1 Gigabyte of graphics memory. Even for main memory, this is significantly more than the standard provision in PCs in 2003/2004. On the other hand, if ×32 modules were used, the total memory would be 128 MByte.

Modules with a large number of I/Os make possible systems with a smaller number of components, and typically a lower total current consumption. The large number of I/O pins means, of course, that this is at the cost of the package size, or demands a more compact and therefore more expensive packaging technology.

In contrast to the historical ×1 organization, nowadays it is normal with every organizational arrangement for the inputs/outputs to be no longer separate. This is actually not a problem, because on a processor bus no distinction is made between input and output. By this point at the latest, the separate inputs and outputs are combined together. However, as far as the chip is concerned, a distinction must be made as to whether it has separate inputs and outputs ("separate I/O") or so-called "common I/O", that is shared pins for input and output. In the latter case, no output may be made whilst writing is in progress.

Scrambling

If one wishes to make a physical analysis of a faulty cell, or to write a particular pattern into the cell array for test purposes, it must be noted that the actual arrangement of the cells on the chip does not conform to one's preconceptions of an ideal matrix. One reason for this is that the chip is usually not square. However, a much more important reason is that the manufacturer is concerned to accommodate as many cells as possible in the smallest possible area, and to address them with the least possible delays. In other words, we can say that the logistics required in order to write data into particular cells, or to read it from them, calls for the smartest possible addressing arrangement.

In order, despite the above, to obtain the physical position of any cell which is being addressed, the address must be "descrambled" either by using a special program or by reference to the address coding. Scrambling means that the address is recalculated so that a defined cell in the chip is addressed. What this means, however, is that the address which must be set on the chip's address line is actually different from the one which one wishes to address logically.

For this purpose, in series production testers specific scramblers have been programmed for each DRAM type, and these are inserted after the address generator to produce the required physical addresses.

Scrambling differs from one manufacturer to another, because each manufacturer has its own philosophy on the addressing of cells. In addition, however, scrambling can vary between the different design variants from one manufacturer. Information about scrambling is not usually included in the specification, but is only provided if specially requested.

Apart from "address scrambling" which we have just described, "data scrambling" is also important. As explained below in the section on "Sensing", depending on the word or bitline addressed, the actual signal is stored in inverted or unchanged form. In order to perform a cell array analysis, it is absolutely essential to know what physical data was actually written into the cell. Using the data scrambler, the data values are set at the input to the chip so that the data held in the cell is that which was actually wanted.

Note must be taken of whether data scrambling is used with address scrambling or without. If the addresses are scrambled before the data items are scrambled, the data scrambler must take this into account differently from when no address scrambling takes place.

Redundancy

Irrespective of the large number of cells which have by now been integrated onto one chip, it is essential that all the cells function under all the foreseeable conditions. Not one single cell may fail. Because the internal structures have dimensions in the submicron range, it is almost impossible to prevent every single fault. This applies particularly to contamination by particles which can affect the processing activities on individual cells, and their subsequent functioning. A financially sensible yield would hardly be possible if only the fault-free chips could be sold.

In the initial stages of the development of a DRAM, the process is generally not yet sophisticated enough for it to be possible to manage without redundancy. Indeed, as much redundancy is integrated on the chip as it is possible to find space for in one way or another. This enables the yield to be raised from virtually zero to 30, 40 or an even greater percentage. Over the course of the life cycle of a DRAM, as the process technology gets run in, less and less redundancy will be incorporated. This may lead to the number of redundant cells being reduced to the benefit of a smaller chip. Another consequence of the chip becoming smaller is that more chips can be accommodated on a wafer.

To enable individual faults to be tolerated, a repair process is set in train. For this purpose additional cells are available, apart from the actual cell array, in the form of redundant bit and wordlines. Because only a restricted number of these redundant cells is present, and they are not available for any arbitrary repair, there are still some chips which are unusable and must be discarded.

To enable faulty cells to be replaced by fault-free redundant cells, the faulty address is written into a fuse bank on a redundant line. Each redundant word or bitline is controlled by such a fuse bank. Technically, the fuse banks are a programmable read-only memory (PROM), which is permanently programmed during manufacture. As this has to be done after the test runs, there are two common techniques: metal bridges, which can be severed using a laser, and electrical fuses the resistance characteristics of which can be changed on a one-off basis by a specified overload. In operation, the addressing logic can then recognize which address must be redirected to which redundant line.

Cells

A cell comprises a capacitor and a selection transistor. The capacitor stores the two states "full" and "empty" as representations of "1" and "0". The capacitive effect is created by a conducting layer which is insulated from the cell plate (shared by all the cells) by a thin oxide. The capacitance is greater, the greater the area of the conducting layer and the thinner the insulating oxide. Adequate cell capacitance is a prerequisite for sensing the cell content and determining whether a "1" or a "0" was stored in the cell.

Up to the 1M generation, it was possible to build up the cell capacitor as a parallel plate on the surface of the silicon. As from the 4M generation however, even in spite of improvements in the dielectric, it has become impossible to use planar cells. A three-dimensional structure, the "trench", has now come into use. As its name indicates, a trench is a 'ditch' or, more accurately, a hole in the silicon. In order to raise the capacitance of the capacitor, the surface of the trench is used as the capacitor. Here, the dielectric is deposited on the walls of the trench, providing high capacitances within a confined surface area.

Because the trench process is difficult to control, various manufacturers have developed the stack cell. In principle, the stack cell is the same as the trench except that, instead of hollowing out a hole, a "hill" is built up increasing the capacitance, too.

Sensing

A significant problem in the development and operation of a DRAM is how to retrieve the data from the cell. On the one hand, the data in a cell is represented by an extremely small capacitance (approx. 25-50 fF), while on the other hand this capacitance is often yet further reduced by various influences. The need is to amplify the magnitude of the electric charge, so that the correct data can be reconstructed.

Each cell capacitor "hangs", so to speak, on a selection transistor, through which the charges are fed in or out. This transistor is switched on or off by the wordline. The data going from and to the cell flows along the bitline to which the selection transistor is connected. During writing there is no problem, because in this case the maximum charge is always offered to the cell by the supply voltage. During reading, the cell is then switched onto the bitline. This changes the charge relationships on this particular bitline. By activating the read amplifier, the so-called "sense amplifier", the change in the charge is amplified so that a "1" or a "0" can be recognized.

A bitline consists of a pair of complementary bitline halves, which are both connected to the same sense amplifier. The two halves are referred to respectively as bitline true and bitline complement (BT, BC). One wordline then always corresponds with a bitline half. Each cell capacitor is connected via the drain/source of its selection transistor to a bitline half, and via the gate to a wordline. When the wordline is activated, all the cells on this wordline are switched through to their associated bitlines. As a result, the voltage on the bitline is affected by the charge which flows from the cell into the bitline. The effect on the bitline voltage is very small, corresponding to the ratio of the capacitances of the cell and the bitline (approx. 1:5). But this small voltage change must be detected during the sensing. For this purpose, use is made of sense amplifiers, which recognize and amplify the correct data from a minimal voltage change on the bitline. Each bitline pair has one p-channel and one n-channel sense amplifier. The function of these is to amplify the cell signal, which is switched through to a bitline half by the activation of the wordline, so that it is unambiguously possible to distinguish a "1" or a "0". The details of how the sensing is effected depend on the chip construction and the sensing concept.

It will be apparent that, due to the dynamic nature of the storage, and the sensing of data when the wordline is activated, there would generally be no timing advantage if, instead of providing multiplexed addresses, all the line and column addresses were made available to the DRAM simultaneously. There is always a time lag until the sensing process has been completed and the data is available, so that the column addresses can then be used internally.

Refresh

Because the cell in a DRAM consists of only a very small capacitor, it is physically inevitable that the charge stored in the cell will be lost over time, due to various leakage mechanisms. Of these, the main one is the temperature, which has the effect that the charges contain higher energy, and can more easily overcome the insulating barriers. The time which elapses before a cell will be incorrectly sensed, because some of the charge has drained away, is called the refresh or retention time. This is dependent on which signal has been stored. If a "0" is stored in a cell and the cell environment, the substrate, is at ground potential, then there is no reason for the cell to lose its charge in the normal way, by leakage currents. However, for a more precise consideration of the refresh needs, one must take into account the fact that the environment of a cell can also consist of neighboring wordlines or bitlines which are conducting "1"s, and also other cells. Nevertheless, if a cell has been optimally processed, and if it has no unintended weak points, it is possible to determine the ultimate refresh time. This time is then referred to as the intrinsic refresh time.

When DRAMs were being developed, it was necessary to take appropriate account of the fact that over time (generally a few milliseconds) the data is lost. In addition, it was necessary to ensure that the enforced pause before the chip is in the ready state, which arises from the refresh mechanism, is kept as short as possible. In order to retain the data in the cells, it must be read out from the cells at certain intervals and written back after amplification. The process takes place each time that a wordline is selected. When a wordline is selected, all the selection transistors along this wordline are connected through to their associated bitline, and the cell signals are sensed by a read amplifier and then written back into the cells again at full signal strength. In order to reduce the number of refresh operations, in general more than just one single wordline is addressed. Furthermore, for the SDRAM it has been specified that for each auto-refresh command one out of 8K wordlines is activated in parallel in each bank, and all the connected cells are sensed and written back.

In order to refresh all the cells in the memory in good time, all the wordlines must be addressed at specific intervals. As such regular addressing practically never happens in the everyday use of a computer, a special refresh command has been defined which, for asynchronous DRAMs was called "CAS before RAS" (CBR) and in the case of an SDRAM is called "Auto Refresh". With these commands, the wordline address is generated internally by means of a counter and appropriate logic, and is not set from outside. "Self Refresh", also called "Sleep Mode" is a continuous refresh, completely without external interaction.

In general, however, it is up to the processor when and in what way the DRAM is refreshed. There are two normal variants: the "Burst Refresh" ensures that all the wordlines for the memory are addressed in a single pass. A "Distributed" or "Steal" refresh ensures that only one or a few refresh commands are ever executed together. The pause between the individual refresh cycles must then be chosen such that all the wordlines are addressed at least once within the maximum refresh cycle.

As a corresponding number of cells must be sensed simultaneously for each wordline, and this is reflected in the current consumption, it is possible to reduce the current consumption by increasing the ratio of the number of line addresses to column addresses, and hence reducing the number of cells to be refreshed simultaneously. If one were to consider only the refresh operation, it makes no difference to the current consumption whether one uses 4096 cycles to refresh 1024 cells, or 2048 cycles to refresh 2048 cells. The difference in the current consumption arises from the fact that each normal activation in a write or read cycle includes the corresponding number of cells. For example, if it is 2048 cells each time, then the current consumed is twice as much as if only 1024 cells are refreshed each time.

This explains why, in current generations of DRAM, symmetric addressing is no longer usual – the number of line addresses is larger than the number of columns. This requires more address pins, but does reduce the active current. For mobile systems, powered by accumulators or batteries, additional power-saving functions are incorporated. Further power savings are possible by means of temperature-dependent refresh control in the sleep mode, or by restricting the memory areas addressed by auto refresh in the active mode, and by special power-saving circuit variants which only permit lower maximum frequencies.

RAS only Refresh

As already mentioned, to refresh all the cells on a wordline it is sufficient to address this wordline. This also means, however, that it is sufficient to simply supply the row addresses, by means of the activate command, in each cycle, followed by a precharge command, without there necessarily being a command containing a column address. It is only necessary to ensure that in each case the correct row address is set for the next wordline to be refreshed. This is either the responsibility of the processor, or requires a special refresh controller.

Self Refresh

Normally, a peripheral circuit determines when and how often a refresh cycle is inserted. With Self Refresh, an internal timer ensures that internal Auto Refresh commands are started at the correct intervals, and in turn these are executed without external commands or addresses. There is a dedicated command for swapping out of the standard operating mode into the Self Refresh mode. In this state, the module will respond only to one command, Self Refresh Exit. This is the signal to reawaken the memory module which appears to be asleep, for which reason the operating state is also called Sleep Mode. Self Refresh is a mode which puts the chip into a state of minimal current consumption while guaranteeing the data content. As the user never knows precisely which part of the chip was the last to be refreshed while in the sleep mode, the first thing that should be done on exiting from self refresh is to completely refresh the chip (Burst Refresh).

The internal timer for the self refresh is subject to typical process variations. By the use of correction circuits, which like the redundancy circuits are programmed by means of fuses, the correct interval specific to the module can be programmed.

Refresh problems

The term refresh normally means the mechanism which handles the refreshing of the cell contents. Refresh time is the term for the time between the writing and subsequent reading of cell data above which the first errors arise due to loss of data. This time is also referred to as the retention time.

As the storage of data is the actual function of a DRAM, this retention time is of particular importance. The retention time depends on various factors, and theoretically it can be different for each cell. In connection with a cell array, the retention time is an excellent indicator of process quality. The retention time should be approximately the same for all the cells, and as large as possible.

The problem, which arises as a result of the large number of interdependent factors, is that the chips which are sold must have a certain minimum retention time under all conditions. This time must be guaranteed by suitable tests. However, one cannot use tests to simulate every possible condition under which a chip will be operated. Apart from which, the test time must be kept within justifiable limits.

The need is, therefore, to develop tests which do not test the possible operating conditions but rather investigate selectively the physically possible malfunctions of the cells and their environment. This means that an appropriate selection of tests must be developed for each chip architecture and technology.

If one is in a position to investigate all the possible fault mechanisms, faulty cells can be localized and repaired in good time. This is beneficial in terms of yield stability, and avoids customer complaints.

When considering refresh times, one must first subdivide them into two categories. The first is the intrinsic refresh time. The retention times achieved for an intrinsic refresh, when the construction of the cell is fault-free, depend only on the physical parameters of the cell structure and the technology. The intrinsic refresh time depends essentially on the strongly temperature-dependent generation and recombination in a semiconductor.

Depending on the technology, the retention time also depends on what data is stored. The reason for this lies in the fact that the main data losses take place via the immediate surrounds of the cell capacitance. For example, the cell capacitance can have two states if the second connection to the memory capacitor is held at ground potential: 1st charged; 2nd uncharged. If in one of these states its potential difference from its environment is equal to 0, then it is clear that no charge will flow in this situation. This means that the data value for which the potential difference to the environment is the smaller must have a longer retention than the data with the inverse value.

One can say that the significance of the intrinsic refresh time is that it is good to know roughly what its value is, and to know that it is long enough to satisfy the specification. However, what is more important for everyday DRAM business is the second group of refresh times. Into this group falls the refresh time which is less than the intrinsic refresh time due to faulty processing. Moreover, with the complexity in manufacturing a DRAM as a mass-production product it is normal that the cells do not all uniformly have the maximum possible retention time. Depending on their magnitude, these faults determine the yield. As absolute freedom from faults cannot be achieved in practice, one must give thought to which faults will be permitted, and which will be regarded as a problem.

At 85°C the intrinsic value is of the order of magnitude of seconds. The specified retention time is generally 64 ms. One cannot now simply say that a chip which is tested and just achieves the 64 ms plus a safety margin of perhaps a further 64 ms is actually good. As a "healthy" cell has a refresh time of seconds, cells which have a just adequate refresh time must be viewed as "sick", and represent a major service life problem. One can make the assumption that these cells are suffering from so much prior damage that they may fail prematurely when in operation with a customer. In assessing which cells are categorized as sick or healthy, a middle way must be adopted between financial and quality considerations. Where this way lies will only be found from experience with the characteristics of the fault mechanisms which are decisive in limiting the service life. So a distinction must be made between faults which are stable over the service life of the chip and those which intensify with time.

Basically, one can assume that there are possibilities of faults everywhere where the contents of the cell produce a potential difference relative to its environment. The latter includes, apart from the substrate, also the driving transistor and the bitlines and wordlines which are routed over or in the neighborhood of the cells. In the case of the bit/word lines this can also be a temporary state which depends on the timing sequence of the drive activities. So, certain addressing might hold a wordline at such a potential over a lengthy period of time that a failure occurs in a cell, which would not occur with well considered driving of the chip. However, as no-one can predict how the cell array will be selected, it is necessary to work on the assumption of the most unfavorable conditions.

In total there is a host of potential leakage paths in a cell's environment, each of which will be activated only under quite specific conditions. The task now is to find the mechanisms which activate the leakage paths, and to convert this into suitable testing programs, which activate these mechanisms in such a way that the potentially weak cells can be found in good time. A weak cell which is found in good time can be replaced, provided that there is still redundant capacity available. Although this repair capacity is limited, it should nevertheless be utilized as well as possible.

Soft Error Rate

The cell data is not only affected by electrical and physical phenomena due to internal conditions. Another fault mechanism is that of soft errors, caused by alpha particles. These alpha particles are produced in the package by unavoidable radioactive contamination, for example with uranium 238. They penetrate the surface of the chip and produce a shorter or longer track of ionized molecules. Their range amounts only to a few µm. But they are sufficient to create or neutralize cell charges along their paths. It is also possible for bitline data to be corrupted during its sensing. These errors are called "soft errors", because they are generated by chance, corresponding to the random process, and are not reproducible. Errors caused by alpha particles do not cause any permanent damage. The next time that the cell is written to, all will be in order again. But even an error of this sort, which occurs temporarily, can in some circumstances cause substantial data loss in a computer system.

Because it is impossible to manufacture a packaging material without impurities, one must live with an SER (Soft Error Rate). The acceptable error rate is specifies as a "FIT" (Failure in Time) value. Here, one FIT means one error in 10^9 hours of operation. 500 FIT is a common level which the customer expects as the maximum value for DRAMs.

The susceptibility of a DRAM to this problem depends mainly on the cell architecture and the cell capacitance. Thus the physical design of a cell offers more or less opportunities for alpha particles to discharge it.

Ultimately, the SER must also be determined. There are two methods for doing so. The accurate but very costly method is field trials. Here, thousands of chips are operated over many hours, and the natural SER is recorded. To achieve as good statistics as possible, long period of operation and large numbers of chips are required.

A second method is an accelerated test, using a radioactive substance. This test is also called the ASER test (Accelerated Soft Error Rate). This involves holding a radioactive probe over the cell array. Using the known radiation from the radioactive probe and a suitable conversion procedure, it is possible to calculate a natural SER. However, the calculations must be performed with great care. There are various calculation procedures, which are often very doubtful, for determining the absolute SER. However, what can be done well is to make comparative measurements.

Interface to the external world (inputs and outputs)

The interface to the outside world is precisely defined by the specification. The definitions are formulated less in the interest of maximizing progress and much more to achieve the greatest possible continuity and compatibility.

However, the specification is binding and, even if there is only one competitor who can achieve this specification to an adequate extent, there is no way round it as a manufacturer, but to accept these requirements. Over and above this, the user demands adequate sureties, some of which go well beyond the specification.

6.2.8 Development and production of a DRAM

- Fundamental development of a processes for the structural dimensions which are being aimed at, and determination of the cell type and architecture of the storage.
- Circuit development (design) for all the required circuits, and their functional testing with the help of simulation.
- Drawing (layout) of the circuits and comparison (verification) with the circuit plans from the design stage.
- Calculation of the individual masks and their production.
- Processing of the raw silicon wafers to finished wafers.
- Test and repair.
- Packaging (package production).
- Testing of the finished chips.
- Burn-in.

- Final test and sorting by speed.
- Sale.
- Customer support (fault analysis) for faulty chips.
- All the steps after the operable wafers are completed are followed up by analytical measures, and the results feed into the processing, design and layout.
- This cycle is repeated many times in the course of one generation of DRAMs, because of the ongoing attempts to make an ever smaller, more reliable and faster chip.

Design/simulation

The development of the electrical circuits which make up an IC is carried out by developers at workstations, with the aid of CAD programs. By this process, circuits are built up graphically on the screen, using the usual electronics symbols. From such a circuit, a netlist is produced, showing an exact description of the contents of the circuit, its elements and their connections to each other. This netlist can now be subject to a check for errors, and can be used as the basis for a simulation. The checks on the netlist can, of course, only detect syntactical errors; that is, errors which infringe the general rules for circuit engineering. This includes such errors as, for example, 4 connections to a circuit which only expects 3, or inconsistent naming. By contrast, semantic errors will not be recognized. These are errors in the circuit concept. An example of such an error would be if an incorrect signal state is set at an output because an inverter has been forgotten. Because whether an inverted signal or an uninverted signal is to be expected at the output is generally something which only the developers themselves will know.

In order to eliminate such semantic errors, simulations are performed using programs which test the logic of a circuit (logic simulation) and with programs which represent the time sequence of the switching activities for each individual transistor, and hence of the complete circuit. "TITAN" is a program of this sort, and is a development of the widespread SPICE analog simulator. With the help if this program, the exact timed behavior of each circuit can be simulated. A prerequisite for doing this is that the appropriate model parameters must be available for the technology used. Depending on the quality of the model parameters, the predictions about the functioning of the circuit can be very accurate. Even though the functioning of a DRAM is largely digital, the use of analog simulation is unavoidable, because the usual logic simulation only makes statements about what happens when a signal changes from "1" to "0". How long the signal requires to change its state, because a driver transistor has a particular relationship to the driven load, is something about which a logic simulation makes no predictions. In order to get out of a DRAM the most, in terms of speed, that is electrically possible, and hence keep the transistors as small as possible because of the space requirements, analog consideration of the switching behavior is unavoidable. Also, an analog simulation is the only way to take correct account of any process dependencies, such as thermal response.

Layout/verification

The symbolic representation of the circuits, which the designer develops, must then be transformed to meet the physical constraints of a real circuit. The first step in doing so is that all the individual parts of the circuit are drawn graphically in accordance with appropriate design rules. So, transistors are drawn exactly, with their diffusion zones, the gate and the connections to the environment. Unlike the design, in which here and there one transistor more or less can be used, as required, in the layout every individual transistor must be haggled over, because space on a DRAM is always very limited.

Just as a netlist can be produced from the symbolic circuit, one can also be generated from the finished layout, which has been tested against process-conformity design rules. This netlist is now compared against the netlist for the symbolic circuit. Differences are indicated, and must be eliminated if the circuit in the layout is to be as it was conceived by the designer.

Apart from verifying a circuit against the design, the layout must be checked for adherence to certain process-related design rules. Thus, the minimum spacing from metal is checked, just as is the size of a contact hole, or the size of a diffusion zone in a transistor. There are numerous rules which must be observed, because the process cannot produce every arbitrary structure.

Process

The process represents the actual production of the chip. In this, the chip is produced from a pure silicon wafer by a wide variety of processing steps, such as exposure, etching, polishing, coating, implantation, oxidation and so on.

Assembly

After the wafer has been processed, and the chips which are deemed good have been selected, the wafer is divided up using diamond-tipped saws. With the TSOP package, which has been dominant up to now, the usable chips are sorted out and attached to the leadframe with a foil, using the LOC technique (Lead on Chip) (see Figure 6.11). Then, by bonding with gold wires, the connections from the pads in the middle of the chip to the leads on the leadframe are established. This entire construction is then cast in the plastic mass. The leadframe is then freed from the residual supports, and the pins are formed. In the classical form of assembly, the connecting pads were on the edge of the chip, requiring more extensive rewiring on the memory module. In this latter case the leadframe takes a much shorter form (see Figures 6.11 and 6.12).

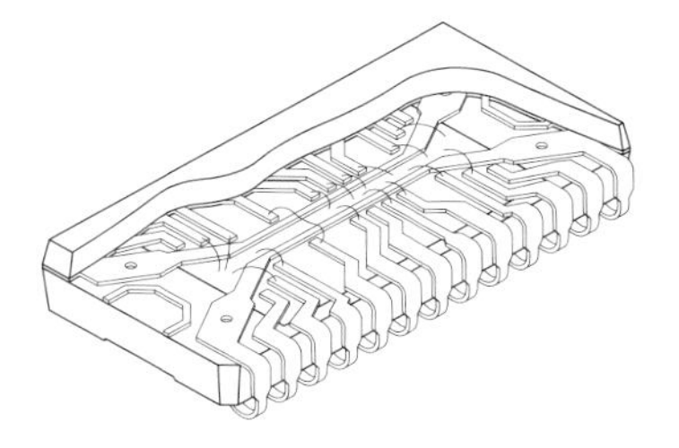

Figure 6.11
LOC (Lead on Chip) assembly technique for DRAM chips

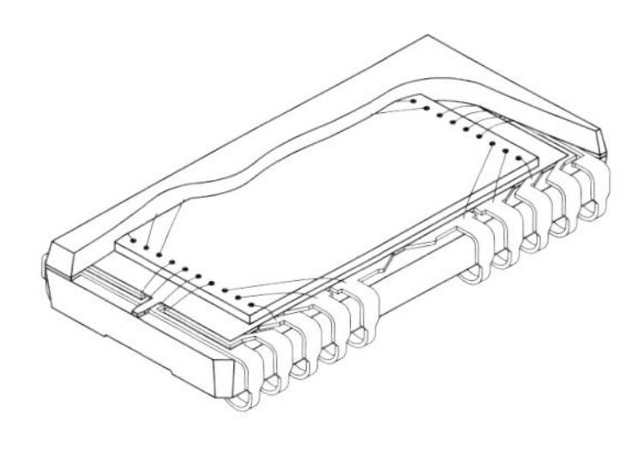

Figure 6.12
Standard assembly technique for DRAM chips

6.2.9 Quality assurance

Essentially, quality assurance is responsible for ensuring not only that the chips which are delivered to customers function, but also that they do not deteriorate markedly in service, or even fail. This involves, particularly, investigations into the service life of the product. The chips are subject to above-normal levels of stress to the point that they fail. Raising the stresses to excessive levels, in terms of supply voltage and temperature, results in an acceleration of the internal disruptive processes, which eventually destroy the chip. As a chip with a predicted service life of 10 years cannot realistically be tested for 10 years, the raised stress is used to achieve a defined shortening of the service life. This makes it possible to decide within days whether a chip would last 10 years in normal operation.

Series testing

All the DRAMs which are manufactured must undergo various tests. The first tests take place on the wafer. These investigate which chips are suitable for further processing, and which chips can be repaired and how, so that they can be further processed. After their assembly into packages, the chips are tested again to detect any faults caused by assembly or by faulty repair work. The chips, which are now functioning, are stressed in a burn-in system for a few hours, with a raised voltage and temperature. After this, the chips which have failed are again selected out by testing, and finally tests are carried out for the purpose of selecting the chips by speed.

The programs required for this purpose must be modified for each DRAM generation and its specific characteristics (new operating modes). The results of the design analysis and quality assurance also provide inputs for this activity. Each design or technology has its own weaknesses, which must be studied and sorted out. As a result, even though the DRAM generations have very similar characteristics, the test programs are very individual.

Design and fault analysis

Unlike series testing technology, by which the chips are automatically measured in their thousands by test robots (handlers), in the case of design and fault analysis each chip is dealt with individually. Here, design analysis investigates the overall functioning of the chip as well as the functioning of the individual circuits. This is possible because, using extremely fine needles, signals can be measured from aluminium tracks within the chip with widths in the submicron region (Figure 6.13). The needles are controlled with the help of micromanipulators, at a microprobe measuring station. Using these manipulators, one can position the needles under the microscope at any required place on the chip. In this way, it is possible to follow the internal time-sequence of the signals. In addition, one can determine the extent to which the circuit corresponds to what is wanted and, if the chip does not function correctly, where a signal is proceeding incorrectly.

In order to be able to use needles to make measurements within a chip, it must be possible to contact the required track with the needle. Normally the circuits are protected by a polyimide, an oxide and a nitride layer, so it is necessary either to

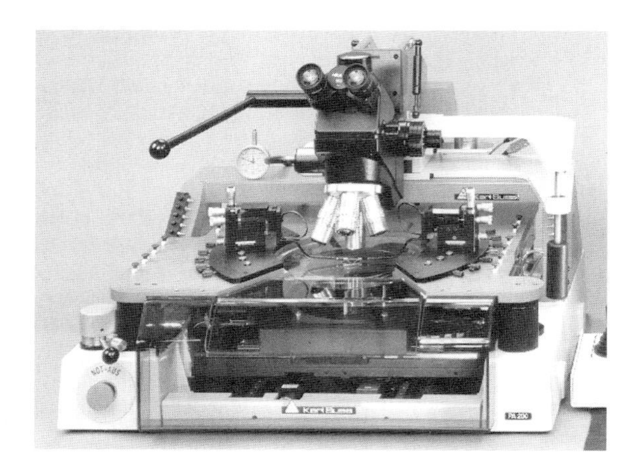

Figure 6.13
Microprobe measurement station with micromanipulators

leave these layers off during the processing or to remove them later by chemical means.

Fault analysis mainly starts from a fault in series testing, for which one is attempting to find the cause. Here, design analysis and fault analysis work closely together, because a fault may well be attributable to a design error. Apart from this, the attention in fault analysis is more on faults with technological causes.

During design analysis matters are relatively simple. One sets a particular test condition and measures what the internal signals do. In doing this, it is generally of no importance whether or not the chip functions. It is often sufficient if the individual circuit functions. In the case of fault analysis on an enclosed chip, it is not possible to satisfy oneself about the quality of the signals. In this case, the chip must be driven in such a way that faulty behavior shows up as a false item of data at an output. With an enclosed chip it is only possible to analyze faults by whether an item of data which one has written into a particular cell is still exactly the same when retrieved. Faults which really are present, but which under the test methods used do not lead to an inversion of the output signal, can only with difficulty be localized. On the other hand, if it is possible to measure the chip internally one can see from the interactions of the individual signals where something is wrong.

6.3 How DRAMs have become faster

EDO DRAMs, and SDRAMs which work synchronously, are examples of how higher data transfer rates can be achieved during memory accesses by circuit-technological measures. Here, experience has shown that it always takes some time before the memory controllers are also able to take advantage of the new features.

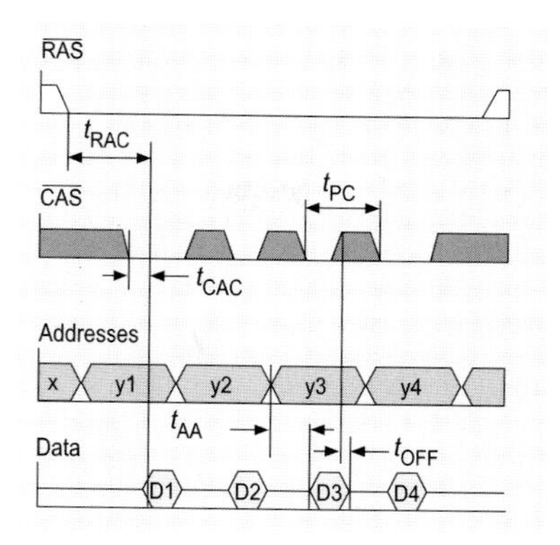

Figure 6.14
In the fast-page mode, when the CAS signal is inactive the data output goes high-ohmic

Until about 1998, "Fast Page Mode" (FPM-DRAM) was the sole standard operating mode for dynamic semiconductor memories. Figure 6.14 shows an example of the main control signals – RAS, CAS – and the addresses, for a fast-page mode DRAM module. The row addresses are transferred first, at the falling edge of the RAS signal, and then the column addresses at the falling edge of the CAS signal. After the time t_{RAC} or t_{CAC}, these denoting respectively the RAS or CAS access times, which are correspondingly measured from the falling edge of the RAS or CAS signal, the first data bit appears at the output for the module. In an FPM module, the data output is switched to be high-ohmic when the CAS signal is set as inactive and the next memory access to the page currently open is started.

6.3.1 EDO DRAMs speed up memory accesses

As from 1995, EDO DRAMs (extended data out) became available on the market. Using these, it was possible to raise the data transfer rate for a memory subsystem without having to give up compatibility with the FPM modules, both in respect of the memory organization and also the forms of packaging used. From a technical point of view, an EDO DRAM offers the advantage of a shorter cycle time t_{PC} in page mode. The main difference lies in the fact that the data values remain set at the data output even when the CAS signal has been set to inactive (Figure 6.15). By means of this circuit measure, it is also possible to save one clock cycle for each page access, and thus to shorten the time for one page-mode cycle. One thereby achieves a higher working speed for the overall system.

With this circuit measure, the cycle time is shortened from 35 ns for a 50-ns FPM module to 20 ns for an EDO module, an increase in the data transfer rate of 75%. However, there is no linked reduction in the access times t_{RAC} and t_{CAC} – the time up to the first access. These times are the same as for the FPM modules.

Most processors request the data from their main memory in a 4-bit burst. An optimal data transfer between the main memory and the processor will be achieved if data can be read or written with each clock cycle. This cannot be achieved with either FPM and EDO memory, because both require a new line address to be set for each new data bit. One must insert one or more waitstates between the clock pulses.

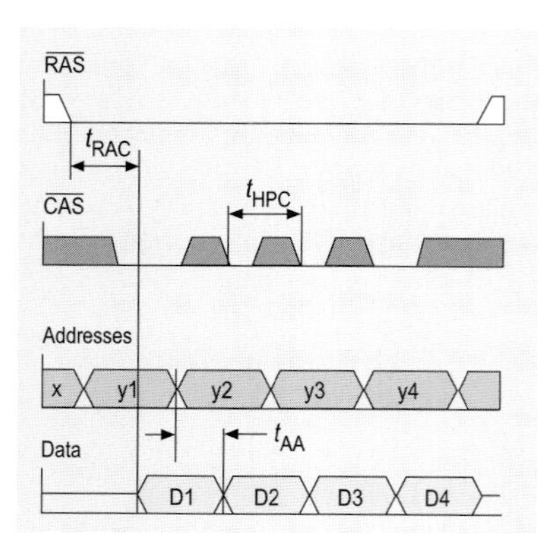

Figure 6.15
With an EDO DRAM, the data value remains at the data output even when the CAS signal is set to inactive

Under the assumption that the data which is being sought is located on a page which is already open, reading a 4-bit burst when the clock rate on the memory bus is 66 MHz requires 5/3/3/3 cycles for FPM modules and 5/2/2/2 cycles for EDO modules. In both cases, the first of the numbers refers to the number of clock cycles until the first data bit is available, the following numbers to the number of clock cycles until the next of the data bits is available.

6.3.2 Synchronous is faster

Synchronous DRAMs (SDRAMs) are memory modules with a circuit architecture which works synchronously with the bus clock. This brings further advantages for the system developer in terms of speed and performance, and at the same time reduces the development cost for the complete system.

To this end, in an SDRAM module all the input and output signals are clocked synchronously with the external bus clock, and the control signals are in each case read into the module for processing at the rising edge of the clock pulse. Using this technique, it is possible to read or write data at each clock pulse. The data items then appear at the data output at 10-ns intervals (at 100 MHz). Furthermore, it only remains to set a start address for the burst. There are no addresses to be set for the subsequent bits in the burst. Even with SDRAMs, however, the time up to the first data bit is no shorter than with the FPM and EDO modules (Figure 6.16). This means that in the most favorable case one now only requires 5/1/1/1 cycles to read a 4-bit burst. Using it, one can now fill the fast SRAM cache with no additional waitstates.

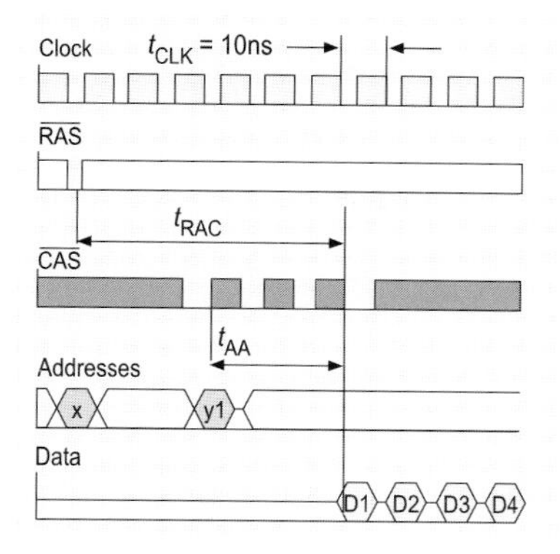

Figure 6.16
With SDRAMs the data appears at the output at intervals of 10 ns

With an SDRAM, both the length of the data burst and also the time up to the first data bit (the "latency") can be programmed with a special command cycle, and thus optimized for every application situation. Furthermore, an SDRAM consists internally of two or more independent memory banks, which can be addressed simultaneously. Memory bits coming from different banks can be interleaved in such a way that the precharge times and waitstates required before the first access to a memory can be hidden. In this way one can achieve a continuous data stream at 133 MHz and more for both write and read operations.

6.3.3 Double data rate

At frequencies above 150 MHz, the I/O-related power loss for the LVTTL interface, considered in relation to the system, is comparable with that of the SSTL interface, and above this frequency it becomes ever more disadvantageous not only in terms of power loss but also in terms of signal integrity. For this reason, the option of an SSTL interface was investigated even for the first SDRAMs, and in some cases was actually offered. On the system side however, only one standard was supported, the LVTTL interface.

For systems with data transfer rates of 200 Megabits or more per second and pin, a search was made for an optimized solution, which should also incorporate the change in the I/O interface definition. To achieve this, it was decided that a higher data throughput rate was required only for the data, and not for the commands and addresses, and so the Double Data Rate (DDR) SDRAM was born. With this, data words were read or written both at the rising and the falling edges of the clock pulse. The clock rate could again be relaxed to 100 MHz, but the data throughput was higher than for 166 MHz SDRAMs.

Before the next specification for SDRAMs is launched on the market as DDR2 in 2004, the first Double Data Rate SDRAM, now designated DDR1, may be reaching its limits in systems with 200 MHz clock rates and 400 Mbit/s/pin. Thus, compared to the simple SDRAM this solution has brought an adequate increase in performance while retaining many proven features. Precisely such evolutionary performance improvements are especially welcome in the field of the mass market for dynamic semiconductor memory.

6.3.4 Modules simplify memory upgrades

Nowadays, memory components are no longer soldered onto the board, but are plugged as memory modules into a socket provided for the purpose. This means that it is also simple for the user, if the need arises, to upgrade the working memory of a PC. All the memory modules are standardized into module families (Figure 6.17) and can be obtained with various memory capacities.

The first modules were called SIMMs (Single In-Line Memory Modules), had 72 connections and could be obtained

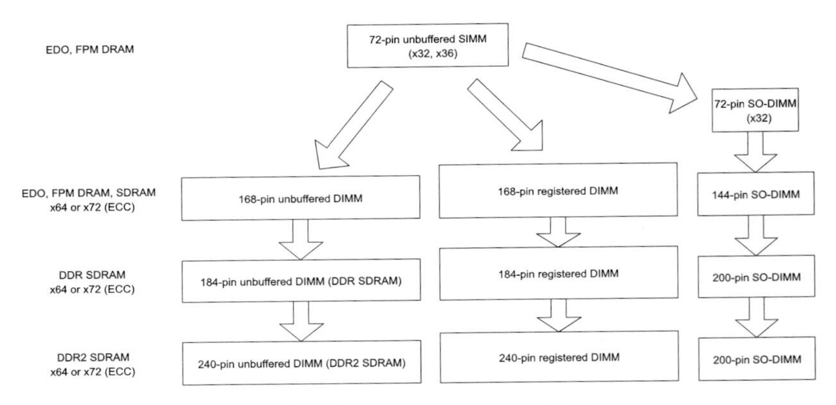

Figure 6.17 Standardized modules for memory components in PCs

with either a 32 bit data width for applications with no parity or with 36 bit data width for applications with parity, and used a 5 V operating voltage. This standard was used for asynchronous memory conforming to the FPM or EDO specification.

DIMMs (Dual In-line Memory Modules) are a development of the SIMM family. They can be regarded as two SIM modules on one memory card. The first modules of this type had 168 connections, 84 on each side of the card, and double the data width of a SIMM. However, the electrical pin dedications on the two sides of the card were different, so that they are not interchangeable. Both asynchronous DRAMs and SDRAMs were used as components, operating at a voltage of 3.3 V.

DIMMs are available as buffered and unbuffered modules. In the case of the buffered DIMMs, most of the input signals pass through a buffer module. This enables the unavoidable input capacitances to be reduced to values around 10 pF. This is mainly of importance if one wishes to implement systems with many sockets for DIMMs, as is the case with servers.

The "unbuffered" DIM modules contain no buffer modules. Instead, these modules could be fitted both with FPM and EDO DRAMs and also with synchronous DRAMs. All these modules had the same connection usages, and were interchangeable on suitable motherboards, and in most cases they could be operated in a mixed combination. This gave the PC manufacturer the maximum possible flexibility for his product up to the point of dispatch. Unbuffered DIMMs were the first to have the feature "serial presence detect" (SPD). For this purpose, the module included in addition a non-volatile E^2PROM memory, containing all the data relating to the module (memory size, organization, electrical characteristics, etc.). These items of data can be read out by the CPU and processed. This enables the memory controller to set up the optimal access, according to the devices fitted in the main memory. With this module it is also possible to implement an optimized "plug & play".

For laptops and notebooks, special small module constructions are available, called SO-DIMMs ("small outline DIMMs"). For portable computers with a 64 bit data width, there was a SO-DIMM family with 144 connections, which can be fitted with either EDO or SDRAMs. Because of their small installed height (67.5 mm length and 25.4 mm height) these modules have spread rapidly for all portable computer applications.

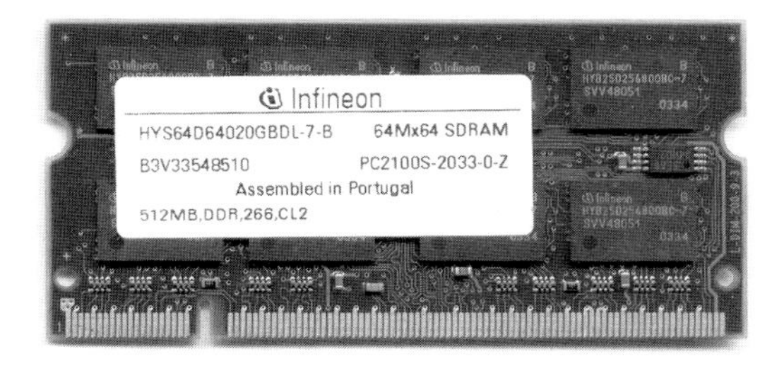

Figure 6.18
Small outline DIMM for portable computer applications: the figure shows a module with 512 MByte

DDR SDRAMs are sold on DIMMs, too. The pin count is 184 for buffered and unbuffered modules and 200 for SO-DIMMs. Figure 6.18 shows an SO-DIMM fitted with DDR-SDRAM for a data transfer rate of 266 Mbit per second and a storage capacity of 512 MByte. The pin count laid down for the following generation, the DDR2, is 200 or 240 respectively. The increasing pin count is caused by the increased number of memory pins for secure data transmission (data strobe signals). Further details can be found in the publicly accessible data sheets from Infineon.

7 Microcontrollers

7.1 Introduction

As early as the seventies, shortly after they had appeared on the market, microprocessors were already being used for automatic control applications. Such control applications included, of course, a microprocessor together with additional circuit elements, such as main memory (RAM), program memory (ROM), and a range of I/O connections, including interfaces such as UART or USART. Initially, these components were arranged together on a controller circuit board, which represented the controller unit.

Attempts were soon made to accommodate as many as possible of these peripheral components on a semiconductor chip, to provide a fully functional single-chip microcontroller. By the end of the 1970s, a range of 8-bit types was already available, equipped with internal RAM, serial connections, timing circuits, and an optional ROM or EPROM. In 1980, Intel's 8051 microcontroller came onto the market, and rapidly became the de-facto standard for 8-bit applications. Even today it is still available in a wide range of configurations (for details, see section 7.2).

Since then, increased demands in terms of speed, complexity and cost-effectiveness have been the driving forces for further developments. Modern high-performance types are microcontrollers with a processor width of 16 bits, and are manufactured in large numbers (see section 7.3). Increasingly, the trend for complex, demanding applications is towards 32 bits (for details see section 7.4).

7.2 8-bit microcontrollers

7.2.1 Introduction

In principle, the members of Infineon's 8-bit C500 microcontroller family are, in terms of architecture and software, fully compatible with the standard 8051 microcontroller family. In particular, they are functionally upward-compatible with the SAB 80C52/80C32 microcontroller. While all the architectural and operating features of the SAB 80C52/80C32 have been retained, the C500 microcontrollers differ in the number and complexity of their peripheral units, which have been adapted for their special application fields.

In this chapter, the basic architecture and functional attributes of the members of the C500 family of microcontrollers are described. More detailed information about the different versions of the C500 microcontroller will be found in the appropriate user manuals and data sheets.

7.2.2 Memory organization

The memory resources of the microcontrollers in the C500 family fall into different memory types (data and program memory), and in addition these can be accommodated either within or outside the microcontroller chip. The subdivision of the memory in C500 microcontrollers is typical of a Harvard architecture, under which the data and programs are stored in separate memory areas. The peripheral units accommodated on the chip can be accessed via an internal special function register memory area.

Table 7.1 C500 address spaces

Memory type	Memory location	Size
Program memory	External	Max. 64 kbytes
	Internal ROM, EPROM	Depending on C500 version, from 2 up to 64 kbytes
Data memory	External	Max. 64 kbytes
	Internal XRAM	Depending on C500 version, 256 bytes up to 3 kbytes
	Internal	128 or 256 bytes
Special function register	Internal	128 / 256 bytes

The available memory areas have different sizes. Table 7.1 shows three types with six memory locations.

Program memory

The program memory in the C500 microcontroller family can consist entirely of external program memory, entirely of internal program memory (integrated ROM/OTP), or of a mixture of internal and external program memory. If the L-signal level is applied to its $\overline{EA}$ pin (EA = **E**xternal **A**ccess), a C500 microcontroller will always execute the program code in the external memory.

C500 variants which have no ROM can only use this program memory. In the case of C500 variants with integral program memory, generally only the internal program memory is used. When the internal program memory is being used, the H-signal level must be applied to the $\overline{EA}$ pin. With the H-signal on the $\overline{EA}$ pin, the microcontroller executes instructions internally, unless the program memory address lies above the upper limit for internal program memory. If the program counter contains an address (e.g. for a jump instruction) which is above the internal program memory, the instructions in the external program memory will be executed. When the instruction address returns below the internal program memory limit again, then the internal program memory will again be accessed.

Figure 7.1 shows the typical program memory configuration of the C500 microcontroller family for the situations $\overline{EA} = 0$ and $\overline{EA} = 1$. The ROM boundary shown in Figure 7.1 applies for a C501 with 8 kbytes of internal ROM. Other microcontrollers in the C500 family with different ROM sizes have other ROM boundaries.

Data memory

The data memory area of the microcontrollers in the C500 family consists of internal and external data memory areas. For internal data memory, 8-bit addressing is used, with 8-bit or 16-bit addressing for the external data memory and internal XRAM data memory.

A reset has no effect on the contents of the internal data memory. After a switch-on, the contents of this memory are not defined, but during and after a reset the contents remain unchanged, provided that the power supply is not switched off. The contents of the XRAM are also retained if the C500 microcontroller is in energy-saving mode.

Internal data memory

The address space of the internal data memory is subdivided into three fundamental, physically separate blocks: the lower 128 bytes of the internal data RAM, the upper 128 bytes of the internal

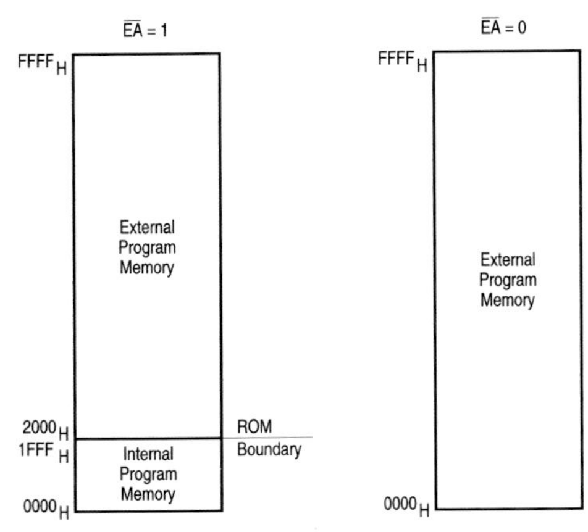

Figure 7.1 Program memory configuration (using the C501 as an example)

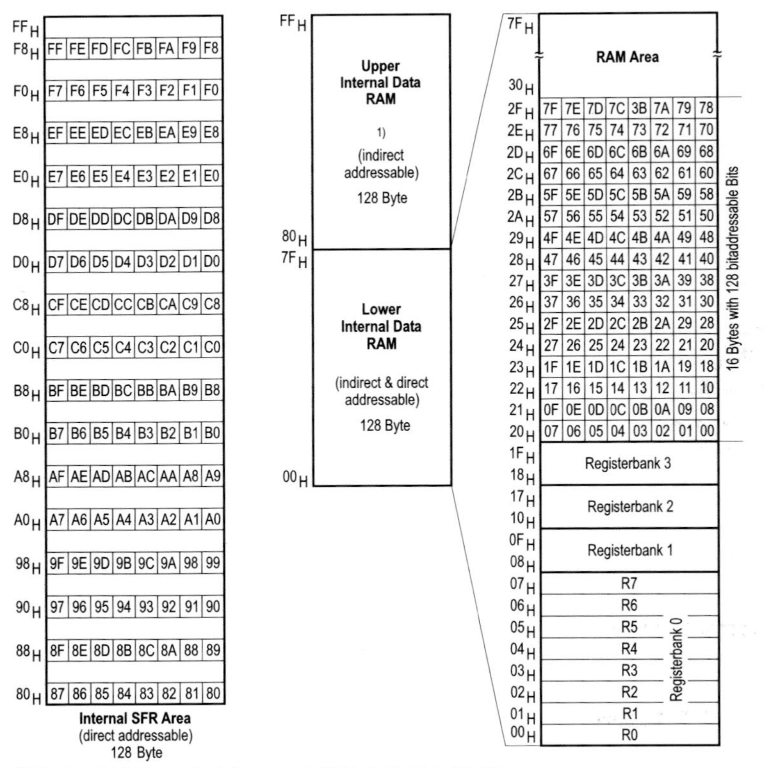

Figure 7.2 Organization of the internal data memory

data RAM and the 128 bytes of the special function register (SFR) area. In addition, the lower data RAM and the SFR area each contain 128 bits which can be changed using special bit manipulation instructions. The position of the ROM boundary depends on the particular version of the C500.

Figure 7.2 shows the configuration of the three basic internal RAM areas. The lower data RAM lies within the address range 00_H - $7F_H$ and can be addressed directly (e.g. MOV A,direct) or indirectly (e.g. MOV A,@R0, where the address is put into R0). Within the byte address range of 20_H - $2F_H$ in the lower data RAM is a bit-addressable area of 128 freely-programmable, directly addressable bits. Bit 0 of the internal data byte at 20_H has the bit address 00_H, bit 7 of the internal data byte at $2F_H$ has the bit address $7F_H$. The lower 32 memory locations of the internal lower data RAM are allocated to four groups with eight general purpose registers (GPRs). Only one of these groups may be enabled for use as general purpose registers at any time.

As the SFR area and the upper internal RAM area use the same addresses (80_H - FF_H), different address modes are required for accessing them. It is only possible to access the upper internal RAM by means of indirect addressing, and on the other hand the special function registers (SFRs) can only be accessed by direct addressing instructions. The SFRs, for which address bits 0 - 2 are equal to 0 (addresses 80_H, 88_H, 90_H, ... $F0_H$), are bit-addressable SFRs.

XRAM internal data memory

Some of the models in the C500 family provide an additional internal data memory area, which is referred to as XRAM. This data memory is logically organized as the upper end of the external data memory area, but is integrated onto the chip. As the XRAM is used in the same way as external data memory, the same instruction types must be used in accessing it.

Figure 7.3 shows a typical 256-byte XRAM address assignment for the C500 microcontroller.

Depending on the C500 variant, the size of the XRAM varies between 128 bytes and 3 kbytes. In addition, the XRAM can be activated or deactivated. If an internal XRAM area is deactivated, it is possible to use the address range of the internal XRAM for addressing the external data memory.

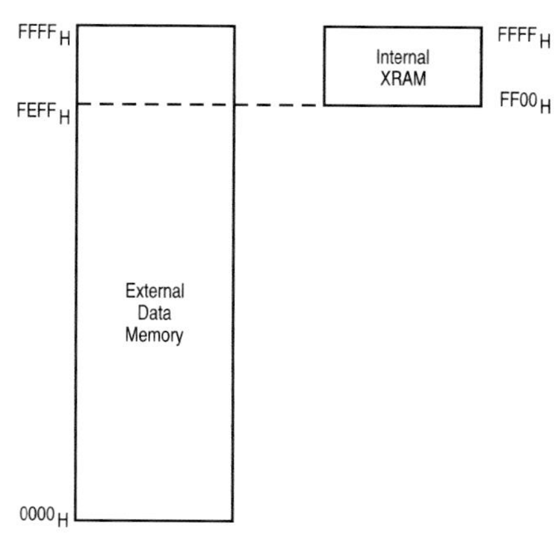

Figure 7.3
XRAM memory assignment (256 bytes)

External data memory

The 64 kbytes of the external data memory can be addressed by instructions which use 8-bit or 16-bit indirect addressing. The 16-bit external memory addressing mode is supported by the MOVX instruction using the 16-bit data pointer, DPTR, for addressing purposes. For 8-bit addressing, MOVX instructions are used in combination with the general registers R0/R1.

7.2.3 Special function register area

With the exception of the program counter and the four general purpose register banks, the registers of a C500 microcontroller are accommodated in the special function register (SFR) area (see Figure 7.2). As a rule, the special function register area provides 128 bytes of directly addressable SFRs. The SFRs, for which address bits 0 - 2 are equal to 0 (addresses 80_H, 88_H, 90_H, ... $F0_H$) are bit-addressable SFRs (see also Figure 7.2). For example, the SFR with a byte address of 80_H has bit memory locations with the bit addresses 80_H to 87_H. The bit-addresses of the SFR bits range from 80_H to $F0_H$.

Because the number of standard SFRs is limited to 128, some variants of the C500 microcontroller family provide an additional SFR area of 128 bytes, the so-called mapped SFR area. This allows the same addressing options (direct addressing, bit addressing) as the standard SFR area. With the indirect addressing modes, the registers R0 and R1 are used as pointers or index registers for the addressing of internal or external memory (e.g. MOV @R0).

7.2.4 CPU architecture

Figure 7.4 shows the typical architecture of a microcontroller in the C500 family. This block diagram contains all the main functional blocks of the C500 microcontroller. The shaded blocks are elementary functional units, which every C500 microcontroller must contain. The other functional blocks, such as the XRAM, peripheral units and ROM/RAM sizes, differ from one variant to another of the C500 microcontroller.

The central block represents the CPU (central processing unit) of a C500 family microcontroller. This CPU consists of the instruction decoder, the arithmetic unit,

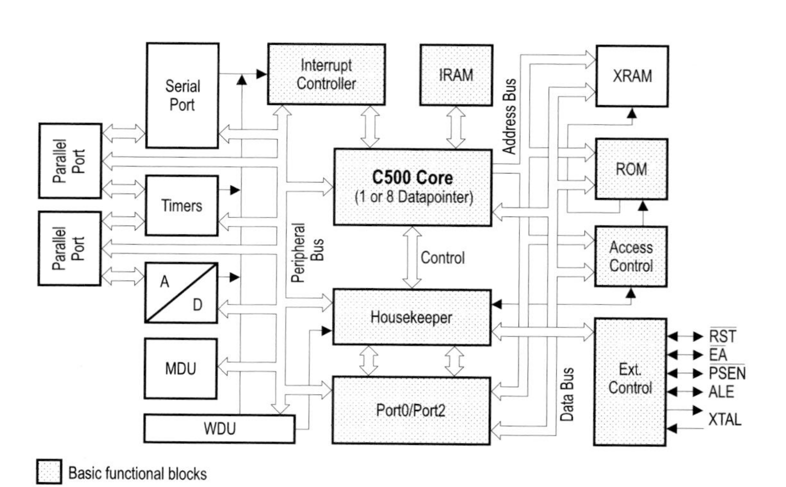

Figure 7.4 Block diagram of the C500 microcontroller architecture

the CPU registers, and the program control unit. The housekeeper unit generates internal signals to control the functions of the individual internal units within the microcontroller. The port 0 and port 2 pins are necessary in order to access external program and data memory, and for emulation purposes. The external control module is responsible for the external control signals and for clock generation. The access control unit is responsible for the selection of the on-chip memory resources. The IRAM contains the internal RAM, including the general registers. The interrupt requests from peripheral units are processed by the interrupt controller. The configuration of the peripheral units integrated onto the chip depends on their particular use. Typical examples of peripheral units integrated onto the chip are serial interfaces, timing circuits, capture/compare units, A/D converters, timeout monitoring units, and a multiplication/division unit. The external signals for these peripheral units can be tapped off at multi-functional parallel I/O pins or at pins specially provided for the purpose.

The arithmetic unit on the main chip carries out extensive data manipulations and consists of the arithmetic/logic unit (ALU), an A register, a B register, and a PSW register. In addition, it provides extensive facilities for binary and BCD arithmetic and powerful bit processing capabilities. Efficient use of the program memory is provided for by an instruction set which consists of 44% single-byte instructions, 41% two-byte instructions, and 15% three-byte instructions. The ALU can process 8-bit data words from one or two sources and, under the control of the instruction decoder, calculates an 8-bit result. The ALU can perform the following arithmetic operations: addition, subtraction, multiplication, division, incrementing, decrementing, BCD decimal addition-conversion, and comparisons as well as the logical operations of AND, OR, exclusive OR, complementing, and rotation (right, left, or exchange half-bytes/nibbles (the left four nibbles)). In addition it contains a boolean processor, which can carry out the following bit operations: set, clear, form complement, jump-if-not-set, jump-if-set-and-clear, and shift to/from carry.

The bit operations ‘logical AND’ and ‘logical OR’ can be performed between any required addressable bit (or its complement if applicable) and the carry flag, with the result being saved back into the carry flag. The program control unit in the main chip controls the sequence in which the instructions, stored in the program memory, are performed. The 16-bit program counter (PC) contains the address of the next instruction to be performed. Using the conditional jump logic, the processor can be instructed to vary the program execution sequence on the basis of internal and external events.

The accumulator

The symbol for the accumulator register is ACC. However, in the pseudo-code for accumulator-specific instructions the accumulator is referred to simply as A.

B register

The B register is used during multiplication and division, and serves both as the source and also as the destination. For other instructions, the register can be used as a scratchpad register.

Program status word

The program status word (PSW) contains several status bits, which specify the current status of the CPU. The bits of the PSW are used for various functions: two register bank selection bits, two carry flags and one overflow flag for arithmetic instructions, one parity bit for the contents of the ACC, and two general purpose flags.

Stack pointer

The stack pointer (SP) register is 8 bits wide. It is incremented before data is saved using a PUSH or CALL, and decremented after data has been retrieved using POP or RET (RETI), i.e. the stack pointer always points to the last valid byte in the stack. The stack can be stored at any required location in the on-chip RAM. However, the stack pointer is initialized to 07_H after a reset. This has the effect that the stack starts at memory location 08_H above the register bank zero. The SP can be read or written under software control.

Data pointer

8-bit accesses to the internal XRAM data memory or the external data memory are executed using the data pointer, DTPR, as a 16-bit address register. Normally, microcomputers in the C500 family have a single data pointer, but some models provide eight data pointers. This provides support, in particular, for programming in higher programming languages, which save data in large external memory blocks.

The enhanced hooks emulation concept

The enhanced hooks emulation concept of the C500 microcontroller family is a new and innovative option for controlling the execution of C500 MCUs and for accessing extensive data about the internal operations in the controllers (Figure 7.5). It is also possible to emulate programs which are based on the on-chip ROM.

Each production chip provides integral logic for the enhanced hooks emulation concept, so that no costly bond-out chips are required for emulation. This also ensures that emulation and production chips are identical.

The Enhanced Hooks Technology™ requires embedded logic in the C500 and ensures that the C500, together with an EH-IC, functions in a similar way to a bond-out chip. This simplifies the design and reduces the costs of an ICE system. ICE systems, which are equipped with an EH-IC and a compatible C500, can emulate all the operating modes of the various versions of the C500, even the modes: ROM, ROM with code rollover, and operation without a ROM. It is also possible to operate in single-step mode and to read the SFRs after an interrupt.

Port 0, port 2, and some of the control lines of the C500-based MCU are used by the Enhanced Hooks Emulation Concept for device control during the emulation, and for transmitting information between the external emulation hardware (ICE system) and the C500 MCU about program execution and data transfer.

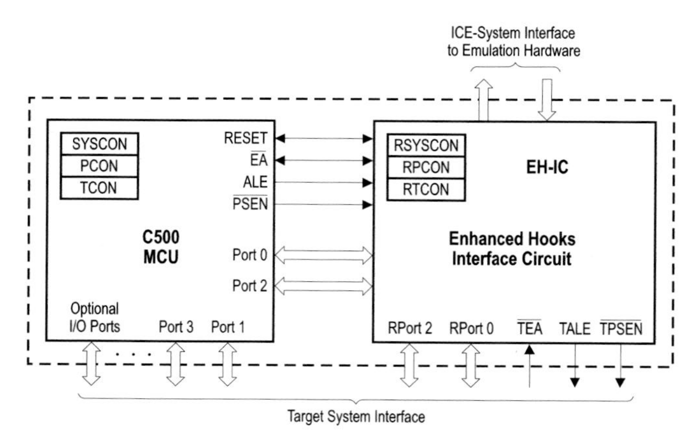

Figure 7.5 Basic configuration of the C500 MCU enhanced hooks

7.2.5 Basic interrupt processing

All the variants of the C500 microcontroller family provide several interrupt sources. These interrupts are generally triggered by external events or by the internal peripheral units. If the CPU receives an interrupt, the microcontroller interrupts any program which is running and continues the execution of the program at a vector address specific to the interrupt source, where the interrupt service routine is stored. After the execution of a RETI (Return from Interrupt) instruction, the program continues from the point at which it was interrupted. Figure 7.6 shows an example of the interrupt vector addresses for a C500 microcontroller (C501). In general, the interrupt vector addresses are located in the program memory area starting at address 0003_H. The minimum separation of two consecutive vector addresses is always 8 bytes, so that interrupt vectors can be assigned to the following addresses: 0003_H, $000B_H$, 0013_H, $001B_H$, 0023_H, $002B_H$, 0033_H ... $00FB_H$.

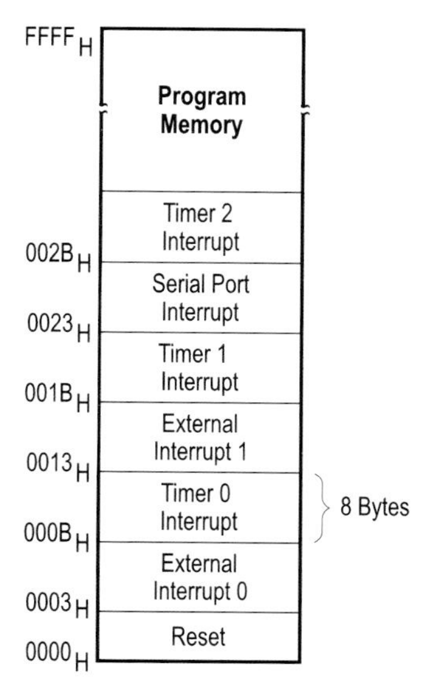

Figure 7.6 Interrupt vector addresses (exemplified by the C501)

An interrupt source indicates to the interrupt controller that there is an interrupt condition by setting an interrupt request flag. The interrupt request flags are sampled in each machine cycle. The sampled flags are polled during the next machine cycle. If one of the flags was in the set state during the preceding cycle, this will be detected in the polling cycle. The interrupt controller will then generate an internal LCALL, to request the CPU to branch to the vector address of the corresponding service routine. This hardware-generated LCALL can be blocked by any of the following conditions:

1. An interrupt with an equal or higher priority is already being processed.
2. The current (polling) cycle is not part of the last cycle for the instruction currently being processed.
3. The instruction currently being processes is a RETI or another write access to interrupt-enable or priority registers.

If at least one of these conditions applies, no LCALL is sent to the interrupt service routine. Condition 2 ensures that the instruction currently being processed is fully completed before a branch is made to a service routine. Condition 3 ensures that if the instruction currently being processed is a RETI or another write access to interrupt-enable or priority registers at least one more instruction will be executed before a branch is made to an interrupt. This delay ensures that changes to the interrupt status can be detected by the interrupt controller.

The polling cycle is repeated with each machine cycle. The values polled are then the values which were set during the previous machine cycle. It should be noted that an interrupt which does not reach the processing stage will not be processed if an interrupt flag is set but there is no response to it because one of the above conditions applies or if, when the blocking condition ceases to apply, the flag is no longer set active. In other words: the sys-

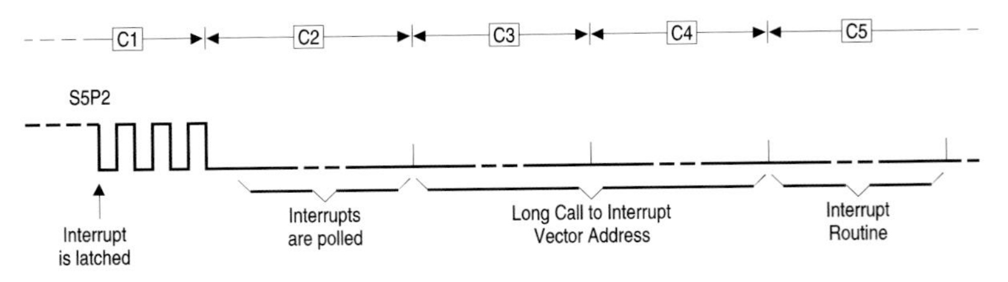

Figure 7.7 Interrupt detection/entry diagram

tem does not remember that the interrupt flag was set active but has not been processed. Every polling cycle interrogates only the pending interrupt requests.

The sequence of actions in the polling cycle/LCALL is shown in Figure 7.7.

It should be noted that, in accordance with the above rules, an interrupt in cycles C5 and C6 which has a higher priority level and becomes active before S5P2 (in the machine cycle labeled C3 in Figure 7.7) will be assigned a destination, without an instruction for the lower priority levels having to be executed.

Consequently, the processor acknowledges an interrupt request by executing a hardware-generated LCALL to the appropriate service routine. In some cases, it also clears the flag which triggered the interrupt. In other cases the flag will not be cleared, and the user software must do this.

Execution of the program is continued at this memory location, until the RETI instruction is encountered. The RETI instruction informs the processor that the interrupt routine is no longer being executed, and fetches the top two bytes from the stack and reloads the program counter. Execution of the interrupted program is resumed at the point where it was halted. It should be noted that the RETI instruction is very important, because it informs the processor that the program has left the current interrupt priority level.

A simple RET instruction would also have resumed the interrupted program, but the interrupt control system would then assume that an interrupt was still being processed. In this case, no interrupt with the same or lower priority would be acknowledged.

Interrupt response time

After an external interrupt has been detected, the corresponding request flag is set at S5P2 in each machine cycle. The value is not polled by the circuitry until the next machine cycle. If the request is active and the conditions permit an acknowledgement, the next instruction executed will be a hardware subroutine call to the requested service routine. The call itself requires two cycles. Thus at least three complete machine cycles will elapse between the activation of the external interrupt request and the beginning of the execution of the first instruction of the service routine.

A longer response time would result if the request were blocked by any of the three conditions listed above. If an interrupt with the same or higher priority is already being processed, the additional delay will obviously depend on the nature of the service routine for the latter interrupt. If the instruction which is currently being executed is not in its last cycle, the additional delay cannot be longer than three cycles, because the longest instructions (MUL and DIV) are only four cycles long, and if the instruction currently being processed is a RETI or a write access to interrupt enable or priority registers, the additional delay cannot be longer than five cycles (at most one further cycle to complete the instruction currently being executed and an additional four cycles to complete the next instruction, if that is a MUL or DIV instruction).

Consequently, if there is a single interrupt system the response time is always longer than three cycles and shorter than nine cycles.

7.2.6 I/O port structures

Digital I/O ports

In general, C500 microcontrollers permit digital I/O operations (input/output) at several pins. These pins are combined in blocks of 8 to form an 8-bit port. Each port consists of a latch, an output driver, and an input buffer. Write and read accesses to the I/O ports are performed through their corresponding special function registers.

Figure 7.8 shows a schematic plan of how a typical bit-latch and I/O buffer (the heart of each I/O port) work. The port-latch (one bit in the special function register for the port) is shown as a D-flipflop, which reads in a value from the internal bus at the clock rate when prompted by a "write-to-latch" signal from the CPU. The Q-output of the flipflop is put on the internal bus when prompted by a "read-pin" signal from the CPU. Some of the instructions to read from a port (i.e. from the corresponding port SFR P0, P2, P3) activate the signal "read-latch", while others activate the signal "read-pin".

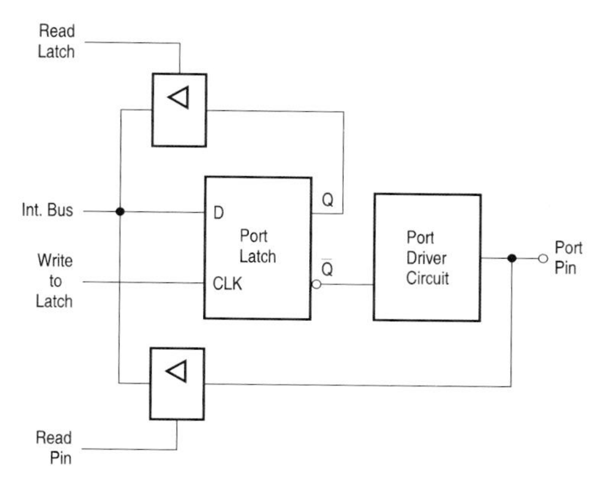

Figure 7.8
Basic structure of the port circuitry for a C500 microcontroller

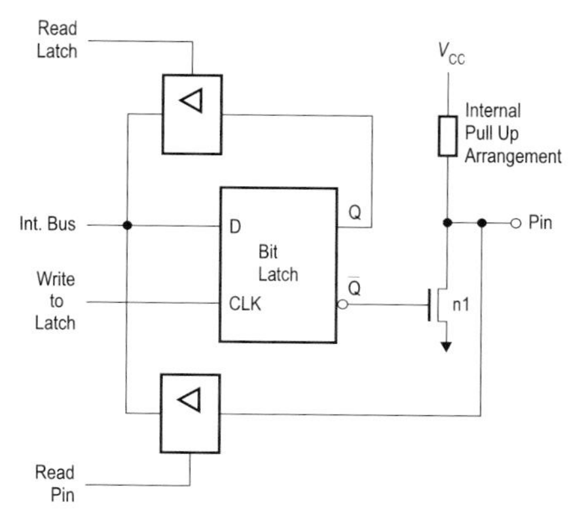

Figure 7.9
Port circuit (all ports apart from 0)

The output drivers for ports 0 and 2, and the input buffer of port 0, are also used for accessing external memory. In this case, the lower-order byte of the address in the external memory is output via port 0, time-multiplexed with the byte which is being written or read. If the address is 16 bits wide, the higher-order byte of the address in the external memory is output via port 2.

The port output drivers (apart from port 0) provide internal pull-up FETs (see Figure 7.9). Each I/O line can be used independently as an input or an output. If used as an input, the port bit must contain a one (1) (i.e., in Figure 7.9 : $Q = 1$ or $\overline{Q} = 0$), which switches off the output driver FET n_1. The pin can then be set to the high potential by the internal pull-ups, or to the low potential by an external source. With a low potential from an external source, a source current will flow. For this reason these ports are sometimes also described to as "quasi-bidirectional".

Some of the C500 microcontrollers also provide a port structure which is arranged for operation either as a quasi-bidirectional port structure (compatible with the standard 8051 family) or as a true bidirectional port structure with CMOS input/output levels.

Analog input ports

Some variants of the C500 microcontroller family are equipped with an integral A/D converter with several analog input lines. These are either dedicated A/D converter inputs or mixed A/D I/O lines (in particular where there is a small number of pins).

Dedicated analog input ports have two functions. When used for analog inputs, the required analog channel is selected using a bit field in the A/D control register. When used as digital inputs, the corresponding port SFR contains the digital value which is present on the port lines. When a digital value is to be read, the levels of the voltages must lie within the prescribed ranges for the input voltage (V_{IL}/V_{IH}). It is therefore possible to use analog input ports for both analog and digital inputs simultaneously.

With mixed digital/analog I/O lines, all analog inputs are blocked after a reset, and the corresponding pins are configured as digital I/O lines. The analog function of these mixed digital/analog I/O port lines is enabled by means of bits in an SFR. A 0 in a bit position of this SFR configures the corresponding pin as an analog input. This enables unused analog inputs to be assigned digital I/O functions.

7.2.7 CPU clock cycles

Basic clock cycles

A machine cycle consists of six states. Each state is subdivided into a phase-1 half, during which the clock generator for phase 1 is active, and a phase-2 half, during which the clock generator for phase 2 is active. Accordingly, one machine cycle consists of the states S1P1 (state 1, phase 1) to S6P2 (state 6, phase 2).

Depending on the type of the C500 microcontroller, each state lasts either one or two periods of the oscillator clock. In general, arithmetic and logical operations are performed in phase 1 and internal transfers between registers in phase 2.

The diagrams in Figure 7.10 show the clock cycles for fetching/executing instructions, in terms of the internal states and phases.

As the user cannot access these internal clock signals directly, the ALE (Address Latch Enable) signal is specified for external references. ALE is normally activated twice during each machine cycle: once during S1P2 and S2P1 and again during S4P2 and S5P1.

The execution of a one-cycle instruction begins at S1P2 with the writing of the operation code into the instruction register.

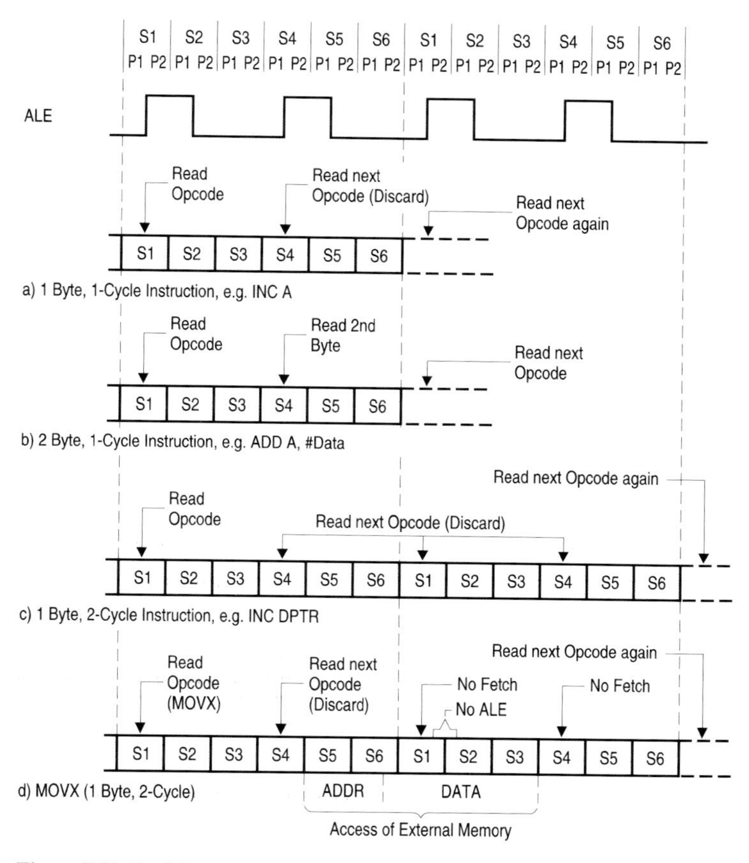

Figure 7.10 Fetch/execute sequence

In the case of a two-byte instruction, a second read is performed in S4 of the same machine cycle. For a one-byte instruction, although data is indeed fetched in S4, the byte which has been read (which would be the next operation code) is ignored (the data fetched is discarded), and the program counter is not incremented. In either case, execution is completed at the end of S6P2.

Figures 7.10a and 7.10b show the clock cycles for a 1-byte one-cycle instruction and for a 2-byte one-cycle instruction.

Most of the C500 instructions are executed in one cycle. The instructions MUL (multiply) and DIV (divide) are the only instructions for which more than two cycles are required, namely four cycles. Normally, two code bytes are fetched from memory in each machine cycle. The only exception is the execution of a MOVX instruction. MOVX is a 1-byte, two-cycle instruction, which accesses the external data memory. During the execution of this instruction, the two data fetch operations in the second cycle are skipped, while the external data memory is addressed and interrogated. Figures 7.10c and 7.10d show the clock cycles for a normal 1-byte two-cycle instruction and for a MOVX instruction.

7.2.8 Accessing the external memory

There are two types of access to the external memory: accesses to external program memory and accesses to external data memory. When accessing external program memory, the signal $\overline{PSEN}$ (Program Store Enable) is used as the read instruction, and for accesses to the external data memory $\overline{RD}$ or $\overline{WR}$ is used (alternative functions of P3.7 and P3.6), as appropriate.

When data is being fetched from the external program memory, a 16-bit address is always used. When accessing external data memory, either a 16-bit address (MOVX @DPTR) or an 8-bit address (MOVX @Ri) can be used.

When a 16-bit address is used, the high-order byte of the address is output at port 2, and is held there for the duration of the read, write or data read cycle.

If an 8-bit address is used (MOVX @Ri), the content of the SFR for port 2 is held at the pins for port 2 throughout the entire external memory cycle. In this case, the pins of port 2 can be used for reading the external data memory.

In either case, the low-order byte of the address is output at port 0, time-multiplexed with the data byte. The ADDRESS/DATA signal controls both FETs in the output buffers for port 0. As a result, in external bus mode the pins of port 0 are not open-drain outputs, and require no external pull-ups. The ALE (Address Latch Enable) signal should be used to prompt the storage of the address byte in an external latch. This address byte is valid when ALE goes low. In the case of a write cycle, the data byte which is to be written then appears at port 0 shortly before the activation of $\overline{WR}$, and is held there until $\overline{WR}$ is deactivated. For a read cycle, the incoming byte is accepted from port 0 shortly before the read instruction ($\overline{RD}$) is deactivated.

When any access is made to the external memory, the CPU writes the value FF_H into the latch for port 0 (special function register), which deletes the information in the SFR for port 0. In addition, the instruction MOV P0 may not be used during an access to the external memory. If the user writes to port 0 during the polling of data from the external memory, the incoming code byte may be corrupted. It is for this reason that no write operations should be performed to port 0 during access to the external memory.

Accessing external program memory

The external program memory can be accessed subject to two conditions:

1. if the $\overline{EA}$ signal is active (low), or
2. if the $\overline{EA}$ signal is inactive (high) and the program counter (PC) contains an address which is larger than the largest internal ROM address (e.g. $1FFF_H$ in the case of an 8-kbyte internal ROM or $3FFF_H$ for a 16-kbyte internal ROM).

This requires, for versions with no ROM, that $\overline{EA}$ always has a conductive connection to V_{SS}, to enable the low-order 8 kbytes, 16 kbytes, or 32 kbytes of program to be fetched from external memory.

When the CPU is executing programs in the external program memory (see clock cycle subdivision diagram in Figure 7.11), all 8 bits of port 2 are reserved for an output function, and cannot be used for general I/O purposes.

During external program calls, the higher-order byte of the PC is output by port 2, using the drivers and strong pull-ups to represent bits which have the value 1.

Accessing external data memory

Throughout the time that they are outputting address bits with the value 1, the drivers for port 2 use strong pull-ups. This is the case when the instruction MOVX @DPTR and external program calls are being executed. During this time, the latch for port 2 (the special function register) must not contain any ones. The con-

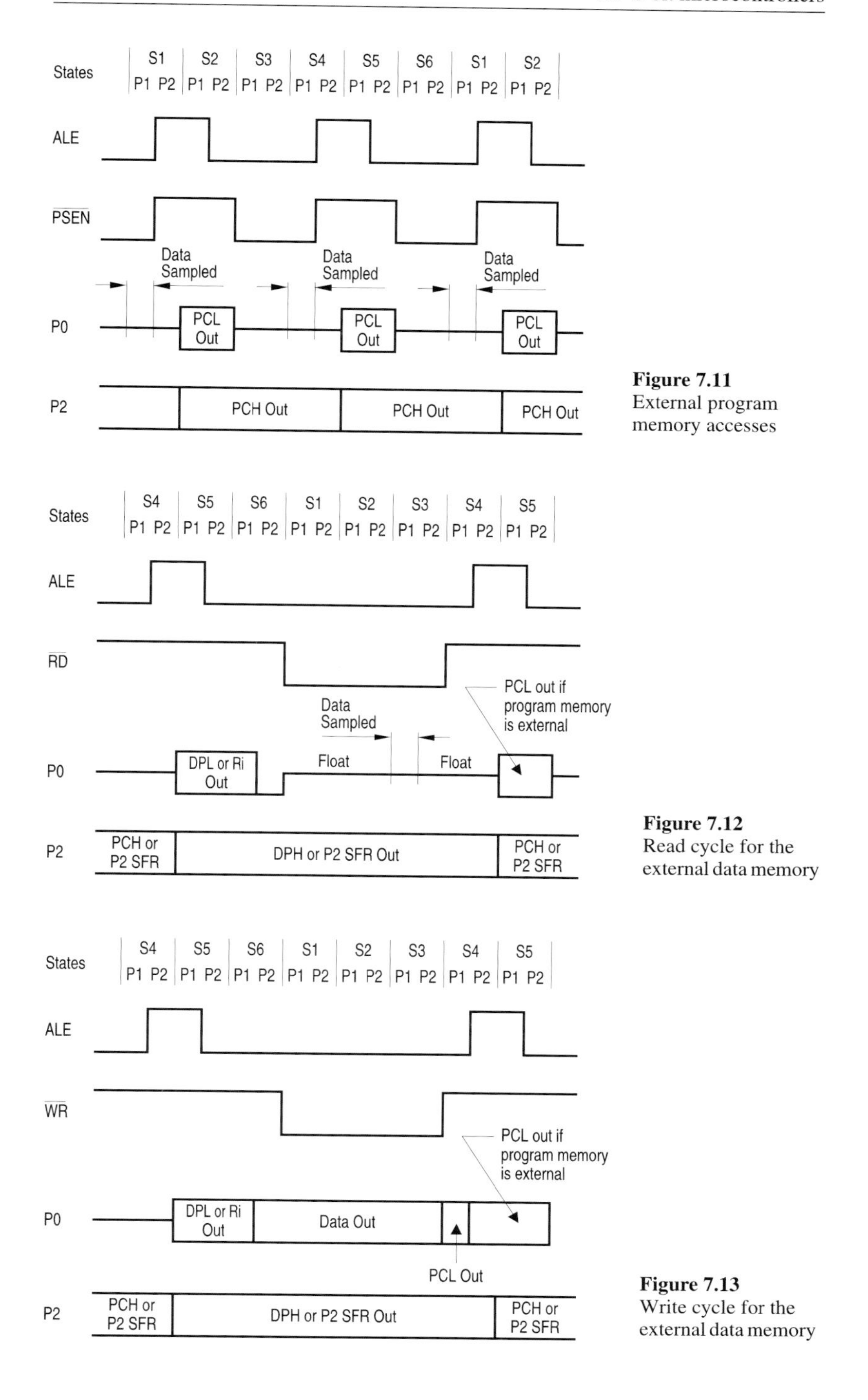

Figure 7.11
External program memory accesses

Figure 7.12
Read cycle for the external data memory

Figure 7.13
Write cycle for the external data memory

tents of the SFR for port 2 is not changed. If the external memory cycle is not immediately followed by another external memory cycle, the content of the SFR for port 2 reappears unchanged in the next cycle.

Figures 7.12 and 7.13 show in detail the clock cycles for the external data memory read and write cycles.

7.2.9 Overview of the instruction set

The instruction set for the 8-bit C500 microcontroller family comprises 111 instructions, of which 49 are single-byte instructions, 45 are two-byte and 17 are three-byte. The format of the operation code for an instruction consists of a function pseudocode followed by an operand field: "destination, source". This field is used to specify the data type and addressing method(s) used.

Like all other members of the 8051 family, the C500 microcontrollers can be programmed using the same instruction set as the SAB 8051, the product which is the ancestor of all other members of the family.

Consequently, the microcontrollers in the C500 family are 100% software compatible with the SAB 8051, and can be programmed either using 8051 Assembler or in higher level programming languages.

Addressing modes

The C500 uses five modes of addressing:

1. register,
2. direct,
3. immediate,
4. indirect addressing via a register,
5. base register plus index register indirect.

Table 7.2 summarizes the memory spaces which can be accessed by each mode of addressing.

Register addressing

Register addressing accesses the eight working registers (R0 - R7) of the selected register bank.

The least significant bits of the instruction operation code specify which register is to be used. ACC, B, DPTR, and CY, the boolean processor accumulator, can also be addressed as registers.

Direct addressing

Direct addressing is the only possible way of accessing the special function registers. The lower 128 bytes of internal RAM are also directly addressable.

Immediate addressing

Immediate addressing enables constants to be accommodated among instructions in the program memory.

Table 7.2 Addressing modes and associated memory areas

Addressing mode	Associated memory area
Register	R0 - R7 in the selected register bank, ACC, B, CY (bit), DPTR
Direct addressing	Lower 128 bytes of the internal RAM, special function registers
Immediate	Program memory
Indirect addressing via a register	Internal RAM (@R1, @R0, SP), external data memory (@R1, @R0, @DPTR)
Base register + index register indirect	Program memory (@A + DPTR, @A + PC)

Indirect addressing via a register

With indirect addressing via a register, the content of R0 or R1 (in the selected register bank) is used as a pointer to memory locations in a 256-byte block, either the 256 bytes of the internal RAM or the lower 256 bytes of the external data memory. It should be noted that the special function registers cannot be accessed using this method.

Only indirect addressing can be used to access the higher half of the internal RAM. Access to the entire 64 kbytes of the external data memory address space is made possible by the use of the 16-bit data pointer.

When the PUSH and POP instructions are executed, indirect addressing via a register is again used. The stack can be located in any required position in the internal RAM.

Base register plus index register addressing

Base register plus index register addressing makes it possible to read from program memory by an indirect path, from the memory location whose address is the sum of a base register (DPTR or PC) and an index register (ACC). This method simplifies access to reference tables.

Boolean processor

The boolean processor is a bit processor which is integrated into the microcontrollers in the C500 family. It has a separate instruction set, accumulator (the carry flag), bit-addressable RAM, and I/O.

The bit manipulation instructions permit:

- a bit to be set
- a bit to be cleared
- the complement of a bit to be formed
- a jump if the bit is set
- a jump if the bit is not set
- a jump if the bit is set and clear the bit
- move a bit to/from the carry flag.

Addressable bits or their complements can be logically ANDed or ORed with the carry flag. The result is saved back into the carry register.

Instruction types

The instruction set is subdivided into four functional groups:

- data transfer,
- arithmetic,
- logic,
- control transfer.

Data transfer instructions

The instructions for data transfer are subdivided into three classes:

- general purpose,
- accumulator-specific,
- address-object.

With the exception of instructions to POP or MOV directly into the PSW, none of these operations affects the PSW flags.

General-purpose transfers

- MOV transfers a bit or a byte from the source operand to the destination operand.
- PUSH increments the SP register and then transfers a byte from the source operand to the memory location in the stack currently addressed by the SP.
- POP transfers a byte operand from the memory location in the stack addressed by the SP to the destination operand and then decrements the SP.

Accumulator-specific transfers

- XCH exchanges the byte source operand with register A (accumulator).
- XCHD exchanges the low-order nibble of the source operand byte with the low-order nibble of A.
- MOVX moves a byte between the external data memory and the accumulator. The external address can be speci-

fied using either the DPTR register (16 bit) or the R1 or R0 register (8 bit).

- MOVC moves a byte from the program memory into the accumulator. The operand in A is used as an index for a 256-byte table, to which the base register (DPTR or PC) points. The byte-operand which is accessed is transferred into the accumulator.

Address object transfers

- MOV DPTR, #data16 loads 16 bits of immediate data into a pair of destination registers (DPH and DPL).

Arithmetic instructions

Microcontrollers in the C500 family use four basic mathematical operations. Only 8-bit operations using unsigned arithmetic are supported directly. However, the overflow flag permits addition and subtraction for both signed and unsigned binary integers. Arithmetic operations can also be performed directly on packed BCD representations.

Addition

- INC (Increment) adds one to the source operand and puts the result into the operand (this does not alter the flags in the PSW).
- ADD adds A to the source operand and puts the result into A.
- ADDC (add with carry) adds A to the source operand and then adds one (1) if CY is set, and puts the result into A.
- DA (decimal-add-adjust for BCD addition) corrects the sum which results from the binary addition of two-digit decimal operands. The packed decimal sum formed by DA is put back into A. CY is set if the BCD result is greater than 99, otherwise it is cleared.

Subtraction

- SUBB (subtract with borrow) subtracts the second source operand from the first operand (the accumulator), subtracts one (1) if CY is set, and puts the result back into A.
- DEC (decrement) subtracts one (1) from the source operand and puts the result back into the operand (this does not alter the flags in the PSW).

Multiplication

- MUL performs an unsigned multiplication of the contents of register A by the contents of register B, and returns a two-byte result. The low-order byte is stored in A, the high-order byte in B. OV is cleared if the more significant part of the result is zero, and is set if it is not zero. CY is cleared, AC is not changed.

Division

- DIV performs an unsigned division of the contents of register A by the contents of register B, stores the integer part of the quotient in register A and the fractional remainder in register B. Division by zero leaves indeterminate data in registers A and B, and OV is set. Otherwise, OV is cleared. CY is cleared. AC is not changed.

Flags

Unless otherwise stated in the above descriptions, the following changes are made to the flags in the PSW:

- CY is set if the operation causes a carry to or borrow from the resulting high-order bit. Otherwise CY is cleared.
- AC is set if the operation causes a carry from the low-order four bits of the result (during addition) or a borrow from the high-order bits to the low-order bits (during subtraction). Otherwise AC is cleared.
- OV is set if the operation results in a carry to the high-order bit but no simultaneous carry from the bit, or vice versa. Otherwise OV is cleared. OV is used during two's-complement arithmetic

because it is set when the signal result cannot be represented using 8 bits.

- P is set if the sum modulo 2 of the eight bits in the accumulator is 1 (odd parity). Otherwise P is cleared (even parity). When a value is written into the PSW register, the P bit remains unchanged because it always indicates the parity of A.

Logic instructions

Microcontrollers in the C500 family carry out basic logic operations on both bit and byte operands.

Operations with a single operand

- CLR sets A or any of the directly addressable bits to zero (0).
- SETB sets any of the directly bit-addressable bits to one (1).
- CPL is used to form the complement of the contents of register A without changing any of the flags or directly addressable bit memory locations.
- RL, RLC, RR, RRC, SWAP are the five operations which can be performed on A. RL (rotate left), RR (rotate right), RLC (rotate left through carry), RRC (rotate right through carry) and SWAP – rotate four bits to the left. For RLC and RRC, the CY flag is assigned the same value as the last bit rotated out of memory. SWAP rotates A by four places to the left, thus exchanging bits 3 to 0 with bits 7 to 4.

Operations with two operands

- ANL performs a bitwise logical AND on two operands (for both bit and byte operands) and puts the result back into the memory location of the first operand.
- ORL performs a bitwise logical OR on two source operands (for both bit and byte operands) and puts the result back into the memory location of the first operand.
- XRL performs a bitwise logical exclusive-OR on two source operands (for both bit and byte operands) and puts the result back into the memory location of the first operand.

Control transfer instructions

There are three classes of control transfer instruction:

- unconditional calls, returns, jumps,
- conditional jumps,
- interrupts.

All the control transfer operations cause the program execution to be continued at a different location in program memory (but in some cases under special conditions).

Unconditional calls, returns and jumps

Unconditional calls, returns and jumps transfer control from the current value in the program counter to the destination address. Both direct and indirect transfers are supported.

- ACALL and LCALL push the address of the next instruction onto the stack and then transfer control to the destination address. ACALL is a 2-byte instruction which is used when the destination address is within the current 2-kbyte page. LCALL is a 3-byte instruction which can address the entire program memory space of 64 kbytes. With ACALL, immediate data (i.e. an 11-bit address field) is concatenated with the five most significant bits of the PC (PC points to the next instruction). If an ACALL is in the last 2 bytes of a 2-kbyte page, the call will be made to the next page, because the PC is incremented before the next instruction is executed.
- RET transfers control to the return address saved in the stack by a previous call operation, and decrements the SP register by two (2), to adjust the SP for the address which has been retrieved.
- AJMP, LJMP and SJMP transfer control to the destination operand. The AJMP and LJMP operations are analogous to ACALL and LCALL. The

SJMP (short jump) instruction permits transfers within a 256-byte range around the start address of the next instruction (– 128 to + 127).

- JMP @A + DPTR performs a jump relative to the DPTR register. The operand in A is used as the offset (0 - 255) to the address in the register. This means that the effective destination address for a jump can be anywhere in the program memory space.

Conditional jumps

Conditional jumps perform a jump depending on a particular condition. The destination address lies within a 256-byte range around the start address of the next instruction (– 128 to + 127).

- JZ performs a jump if the accumulator is zero.
- JNZ performs a jump if the accumulator is not zero.
- JC performs a jump if the carry flag is set.
- JNC performs a jump if the carry flag is not set.
- JB performs a jump if the directly addressed bit is set.
- JNB performs a jump if the directly addressed bit is not set.
- JBC performs a jump if the directly addressed bit is set, and then clears the directly addressed bit.
- CJNE compares the first operand with the second operand and performs a jump if the two are not equal. CY is set if the first operand is less than the second operand, otherwise CY is cleared. Comparisons can be made between A and directly addressed bytes in the internal data memory; or between an immediate value and A, or a register in the selected register bank, or a byte in the internal RAM which can be indirectly addressed via the register.
- DJNZ decrements the source operand and returns the result to the operand. A jump is performed if the result is not zero. The source operand for the DJNZ instruction must not be a directly addressable byte in the internal data memory. Either direct or register addressing may be used to address the source operand.

Interrupt returns

RETI transfers control in the same way as RET does, but in addition it enables the interrupts at the current priority level.

7.2.10 Block diagrams of C500 microcontrollers

Figures 7.14 to 7.22 show the functional units of a number of C500 microcontrollers. Further information on the more exact details of the different versions are given in the user manuals and data sheets for each of them.

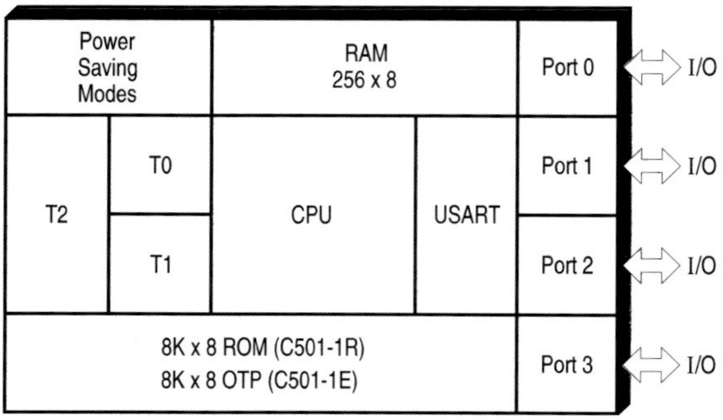

Power Saving Modes
RAM 256 x 8
Port 0
I/O
T0
T2
CPU
USART
Port 1
I/O
T1
Port 2
I/O
8K x 8 ROM (C501-1R)
8K x 8 OTP (C501-1E)
Port 3
I/O

Figure 7.14
C501-F functional units

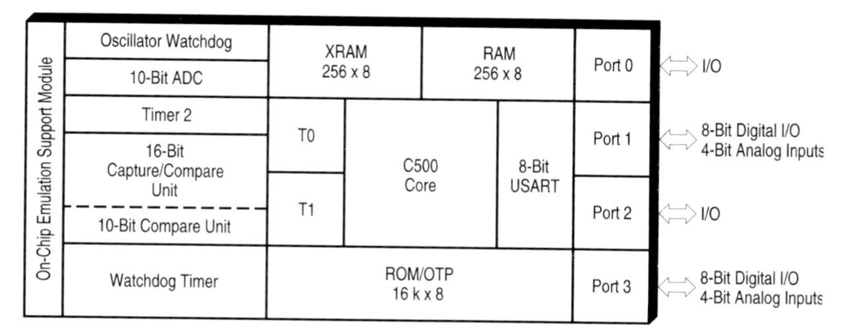

Figure 7.15 C504 functional units

Figure 7.16 C505/C505C/C505A/C505CA functional units

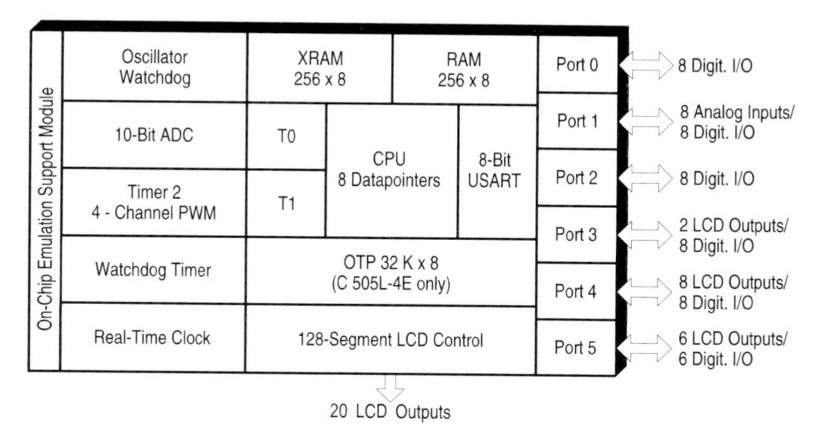

Figure 7.17 C505L functional units

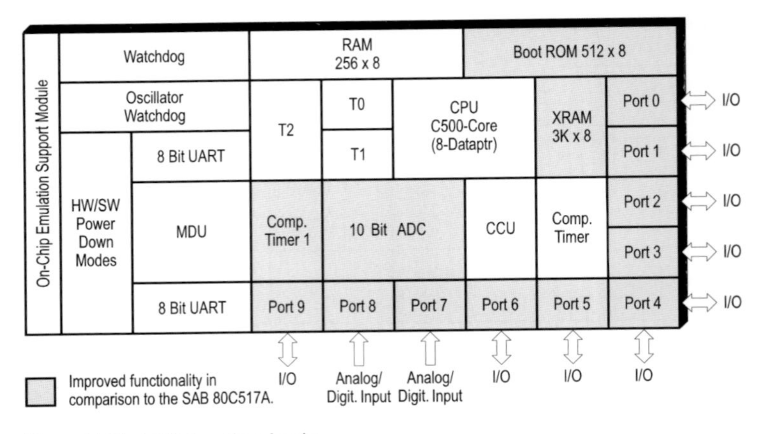

Figure 7.18 C509 functional units

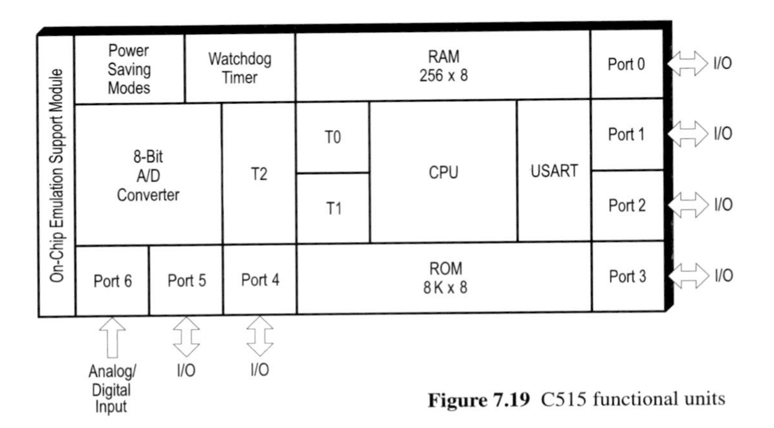

Figure 7.19 C515 functional units

Figure 7.20 C515C functional units

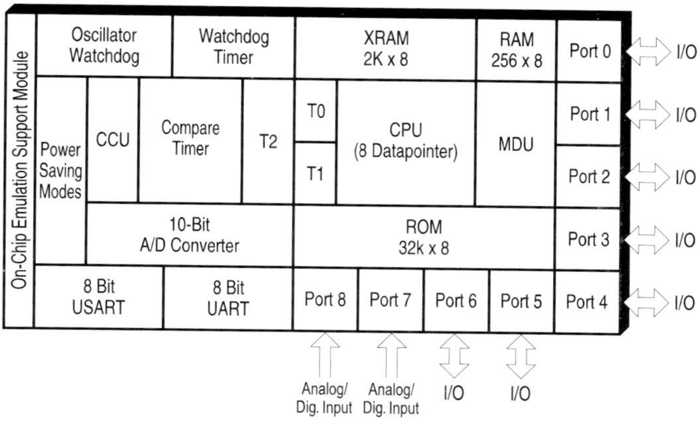

Figure 7.21 C517 functional units

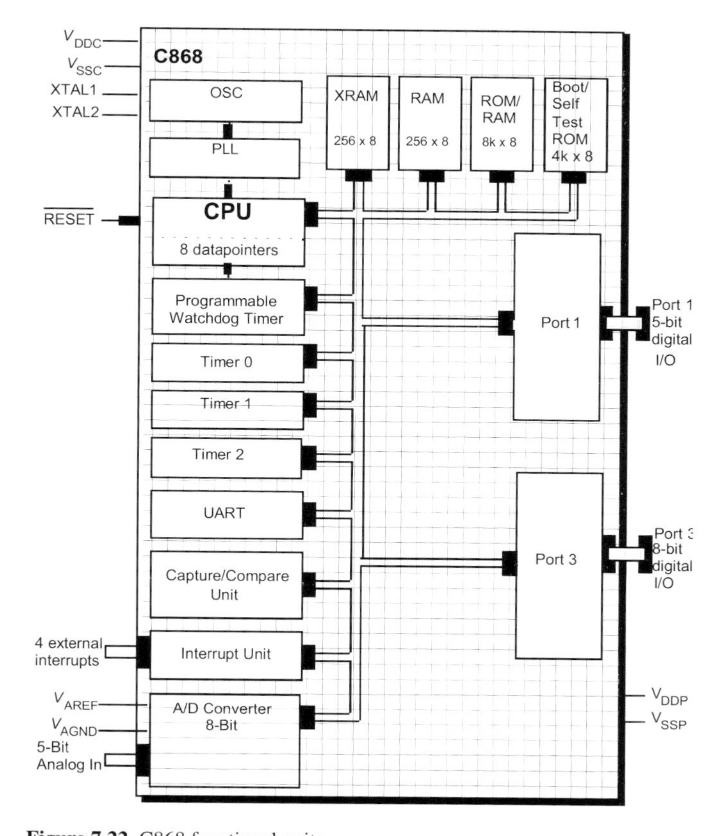

Figure 7.22 C868 functional units

7.3 16-bit microcontrollers

7.3.1 Introduction

The exceptionally fast-growing field of embedded controller applications represents one of the operating environments for today's microcontrollers in which the time factor is particularly critical. Complex control algorithms must be processed on the basis of a large number of digital and analog input signals, and the corresponding output signals must be generated with a defined maximum response time. Embedded controller applications are often sensitive in terms of circuit board space, current consumption and the total cost of the system.

Embedded controller applications thus call for microcontrollers which

- offer a high level of system integration,
- eliminate the need for further peripheral devices, and the associated software costs,
- ensure system security and fail-safe mechanisms,
- provide effective means for controlling (and reducing) the current consumption of a device.

With the growing complexity of embedded controller applications, the microcontrollers require high quality embedded control systems with a significant increase in CPU power and peripheral functionality compared to conventional 8-bit controllers. In order to achieve this performance objective, Infineon made a decision to develop the family of 16-bit CMOS microcontrollers without the limitations of downward compatibility. Naturally, the architecture of the 16-bit microcontroller range follows the successful hardware and software concepts which have already been developed by Infineon in the context of their popular 8-bit controllers.

7.3.2 The members of the 16-bit microcontroller family

The microcontrollers in Infineon's 16-bit range have been developed to satisfy the high performance demands of embedded real-time controller applications. The architecture of this range has been optimized for high instruction throughput and minimum response times to external stimuli (interrupts). Intelligent peripheral subsystems have been integrated to restrict the need for intervention by the CPU to a minimum. This has the additional effect of minimizing the need for communication via the external bus interface. The high flexibility of this architecture makes it possible to satisfy the differing requirements of the various application areas, whether it be in the automotive field, industrial controllers, or in terms of data communication.

The core of the 16-bit range was developed with a modular concept in view. All the systems in this range execute an efficient instruction set, optimized for control purposes (with additional instructions for second generation systems). This permits the simple and rapid implementation of new systems, with their different internal memory sizes and technologies, different on-chip peripheral devices and/or a different number of input/output pins.

Apart from the standard on-chip peripheral devices, the XBUS concept opens up in addition a direct path for the integration of user-specific peripheral modules, to enable application-specific derivatives to be produced. As the programs for embedded controller applications are becoming ever more extensive, programmers prefer higher level programming languages because this makes it easier to write the programs, to debug them and to maintain them.

Microcontrollers of the 80C166 type belong to the first generation of the 16-bit controller family. The C166 architecture is based on these systems. The types C165 and C167 belong to the second generation of this device family. This generation is indeed even more powerful, due to the additional instructions to support higher programming languages, the enlarged address space, larger internal RAM, and exceptionally efficient management of the various resources on the external bus.

Improved derivatives of this second generation incorporate further functional features, such as, for example, extremely fast internal RAM, an integrated CAN module, an on-chip PLL, etc.

The design of efficient systems by integration can necessitate the incorporation of user-specific peripheral devices, in order to raise the performance of the system and to minimize the number of its components. These endeavors are supported by the so-called XBUS, which was defined for Infineon's (second generation) 16-bit microcontrollers. The XBUS is an internal representation of the external bus interface which permits and simplifies the integration of peripheral devices by standardizing the required interface. A characteristic advantage of this technology is the on-chip CAN module. The C165 type systems are 'down-specified' versions of the C167, and have a smaller housing and a lower current consumption at the expense of the A/D converter, the CAPCOM units, and the PWM module.

The C164 type systems and some of the C161 types have been further improved with the help of a flexible energy monitoring system, and form the third generation of the 16-bit controller family. This energy monitoring mechanism represents an effective means for controlling the current consumption for a particular controller status, enabling the total consumption for a specific application to be minimized.

The types XC161, XC164 and XC167 continue this range, and collectively form the XC166 family. These modules are based on the C166SV2 CPU core which, together with the multiply/accumulate unit (MAC unit) makes DSP-oriented instructions possible. This enables the execution times for filter calculations or FFTs to be drastically reduced. The performance of the peripheral system has also been increased, by the use of some new peripheral modules with enhanced functions. Extensive fast memory modules provide the prerequisites for compact high-performance systems for little additional cost.

A range of different versions is available, offering different types of on-chip program memories:

- mask-programmable ROM,
- Flash memory,
- OTP memory,
- ROMless, with no non-volatile memory.

Apart from this, systems are offered with specific functional units. Some of these may be offered with different housings, for different temperature ranges and speed classes. Further standard specifications and derivatives for specific applications are currently in the planning and development stage. Not all of the derivatives are offered for all the variants of temperature range, speed class, housing, or program memory.

7.3.3 Architectural overview of the C166 family

The architecture of the C166 family of microcontrollers combines the advantages of both RISC and CISC processors in an exceptionally well-balanced way (Figure 7.23). The sum of the combined features results in a microcontroller with a high performance, which not only represents the right choice in terms of modern applications, but is also equipped to meet future technical challenges. These microcontrollers contain a powerful CPU core and a group of peripheral units on one chip, and furthermore they link these units together in a very efficient manner.

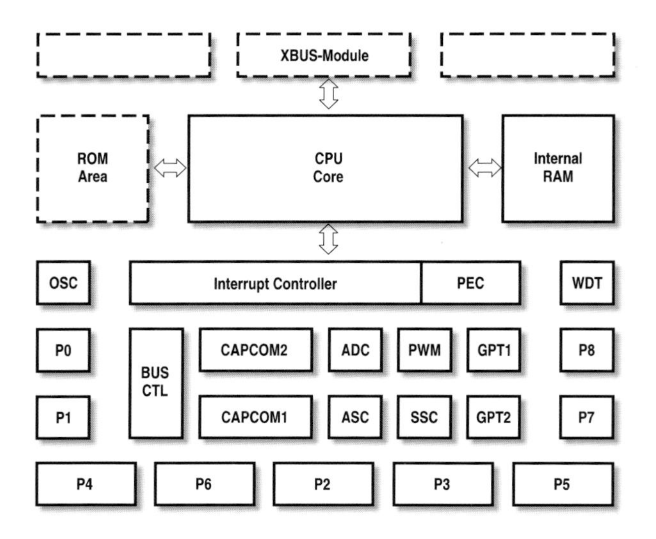

Figure 7.23
Typical, functional block diagram

One of the four buses which are used simultaneously is the XBUS, an internal representation of the external bus interface. This bus offers a standardized method of integrating application-specific peripherals, to create application-specific derivatives of the C166 family.

7.3.4 Memory organization

The 16 Mbytes of the total address space of the C166 architecture is subdivided into 256 segments of 64 kbytes or 1024 data pages of 16 kbytes (Figure 7.24). The external addressing capability de-

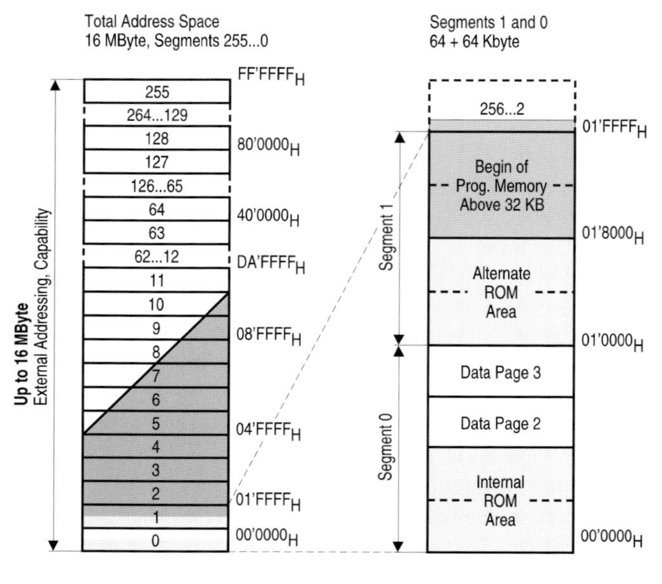

Figure 7.24
Overview of the address space

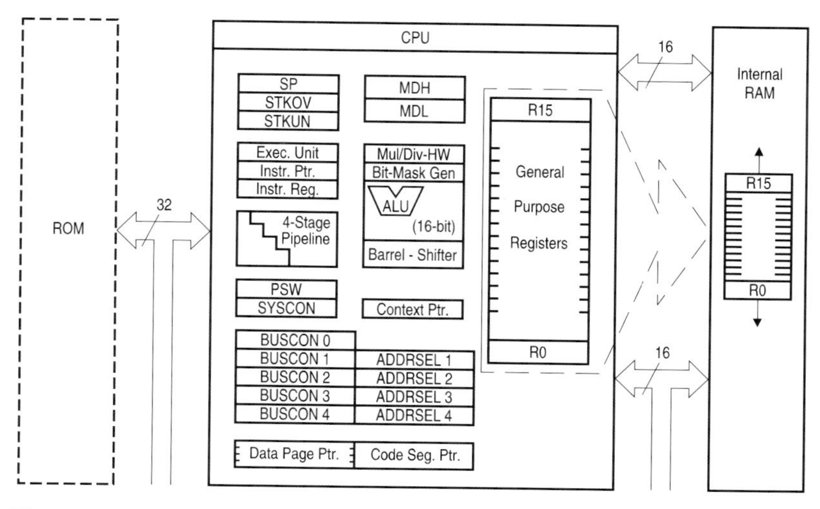

Figure 7.25 Block diagram of the CPU

pends on the configured width of the segment address and – of course – on the number of available pins for port 4. Data page 3 in segment 0 enables most of the on-chip resources to be accessed, for example the peripheral registers, IRAM, XRAM, etc.

7.3.5 Fundamental CPU concepts and optimization measures

The main core of the CPU consists of a four-stage instruction pipeline, a 16-bit arithmetic-logic unit (ALU), and reserved SFRs (Figure 7.25). Additional hardware is available for a separate multiplication and division unit, a bit-mask generator, and a cyclic (barrel) shifter.

Various areas of the processor core have been optimized in order to satisfy the demands for higher performance and greater flexibility. Functional blocks in the CPU core are controlled by signals from the instruction decoder logic. These are summarized below, and are described in detail in the appropriate sections:

1. High instruction bandwidth, fast execution.
2. Exceptionally functional arithmetic-logic unit with 8-bit and 16-bit facilities.
3. Extended bit processing and input/output control.
4. Powerful branching, call and loop processing.
5. Consistent and optimized instruction formats.
6. Programmable interrupt structure for multiple priorities.

High instruction bandwidth / fast execution

Thanks to the hardware devices, most of the instructions can be executed in a single machine cycle, which requires 2 CPU clock cycles ($2 \cdot 1/f_{CPU} = 4$ TCL). So for example, shift and rotation instructions are always processed in one machine cycle, regardless of the number of bits to be shifted.

Branching, multiplication, and division instructions normally require more than one machine cycle. However, these instructions have also been optimized. Branching instructions, for example, only require one additional machine cycle if a branch is actually carried out, and most of the branching operations performed in program loops do not require any further machine cycles, due to the so-called "jump cache".

A 32-bit / 16-bit division occupies 20 CPU clock cycles, and a 16-bit × 16-bit multiplication 10 CPU clock cycles. The instruction execution time has been significantly reduced by the use of instruction pipelining. This technique permits the CPU core to process parts of several sequential instruction sections in parallel. The following four-stage pipeline offers optimal balancing of the CPU core:

FETCH: in this phase, an instruction is fetched from the internal ROM or RAM, or from external memory, depending on the current IP value.

DECODE: in this phase, the instruction previously fetched is decoded and the necessary operands are fetched.

EXECUTE: in this phase, the specified operation is performed using the operands which had been fetched.

WRITE-BACK: in this phase, the result is written into the specified memory area.

Without this special technique, each of these instructions would require four machine cycles. This increased performance permits a larger number of tasks and interrupts to be processed.

Instruction decoder

Instruction decoding starts in the first instance from the PLA outputs, on the basis of the selected operation code. No microcode is used, and each pipeline phase receives control signals which are stored in control registers of the decoding phase PLAs. Pipeline wait cycles are caused mainly by wait states, the waiting phases due to external memory accesses, and lead to holding of signals in the control registers. Multi-cycle instructions are executed with the help of instruction injection and simple internal status mechanisms which modify the necessary control signals.

Exceptionally functional 8-bit and 16-bit arithmetic-logic unit

All the standard arithmetic and logical operations are performed by a 16-bit arithmetic and logic unit (ALU). In addition, for byte operations, signals are supplied by the ALU for bits six and seven, so that the condition flags will be correctly set. Multiple precision arithmetic is achieved by the transmission of a "CARRY-IN" signal to the ALU from previously calculated portions of the desired operation. Most of the internal execution blocks have been optimized for the performance of both 8-bit and also 16-bit operations. As soon as the pipeline has been filled, one instruction is executed per machine cycle, except for multiplications and divisions. An advanced Booth algorithm has been incorporated, to allow four bits to be multiplied and two to be divided per machine cycle. For this reason, these operations use two combined 16-bit registers, MDL and MDH, and require respectively four and nine machine cycles to perform a 16-bit × 16-bit (or 32-bit / 16-bit) calculation, plus one machine cycle to set up and adjust the operands and the result. Even these longer multiplication and division instructions can be interrupted during their execution, to enable a very fast interrupt response to be achieved. In addition, instructions are provided for increasing the packing density of the bytes in memory while at the same time providing sign extension of bytes for arithmetic operations with a large word width. The internal bus structure also permits, depending on the requirements, the transmission of bytes or words to or from peripheral devices. A set of corresponding flags is automatically updated in the PSW after each arithmetic, logical, shift, or move operation. These flags permit branches to be made under specific conditions. Support is provided for both signed and unsigned arithmetic by branch tests, which the user can specify. These flags are automatically saved by the CPU on entry into an interrupt or trap routine. All the destinations for branch operations are also calculated in the central ALU.

A 16-bit cyclic shifter makes repeated bit injection possible in a single cycle. Rotations and arithmetic shifts are also supported.

Extended bit processing and input/output control

A large number of instructions have been dedicated to bit processing. These instructions provide for efficient control and checking of peripheral devices, while improving data manipulation. In contrast to other microcontrollers, these instructions offer direct access to two operands in the bit address space without requiring to move them into temporary flags.

The same logical instructions which are available for words and bytes are also supported for bits. This allows the user to compare and modify a control bit for a peripheral, using one instruction. Several bit shift instructions have been included, to avoid long sequences of instructions for single bit-shift operations. These are also executed in a single machine cycle. In addition, there are bit field instructions which permit several bits of an operand to be modified by a single instruction.

Powerful branch, call, and loop processing

Because of the high proportion of branches made in controller applications, the branching instructions have been optimized and require an additional machine cycle only when a branch is actually made. This is achieved by precalculating the destination address while the instruction is still being decoded. In order to reduce the overhead involved in executing program loops, three improved solutions have been developed:

- The first provides for a jump to take a single cycle after the first iteration of a program loop. This means that during the execution of the entire loop, only a single machine cycle is lost. In program loops which are exited from their end upon completion, not a single machine cycle is lost when the loop terminates. No special instructions are required for the execution of program loops, and branching instructions are automatically recognized when they are executed.
- The second improvement in relation to program loops allows the end of a table to be recognized, and avoids the use of two comparison instructions embedded in loops. The lowest negative number is simply placed at the end of the specific table, and a branch is performed if neither this value nor the compared value have been found. If one of the conditions is met, the program loop is terminated. The terminating condition can then be tested.
- The third improvement offers a more flexible solution than the usual decrement and skip-on-zero instruction, used in other microcontrollers. By using compare and increment or decrement instructions, the user can perform comparisons with any required value. This allows loop counters to cover any range. This is particularly advantageous in table searches.

Details of the system state are saved automatically on the internal system stack, which avoids the use of instructions which preserve the state when entering or returning from interrupt or trap routines. Call instructions push the value of the IP onto the system stack, and require the same execution time as branch instructions. In addition, there are instructions to support indirect branch and call instructions. These support the execution of multiple CASE statement branches in assembler macros and higher level programming languages.

Consistent and optimized instruction formats

To achieve optimal performance with a pipelined design, an instruction set has been developed which also incorporates elements from reduced instruction set computing (RISC). Above all, these permit fast decoding of the instructions and operand, while at the same time reducing the pipeline waitstates times. The RISC concepts in no way exclude the use of the complex instructions which are demanded by microcontroller users. In developing the instruction set, the following objectives were pursued:

1. To provide powerful instructions for carrying out operations which currently require complete instruction sequences and which are used regularly. To avoid transfers in and out of temporary registers such as accumulators and carry bits. In addition, to perform tasks in parallel, such as saving the system state on entry into interrupt routines or subroutines.

2. To avoid complex encoding schemes by placing the operands for each instruction in consistent fields. In addition, to avoid complex addressing modes which are only used infrequently. This reduces the instruction execution time and at the same time simplifies the development of compilers and assemblers.

3. To provide the most frequently used instructions in 1-word instruction formats. All other instructions have a 2-word format. This permits all the instructions to be aligned to word boundaries, which reduces the need for complex alignment hardware. This offers the additional advantage of increasing the operating range for relative branching instructions.

The high performance offered by the hardware implementation of the CPU can be efficiently utilized by programmers using the exceptionally functional instruction set, which includes the following classes of instruction:

- Arithmetic instructions
- Logical instructions
- Boolean bit manipulation instructions
- Comparison and loop control instructions
- Shift and rotation instructions
- Priority instructions
- Data-flow instructions
- System stack instructions
- Jump and call instructions
- Return instructions
- System control instructions
- Miscellaneous instructions.

The possible operand types include bits, bytes, and words. Certain instructions support the conversion (extension) of bytes to words. Various direct, indirect, and immediate addressing modes are available for specifying the required operands.

Programmable multiple priority interrupt structure

The following improvements have been made, to permit the processing of a large number of interrupt sources:

1. Peripheral event controller (PEC): this processor is used to off-load many interrupt requests from the CPU. The overhead of entry to and return from interrupt and trap routines is avoided by performing interrupt-controlled data transfers of bytes or words between any two memory locations in segment 0 as single-cycle operations, with optional incrementing of the PEC source or the destination pointer. Only one single cycle is ‘stolen’ from the current CPU activity to perform a PEC task.

2. Multiple priority interrupt controller: this controller allows any required priority to be assigned to all the interrupts. Interrupts can also be grouped together, allowing the user to prevent tasks with a similar priority from interrupting each other. For each possible interrupt source, there is a separate control register, containing an interrupt request flag and an interrupt-enable flag plus an interrupt priority bit field. As soon as an interrupt task has been accepted by the CPU, it can only be interrupted by an interrupt request with a higher priority. For standard interrupt processing, each possible interrupt source has a reserved vector location.
3. Multiple register banks: this feature permits the user to locate up to 16 general purpose registers in any required position in the internal RAM. A single instruction, requiring only one machine cycle, allows register banks to be switched from one task to the next.
4. Interruptible multiple cycle instructions: A reduction in the interrupt latency time has been achieved by allowing multiple cycle instructions (multiplication, division) to be interrupted. With an interrupt response time in the range of just 5 to 10 CPU cycles (for the execution of internal programs) the microcontroller can react exceptionally quickly to non-deterministic events. Its fast external interrupt inputs are polled every CPU cycle, and this allows even very brief external signals to be detected.

The C166 series also offers an excellent mechanism, the so-called "hardware trap", for identifying and processing exception or error conditions which arise during run-time. Hardware traps prompt an immediate non-maskable system response, which is similar to a standard interrupt task (branching to a reserved vector location). The occurrence of a hardware trap is also indicated by a single bit in the trap flag register (TFR). With the exception of a trap task with a higher priority which is already running, a hardware trap will interrupt any current program execution. On the other hand, tasks for hardware traps cannot normally be interrupted by standard or PEC interrupts. Software interrupts are supported by the "Trap" instruction in combination with an individual trap number (interrupt number).

7.3.6 The on-chip system resources

The C166 series of microcontrollers from Infineon have a range of powerful system resources, which have been developed around the CPU. The combination of the CPU and these resources is the basis for the high performance of the systems in this family of controllers.

Note: the following sections provide a description of the resources presently available. However, it is possible that not all of these resources are available for any particular product of the C166 series.

Peripheral event controller (PEC) and interrupt control

The peripheral event controller makes it possible to respond to an interrupt request with a single data transfer (word or byte), so that only a single instruction cycle is required and it is not necessary to save and restore the machine state. During each machine cycle, all the interrupt sources are assigned a priority in the interrupt control block. If the PEC task option is selected, a PEC transfer is started. If a CPU interrupt task is requested, the current CPU priority level, stored in the PSW register, is checked to determine whether an interrupt with a higher priority is currently being processed. When an interrupt is accepted, the current machine state is saved on the internal system stack, and the CPU branches to the system-specific vector for the peripheral device.

The PEC contains a set of SFRs which store the transfer length and control bits for eight data transfer channels. In addition, the PEC uses a reserved area of RAM which contains the source and destination addresses. The PEC is controlled in a similar way to any other peripheral device, i.e. through SFRs which show the desired configuration of each channel.

A PEC transfer counter is decremented in the background for each PEC task except for the continuous transfer mode. When this counter reaches the value zero, a standard interrupt is performed to the vector location associated with the source concerned. PEC tasks are, for example, very well suited for moving register contents into or out of memory tables. 8 PEC channels are available, each of them offering such fast, interrupt-controlled data transfer capabilities.

Memory areas

The memory space of the microcontrollers in the C166 range is configured in accordance with the von Neumann architecture, which means that code memory, data memory, registers and input/output ports are organized within the same linear address space, which covers up to 16 Mbytes. The entire memory space can be accessed bytewise or wordwise. Particular areas of the on-chip memory can also be addressed directly as bits.

An internal 16-bit RAM (IRAM) allows fast access to general purpose registers (GPRs), user data (variables), and the system stack. This internal RAM can also be used for code. A unique decoding scheme provides flexible user register banks in the internal memory and optimizes the remaining RAM for user data.

The CPU provides a current register set consisting of up to 16 GPRs aligned to word and/or byte boundaries, which from a physical point of view are located within the on-chip RAM area. One register contains the context pointer (CP), which specifies the base address of the register bank currently active, which the CPU will access when required. The number of register banks is only limited by the available internal RAM space. To simplify parameter passing, one register bank may overlap others.

A system stack for up to 512/1024/1536 words (depending on the size of the IRAM) is available for the storage of temporary data. The system stack is also located within the on-chip RAM area, and the CPU accesses it via the register for the stack pointer (SP). Each time the stack is accessed, two separate SFRs (STKOV and STKUN) are compared in the background with the value of the stack pointer, to detect any stack overflow or underflow. Hardware identification for the selected memory space takes place in the internal memory decoders, and enables the user to specify any address directly or indirectly, and to obtain the desired data without having to use temporary registers or special instructions.

A 16-bit on-chip XRAM offers fast access to user data (variables), user stacks, and code. The on-chip XRAM is treated as an X-peripheral, and appears to the software as an external RAM. For this reason, it cannot store register banks and cannot be bit-addressed. The XRAM permits 16-bit accesses at maximum speed.

An optional internal ROM/OTP/flash memory handles the storage of code and also constants. This memory area is connected to the CPU via a 32-bit bus. This means that in just one machine cycle it is possible to fetch a complete 2-word instruction. The ROM is mask-programmed in the factory, while the OTP/Flash memory can also be programmed as part of the application. The execution of programs using the on-chip program memory is the fastest of all the possible alternatives.

1024 bytes of the address space are reserved for Special Function Registers. The standard area for special function registers (SFRs) uses 512 bytes, while the extended area for special function registers (ESFRs) utilizes the other 512 bytes. (E)SFRs are word-wide registers and are used for controlling and monitoring the functions of the various on-chip units. Unused (E)SFR addresses are reserved for future C166 microcontrollers with enhanced functions.

7.3.7 External bus interface

In order to meet the requirements of designs which call for more memory than is available on the chip, up to 16 Mbytes of external RAM and/or ROM can be connected to the microprocessor via its external bus interface. The integral external bus controller (EBC) permits exceptionally flexible access to external memory and/or peripheral resources. For up to five address ranges, the bus mode (multiplex/demultiplex), the databus width (8-bit/16-bit), and even the bus cycle length (wait states, signal delays) can be selected independently of one another. This allows various memory and peripheral components to be accessed directly and with maximum efficiency. If the module is not running in single-chip mode, for which no external memory is required, the EBC can control external accesses in one of the following external access modes:

- 16-/18-/20-/24-bit addresses, 16-bit data, demultiplexed
- 16-/18-/20-/24-bit addresses, 8-bit data, demultiplexed
- 16-/18-/20-/24-bit addresses, 16-bit data, multiplexed
- 16-/18-/20-/24-bit addresses, 8-bit data, multiplexed

The demultiplexed bus modes use port 1 for addresses and port 0 for data input/output. The multiplexed bus modes use port 0 both for addresses and also for data input/output. Port 4 is used for the upper address signals (A16 ...), if these have been selected.

Important timing characteristics of the external bus interface (wait states, ALE length, and read/write delay) are now programmable, to permit the user to make adaptations for a large range of different types of memory and peripheral devices. Access to very slow memory or peripherals is supported by a special "Ready" function.

For applications which require an address space of less than 64 kbytes, a non-segmented memory model can be selected, with which all the memory locations can be addressed using 16-bit addresses, so that port 4 is not required for the output of the upper address bits (Axx ... A16), as is required, for example, with a segmented memory model.

The on-chip XBUS is an internal representation of the external bus, which permits integrated application-specific peripherals/modules to be accessed in the same way as external components. It provides a defined interface for these customized peripherals. The on-chip XRAM and the on-chip "I^2C" or CAN module are examples of such X-peripherals.

7.3.8 The on-chip peripheral blocks

With the C166 range, the peripherals are clearly separated from the core. This structure permits a maximum number of operations to be executed in parallel, and allows peripherals to be added to or removed from any of the systems in the range, without having to make modifications to the core. Each functional block processes data independently, and communicates data over the shared buses. Peripherals are controlled by data written to the appropriate special function registers (SFRs). These SFRs are located either in the standard area for SFRs ($00'FE00_H$ ... $00'FFFF_H$) or in the extended ESFR area ($00'F000_H$ ... $00'F1FF_H$). Peripherals which are linked to the XBUS (X-peripherals) occupy separate address windows, in which their register and memory segments are located. These built-in peripherals enable the CPU to establish a link to the "external world", or to integrate on the chip functions which would otherwise have to be added externally to the system concerned.

The general peripherals of the C166 family include:

- input/output ports with various alternative functions,
- two general purpose timer blocks (GPT1 and GPT2),
- a watchdog timer,
- serial interfaces (ASC, SSC),
- capture/compare units (CAPCOM1, CAPCOM2, CAPCOM6),
- a 4-channel pulse width modulation unit,
- an analog-digital converter,
- a real time clock.
- The XBUS peripherals include:
- a high-speed SSP (synchronous serial port),
- an additional USART (ASC1),
- I^2C bus interface (400 kbit/s, 10-bit addressing),
- CAN interface (Rev. 2.0B active, full-CAN/basic-CAN),
- USB interface (maximum speed, 8 end-points),
- serial data transfer module (SDLM), conforming to J1850 Class 2.

Note: it is possible that not all of the peripherals mentioned are available for any specific derivative. Where necessary, there may also be different versions of a peripheral.

Peripheral interfaces

The on-chip peripherals generally have two different types of interface, an interface to the CPU and an interface to external hardware. Communication between CPU and peripheral devices is performed through Special Function Registers (SFRs) and interrupts. The SFRs serve as control/status and data registers for the peripherals. Interrupt requests are generated by the peripherals based on specific events which occur during their operation (e.g. operation complete, error, etc.).

For interfacing with external hardware, specific pins of the parallel ports are used, when an input or output function has been selected for a peripheral device. During this time, the port pins are controlled by the peripheral (when used as outputs) or by the external hardware which controls the peripheral (when used as inputs). This is called the 'alternate (input or output) function' of a port pin, in contrast to its function as a general purpose input/output pin.

Peripheral timing

Internal operation of CPU and peripherals is based on the CPU clock (f_{CPU}). The on-chip oscillator derives the CPU clock from the crystal or from the external clock signal. The clock signal which is transmitted to the peripherals is independent of the clock signal which synchronizes the CPU. In the idle mode, the CPU's clock is stopped while the peripherals continue their operation. The energy monitoring facilities can also switch off peripheral groups temporarily, by shutting down each of their clock signals. Peripheral SFRs may be accessed by the CPU once per state. If software writes to an SFR in the same state as that in which it is also to be modified by the peripheral, the software write operation is given priority. Further details on peripheral timing are included in the appropriate sections for the peripherals concerned.

Parallel ports

The input/output links of the microcontrollers in the C166 range are organized into (input/output) ports. All port lines are bit-addressable, and all input/output lines are individually (bit-wise) programmable as inputs or outputs via direction registers. The input/output ports are true bidirectional ports which are switched to a high impedance state when configured as inputs. The output drivers of some input/output ports can be configured (pin by pin) via control registers for push/pull or open-drain operation. During an internal reset, all port pins are configured as input lines. The input stages of some input/output ports can be configured for TTL or CMOS compatible input thresholds, via a control register.

Most of the port lines have programmable alternate input or output functions assigned to them. This comprises the address and data lines in the case of an access to external memory, chip-select signals, fast external interrupt inputs, timer inputs/outputs, serial interfaces, a system clock, and analog inputs for the A/D converter. All the port lines which are not used for these alternate functions can be used as general purpose input/output lines.

General purpose timers (GPT)

The GPT units represent a very flexible multifunctional timer/counter structure which may be used for many different time-related tasks, such as event timing and counting, pulse width and duty cycle measurements, pulse generation or multiplication. Each timer can operate independently in a number of different modes, or may be linked with another timer of the same module. Each timer can be configured individually for one of four basic modes of operation, namely as a timer, a gated timer, in counter mode, and in incremental interface mode (GPT1 timers). In timer mode, the input clock for a timer is derived from the internal CPU clock divided by a programmable factor, while counter mode allows a timer to be clocked taking account of external events (via TxIN). Pulse width or duty cycle measurement is supported in gated timer mode, in which the operation of the timer is controlled by the 'gate' signal on its external input pin TxIN.

In the incremental interface mode, the GPT1 timers can be directly connected to the incremental position sensor signals A and B via the corresponding input lines, TxIN and TxEUD. Direction and count signals are derived internally from these two input signals, so that the content of timer Tx corresponds to the sensor position. The third position sensor signal TOP0 can be connected to an interrupt input.

The count direction (up/down) is software-programmable for each timer, or can also be altered dynamically by an external signal (TxEUD), for example to simplify position synchronization.

The core timers T3 and T6 have output toggle latches (TxOTL) which change their state on each timer overflow/underflow. The state of these latches may be used internally to combine the core timers with the appropriate auxiliary timers, to give 32/33-bit timers/counters for measuring long time periods with high accuracy.

There is a choice of various capture or reload functions, for reloading timers or for capturing the content of a timer triggered by an external signal or a selectable transition of the toggle latch TxOTL.

The maximum resolution of the timers in the GPT1 module is 8 CPU clock cycles (= 16 TCL). With their maximum resolution of 4 CPU clock cycles (= 8 TCL), the GPT2 timer ensure precise event control and time measurement.

Watchdog timer

The watchdog timer represents one of the fail-safe mechanisms which have been incorporated to prevent the controller from malfunctioning over longer periods of time. The watchdog timer is always enabled after a reset of the chip, and can only be disabled in the time interval until the EINIT (end of initialization) instruction is executed. Thus, the chip's start-up procedure is always monitored. The software must be designed so that it services the watchdog timer before it overflows.

If, due to hardware or software related failures, the software does not carry out this task, the watchdog timer will overflow and generate an internal reset, and will set the $\overline{\text{RSTOUT}}$ pin to the low signal level, to allow external hardware components to reset.

The watchdog timer is a 16-bit timer, clocked with the CPU clock divided by a factor of 2/4 ... /128/256. The high byte of the watchdog timer register can be set to a prespecified reload value (stored in WDTREL) in order to allow further variation of the monitored time interval. Each time it is addressed by the application software, the high byte of the watchdog timer is reloaded.

Serial channels

Serial communication with other microcontrollers, processors, terminals or external peripheral components takes place through several serial interfaces with different functional characteristics. The ASC interface is upward compatible with the serial ports of Infineon's 8-bit microcontroller family, and supports full-duplex asynchronous communication at up to 625 kbit/s and half-duplex synchronous communication at up to 2.5 Mbit/s with a CPU clock rate of 20 MHz. A dedicated baud rate generator is provided to allow all standard baud rates to be set up without oscillator tuning.

Separate interrupt vectors are provided for transmission, reception, and error handling. In asynchronous mode, 8-bit or 9-bit data packets are transmitted or received, preceded by a start bit and followed by one or two stop bits. For multiprocessor communication, there is an integral mechanism to distinguish address bytes from data bytes (8-bit data plus wake-up bit mode).

In synchronous mode, the ASC interface transmits or receives bytes (8 bits) in synchrony with a shift clock which is generated by the ASC. The ASC always shifts the least significant bit first. A test loop option is available for testing purposes. In addition, a range of optional hardware error detection capabilities has been incorporated, to increase the reliability of data transfers. A parity bit can be automatically generated on transmission or checked on reception. The error detection facilities enable data frames with missing stop bits to be recognized. An overflow error will be generated, if the last character received has not been read out of the receive buffer register when the reception of a new character is complete.

The SSC interface supports full-duplex synchronous communication at up to 5 Mbit/s at a CPU clock rate of 20 MHz. It can be configured to permit the connection of serially linked peripheral components. The baud rate generator reserved for this purpose permits all standard baud rates to be set up without oscillator tuning. Separate interrupt vectors are provided for transmission, reception, and error handling.

The SSC interface transmits or receives characters of 2 ... 16 bits length in synchrony with a shift clock which can be generated either by the SSC (master mode) or by an external master (slave mode). The SSC can start a shift operation with the LSB or with the MSB, and allows the receiving and transmitting clock pulse edges to be selected, as well as the clock polarity.

In addition, a range of optional hardware error detection capabilities has been incorporated to increase the reliability of data transfers. The checks on transmit and receive errors control the correct handling of the data buffer. Phase and baud rate error checks detect incorrect serial data.

The SSP interface supports half-duplex synchronous communication at up to 10 Mbit/s at a CPU clock rate of 20 MHz. It can be configured to permit the connection of serial peripheral components. The baud rate generator provided for this purpose permits all standard baud rates to be set up without oscillator tuning. A separate general purpose interrupt vector is provided for transmission, reception, and error handling.

The SSP transmits 1 ... 3 bytes, or receives one byte after sending 1 ... 3 bytes, in synchrony with a shift clock which is generated by the SSP (master mode). The SSP can start a shift operation with the LSB or with the MSB, and allows the receiving and transmitting clock pulse edges to be selected, as well as the clock polarity.

In continuous transfer mode, a group of data bytes can be transferred consecutively with no additional address or status information. Up to two chip control signals can be activated, in order to direct data transfers to one or both of two peripheral devices.

Capture/compare units, CAPCOM1/2

The two CAPCOM units support the generation and control of timing sequences on up to 32 channels with a maximum resolution of 8 TCL. The CAPCOM units are normally used to process high speed inputs/outputs, such as pulse or waveform generation, pulse width modulation (PWM), digital-analog conversion, software timing or clock recording in relation to external events.

Two 16-bit timers per unit, with reload registers, provide two independent time bases for the capture/compare register system.

The input clock rate can be programmed as various fractions of the internal CPU clock rate, or can be derived from an overflow/underflow of timer T6 in the GPT2 module. This provides a wide range of variation for the timer period and resolution, and enables them to be precisely matched to the application-specific requirements. In addition, external counter inputs for one CAPCOM timer per unit permit the capture/compare registers to be event-controlled in relation to external events.

Each of the two capture/compare register systems contains 16 combined capture/compare registers, and each of these can be assigned individually to one of the two CAPCOM timers and programmed for a capture or compare function. Each register is assigned to a port pin, which serves as the input pin for activating the capture function or as the output pin for indicating a comparison event.

If the capture mode has been selected for a capture/compare register, then the current contents of the assigned timer will be saved (captured) in the capture/compare register in response to an external event on the port pin which is assigned to this register. In addition, a specific interrupt request will be generated for this capture/compare register. The triggering event at the pin can be chosen as either a positive external signal transition, a negative one, or both positive and negative transitions.

The contents of all the registers selected for one of the five comparison methods will be continuously compared against the contents of the assigned timer. If the timer value corresponds to the value of a capture/compare register then specific actions will be taken, depending on the selected compare mode.

Capture/compare unit CAPCOM6

The CAPCOM6 unit provides 3 capture/compare channels together with an additional compare channel. Each of the 3 capture/compare channels can control two output lines, which can be programmed to generate non-overlapping pulse patterns. The additional compare channel can either generate a separate output signal or can modulate the output signals of the other three channels.

Versatile multi-channel PWM signals can be generated, either under the internal control of a timer or under external control, e.g. by Hall sensors. The active signal level can be selected individually for each output signal.

The trap function makes it possible to set the output signals to a defined level in response to an external signal.

Pulse width modulation unit

The pulse width modulation (PWM) unit supports the generation of up to four independent high speed PWM signals. It can be used to generate both standard PWM signals (edge aligned) and also symmetrical PWM signals (center aligned). In burst mode, two channels can be combined so that their signals are AND-linked and one channel triggers the output signal of the other channel. The single-slot mode permits a single output pulse (re-triggerable) to be generated under software control. Each PWM channel is controlled by a forward/backward counter with associated reload and compare registers. The polarity of the PWM output signals can be controlled by the corresponding port output latch (EXOR combination).

A/D converter

For the purposes of analog signal measurement, an A/D converter with multiplexed input channels and a sample-and-hold circuit has been incorporated on the chip. This uses the method of successive approximation. The sample time (for loading the capacitors) and the conversion time are programmable, and so can be matched to the external circuitry.

Detection and protection against overflow errors is provided for the converter result register (ADDAT): either an interrupt request is generated if the result of a previous conversion has not been read out of the result register before the next conversion is completed, or the next conversion will be suspended until the previous result has been read.

With applications which require less than 16 analog input channels, the remaining channel inputs can be used as digital input port pins.

The A/D converter supports four different types of conversion. In the standard single-channel conversion mode, the analog signal level on a specific channel is sampled once and converted to a digital result. In the standard single-channel continuous mode, the analog signal level on a specific channel is sampled repeatedly and converted without software intervention. In the auto-scan mode, the analog signal levels on a predetermined number of channels are sampled sequentially and converted. In the continuous auto-scan mode, the analog signal levels on the predetermined number of channels are sampled repeatedly and converted. Furthermore, the converted values for a specific channel can be inserted (injected) into a running sequence without interrupting the latter. For this reason the mode is referred to as channel injection mode.

The peripheral event controller (PEC) can be used to store the conversion results automatically in a table in memory, for later evaluation, without incurring the overhead of entering and returning from interrupt routines for each data transfer.

Real time clock

The real time clock (RTC) serves various purposes:

- System clock to determine the current time and date, even during idle mode and power-down mode (optionally).
- Cyclic interrupt, e.g. to supply a system time signal independent of the CPU frequency, without loading the general purpose timers, or to wake up from idle mode at regular intervals.
- 48-bit timer for measuring long times, with a maximum usable timespan of over 100 years.

The RTC module consists of a chain of 3 divider blocks, a fixed 8:1 divider, the reloadable 16-bit timer T14, and the 32-bit RTC timer (accessible via registers RTCH and RTCL). Both timers count forward.

The on-chip I^2C bus module

The integral I^2C module controls the transmission and reception of data frames over the two-line I^2C bus in accordance with the specification. The on-chip I^2C module can receive and transmit data with 7-bit or 10-bit addresses, operating in slave mode, master mode, or multi-master mode.

Several physical interfaces (port pins) can be set up via software. Data can be transferred at a rate of up to 400 kbit/s. Two interrupt nodes which are reserved for the I^2C module permit efficient interrupt services, and in addition support operation via PEC transfers.

Note: the port pins assigned to the I^2C interface only provide open-drain drivers, as required by the I^2C specification.

The on-chip CAN module

The integral CAN module controls the completely autonomous transmission and reception of CAN data frames in accordance with the CAN specification V2.0 Part B (active), i.e. the on-chip CAN module can receive and transmit standard data frames with 11-bit identifiers and extended data frames with 29-bit identifiers.

The module offers full-CAN functionality for up to 15 message objects. Message object 15 can be configured for basic-CAN functionality. Both modes have separate masks for acceptance filtering, which permit a number of identifiers to be accepted in full-CAN mode and a number of identifiers to be ignored in basic-CAN mode. All message objects can be updated independently of the other objects, and are suitable for the maximum message length of 8 bytes.

The bit timing is derived from XCLK, and is programmable for a data transfer rate of up to 1 Mbit/s. The CAN module uses two pins as the interface to a bus driver.

Universal serial bus (USB) interface

The USB module controls all transactions between the serial USB bus and the internal (parallel) bus in the microcontroller. The USB module contains various units which are required to support data handling on the USB:

- on-chip USB bus transceiver,
- USB memory with two pages of memory, each of 128 bytes,
- memory management unit (MMU), for USB and CPU memory access control,
- UDC device core for USB protocol handling,
- microcontroller interface with the USB-specific special function registers,
- interrupt control logic,
- clock generator unit to supply the clock signal for the USB module for higher and lower speed operation of the USB.

SDLM interface

Together with an external J1850 bus transceiver, the serial data link module (SDLM) is responsible for serial communication via a multiplexed serial bus of type J1850. The module conforms to the SAE Class B J1850 specifications for variable pulse width modulation (VPW). The SDLM is incorporated as an on-chip peripheral, and is linked to the CPU via the XBUS.

General SDLM equipment features:

- conforms to the SAE Class B specifications for J1850 (VPW),
- class 2 protocol is fully supported,
- operation with variable pulse width (VPW) at 10.4 kbit/s,
- high speed 4X operation at 41.6 kbit/s,
- programmable normalization bit,

- programmable delay for the transceiver interface,
- digital noise filter,
- power saving mode with automatic reactivation mechanism when bus activities occur,
- support for a single-byte header and compact header,
- CRC generation and checking,
- block mode for receiving and transmitting.

Data connection features:

- 11-byte transmit buffer,
- double-buffered 11-byte receiving buffer (optional overwrite enable),
- support for IFR types (in-frame response) 1, 2, and 3,
- transmit and receive message buffer, configurable for FIFO or byte mode,
- advanced interrupt management with 8 sources which can be separately activated,
- automatic IFR transmission (types 1 and 2) for 3-byte consolidated headers,
- configurable clock divider,
- bus status flags (IDLE/stand-by, EOF, EOD, SOF, Tx, and Rx in progress).

7.3.9 Power management monitoring features

The familiar basic methods of power saving (stand-by and switch off) have been enhanced with a range of additional energy monitoring functions (see below). These functions can be combined to reduce to a minimum the energy consumption of the controller for the application concerned:

- Flexible clock generation.
- Flexible peripheral management (peripherals can be switched on or off separately or in groups).
- Periodic reactivation in stand-by (idle) mode, using the RTC timer.

The features listed represent effective means of realizing stand-by states for the system, with an optimal balance between power reduction (i.e. time on stand-by) and peripheral operation (i.e. system functionality).

Flexible clock generation

The system for flexible clock generation combines a host of improved mechanisms (some of them user-defined) to supply the on-chip modules with clock signals. This is mainly of importance with energy-sensitive modes of operation, such as the idle (stand-by) function.

The energy-optimized oscillator reduces the amount of energy consumed in generating the clock signal.

The clock pulse supply system exercises efficient control over the amount of energy consumed in distributing the clock signal.

A reduction in the operating rate is achieved by means of a programmable factor (1 ... 32), by which the oscillator clock rate is divided, which results in lower frequency operation of the devices and hence a significant overall reduction in energy consumption.

Flexible peripheral management

The flexible peripheral management facilities offer a mechanism for switching each peripheral module on and off separately. In any situation (e.g. various system operating modes, standby, etc.) only those peripherals required for the function concerned need be operational. All the other ones can be switched off. This also permits operational controls of whole groups of peripherals, including the energy required for the generation and distribution of their clock input signals. Other peripherals can remain switched on, for example to maintain orderly communication channels. It remains possible to access the registers for peripherals which have been switched off separately (not within a group which has been switched off).

Periodic wakeup from idle mode

Periodic reactivation in stand-by mode combines the drastically reduced energy consumption of the stand-by mode (in combination with the supplementary energy monitoring functions) with a high level of system availability. External signals and events can be polled (at a lower speed) by activating the CPU and selected peripheral devices at periodic intervals and then returning them to the energy saving mode after a short time. This reduces the average energy consumption of the systems substantially.

7.3.10 Special features of the XC166 family

The basic architectural features of the C166 family have been retained. These include, in the first place, the instruction set which both supports the reuse of existing software and also preserves the knowledge gained. Additional DSP-oriented instructions enormously increase the performance of the CPU in the appropriate applications (e.g. filter calculations).

The fundamental concept of using registers for controlling peripherals has also been retained. The register address space has been extended, to enable new modules to be supported. Various optimized memory modules permit efficient access to registers, data, and code.

The communication capabilities of the devices in the XC166 family have been improved by the fact, on the one hand, that it is possible to incorporate several copies of some familiar interfaces (ASC, SSC), and on the other hand that powerful new modules (e.g. TwinCAN) are being used.

7.3.11 Summary of the instruction set

The following section summarizes briefly the instructions for the microcontrollers in the C166 family, arranged by class of instruction. This will give a basic understanding of the instruction set, the power and versatility of the instructions and their general usage.

Arithmetic instructions

- Addition of two words or bytes: ADD, ADDB
- Addition with carry of two words or bytes: ADDC, ADDCB
- Subtraction of two words or bytes: SUB, SUBB
- Subtraction with carry of two words or bytes: SUBC, SUBCB
- 16 × 16-bit signed or unsigned multiplication: MUL, MULU
- 16 / 16-bit signed or unsigned division: DIV, DIVU
- 32 / 16-bit signed or unsigned division: DIVL, DIVLU
- 1's complement of a word or byte: CPL, CPLB
- 2's complement (negation) of a word or byte: NEG, NEGB

Logical instructions

- Bitwise ANDing of two words or bytes: AND, ANDB
- Bitwise ORing of two words or bytes: OR, ORB
- Bitwise, exclusive ORing of two words or bytes: XOR, XORB

Compare and loop control instructions

- Comparison of two words or bytes: CMP, CMPB
- Comparison of two words followed by incrementing by 1 or 2: CMPI1, CMPI2

- Comparison of two words followed by decrementing by 1 or 2: CMPD1, CMPD2

Boolean bit manipulation instructions

- Manipulation of a maskable bit field in either the high or low order byte of a word: BFLDH, BFLDL
- Setting a single bit (to "1"): BSET
- Clearing a single bit (to "0"): BCLR
- Moving a single bit: BMOV
- Moving a negated bit: BMOVN
- ANDing of two bits: BAND
- ORing of two bits: BOR
- Exclusive ORing of two bits: BXOR
- Comparison of two bits: BCMP

Shift and rotate instructions

- Shifting a word to the right: SHR
- Shifting a word to the left: SHL
- Rotating a word to the right: ROR
- Rotating a word to the left: ROL
- Shifting a word to the right arithmetically (shift the sign bit): ASHR

Prioritization instructions

- Determining the number of shift cycles required to normalize a word operand (floating point support): PRIOR

Data movement instructions

- Standard movement of a word or byte: MOV, MOVB
- Movement of a byte to a word location in memory, with sign or zero byte extension: MOVBS, MOVBZ

Note: the data movement instructions can be used with numerous different addressing options, including also indirect addressing and automatic incrementing/decrementing of the pointer.

System stack instructions

- Storing a word temporarily on the system stack: PUSH
- Extracting a word from the system stack: POP
- Saving a word on the system stack and then updating the old word with a new value (for register bank control): SCXT

Jump instructions

- Conditional jump to a destination instruction within the current code segment, addressed either in absolute, indirect or relative form: JMPA, JMPI, JMPR
- Unconditional jump to an absolutely addressed destination instruction within any required code segment: JMPS
- Conditional jump to a relatively addressed destination instruction within the current code segment, depending on the value of a selectable bit: JB, JNB
- Conditional jump to a relatively addressed destination instruction within the current code segment, depending on the value of a selectable bit, followed by inversion of the tested bit if a jump is made (semaphore support): JBC, JNBS

Call instructions

- Conditional call of a subroutine within the current code segment, addressed either absolutely or indirectly: CALLA, CALLI
- Unconditional call of a relatively addressed subroutine within the current code segment: CALLR
- Unconditional call of an absolutely addressed subroutine within any required code segment: CALLS
- Unconditional call of an absolutely addressed subroutine within the current code segment plus the additional temporary storage (pushing) of a selectable register on the system stack: PCALL

- Unconditional branching to the interrupt or trap vector table in code segment 0: TRAP

Return instructions

- Returning from a subroutine within the current code segment: RET
- Returning from a subroutine within any required code segment: RETS
- Returning from a subroutine within the current code segment plus the additional retrieval (popping) of a selectable register from the system stack: RETP
- Returning from an interrupt control program: RETI

System control instructions

- Resetting the microcontroller by software: SRST
- Activation of the stand-by mode: IDLE
- Activation of the power reduction mode: PWRDN
- Servicing the watchdog timer: SRVWDT
- Disabling the watchdog timer: DISWDT
- Indicating the end of the initialization routine (sets the $\overline{\text{RSTOUT}}$ pin to the high level and disables the effects of any subsequent execution of the DISWDT instruction): EINIT

Miscellaneous instructions

- No operation, which requires 2 bytes for storage and the minimum time for execution: NOP
- Defining an uninterruptible instruction sequence: ATOMIC
- Switching the addressing modes "reg", "bitoff", and "bitaddr" over to the extended SFR area: EXTR
- Disabling the DPP address schema by the use of a specific data page instead of the DPPs, and optionally switch over to the ESFR area: EXTP, EXTPR
- Disabling the DPP address schema by the use of a specific segment instead of the DPPs, and optionally switch over to the ESFR area: EXTS, EXTSR

Note: the instructions ATOMIC and EXT* are used to support uninterruptible code sequences, e.g. for semaphore operations. In addition, they support data addressing beyond the limits of the current DPPs (except for ATOMIC), which is advantageous for models with larger memories in higher programming languages.

DSP-oriented instructions

The derivates of the XC166 family are based on the C166SV2 CPU core. Its multiply/accumulate unit can execute a range of additional instructions.

- Data movement: CoMOV, CoLOAD(2), CoSTORE
- Multiplication of two 16-bit values: CoMUL
- Multiplication of two 16-bit values followed by 40-bit addition: CoMAC(R)
- Multiplication followed by addition and simultaneous data movement: CoMACM(R)
- 31-/40-bit arithmetic instructions: CoADD(2), CoSUB(2), CoASHR, CoSHR, CoSHL, CoCMP
- Special arithmetic functions: CoABS, CoMIN, CoMAX, CoNEG, CoRND
- Idle instruction (for modifying address pointers): CoNOP

Protected instructions

Some instructions which are critical for the functioning of the microcontroller are provided as so-called protected instructions. These protected instructions use the maximum instruction format of 32 bits for decoding, while normal instructions only use a part of it (e.g. the lower 8 bits) with the other bits providing additional information, such as the registers involved. Decoding all 32 bits of a protected double word instruction increases the security against corruption of the data while it is being fetched. Critical operations like a software reset are therefore only executed if the complete instruction is decoded without error. This raises the security and reliability of a microcontroller system.

7.3.12 Block diagrams of the 16-bit microcontrollers

The block diagrams in Figures 7.26 to 7.34 on the following pages emphasize the diversity of the 16-bit microcontrollers in the C166 family. Please note that not all of the available derivatives are listed, and that new ones are continually being developed.

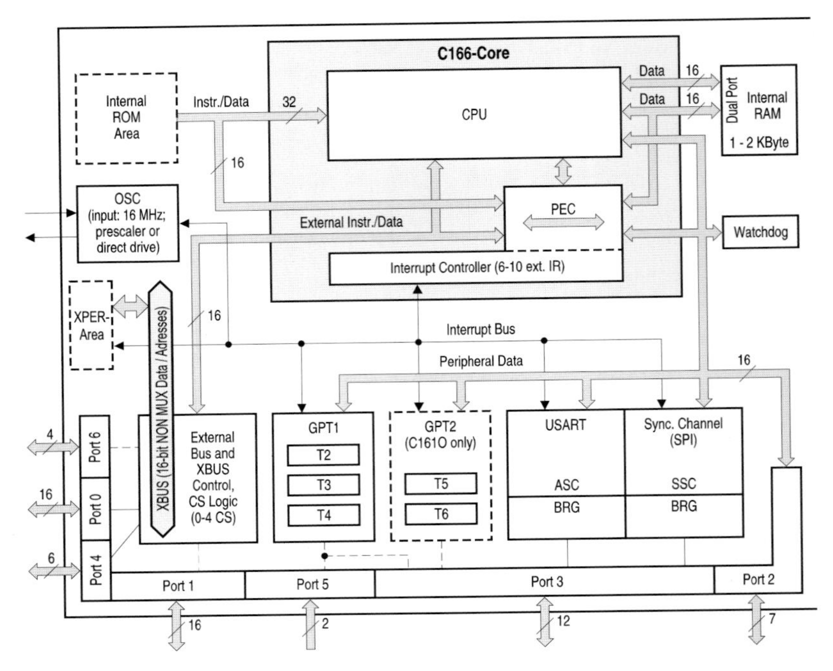

Figure 7.26 Block diagram of the C161/K/O

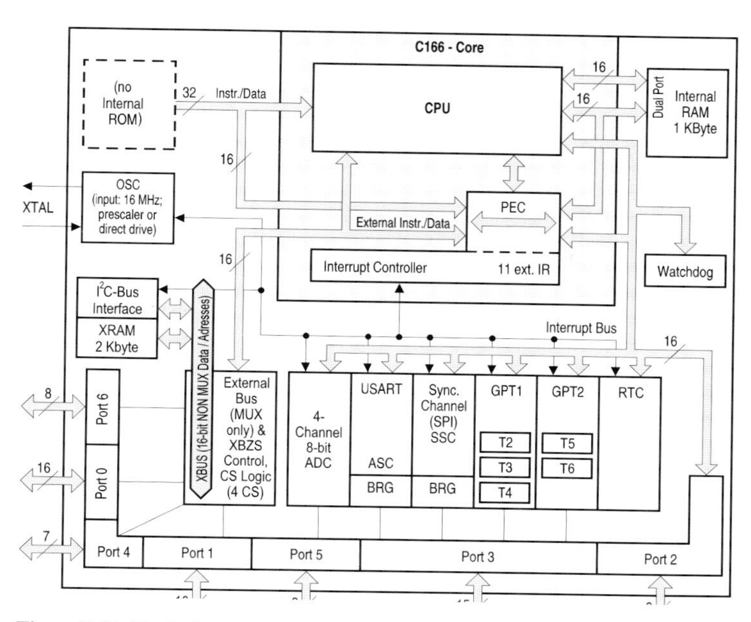

Figure 7.27 Block diagram of the C161PI

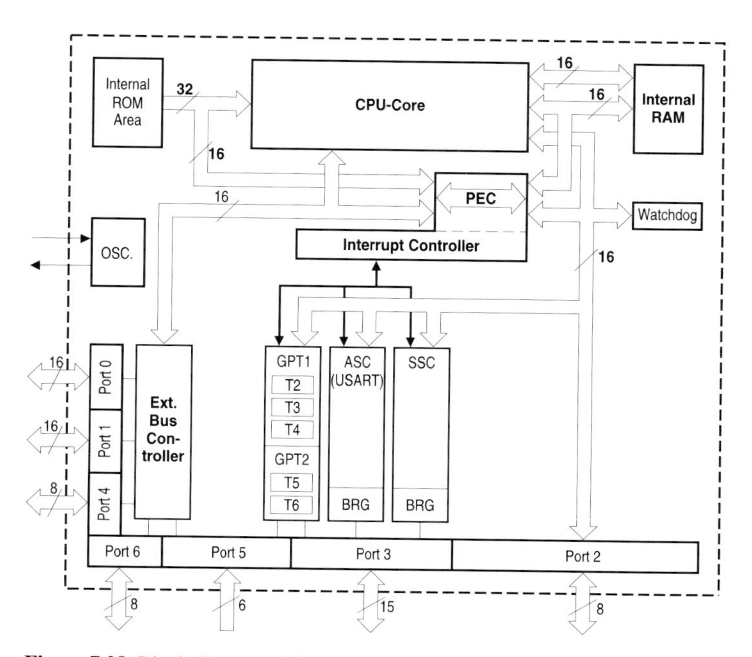

Figure 7.28 Block diagram of the C165

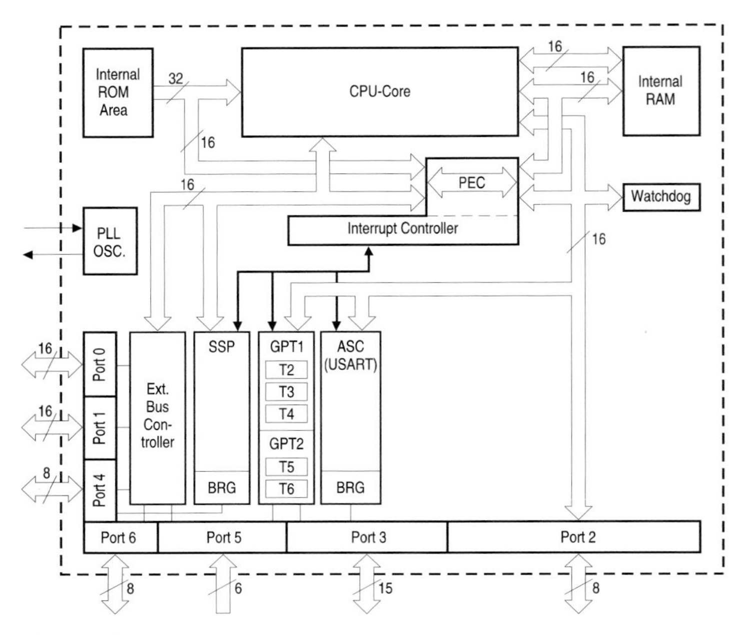

Figure 7.29 Block diagram of the C163

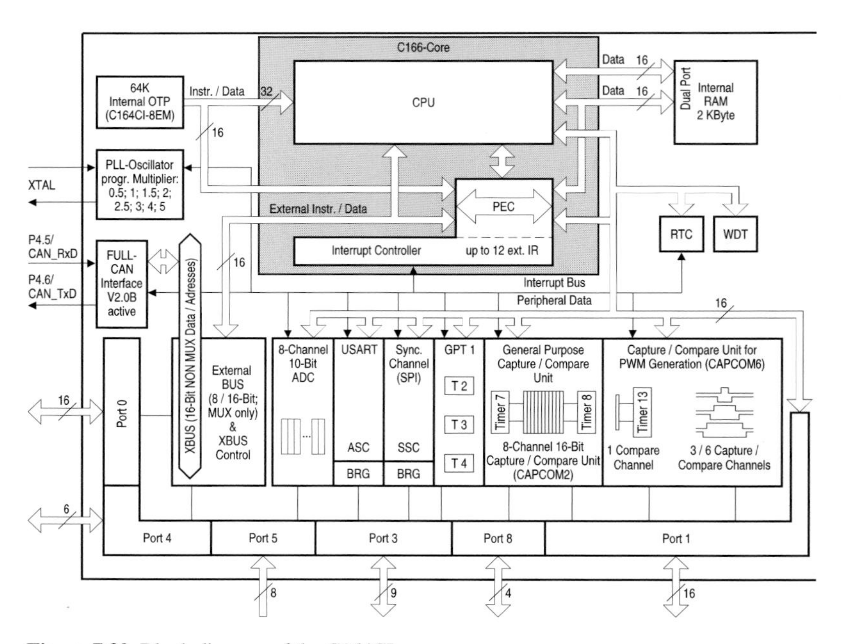

Figure 7.30 Block diagram of the C164CI

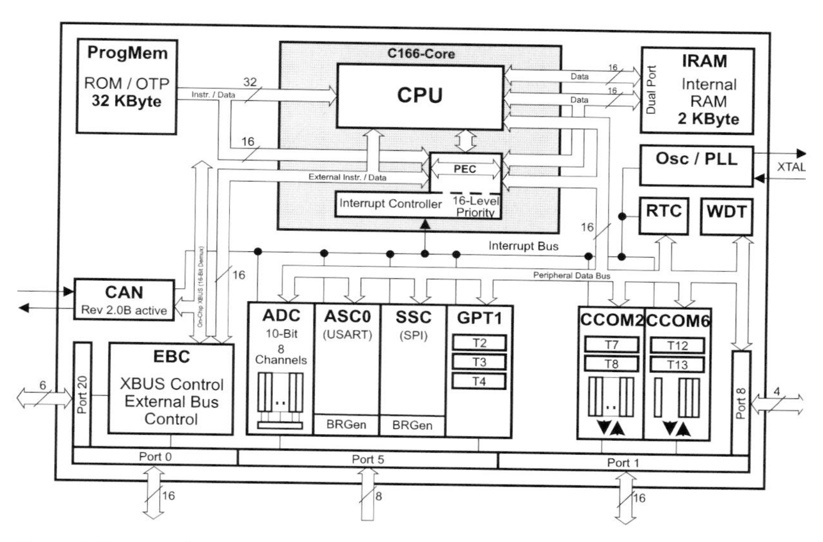

Figure 7.31 Block diagram of the C164CM

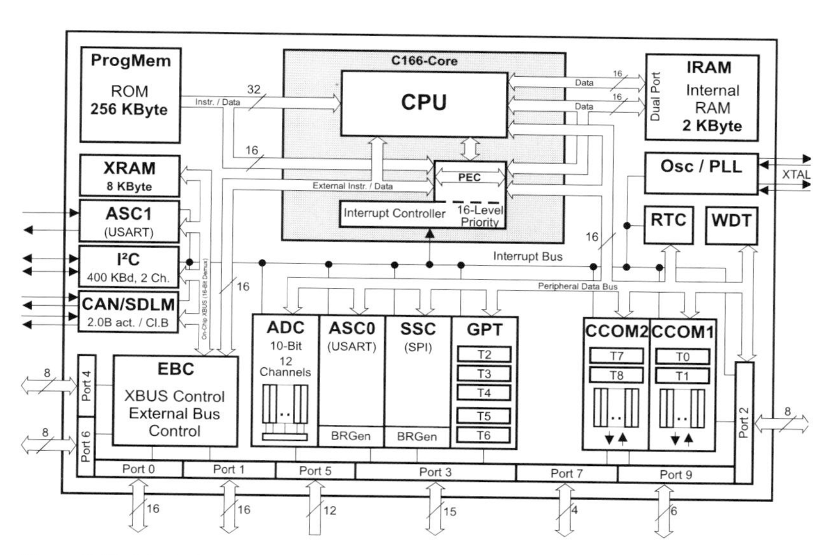

Figure 7.32 Block diagram of the C161CS

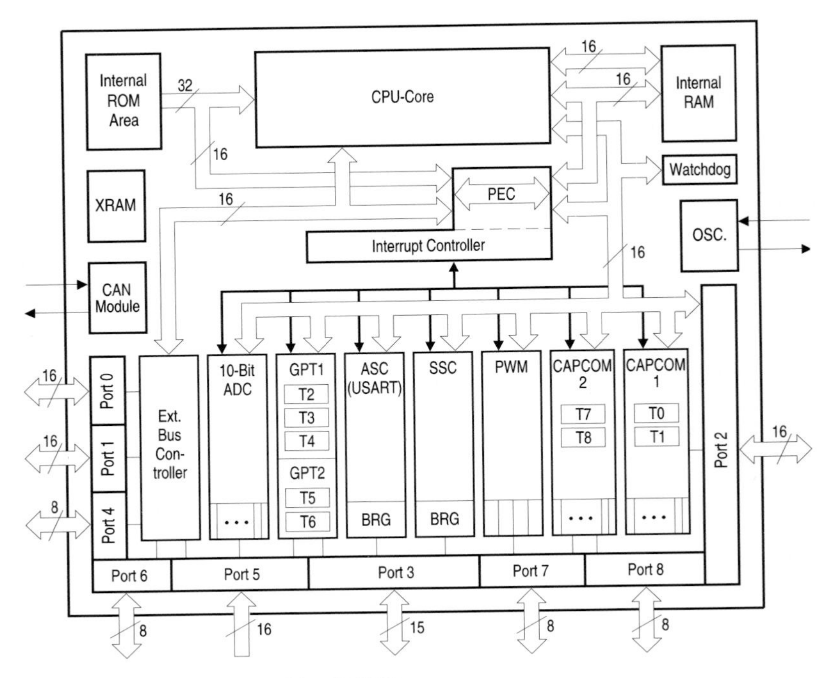

Figure 7.33 Block diagram of the C167CR

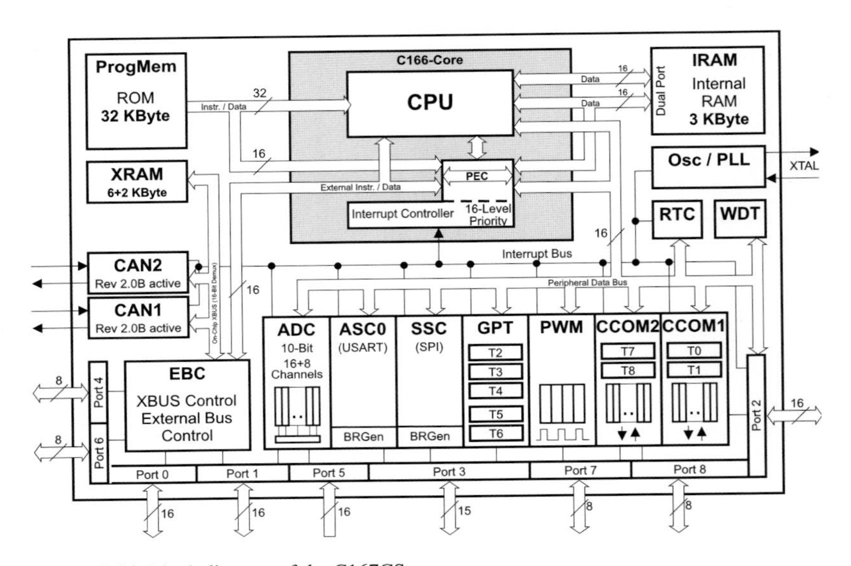

Figure 7.34 Block diagram of the C167CS

7.4 32-bit TriCore architecture

TriCore is the first combined 32-bit microcontroller/DSP architecture optimized for "integrated real-time systems". It combines the outstanding characteristics of three different areas: the signal processing of DSP, real-time microcontrollers, and RISC processing power; and it permits the implementation of RISC load-store architectures. It offers an ideal price/performance relationship, because a system requires fewer modules because more functions are integrated on the chip. This family of controllers is equipped with an optimal range of powerful peripherals for a wide spectrum of applications in power train, safety, and vehicle dynamics, driver information and entertainment electronics, and for body and convenience applications.

Figure 7.35 shows a diagram summarizing the TriCore architecture.

The instruction set architecture (ISA) supports a global linear 32-bit address space with memory-oriented I/O. The operation of the core is superscalar, i.e. it can execute simultaneously up to three instructions with up to four operations. Furthermore, the ISA can work in conjunction with different system architectures, also with multi-processing architectures. This flexibility at the implementation and system level permits different cost/performance combinations to be created whenever required.

TriCore contains a mixed 16-bit and 32-bit instruction set. Instructions with different instruction lengths can be used alongside each other without changing the operating mode. This substantially reduces the volume of code, so that even faster execution is combined with a reduction in the memory space requirement, system costs and energy consumption.

The real-time capability is essentially determined by the interrupt wait and context switching times. Here, the high-performance architecture reduces response times to a minimum by avoiding long multi-cycle instructions and by providing a flexible hardware-supported interrupt scheme. In addition, the architecture supports rapid context switching.

Detailed information about the TriCore architecture with the complete instruction set is contained in the "TriCore Architecture Manual".

7.4.1 Overview of the features of TriCore architecture

The list below summarizes the basic features of the TriCore architecture:

- 32-bit architecture,
- unified 4-Gbyte data, program, and input/output address space,
- 16-bit/32-bit instructions to reduce code volume,
- low interrupt response times,
- fast, automatic HW context switching,

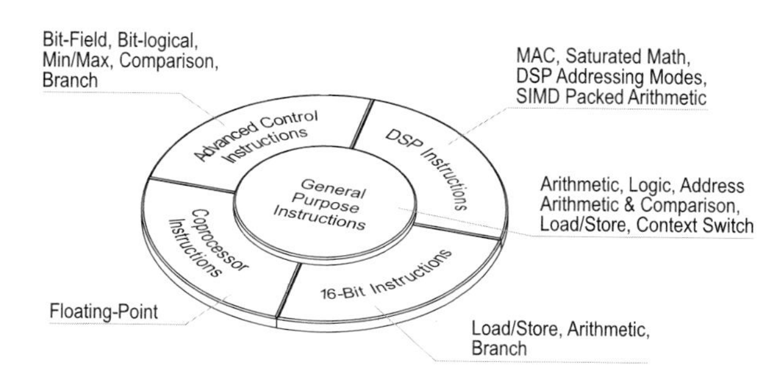

Figure 7.35
TriCore: one architecture with a modular instruction set

- multiplication-accumulation unit,
- saturation integer arithmetic,
- bit-operations and bit addressing supported by the architecture and instruction set,
- packed data operations (single instruction multiple data, SIMD),
- zero overhead loop for DSP applications,
- flexible power management,
- byte and bit addressing,
- little endian byte order,
- support for big and little endian byte ordering on the bus interface,
- precise exception states,
- flexible, configurable interrupt management with up to 256 levels.

7.4.2 Program status registers

As shown in Figure 7.36, the TriCore register sets consist of 32 general purpose registers (GPRs), two 32-bit registers with program status information (PCXI and PSW) plus one program counter (PC). PCXI, PSW, and PC are core special function registers (CSFRs).

The 32 general purpose registers are subdivided into sixteen 32-bit data registers (D0 to D15) and sixteen 32-bit address registers (A0 to A15). Four GPRs perform special functions: D15 serves as an implicit data register, A10 is the stack pointer (SP), A11 the return address register, and A15 the implicit base address register.

Registers A0 and A1 in the lower address registers, together with A8 and A9 in the upper address registers, are defined as system global registers. These registers are not included in any context partition and are not automatically saved or restored when there is a HW context switch. The operating system uses them, for example, to reduce the system overhead.

The PCXI and PSW registers contain status flags, information about instructions which have been executed, plus protection information.

Important: the register database is split into a lower and an upper context; when there is an automatic switch of one half: saved in a HW-controlled double chained list.

7.4.3 Data types

The TriCore instruction set supports boolean operations, bit sequences, characters, fixed point, addresses, signed and unsigned integers, and single-precision floating point numbers.

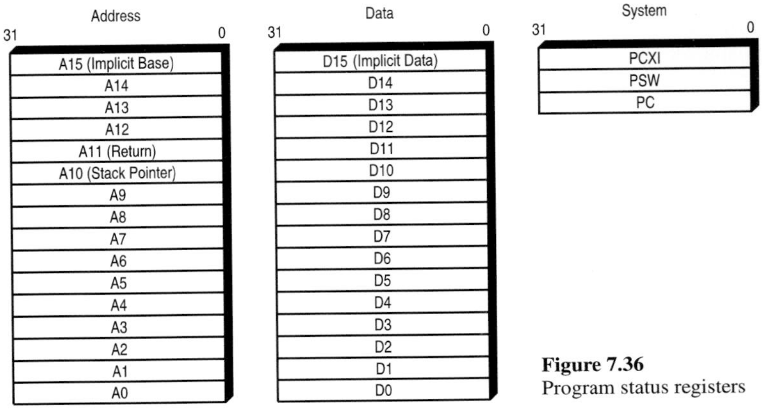

Figure 7.36
Program status registers

Most of the instructions process specific data types, while there are others which are suitable for manipulating various data types.

7.4.4 Addressing modes

The addressing modes enable load and store instructions to effect efficient access to simple data elements within data structures such as records, arrays with direct or sequential access, stacks and circular buffers. Simple data elements have a width of 1, 8, 16, 32, or 64 bits.

The addressing modes also ensure the efficient compilation of C, easy access to peripheral registers or the efficient implementation of standard DSP data structures (DSP addressing modes such as circular buffers for filters, and bit-reverse addressing for FFTs). The following seven addressing modes are supported by the TriCore architecture:

- Absolute
- Base + short offset
- Base + long offset
- Pre-increment or pre-decrement
- Post-increment or post-decrement
- Circular
- Bit-reverse

7.4.5 Instruction formats

The TriCore architecture supports both 16-bit and 32-bit instruction formats. All instructions have a 32-bit format. The 16-bit instructions form a subset of them, chosen because of the frequency with which they occur, to reduce the volume of code. Instructions are selected by the compiler, and can be used in parallel alongside one another with no mode change etc.

7.4.6 Tasks and contexts

In this book, the term "task" is used for an independent control procedure. Two types of task must be distinguished: software-managed user tasks (SMTs) and interrupt service routines (ISRs). Software-managed tasks are produced by the services of a real-time kernel or operating system, and are chosen for processing under the control of scheduling software. ISRs are chosen for processing by the hardware, as a response to an interrupt. In this architecture, ISR refers only to the code which is called directly by the hardware. Software managed tasks are sometimes referred to as user tasks, on the basis that they are executed in user mode.

Each task can be assigned its own authorization level. These individual rights are mainly enabled/disabled by I/O mode bits in the program status word (PSW). Associated with each task is a set of state elements, known collectively as the task's context. The term context is used for everything which the processor requires in order to determine the state of the corresponding task and to enable its further execution. This includes the CPU general registers used by the task, the task's program counter (PC), and its program status information (PCXI and PSW). The TriCore architecture exercises efficient management of the tasks' contexts by means of the hardware.

Upper and lower contexts

As shown in Figure 7.37, the context is subdivided into the upper context and the lower context.

The upper context comprises the upper address registers A10 - A15 and the upper data registers D8 to D15. These registers are designated as non-volatile for the purpose of function calls. The upper context also includes the PCXI and PSW registers.

The lower context comprises the lower address registers A2 to A7, the lower data registers D0 to D7 and the PC.

Both the upper and the lower context are associated with a LINK WORD. The con-

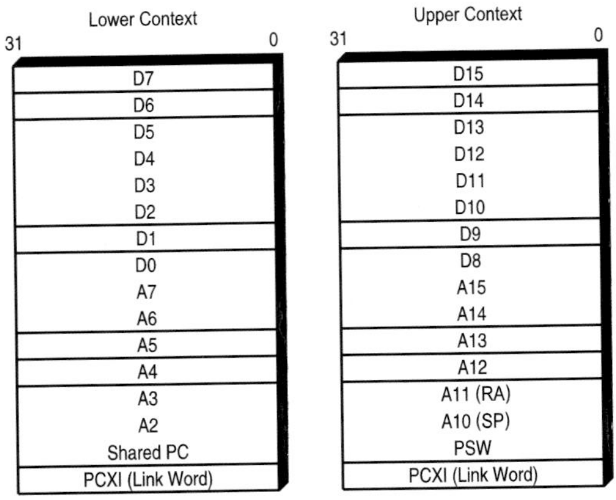

Lower Context (31 … 0)	Upper Context (31 … 0)
D7	D15
D6	D14
D5	D13
D4	D12
D3	D11
D2	D10
D1	D9
D0	D8
A7	A15
A6	A14
A5	A13
A4	A12
A3	A11 (RA)
A2	A10 (SP)
Shared PC	PSW
PCXI (Link Word)	PCXI (Link Word)

Figure 7.37
Upper and lower context

texts are stored in areas with a fixed size, which are linked together by the link word (see following section).

When an interrupt occurs, the upper context is automatically saved and is restored when the return occurs. If the interrupt service routine (ISR) needs to use more registers than are available in the upper context, then the lower context will be explicitly saved and restored by the ISR.

Context save areas

The TriCore architecture makes use of linked lists of context save areas (CSAs) with fixed sizes, supporting systems with multiple, linked control threads. A CSA consists of 16 words of on-chip storage facilities, aligned to a 16-word boundary. One individual CSA can hold exactly one upper or lower context. CSAs which are not used are linked by an 'unused' list. They are assigned from this unused list as necessary and, when they are no longer required, are returned to the list. The processor hardware controls their assignment and release. They are transparent to the application code. Only the system start code and certain of the operating system's exception handling routines need to explicitly access the CSA lists and the memory device.

Fast context switching

To increase its performance capability, the TriCore architecture has a uniform context switching mechanism for function calls, interrupts, and traps. In all cases, the upper context is automatically saved and retrieved by the hardware, with the saving and retrieval of the lower context being open to the new task as an option. As a result of the unique memory subsystem design of TriCore, which permits the transfer of up to 16 data words between the processor registers and memory, so the entire context can be saved in a single operation, fast context switching is speeded up even more.

7.4.7 Interrupt system

A service request can be defined as an interrupt request from a peripheral device, a DMA request, or an external interrupt. For the sake of simplicity, a service request will be described simply as an interruption. The entry code for the ISR consists of a block within a vector of code blocks. Each code block represents the entry for an interrupt source. A priority number is assigned to each source. All the priority numbers are programmable. The service program uses the priority number to determine the memory location for the entry code block. This prioritization of the service programs permits nested interrupts. A service request can interrupt the processing of an interrupt with lower priority. Interrupt sources with the same priority cannot mutually interrupt each other.

7.4.8 Trap system

A trap is a special form of interrupt for error handling, and is initiated in the event of an exception which falls into one of the eight classes identified below:

- reset
- internal protection
- instruction errors
- context management
- internal bus and peripheral errors
- logically 'true' signal state for L signal level
- system call
- non-maskable interrupt

The entry code for the trap processing routines comprises a vector of code blocks. Each code block contains the entry point address for one trap. When a trap is triggered, the trap's identification number (TIN) is stored in data register D15. The trap processing routine uses this TIN to identify the precise reason for the trap. During arbitration the trap with the lowest TIN number takes precedence.

7.4.9 Protection system

The protection system gives the programmer the option to assign access permissions to memory regions, for both data and code. The ability can be used to protect the core system functionality from bugs that may have slipped through testing and from transient hardware faults. In addition, TriCore's protection system contains the critical features for isolating errors, thus facilitating debugging.

Permission levels

TriCore's embedded architecture allows each task to be assigned the specific permission level it requires to perform its function. Individual permissions are enabled by means of the I/O mode bits in the program status word (PSW). The three authorization levels are called User-0, User-1, and Supervisor:

- User-0 mode is used for tasks which do not access peripheral devices. Tasks at this level do not have permission to enable or disable interrupts.
- User-1 mode is used for tasks which access common unprotected peripheral devices. These accesses usually include read/write accesses to SIO ports and read accesses to timers and most of the I/O status registers. Tasks at this level can disable interrupts.
- Supervisor mode permits read/write accesses to the system registers and all peripheral devices. Tasks at this level can disable interrupts.

Protection model

The memory protection model for the TriCore architecture is based on address ranges, with each address range having its own permission setting. The address ranges and the associated permissions are defined in two to four identical sets of tables, which are stored in the core SFR (CSFR) space. Each set is referred to as a protection register set (PRS). When the protection system is active, TriCore checks the legality of every load, store, or instruction fetch address before performing the access. In order to be legal, the address must fall within one of the ranges specified in the currently selected PRS and permission for that type of access concerned must be available in the matching range.

7.4.10 Reset system

Various events can force a reset of the TriCore device:

- Power-on reset: triggered through an external pin when the power supply for the device is switched on (cold start).
- Hard reset: triggered through an external pin during operation (warm start).
- Soft reset: triggered by a software write into a reset request register. This register has a special protection mechanism to prevent unintended accesses. Implementation-specific controls in this register effect either a partial or a complete reset of the device.
- Watchdog timer reset: triggered by an error state recognized by a watchdog timer.
- Wake-up reset: triggered through an external pin when the device is reactivated from the energy-saving mode.

The core can check, using a reset status register, which of the various triggers has invoked the reset.

7.4.11 Debugging system

TriCore contains mechanisms and resources to support on-chip debugging, which are used by the Debug Control Unit, a module which is located outside the core. Most of the functions and details of the Debug Control Unit are implementation-specific. For this reason, no further description of the unit and its registers is included here. The details will be found in the documentation for the products concerned.

7.4.12 Programming model

This section discusses the following aspects of the TriCore architecture which are of relevance for the software: the data types supported, the formats of the data types in registers and memories, the various addressing modes provided by the architecture, and the memory model.

Data types

The TriCore instruction set supports boolean operations, bit strings, characters, signed fractions, addresses, signed and unsigned integers, and single-precision floating point numbers. Most of the instructions process specific data types, while there are others which are suitable for manipulating various data types.

Boolean expressions

A boolean expression is either TRUE or FALSE. TRUE corresponds to a value of one (1) when it is computed, or non-zero when it is checked; FALSE corresponds to a value of zero (0). Boolean expressions are generated as the result of comparison and logical instructions, and are used as source operands in logical and conditional jump instructions.

Bit strings

A bit string is a packed bit field. Bit strings are generated and used by logical, shift, and bit field instructions.

Characters

A character is an eight-bit value which corresponds to a very short unsigned integer. It does not assume any specific coding.

Signed fractions

The TriCore architecture supports signed fractional 16-bit data for DSP arithmetic. Data values in this format have a single leading sign bit with a value of 0 or -1, followed by an implicit binary point and a fraction. Their values thus lie in the range [-1,1]. When they are saved into registers, 16-bit fractional data occupies the 16 most significant bits, and the 16 least significant bits are set to zero.

Addresses

An address is a 32-bit unsigned value.

Signed / unsigned integers

Signed and unsigned integers normally comprise 32 bits. When smaller signed and unsigned integers are loaded from memory into a register they are extended to 32 bits with a sign or zeros. Multiple precision integers are supported by addition and subtraction with carry. During move and masking operations, integers are regarded as bit strings. Multi-precision shifts can be done using a combination of single precision shifts and bit field extracts.

Single precision IEEE-754 floating point numbers

Depending on the particular implementation of the core architecture, IEEE-754 floating point numbers are supported by direct hardware instructions or software emulation.

Data formats

All the general purpose registers have a width of 32 bits, and most of the instructions operate on word values (32 bits). If data containing less bits than a word are read out of memory, they must be sign-extended or zero-extended before any operations can be applied to the whole word. The required alignments are different for addresses and data. To permit transfers between address registers and memory, addresses (32-bit) must be aligned to a word boundary. For transfers between data registers and memory, a data item can be aligned to any arbitrary half-word boundary, regardless of its size;

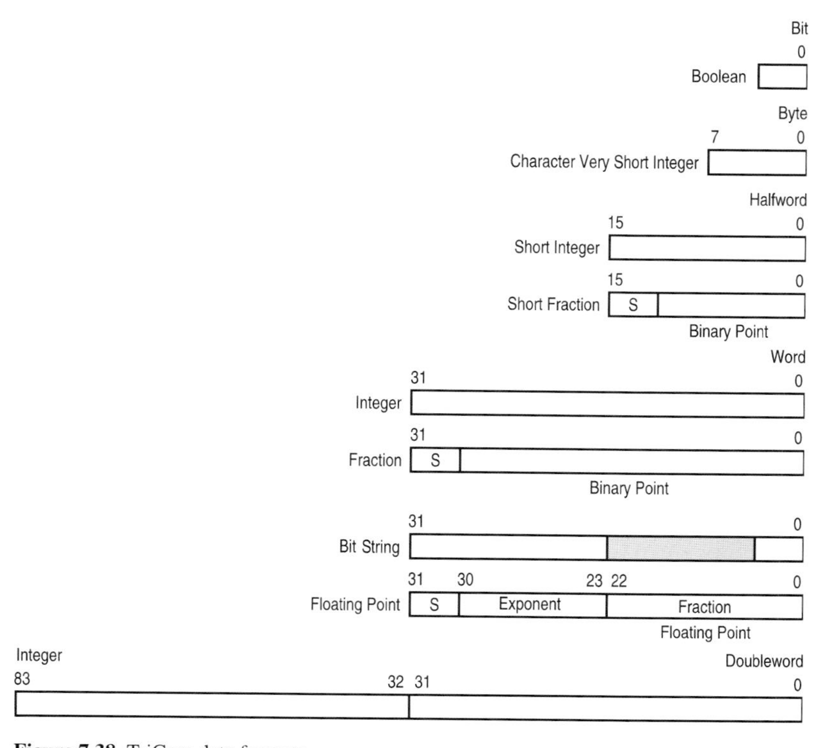

Figure 7.38 TriCore data formats

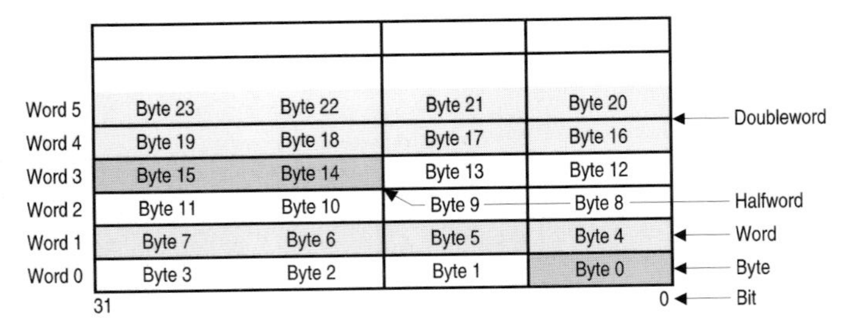

Figure 7.39 Byte order

bytes can be accessed using any valid byte address. Figure 7.38 shows the data formats which are supported.

The data memory and the CPU registers save data in 'little endian' byte order (i.e. the least significant bytes are stored in the lower addresses). Figure 7.39 shows the byte order. Little endian memory referencing is used consistently for data and instructions.

If the TriCore system is connected to an external big endian device, the bus interface carries out the translation between the big and little endian formats. As already mentioned, bytes must be saved on byte boundaries, and half-words, words and double-words on half-word boundaries.

7.4.13 The memory model

The TriCore architecture permits up to 4 Gbytes of memory to be accessed. The address width is 32 bits. The address space is divided into 16 regions or segments (0 to 15). Each segment comprises 256 Mbytes.

The upper four bits of an address are used for selecting a specific segment. The first 16 kbytes of each segment are accessed using either absolute addressing or absolute bit addressing.

Speculative read accesses are not supported for segments 14 and 15. Accesses to this memory area are only initiated if the core is sure that the access will be successfully completed. These segments are used for peripheral registers (PSFRs) or for external peripheral devices. FIFOs, peripheral devices with status registers and other devices should be arranged in this address segment, so that they cannot be the subject of speculative read operations which could result in the deletion of data. Accesses cannot be made in the User-0 mode.

Segments 0 to 7 are reserved. Any access to them triggers a trap. For segments 8 to 13 there may be further restrictions in the case of certain products; these are noted in the appropriate product documentation.

Many data accesses are made using addresses which are formed by the addition of an offset value to the contents of a base address register. In these cases, however, the offset value used must not produce an address beyond the segment boundary, as this would trigger a trap. This restriction means that the base address can be used at any time to determine which segment may be accessed.

The core special function registers (CSFRs) are arranged in one 64-kbyte area. The base address of this area is product-specific, and is stated in the appropriate product documentation.

7.4.14 Addressing model

Apart from the addressing modes used in the instruction set, the TriCore architecture allows extended addressing modes to be realized using short instruction sequences.

Built-in addressing modes

These addressing modes allow load and store instructions to efficiently access simple data elements with data structures such as records, arrays with direct or sequential access, stack memories, and circular buffers. Simple data elements have a width of 1, 8, 16, 32, or 64 bits. Table 7.3 gives an overview of the addressing modes which are supported.

The addressing modes have been chosen to permit the efficient compilation of C, easy access to peripheral registers and the efficient implementation of typical DSP data structures (circular buffers for filters, and bit-reversed indexing for FFTs).

The instruction formats have been chosen so that they provide the greatest possible space for direct addresses, or offsets in the case of indirect addressing.

Absolute addressing

Absolute addressing is suitable for addressing peripheral registers or global data. It uses an 18-bit constant for the memory address, this being specified in the instruction itself. The complete 32-bit address is formed by using the top 4 bits of the 18-bit constant as the top 4 bits of the 32-bit address, the lower 14 bits being directly copied into the address, and inserting 14 zero-bits between them.

Base+offset addressing

Addressing by means of a base + offset is useful for accesses to structural elements, local variables (with the stack pointer as the base), and static data (with an address register as the base pointer). Here, the final address is the sum of an address register and an offset value with sign extension. The width of the offset is 10 bits, or for some instructions 16 bits. This allows any required memory location to be addressed.

Addressing with post-modification

This mode of addressing uses the value of an address register as the final address and, after the access, changes the address by the addition of a signed 10-bit offset. The sign can be chosen to move the pointer forward or backward. Both options can be used, for example, for sequential accesses to arrays or to remove (POP) data from a stack memory.

Addressing with pre-modification

This mode of addressing adds a signed 10-bit offset to the value of an address register and uses the result as the final address. The address register is overwritten with the sum which has been formed. The sign can be chosen to move the pointer forward or backward. Both options can be used, for example, for sequential accesses to arrays or to store (PUSH) data into a stack memory.

Table 7.3 Built-in addressing modes

Addressing mode	Address register used	Offset width (bits)
Absolute	None	18
Base + offset	Address register	10/16 (short/long)
Post-op change	Address register	10
Pre-op change	Address register	10
Circular addressing	Address register pair	10
Bit reversed	Address register pair	–

Circular addressing

The main application of circular addressing is to access data in a circular buffer during filter calculations. In the case of circular addressing, the current state is stored in an address register pair. The even-numbered register contains the base address, the upper half of the odd-numbered register contains the buffer size, and the lower half the buffer index. Here, the final address is the sum of the base address and the index.

After the access has been made, the index is modified by a signed 10-bit offset (contained in the instruction). Provided that this offset is less than the buffer size, the index will automatically wrap-round from one end of the buffer to the other. For example, if the buffer has a size of 50 and the index of 48 is increased by an offset of 4, the index will be given the value 2 (48 + 4 – 50).

A circular buffer is subject to the following restrictions:

– the start of a circular buffer must be aligned to a 64-bit boundary
– the buffer size must be a multiple of the data size, which is implicitly determined by the access instruction. For example, when using a "load word" instruction for the access, the buffer size must be a multiple of 4 bytes, and with "load double word" instructions a multiple of 8 bytes.

Bit-reverse addressing

This mode of addressing is used for FFT algorithms, because with the usual implementations of the FFT the results appear in bit-reversed sequence. For bit-reversed addressing, an address register pair stores the current status. The even-numbered register contains the base address, the lower half of the odd-numbered register contains the array index, and the upper half the modification value. Here, the final address is the sum of the base address and the index. After each access, the modification value is added to the reversed index and the result is again reversed. The result of this with a modification value of 1024, for example, is the following sequence of indices: 0, 1024, 512, 1536, 256, etc.

As a rule, the modification value represents the reversed value of half the array size.

As bit reversal does not represent a bit field operation, and is therefore only programmed with difficulty, this addressing mode permits simple programming and rapid processing.

Extended addressing modes

Special addressing modes which are not directly supported by the architecture can be implemented by short instruction sequences.

Indexed addressing

Using the ADDSC.A instructing, which adds a scaled value to an address register, indexed addressing of byte, half-word, word, or double-word arrays can be realized (scaling by 1, 2, 4 or 8).

To allow indexed addressing of bit fields, the instruction ADDSC.AT determines the word in which the required bit or bit field is located. Bits are extracted using the EXTR.U instruction, and are stored using the instruction LDMST (load / modify / store).

PC-relative addressing

For branching and subroutine calls, it is usual to use PC-relative addressing. However, as this mode of addressing for data accesses would reduce the performance of the memory system, the TriCore architecture does not directly support such data access. If PC-relative data access is required, the address of a neighboring label can be loaded into an address register and used as the base address.

In the case of code which is loaded dynamically, the current value of the PC can be determined with the instruction JL (jump and link), which stores the address of the subsequent location in register A11. The return address for the current routine must be saved beforehand.

Extended absolute addressing

Extended absolute addressing is effected by combining two instructions. The instruction LEA (load effective address) loads a 32-bit address into an address register. After execution of the instruction MOVH.A (move high word), the data is addressed using the base address + 16-bit offset.

7.4.15 Core registers

In the TriCore architecture, a set of core special function registers (CSFRs) is defined. These CSFRs control the operation of the core, and provide status information about the core's operation. The CSFRs are split into the following groups:

- Program state information
- Stack management
- Context management
- Interrupt and trap control
- System control
- Memory protection
- Debug control

Table 7.4 Core register matrix

Register name	Description
D0 – D15	Data registers
A0 – A15	Address registers
PSW	Program status word
PCXI	Previous context information
PC	Program counter (read only)
FCX	Pointer to first free location in list
LCX	Pointer to last free location in list
ISP	Interrupt stack pointer
ICR	Interrupt control register
BIV	Base address of the interrupt vector table
BTV	Base address of the trap vector table
SYSCON	System configuration register
DPRx_0 – DPRx_3	Data segment protection register sets (x = 0 – 3)
CPRx_0 – CPRx_3	Code segment protection register sets (x = 0 – 3)
DPMx_0 – DPMx_3	Data protection mode register sets (x = 0 – 3)
CPMx_0 – CPMx_3	Code protection mode register sets (x = 0 – 3)
DBGSR	Debug status register
EXEVT	External break input event specifier
SWEVT	Software break event specifier
CREVT	CSFR access event specifier
TRnEVT	Specifier for trigger event n (n = 0, 1)

The sections below present a summary of these registers. The CSFRs are supplemented by a set of general purpose registers (GPRs). Table 7.4 includes all the CSFRs and GPRs.

Accessing the core registers

The core uses the two instructions MFCR and MTCR to access the CSFRs. The instruction MFCR (move from core register) moves the contents of the CSFR which is addressed into a data register. MFCR can be executed on any privilege level. The instruction MTCR (move to core register) moves the contents of a data register into the CSFR which is addressed. To prevent unauthorized write accesses to the CSFRs, the MTCR instruction can only be executed at the Supervisor privilege level.

The CSFRs are also mapped into the top of the local code segment in the memory address space. This assignment makes the complete architectural state of the core visible in the address map. This feature ensures efficient support for the debugger and the emulator.

Note: the core may not use this mechanism to access the CSFRs, but must use the instructions MFCR and MTCR for that purpose.

There are no instruction which allow bit, bit field, or load-modify-store accesses to the CSFRs. The instruction RSTV (reset overflow flags) only resets the overflow flags in the PSW, without changing any other PSW bits. This instruction can be executed at any privilege level.

7.4.16 General purpose registers (GPRs)

Figure 7.40 shows the general purpose registers. The 32-bit general purpose registers are split evenly into 16 data registers or DGPRs (D0 to D15) and 16 address registers or AGPRs (A0 to A15). The separation of data and address registers facilitates efficient implementations, in which arithmetic and memory operations are performed in parallel. A range of instructions can be used for the interchange of information between data and address registers, in order to generate or

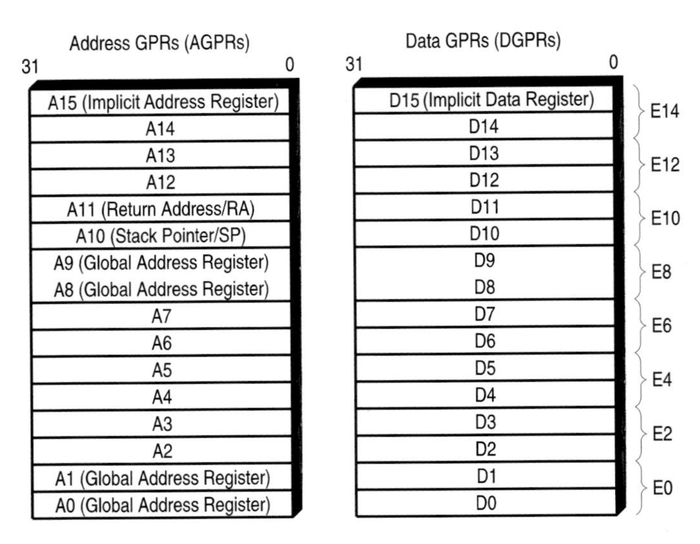

Figure 7.40 General purpose registers (GPRs)

derive table indices etc. Two consecutive data registers can be concatenated to form eight extended size registers (E0, E2, E4, E6, E8, E10, E12 and E14), in order to support 64-bit values (see Figure 7.40).

Registers A0, A1, A8, and A9 are defined as system global registers. Their contents are not saved and restored when calls, traps, or interrupts occur. Register A10 is used as the stack pointer (SP); Register A11 is used for saving the return address (RA) for calls and linked jumps, and for saving the return program counter (PC) value during interrupts and traps.

While the 32-bit instructions have unrestricted access to the GPRs, many of the 16-bit instructions implicitly use A15 as their address register and D15 as their data register. This implicit use facilitates the coding of these instructions in 16 bits. To support 64-bit data, these values are stored in an even/odd register pair. In Assembler syntax, these register pairs are designated either as one pair of 32-bit registers (for example D9/D8) or as one extended 64-bit register (thus, for example, E8 is the concatenation of D9 and D8, where D8 represents the less significant word of E8).

It should be noted that there are no separate floating point registers, but that the data registers are used in performing floating point operations. The saving and retrieval of floating point data is effected automatically, making use of the fast context switching. The GPRs represent an important element in the context of a task. When the context for a task is being saved to or restored from memory, the context is divided into an upper and a lower context. Registers A2 to A7 and D0 to D7 are part of the lower context. Registers A10 to A15 and D8 to D15 belong to the upper context.

Registers for program state information

The PC, PSW, and PCXI registers store and reflect the program state information. When the context for a task is being saved or restored, the contents of these registers form an important part of the procedure, and are saved/restored or modified during this process.

The program counter (PC) contains the address of the instruction which is currently being executed.

The five most significant bits of the PSW contain ALU status flags, which are set or cleared by arithmetic instructions. The remaining bits of the PSW control the permission levels, the protection register sets, and the call depth counter. The PCXI register contains the link to the previous execution context and supports fast interrupts and automatic context switching.

Context management registers

The context management registers comprise three pointers: FCX, PCX, and LCX. These pointers handle context management and are used during the operations to save and restore the context.

Each pointer consists of two fields: a 16-bit offset and a 4-bit segment specifier. A context save area (CSA) is an address range which contains 16 word locations (64 bytes), which is the memory space required to save one upper or one lower context. By incrementing the offset value by one, the effective address is always incremented to the address 16 word locations above the previous address (offset moved 6 bits to the left). The total usable area in each address segment for CSAs amounts to 4 Mbytes, which gives memory space for 64 K of context save areas.

The effective address must point to memory which exists. Otherwise the system behavior is undefined.

The FCX pointer register contains the pointer to the start of the free locations list, which always points to an available CSA.

The previous context pointer (PCX) stores the address of the CSA for the previous task. PCX is part of the PCXI register.

The LCX pointer register contains the pointer to the end of the free locations list, and is used to recognize impending CSA list range underflows. If the value of FCX resulting after for an interrupt or a CALL corresponds to the limit value then, although the context save operation will be completed, the destination address written into the trap vector address will be forced to that for the emptying of the CSA list.

Stack management register

Stack management in the TriCore architecture supports a user stack and an interrupt stack. The management of the stack involves the address register A10, the interrupt stack pointer (ISP), and one PSW bit. The general purpose address register A10 is used as the stack pointer. The initial contents of this register are usually defined by an operating system when a task is generated, where individual tasks can be assigned a private stack area. The interrupt stack pointer (ISP) prevents the interrupt service routines (ISRs) from accessing the private stack areas and possibly interfering with the context of the software-administered tasks. The TriCore architecture implements an automatic switch to the use of the interrupt stack pointer instead of the private stack pointer.

Interrupt and trap control registers

Three CSFRs support interrupt and trap handling: the interrupt control register (ICR), the interrupt vector table pointer (BIV), and the trap vector table pointer (BTV).

The interrupt control register (ICR) contains the current CPU priority number (CCPN), the enable/disable bit for the interrupt system, the pending interrupt priority number (PIPN), and an implementation-specific control for the interrupt arbitration schema. The other two registers contain the base addresses for the interrupt (BIV) and trap vector tables (BTV).

When an interrupt is accepted, or when a trap occurs, the pointer to the address in the interrupt/trap vector table is formed by shifting the value of the priority number / trap class 5 bits to the left and then ORing it with the value of the BIV/BTV register. The shift to the left results in a step spacing of 8 words (32 bytes) between the individual entries in the vector tables. The base addresses must be carefully aligned. On the one hand, the addresses must represent even byte addresses (half-word addresses) and on the other hand the base addresses must be aligned to a boundary which is a power of two, because of the simple ORing with the shifted priority values. The appropriate power of two is determined by the number of entries used. For example, the full extent of 256 interrupt entries requires alignment to an 8-kbyte boundary. For the 8 trap classes (0 to 7), alignment to a 256 byte boundary is sufficient.

System control registers

Three registers provide system control: the system configuration control register (SYSCON), the local program memory unit control register (PMUCON), and the local data memory unit control register (DMUCON).

Memory protection registers

The TriCore architecture incorporates hardware mechanisms which protect memory areas specified by the user from unauthorized accesses by read, write, or call instructions. Furthermore, the protection hardware can be used to generate signals to the debug unit. TriCore includes register sets in which the address ranges and access rights for a series of memory areas are specified. There are separate register sets for code and data memory.

The 2-bit PRS field in the PSW allows up to four such register sets to be selected in each case (four for data and four for code). The number of register sets provided for memory protection is defined specifically for each implementation of the TriCore architecture.

Data and code segment protection registers

The register pairs DPRx_n/CPRx_n comprise the two-word registers which define the lower and upper boundary addresses of the corresponding memory area. If the lower boundary is greater than the upper boundary, then no range checks will be carried out. If the lower boundary is the same as the upper boundary, then the area is considered to be empty.

When debug signals are being generated, the values in DPRx_n/CPRx_n are regarded as individual addresses, instead of defining a range. Signals are then generated to the debug unit if the address of a memory access corresponds with the contents of one or more of DPRx_n/CPRx_n. For this purpose, a comparison is made with the contents of the upper boundary register.

The 8-bit data/code protection mode registers determine the access permissions and debug signal generation for the data/code protection areas defined by the respective registers.

Debug registers

Seven registers have been implemented to support debugging. These registers define the conditions under which a debug event is generated, or what actions are to be initiated when a debug event occurs, and they also supply information about the state of the debug unit. The precise functions of the debug unit depend on the particular implementation. A description of the functions and the associated registers will be found in the appropriate product descriptions.

7.4.17 Block diagrams of 32-bit microcontrollers

Figures 7.41 to 7.43 show block diagrams of three selected 32-bit microcontrollers.

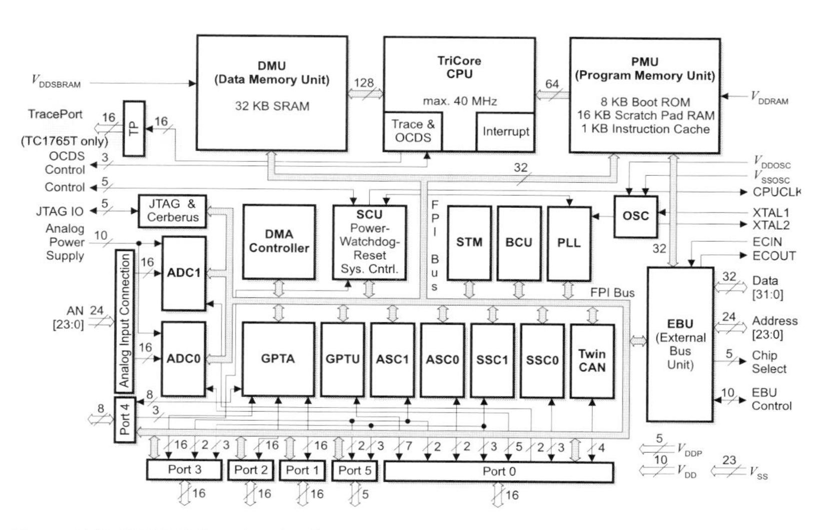

Figure 7.41 TC1765 functional units

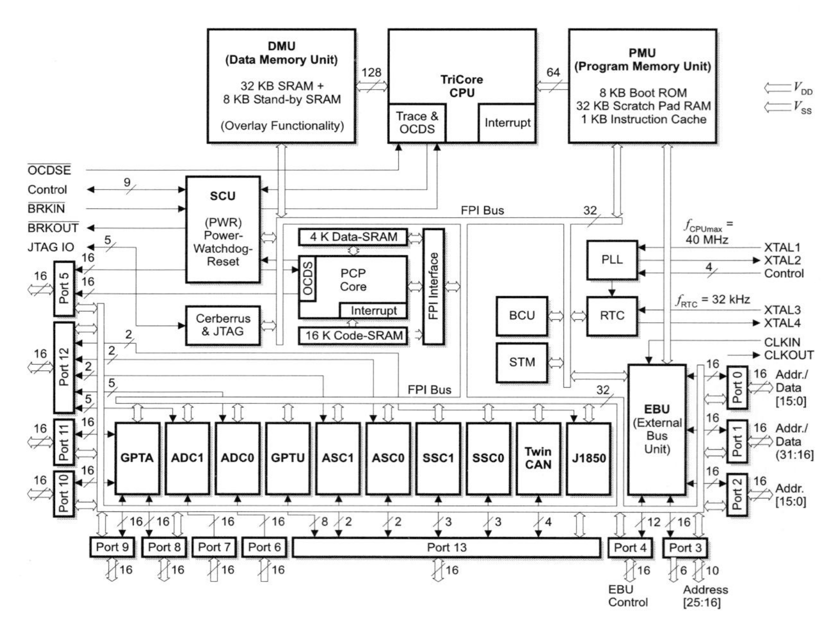

Figure 7.42 TC1775 functional units

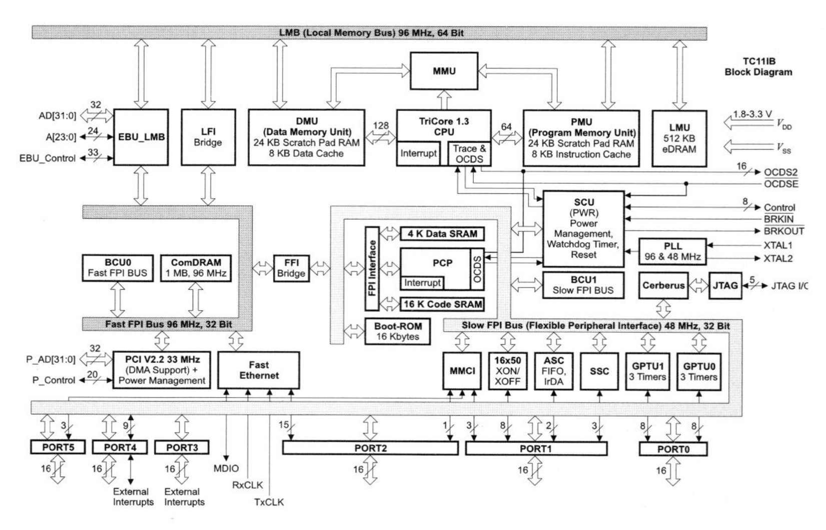

Figure 7.43 TC11IB functional units

8 Smart cards

8.1 Overview

ICs for smart cards (also called chip cards) have an exciting future ahead of them: the most advanced semiconductor technologies are making ever smaller chips possible, with every higher performance, and together with innovative production concepts they provide the prerequisites for unimagined applications on smart cards.

Smart cards will in future play a central role in the global information and services society.

8.2 Introduction

Communication is the basis of our social life and is also a normal feature of business life. Companies which communicate rapidly and efficiently will be the winners in any market. Speed is a critical competitive advantage, and it is the growing communication networks which enable information to be exchanged quickly and conveniently.

In the age of global and transparent communication networks this means not only the physical networking of telecommunication systems but also an ever more intensive "networking of minds" around the world: intellectual property – previously protected by strongrooms and only passed on personally or in writing – is today, and will in future be, increasingly exchanged over open communication networks and "held in safekeeping" in these networks.

This produces new security requirements which must be met by the communication networks: in future a network security infrastructure must replace the strongrooms of the past. This network security infrastructure will in turn make possible new international services, in such areas as telecommunication, multimedia, payment transactions, health management, and completely new future applications.

In this context, smart cards are the ideal medium for supporting and marketing such services.

8.3 The market

The smartcard industry is one of the driving forces in the development of national economies. Starting from today's predominantly European market, by the year 2010 a genuine world market will come into being (in the sense of market penetration and "local content") with annual growth rates of over 25%. This will be brought about by the globalization of services, with increasing security demands from all the participants, the low levels of investment (software, hardware) required to gain access to the high-technology smart card systems, and an infrastructure which is simple to provide (communication networks).

8.3.1 The market for smart card ICs by application

The market for smart card ICs, which today still makes up less than 1% of the total IC sales of US$118.5bn p.a., is gaining ever more importance and, with an annual growth rate of more than 25% per annum, can be expected to achieve a dominant role in the early years of this millennium.

The growth impulses for the highly-developed economies are coming increasingly from the service companies, and the functions and capabilities of smart card systems as the ideal medium for the distribution and marketing of services are mainly determined by the IC on the card.

With an average share of sales at around 45%, the mobile communications area of application is likely to remain the strongest segment. After this come the segments: "Payments" (15% in 2003), "Identification" (14% in 2003), PayTV (12% in 2003) followed by "Telecommunication services" with an expected average of 10%.

At 45% and 28% respectively, the growth rates in the areas of "Identification" and "Transport" are the strongest. The application segment "Prepay phones" has the need for the greatest number of units, at 50% of the total (security storage).

8.3.2 Requirements of the market

In terms of market requirements, these are the typical constellation for semiconductors which, in the case of the smart card business, are particularly characterized by the following demands:

- low price,
- high security,
- low power consumption,
- large program and data memories ,
- optimize packaging technology.

The smart card IC segment is distinguished by price falls of the order of 15 - 25% per annum, which puts a considerable strain on the semiconductor industry in justifying financially any participation at all in the market, and in making the enormous technical outlays possible. These demanding requirements cannot be met using standard modules from the commodity portfolio of non-volatile semiconductor memories and microcontrollers.

8.4 Applications

Smart card applications are volume-driven applications, the success of which depends greatly on the successful mastery of numerous factors. The more efficiently the semiconductor manufacturer can meet the demands of this new technology, the greater will the success be in realizing the applications concerned, and the associated financial success. It is up to governments to establish a legal framework for the use of smart card solutions (e.g. the signature law), so that they are based on secure foundations, and to speed up their introduction and dissemination. An important prerequisite, however, is that the service provider, system provider, card manufacturer, and others involved in the realization of smart card systems, are clear about what the possibilities actually are for the semiconductor industry. An understanding of these mechanisms is a prerequisite if this example of a "service technology" with a secure future is to make the breakthrough.

As representatives of a practically unlimited number of applications, just a few will be mentioned here: telecommunication (card telephone, mobile communications) payment transactions (electronic purse, credit cards), pay-TV, transport (electronic ticket, ÖPNV), governmental applications (electronic ID Cards), health management (patient's data card), access control.

8.4.1 Digital signature – the signature of the future

Electronic business transactions are subject to incalculable risks if the recipient of a message (purchase order, payment order, ...) cannot really be sure whether it actually originated from the claimed sender. However, transmission procedures such as those used in the Internet really put just as few obstacles in the way of the creation of forged sender identities as there are against tampering with existing messages.

Only a digital signature gives to "electronic commerce", and to other electronically supported legal business transactions, the indispensable legal validity. From the present-day point of view it is the only means by which electronic documents can be signed – in an analogous way to a conventional signature – with computerized support, authentically and with security against forgery.

A specific and notable step on the way towards the ultimate objective, of the complete legal equality of a signature effected electronically with a manual one, was the German Signature Law, passed into law by the Bundestag in August 1997.

A prerequisite for the digital signature to have legally binding force is, of course, that forging it should be more difficult than a manual one by at least as much as it is easier for the swindler to generate digital messages rather than paper ones (in reality, what is demanded is absolute unforgeability, at least with the technology available to a hacker over the next few years). This imposes demanding requirements on the system components used.

Technically, the testing of a signature on a message consists in proving that sender-specific secret data (the so-called secret key) is incorporated into the signature. This proof is made by using the associated public key. Infrastructure measures must put the recipient of a message in a position to determine without any doubt that the public key is associated with the identity of the message-sending partner.

The algorithms used, that is the computing rules by which the digital signature is retrieved from the message and the secret key, must ensure that it is impossible without knowing the partner's secret key to calculate a signature which appears to be valid (and, indeed, impossible to determine the key itself).

The disclosure of the secret key in other ways must be prevented. In this context, not only should theft and the analysis of any component of a personal signature be considered, but also scenarios such as when a partner makes a signature on a terminal which is not his and may have been tampered with.

The agreement between a digitally signed document and the one which the signer is looking at must be ensured by the system creating the signature (which is in no way as simple as it sounds – just think of the numerous Windows documents which are open at the same time on a PC...). It is also necessary to avoid the signer producing a signature unknowingly.

Modern smart cards offer the necessary security against spying on the signature key, and thus permit signatures to be applied even at terminals other than the user's, that is they support user mobility.

A direct readout of the key is prevented by a combination of physical and logical security measures.

In spite of the high computational complexity of the algorithms used, they can be executed on the smart card: the danger associated with transferring the key onto a terminal which may have been tampered with is thereby eliminated.

A true random source on the card even permits the key to be generated "on-card": this means that the risk of the card issuer (or the system used by the issuer) disclosing the key is eliminated.

In the final analysis, all the capabilities of the smart card are, by comparison with a PC, say, still a manageable system. This manageability is, however, a prerequisite for an adequate security evaluation, such as the evaluation in accordance with ITSEC E4 required by the German Signature Law.

8.4.2 Electronic commerce – the world economy on the Internet

The Internet is ever more becoming an all-embracing data network, which is also even being used by the private consumer who will, in future, inquire via the web about current offerings from around the world and will then want to place an order on-line. An important point in this connection is the question of payment, for which – regardless of whether it is by credit card or freely-circulating "cyber money" – the problem of security must be solved.

Internet payment systems can take the most varied of forms. They can be based on bank accounts (money transfers using home banking, "electronic check"), linked to credit cards or use freely-circulating electronic money (E-Cash, Net-Cash).

In all these cases, the security which can be achieved is one of the determining factors. Here, purely software solutions are ultimately not adequate, because of the susceptibility of a PC to harmful or criminal software, such as viruses or Trojan horses. Although additional hardware such as smart cards and smart card readers do involve costs, they do also permit higher levels of security if used with cryptographic protocols.

In France, the SET standard used by Visa and MasterCard has now been enhanced with a smart card protocol. This enables smart cards now to be used as credit cards which confirm their authenticity themselves.

Electronic money purses, in particular, which are finding ever more applications in Europe, are very well suited as an Internet means of payment. In this context, preparations are already being made for the use of the German "GeldKarte" (cash card) in the Internet. Similar efforts are also be in made for other systems.

8.4.3 Home banking

In the field of home banking there is also a security problem to be solved: because ever more bank customers want on-line access to their accounts. On the other hand, up until now home banking has been a service with an exceptionally poor security concept:

Apart from unreliable TAN lists (transaction numbers), which one puts away of necessity in a writing desk drawer which cannot always be regarded as secure, or even saves in the home banking program, there is at least one major security gap, in that the data is transmitted over the net in plain text form.

The use of smart cards, the replacement of the TANs as the security instrument, and a further application of appropriate cryptographic methods, e.g. for data encryption, are the developments which will characterize home banking in the immediate future.

8.5 The business relationships network

The network of relationships specific to today's smart card, between the chip manufacturer, card manufacturer, system integrator and service provider, will in future be determined by the following trends:

- dominance of multinational groups,
- vertical integration (both forward and backward integration),
- security as the basis and framework for the entire business.

The business for smart card ICs is mainly determined by major projects. At the center of the decision-making and implementation process stands the service provider (e.g. Internet supplier, health insurer, bank). In addition there is a complicated project-specific network of relationships between the service provider, system supplier, chip manufacturer and card manufacturer. External determining factors, such as government policy, laws, regulations, institutions are here typical of the smart card business. A critical prerequisite for the success of a project is that these factors should constitute a positive environment or even promote particular applications (e.g. the health policyholder's card).

The basis and framework for all business links in the field of smart card systems is a security philosophy which runs through all the processes. This concerns security in, for example:

- the hardware components (chips, terminals),
- the interrelationships (e.g. certificated customers and manufacturers),
- the procedures (e.g. for the exchange of datasheets and other technical documentation),
- communication (secure data transmission),

the working environment, and the building and network infrastructure, to provide protection against attacks, from outside and also from inside, because smart card ICs are electronic money or contain similarly valuable data. The security of the building and network infrastructures can be achieved, for example, by building design to provide secure areas with personnel movement locks and access control using biometric features, monitoring of the building or rooms by cameras and movement sensors, separate PC and workstation networks with firewalls, and similar measures.

8.6 Products

8.6.1 "Chip on card" – state of the art

Today's microcontroller ICs for smart card applications, such as the typical example in Figure 8.1, are characterized by an 8-bit, 16-bit or 32bit security CPU, volatile and non-volatile memory with sizes of up to 2 KByte RAM, 32 KByte EEPROM, 32 KByte ROM and options for a hardware-optimize crypto-coprocessor (512 and 1024 bit) and a contactless interface with a transmission frequency of 13.56 MHz. Using this last, data and power can be transmitted over distances ranging from a few millimeters up to 10 cm.

The various requirements of the market are satisfied by Infineon's "family concept" of security controllers for smart

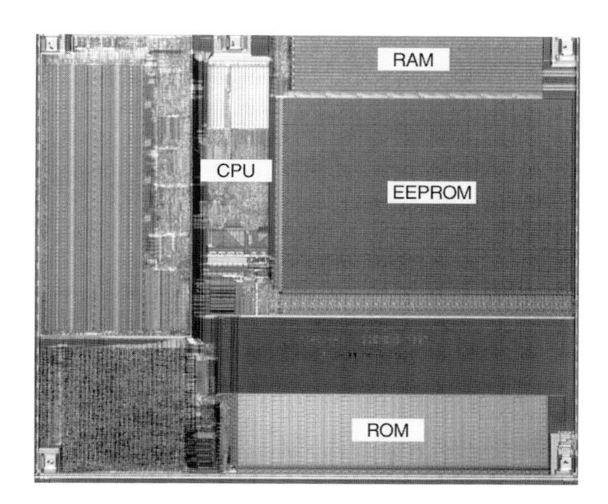

Figure 8.1
Layout of a modern "high-end" security controller for smart cards, with 16 kByte EEPROM and integral crypto-coprocessor

cards, which fall into bands in terms of their prices, performance and options.

The most successful controller family (SLE 66CxxP) is based on a 16-bit CPU core specially developed for smart card applications, which can be fitted with 64 to 136 KBytes ROM, 2 to 4 KBytes RAM and 8 to 64 KBytes EEPROM or flash memory. Advanced special hardware, such as for example the "Advanced Crypto Engine" (crypto-coprocessor), which is rapid and efficient in executing the calculations for cryptographic schemes such as the RSA (up to 2048 bit key length) or elliptical curve, together with a fully hardware-implemented "true" random number generator, increase the performance. Sensors realized in hardware form protect the chip in every operating state. This controller family can today already offer a platform for multifunctional smart cards. Notwithstanding the high performance demands, all the chips are optimized for the lowest possible power consumption, so that they can also be used in contactless applications.

8.6.2 "System on card" – the challenge of the future

For the realization of the high-security pocket computer of the future ("system on card"), assuming the core of the system is a special smart-card-specific security CPU, the following points must be taken into account:

- In the field of crypto-controllers, there are salient application requirements which call for the development of dedicated, scalable high-performance controllers, which must satisfy the highest security requirements in the context of the several applications implemented alongside each other on one chip.
- For the chip's peripherals, dual-interface chip modules will use both a physical contact and a contactless electrical interface. Input/output modules integrated on the card, together for example with keyboards, finger sensors, flexible LCD displays, solar cells, acoustic receivers and sources, will form the "human interface".
- Rather as with silicon technology, the integration of the components onto the card will result in completely new approaches to how the packaging technology is realized. Only by using intelligent security concepts, which take into account the whole system comprising the chip, package and card, will it be possible to satisfy the high hardware security requirements of the future.
- New security characteristics, based on sensor techniques, coprocessors and additional physical security structures with self-destruction features, represent the development on which work has already started for future generations of smart card ICs.

8.7 Cryptographic expertise

In the field of asymmetric or public key cryptography, a range of procedures have been developed over the last two decades, which mainly exploit two mathematical problems: the factorization of large numbers into their prime factors, and the calculation of a discrete logarithm. Both of these tasks are, for sufficiently large numbers, effectively insoluble with the available computer power. This can be used as the basis for constructing information-technology protocols for the authentic and confidential transmission of electronic data, such that in order to hack into a message protected in this way it would be necessary to solve the mathematical problems mentioned.

In order to apply these algorithms one requires a specialized arithmetic processor, capable of performing modular computational operations on extremely large numbers. The length of the numbers used is critical in determining the security of the protocol, and is the subject of continual discussion. After it was initially held that numbers with 150 decimal digits were unbreakable, nowadays the numbers used typically have 200 to 300 decimal digits, to take account of the progress with analytical algorithms and the generally available computer power.

To meet these requirements, the semiconductor area of Infineon AG has a crypto-coprocessor for smart card applications which, since its introduction five years ago, represents the best compromise between computational speed, chip area and power consumption. It has been designed to be scalable, i.e. depending on the application, a module can be generated which is optimized either in respect of a single one of the above parameters or in respect of an appropriate combination of parameters.

Unlike its competitors, the crypto-coprocessor from Infineon works as an independent processor, i.e. it has an accumulator, an arithmetic logic unit and registers. Together with a separate instruction set, this provides a freely-programmable coprocessor which can, depending on the implementation, even process complete tasks, and hence noticeably relieves the load on the smart card controller's CPU.

Its optimized full-custom design is the smallest and most compact on the market. For the technology concerned, it is distinguished by having a transistor density which is higher than usual by a factor of up to 3, i.e. in comparison with a "standard full-custom design" either three times as much functionality can be accommodated on the same chip area, or for the same functionality one requires only around one third of the area.

The first generation of this arithmetic processor was optimized for up to 300 decimal digits, the current generation has been developed for protocols using 600 decimal digits and more. The calculations are performed in a time which is significantly less that the definitive "human response time" of one second – and that on a chip which is small enough to be integrated without difficulty as a coprocessor on a smart card.

Because this performance has been achieved without the need for frequency multipliers and at an extremely low power consumption, the architecture still has adequate reserves to process even larger numbers with appropriate speed for more demanding requirements.

As an alternative to the factorization of numbers or the use of the discrete logarithm, attention is now also turning to the use of a third mathematical problem: the calculation of the logarithm on an elliptical curve. As the basic operations in this system of calculation require several times more resources than conventional arithmetic, the same level of security can be achieved with significantly shorter numbers; thus 50 decimal digits are currently sufficient. Because of their lower demands on memory capacity and transmission bandwidth, increasing commercial use is presently being made of procedures based on elliptical curves.

As the architecture of Infineon's crypto-coprocessor is equally suitable for these algorithms, here again powerful hardware is available today and for the future.

Smart cards combined with asymmetric cryptography will be a portable, personal security tool for the user, allowing him/her access to any desired items which require protecting or goods and services which must be paid for. This access will be secure, authentic and, if necessary, even anonymous.

Apart from this, however, the availability of asymmetric cryptography is a necessary prerequisite for the practically unlimited enhancement of smart cards, which today are still extremely purpose-specific. Only by using algorithms such as those mentioned above will it be possible to change or enhance the capabilities of a smart card securely, even after it has been issued to the user.

8.8 Chips for multifunctional cards

Smart cards are nowadays predominantly aimed at individual applications (credit card, SIM card, ...). Increasingly however, a need is arising for several applications to coexist on one card. For example chains of traders want a 'reward' program, which permits electronic "reward points" to be collected when a cash card is used. The problem with this requirement is that the cash card implementation, checked by an expert, is to be enhanced by adding a function which, on grounds of cost and flexibility, will not also be checked.

A remedy is provided here by multifunctional cards, which guarantee a secure separation of the applications. They permit applications which have been subject to checking to be loaded onto the card alongside other unchecked applications, because the unchecked applications cannot influence the checked ones.

A secure separation of applications can be achieved in various ways. One very promising approach is to use an operating system with an integral interpreter.

From the programmer's point of view, an interpreter creates a machine which does not actually exist, hence is a virtual machine. This virtual machine provides similar instructions (called "byte codes") to a real CPU, but is in general not implemented in the hardware but in the form of machine code sequences (software) which are executed on a real CPU.

The applications on the card are implemented in these byte codes, and are executed by the interpreter. The advantage of this method is that the interpreter can limit the access rights of the interpreted code to data for the associated application. This insures, even on chips for which the machine code permits free access to all memory areas, that the interpreted code can process solely the data assigned to it, while the available byte codes do not allow the data and programs for other applications to be read or manipulated.

This makes it possible to add further applications to an existing security application (e.g. cash card) without endangering the integrity of the security application. In the example described above, the trading chain can commission its reward points system from any software house it chooses, and load the system onto its customers' cash cards if they wish to participate in it.

The secure separation of applications permits new applications to be loaded retrospectively in the field. The card owners can themselves determine the applications in which they wish to participate. On the assumption that it is in widespread use, the implementation of an interpreter on the smart card offers the possibility of developing applications on an open platform, rapidly, efficiently and in a standardized form, that is at low cost. In future, the base of existing smart cards will reduce to virtually zero the initial costs, which today are substantial, for the manufacture and distribution of smart cards, because only the infrastructure for loading the application onto the card will need to be provided.

Today, three interpreter solutions are competing for widespread introduction into the smart card market:

- JavaCard® V2.1, a virtual machine from SUN, based on Java
- MultOS, a virtual machine derived from the Mondex payment system, and
- WinSC, an interpreter solution from Microsoft.

There are various possible methods for implementing the interpreter on the smart card, from a complete software solution through to an (almost) complete hardware solution. Each variant has specific advantages and disadvantages in terms of execution speed, memory requirement, power consumption and flexibility.

8.8.1 Interpreter support in the high-end microcontroller family from Infineon

Infineon is offering a 32-bit CPU with dedicated hardware support for interpreters, which takes into account the prevailing smart-card-specific restrictions and requirements. The SLE88Cxx family offers a special smart card oriented CPU core, which implements in hardware the frequently used functions of the interpreters identified above. The objective of optimizing performance and minimizing the program code is pursued by a consistent use of parallelism for operations which are executed in the interpreter when the byte codes are being processed. By limiting the hardware costs to the functions which already exist in the processor, the power consumption of the CPU can at the same time be kept low. Focusing on the basic functions of the interpreter gives the programmer the flexibility to react to future enhancements and new interpreter concepts.

The result is an optimized multi-application smart card, which permits significantly accelerated execution of the byte codes which are to be interpreted, and thereby offers higher performance and faster transactions. Further advantages are that operational security is higher due to the hardware support, and that it is realized on a minimal silicon area. As a result, sufficient memory resources are available on the chip for diverse applications with a minimal power consumption, thus also making contactless card applications possible.

The combination of a smart card controller from Infineon which supports interpreters, and the operating system or virtual machine, as applicable, thus provides a high performance and secure multifunctional smart card.

8.9 "Human interfaces" – a new peripheral

In the implementation of a "system on card", its interfaces to the human will assume a critical role, and hence substantially raise the acceptance of smart cards.

From a technical point of view, the purpose of all these interfaces is to communicate the information provided by card owners to the computer on the card, and conversely to indicate to card owners the status of their cards. To do so, use is made of display and input units (display, loudspeaker, keyboard) together with biometric sensors (e.g. finger sensors).

It should be particularly emphasized that the processing of data takes place on the personal smart card, and hence is available solely to the card owner. If the interface units were accommodated into a terminal, this could lead to individual users having reservations, because they do not know what might be done with the input and display data.

An important example of the advantages of the "systems on card" is shown by its use as a replacement for the so-called PIN (personal identification number). Today, input of the PIN is used to check whether the legal owner of the card is using it. So-called biometric methods are being considered as a possible replacement for the PIN. If the card owners could use an identifying attribute which always goes with them, this would permit really simple and yet also secure use of the smart card.

8.10 Technology and production

The most advanced semiconductor technologies make possible ever smaller chips with higher performance and, together with innovative production concepts (e.g. 3D/multi-chip modules) provide the prerequisites for undreamt of applications on smart cards.

The chip on the card is the most important core item in a smart card system. The restriction on its area, of around 25 mm^2, due to the potential danger of breakage of the inherently brittle silicon in a card which is required to be flexible, is a substantial determinant of the performance, security and cost-effectiveness of the entire system.

This restriction on the area, at the same time as the increasing requirements for the features of smart card ICs, always calls for the use of the latest semiconductor technology. The performance demanded in the fields of computational speed (e.g. cryptographic applications) security (e.g. electronic money purse) and reliability (e.g. over 100 million health policyholders' cards) call for action at the "front line" of semiconductor technology. The current latest semiconductor technology must be rapidly adapted for the special smart card requirements.

Incorporating the ICs into a flexible card medium requires the highest capabilities in the area of packaging technology.

8.10.1 "Leading edge" technology

Apart from having a leading position for innovative smart card IC architectures with a "multi-lingual" CPU core and high performance peripherals, the Security und Chip Card ICs Division of Infineon AG considers the further development of its expertise in modern silicon technology, especially non-volatile memory technologies, as an important core competence for the production of high-end smart card ICs. The current basic process is a 0.2 µm embedded flash technology, which permits security controllers to be produced, for example for GSM applications, with an area of < 10 mm^2 and a 32 KByte EEPROM. The introduction of a 0.13 µm process in 2003 was an essential step towards satisfying the requirements for ever more chip functionality subject to simultaneous cost pressures.

Infineon meets the requirements of the market, and attempts to consolidate them in the market, by trend-setting (in the form of higher functionality for new innovative applications) for the hardware platforms for smart cards. This includes the 0.6 µm CMOS technology as the present basic process for the production of controller ICs (which means, for example, that a "standard" security controller with 8 KBytes of EEPROM, such as is commonly introduced into GSM applications, requires less than 8 mm^2 of silicon surface), the introduction of a "quarter micron" processes (0.25 µm) in 1998, which meant a significant step forwards in raising functionality and "on-chip" memory resources, the use of flash memory technology for faster personalization of program and data memory areas, to the benefit of the customer, 8′-wafer technology, to achieve further improvements towards more cost-effective high-volume production.

The next step will be the integration of the FRAM memory technology into the smart card IC roadmap for even further reductions in the power consumption, and faster memory access.

8.10.2 Requirements to be met by the technologies, products and design

Because smart card ICs can only be developed at the "technological front line", and the performance demands are continually rising, the product lifetime is limited. The required changes in technology, at intervals of about 1 1/4 years, call for flexibility and innovative abilities from the developer.

The form in which a smart card is made subjects it in use to high mechanical stress and corrosive influences. This means that appropriate packaging is an essential component of the overall concept, because of the greater flexibility of the extremely thin chips. This leads distinctly away from established technologies – new processes today already permit chip thicknesses of around 120 µm to be realized, future processes such as 3D integration require values of between 5 and 10 µm.

The "stacking" of the chip layout in the third dimension permits a multiple of the electrically active integration layers on the same area (with smart card ICs, the maximum area which can be justified, for mechanical reasons, is 25 mm^2). The mutual "shielding" of the layers which this produces achieves another significant increase in security.

In sum, the high rate of innovation and the special forms taken by the ICs call for specially trained designers with a marked security consciousness and the ability to produce optimal and minimum-area mappings of mathematical algorithms "onto silicon".

8.10.3 Requirements to be met by production

The production of smart cards is characterized by product run-ups which follow rapidly one after another, sometimes using different generations of technology. This and the greatly fluctuating demands of customers call for a maximum of flexibility throughout the supply chain.

The complexity and the mixed occurrence of possible design and production process errors represent a major challenge. To minimize these risks, the prerequisites are effectively functioning technologies and analysis for the product and process.

8.11 Security

The smart card security system with the chip as "security in silicon" guarantees the confidentiality, freedom from forgery and availability of data.

8.11.1 The smart card as a security system

Today's outstanding market growth rates, both in security applications and also for crypto-controller ICs, reflect the importance of security considerations in modern information technology.

Because of their superior security functions, smart cards have become the de facto security tool in information processing.

The basis for effective protection against the manipulation of a smart card system is always an interplay of

- hardware security (chip),
- software security (smart card operating system and application software), and
- system security (in the communications between card, reader and background system).

The highest level of security can be realized if the semiconductor manufacturer, card manufacturer, system vendor and system operator constantly push forward in close teamwork, each in their own area of core competence, with developments which are effectively matched to each other in respect of security.

8.11.2 Hardware security

To guarantee the highest possible hardware security, the entire life cycle of an IC must be taken into consideration. In doing so, security in the areas of architecture, design and technology lie within the hardware manufacturer's responsibility, together with the effectiveness of the security mechanisms on the chip (security philosophy) and security in the development and production processes.

8.11.3 Security pyramid

Because different applications call for different security levels from the chip, the semiconductor industry must provide a differentiated range of smart card ICs, tailored to the security needs of the intended applications ("security pyramid").

Starting from the basic level of secure (non-volatile) memory technologies, hardware-implemented security and high-security mechanisms (such as cryptographic coprocessors and special design techniques) form the next level.

The highest level of security, which is essential for many smart card applications, can be provided by public key cryptography methods. Digital signatures, key-splitting and advanced key management schemes are only feasible with the introduction of public key encryption systems. The advantage of the logical "uniqueness" this gives the cards is that it makes any hacking extremely ineffective and hence, ultimately, pointless.

In order to achieve extremely high performance in the execution of the asymmetric algorithms on the chip, "high-end" security controllers for smart cards are equipped, in addition to the main processor, with a special mathematical crypto-coprocessor, which is optimized for the calculation of public key algorithms.

The combination of the algorithmic security of public key cryptography with the physical and logical security of the smart card ICs results in the smart card being a handy, universally applicable security tool.

This means that the manufacturer of the smart card IC bears a heavy responsibility, because the security of the entire system is critically dependent on the security performance of the chip on the card.

8.11.4 Security as a technical and organizational challenge

Even using the latest development and production methods, total and permanent security against successful hacking cannot be guaranteed. Security is much more a dynamic process, which must be so structured that the know-how gap between the chip developer and the hacker is kept constant or, if at all possible, increased. As this process is in constant flux, it has implications for the product life cycle and calls for the greatest care in the technical and organizational arrangements for smart card systems.

From today's perspective, a smart card IC needs to be replaced by a more powerful successor after at most two or three years. This competition between the semiconductor industry and potential hackers demands major innovative efforts, outside the concept and development structures which are typical of the semiconductor industry. This raises the complexity level, and with it the difficulties, of the smart card business, by at least one more dimension compared to the "standard" semiconductor business.

8.12 Outlook

In future, smart cards will play a central role in the global information and service society, because:

The smart card is the key to the information society of the 21st century.

Even today, smart cards already have their place in all applications in which security, identification and the storage of personal data play a role. This will increase as the scenario addressed above unfolds.

Smart cards are a secure basis for tomorrow's financial processes

Smart cards incorporate – as if protected in a strongroom – keys which permit cryptographically secured communication in any form. The driving force here is the electronic signature, which is an adequate substitute for a "manual" signature in the networks, and hence just meets the minimum requirements for having legal force on the network.

The smart card is a guarantor of the individual's quality of life (private sphere, mobility, services).

The smart card makes it possible, for the first time in history, to store and manage personal data independently of central archives or databases. The smart card thereby protects the individual's private sphere, in that he/she can permit or refuse access to his/her data: The individual holds the "trump card".

The smart card enhances mobility, because individuals can conveniently take the required data with them, and thus take advantage everywhere of numerous services.

Smart cards are the money of tomorrow.

Whether at a kiosk, in the supermarket, the subway, bus or when buying on the Internet: it will be possible to pay everywhere by smart card, conveniently and securely.

In 2010 every human will have an average of 3 to 5 smart cards in their pocket (making a worldwide total of 21 - 35 billion cards).

A wide diversity of card types will spread around the whole world: telephone cards, Eurocheck cards, credit cards, patient data cards, pay-TV cards, cards as electronic money purses, identity cards, auto keys, access cards in the most diverse forms (office, PC, security zones) etc. In this context, there will be no difference between countries which are highly-industrialized and those which are less so.

As a result of its prominent position in the field of ICs for both physical contact and contactless smart cards, Infineon has all the technological prerequisites for the realization of powerful smart card ICs, which will also keep up with the requirements of the future.

9 Automotive Silicon Solutions

From the time when automobiles first appeared, electrics have always been a component of them, initially in the form of electric ignition, then later in the lighting, and finally also the horn. By the time that the lighting was electrified, the battery had already made its appearance in automobiles, supplied by an electric generator. This very simple vehicle electric power supply was retained, with only a few very limited changes, right into the 60s. One of the greatest alterations over this period was the change-over from a battery voltage of 6 V to the level which still applies today, 12 V.

9.1 Electronics in the automobile

In the 60s and 70s, the first attempts were made to integrate electronics into automobiles. Mention should be made of such applications as simple engine controllers, lighting dynamo regulators, flasher relays and indicator instruments, together with the radio.

From this point on, the development of electronics in vehicles proceeded apace. When microcontrollers were used for the first time, at the end of the 70s, it was not yet possible to forsee that by today up to 80 microcontrollers and more than 100 electric motors would be used in vehicles, together with countless other components, such as voltage regulators, power MOSFETs, integrated circuits, relays, sensors, capacitors and inductances, arranged in individual controllers which often number more than 50. The state of the art is today represented by such applications as electronic engine controllers, (e.g. fuel injection, lamda regulation, electric throttle valve control), electronic gearbox control, ABS (anti-lock braking system), airbags, air conditioning, light control, window and mirror adjusters, rain sensors, proximity warning, on-board diagnostics, menu prompted operating units, together with multifunctional display devices, down to Internet access and much more beside.

Military level requirements at entertainment industry prices would be one way of describing the challenge of automotive electronics. This statement sumarizes the enormous temperature requirements, from –40°C to 150°C and in some cases right up to 200°C, the high mechanical stresses (vibration) plus the aggressive chemical environment (salt water, oil). The electromagnetic compatibility, in terms of both emmision and immision, calls for expensive detailed solutions to ensure, for example, that there is absolutely no degradation in the functionality of safety-related applications up to the required immision limits. To all this must then be added the highest reliability and a minimal failure rate over more than 6,000 operating hours, combined with a long-term supply of the components for at least 15 years after the end of their production. The technologies used are subject to demanding requirements in terms also of the breakdown voltages and robustness to ESD (electro-static discharge) which are required.

In spite of all the progress which there has been in automotive electronics, developments are only at their beginning. Today's visions of the future start with the assumption that modern vehicles will undergo further electrification and use of electronics. According to experts, in future 90% of the innovations in a car will be made possible by electronics. This applies particularly to the areas of infotainment and telematics, which are all a matter of the communication of individual controllers with each other and with the overall vehicle, between the vehicle and the occupants or between the vehicle/occupants and the external world.

Electrification is often summarized by the term x-by-wire. This refers to the transition from components which are today actuated and controlled mechanically to units which are actuated and controlled electrically. Thus, brake-by-wire stands for an electrically actuated brake unit, which is connected by an electrical link to the electric brake pedal. The fail-safe functioning of all three components, under all conditions, is here absolutely vital. Other applications are referred to as steer-by-wire, shift-by-wire (gear-change), throttle-by-wire, valve-by-wire (valve actuation), wipe-by-wire or start-by-wire (engine starter). All these applications have one thing in common: they require high currrents and high powers, which are difficult to manage in a 12 V system. For this reason, there is currently intense discussion on the introduction of an additional operating voltage of 42 V in vehicles.

Behind all these efforts towards more electronics in cars there are a number of driving factors, such as the environmental compatibility of cars themselves, and cost reduction, as well as the constant pressure for increased convenience and safety, but also for greater competitiveness.

Today it remains as always very difficult to assess the future development of automotive electronics. The electronic element in a car today already accounts for nearly 20% of the total costs, and in ten to fifteen years could well have grown steadily to around 30-40%, perhaps even as high as 50-60%. Over the same period, the proportion of these electronic systems which are semiconductor in nature will go from less than 50% today towards 50-70%.

9.2 Body and convenience electronics

Applications such as electric mirror adjustment, air conditioning or seat heating, have now become a permanent feature of all vehicle classes. Here, the most important applications in the field of body and convenience electronics are:

- Body control modules/light-control modules, with just under 50% market share
- Door and seat control modules with around 15% market share
- Air conditioning systems with around 11% market share

The greatest technical challenge in the field of body electronics is to achieve a low quiescent current consumption, because the controllers are generally connected directly to the battery, rather than being activated when the ignition is turned on. On the other hand, the climatic conditions, from −40°C to +105°C are less critical than in the field of engine management and gearbox control.

9.2.1 Vehicle power supply controllers and lighting modules

These devices and modules represent the central controllers in the body area. The scope of their function is very diverse and there are numerous variants.

The most important element in these controllers is the switching of the vehicle's lights, that is the interior lights and all the external lights such as the full beam headlights, dipped beam, flashers, brake lights and parking lights. In addition to these functions, which are prescribed by law, there are also system diagnostics, i.e. the controller can diagnose the failure of a lamp, a broken wire or disruption of a lighting function. The data is then stored in a fault memory. In addition, the driver is informed that the lighting function has a fault. Apart from purely providing the information, some systems even offer alternatives for the function which has failed. So, for example, a rear light which has failed can be replaced by the rear fog lamp with a reduced brightness. This reduction in brightness is achieved by using pulse-width modulation (PWM) with a frequency of around 100 Hz. This frequency cannot be achieved with a relay.

Light-control modules often switch up to 600 W of light power. By using ever more efficient semiconductors, the power loss can be reduced to 7-10 W, thus eliminating the need for expensive and heavy heat sinks. In addition, it is becoming more and more common to use multi-channel solutions, that is modules with four independent channels. This makes it possible to achieve miniaturization, savings on component fitting costs, and a reduction in space requirement. As a result, controllers to provide comparable functionality are becoming ever smaller and more maintenance-friendly. In the latest generation of lighting modules (see Figure 9.1), there are no longer any fuses. Consequently, the power semiconductors are also inaccessible, being located directly behind the rotary light switch on the dashboard, and there are no longer any wires between the operating element and the power module.

A critical advantage which Infineon Technologies can offer in the field of body electronics is its complete product portfolio. Thus, the product spectrum ranges from low-side switches (HITFETs) through relay drivers and LED drivers (multi-channel HITFETs), intelligent high-side switches (PROFET, HiC-PROFET) up to complex transceiver products for communications using the CAN or LIN protocol. Powerful microcontrollers with 16-bit or 32-bit cores round off the portfolio (see Figure 9.2).

Figure 9.1 The development of controllers

The use of PROFETs offers a range of advantages, of which the greatest results from the high-side concept itself: one side of the load is at ground potential, and the power supply is switched through to the load by a high-side switch. If the load has direct access to the ground, in the form of the chassis, only one wire is required to feed the load with power. This saves on the system costs!

Because the major lamp loads, such as dipped (low) and main (high) beam headlights and fog lights, use 55 W, 65 W or even HIDL lamps (xenon), only very low resistance high-side switches are used (HiC PROFETs). The HiC PROFETs thus offer a very space saving and efficient solution.

The component and the application are protected against overtemperature and overload by integrated logic combined with a protection circuit. If a short circuit should occur in the feeder wire or in the load, the PROFET will switch off safely. No additional fusing is required. The maximum short circuit current which can occur is also limited. Consequently, wires with smaller diameter can be used, which in turn saves on cost and weight.

The diagnostic concept covers not only overload and overtemperature, but also wire breakages. In addition, depending on the product feature, a signal proportional to the load current (Intelli-Sense) can be supplied. This means that there is no need for a shunt resistance with evaluation circuit. Further advantages are the CMOS-/TTL-compatible inputs and status outputs. The smart switch can be driven directly from a microcontroller port, whereas a relay would require a driver (see Figure 9.3).

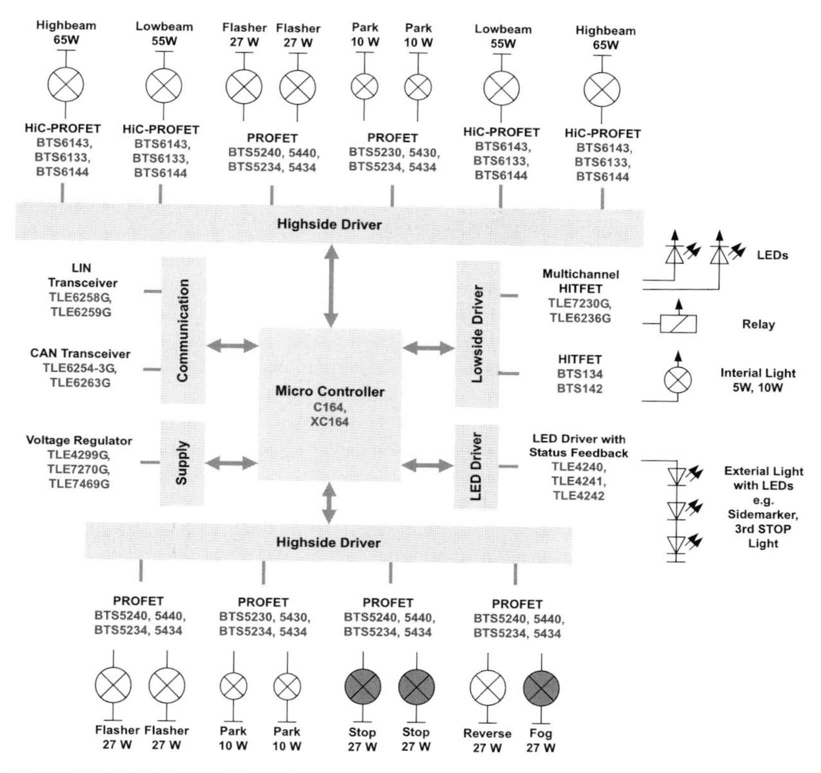

Figure 9.2 An Infineon chipset for the lighting module

Again, there is no problem in switching inductive loads. An integrated Zener diode chain protects the module against overvoltages by active clamping. For this, a relay would require an external freewheeling diode.

With the HiC PROFETs which have already been referred to, use is made of the chip-on-chip technology with which a top-chip is bonded onto a base-chip, and they are then mounted in a package. The logic and protective circuitry described above are integrated into the top-chip. In providing its functions, use is made of Infineon's SMART technology, which offers CMOS and DMOS circuit elements.

For the base-chip, a MOSFET is used. Although this does not offer the lowest on resistance per unit area, its lower complexity means that it is even more cost-effective than the SMART process. Consequently, the product portfolio offers not only modules for a 12 V vehicle power supply, but also solutions for truck applications or for a 42 V vehicle power supply (see Table 9.1).

For the other lights in an automobile, such as a 21 W flasher, 10 W and 5 W lights for parking lights etc., Infineon offers a very wide and powerful product portfolio in its PROFET family (Table 9.2).

In future, increasing use will be made of LEDs for the lights. In spite of having a higher cost, they are today already being used as the third brake light. The reason

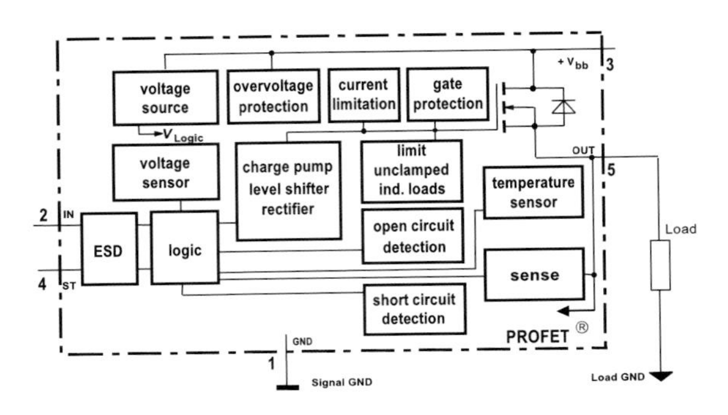

Figure 9.3
PROFET block diagram

Table 9.1 Extract from the HiC PROFET product portfolio

Type	$V_{DS(AZ)}$ [V]	$R_{on(max)}$ [mOhm]	$I_{L\text{-}SC(typ)}$ [A]	Package
BTS 443 P	42	16.0	50	DPAK5
BTS 6143D	42	10.0	50	DPAK5
BTS 6144P	42	8.0	65	TO220/7
BTS 650 P	42	6.0	130	TO220/7
BTS 6510	42	6.0	130	TO220/7
BTS 550 P	42	3.6	180	TO218/5
BTS 555	42	2.5	300	TO218/5
BTS 6163D	60	21.0	45	DPAK5
BTS 660 P	60	9.0	145	TO220/7

Table 9.2 Extract from the product portfolio

Type	R_{ON} [mOhm]	I_{NOM}[A]	Diagnostics	Package
BTS 5240L	2 × 25	5.7	Intelli Sense	Power SO 12
BTS 5240G	2 × 25	5.7	Intelli Sense	P-DSO 20
BTS 5440G	4 × 25	5.7	Intelli Sense	P-DSO 28
BTS 5234L	2 × 60	2.8	Intelli Sense	Power SO 12
BTS 5234G	2 × 60	2.8	Intelli Sense	P-DSO 20
BTS 5434G	4 × 60	2.8	Intelli Sense	P-DSO 28

for this is their faster response time which, particularly in the case of a brake light, can mean critical tenths of a second for the driver behind. In addition, they offer new levels of design freedom (shape and smaller installed depth). The reliability of an LED is also significantly better than that of a halogen lamp. In order to drive an LED optimally, a driver IC such as the TLE4240G is required.

The TLE4240G acts as a constant current source for up to 60 mA, and hence protects the LEDs against overvoltage. This module provides protection against overload, short circuit of the output to ground and to the supply voltage, overtemperature and polarity reversal, and an open-load diagnosis which reports a faulty LED or a broken wire.

The TLE4241 represents a development of this. This IC provides an adjustable current source in the range 8-75 mA. In addition, it is possible to select either of two current values, a low one and a high one. This makes it possible to use the LED chains at the lower current threshold as a rear light and at the higher current threshold as a brake light, because of its greater brightness. This also saves space and system costs.

9.2.2 Door control modules

The doors on modern vehicles also conceal ever more convenience electronics. Central locking, actuated remotely by radio, has now become standard, as have electric mirror adjustment and heating, or electric window lifts.

It is easy to imagine that the wiring harness would indeed have a considerable thickness if these loads were all to be actuated from a central vehicle power supply controller. It is for precisely this reason that the trend to more convenience and greater electrification has seen a migration of the electronics into the doors. The intelligent sub-modules which are to be found there are called door controllers. Communication with the other controllers uses bus communication; in general according to the CAN protocol (see Figure 9.4).

The functions of a door controller can vary to a greater or lesser extent, depending on the level of equipment fitted. Intelligent power semiconductors control, protect and diagnose the actuators, such as door-lock motors, mirror adjustment/fold-in and heating, lights and window lifts (Figure 9.5).

It is generally true that, apart from the microcontroller there are always functions which can be integrated into a power system IC. These include the power supply for the microcontroller and any sensors, the drivers for the "physical layer" of the

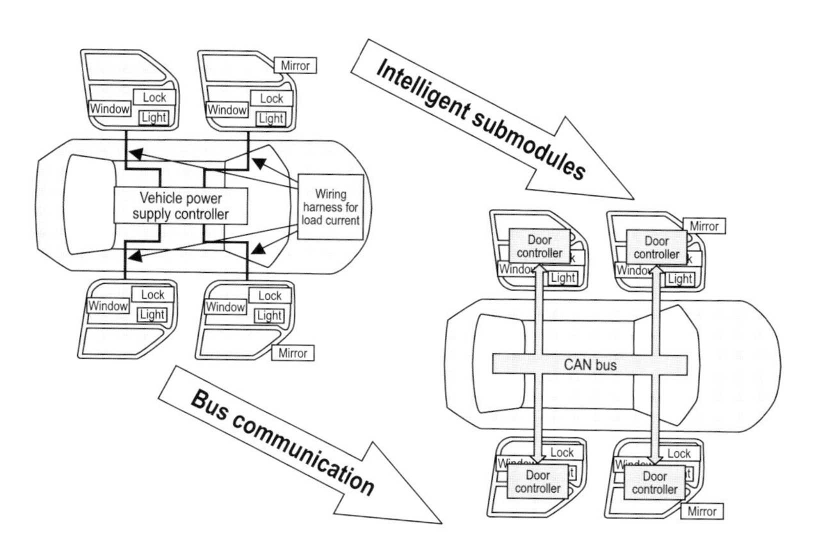

Figure 9.4 Trends towards local door controllers and bus communication

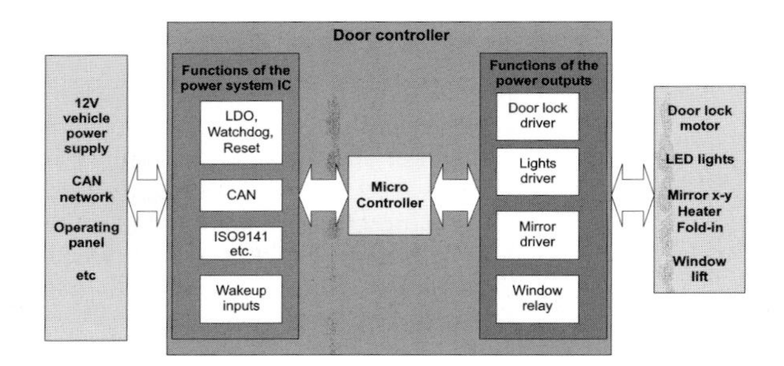

Figure 9.5
Block diagram of a door controller

bus communication, sampling of the operating panel and monitoring of the microcontroller by means of watchdog and reset functions.

For all these functions, Infineon Technologies offers optimized chipsets.

For use as power system ICs, products such as the TLE 6263 or TLE 6266 are the best-known of a whole family of so-called system base chips (SBCs). These are controlled via an SPI interface, and offer extensive diagnostics, including for all bus errors. In addition, the SBCs provide various operating modes: normal, silent, sleep and stand-by. They thus make a critical contribution to enabling the stringent requirements, for quiescent currents of less than 100 μA per controller, to be satisfied. This makes it possible to reduce the components significantly compared to a discrete implementation (Table 9.3).

Infineon's product range for power drivers extends from specialized drivers for dedicated applications to highly-integrated solutions.

For example, the TLE 7201R combines the drivers for door locks, mirror adjust-

Table 9.3 Chipsets for door modules

Function	Product	
Door lock	BTS 7740G, BTS 7741G, BTS 7750G, BTS 7751G	or TLE7201R
Mirror adjustment	TLE 6208-3G	or TLE 7201R
Mirror heating	BSP 772	or TLE 7201R
Mirror fold-in	BTS 7740G, BTS 7741G, TLE 6208-3G	or TLE 7201R
Lights driver	BTS 5210G, BTS 5215L	or TLE 7201R
Window lift	BTS 781GP	
LED driver	TLE 4240G, TLE 4241G	
LIN transceiver	TLE 6258G, TLE 6259G	
CAN transceiver	TLE 6254G	or SBC: TLE 6263G, TLE6266G
Voltage regulator	TLE 4299G, TLE4278G	or SBC: TLE 6263G, TLE6266G
Operating panel sampling	TLE 6263G, TLE6266G together with μC	
Microcontroller	XC 164	
Hall sensor	TLE 49x5, TLE 49x6, TLE4966	

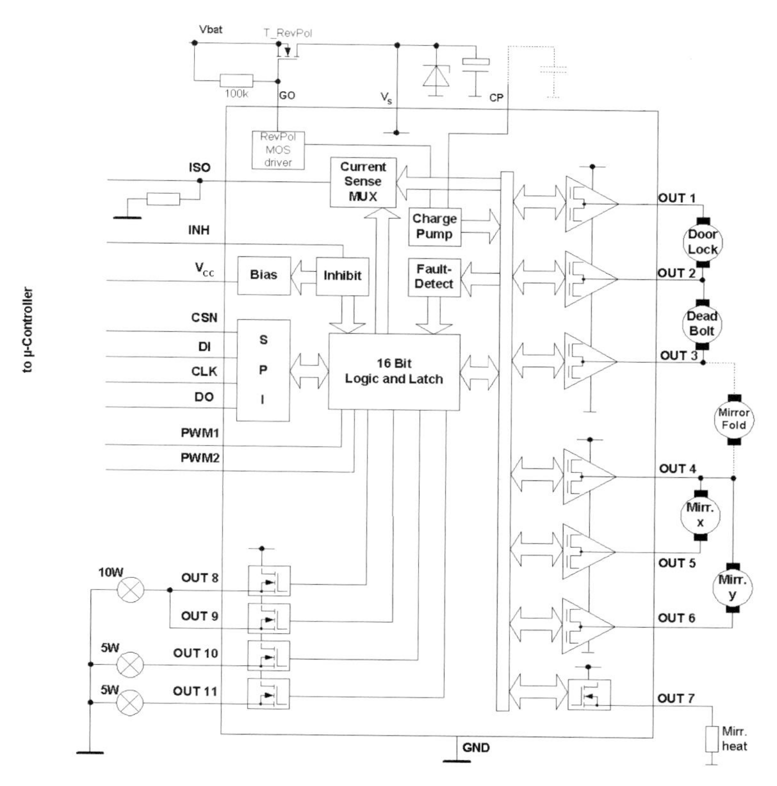

Figure 9.6 Block diagram of the TLE7201R in an application

ment, mirror heating and fold-in, and for four 5 W lamps. All the switches are protected against overload, overtemperature and short circuit. In addition, it can diagnose short circuits in the load or to ground, and wire breakages. The module also provides analog current sensing which can be used, for example, to recognize the end position during mirror folding or in the door locking system. An important objective during development was to massively reduce the quiescent current consumption. Here, the TLE 7201R stands out clearly from the comparable competitive products, with its value of a mere 6 µA (Figure 9.6). If the operating voltage is connected with the incorrect polarity, the module protects itself by switching off the polarity reversal MOSFET in the power feed path.

9.2.3 Air conditioning

In progressing from a purely mechanical "air conditioning" system to a semi- or fully-automatic air conditioning system, it has been necessary to achieve a clear reduction in the weight and installed space (Figure 9.7).

Whereas, with a mechanical air conditioning system, the valves controlling the airflow were actuated by rods or Bowden cables, small electric motors are now used. For this purpose, bipolar stepper motors, unipolar stepper motors or DC motors are

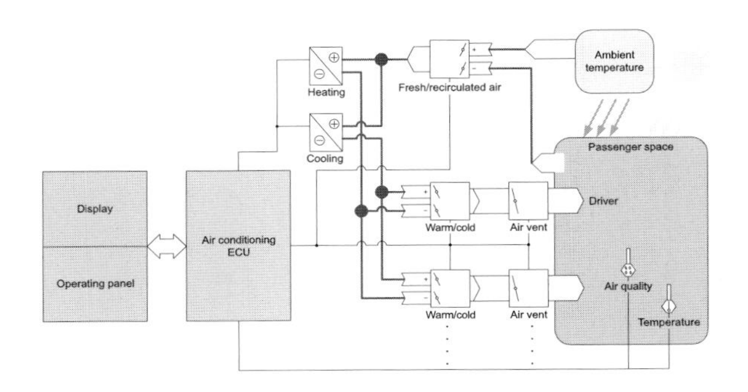

Figure 9.7
Block diagram of an air conditioning system

used. Presently, Infineon Automotive Power can already offer two outstanding ICs for the control of DC motors: the TLE 6208-6 and the TLE 6208-3.

These two ICs use the SPT technology, which provides bipolar devices for analog circuit elements, with C-MOS logic and D-MOS output stages. Thus the TLE 6208-6 provides a total of six high-side and six low-side switches which, if connected to form half-bridges, can drive up to five motors in cascade. The output stages have an $R_{DS\text{-}ON}$ of only 800 mOhm with a current limitation of 1.2 A. If more power is required, there is no problem in connecting the output stages in parallel. In addition, the module provides an internal freewheeling diode, so that no other external circuitry is required for driving motors. The output stages offer overload and short circuit protection together with temperature sensors which protect not only the module but also the application. As a result is no longer necessary to protect the motor itself against overload, and the motor does not need a PTC or bimetallic switch.

To control these high-side and low-side switches, a modern SPI interface has been used, rather than direct I/O logic, in order to save on the connections to the microcontroller. However, this SPI interface also offers advantages in relation to diagnosis. The diagnostic word, with a total of 16 bits, informs the microcontroller about the status of each switch. This enables any overload or overtemperature situation to be recognized, and appropriate countermeasures taken. This greatly simplifies diagnosis in a technical workshop.

The wide range of the supply voltage, from 7 V up to 40 V, the particularly low quiescent current consumption of 10 µA, and the compact P-DSO-28 package, make the TLE6208-6 a smart power device which is outstandingly suitable for use not only in automobiles but also in industrial electronics. The TLE6208-3 offers three half-bridges in a P-DSO-14 package, and is at the same time fully compatible with its “big brother”, the TLE 6208-6.

For unipolar stepper motors too, Infineon Technologies has a new solution to offer. The BTS 3408G is a two-channel HITFET, that is a fully-protected low-side switch.

Apart from the protective functions against overload and overtemperature, and for current limitation, the BTS3408 also offers diagnosis of short circuit and broken wire. It is also specially suitable for use as a driver for stepper motors. In a 2-zone air conditioning system there are five stepper motors. For each unipolar stepper motor one requires four low-side switches, that is two BTS 3408Gs (Figure 9.8).

An additional challenge for a modern automobile air conditioning system is to reduce the time required to heat up the pas-

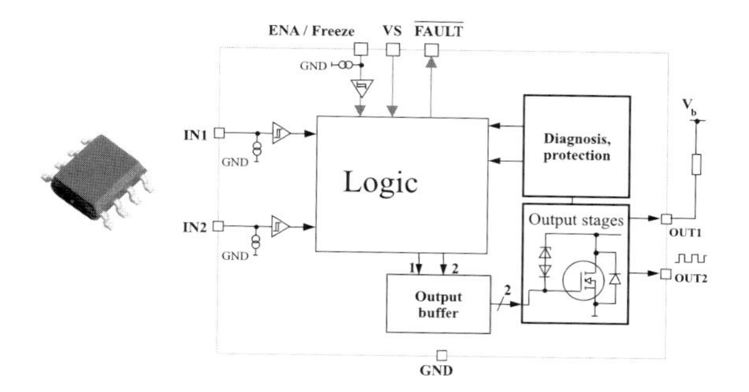

Figure 9.8
Block diagram of the BTS 3408 in a P-DSO-8

senger space to the desired temperature. The efficiency of modern combustion engines has now become so good that the engine alone no longer provides sufficient heat for the passenger space. The only way to get sufficient heat to the occupants within a few seconds, even in winter, is by supplementary electrical heating or seat heaters. These heating systems consume some hundreds of watts. With the current 12 V power supply, such supplementary heating systems represent genuine high-current applications. With respect to the supplementary heaters, a distinction is made between systems which heat water or heat air.

A coolant heater has the advantage that the engine is also brought up to its operating temperature more quickly. However, with this method, a passenger must wait longer for the heat to reach the passenger compartment, because the air is only heated up by the coolant liquid via a heat exchanger. Supplementary electric air heaters are becoming increasingly commonplace (see Figure 9.9). A supplementary electric heater also has clear advantages over an independent vehicle heater: it is lighter, cheaper and more reliable.

For the supplementary heater, one can use a PTC element which is controlled by a high-current switch. The heat can then be regulated very effectively by means of pulse-width modulation (PWM). Because of the high currents involved, the switching is at frequencies above 100 Hz, so that it represents a quasi-constant load for the vehicle power supply and generator. For this application, Infineon Technologies' HiC PROFET offers a wide product portfolio with modules from 16 mOhm right down to 2.5 mOhm (see also Table 9.1).

For this application, it is also possible to use a semiconductor in linear mode for heating purposes. The hotter the semiconductor becomes, the more efficient is the

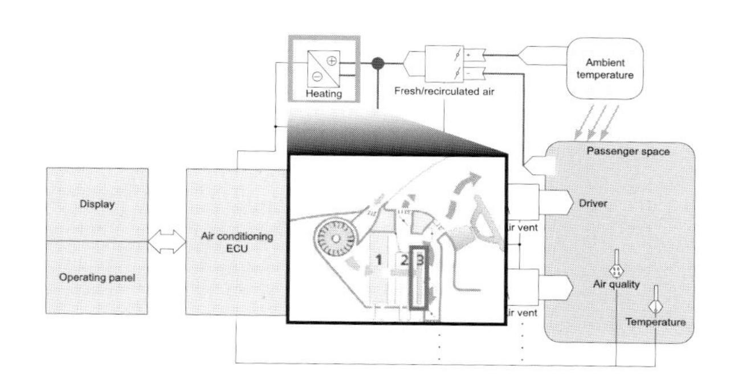

Figure 9.9
Installation location of the supplementary electrical heater

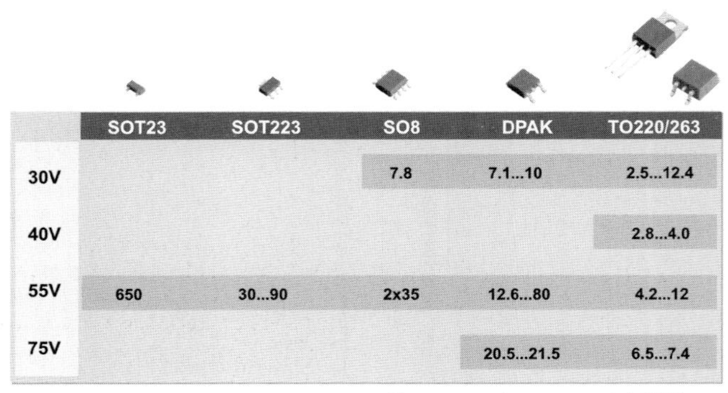

	SOT23	SOT223	SO8	DPAK	TO220/263
30V			7.8	7.1...10	2.5...12.4
40V					2.8...4.0
55V	650	30...90	2x35	12.6...80	4.2...12
75V				20.5...21.5	6.5...7.4

$R_{DS(on)}$ max. values measured at 10V V_{GS}

Figure 9.10 Extract from the OptiMOS™ product portfolio

heating. However, in this mode one is operating very close to the destruction limits of the semiconductor. In order to remain reliably below the destruction temperature, a temperature sensor is required. This is integral in the case of the products in the Speed-TempFET family. These are MOSFETs, onto the chip for which a temperature sensor is attached. The HiC-PROFETs and Speed-TempFETs are based on Infineon's new generation of MOSFETs, which go under the name of OptiMOS™ (Figure 9.10). In developing these, two objectives were pursued: on the one hand efforts were made to reduce significantly the on-state resistance for the voltage classes 30 V, 40 V, 55 V and 75 V, this being achieved by a new DMOS cell concept. Thus, for the 55 V class in a D²Pak (TO-220), a $R_{DS\text{-}ON}$ of a mere 4.2 mOhm has been achieved.

The other development objective was to make the modules sufficiently rugged that they would have no problem in withstanding the harsh conditions in an automobile. Hence, the avalanche energy (E_{AS}) is 810 mJ. In terms of the ability to stand high temperatures, the aim was again to make significant advances. Thus, the OptiMOS technology is approved for a junction temperature of 175°C over its entire service life in the vehicle. To enable the heat to be dissipated effectively, the chip thickness was reduced to 175 μm, which has resulted in a reduction in the thermal resistance $R_{thJC(max)}$ to 0.5 K/W in the D²Pak.

Air conditioning systems, door controllers and lighting are only three examples of the applications in the automotive area. Infineon offers a range of solutions which are precisely tailored to the particular requirements of these applications.

9.3 Safety electronics

The focus of the developments in automotive electronics has differed from one cycle of development to the next. The result has been the following major development steps:

Complex control electronics were originally introduced in the area of engine management, and it was the requirements of this area which initially dominated the developments and technologies.

During this wave of development, bus systems such as CAN became popular. With increasingly more powerful vehicles and increasing numbers of them, the safety requirements then increased. Applications such as ABS and airbags were introduced.

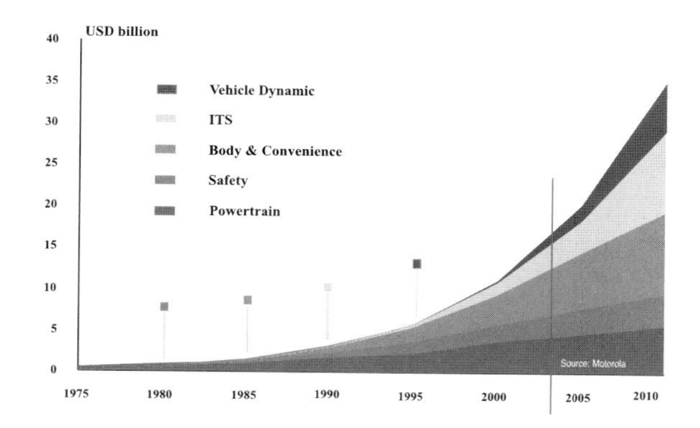

Figure 9.11
Technology developments in five phases

After performance and safety, the need for convenience increased, with the result that door openers, and electric mirror and seat adjustment, became popular. This prompted more intensive networking of the electronic units.

The next wave for automotive electronics was driven by new requirements from the field of communication and information electronics in the vehicle, by telematic services and also increasingly, as developments unfolded, by multimedia.

Finally, the fifth wave of technological development focused on total vehicle dynamics. This includes, on the one hand, the rapid and comprehensive networking of all the control devices involved in the vehicle kinetics, the development of new types of safety buses with higher bandwidths, the combining of individual regulators into one common unit and, in the course of this, a change over from local to global regulation strategies together with greater use of x-by-wire technologies. These five waves of development are shown graphically against an approximate time axis in Figure 9.11.

The market, as represented by the number of systems over the period, is shown in Figure 9.12.

One can see that the airbag and braking systems dominate, with more than 75% of the volume. They are followed by the steering system, which is a rapidly growing (emerging) application. The dark bars

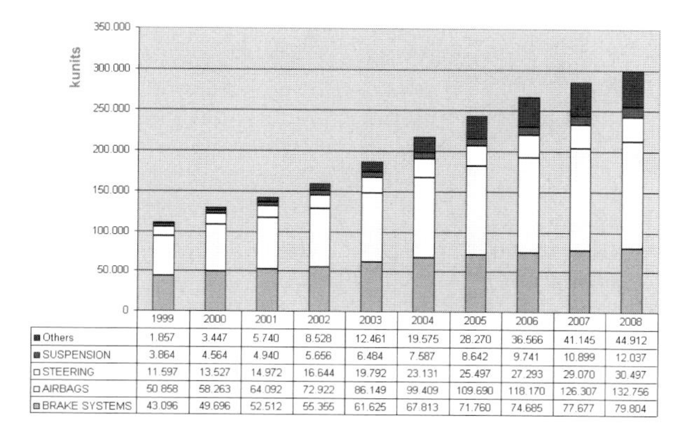

	1999	2000	2001	2002	2003	2004	2005	2006	2007	2008
■ Others	1.857	3.447	5.740	8.528	12.461	19.575	28.270	36.566	41.145	44.912
■ SUSPENSION	3.864	4.564	4.940	5.656	6.484	7.587	8.642	9.741	10.899	12.037
□ STEERING	11.597	13.527	14.972	16.644	19.792	23.131	25.497	27.293	29.070	30.497
□ AIRBAGS	50.858	58.263	64.092	72.922	86.149	99.409	109.690	118.170	126.307	132.756
▩ BRAKE SYSTEMS	43.096	49.696	52.512	55.355	61.625	67.813	71.760	74.685	77.677	79.804

Figure 9.12
Worldwide automotive safety market in k-units (1000s of devices) versus time

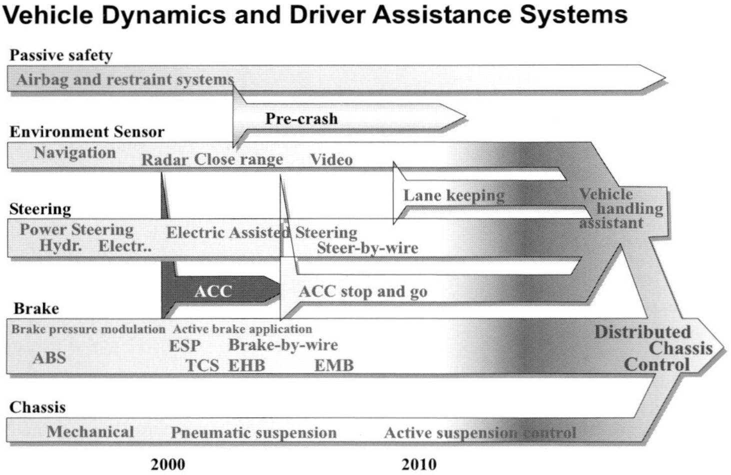

Figure 9.13 Overview and development of the vehicle dynamics and safety applications

include the new systems, such as a tire pressure monitoring system.

Broadly, the field of safety electronics can be subdivided into two areas, active safety (vehicle dynamics) and passive safety (driver assistance). Figure 9.13 shows the sub-applications and their interrelationships, over time.

Whereas active safety systems are primarily used to prevent accidents, many passive systems attempt to reduce the possible consequences of an accident to a minimum. The active systems therefore include mainly the applications for braking, steering and damping/suspension, while the passive systems include the airbags and restraint systems. Over and above these there are the fields of driver assistance systems which can include, on the passive side, such features as a parking assistant (proximity warning) or pedestrian recognition and, on the active side, emergency braking systems or automatic collision avoidance.

The boundaries between these systems are fluid, and can even move over the years. For example, parking aids are currently limited to the recognition of the proximity of objects in the immediate vicinity of the vehicle, and appropriate auditory or optical indications for the driver. New systems make it possible in addition, for example, to determine the length of a parking space as it is driven past, and possibly even to make recommendations to the driver as to the optimal rotation of the steering wheel for parking in the space. For future systems, active electronic intervention may also be conceivable, by which braking is enforced before a collision can occur, or perhaps even automatic steering into a parking space. Presently however, such systems are tightly limited because of legal restrictions, their high costs and the question of general liability.

Apart from drawing a line between active and passive systems, it is also possible to see a spectrum from convenience through to safety. Systems such as adaptive speed control with enhancement for stop-and-go functionality are currently viewed as convenience features, by contrast with which collision avoidance systems such as automatic emergency braking are categorized as safety functions. Pedestrian protection can, on the other hand, be supported by both active and passive systems: an active system might, for example, automatically brake the speed when a pedestrian is recognized, whereas a passive system could implement such functions as pedestrian airbags, or mechanisms which automatically tilt the engine hood in the event of a collision, to cushion the impact energy.

In the context of safety electronics, safety comprises at least two different nuances of the term: on the one hand, safety in the sense of traffic safety, measured rather by the potential injury risk or the severity of a possible injury; but on the other hand safety in the sense of guaranteed functionality, fault tolerance, reliability of the systems used. Safety in the sense of "security" – protection against theft – is not within the domain of safety systems, but falls under the heading of body electronics.

The next section uses the example of the development of braking systems to elucidate a typical evolution path for active safety systems and the safety considerations which underlie it. This is followed by descriptions of passive systems for airbags and tire pressure measuring systems.

9.3.1 Active safety systems

Applications in the field of active safety systems make the major contribution to the control of vehicle dynamics. Practically all the functions which directly affect the vehicle dynamics (with the exception, at this point, of the power train) are highly safety-critical, i.e. a failure of any such system could have the most serious consequences. Because of the close link between vehicle dynamics systems and safety, the application segment is sometimes referred to as "Vehicle Dynamics and Safety".

The development of braking as an application is shown schematically in Figure 9.14. This shows the development steps from a simple hydraulic brake through electronic enhancements such as antilock systems, limited slip control, electronic stability program (ESP), newer supplementary functions such as brake force distributors and emergency braking assistance. After the introduction of hydraulic brakes, the first supplementary electronic functions were introduced, affecting the braking force in certain danger situations regardless of the driver's wish-

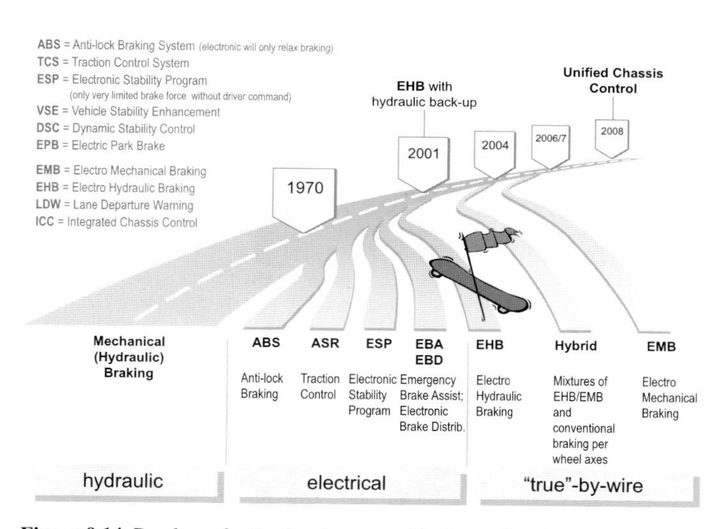

Figure 9.14 Roadmap for the development of brake applications

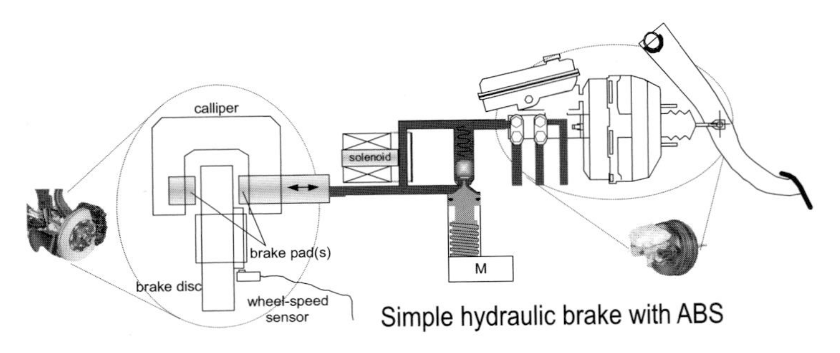

Figure 9.15 Principle of a hydraulic brake with ABS functionality

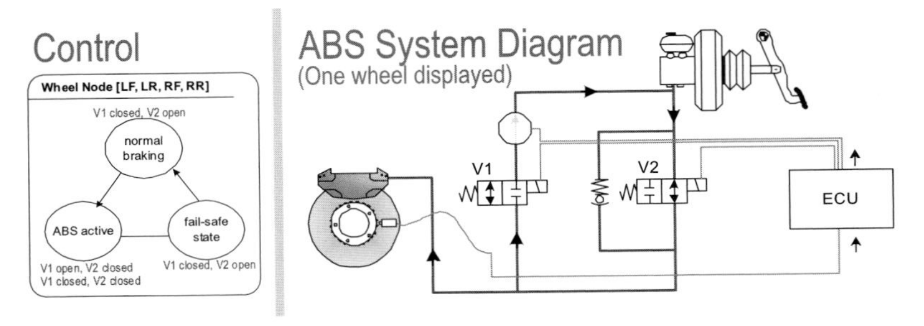

Figure 9.16 Equivalent circuit for a simple ABS system (circuit for one wheel)

es. The next generation of brakes covers various further developments in the field of genuine "by-wire" braking systems.

The anti-lock braking system (ABS) can cancel the braking force briefly when a wheel locks up. The result is that the vehicle remains steerable even on surfaces where there is inadequate grip, because the wheels continue to turn and do not slip, as would happen if the wheel locks up due to excessive braking force. The principle of a hydraulic brake with ABS is shown in Figure 9.15.

In Figure 9.16, the functional principle of ABS can be clearly seen: the brake pedal builds up a braking pressure in the hydraulic system, and in the initial situation this is transmitted directly through the open magnetic valve (V2) to the brake, pressing the brake pads against the brake disk. The rotational speed of the wheel is detected by a sensor, and is evaluated electronically in the ABS controller (ECU). If a wheel locks up, the pressure applied to the brake cylinder can, by opening a return valve (V1) and closing another (V2), be diverted and modulated, i.e. the electronics can override the braking command from the driver and reduce the braking force briefly, but cannot by itself build up a braking force.

An enhancement, for which the sign is reversed, is anti-slip regulation (ASR), also know as traction control (TC). This system is used to reduce the torque applied to a wheel when the ground adhesion is no longer sufficient to transmit the entire drive force to the road and the wheel begins to slip. This is counteracted by feedback to the engine control unit, or by a limited build up of an electronically initiated braking force.

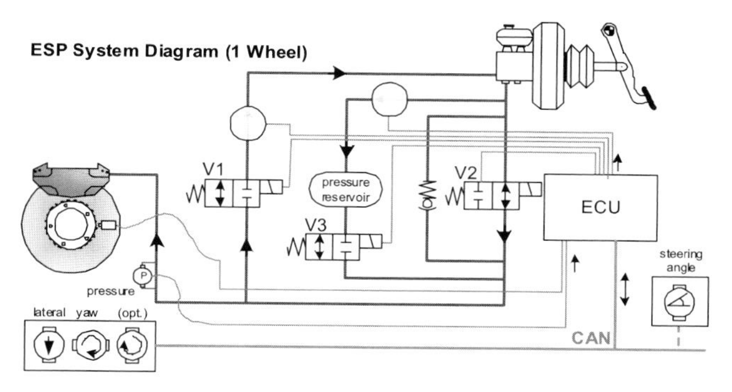

Figure 9.17 Equivalent circuit for a hydraulic ESP

The circuit diagram for an ASR corresponds essentially to that of an ABS, with an additional valve which permits a build up of pressure on the brake in parallel to the regular hydraulic path. In addition, in the case of an ASR system a communication link is required to the engine control unit – generally a CAN link. In the case of an ABS, only wheel speed sensors are used.

After ABS became a pseudo-standard for all vehicles in Europe, ESP was developed, and is predicted to have good growth prospects for the first half of this decade. The system in the case of an electronic stability program (ESP) is comparably expensive (Figure 9.17). This development of ABS and ASR provides additional sensors for the vehicle dynamics, namely longitudinal acceleration and ro-

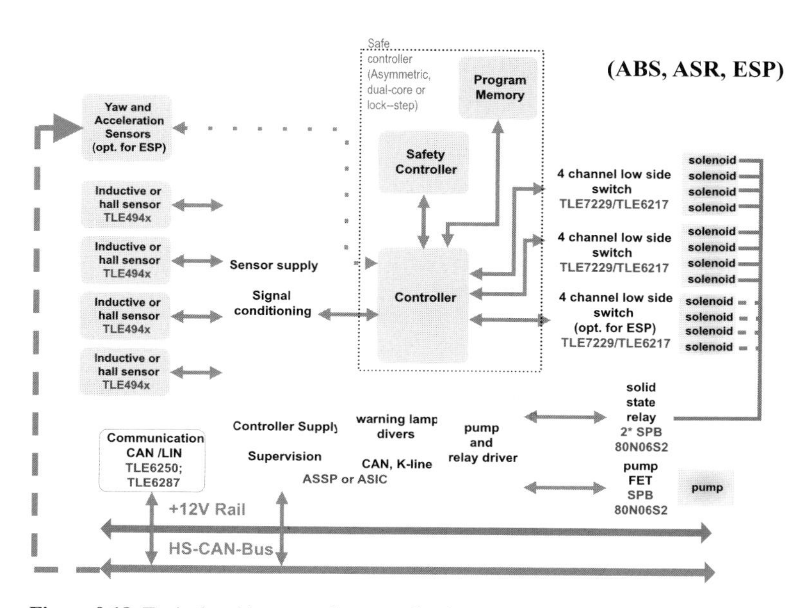

Figure 9.18 Typical architecture of a controller for ABS, ASR or ESP systems

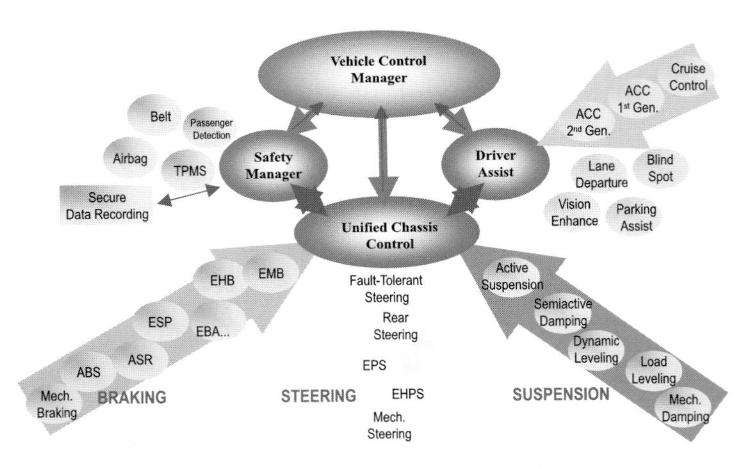

Figure 9.19 Possible scenario for a combined chassis control system

tation rate sensors. Depending on the system, there may also be an integral steering angle sensor. From these items of data, the ESP controller can compute when the actual movement of the vehicle is not following the trajectory defined by the steering angle, and counteract this by appropriate over- or under-steering, by selective application of the brakes.

The block diagram for a typical ABS/ESP controller can be seen in Figure 9.18. This shows almost complete coverage of the functions by Infineon components.

Further developments for braking systems have arisen from the requirements for improved vehicle stability, braking performance (shorter braking distances), elimination of the highly toxic hydraulic brake fluid, and other cost reductions along the development chain. Accident statistics confirm greater traffic safety with every meter by which the braking distance is shortened, and every half-millisecond saving in reaction time, together with predictive proximity warning systems.

With the standards which have already been reached, maximum braking force alone is no longer sufficient to achieve an optimal braking rate. Instead, as well as the braking force, the road adhesion must be optimized, in particular on uneven surfaces. To an ever greater extent, this is being achieved by an interaction between brakes, shock absorbers, steering and the drive train, which demands further interlinkage of the individual components. Harmoniously matched control strategies permit control of the vehicle dynamics at the whole-chassis level, with local control algorithms for individual systems being replaced by a common analysis of the vehicle status. From this, harmonized control requirements are then derived, for application locally in the individual systems. A broad overview of how the individual applications develop (from outside inwards) and how they ultimately grow together into one “Unified Chassis Control” system, can be seen in Figure 9.19.

There is a range of other motivations for the further development of brakes, in particular with the use of the “by-wire” technology.

The first electrohydraulic systems were introduced back in 2001 by DaimlerChrysler, using the SensoTronic system developed by Bosch. The second generation of this system promises further improved response times. Further develop-

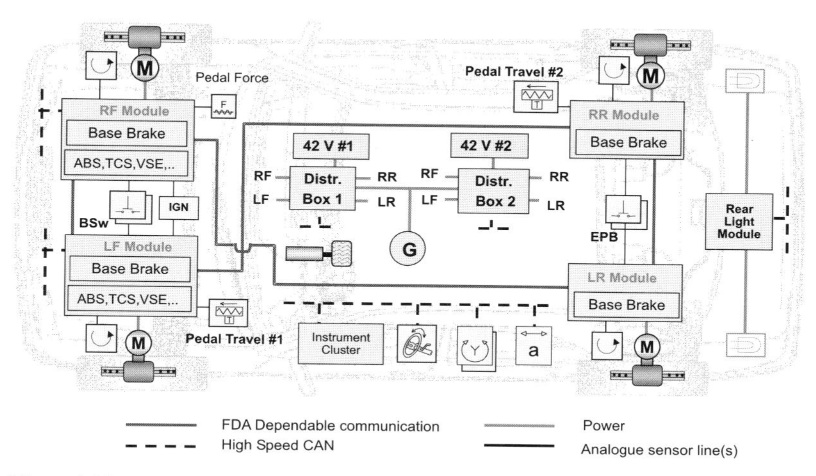

Figure 9.20 Example of a distributed, local safety architecture for electrical brakes

ments, with the objective of extending the use of this braking system to vehicles with larger volumes, are also pursuing hybrid braking concepts. This term refers to the conversion of one axle to an electric brake, while the others continue to be braked hydraulically or electro-hydraulically. This enables production costs to be saved, because the hydraulic system can be filled and bled of air even before the axle module is assembled, with good safety levels being provided nevertheless by the fully independent subsystems. This means that it is not necessary to guarantee total fault tolerance in the electronics, because if there is a failure of the electronics it remains possible to brake the hydraulic axles. In the somewhat longer term, it will be possible to implement electrical brakes on all four wheels (Figure 9.20).

However, this step will call for complete fail-safety of the electronic system, including the power supply. Apart from this, any such system will be dependent on the availability of higher voltages, such as 42 V, because of the extreme power peaks, currents, wiring cross-section and hence also the weight. Depending on the system architecture, another absolute requirement for such an implementation will be a new, more deterministic and fault-tolerant safety bus standard. Various buses, such as TTP/C (time triggered protocol) and FlexRay, are currently under development for such requirements. Ultimately, a cost-efficient implementation will require the formulation of a unified standard common to all manufacturers. And in the final analysis, such a system also calls for new model-based development methods, ensuring high levels of correctness for the prevailing complexity. There are already various development vehicles, from many of the vehicle manufacturers.

One of these has been developed within the framework of a common research project – "BRAKE" – promoted by the European Union in partnership with Delphi Systems, Infineon Technologies, Volvo Cars and Wind River. In the context of this cooperative research, the technologies mentioned above have been investigated and developed, and various safety concepts investigated. The overall system has been implemented using a redundant 42 V power supply, local (distributed) redundancy which utilizes resources inherent in the system, rather than a central, redundant controller. The special feature of this development is the safety concept, which is based on completely distributed redundancy, i.e. in this implementation there is no central braking controller, but all the calculations and regulatory actions are performed locally on the brakes, and are mutually checked and synchronized via a fail-safe, deterministic and fault-tolerant bus system. In addition, the complete system including the engine management, vehicle dynamics and all the components involved on the brakes, such as the algorithms, buses etc., has been fully simulated on a PC, and using error injection has been validated for correct regulatory behavior even when errors occur.

Due to the high levels of technological interdependence, it is to be expected that purely electronic braking systems will first be launched on the market in about five years time.

9.3.2 Passive safety systems

Passive safety systems help to protect the vehicle occupants in the event of an accident, and supply diagnostic data about the environment and critical components of the vehicle. One thing common to all passive applications is that they do not at present intervene in the transmission path (clutch/gearbox), the engine management system, the steering system, the braking system or the suspension system. The following applications count as passive safety systems:

- Airbag and personal restraint systems
- Occupant detection
- Tire pressure monitoring system (TPMS)
- Proximity checking
- Driver assistance systems, e. g. parking assistant (proximity warning) or pedestrian detection

By far the most important passive safety system is the airbag and restraint system. Table 9.4 summarizes the names of the various restraint systems, their abbreviations and a short functional description.

Figure 9.21 gives an overview of the locations of the various restraint components in an automobile.

A large number of the sensors around the passenger compartment are located in the external shell of the chassis. Then there is the airbag ring directly around the passenger space concerned, combined with the restraint system (buckle-switch; belt pretensioner). Located in the center of the

Table 9.4 Restraint systems in an automobile

Brief description	Acronym	Details
Airbag centrally fired system	ACFS	Front driver and/or passenger airbag system; controlled/fired centrally from one ECU
Multiple airbag centrally fired system	MCFS	Front driver and/or passenger airbag system; controlled/fired semi-centrally from more than one ECU
Side impact (front)	SIF	Side-impact bags to protect front passengers
Side impact (rear)	SIR	Side-impact bags to protect rear passengers
Airbag distributed firing system (firing bus)	ADFS	Driver and/or front passenger airbag system; bus structure
Other (knee bags, head bags…)	Other	Detects the presence of a passenger or child seat (to deactivate airbag deployment accordingly)
Occupant detection	OcD	Detects the presence of a passenger or child seat (to deactivate airbag deployment accordingly)

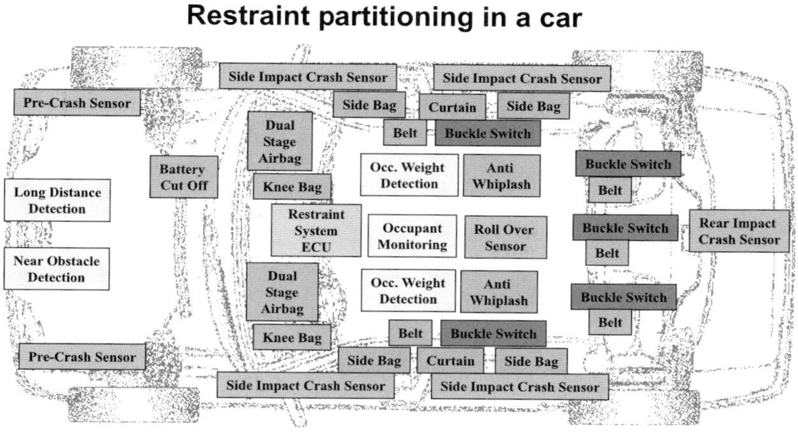

Figure 9.21 Restraint systems and their associated sensors in an automobile

chassis, optimally protected against any form of external influence (e.g. including against EM immission), is the system electronic control unit (ECU). A major part of the complete system electronics is brought together in this controller.

The "state of the art" device, which today is produced in very high volume, has already been described in Chapter 3 "Power semiconductors". The further development of this standard ECU, with its most important sub-functions, is shown in Figure 9.22.

A central airbag ECU can be broken down into various functional units:

The power supply unit

This supplies power to the sub-functional units of the ECU. In an accident, it is possible for the battery leads to be torn off, and consequently there must be a buffered supply of a finite quantity of energy. For this purpose, large capacitors are charged up to the highest possible voltage by a boost circuit ($W = 0.5 \cdot C \cdot U^2$). This energy must be sufficient, on the one hand, to fire the so-called squibs (gas generators for the airbags) and, on the other hand, for controlling the sensors, the microcontroller and of course diagnostics (firing protocol). In addition, this block undertakes the so-called supervision functions. A reset monitoring facility checks the power supply to the microcontroller and if necessary resets the system. A window watchdog monitors the processor and releases it if, for example, it gets trapped in an endless software loop (dead-loop). In the next generation all these functions, plus the CAN transceiver and a safety controller, could be integrated into a so-called system-base chip (SBC). This was already discussed for body electronic in Figure 9.5.

The communication unit

This comprises a high-speed CAN interface and special interfaces to the externally-located sensors. The latter are usually current-controlled serial interfaces. Some of the sensors are located on the ECU itself, and can be controlled directly by the processor.

The microcontroller

In general, this is nowadays a 16-bit microcontroller with extensive analog (A/D converter) and digital (memory) peripher-

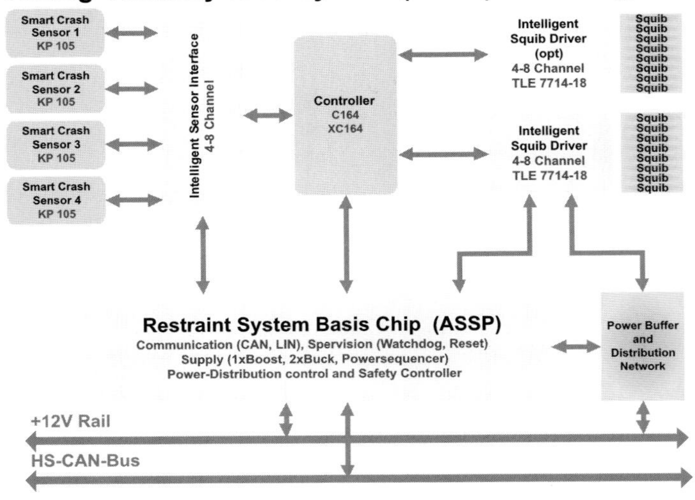

Figure 9.22
System block diagram and system coverage by components for a standard central airbag ECU of the next generation

als. The demanding diagnostic requirements in the safety field, and the real-time functionality, call for a relatively high computational power in the ECU. For this reason, the next generation will be fitted with 32-bit controllers. In order to increase the safety, a small second controller is used to provide redundancy. In the near future, this function will be integrated into the system-base chip.

The firing units

These smart power ICs fire the airbags by sending currents of 1 to 2 A through the gas generator for a few 10-ms, heating them up and thus bringing them to the point of "explosion". A well-honed diagnostic system and safety firing controller prevent failures and malfunctioning of the system with a very high degree of reliability. Today, a standard airbag ECU already contains 8-12 such firing units. In future there will be further significant increases in this number.

As in the case of other automobile applications, Infineon has developed a complete system chipset, and is supplying this to many of the world's ECU manufacturers.

For a large number of peripheral devices (sensors and firing units) in the restraint system it would be logical to link the satellite ECUs by bus wiring. For the sensors, this is today already a reality. The way that these controllers are constructed, and their individual peculiarities, will be explained below by reference to a side-airbag sensor system.

Pressure-based side-airbag systems

Side-airbags in the front and rear side doors are becoming ever more a standard fitment on modern vehicles. Until now, the sensors used to recognize a side-on crash have been mainly acceleration sensors. In recent years, some automobile manufacturers have implemented a new concept for detecting a side impact; detection of the pressure pulse generated in the side door.

By comparison with the acceleration-based systems, the pressure detection system offers a number of advantages: the triggering (firing) of the airbag is effected very rapidly and reliably. In addition, this concept is distinguished by the fact that it can make very good decisions as to whether the airbag must be triggered in a side impact, or its firing is unnecessary.

For these controllers too, Infineon offers all the necessary ICs, such as for example the pressure sensor, the matching interface IC and supply ICs optimized for the application.

Restraint system with the functions of a side-airbag

Figure 9.23 shows a block diagram of an airbag system with the functions of a side-airbag. The side-airbag module consists of the pressure sensor IC, a voltage regulation IC and a few discrete components, for example a transistor for current modulation. No quartz crystal is necessary, because the sensor IC is equipped with a digitally tuned chip oscillator, to satisfy the timing requirements of the Manchester output signal.

The central airbag ECU consists of the sensor interface IC, a power supply IC, a microcontroller and a firing driver IC for triggering the airbag.

The decision as to whether or not to fire the airbag is made by a special algorithm, implemented in the microcontroller. This algorithm analyses the pressure data received from the sensor IC. The important criteria here are the level of the pressure pulse, and its rate of change. In systems which have been further developed, account is taken not only of the data from the pressure sensor, but also that from a central storage sensor or from supplementary acceleration sensors located, for example, in the rigid side pillars of the passenger cell. By combining acceleration and pressure sensors, the performance of the side-airbag system is substantially improved, and a better distinction can be made as to whether or not the airbag should be triggered.

In order to implement combined systems of this type, acceleration sensors can also be linked to the receiving channels on the sensor interface IC.

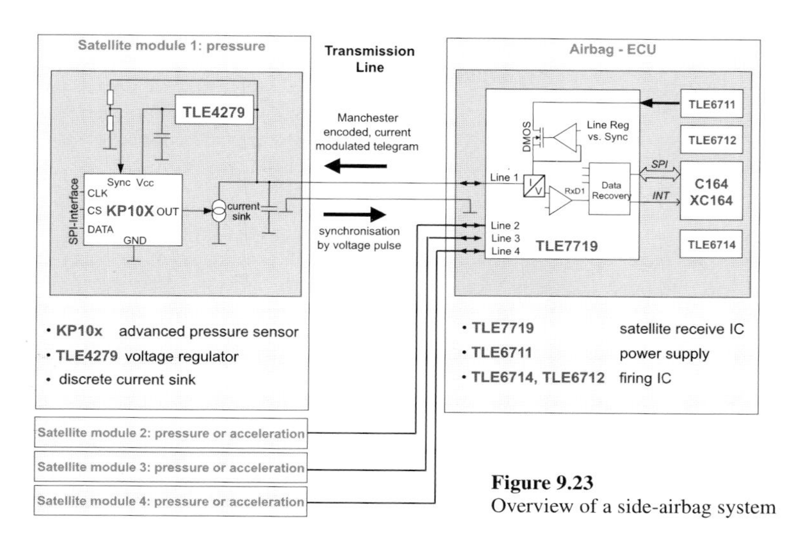

Figure 9.23
Overview of a side-airbag system

Safety functions

For safety systems, self-diagnostic functions are absolutely essential. For this reason, the sensor ICs provide various diagnostic functions to warn the driver if the airbag system is defective or no longer active.

For the reason of traceability, the latest generation of pressure sensor ICs has a unique identifying number. This number is copied into a start sequence after the chip has been commissioned. In addition, after each switch-on operation a diagnostic value is also transmitted to the ECU. This value shows either that the sensor is functioning correctly, or it may indicate any malfunction of the sensor cells. Other self-diagnostic functions carry out constant checks on parts of the electronic circuits. If a fault is detected, a fault code is transmitted to the ECU.

The link to the central airbag unit is implemented by special communication ICs. These ICs are able to received Manchester-coded current-modulated telegrams, sent from the pressure sensor over a two-wire line. For example, the satellite interface module TLE 7719 can service four two-wire interfaces simultaneously. The data received is made available to the central microcontroller via a standard SPI interface (serial interface). This enables substantial savings to be made on computing power.

The heart of the satellite system is the sensor module, which will now be considered in more detail.

Pressure sensor in the sensor module

The core of a pressure sensor IC comprises capacitive sensing cells, in surface mounted micromechanical form, with a monolithically integrated A/D converter and circuitry for digital signal processing. The pressure sensor ICs have a digital interface, which is compatible with standard µCs. Over and above this, there are implementations of the self-diagnostic functions which are indispensable for safety-related applications.

Two types of pressure sensor are available: a simple type, which supplies an absolute pressure signal via a SPI, and a more-developed type which supplies a pre-processed relative pressure signal in Manchester code. The more developed type enables the number of devices in the module to be reduced, and hence lower cost door modules to be manufactured. Both types have a proven small SMD (surface-mounted device) package which fully satisfies the requirements in vehicles.

Technology and signal processing

In the vehicle safety market, the demands are for low-cost and highly-specialized semiconductor modules, which satisfy high quality standards.

The process which is used in the manufacture of the pressure sensor ICs which can currently be obtained is a 0.8-µm BICMOS process, which is performed on standard equipment in order to achieve high process stability.

The pressure sensing cell forms the basic element of the sensor IC. Unlike when so-called bulk micromachining is used to create a pressure-sensitive membrane on the rear of a wafer by anisotropic etching, in the present case so-called surface micromachining is used to manufacture sensor cells using the standard process steps of layer deposition and etching.

Figure 9.24 shows the typical cross-section of a sensor cell, which consists of an elastic membrane made of conductive polycrystalline silicon (polysilicon), a sealed space (cavity) together with a conductive electrode on the surface of the substrate (drain).

The intermediate cavity is produced by a special sacrificial layer process. During this process step, a field oxide which has previously been deposited is removed locally through small holes in the polysili-

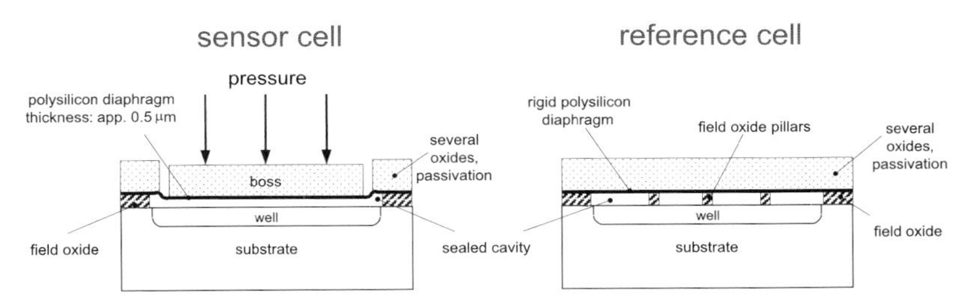

Figure 9.24
Cross-section through the capacitive sensor and a reference cell made by surface micromachining (not to scale)

con membrane. The empty space which results is then sealed off by an oxide deposition process. To achieve the required pressure sensitivity, a last process step etches a trough in the oxide layer, down as far as the polysilicon. This produces a square shaped connecting surface in the middle of the membrane.

Reference cells which are almost identical are also created. These contain small pillars of field oxide, which prevent deflection of the membrane. No troughs are etched. For this reason, the reference cells are not pressure sensitive, but their capacitance is comparable and temperature characteristics similar to the pressure-dependent cells. For this reason, reference cells are specially suitable for compensation purposes, e.g. in a bridge circuit.

Figure 9.25
SMD package for pressures sensor ICs

Packages

The surface of the chip must be exposed to the environmental pressure, so that a special package is required. The sensor ICs are finished in a specially-developed SMD package, which enables small modules to be constructed. The chip is not covered over with a rigid material (molding compound), but instead is protected by a flexible gel. This gel is resistant to the chemicals present in the vehicle environment. In addition, it prevents fluids from condensing on the surface of the chip.

The devices are supplied with a protective cap, which is removed after the chip has been soldered into position. Figure 9.25 shows the SMD package, omitting the protective cap.

Further development of the airbag systems: the firing bus

It is an obvious idea to locate the firing electronics at the place where the airbag must expand. This would significantly simplify the star-shaped wiring of today's "centrally fired airbag system". In addition, the interference from the parasitic characteristics of the firing lines would be eliminated. Hence, the next evolutionary step which is under consideration for airbag controllers is to localize the firing stage. The firing bus for this purpose is in the definition stage. Before this step is

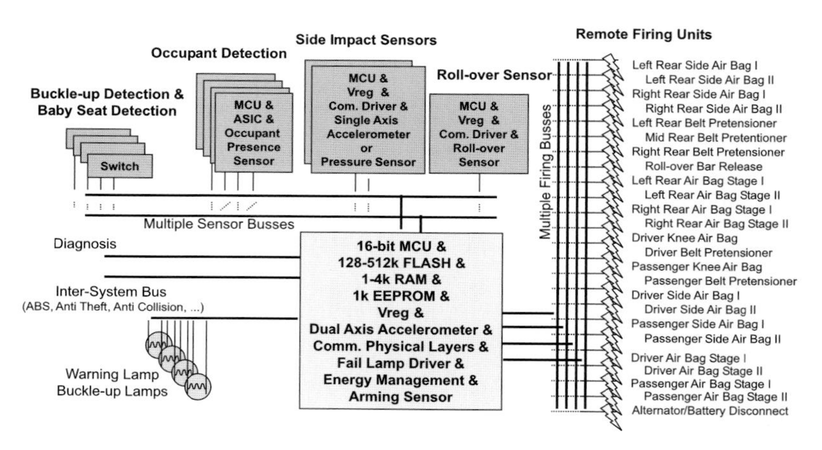

Figure 9.26 Restraint system with sensor and firing bus system to simplify the wiring

ready for series production, there are still many reliability requirements which must be satisfied. In this connection, the possibility of a rupture in the bus during an accident has been a source of many headaches for developers. The structural principle for such a firing bus system is shown in Figure 9.26.

Figure 9.26 shows that it is possible to construct a multiple-bus system, with several bus wires and intelligent satellite triggers, which enables substantial savings to be made in the wiring. However, it will certainly be a few years yet before these systems are constructed in numbers worth mentioning.

The safety system described immediately below, for the measurement of tire pressures, will on the other hand appear in automobiles significantly more rapidly:

System for monitoring tire pressure using a micromechanical pressure sensor

Following a series of fatal accidents caused by a loss of tire pressure, a system for the measurement of tire pressures is to be introduced. In the USA for example, legislators have enacted relevant regulations to this end. The increase in safety is especially relevant at higher speeds and with greater traffic densities. For this purpose too, a pressure sensor from Infineon, the KP500, plays an essential role. We explain below the interactions of the electrical and mechanical components required for this micro-mechatronic system.

System aspects

Each direct TPMS consists of four or five pressure-sensitive tire modules (one for each tire) and a central unit. The tire modules measure the pressure and the temperature in the tire cavity. They are battery-powered and transmit the sensor data at HF to a central receiver. To save on current, the interval between the transmissions must be matched to the situation.

Consequently, the module has the following three basic functions:

- measurement,
- control,
- transmission.

These functions can be realized by various system approaches and different solutions for the split between sensor, controller and transmitter. For example, the new Infineon transmitter, PMA5100, has microcontroller functions and is precisely matched to the KP505, one of the tire pressure sensors in the KP500 series. As an option, additional controller functions can be implemented in the sensor. The KP510 will be such an intelligent sensor. It can be connected directly to a standard RF transmitter chip, such as the TDA 5100. This transmitter can be operated in either ASK or FSK mode (i.e. using amplitude sampling or frequency sampling) at 434 or 869 MHz.

The development of the KP500 series thus provides a platform of sensors which can be flexibly adapted for various system specifications.

KP500 series: a sensor platform for TPMS

Infineon's KP500 intelligent tire pressure sensor offers all the functions required for tire module sensors, without the need to use any further external sensor elements. It measures the relevant physical parameters of pressure, temperature and battery voltage.

Figure 9.27 shows a schematic block diagram. The sensor elements are monolithically integrated with an A/D converter, signal processing, calibration data memory, power management and an SSI communication interface, to offer simple and flexible system integration.

In order to achieve a long service life, highly-developed energy management is indispensable. The KP500 sensors are arranged in such a way that with a single battery they will have a service life of more than 10 years (including the power consumption of a transmitter and of a microcontroller optimized for this application).

Most of the circuits need only be active during the brief time intervals when measurements are being made. A state machine (hard wired controller) controls the measurement cycle by issuing commands to the energy management system. A measurement cycle can be initiated either by an external command or by the internal wake-up algorithm. In general measurements will be made once or twice per second. This is an adequate frequency because the pressure changes which are to be detected take place only slowly.

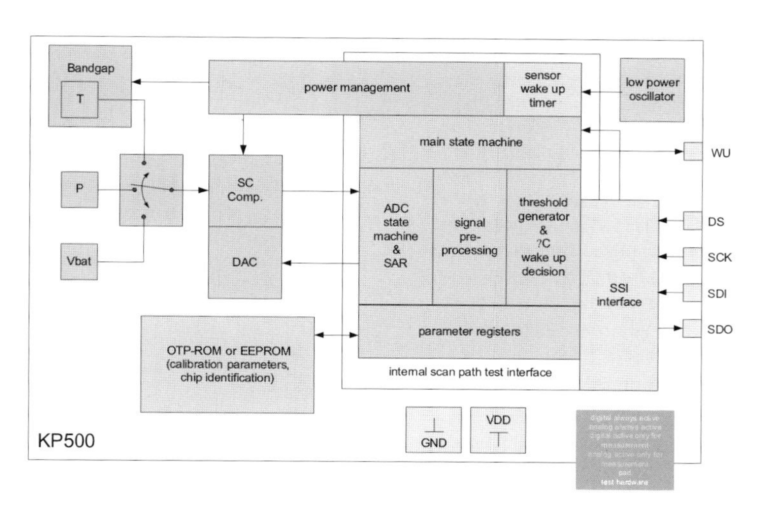

Figure 9.27 Block diagram of the KP500 pressure sensor

Pressure sensors in the KP500 series

The pressure sensor consists of capacitive sensor cells made by surface micromachining. The sensor cell is designed for a pressure range of up to 650 kPa (6.5 bar). Two sensor arrays and two reference arrays form one capacitive bridge circuit. Any rise in pressure alters the capacitance of the sensor arrays, and thus disturbs the bridge from its balance point. The bridge voltage which results is fed into a full differential A/D converter. The conversion is performed by applying the SAR method (successive approximation register, incremental step converter procedure).

The sensor technology is based on the SMM technology (surface micromachining), which has now become common and has been repeatedly proven. Infineon's surface micromachining integrates a pressure sensor cell into a BiCMOS process. As BiCMOS supports both bipolar and also CMOS devices, this technology is the most advanced micro-system technology currently available.

In 1998, Infineon started with mass production of this new type of micromachining process for pressure sensors. SMM sensors with analog and digital signal outputs are today manufactured in large numbers. It is because only the front side of the wafer together with the integrated circuits (ICs) is processed that this process is referred to a surface micromachining.

With surface micromachining, sensors are manufactured using standard semiconductor processes. Currently, a 0.8-μm BiCMOS process is used for SMM production. Work is currently in progress on reducing the minimum feature size down to 0.5 μm, which will give surface micromachining even further potential.

The pressure sensor itself is based, as shown in Figure 9.28, on a pressure sensor cell which consists of a thin, flexible membrane, the silicon substrate, and a sealed cavity. The membrane and the substrate represent the two plates of a capacitor. Under pressure, the membrane bends inward, leading to a change in the capacitance of the sensor cell.

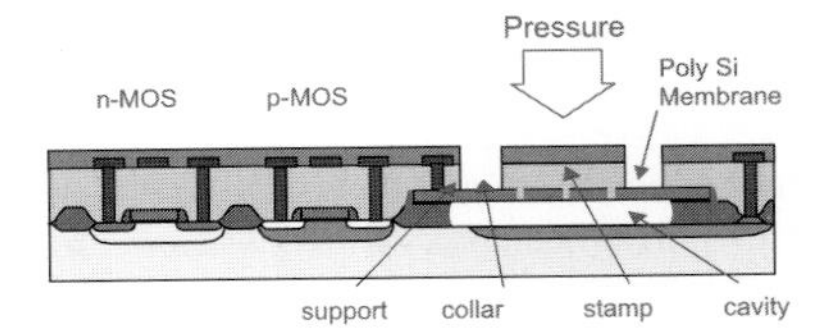

Figure 9.28
Schematic cross-section through a pressure sensor device

The membrane consists of polysilicon. This layer is a part of the BiCMOS process, and normally serves as a resistive layer and as a component of a linear capacitor. The thermic oxide, used in the BiCMOS process as an insulating layer, is used as a sacrificial oxide in the micromachined component. In one of the few additional steps, this sacrificial oxide is removed by selective wet-chemical etching under the polysilicon membrane (i.e. the polysilicon of the membrane is unaffected). The etching of the sacrificial layer is effected through 1 μm holes in the sensor membrane. The dimensions of the cavity are determined by the lateral isotropic etching rate.

Particularly for applications of this type, this unique technology offers cost-effective solutions in which the overall system is optimized by immediate digital processing of the pressure signals. At the same time, its integration into a standard BiCMOS process offers an economically viable way of implementing extremely reliable intelligent sensors for use in vehicles. Surface micromachining is especially suitable for tire pressure monitoring systems, with their requirements for a low total energy consumption and the highest levels of accuracy.

Outlook

The development of advanced micro-system technologies enables intelligent sensors to be designed. Battery-powered systems will, because of their simple and robust design, predominate for quite some time yet in tire pressure monitoring systems. Nevertheless, systems are currently being developed which have no battery, and direct integration of the sensors into the tires is under consideration. This represents a further challenge for sensors and electronics, and it is to be expected that the development of micro-systems is very far from exhausted.

The automobile of the future will contain a host of additional safety functions. Here, the applications will grow together into a powerful computing system, in order to satisfy the requirements with the reliability demanded. The necessity for particularly high redundancy requirements can easily be imagined in the case of steering or an electrical braking system. Here, one can foresee the use of highly-complex controllers, as used nowadays in aviation.

The development of such electronic units is only possible with a general cooperation: automobile manufacturer, supplier and semiconductor manufacturer. This is particularly true for the highly-complex applications in section 9.4 "Powertrain electronics", with its sub-applications of engine and gearbox control.

9.4 Powertrain electronics

Exhaust gas standards and the steady rise in fuel prices make the use of electronics in the regulation of combustion engines unavoidable. Electronic controllers measure a host of signals, calculate the fuel quantities to inject and the ignition timing, and control the appropriate actuators, such as the throttle valve, by means of power output stages. Here, semiconductors are the key components in the electronic control units (ECUs). The challenge today no longer lies solely in designing a perfect engine, but also in developing the right electronics and software, and hence also the semiconductor devices. It is under the engine hood that the most demanding requirements are posed under the most severe conditions of use. A complete portfolio of technologies and products for the entire application is the ideal prerequisite for enabling one to determine the optimum performance and cost at system level.

9.4.1 Semiconductor technologies for the powertrain control loop

Today, semiconductor solutions already cover the entire control loop for a powertrain system (Figure 9.29).

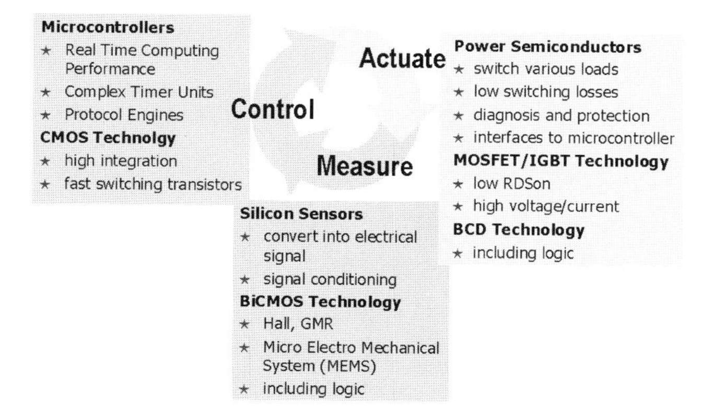

Figure 9.29 Semiconductor products and technologies for the powertrain control loop

Powertrain applications call for a whole range of sensors for converting the measured physical quantities into electronic values. There are semiconductor sensors for capturing data on rotation speed, position, temperature and pressure. As described in section 9.3, the trend is from passive sensors to active ones, in that the actual sensing element is combined with signal processing facilities.

The growing level of integration for CMOS technology permits the implementation of powerful processor and peripheral functions in microcontrollers. The computer architecture for powertrain systems must meet the needs for real time capability, because otherwise exact regulation of the combustion process is not guaranteed. Here, the architecture differs from that developed for personal computers, for example. Ever-increasing integration densities also permit the incorporation of functions for digital signal processing (DSP) into microcontrollers. This makes it possible to realize functions in software form which were previously implemented in hardware, for example in ASICs.

In recent years, the development of semiconductor technologies has been mainly concerned with optimizing existing systems in respect of costs and functionality, by appropriate integration. The results of these efforts are mixed technologies (BCD technologies), which permit the integration of power output stages (DMOS), analog circuits and digital logic into highly-integrated products. This means that products based on this technology can significantly reduce the expense of modules in the controllers. The MOSFET technologies are following the trend of minimizing the cost for unchanged functionality. Infineon has developed a family of power MOSFETs – the OptiMOS™ range. The advantages of these products are that they have significantly lower switching losses while maintaining the required ruggedness (see also Chapter 3).

9.4.2 Powertrain applications – system overview

Modern engine and gearbox controllers are today dominated by active semiconductor components, comprising microcontrollers, power semiconductors and sensors. Figure 9.30 shows a typical

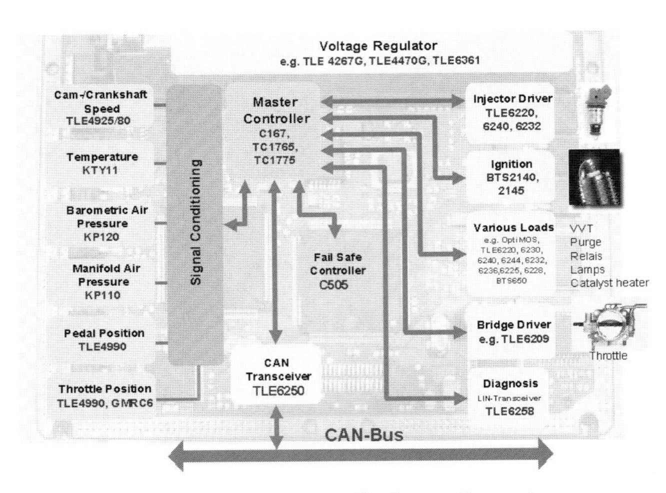

Figure 9.30 Block diagram of a controller for gasoline engines

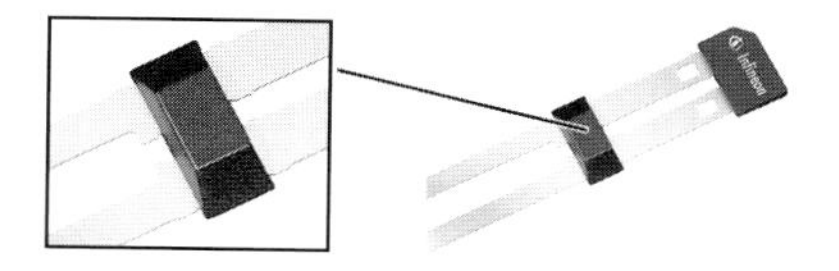

Figure 9.31
Hall sensor IC with an integral capacitor in one package

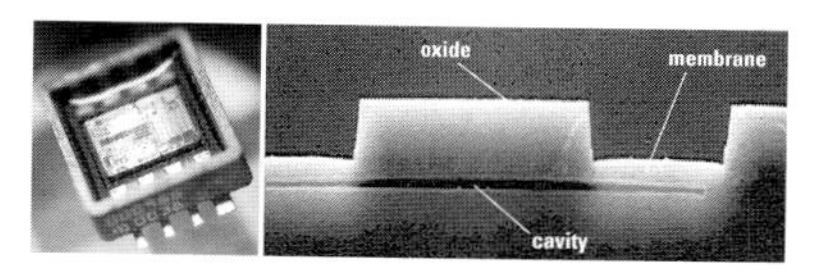

Figure 9.32
Technology for silicon membranes in the KP120 (BAP sensor)

block diagram of an engine management controller.

Semiconductor sensors for powertrain applications

Dynamic differential Hall ICs such as Infineon's TLE 4925 or TLE 4980 are suitable for capturing rotation speed data from the crankshaft or camshaft, the gearbox or the wheels.

By integrating components such as capacitors with the Hall ICs in a single package, it is possible to pre-connect the devices in one sensor package with no external circuit board and components (Figure 9.31).

Apart from the classical temperature and Hall sensors, micromachining technologies are opening up a completely new world of pressure, acceleration and rotation speed detectors.

In the case of analog pressure sensors, which are intended to detect the absolute pressure in the inlet manifold (MAP) and the absolute atmospheric pressure (BAP), the use of a conventional BiCMOS technology offers the option of integrating logic (Figure 9.32).

Power semiconductors for powertrain applications

Smart power output stages

Engine and gearbox controllers are required to switch such consumer elements as injector nozzles, heating elements for lambda probes, ignition coils, fans and

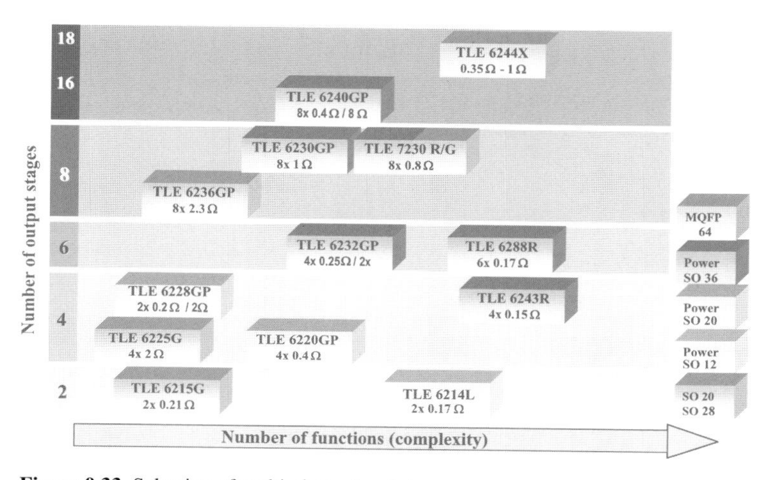

Figure 9.33 Selection of multi-channel switches

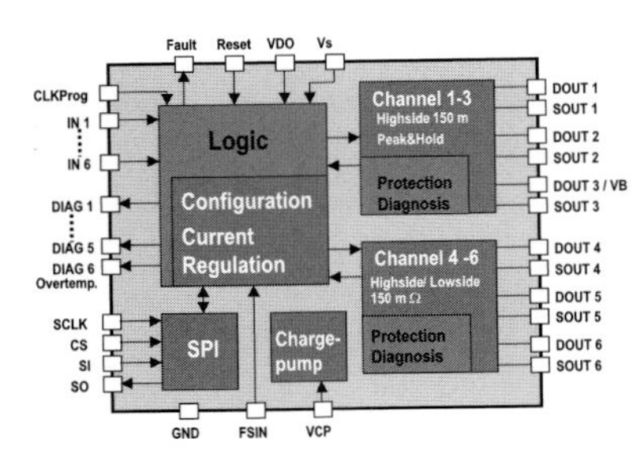

Figure 9.34
Block diagram of a multi-channel switch (TLE6288R) for gearbox applications

various relays. For these applications, Infineon offers a wide range of multi-channel switches with 2 to 18 channels which, depending on the application concerned, can switch currents of between 50 mA and 10 A (Figure 9.33). Modern multichannel switches are equipped with special circuits which can perform these demanding functions autonomously and with no further external involvement (Figure 9.34).

Apart from this control functionality, these products also contain protection and diagnostic functions for recognizing short circuits, overload, excessive temperatures and any breakage in the wire to the consumer device. In addition, such safety functions as a shutdown after overloading or an emergency running mode (limp home) can also be implemented. A fast serial interface will soon be available, for use in real time control (current regulation by PWM) and for communicating the diagnostic data from the switch.

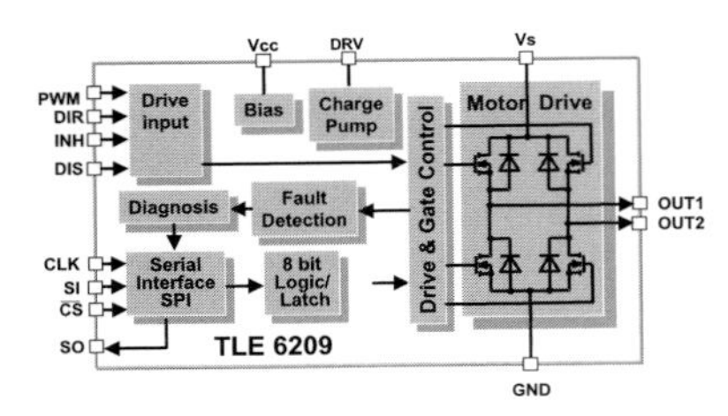

Figure 9.35
The TLE 6209 motor bridge for driving an electronic throttle flap (e-gas)

Motor bridges:

Powertrain applications involve mainly the driving of DC motors by means of motor bridges. These motors move valves and throttles in the inlet manifolds of gasoline engines, and valves for exhaust gas feedback systems for Diesel engines. Today's motor bridges, such as Infineon's TLE 6209, include all the functions required for controlling DC motors. The signal used for control purposes is a pulse-width modulated signal at frequencies above 20 kHz. Here again, the integration of logic functions enables the H-bridges to be precisely controlled, together with protective and diagnostic functions (Figure 9.35).

The serial interface can be used for transmitting to the microcontroller diagnostic and error reports, such as short circuits. A three-stage temperature monitoring arrangement (pre-warning, warning, shutdown) represents the safety function required for the safety-critical function of the throttle flap.

MOSFETs and IGBTs:

There is a wide offering of MOSFETs and IGBTs from which the appropriate device can be selected for the application concerned. Nowadays, the main applications in this area are, for example, the driving of the injector nozzles in gasoline or Diesel engines, fuel and water pumps, and valves in gearbox controllers.

IGBTs are ideally suitable for ignition systems under the extreme conditions involved (high temperatures, vibrations, electromagnetic interference, etc.). The advantages of IGBTs are that they can be driven using logic signals, their low saturation voltage, integral ESD protection and the active Zener clamping. Apart from satisfying all these requirements,

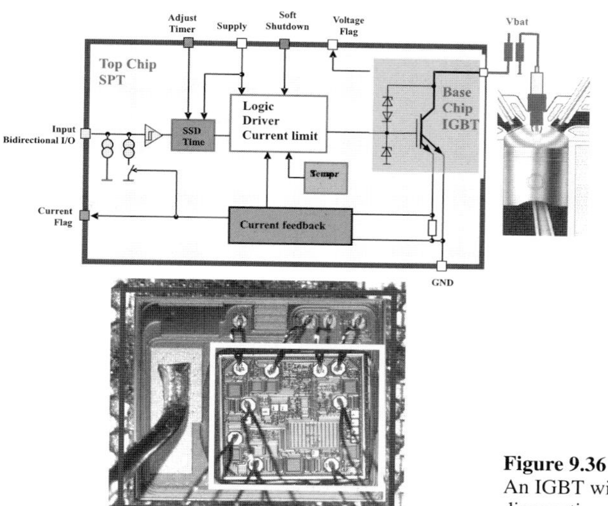

Figure 9.36
An IGBT with a top-chip for diagnostic and safety functions

modern ignition systems require further safety and diagnostic functions. These are provided by Infineon's SMART IGBTs, which are based on the chip-on-chip technology (Figure 9.36). A sense IGBT for the power functions acts as the base chip, with a second chip in BCD technology being arranged on top of this to provide the intelligence. This top-chip can include as monolithic elements such functions as current limitation, current and voltage level reporting or a "soft shutdown".

In future, piezo-electric injectors will represent another application for IGBTs. Here particularly, Infineon already offers a fast-switching IGBT, the so-called Fast IGBT.

In high current applications (up to 1000 A), Infineon's PROFET™ family offers an option for drastically reducing the components required and raising the performance of the drive train. By contrast with obsolescent solutions using relays, these devices offer protective and diagnostic functions such as current limitation, overvoltage and overtemperature protection, recognition of a broken wire, and current sampling. Typical applications include control of the engine starter, the radiator fan, the pressure pumps and the glow plugs for Diesel engines.

Transceivers:

Transceivers transmit the signals from the microcontroller over the vehicle's own communication lines. In powertrain applications, the bus protocol used as standard is high-speed CAN. Infineon has developed the TLE 6250G high-speed transceiver to have the best characteristics in terms of EMC. In addition, it incorporates functions which ensure secure transmission of data between the various electronic controllers.

Power supplies:

The power supply for semiconductor components requires new power supply architectures, which differ fundamentally from traditional approaches. Whereas in earlier systems individual linear regulators would suffice, it is now sometimes essential to have power supply units with several outputs, because of the reduced operating voltages, the higher current levels and the special requirements to be met in terms of the switch-on and switch-off processes.

Infineon has therefore developed a new power supply system IC. This consists of a switched-mode buck regulator, downstream from which are three linear low dropout (LDO) voltage regulators on one chip. The asynchronous switched-mode buck regulator provides a regulated voltage, which feeds the three linear controllers. With this buck-and-linear concept the efficiency is significant improved. If one is considering a powertrain system, with a total current requirement of more than 1 A, then this concept is already very attractive for 12 V systems. For the future 42 V vehicle power system, it will be indispensable. In addition to a comprehensive portfolio of linear regulators, Infineon offers the TLE 6361 smart power supply IC, which incorporates the power supply concept just described (see Chapter 3).

Microcontrollers in powertrain applications

A microcontroller plays an essential role in the system when to comes to the reduction in exhaust gas emissions and fuel consumption (Figure 9.37).

Complex computational algorithms must be processed in real time to ensure that the engine and power transmission are controlled under all operating conditions. Apart from the microcontroller's CPU, the entire system architecture is critical in determining the performance of the system under the adverse environmental conditions in the engine compartment.

8-bit architectures:

8-bit systems such as Infineon's C505 may by nowadays be reserved mainly for use in basic models of vehicles, but are also attractive in addition as the controllers for motorcycles. As electric motors make increasing inroads into powertrain applications, new possibilities continue to appear for the use of 8-bit microcontrollers. Infineon has launched a new type C868 on the market for the control of electric motors, such as will be used in future e.g. for turbochargers.

16-bit architectures:

Although the market share of 32-bit microcontrollers is steadily growing, 16-bit products continue to be installed in vehicles on a large scale. The C167 16-bit family from Infineon has been designed to satisfy the high performance requirements of embedded control applications

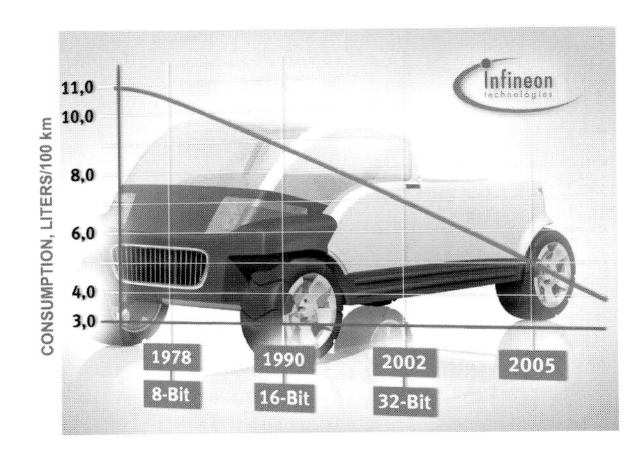

Figure 9.37
Fuel consumption and microcontroller architectures

with real-time capability, and today represents a standard in the powertrain application area. The success of the C167 family is based on its peripherals, which are ideally adapted for the requirements of powertrain applications.

In addition to asynchronous, synchronous and CAN interfaces, the C167 offers a range of special peripheral functions, which increase the performance of powertrain applications.

As a new generation of the C166 core, the C166v2 attains a doubling of the processing power by the clocked processing of commands. An integral highly-developed MAC unit (multiply-accumulate) gives a dramatic increase in the DSP performance. Products based on this CPU, such as the XC164, are now beginning to appear in controllers for electronic turbochargers, starter-generators and gearbox controllers.

32-bit architectures:

Today, the most powerful microcontrollers are made with 32-bit CPUs. Some of these CPUs are derived from microprocessor architectures. CPUs such as the TriCore from Infineon are, on the other hand, conceived for embedded control applications with real-time capabilities,

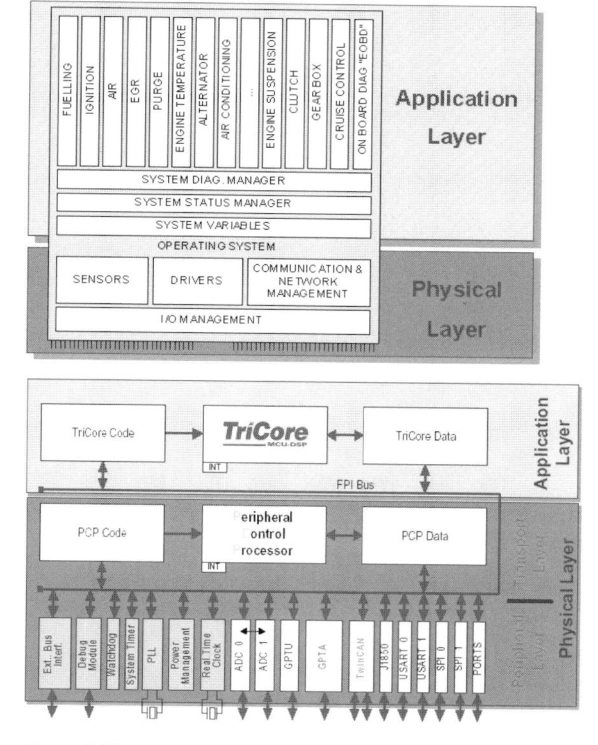

Figure 9.38
Software structure in accordance with OSEK, and its implementation in the AUDO architecture

and consequently are ideal for applications in the powertrain. The family of microcontrollers offered by Infineon under the name AUDO comprises, in addition to the TriCore CPU, special peripherals, which have been developed as dedicated peripherals for powertrain applications.

TriCore has a RISC architecture, and provides special instructions for the common DSP operations, which permit complex signals to be efficiently analyzed. The interrupt processing architecture of the CPU combines the typically short response time of a microcontroller with a high degree of flexibility.

TriCore supports both a 16-bit and a 32-bit instruction format. This reduces the length of the code, and ensures a higher code bandwidth. Hence the need for program memory is reduced, and with it the system costs.

Another outstanding element in the AUDO architecture is the Peripheral Control Processor (PCP), which enables the application software running on the TriCore CPU to be clearly delineated from the driver software processed by the PCP. The software modules are linked by a standard operating system, such as for example OSEK (Figure 9.38).

The so-called General Purpose Timer Array (GPTA) provides autonomous and complex functions, because the timer functions can be combined using software (Figure 9.39). The architecture thus supports the user in the use of software to implement functions for fuel injection, ignition systems (including knock-control) or air-bypass management.

The A/D converter in the AUDO architecture can also take some of the load off the processor by such autonomous functions as the recognition of short circuits or wire breakages.

Infineon's AUDO architecture is presently available as the products TC1765 and TC1775. Further derivates such as TC1766 and TC1796 (AUDO Next Generation) are under development. In order to reduce the system costs, the facilities integrated into these are to include flash memory of up to 2 Mbytes and a time-triggered CAN module (TTCAN). The AUDO NG products will provide the power for complex powertrain applications such as electromagnetic valve operation.

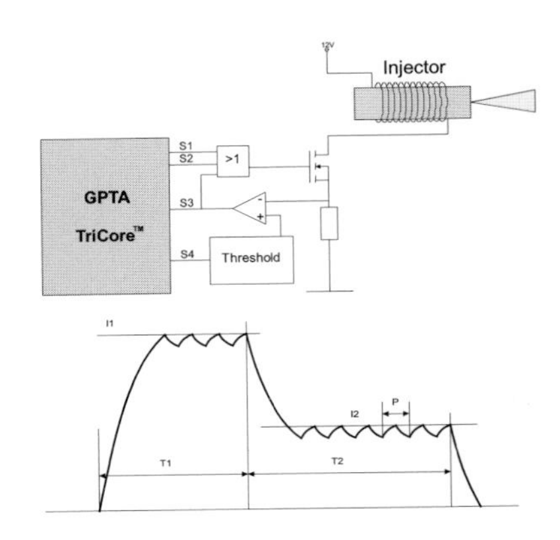

Figure 9.39
Possible connection of the GPTA for driving an injector valve, and the "Peak and Hold" signal which is generated

Further details on the topic of microcontrollers will be found in the relevant chapters of this book.

9.4.3 The future allocation of powertrain applications

Future controllers in the powertrain field will be characterized by increasing power requirements together with optimization of the system costs. Today, Infineon is able to offer complete chipsets, and in future to optimize them for cost and rising system performance.

One reason is that the increasing logic density of the power technologies permits control functions to be further integrated with the power semiconductors. Another is that by using the BiCMOS technologies it is possible to integrate ever more functions for signal capture, processing and transmission alongside the sensor element. Apart from this, it is also possible to realize a non-integrated solution by using multi-chip technology to produce either chip-by-chip or chip-on-chip products, as with the smart IGBT.

The semiconductor devices of the future must satisfy the requirements, in terms of quality, reliability and performance at an acceptable price. By working in close cooperation with the automotive industry, it is possible to achieve cost-optimization of the semiconductors within the overall powertrain system by partitioning the functions in an intelligent way.

9.5 Infotainment electronics

In the field of automotive electronics, the general term of infotainment traditionally covers those systems which are used for displaying vehicle-specific information and for the entertainment of the occupants. This includes, in addition to the display instruments in the so-called instrument cluster, the integral audio systems but also, more recently to an increasing extent, any telematic systems, navigation and multi-media systems.

Even where the developments, particularly in telematic, navigation and multi-media applications, tend to be based on the consumer area, these systems are nevertheless integrated into the vehicle environment, and the quality requirements are therefore based primarily on those for automotive electronics.

9.5.1 Dashboard/instrument cluster

This application segment includes the primary instrumentation, the so-called dashboard and instrument cluster, with mechanical, electro-mechanical and, increasingly, also electronic displays for vehicle-specific data including such items as speed and engine revs. The secondary instrumentation covers such applications as the trip computer.

Whereas, in the past, purely mechanical instruments were used for dashboard applications, these being linked to the drive by a shaft, nowadays 8-bit and 16-bit microcontrollers are used, and to an increasing extent even 32-bit controllers, which use PWM to drive the stepper motors for the individual instruments. In order to show the information on displays, such processors also have appropriate controllers and drivers integrated with them.

9.5.2 Car audio

The car audio segment covers the high-volume car radio application, with its diverse manifestations: simple AM/FM radios, radio systems with cassette players, and to an ever greater extent with an integral CD player, some with support for such audio formats as MP3.

Standard functions such as RDS permit the transmission of a radio station identifier and program type, and enable whichever transmitter has the strongest signal for the chosen program to be selected. The TMC service (Traffic Message Channel) permits the terrestrial transmission of traffic information in a standardized format, which is also used in navigation systems when calculating the most advantageous route for the journey.

Many audio systems use two tuners to support the RDS function. While one of the tuners is used to play the signal from the radio station which is currently being listened to, the second tuner is constantly searching the spectrum for stronger transmitters of the same radio station, so that it can always guarantee the best possible reception.

New system architectures also provide for the flexible implementation, of functions in software form. This then poses particular requirements for the processor which is used; these requirements include not only the functions of a microcontroller for control purposes, but also DSP functions for processing audio signals (e.g. equalizers, MP3). In addition there are such requirements as voice control and speech output (text-to-speech), the integration of hands-free mobile communication systems (including, in future, by Bluetooth connection to a mobile phone), etc.

9.5.3 Telematic systems

Telematic systems represent a relatively new class of infotainment systems. These systems use an integrated mobile communication link (e.g. a GSM/GPRS module in Europe, CDMA/AMPS module in the USA) to provide basic vehicle-specific and communication services. These include emergency services (e.g. after an airbag has been triggered), remote diagnosis for determining malfunctions, remote maintenance, location-dependent information services (location-based services/point-of-interest services), but also the familiar mobile phone (Figure 9.40).

On the hardware side, the main challenge in this application segment consists in the optimization of the system, and in the cost-effective provision of terminals for high-volume implementation in vehicles.

In this and other infotainment applications, particular advantages are offered by the TriCore architecture which is already widespread in the automotive field. This processor architecture combines the functionality of CISC, RISC and DSP in one processor core, and thus permits the use of a single development environment. The system-on-chips which have been specially developed for telematic applications on the basis of this core combine the processor functionality required for this application with specially customized application-specific peripherals.

Another important module in a telematic system is the GPS receiver which, in combination with other signals supplied by the vehicle or the telematic system (e.g. a directional sensor), make it possible to determine the vehicle's location.

In order to support vehicle-specific services, the telematic system must be integrated into the vehicle network via a range of vehicle buses: in Europe the

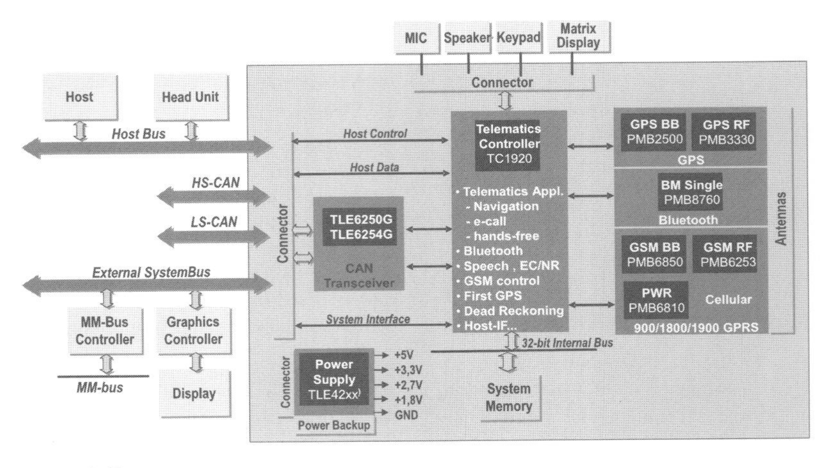

Figure 9.40
Example of a highly-integrated telematic platform, based on Infineon's semiconductor components

most common bus is the CAN bus, the bus used in the USA is the J1850.

9.5.4 Navigation systems

In the case of navigation systems, there are two different groups of systems, which are distinguished mainly by where the cartographic data is stored: in the case of on-board navigation, the geodata is stored on a CD-ROM or DVD supplied with the system, and is made available locally in the vehicle by means of an appropriate drive.

In the case of off-board navigation, the map data is provided on a server by a service provider, and is transmitted to the vehicle via a mobile communication link (e.g. using a GSM/GPRS module). Preprocessing of the data on a powerful server is intended to reduce the computing effort required and hence the system costs in the vehicle. The so-called Internet radios can be counted as one of this group of systems; these implement navigation as an Internet-based application.

Hybrid navigation represents a mixture of the two methods described above, and is intended to ensure the availability of the latest street data and traffic information combined with a reduced data communication load. With this, traffic information such as TMC is also used to give the best possible choice of route.

As in the case of the telematic system, an essential element of a navigation system is again the determination of the vehicle location by means of GPS. Determination of the location can be further improved by using dead-reckoning algorithms together with tachometer impulses and a gyroscope (angular sensor). Map-matching algorithms then make it possible to synchronize the map data with the vehicle's position. In this application, high-performance navigation computers are required for route calculation and location determination.

The navigation system can also call on any telematic system which is integrated into the vehicle, for use as a communication infrastructure, or the function of the telematic system can be integrated as a permanent element in it.

Increasingly these systems are using high-speed buses, based on POF (plastic optical fiber), which have been specially developed for use in vehicles and which can, for example like the MOST bus (Media Oriented Systems Transport), support data transfer rates of up to 24.8 Mbps. These buses use a ring structure and, for example, link the CD changer to the navigation system.

9.5.5 Multimedia systems

Multimedia systems represent a combination of a top-class audio system, a navigation system and an entertainment unit. Apart from TV receivers and DVD players, these systems can provide an integral games console, in some cases using displays built into the headrests of the front seats.

Multimedia systems are based on high performance processor platforms, and provide expensive graphics facilities.

9.5.6 Cross-application technologies

GPS (Global Positioning System)

In the infotainment field, GPS receivers are used in very large numbers for the determination the vehicle's location. These solutions are capable of searching for and tracking between eight and 16 different satellites in parallel.

Because of the need to service tasks with high priorities (interrupts at 1 ms and 20 ms), a 32-bit controller is used in a GPS receiver, with appropriate memory resources. In a so-called autonomous GPS implementation this processor is integrated into the GPS baseband IC, the neces-

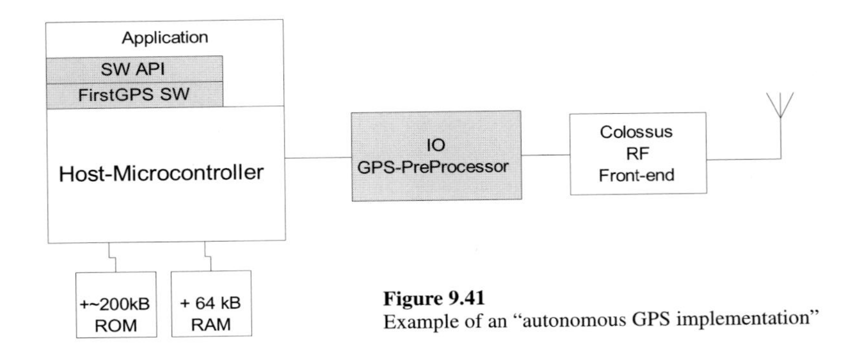

Figure 9.41
Example of an "autonomous GPS implementation"

sary memory must be provided externally (Figure 9.41).

Especially in higher-volume applications, particular advantages are offered by the so-called host-based GPS solution (example: the FirstGPS solution from Infineon/Trimble). With this implementation, all the time-critical functions are realized in hardware on the base-band IC, with the less time-critical tasks being implemented in software form on the host microcontroller (e.g. the navigation processor) which is anyway generally available in the system. The software is optimized so that it only generates a small additional processor load (<= 4 MIPS) on the host controller. Furthermore, the RAM/ROM resources which are available there are also used, which further reduces the number of dedicated components required for GPS. This also makes it possible to dispense with the 32-bit controller integrated into the autonomous architecture in the baseband IC/correlator.

Bluetooth

With the increasing spread of Bluetooth-based consumer devices such as PDAs, mobile telephones and laptops, the requirement for Bluetooth support in vehicles is also growing, because it produces a host of new application possibilities. Bluetooth provides a standardized interface to consumer devices from various manufacturers, and offers the potential to provide a bridge between the comparatively long-lived infotainment systems and short-life consumer products.

Bluetooth functionality is often integrated in the form of modules, on which the latest current Bluetooth ICs are used. Bluetooth ICs are primarily developed for high-volume consumer applications, and thus have a very short product life span (1-2 years).

Today, the most advanced semiconductor technologies make it possible to manufacture single-chip solutions with integral baseband and high-frequency circuits and a minimal number of external components (Figure 9.42). These ICs support the enhanced requirements of the automotive industry in respect of the temperature range.

Apart from the availability of certified Bluetooth ICs, particular importance also attaches to the Bluetooth software.

Bluetooth profiles and the Bluetooth protocol stack are specified for the various application situations. To confirm conformity with the Bluetooth standard, each system integrator must demonstrate that the profiles implemented in his system are interoperable with other systems.

Apart from the standard profiles developed for other markets, special application profiles are also developed for Bluetooth applications in cars. These include

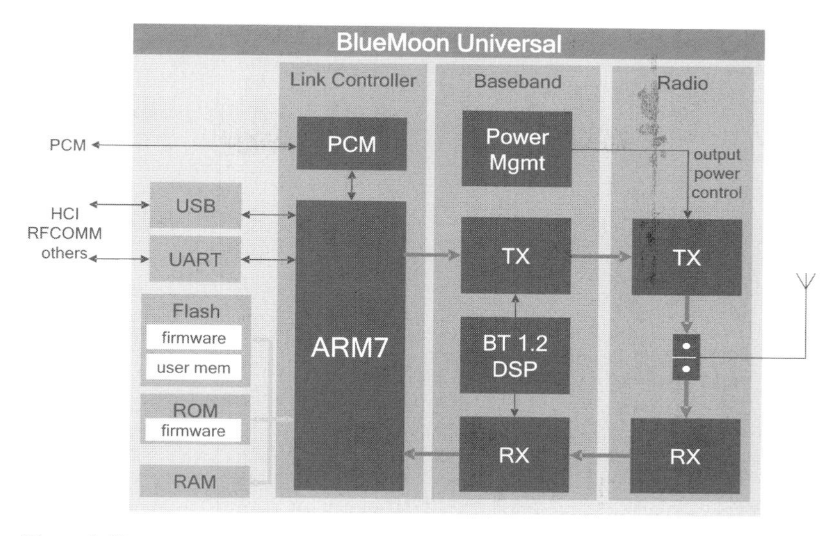

Figure 9.42
Block diagram of a fully-integrated Bluetooth single-chip solution, BlueMoon Universal (PMB8754), with integral baseband and HF functionality

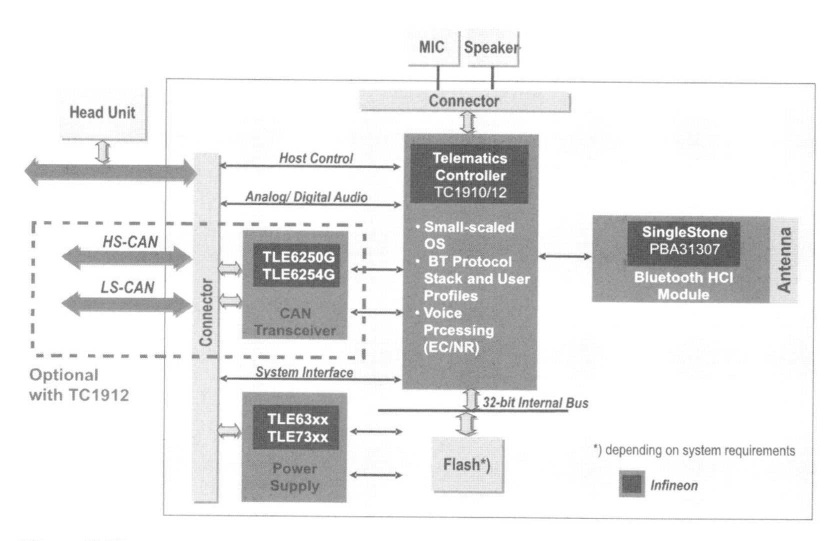

Figure 9.43
Bluetooth communication gateway implementation for use in infotainment systems

the hands-free profile, which supports the implementation of hands-free functions for mobile telephones of an appropriate design, plus the SIM Access and Phone Access profile to support enhanced telephony functions (Figure 9.43).

9.6 The new 42 V vehicle power supply system

Since 1995 there has been increasingly intense discussion about the introduction of a 42V vehicle power system (42 V PowerNet) in international forums. Now that the major technical problems had been largely identified, it has been possible to draw up a usable and internationally agreed draft specification, which gives the system suppliers a certain degree of development confidence.

More controversial than the technical details was the discussion held with the individual automobile manufacturers about the cost-benefit ratio. The first starter-generators with a very simple 42 V vehicle power supply have gone into series production in 2003, and from 2005 the first more complex 42 V/12 V dual-voltage system will appear on the market (Mercedes S Class). However, many projects have been either delayed until 2005 or halted.

As early as 2001, Toyota had already announced a simple hybrid vehicle (THS-M, Toyota Hybrid System Mild), which is equipped with a belt-driven 42 V starter-generator and permits braking energy to be recovered and stored in lead-acid batteries.

Even with this modest equipment it is becoming clear that the general introduction and spread of a 42 V vehicle power supply system can no longer be halted.

9.6.1 Definition of the terms 12 V and 42 V

In today's vehicle power supply systems, the value of the operating voltage is specified as 12 V. This value refers to the open-circuit voltage of the lead-acid batteries which are almost exclusively used, and which consist of six battery cells each with a cell voltage of 2 V. These batteries permit a very simple charging process, in that when the combustion engine is running a constant charging voltage is applied to the battery. At room temperature, this charging voltage is normally approximately 13.8 V. Because of this fact, one speaks of a 14 V vehicle power supply. The terms 12 V and 14 V vehicle power supply thus refer to the same system, but in the two different operating states of open-circuit (vehicle switched off) and charging (vehicle engine running).

In defining the 42 V vehicle power supply on the other hand, reference is made only to the voltage level for the vehicle in nominal operation, that is to say in charging mode at room temperature. The open-circuit voltage has deliberately not been used in its name, because the definition is intended to remain independent of the type of battery used. If one were to construct a 42 V vehicle power supply using lead batteries, the open-circuit voltage would presumably be 36 V, while with other types of battery such as nickel-cadmium the open-circuit voltage would be different. Nevertheless, regardless of the battery used we speak of the actual charging voltage as 42 V.

For today's vehicle power supply, the term 12 V power supply has become established, so that when comparing the two supply systems we speak of a 12 V or 42 V vehicle power supply respectively. However, when the charging voltages are being compared, the two systems are also often designated as 14 V and 42 V vehicle power supplies. When the comparison is made in these terms, it is clear that the new 42 V charging voltage is exactly three times as large as today's 14 V charging voltage. The term 36 V vehicle power supply should not be used.

9.6.2 The 42 V PowerNet for new solution approaches

Time and again, the central question in all discussions about the 42 V vehicle power supply is "why".

- Lower currents	e.g. a factor of 3
- Reduction of power semiconductor costs	e.g. 20% chip size
- Cable cross section reduction	e.g. 12kg to 6kg mass
- Efficiency increase Alternator, distribution, switching	e.g. from 40% to 85%
- Cost reduction due to new specifications Overvoltage, reverse battery load-dump, jump-start,	e.g. -12V to -2V
- New power application can be realized	e.g. EVT, Mild Hybrid
- Enables reduction of fuel and emission	e.g. start/stop, recup.
- Enables electrification of accessory drives	e.g. hydraulic pumps

Figure 9.44 Advantages from the use of the new 42 V PowerNet

Figure 9.44 shows a few reasons for the use of 42 V. In the first place, the higher voltages reduce the current levels. This makes it possible to reduce the power semiconductors down to a smaller chip area, or for the average cable harness to be made lighter. In addition, the changed voltage also permits significant increases in the efficiency of the electrical power generation, distribution and switching.

In addition, it is only as a result of the reduced currents that certain new power applications, such as electromagnetic valve operation (EVT) or a so-called mild-hybrid system with a starter-generator, become realizable and economic. Provided that these conditions are met, then a start/stop function for the internal combustion engine combined with a low-emission engine start and the recovery of braking energy provide the best foundations for a reduction in the average emissions and gasoline consumption. The ongoing electrification of auxiliary units which are today either mechanical or mechanical/hydraulic, combined with control systems appropriate for the consumption, will enable this effect on the savings of energy and fuel to be continued further yet.

Similar remarks apply to the air-conditioning compressor, ABS system, fuel pump or water pump. On the other hand, more far-reaching considerations are involved in saving up to 5% on fuel usage by thermally optimal operation of the combustion engine. The key to success in the case of engines operated in a thermally optimal manner is again the electrified auxiliary units such as the cooling fan, but also electrical analog valves in place of the classical coolant thermostat. The basis of this thermally optimized management will be the 42 V vehicle power supply.

This 42 V vehicle power supply will still generally be used in conjunction with classical vehicles, with combustion engines and starter-generators; it is common to speak of these as mild-hybrid or soft-hybrid vehicles (see Table 9.5).

As the first step, the 42 V power supply will be used in these vehicles for the start/stop function, boost function and the function of recovering braking energy, with the remaining vehicle power supply continuing at 14 V. In the long term, both in these vehicles and in classical vehicles the only supply will be 42 V. This presumes that 42 V controllers will be available, for example, for electromechanical steering and braking, electric fans, door and seat controllers, but also lighting modules etc. Subject to this assumption, the 42 V vehicle power supply will also make its appearance in other types of vehicle, such as the parallel hybrid, serial-hybrid, fuel cell or purely electrical vehi-

Table 9.5
Oveview of different vehicle types and their operating voltages, today and in the year 2010

Car Type	Standard	Mild Hybrid Toyota Crown, 2001 GM, 2004 ? PSA, 2003 ?	Parallel Hybrid Toyota Prius, '97 Honda Insight, '98 Dodge Durango 03	Serial Hybrid	Fuel Cell DC Necar, 2004 Honda FCX, '03 GM HydroGen1	Electr. Car Honda EV+
Energy	Fuel	Fuel + Battery	Fuel + Battery	Fuel, Propane + Battery	Hydrogen, Methanol, Fuel + Battery	Battery
IC-Engine	100kW	70kW	60kW	40kW		
E-Motor		5kW	30kW	80kW	80kW	80kW
Drive	Engine	Engine + E-Boost + Recuperate	Engine or/and E-Traction + Recuperate	E-Traction + Recuperate	E-Traction + Recuperate	E-Traction + Recuperate
Today						
Powernet:	14V, 1.5kW	14V, 1.5kW	14V, 1.5kW	14V, 1.5kW	14V, 1.5kW	14V, 1.5kW
E-Traction:		42V, 10kW	288V, 30kW	400V, 80kW	400V, 80kW	400V, 80kW
2010						
Powernet:	(14V), 42V, 5kW	(14V), 42V, 5kW	(14V), 42V, 5kW	(14V), 42V, 5kW	(14V), 42V, 5kW	(14V), 42V, 5kW
E-Traction:		42V, 10kW	288V?, 30kW	400V, 80kW	400V, 80kW	400V, 80kW

cle, because in principle the same convenience and safety functions must also be fulfilled in these vehicles. However, as powers of greater than 30 kW are generally required in these vehicles to operate their electrical drives, a significantly higher voltage must be chosen for these electrical drive units, generally somewhere in the range between 200 V and 400 V. The introduction of a 42 V vehicle power supply will make no difference here.

The so-called environmental policy obligations accepted by the European automobile manufacturers under the ACEA organization (Association des Constructeurs Européens d'Automobiles) call for a step-by-step reduction of 25% in the average CO_2 emissions from new vehicles by the year 2008, down to 140 g/km, starting from 186 g/km in 1995. As a further objective for 2012, the aim is to achieve 120 g/km. Experts are agreed that future reductions can only be achieved by drastic measures such as electrification and the steady conversion of high-power auxiliary units to 42 V. In the USA, similar values are called for by the CAFÉ program (Corporate Average Fuel Economy). This program for gasoline saving, which has been in existence since 1975, today prescribes an average gasoline consumption of 27.5 mpg (miles per gallon) for new cars, and 20.7 mpg for new off-road vehicles and so-called SUVs (sport utility vehicles).

In the long term, the 14 V operating voltage will disappear from all vehicles types. This is the only way to ensure that common parts are used and that the production volumes of the individual components are increased.

The same considerations apply to the use of the 24 V vehicle power supply in the European truck market and the 12 V supply in the American truck market. From about 2010, the 42 V vehicle power supply will be used in all newly-developed commercial vehicles.

9.6.3 42 V and its effect on power semiconductors

In today's 12 V vehicle power supply, power semiconductors require a withstand voltage significantly higher than that. To achieve it, protected power semiconductors commonly have active Zener protection in the range from 45 V to 60 V. In the case of semiconductors for a 42 V vehicle power supply, this active Zener

application	*supply*	*V_{AZ}*
12V automotive el. power net power switching	generator	>45/60V
42V automotive el. power net starter/generator, power switching	generator	>60/70V
24V truck el. power net power switching	generator	>65V
80V e.g. local high voltage fuel direct injection	DC/DC conv.	>80V
60-80V active zener clamp fast inductance de-excitation	--	>60/80V
12/24/48V industry application power switching	power supply	>65V

Figure 9.45
Voltage classes for various automotive and industrial applications

protection must be at between 60 V and 70 V.

Figure 9.45 shows this relationship, together with the Zener voltages commonly applying for other automotive and industrial applications. In the case of industrial applications or 24 V truck applications the need is commonly for a 65 V Zener voltage, while for rapid demagnetization of inductances the Zener voltage desired and used is often in the range 60 V to 80 V.

It can be clearly seen from Figure 9.45 that the requirements for 42 V semiconductors match today's existing applications very well, and indeed it is possible here to determine uniform requirements to be met.

The upper part of Figure 9.46 shows the voltage levels, as planned in the current proposal for the DIN and ISO norm. Nominal operation ranges from 30 V to 48 V. Superimposed on the effective upper limit of 48 V_{rms} there may be a generator ripple, giving a peak value of at most 50 V. In addition, an overvoltage up to 58 V may occur for a maximum of 400 ms.

Voltage dips to 18 V for 15 ms and to 21 V for a maximum of 20 s are specified by a special start profile. Central polarity reversal protection limits the load for electronic systems to -2 V for a maximum

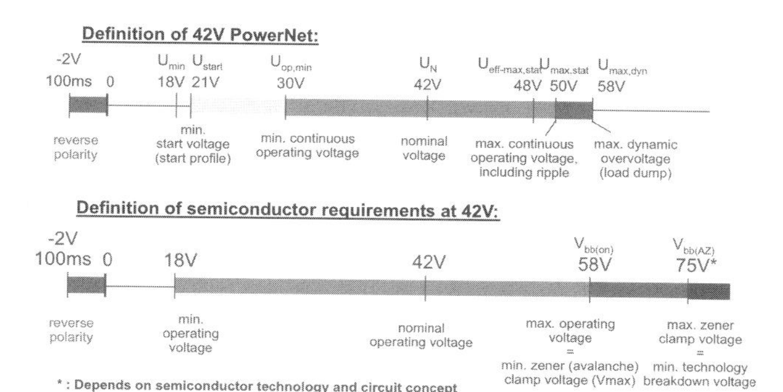

Figure 9.46
Voltage ranges for the 42V PowerNet, and the resulting voltage requirements for 42 V power semiconductors

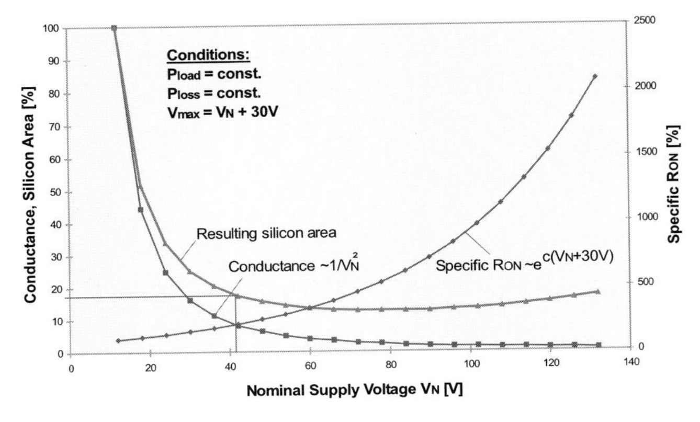

Figure 9.47
Drastic reduction in the chip area for power semiconductors when using the 42 V PowerNet (same operating conditions as for 14 V)

of 100 ms if a connection is made with the wrong polarity.

It is convenient in this connection that intelligent power semiconductors have a nominal operating range of 18 V to 58 V. Protective functions for short-term voltage peaks in the µs region thus lie above 58 V. In order to achieve this performance, Infineon has chosen breakdown voltages of 75 V to 90 V for the various semiconductor technologies (cf. also Figure 9.51). In the event of polarity reversal, a semiconductor module should withstand –2 V for 100 ms with no supplementary circuitry.

It has already been mentioned that, because of the lower currents, the chip areas for the semiconductors are reduced, and hence also their costs.

Figure 9.47 shows the chip area for a power semiconductor for different operating voltages, assuming that the power in the load and the power loss in the switch are the same. The chip area is proportional to the conductivity of the semiconductor switch multiplied by its area-specific on-state resistance. This area-specific on-state resistance is here dependent not only on the nominal operating voltage, but also on the withstand voltage for possible overvoltages. An overvoltage robustness of 30 V has been assumed in Figure 9.47.

The chip area thus drops very dramatically as the operating voltage rises, due to the higher permissible conductivity of the switch, after which the graph changes to a flatter and slowly rising curve. By comparison with 14 V, the required chip area at 42 V is down to a mere 20% from this point of view.

Under constant conditions, therefore, power switches in a 42 V vehicle power supply system can have a significantly smaller chip area. These smaller chip areas can then also be mounted in smaller packages (Figure 9.48).

This shows a comparison of a switch for operating a load of 280 W at 14 V or at 42 V, e.g. an electrically heated rear screen. At 14 V, this function can be performed by a switch with a resistance of 2.9 mΩ in a TO-218 package, while by comparison at 42 V a mere 18 mΩ in a D-Pak package is sufficient, and indeed with reduced power loss. Understandably, the smaller chip area with the smaller package and reduced power loss in this example result in drastic cost advantages.

Figure 9.49 shows this cost advantage in somewhat more detail. On the basis that

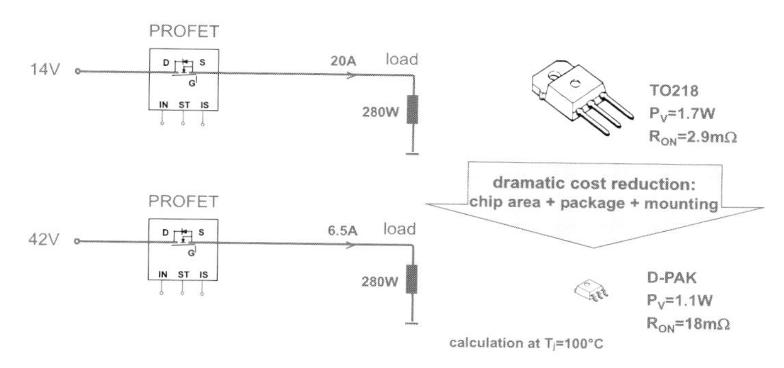

Figure 9.48
Smaller chips in smaller packages reduce the total costs of power semiconductors

the chip costs at 14 V are 100%, cost advantages in the system can be realized in various ways. If the route of chip cost optimization is used, the chip area can be reduced to about 20%, with the power loss in the switch being unaffected by this, and the on-state resistance is increased by a factor of 9. The cost advantage results from the reductions in chip and package costs.

On the other hand the chip area, and with it to a first approximation the chip costs, can be kept unchanged. This increases the on-state resistance by a factor of only 1.4, but the power loss is reduced to 16%. There is a cost advantage here from the drastically reduced cooling requirements. Thus, depending on the point of view and application it is possible to achieve greater cost optimization either by reducing chip costs or by reducing the cooling provisions. By way of example, this is shown for the power stages of an electromechanical steering system EPS. A solution at 14 V using six power MOS transistors at 4.5 mΩ in the TO220 package with a total power loss of 300 W can now, at 42 V, be represented either with six devices at 6.5 mΩ in the TO220 package with 50 W power loss, or alternatively using six times at 20 mΩ in the D-Pak package with a power loss of 150 W.

Figure 9.50 shows a complete electrical system. It can be clearly seen that at 42 V the output stages can be adjusted as required in terms of size and power loss. On

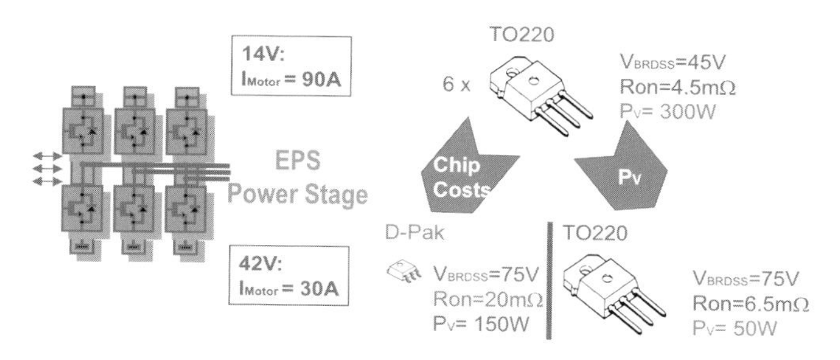

Figure 9.49
Cost advantages for the typical application of an electromechanical steering system EPS

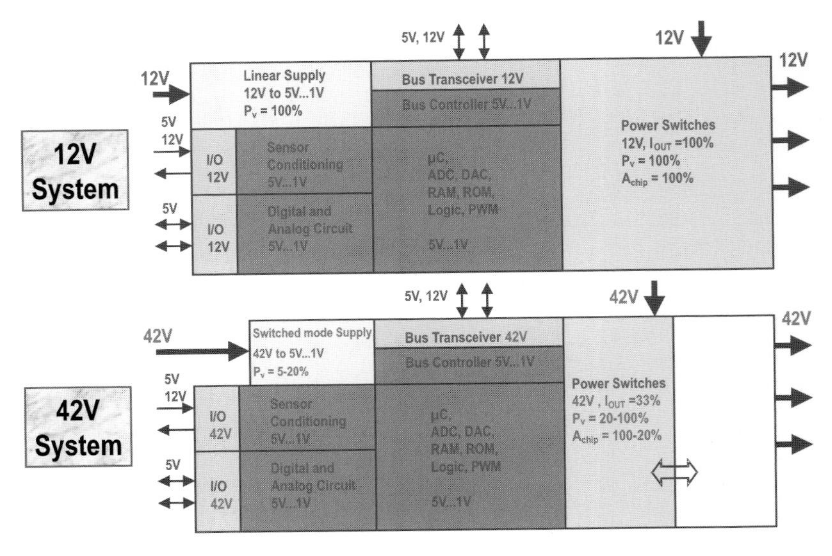

Figure 9.50
The changes in principle when an electronic system is changed from 12 V to 42 V

the other hand all the other subsystems, such as for example any microcontrollers and memory which are operated at 5 V or a lower voltage, are unchanged.

Only the power supply for these systems needs to be changed. Because of their high power losses, linear voltage regulators can no longer be used. The switched mode power supplies which must be used offer the advantage of a clearly reduced power loss (increased efficiency) in the voltage regulator, although this must be paid for by a more expensive inductance and the cost of the filtering (cf. also Figure 9.53). In addition, throughout the system all the input and output channels for sensor connection, transmission links or communication lines, such as CAN or LIN, must be designed to be short-circuit resistant at 42 V. This is technically feasible and has no significant effect on the total costs, because only the output stages concerned need to be modified for the higher voltage. The new communication modules could, however, then be used for any remaining 12 V systems.

The technologies which Infineon Technologies plans to use in 42 V vehicle power supply systems are shown in Figure 9.51. MOSFETs have been realized for use in starter-generators or in DC/DC converters, using the OptiMOS technology with a 75 V breakdown voltage. On the right of the diagram are shown the performance and availability of each of the products.

Infineon is the only manufacturer in the world which today already offers a wide spectrum of 42V compatible products.

Intelligent power switches, on the other hand, have already been implemented with an 80 V breakdown voltage, using the S-Smart technology. Products incorporating this type of technology have been in use in 24 V truck applications for over 10 years.

Finally, using the SPT4/90V BCD technology it is possible to realize more complex functions. Examples which can be cited here are system ICs for direct gaso-

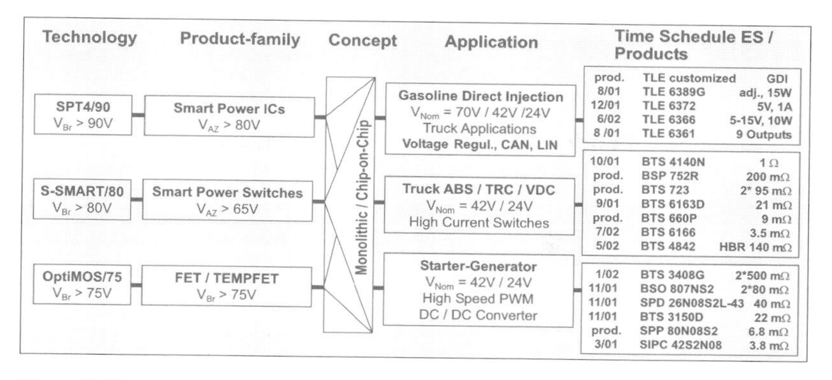

Figure 9.51
Selection of the power semiconductors presently available for 42 V and their technologies

line injection, communication modules and voltage regulators.

With the help of the three basic technologies mentioned, it is possible to provide for all the applications in a 42 V vehicle power supply system. On cost grounds, these technologies are often combined in so-called chip-on-chip modules, in one package.

Quiescent current requirements and the power supply in the 42V PowerNet

In 12 V systems, the quiescent current requirements when the vehicle is switched off play a very important role in ensuring that the battery remains capable of starting the vehicle after several weeks. In the future 42 V systems, the quiescent current problem will be aggravated for two reasons, see Figure 9.52: assuming a battery of the same size and a fixed energy at 14 V and at 42 V, then at 12 V this might require, say, a 66 Ah battery, while at 36 V the corresponding battery would be a 22 Ah one. Presently, allowance is made for a quiescent current of, say, 18 mA for the entire vehicle at 12 V. This represents about 300 µA per electronic module. At 42 V the corresponding figures are 6 mA per vehicle and 100 µA, or even significantly less, per electronic module. The quiescent current requirements are thus more demanding by a factor of at least three.

The second reason which aggravates matters is that in future there will be ever more consumers with separate semicon-

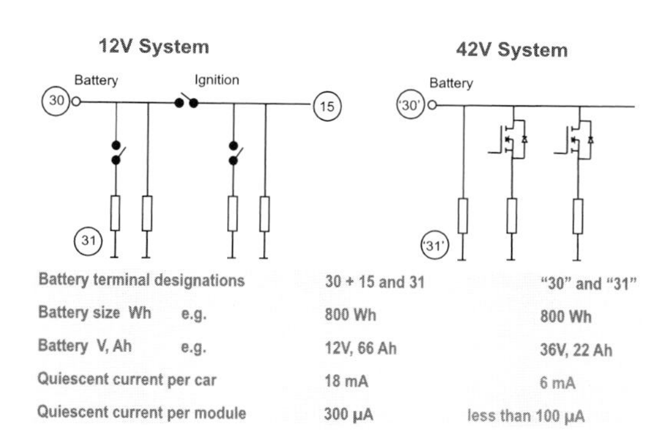

Figure 9.52
Illustrating the quiescent current problems in future 42 V systems

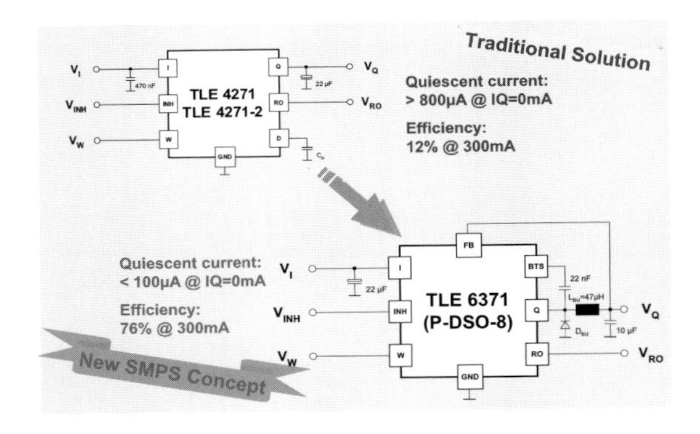

Figure 9.53
Proposal for a universal switched mode voltage regulator in a 42 V PowerNet

ductor switches which are operated directly from the battery (terminal 30). The classical ignition key relay (terminal 15) is more and more losing its importance, its switch-off function being realized by the existing electrical switch and a software function. This implies a substantial increase in the potential quiescent current consumers. In plain English, this means that power semiconductors for use with a 42 V battery must be optimized in terms of their quiescent current. The same applies too for the voltage regulator. Infineon will meet this challenge with various technical solutions.

Linear voltage regulators will hardly have any further use in a 42 V vehicle power system because of their high power losses. The changeover of the logic voltage from 5 V to, say, 2.5 V particularly intensifies this problem. Thus, a 2.5 V 200 mA linear voltage regulator with an input voltage of 50 V would dissipate 9.5 W. Switched mode voltage regulators perform this function with a power loss of a mere 200 mW, for example, which is significantly less than with today's 12 V supply. This very low power loss opens up new possibilities for the integration of voltage regulators into integrated system ICs.

Figure 9.53 shows a proposal for a switched mode voltage regulator for a 42 V vehicle power supply with the functions of the well known standard regulator TLE 4271. The particular development objectives of this voltage regulator are its minimized quiescent current, high efficiency in nominal mode, and low interference operation with minimal filtering costs.

Of the use of switching regulators in a 42 V vehicle power system it can be said in general that, by comparison with a linear regulator at 14 V, the semiconductor costs tend to drop or remain the same, but there is a certain additional cost due to the use of a coil and additional filtering components. In many cases it is certainly possible to compensate for this by a efficient design with lower power loss and fewer heat dissipation problems.

The problems of short circuits in dual-voltage vehicle power supply systems

One of the major challenges in a dual-voltage vehicle power supply is the problem of short circuits between 42 V and 14 V. To anticipate a little, a hundred-percent solution to this problem using simple means is not presently in sight.

The possible short-circuit paths in such a system are shown in Figure 9.54. It is virtually impossible to exclude a short circuit directly between the batteries (1,I), and this can only be prevented by constructional measures. The same applies to the short-circuit path 2,II, even if in this case the intelligent (smart) battery termi-

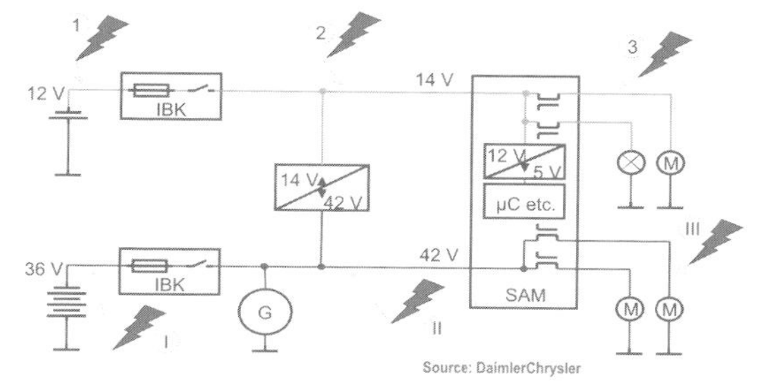

Figure 9.54
Possible short-circuit paths in a dual-voltage vehicle power system

nal SBT can ensure electrical protection under certain conditions.

The most frequent short circuit which will occur is at the end consumers at level 3,III, and this should as far as possible be safeguarded electrically. Figure 9.55 shows the problems in detail. In broad terms, a distinction should be made here between case A, with a high-impedance short circuit at, for example, 5 A between 42 V and 14 V, and on the other hand case B with a short-circuit current of 100 A, say.

In case A, the 42 V switch which is designed for a nominal current of 10 A will not recognize any problem. In case A therefore, the short circuit must be recognized in the 14 V controller, and must be withstood without damage until the problem has been eliminated. Initially, the short-circuit current will flow through the 14 V controller via the reverse diode of the load transistor and the 14 V fuse to the 14 V battery. The 14 V fuse may be blown. The problem can be recognized either by the polarity reversal of the 14 load transistor ($V_{DS} < 0$) or from the rise in the 14 V operating voltage, which is then limited by a Zener diode, for example at $V_{bat} > 16$ V. A so-called switch-off search routine could now trace the 42 V switch which is the cause of the problem, and the short-circuit current could be switched off. During this switch-off time, with a duration of a few seconds, the 14 V controller should remain undamaged, although the connected load might be destroyed.

In case B, the conditions are rather clearer. The overcurrent must be recognized by

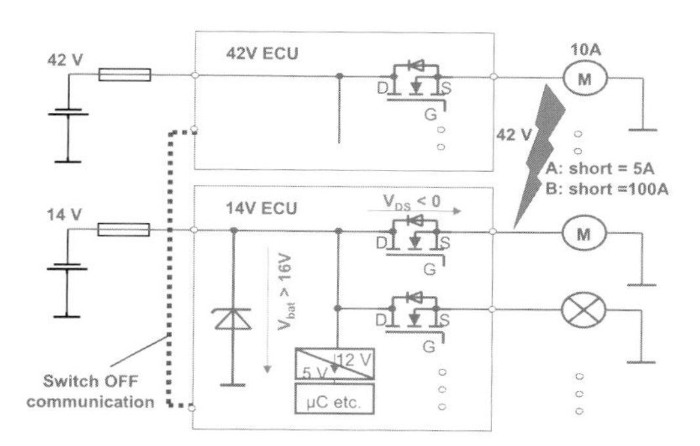

Figure 9.55
The problems and possible solutions for short circuits from 42 V to 12 V

the 42 V controller and switched off within a few µs. Measurements have shown that during this short time even 14 V smart power switches with relatively small and sensitive power switches in SO packages are not damaged.

The handling of short circuits on the borderline between cases A and B are particularly difficult. Here, one hundred percent protection is exceptionally costly or even impossible. It is possible that this problem can be solved with a statistical approach.

The implementation of a 42V vehicle power supply in a car offers, on the one hand, countless opportunities and advantages, but on the other hand there are currently still many detailed problems to be solved.

Outlook

Just as the electrification of vehicles will contribute to the optimization of individual functions, electronics can and must be consistently called on to solve the various detailed problems. Here, it is not always a matter of new additional semiconductor components, but what is required instead is attempts to make use of the available functions of existing components, such as for example PWM, to avoid possible problem situations in a cost-effective way. The future of automotive technology lies in electronics.

9.7 The challenges and opportunities of x-by-wire

The application of x-by-wire as a technology in cars could be of benefit not only for the much-vaunted applications of electric steering and brakes, E-gas or electronic window lifts but also for a host of other applications. Growing demands in terms of functionality, safety and cost are coming up against the limits of physics and development methodology (Figure 9.56).

Various studies have all shown that the increasing individualization and new functions are mainly being realized by electronics and software, and indeed to a certain extent are replacing hydraulic or mechanical linkages in vehicles. With the growing proportion of electronics in cars, it is inevitable that the complexity of the systems is increasing, as too in the long term are the safety requirements to be met by the overall electronic system.

9.7.1 System and design requirements

The driving forces when it comes to the vehicle requirements are:

- environmental compatibility and legal requirements,
- increasing traffic safety,

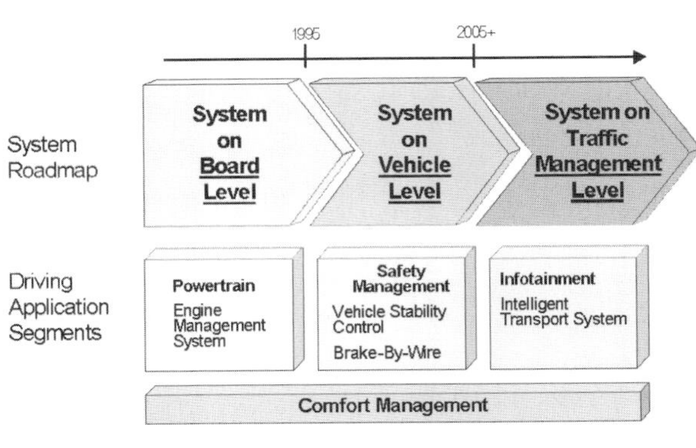

Figure 9.56
How the system perspectives are changing for driver applications

- enjoyment of travel and manufacturer differentiation,
- cost savings in the manufacturing chain.

Environmental compatibility is increasingly being considered in terms of the total energy balance over the life cycle of the vehicle, including its manufacture and final disposal. Safety is no longer defined mainly by the risk of injury for the occupants of the vehicle, but to an increasing extent also by accident avoidance. Enjoyment of travel is a significant feature in differentiating between different vehicles, as also are customer-specific and individually adapted equipment features. In a similar vein to the characteristics mentioned, costs are also increasingly being assessed not in isolation for individual components or subsystems, but taking into account also manufacturing factors, maintenance costs and recycling.

Hardware and software design must be scalable and, in spite of being cost-optimized as far as possible, must incorporate adequate design flexibility in respect of choices such as the specific sensors or actuators used in the system. Reusability is here only one of many criteria.

Reliable solutions for fault recognition, fault limitation and fault handling are indispensable. There remains the challenge of finding the right balance between costs and benefits, and the timing and extent of fault handling and informing the user.

In summary, for systems which have the same functional definition, the safety, quality, reliability and durability must remain at least the same as for a conventional mechanical solution, or if possible be improved. The same applies for the overall costs, which must under no circumstances increase for the same functionality. New features will only take hold in a sustained way if they bring visible advantages for the customer, in terms of functionality and/or cost.

9.7.2 The possibilities of x-by-wire

From a system perspective, the first point about x-by-wire is that it decouples the driver's wishes from the actual actuators. The mechanical or hydraulic link between the operating elements and the throttle valve or brake is broken, and replaced by electronics. From the point of view of system architecture, x-by-wire allows information, current (consumption) and sensors, computational capacity and actuators to be distributed. Items of information can be distributed with regard for real time performance and safety (redundancy). Current consumption can be distributed, and the line losses and electromagnetic emissions reduced.

Examples of x-by-wire applications

A number of examples from the various established application segments in vehicles will be used to illustrate the possible benefits and advances when x-by-wire is introduced. Classically, vehicles and their architecture have been dominated by mechanical linkages. Correspondingly, the exterior of the vehicle is determined by the energy converters used, and is constructed around them. Logically, the engine and power transmission are central to the design (powertrain). The wheels and suspension transmit the drive energy to the road, and are therefore have a critical role in the vehicle's stability and hence its safety (safety & vehicle dynamics). Convenience features and miscellaneous elementary functions such as lighting and visibility are combined in the body and convenience application segment. Functions for information, communication and entertainment belong under infotainment.

The driving motivation in the powertrain segment is the harmonization of engine control and the drive train to minimize emissions while at the same time giving the ideal responsiveness, which promises driving pleasure. The "by-wire" application fields within the powertrain segment

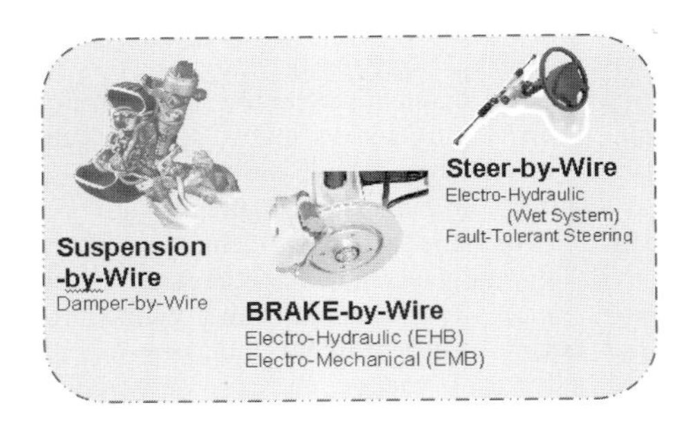

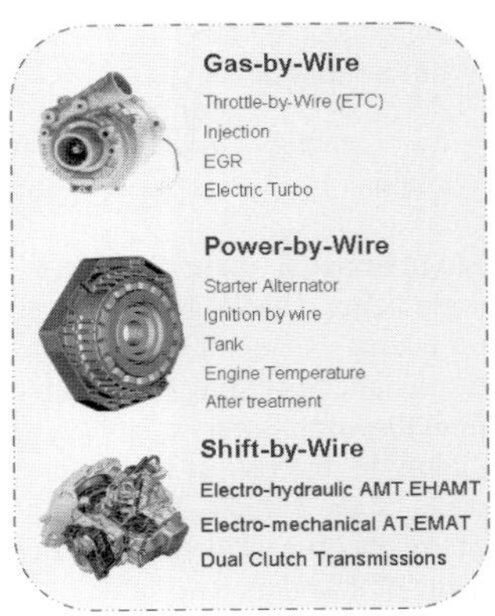

Figure 9.57
Application fields within the powertrain and safety & vehicle dynamics segments

can be grouped into gas-by-wire, power-by-wire and shift-by-wire (Figure 9.57).

The gas-by-wire group includes such individual functions as electronic throttle valve, electronic fuel injection, exhaust gas feedback and an electrical turbocharger. The advantages here lie in improved control of the combustion mixture. In order to reduce the CO_2 emissions, one makes use of engine downsizing, e.g. by using exhaust gas feedback systems, which enable dynamics to be improved by an electrical turbocharger or booster. In addition, the losses in the inlet manifold can be reduced by more flexible valve control, e.g. by electro-mechanical, electro-hydraulic or electro-magnetic valve operation.

Power-by-wire brings together a range of engine functions, such as engine temperature management, tank management, ignition, exhaust gas reprocessing and so on. To achieve further reductions in the CO_2 output, new concepts are appearing, such as hybrid drives, with first implementations in the form of starter-generators. These enable such functions as stop-go, regenerative braking and such like to be realized.

Shift-by-wire includes the various forms of automated gearboxes, such as electro-hydraulic automatic indexing gearboxes, electromechanical automatic indexing gearboxes, continuously variable gearboxes and dual clutches and all-wheel drives. This allows the gear ratios to be better matched to the engine controller, and hence the CO_2 exhaust gases to be reduced as well as increasing driving enjoyment.

In the safety & vehicle dynamics segment too, the focus of interest is on unifying and combining the individual systems for braking, steering and damping, in order to maximize vehicle stability and road contact by appropriate behavior. In this segment, the central applications for "by-wire" are steering, braking and shock absorption (dampers).

Particularly in relation to electrical braking, there is a host of additional motivating factors from the fields of safety, environmental compatibility, convenience and cost. Just a few of these will be mentioned here: electrical actuators on the brakes permit precise control of the braking force, improved response times, drying of wet brake disks, and fade checking, for example for lengthy braking sessions when descending hills. In this field again, the introduction of x-by-wire will not be overnight, but in steps. In general, the path from traditional, hydraulic or mechanical systems goes via systems which are electrically influenced (including ABS, ESP, emergency stop), through hybrid forms down to genuine by-wire systems, in which there is no longer any mechanical link between the driver and the actuator. Hybrid forms in the case of the brakes could, for example, be hydraulic backup variants or hybrid approaches in which, to start with, only one axle is braked electrically.

Dynamic shock absorber adjustment and active level control permit modifications which depend on the driving conditions, for the purposes of reducing consumption levels and improving road contact. In combination with appropriate regulation strategies for braking, steering and shock absorption, this can make a good contribution to vehicle stability.

In general, steering is regarded as a safety-critical application, for which it is difficult to realize even a mechanical backup solution,.

The main interest in the body & convenience application segment (HVAC-by-wire, wiper-by-wire, lighting-by-wire, door-by-wire) consists mainly in the local distribution of the functionality, where the cost sensitivity is very high but the requirements in terms of miniaturization and safety are many times less demanding. Clever distribution of the subfunctions must be used to minimize the overall costs and wiring, optimize the scalability and limit the effects of local failures, e.g. due to minor accidents. For example an collision causing local damage should not result in a total failure of all the lights for the entire vehicle. On the other hand it is undesirable, for example, to integrate the electronics for a headlight into the latter, because this could drive the repair costs sky high, and with them the insurance premia.

An example of the beneficial use of by-wire technology might be the mechatronic integration of air-conditioning flap actuators by using simple bus systems, to maximize scalability and design flexibility in terms of their number and location, and to minimize the wiring costs. Similarly, replacement of the mechanical synchronizing links between the two front windscreen wipers by an electrical link can give flexibility in the design of the vehicle.

9.7.3 Semiconductor concepts for x-by-wire systems

Gateway/backbone network architecture

Nowadays, most vehicles use a central gateway topology to link together the vehicle's different subsystems. Gateways permit the simple interchange of data between the individual segments, with the option for filtering, conversion and encapsulation of the subsystems. On the other hand, gateways are also relatively expensive, because they provide something which is necessary but brings no visible added value, and especially when messages are being exchanged via several gateways they make network planning more difficult and restrict transparency. This is even more true if an existing network needs to be extended. The propagation of what appear initially to be local effects throughout the entire system can often only be determined with difficulty when changes are made, and in many cases calls for an expensive revalidation of the overall system. On the other hand, the alternative topology of a common backbone has certain advantages provided that certain prerequisites are satisfied (adequately fast and secure communication, with transparent and deterministic behavior with respect to the application software which uses it, well-defined and standardized programming interfaces (APIs), transparent and hardware-/resource-independent software functions). The separation of the data transport from the functional design permits better testability and reusability, and hence additional flexibility in the representation of the available resources.

Intelligent partitioning / embedded power

In mapping functions onto technologies, technologies onto chips and chips into packages, there are basically three main families of solutions: monolithic single-chip solutions, multi-chip approaches and discrete circuit designs.

Intelligent I/O interfaces / companion ICs

Modern central controller architectures call for the use of components which are optimally matched to the application, to keep down their numbers and the overall chipset costs. Costs can be further reduced if it is possible to use standard products (ASSPs) matched to the application, instead of application-specific ICs (ASIC), developed specially for the individual situation.

Inter-chip communication, multi-processor, multi-chip links

Controllers continue to be under pressure for reduced development times and for cost efficiency. By comparison with ASICs, ASSP are generic and hence permit lower costs combined with better quality and reliability by their higher production volumes. However, a prerequisite for the success of this ASSP approach is their linkage to the rest of the system. One of today's common interfaces, the SPI (Serial Peripheral Interface) is restricted to some five megabits per second, whereas the future requirements tend to lie in the region of 30 Mbit/s. A new interface from Infineon permits processors or other memory-based modules to be linked via a fast serial interface. In conjunction with the architecture of the Tri-Core-based powertrain controller, resources can be cost-efficiently exchanged via the Micro-Link-Interface (MLI), and different modules linked to each other (Figure 9.58).

On-chip communication

The prerequisites for good partitioning are clearly defined interfaces and separation of the analysis of data and the control flow in the system. Consequently, Infineon has provided a hierarchical system structure in its 32-bit microcontrollers, which permits the application software, the status/network management and the peripheral level to be separated from each other and hierarchically arranged. For the application software, the TriCore 32-bit RISC DSP processor is available. Typical tasks in network management, status and diagnosis or peripheral interactions can optionally be handled either by DMA (Direct Memory Access) or by a second independent 32-bit peripheral processor (PCP). At this level too, the AUDO-NG

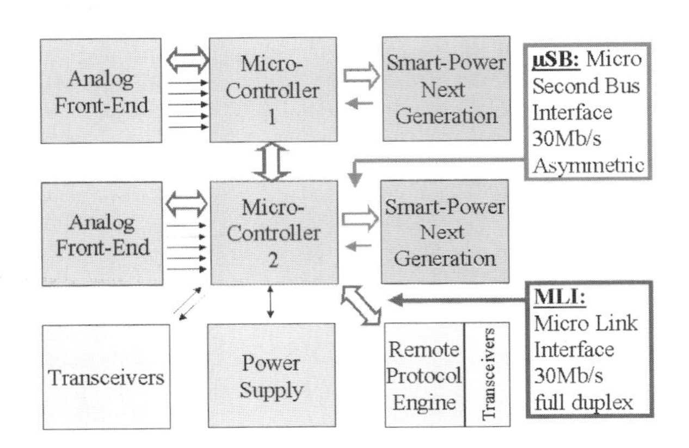

Figure 9.58
New communication paths within controllers

family of microcontrollers permits scalability in respect of the requirements for incoming data volumes, with various levels of expansion in terms PCP and DMA, and appropriate bus connections. This makes it possible for up to three processes to be executed simultaneously within one microcontroller, in addition to the operations performed by the peripheral modules. Its additional strengths are in the diversity of further communication interfaces, such as SPI, K-Line (ISO9141), LIN, CAN, TTCAN, MLI, MSB, and other points to be highlighted are the fast analog-digital converters and freely-configurable timers.

9.8 The future of automobile electronics

In spite of the enormous challenges of automobile electronics, this field is fiercely fought over. The reasons are firstly the relatively high production volumes, worldwide around 55 million cars are manufactured annually, and secondly the relatively constant and crisis-resistant growth.

Electrification and electronification will undoubtedly continue. Development must be aimed at finding a definitive concept for central and local software and hardware in vehicles, to ensure the problem-free interaction of all the components. Without such concepts, improvements in cost and functionality will in future only be achieved with difficulty.

Standardization could be the key to success here. Thus it would be entirely desirable for established applications, for example door or airbag electronics, to be realized using standardized or uniform software and hardware components. This approach would substantially increase the certainty and speed of development. Production volumes would increase, while at the same time the product variety would be reduced.

The pressure for unification or standardization of hardware components would redirect the trend in the case of most semiconductor components away from standard devices and customer-specific ASIC devices to ASSPs specially tailored for a particular application. In future, simulation of developments will have a higher importance, in order to improve both the development time and also the quality of development.

The automobile industry is subject to constant change. Alternative concepts such as electric drives (electric propulsive energy from batteries), hybrid drives (driven by combustion engines and/or electric motors) or fuel-cell drives (electrical propulsive energy from hydrogen or methanol) are being developed at great cost, and today show great potential. But the challenges of automobile electronics will not change significantly from the use of these alternative drive technologies. The application areas of body, safety and infotainment electronics will hardly be affected at all, while only in powertrain electronics will there be small changes.

With the changes in automobile electronics, there are still many detailed problems to be solved. The rapid changes in technology mean that it will very soon become impossible to make replacements for the semiconductor components used and it will be difficult to adapt the new semiconductor solutions for the old applications. One possible and pragmatic solution could be the recycling of used ECUs from scrap vehicles.

10 Entertainment electronics

10.1 Broadband communication take-offs

The digitalization of broadcasting and television (whether it be cable, satellite or terrestrial), not only brings a significant number of additional channels with better quality to private households, but rather also provides online access to global services, such as the Internet or regional information channels and other interactive services with high transmission speeds (Figure 10.1).

The digitalization of consumer networks and devices is picking up speed and quickly moving via digital television and interactive set-top boxes to incorporate multi-purpose and high-performance broadband communication.

With digitalization, however, not just a greater number of radio and television channels can be broadcast than was previously possible, but data can also be transmitted and better picture quality is provided. The future LANs to which households will be connected open up an immense field for new data services, which range from regional information (e.g. weather, travel, sport and community affairs) to Internet access and telephone communication via TV cable and video conferencing, extending on through to high-speed Internet accesses and telephony and video-telephony, all of which can be implemented in broadband cable networks.

A specific interactivity potential will be indispensable for the majority of these new services. The cable network serves here not just as a medium for the rapid downstream to the customer, but rather it also transfers the upstream in exactly the same way, that is, the feedbacks to the service provider. Currently, however, only

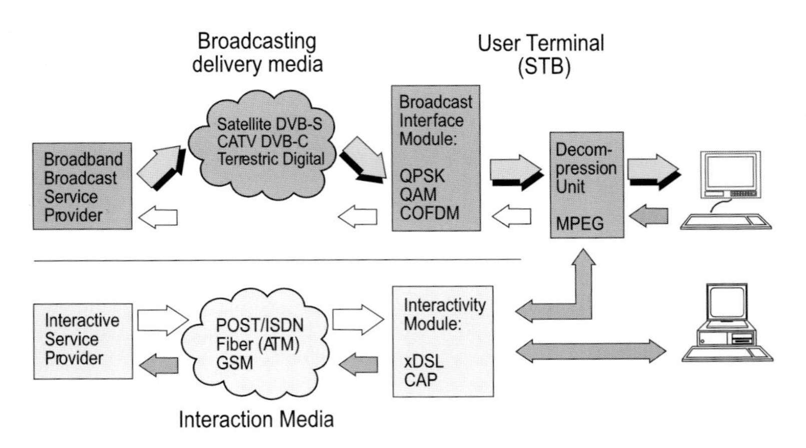

Figure 10.1
More than just television – Digital television opens up broadband communication for private households

a small part of the cable networks is "return channel-capable", a technical upgrading is all but inevitable for using interactive services. At the moment there are two competing procedures available in this respect: MCNS and DVB C-RC. Alternatively, the public telephone network can also be used for feedback via analog modems or optionally via ISDN. This is the standard return channel procedure with digital satellite broadcasting (DSB – Digital Satellite Broadcast), which can also be used in the future with digital terrestrial broadcasting (DTTV – Digital Terrestrial Television).

As the interactive potential of digital cable television will soon also enable telephone communication per cable, the digitalization of the cable signifies competition for the public telephone network, because even the increased bandwidth of ISDN in the long term will no longer be able to keep up with the greater performance capacity of the TV cable. For this reason, operators of conventional telephone networks are setting their sights on technologies like DSL (= Digital Subscriber Line) and HFC (Hybrid Fiber Coax) so that increased bandwidths will help them to remain competitive.

10.1.1 Digitalization of cable television

Cable television networks must be digitalized for two reasons:

1. Competition from digital satellite television: Digital satellite television is a market runner and has become so popular in the USA that substantial numbers of subscribers have been terminating their cable television connection and changing over to satellite television. Cable television operators can only retain or win back their customers if they provide a similar number of channels and additional services.
2. The limited bandwidth: Today's analog cable networks can transmit approx. 30 television channels. Bearing the current technology and consumer expectations in mind, this is clearly not sufficient. The easiest way to sharply increase the capacity of cable TV is to digitalize it.

A conventional analog cable television network with coaxial cables transmits 25 to 30 channels with a bandwidth of 6 to 8 MHz respectively. When digitalized, each channel can deliver 10 to 30 MBit/s to the user and can send 3 to 10 MBit/s in the reverse direction (return channel – Figure 10.2). Approx. 5 - 6 digitalized television channels could consequently be transmitted via each cable conduit, which corresponds with a capacity of some 180 television channels. If some of these digital channels are used for data services, the entire available data capacity (number of channels provided × 30 MBit/s) can be used with 'time-sliced' multiple access of the connected households. However, in order to ensure a rapid access to data services, even at peak times, the number of subscribers per network segment can be limited. For this reason, digital cable television networks are divided into sections with their own server. Estimations on the basis of statistics and network usage show that a digital cable conduit can supply 100 to 150 households with an average data rate of 3 MBit/s. The resulting section size is consequently 150 times greater than the number of reserved channels. A return channel designated within the digital cable television network is used collectively by the users according to the same model.

The section servers are connected with one another via HFC networks and have access to Internet service providers, commercial online service providers and also to company systems. The section structure enables a very efficient integration of global and local content.

The transmission technology for digital television cable networks in most countries is QAM (Quadrature Amplitude Modulation) in accordance with the DVB-C Standard or the MCNS Standard in North America With QAM, both the phase and the amplitude are modulated

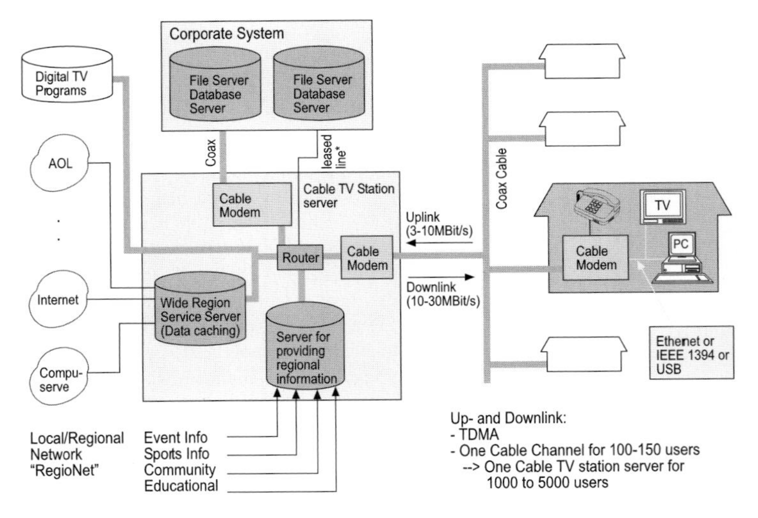

Figure 10.2
Program and information sources with feedback in a digital cable television network

for highly efficient bandwidth usage. Infineon has various 'Superhet Tuner ICs' (TUA 6020, TUA 6030) for using in set-top boxes (STB) for digital television cable networks, which are also still reverse compatible for analog reception.

10.1.2 Digital terrestrial broadcasting is picking up speed

Digital terrestrial TV reception (DTTV) can be considered introduced in the UK, Skandinavia, Singapore and Australia, and this is with the DVB-T Standard. In other European countries as well as in the USA and South Korea (ATSC Standard) and various Asian countries (India, Malaysia and Thailand with DVB-T Standard), and Japan (ISDB-T Standard) the introduction of DTTV is imminent or already underway. In Germany, Berlin has already been completely converted to DVB-T. The regions in northern Germany, such as Hamburg / Lübeck/ Kiel, Hanover / Bremen and Cologne / Düsseldorf and the Ruhr area will follow suit during 2004. With DTTV with a telephone line to the service provider (Figure 10.3), compression algorithms and efficient channel coding will also be used for far more intensive utilization of the available transmission bandwidths. Depending on the modulation technology, simultaneous broadcasting network and even mobile reception can be implemented as additional benefits.

The bandwidth increased as a result of digitalization can be used for higher resolution, additional television channels or new data services. These new services can range from school and university transmissions to community information, on through to services that may include news, advertising, pay-TV and travel information. With digital technology, DTTV would consequently become just as attractive as DSB and public television broadcasters, many of which cannot be received via satellite. With digitalization they could, however, keep up with the growing number of private competitors.

The UK, Skandinavia, Germany, Italy, France, Spain, Australia, Singapore, South Africa, Malaysia and other Europe-

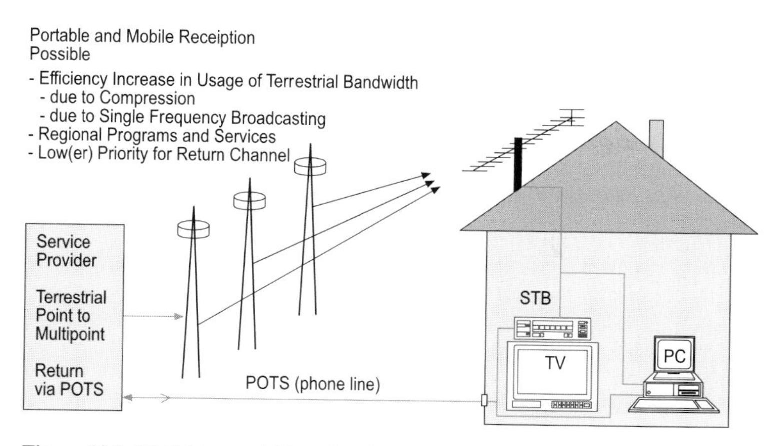

Figure 10.3 Digital terrestrial broadcasting

an countries have decided for the DVB-T standard with DTTV. This brings the use of coherent 'Coded Orthogonal Frequency Division Multiplexing' (COFDM) into operation, which divides the transmitted symbols among a large number of carriers and then transmits for a relatively long period of several milliseconds. This results in the option of the very effective echo suppressor in the receiver, which allows the simultaneous broadcasting network to be set up.

A common frequency network transmits the same channel on the same frequency of several neighboring broadcasters. The receiver takes up the same signal (with phase distortion) of several broadcasters at the same time. The distortions are removed in the receiver using echo suppression, so that a completely self-correcting signal can be decoded.

This principle functions with specific restrictions, even with mobile reception in cars, trains, etc. A restriction is determined by the broadcaster network that has used the analog standard until now. Because of the wide distances between broadcasters/transmitters and the very restricted antenna sizes with the mobile reception, a full-coverage supply with sufficient received field strength is not provided. Additional gap-filling transmitters are therefore required (see below). As neighboring transmitters must no longer use different frequencies, simultaneous broadcasting network provide the advantage of being able to free up a significant number of bandwidths.

An expansion of the current digital standards for full-coverage mobile reception is currently being worked on in various countries. In Europe the standardization work for determining the DVB-H standard (-H for 'Handheld') is just about to be completed. DVB-H is based on DVB-T, as the overall source coding remains the same. Infineon is actively working together with others on the standardization work for DVB-H. There are other activities ongoing within Europa in the 'Digital Video Broadcasting' group (DVB) to complement the use of the convergence between DVB and cellular networks, terminals and platforms. This is a step in the direction of combining the advantages of broadcast ("one-to-many" as cost-effective distribution) and cellular ("one-to-one" with interactivity) with one another. As an essential difference between DVB-H and DVB-T, the data flow per channel will, however, no longer be continuously transferred, but rather in time slots (TD-

MA). Additionally transferred data enables the receiver to connect in cycles only when the service desired (TV program) is being transmitted. With the 'power down' times possible here, in comparison with DVB-T, energy savings of more than 80% are possible at the receiver side. With mobile receivers operated with small batteries, such as mobile phones or PDAs, this is both desirable and necessary in order to receive sufficient operating time and in order to ensure sufficient stand-by time with, for example, mobile telephones. With channel decoders, a memory is additionally required for DVB-T, which re-generates a continuous data stream from the received data packets.

This new transmission standard does not yet change anything with regard to the problematic of the full-coverage supply described further above. DVB-H here uses the available or still existing DVB-T transmitters, which in turn replace the existing analog television transmitters. As a basic requirement for DVB-H, the DVB-H standard must first be transmitted from existing large transmitters. One significant advantage of the simultaneous broadcasting network also used here is that reception gaps caused by gap-filling transmitters of lesser capacity that are independent and that function without any control from control stations can be closed.

Infineon is now producing the currently sole digitally-capable TV tuner, IC TAIFUN, alias TUA6034, which enables the implementation of combi tuners that allow the reception of both digital and analog TV. Infineon also plans to provide solutions for the future DVB-H systems.

10.1.3 Improved feedback for digital satellite broadcasting

Digital satellite television programs have been transmitted since 1994. The first

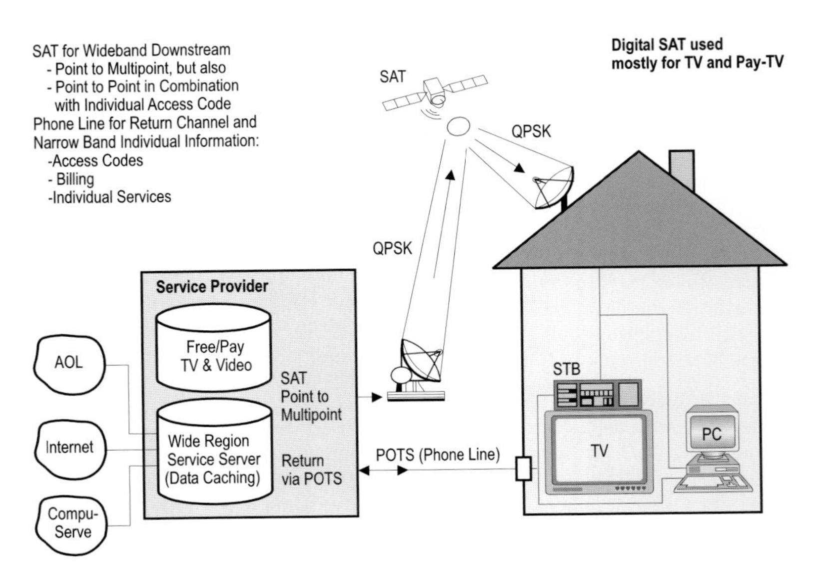

Figure 10.4
Digital satellite television with feedback either directly from orbit or via the telephone network

digital data services were used in the USA in 1996. As digital SAT-TV does not have a return channel, the transmission of television channels will remain its main application. As already mentioned, an interactivity potential for pay-TV, home shopping, home banking and Internet access can only be offered through the use of the standard public telephone network or by ISDN (Figure 10.4).

With this arrangement, a service provider operates a local server with a proxy server at the link station. The proxy server saves the most frequently requested Internet content and minimizes the response times. The server is connected with different information sources, not just with Internet websites, but rather also with other local providers and private companies. The most frequently requested data in the proxy server can even be continuously transferred via satellite (similar to teletext with analog television). Less frequently requested data must be retrieved via the telephone return channel. Depending on the coverage, this data is then transferred to the subscriber directly via satellite or indirectly via the telephone network. For sensitive applications such as home banking or downloading software purchased online, the access control is system-immanent.

A satellite transponder enables a data transfer rate of approx. 38 Mbit/s, which is sufficient for supplying a large number of users with a high average data access time.

A satellite transponder used for digital television can consequently transmit six programs in PAL quality, or even more, should the compression rate be increased, as for example is usual with news broadcasts or cartoons, as these programs do not require the highest picture quality.

With the change over from analog to digital technology, the useable transponder capacity increases six-fold on average. As soon as the supply for households with DSB has reached a critical point, the prospects for the placement of more advertising time alone will have a major influence on the change from analog to digital transponder use. In North America and in Europe, a total of some 1,000 digital channels are already transmitted today via satellite.

DVB-S is used today in most countries as standard for channel modulation with digital satellite television. With DVB-S and other standards, phase shift keying modulation of the 'Quadrature Phase Shift Keying (QPSK)' is used. With regard to the spread in energy and the comprehensive error correction, QPSK is the most suitable modulation technology for DVB-S and its typical interference forms.

With the TORNADO, alias TUA 6120, Infineon has developed a receiver chip that works as a silicon tuner according to the 'Direct Conversion Receiver (DCR)' principle and which has an I/Q output interface, which enables the use of different QPSK channel decoder ICs. 'Silicon tuner' means that external RF components such as coils, varactors or RF diodes or transistors are no longer required. All relevant functions such as controlled LNA, VCOs etc. are fully integrated on the chip; the system is fully alignment-free. The RF signal connected on the satellite receiver is fed directly into the IC. Integrated low pass filters in the I/Q output circuitry with switchable cut-off frequen-

Figure 10.5
QPSK evaluation board, DiRec-Sat2TS

cies enable the adaption to different QPSK data ratings and codings.

10.2 Multimedia card – Ideal mass storage for mobile terminal equipment

With its extremely small dimensions, the MultiMediaCard is a very compact mass storage media, particularly for mobile terminal equipment (Figure 10.7). A 3D memory technology specially developed for this application brings the highest storage densities to the smallest chip.

In contrast to its competitive products, the MultiMediaCard from Infineon, with 24 mm × 32 mm × 1.4 mm, is currently the smallest exchangeable mass storage media on the market (Figure 10.6). It is consequently especially well-suited for handheld devices with which there is a minimum amount of installation space and where the weight of the storage media and the extent of the recording system (connector and control electronics) are an important factor.

With the development of a 3D memory technology and a new packaging procedure (see Table 10.1, Figure 10.9), the relationship of storage capacity to manufacturing costs for electronic storage has

Figure 10.6
MultiMediaCard as the currently smallest and cheapest storage medium of its kind

Figure 10.7
The MultiMediaCard is used in mobile devices of the most varied markets

been optimized, with the result that silicon now for the first time presents a real alternative to optical and magnetic storage media and demonstrates enormous advantages over these in the respective areas of application.

The solid plug-in contact between card and play-back device consequently enables a vibration-insensitive data transmission. And as no moving parts are required in the play-back devices, the MultiMediaCard guarantees (in comparison to CD-ROM or floppy disks) the highest possible robustness with regard to external mechanical, thermal and mositure related influences. Additionally, the MultiMediaCard also provides a higher data transmission rate and requires considerably lower power consumption. This has a particularly positive effect on the operating time of battery-operated play-back equipment.

Leading mobile phone and PDA (Personal Digital Assistant) manufacturers have therefore already equipped themselves with a MultiMediaCard recording system. Further applications, such as digital cameras, navigation equipment, electronic games or purely play-back equipment will follow later.

10.2.1 Diverse applications

One of the numerous MMC applications is the integration of software in handheld devices, whereby end users can enhance their equipment by, for example, retroactively loading programs or data from the card onto the device. If the data to be transferred is available in a specific format, any kind of exchange of cards with different content is possible between different play-back equipment. For example, the owner of a mobile phone can use a MultiMediaCard with telephone numbers in their desk phone and simultaneously in their mobile phone when they leave their office. Further applications in mobile phones are voice recording for use as dictation, saving faxes and e-mails or importing updates of operating sofware.

Another scenario regarding MultiMediaCard use is the downloading of every kind of data from central servers onto PCs and subsequent use in mobile MultiMediaCard players. On a MultiMediaCard with 32 Mbyte storage capactiy, it is now possible using the appropriate compression algorithms, e.g. MPEG2 layer 3, to digitally save half an hour of music in stereo quality. Via the Internet, music titles and other entertainment forms will be offered in the future subject to a charge. The songs etc. can be loaded onto a MultiMediaCard and can then be played while jogging for example in an audio player the size of a matchbox with absolutely not sensitivity to shaking.

In addition to mobile phones and audio players, games are also a mass application for the MultiMediaCard. This is where primarily ROM cards are used, on which audio books/radio plays or children's songs can be digitally saved.

10.2.2 Standardization gets up and running

In order to guarantee the compatibility of MultiMediaCards with different content and the most diverse recording systems, Infineon naturally chooses to follow the path of standardization. Strategic partnerships with international companies from the semiconductor, mobile phone and personal digital assistant industries have already been set up for this very purpose. The objective is to set up a "MultiMediaCard association" that actively promotes the MultiMediaCard standard in the public sphere and ensures compliance with specifications.

For a rapid distribution of the MultiMediaCard, the product spectrum, which originally consisted of purely read-only memories, has also been supplemented by the re-writeable flash cards. Through the cooperation with leading flash companies re-writeable products that correspond with the MultiMediaCard standard can now be offered in addition to read-only MultiMediaCards.

10.2.3 Flexible interface

Because of the low power consumption (1 mW with 100 KHz) and the variable voltage supply (in the 2 to 3.6 V range), the MultiMediaCard is optimally suited for battery operated devices. The data rate with reading is variable from 0 to maximum 20 Mbit/s, so that the requirements of the most diverse applications are covered.

The MultiMediaCard interface is serial, synchronous and has a single master bus architecture. Several cards can consequently be operated at one controller simultaneously by simply through-connecting only the pads of the cards. With only three signal lines (CMD, DAT, CLK), the interface is optimized for secure but nonetheless rapid data transmission.

With the flexibility of the interface with regard to clock frequency and voltage supply, the MultiMediaCard is ideal for use in the most diverse devices, from low-cost play-back equipment to Pentium PCs in which the full transmission speed is fully utilized.

A very simple and therefore cost-effective option of utilizing the MultiMediaCard is the emulation of the three signal lines with software. Here only three free I/O ports are required in the terminal device. With the software emulation (depending on the available processor capacity) data rates of between 100 kbit/s (8051, 16 MHz) and 2 Mbit/s (RISC processors) can be attained. For high-end applications that require the full speed, using a VHDL model, the interface functionality can be implemented in the hardware. The customer can then integrate the MultiMediaCard interface together with other periphery functionalities in a shared ASIC.

Further available system components for the integration of the MultiMediaCard are the necessary plug-in connections and PC card adapter for connecting to devices that do not at the time have dedicated MultiMediaCard slots, but do however have PC card slots.

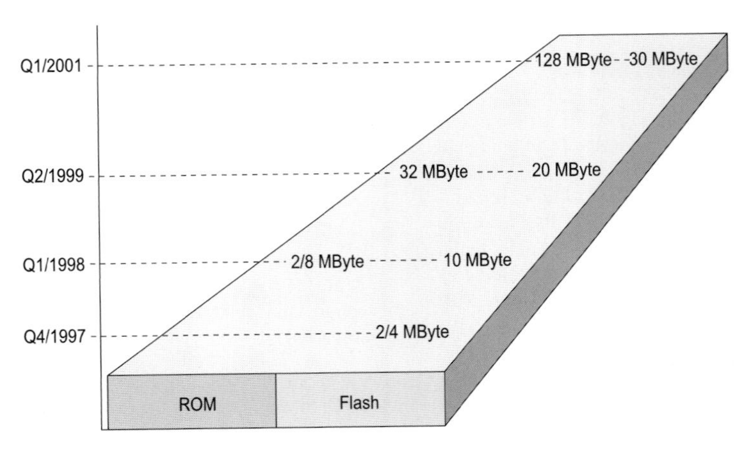

Figure 10.8
MultiMediaCards area available in ROM technology with 128 Mbyte and as flash memory with 30 Mbyte

10.2.4 128 Mbyte in 2001

Since 1998, the MultiMediaCard has been on the market in ROM technology with a storage capacity of between 2 and 8 Mbyte (Figure 10.8 and Table 10.1). Just one year later, the 32 Mbyte card was to follow, and a 128 Mbyte variant was planned for 2001. Likewise the end of the year was to see MultiMediaCards with re-writeable flash memory of 2 and 4 Mbytes available on the market.

The patented cell architecture, as used with the MultiMediaCard, is breaking new ground in semiconductor technology. In comparison with the previous cell architecture that only allows for the arrangement of the storage transistors in two dimensions (2D-ROM), the three dimensional arrangement of memory cells has now been successfully completed.

Without having to draw on smaller structure widths, it is consequently possible with the same storage capacity to increase the packing density of the memory cells, which results in a halving of the original chip surface (Figure 10.9). The required surface of a 3D cell of the MultiMediaCard can be additionally reduced as a result of ‘shrinking’.

Most MMC plugs will be customer-specific designs, with which different plug variants bring advantages (Figure 10.10).

Table 10.1 MultiMediaCard – Technical data

Electrical	
Operating voltage range	2.0 bis 3.6 V
Power consumption	< 20 mW @ 20 MHz operation,
	< 1.0 mW @ 100 KHz operation, < 0.1 mW stand-by
Serial transport data rate	up to 20 MBit/s
Access time	< 3 µs
ESD protection	±4 kV
Logical	
Storage capacity	2/4/8/10 MByte
Operating modes	card identification mode to identify cards,
	data transfer mode for high speed data transport
Data Rate	Read: 20 MBit/s Write: 1.6 MBit/s
Stackability	The MMC bus supports up to 30 cards on a single bus
Mechanical	
Card sizes	24 mm × 32 mm × 1.4 mm
Number of pads	7
Insertion/removal endurance	10 000 cycles
Ambient operating temperature range	–20 bis +85°C
Ambient storage temperature range	–40 bis +85°C
Technology	ROM/Flash

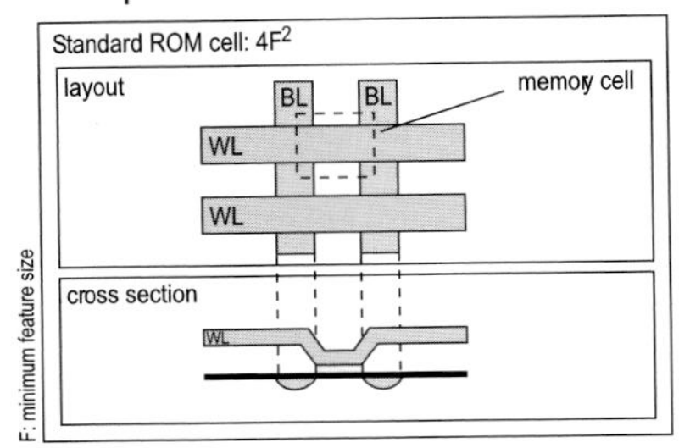

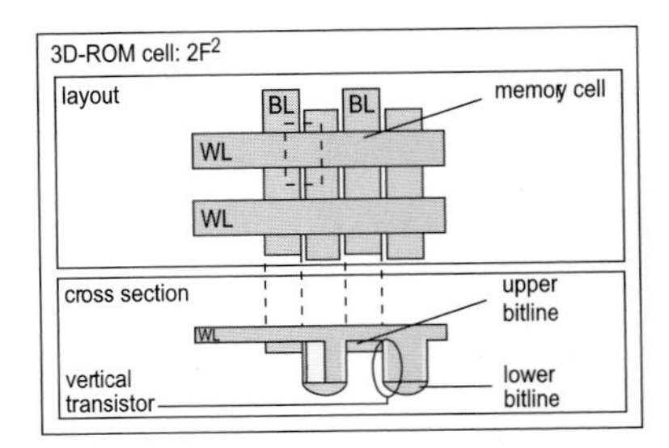

Figure 10.9 The chip surface of a storage cell is halved by integrating the 3rd dimension

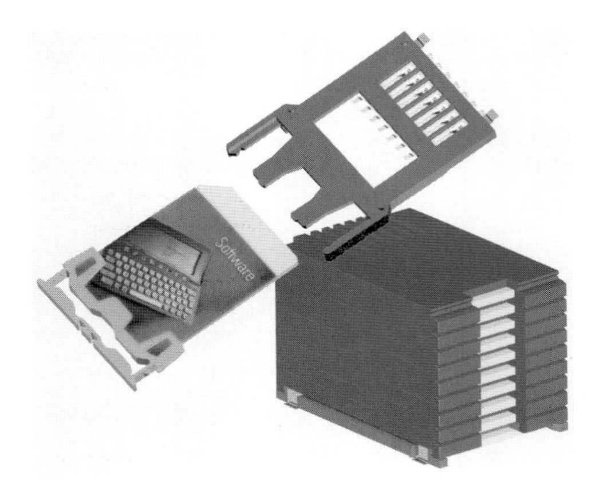

Figure 10.10
Standard plug (left) for inserting the MMC and "flip" multiple plug in SMT version

From the system side, several plug variants, from standard to multiple contact in, plug-in, "flip" or clip variants, are possible for mounting and contacting the MMC. On the basis of the first applications of the MMC on the handheld market, the design of the plug-in connector is principally determined by the space that has been provided. Therefore the design of the MMC contact system is intended to allow up to 10,000 plug cycles of the operation of one or more MMCs to be implemented in the most confined space on the bus. At the same time, in comparison with standard chip cards, the contacts have been placed on the card edge.

11 Communication modules

11.1 Overview and trends

Today, there are more than 600 million telephone connections worldwide. Microelectronic innovations make it possible to utilize this globally available communications network for even more services: as access to the Internet; for video conferencing; or for exchanging information with subscribers in mobile networks, for example.

It is now crystal clear that global communication will be performed via the Internet. It is important in this respect that users are accommodated with as convenient an access to these networks as possible. ISDN (Integrated Services Digital Network) is an important milestone in achieving this goal, and the rapid global market development for ISDN provides the respective confirmation.

The growing availability of multimedia information creates ever increasing requirements for rapid access to this information. This necessitates innovation of the networks themselves. Instead of line switched telephone connections, new technologies, such as Asynchronous Transfer Mode (ATM) or Fast Ethernet provide very fast connections of up to one gigabyte per second, which are dynamically adjusted to the user's requirements. And, of course, innovative semiconductor products are also deployed in this area.

The future has already begun for high performance networks via which data and voice can be transmitted. ADSL, VDSL and SHDSL are some of the promising technologies on offer here.

Focusing on standards and core competencies

One interesting area of application is presented by 'Solutions on a Chip' for market-specific applications in telecommunications, which enables:

- ICs for digital and analog switching and transmission technology, such as 'Linecards for Central Office', 'PBX', 'Radio in the Loop (RITL)' and 'Integrated Access Devices (IADs)'
- IC solutions for ATM systems and other high-speed data networks and
- ICs for digital voice, video and data terminals

'Telecom System Solutions on a Chip' means that increasingly more telecommunication system functions are combined on one highly integrated chip. This reduces the costs of these systems for our customers and at the same time enables the continuous expansion of the functionality. A good example of this is the DOC module presented three years ago, which provides the essential functions of a PBX on a chip.

The necessary investment for such high integration calls for industry standards in order to ensure as wide a global market for these products as possible. One of the best examples in this respect is ISDN, for which the first complete chip solutions were developed some 10 years ago.

11.1.1 Strategic objectives

Our customers are in constant competition, competition in which we can only make them successful through ever increasing performance with falling costs.

Technologically advanced products are just as important with the design of the respective telecom systems as modularity and flexibility are. A hardware package and numerous firmware modules provide the customer with the option of setting up costs-optimized solutions. Infineon ICs provide standard interfaces that, with flexible combinations, help to design tailor-made systems. This reduces the design-in investment, accelerates the market launch and brings system costs down.

Integrated, customer-specific solutions

The majority of these solutions are composed of system solutions based on standards. However, our developers also use our modular architecture to combine customer-specific design with application-specific standard products. This accommodates an early presence with trendsetter customers.

11.1.2 High rates of innovation

Production technology is also decisive for long-term market success with telecom ICs. The rate of innovation with memory modules is extremely high. Performance capacity doubles every 18 months. The logic modules follow in ever increasingly shorter intervals and production processes for the 'Mixed Signal' modules are facilitated by CMOS technology. This has already resulted in telecom IC production being raised to a high technological plateau.

Infineon's chip production in München-Perlach (Munich) avails of the most modern, complex technologies one could possibly require to create high-grade technical product properties. With direct access to this chip production plant, rapid and secure product start-ups are attained and the requirements of the customer are satisfied as quickly as possible. Another plant in Regensburg represents a production cluster that enables flexible reactions to market changes.

11.1.3 Switching ICs

Over 25% of cable-connected global communication currently runs via Infineon switching ICs. In the ISDN area, ICs are supplied for more than 50% of all connections installed worldwide. The switching ICs business division is consequently the global leader in its target markets. In order that our customers can provide functions and costs-optimized solutions for the current and future communications requirements, we act as their partner with the optimization of existing systems or the development of new systems.

The ever-increasing miniaturization of switching equipment with simultaneously higher functionality demands increasingly more cost-effective transmission of increasingly greater data volumes. This is attained through the continuous innovation of the respective technologies, constant expansion of core competence and, of course, through close cooperation with customers all over the world.

11.1.4 Network ICs

The Network ICs business division provides circuits for high-speed networks, including Local Area Networks (LANs) and Wide Area Networks (WANs). Our MUNICH32X is the fastest multi-channel interface controller for transmission networks and is used in almost every router in the world.

The communications networks of the future will integrate the data and telephone infrastructure using ATM and other high-speed systems. We are working here on solutions with which a switching system can be set up with just a couple of hundred ICs, a system which would meet the communications requirements of several million people. The merging of telecommunications and data networks becomes particularly evident with internetworking. System competence is a central success factor in this area. Our simulation tools are one of the results of this system competence. Using these tools, the customer can start the system development even before the silicon is available.

11.1.5 Communication Terminal ICs

The Communication Terminal ICs business division provides a wide range of ICs for voice, data and video communication. The main applications for these products are digital telephones and answering machines, base stations for cordless telephones and PC and video communication. Included among the latest developments are the single-chip telephone module, INCA, and the diverse 'Linecard Transceiver' family.

The information exchange via the Internet and teleworking significantly contributes to a rapidly growing requirement for transmission bandwidths. Analog modems can only meet these requirements under certain conditions. A step in this direction is provided by ISDN with the corresponding bandwidth at affordable prices. Infineon is one of only a few telecom IC manufacturers that have invested right from the outset in the development of ISDN technology, and who today provide an extensive range of modules. Modular products with standardized interfaces provide our customers with optimal flexibility to forge ahead of the competition with their products.

One of the basic elements of this strategy is the IOM-2 architecture developed by Infineon. Together with a number of highly integrated modules, passive or active ISDN applications with integrated voice, video, modem or fax function can be developed with this architecture. The emphasis of our work is on complete system solutions that are successfully implemented in hundreds of designs around the world.

You will find a selection from the very wide-ranging product spectrum in the following sections.

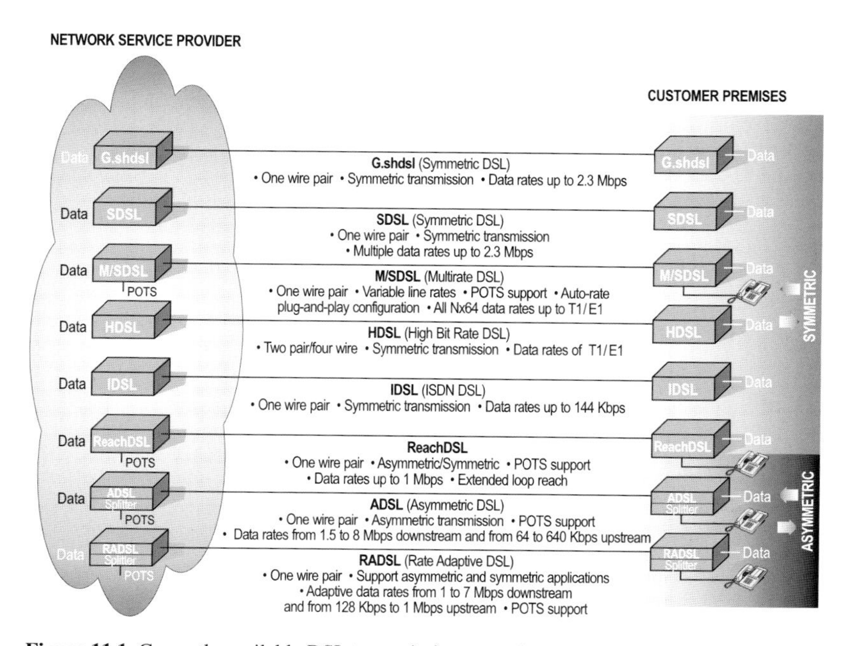

Figure 11.1 Currently available DSL transmission procedures

A high-speed connection to the Internet can, however, only be realized via a DSL connection. DSL stands here for 'Digital Subscriber Line'. In recent years numerous DSL procedures have been established and in the future further procedures will also be developed, which will result in even higher data rates. Figure 11.1 provides an overview of the currently available DSL transmission procedures.

These transmission procedures are explained in section 11.10. ADSL2, the recently standardized successor to ADSL, will be discussed in detail.

11.2 ISDN: From the exchange to the subscriber

An abundance of highly-integrated components guarantee a completely fault-free system solution in ISDN. With intelligent network terminations, analog and digital terminal equipment can be operated alongside one another.

Thanks to the constantly increasing integration density, in the second generation, ISDN chips with a minimum number of components and additional features are now possible at Infineon. The ISDN-S Interface Feeder (ISFC), for example, a feeder circuit for the S-interface, replaces large relays and transformers. Or the 'ISDN Network Termination Controller' for the network termination in ISDN, which presents a solution with a single chip that contains the transceiver for the U interface, the transceiver for the S interface, the HDLC controller and an IOM-2 port so that a codec filter can be directly connected.

With digital linecards, an eight-way U transceiver reduces the space requirement on the circuit board and consequently also the costs. A DSP-based controller module can serve up to three of these eight-way U transceivers. In addition to the provision of 24 channels, it also provides additional functions, such as conferencing and music-on-hold. An integrated power controller feeds the battery voltage into the U line while it simultaneously takes over important control functions such as current limitation and overvoltage protection. The new chip replaces large discrete circuits on the digital linecard.

11.2.1 Functional blocks in ISDN

ISDN is a network with digital end-to-end connection for supporting a wide range of services. Including among these are voice and data services. All subsrciber devices receive and send information in the form of bit streams to and from telecommunications networks and this is actually carried out regardless of whether or a telephone, a computer, a fax machine or a television is being used. In this procedure, different devices use the same network via the same subscriber access line.

Four different functional blocks can be combined between the terminal device of the customer and the network as ACE – Access Connection Element (Figure 11.2)

The local ISDN exchange, known as LE (Local Exchange), is set up using the elements, LT (Local Termination) and ET (Exchange Termination). LT here terminates the local loop, while ET takes over the switching functions.

The NT1 (Network Termination Type 1) physically terminates the connection between the customer side and the LT. It manages the voltage supply, takes over the monitoring of the line, the physical conversion of the signaling protocol and it multiplexes the B and D channels. The NT2 (Network Termination Type 2) is optional and is designed for PBXs or LANs and is not normally planned for home use applications.

Terminal Equipment (TE1) are end user devices that are compatible with the ISDN protocols – digitals telephones, for example. TE2 are terminal equipment

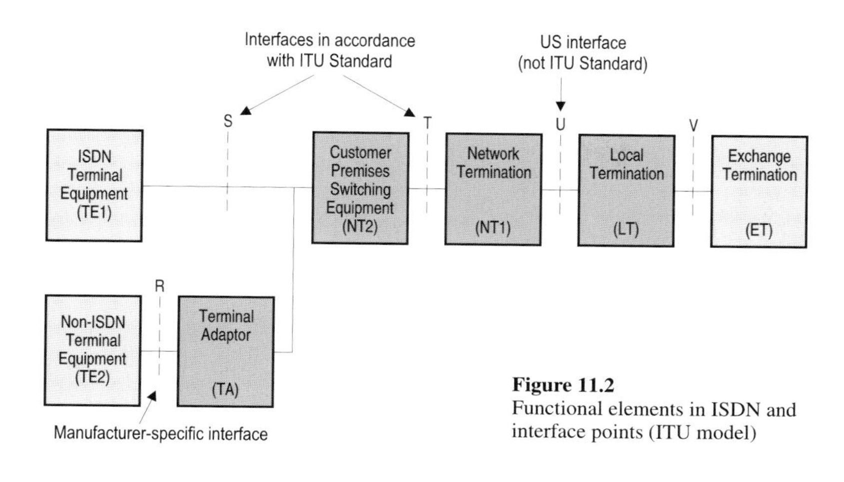

Figure 11.2
Functional elements in ISDN and interface points (ITU model)

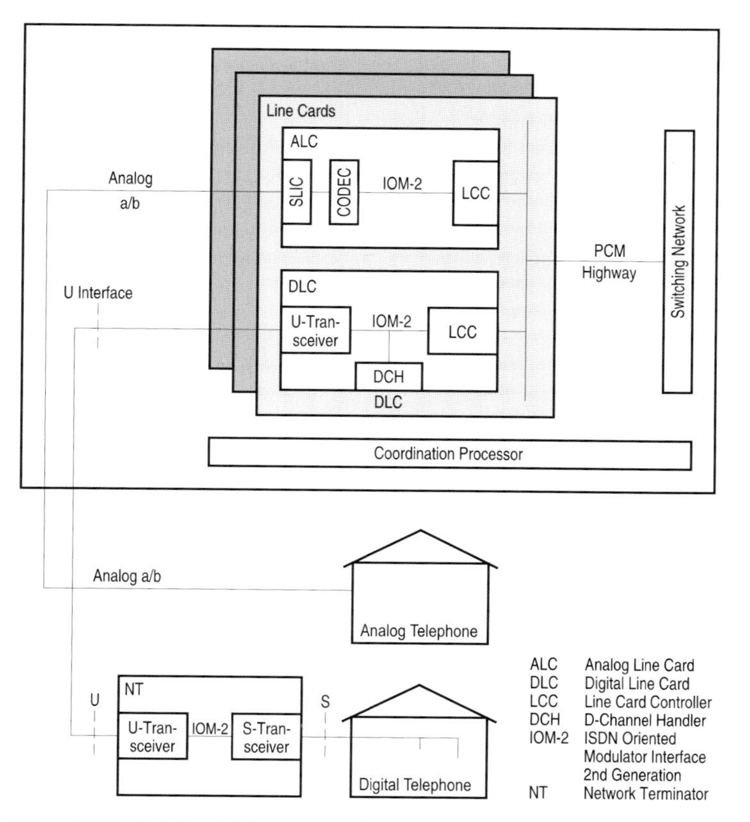

Figure 11.3 Digital switching system with digital and analog lines

that do not support ISDN. Included among these are analog telephones, for example. It is possible to connect these devices to ISDN with a terminal adapter (TA). An "intelligent NT" unifies the two blocks, NT1 and TA in one device.

The reference points correspond with different protocols that are used at different interfaces. If no devices of type NT2 are used, for example, with applications in private households, then S and T interfaces are identical. A more detailed consideration of the blocks ET, LT and NT1 results in the block diagram, which is shown in Figure 11.3.

The IOM-2 interface (IOM: ISDN – Oriented Modular) – a wired interface – can connect all transmission ICs, apart from the primary connection ICs. It consists of two unidirectional data lines, a frame alignment signal and a data clock. With one interface, up to eight digital transceiver modules can be operated and programmed via one controller chip. Figure 11.4 shows a sub-channel structure of the IOM-2 interface.

The advantage: The IOM-2 interface significantly reduces the number of lines, as each transceiver is programmed via the monitor channel and no fixed connection to the microcontroller is required. Each IOM-2 interface enables, for example, up to eight U lines to be connected. Four IOM-2 interfaces can be implemented per linecard controller (ELIC), so that a total of 32 ISDN subscribers can be connected to one digital linecard.

11.2.2 Digital linecard

The digital linecard acts in the ITU model (Figure 11.2) as an LT block. The linecard switches between the PCM highway and the subscriber lines (local loop), accepts the D-channel protocol, if this involves a decentralized D-channel architecture, and supplies the U line with power (Figure 11.5).

11.2.3 Extended Linecard Controller (ELIC)

In order to connect the transceiver, the ELIC (PEB 20550H), equipped with a multiplex function for the D-channel, has a configurable interface, CFI (Config-

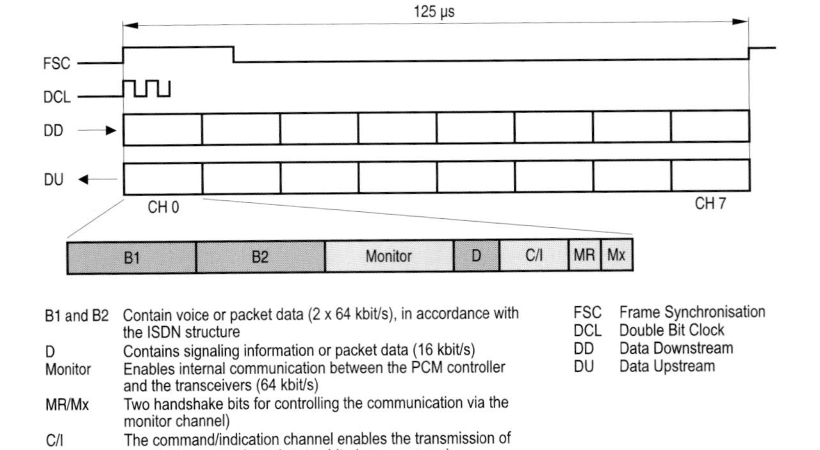
125 µs
FSC
DCL
DD
DU
CH 0
CH 7
B1
B2
Monitor
D
C/I
MR
Mx
B1 and B2 Contain voice or packet data (2 x 64 kbit/s), in accordance with the ISDN structure
D Contains signaling information or packet data (16 kbit/s)
Monitor Enables internal communication between the PCM controller and the transceivers (64 kbit/s)
MR/Mx Two handshake bits for controlling the communication via the monitor channel)
C/I The command/indication channel enables the transmission of specified commands and status bits (e.g. power-up)
FSC Frame Synchronisation
DCL Double Bit Clock
DD Data Downstream
DU Data Upstream

Figure 11.4 IOM-2 interface with up to eight transceiver modules on the controller chip

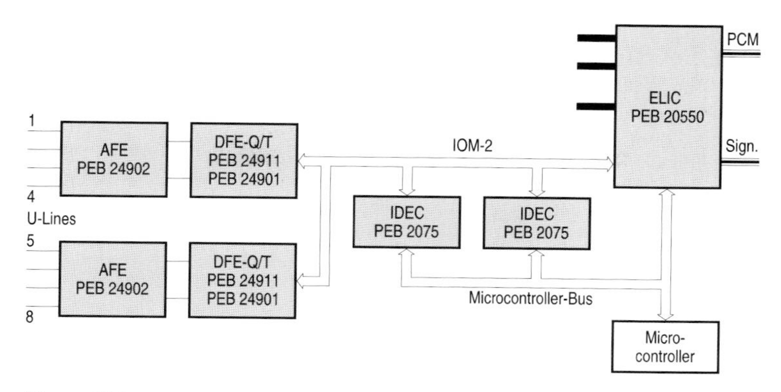

Figure 11.5
Digital linecard (2B1Q/4B3T) switches between PCM highway and the subscriber lines (local loop)

urable Interface). This can be operated in modes 4_PCM and 4_IOM-2. Various clock and data rates can consequently be programmed and up to 32 digital or 64 analog lines can be connected in the IOM-2 mode. Additionally, depending on the data rate, up to four more programmable PCM interfaces are available to set up the IC connection with a PCM highway. The maximum data rates thereby are, 4 × 2 MBit/s, 2 × 4 MBit/s or 1 × 8 MBit/s. With the help of its 128 × 128 matrix, the ELIC is able to connect any data or voice channel (coded in time slots with 8 bits each) between the CFI and the PCM interfaces. The ELIC supports the monitor channel and the C/I codes in accordance with the protocols for the IOM-2 interface. The chip contains a HDLC controller for the signaling information between the card and the combinatorial circuit. As an alternative to the ELIC, there is the EPIC, which apart from the HDLC controller provides the same functions.

11.2.4 ISDN-D Channel Exchange Controller (IDEC)

The IDEC (PEB 2075) has been optimzed for processing the D channel signaling information on digital linecards. Four HDLC controllers, which are independent of one another, have been integrated into one single IC for this purpose. Therefore an IDEC is capable of processing the D channels of four subscriber lines simultaneously. Each channel contains an LAPD controller and, per direction, a 64-byte high-performance FIFO memory. To significantly reduce software expense, emphasis was primarily concentrated on the support of layer-2 functions in accordance with the OSI/ISO model with the switching definition. These and other characteristics significantly reduce the dynamic stress of the microprocessor system. Via the IOM-2 monitor and the C/I channel protocols, the IDEC provides control options for periphery equipment. Because each IOM-2 interface can connect up to eight subscriber lines, two IDECs are used per interface. A total of eight ICs process the D-channel information of all 32 subscriber lines, which are served by one digital linecard.

11.2.5 U transceiver for the analog frontend

Analog FrontEnd (AFE) and the Digital FrontEnd (DFE) (PEB 24902 and PEB 24911/01) form a four-way U transceiver for digital linecards. Depending on the line code used for the U line, 2B1Q or 4B3T, either the PEB 24911 DFE-Q or the PEB 24901 DFE-T is required. The analog AFE can process both line codes. With the quad IEC transceiver, (AFE+DFE) this involves a fullduplex, four-way U transceiver based on the single chip transceiver, PEB 2091 IEC Q, for the U interface in ISDN. The set complies with the latest specifications of ANSI, ETSI and CNET. If the 'echo cancellation' method is used, then transmission can be made fault-free on a line length of up to 8 km.

Included among the main features of the set are:

- A built-in wake-up unit, that activates the IC should it be in power-down mode
- Adaptive echo cancellation
- Adaptive equalizing
- Automatic adjustment to the polarity
- Automatic gain control
- Clock recovery
- Low power consumption

As each chipset supports up to four subscriber lines, two sets are required per IOM-2 interface. On a digital linecard, that uses an ELIC with four IOM-2 interfaces, eight four-way IEC chip sets can be accommodated, which then serve 32 subscriber lines.

11.2.6 ISDN High Voltage Power Controller (IHPC)

The IHPC (PEB 2026) is shown in the detailed illustration of the linecard (Figure 11.6). This IC enables the power supply of the U line from the CO. The component can consequently tolerate voltages of up to 130 V and takes on critical control functions, such as current limitation and overvoltage protection. So that the IC can also withstand extreme conditions undamaged, an internal excess temperature control switch is implemented. The IHPC replaces circuits that are made up of relays and discrete components.

11.2.7 Network termination

Two different types of network terminations are currently in use today. The standard NT1 converts the two-wire U interface into the four-wire S_0 interface. It is a layer-1 component that channels the information of the B and D channels between the exchange and terminal equipment (TE). Furthermore, the standard NT1 supplies the TEs with voltage and carries out the collision management of the D channel should more than one subscriber try to seize the D channel. The incoming signal on the U interface is coded either with 2B1Q or 4B3T. Infineon Technologies provides complete IC systems for both coding types.

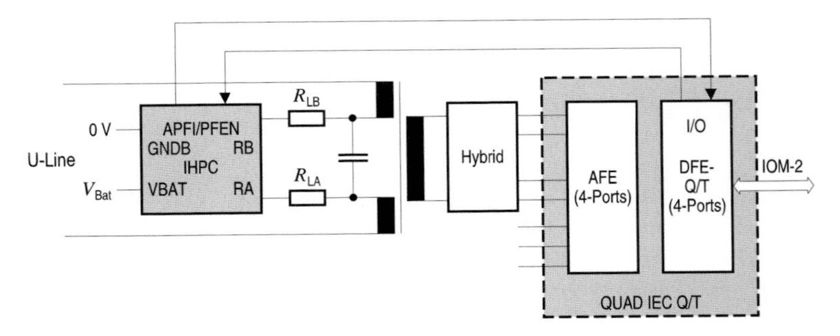

Figure 11.6 The IHPC enables the power supply of the U lines from the CO.

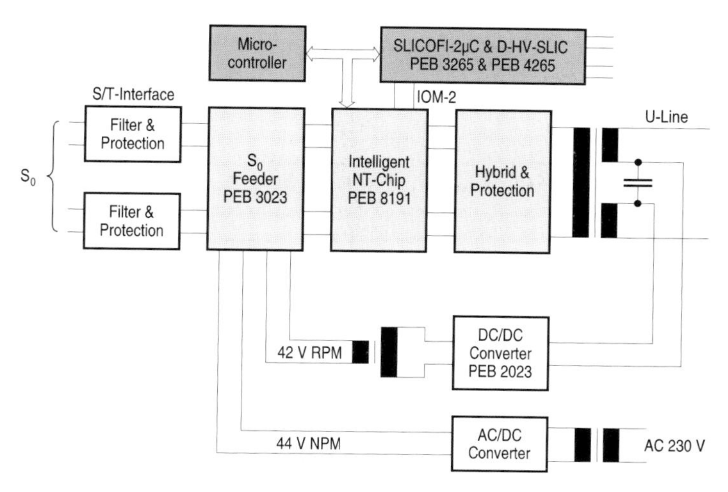

Figure 11.7 The intelligent NT is an expanded standard NT1

The second type of network termination is described as an 'intelligent NT' (Figure 11.7). Supplementary to the standard NT1 functions, this NT type provides a connection to the analog lines of those users that actually want to retain their analog terminal devices, but also want an ISDN access. The intelligent NT corresponds with a standard NT1 that is supplemented with a microprocessor, a codec filter and an SLIC.

11.2.8 Intelligent Network Termination Controller (INTC)

This component (PEB 8191H) (Figure 11.8) is the integration of S and U transceivers of the previous generation including a HDLC controller. It processes the line code 2B1Q in accordance with the guidelines, ETSI ETR 080 1995 and ANSI T1E1.601 1992 together with the S/T interface in accordance with ITU Recommondation, I.430. Together with a parallel or serial microcontroller interface, the integrated HDLC controller enables the access to the B channel, the D channel and the communications channels (inter communication channels) for communication between the units. Layer-2 functions of the LAPD protocol are supported by the hardware at a higher level. In addition to the conversion from U to S, thanks to its IOM-2 interface, INTC also offers a direct connection to the CODEC.

Two additional versions of the PEB 8191 complete the family of network termination controllers (NT Controllers). The PEB 8091 is identical to the PEB 8191, however, it contains neither the HDLC controller nor the IOM-2 interface. It is designed for standard NT1 applications with which no additional analog lines are

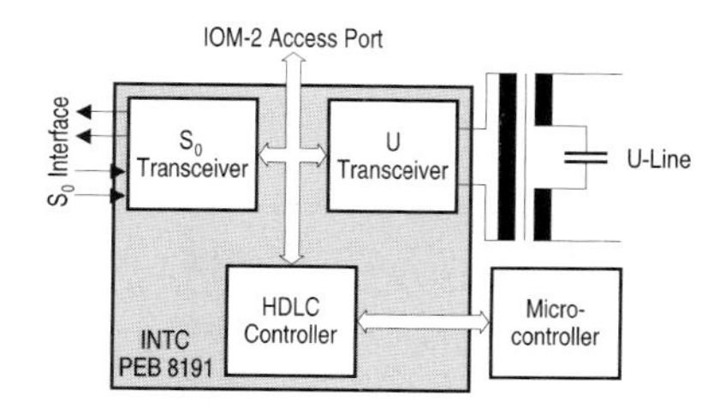

Figure 11.8
The INTC (PEB 8191) integrates an S_0 and a U transceiver

required. The PEB 8090, a network termination controller for standard NTs that use the line code 4B3T, is a similar product.

11.2.9 ISDN DC-DC Converter (IDDC)

The IDDC (PEB 2023), a DC-DC convertor, works according to the switch-mode power supply principle and uses PWM technology for regulation. It was conceived for applications in the telephone and ISDN area and fully complies with the ITU recommendations for the power supply of S interfaces. Together with just a few external components, the IDCC provides a stable direct current source with 5 V. It can also be programmed for higher output voltages to supply the S interface and it can be adjusted to different voltage ranges: e.g. 12 to 80 V or 20 to 120 V. A programmable overcurrent protection, an overvoltage detection and a softstart function have also been integrated onto the chip. Its low power dissipation and the highly precise voltage reference ensure that the IC is ideally suited for NTs that have very high requirements with regard to power dissipation.

11.2.10 ISDN-S Interface Feeder Circuit (ISFC)

Additional circuits are inevitably required in order to route user data from the NT1 via the S interface to the terminal equipment (Figure 11.9). The direct voltage feed of the AC signal coming from the S interface therefore requires two transformers. A relay is required for switching between the normal power mode, NPM, and the restricted power mode, RPM. Switching depends on whether the power feed is local or comes from the U interface. Circuits are implemented for current limiting for each power mode. A reset generator has also been implemented. It ensures a reset of all ICs in the NT1 after the supply voltage has reached a prescribed value.

The ISFC (PEB 3023) reduces the number of discrete components that are required for an NT. The component contains a reset generator on the chip and an individually programmable line current protective device for both power modes (NPM and RPM). It replaces both S transformers and automatically switches between NPM and RPM. It additionally absorbs only very low power dissipation. It is not only compatible with INTC, but rather also with previous component generations of the ISDN product family. Examples in this respect are the components, SBC (PEB 2080), SBCX (PEB 2081) and ISAC-S (PEB 2085 and PEB 2086).

11.2.11 Dual Signal Processing Codec Filter

The dual channel combination, DuSLIC, consisting of SLICO-FI2 (PEB 3265) and Dual-HVSLIC (PEB 4265), provides an optimal solution for the intelligent NT. It delivers the complete connection from the IOM-2 interface of the INTC to two analog telephone terminal devices. Furthermore, the specifications of ITU, EIA and LSSGR are also met. Independent programmable digital filters for country-specific applications are integrated. The DuSLIC family contains a DTMF generator and a microprocessor interface, and it additionally enables the polarity inversion and data transmission with on-hook handsets.

11.3 ISDN terminal equipment: The subscriber end

Users are recognizing the advantages of ISDN at an ever-increasing rate. In addition to the beneficial features of the bandwidth, such as channel packing, the fast connection set up and regionally varying added functions, higher flexibility and lower operating costs are also attractive factors.

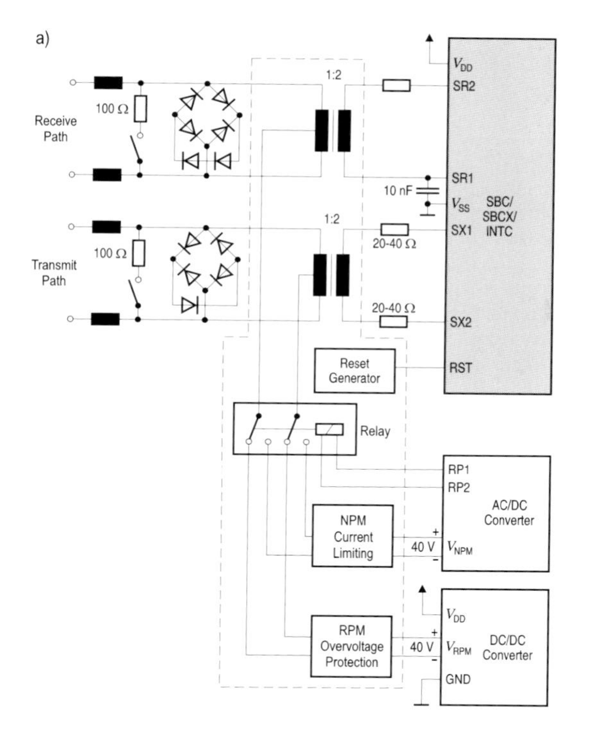

Figure 11.9
The ISFC (PEB 3023) reduces the number of required components:
a) discrete circuit,
b) set-up with ISFC

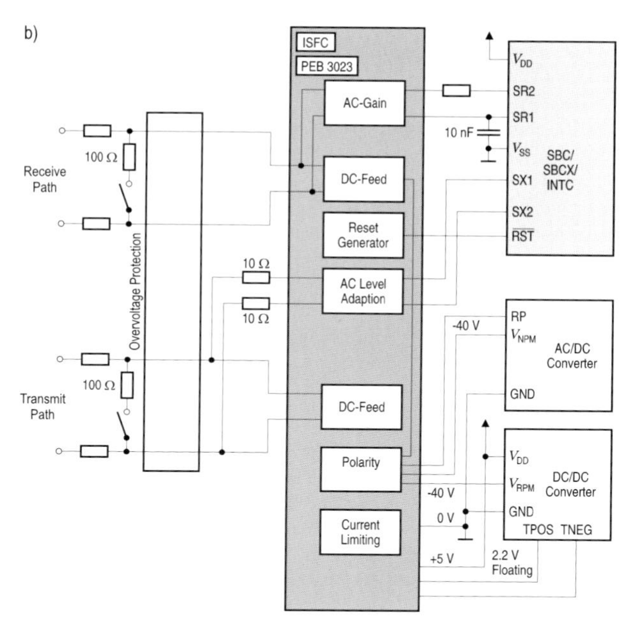

A full range of ISDN transceivers is available:

- Basic Rate Interface (BRI)
- Primary Rate Interface (PRI)
- U-Interface (2B1Q)

ICs are likewise on offer for PC plug-in cards or ISDN terminal adapters. They are designed for purely digital solutions or hybrid solutions, that is, for the ISDN network combined with modem and fax. They are also suitable for video communication systems or voice-over data solutions.

The developer can consequently design the most varied terminal equipment in the telephone, data and video communications area.

The classic terminal device at the subscriber end is, of course, the telephone set. The choice ranges here from standard one-hand through to conference telephone with full duplex handsfree telephoning.

11.3.1 Telephone

For cost-effective telephone set-ups, Infineon Technologies provides the SCOUT-S (PSB 2181/PSB2182) and SCOUT-SX (PSB 2183/PSB 2184) with serial or parallel µC interface. While the SCOUT-S contains all the necessary functional blocks, such as, Analog FrontEnd, HDLC Controller and S_0 Transceiver, the SCOUT-SX additionally provides a half duplex handsfree algorithm in accordance with the 'Stronger Wins' procedure. The fully digital handsfree function implemented in the SCOUT-SX provides the highest performance at a low price. To suppress the echo and feedback that occurs between loudspeaker and microphone, the louder speaking subscriber is always patched through non-attenuated.

The analog frontend provides direct connection option for two microphones and two loudspeakers (Figure 11.10). These can be used for both handset and hands-

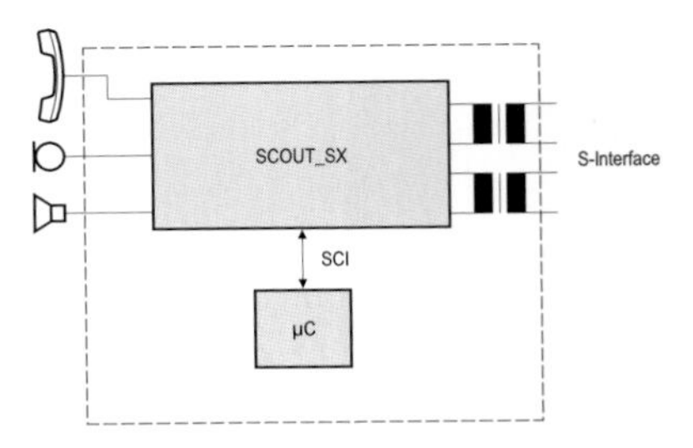

Figure 11.10 Scout-SX telephone

free operation. A third input can be used for an optional external handsfree microphone. This codec provides the following functions:

- Complete compatibility with ITU-T G.712 and ETSI-(NET33) specification
- PCM A-Law / µ-Law (ITU-T G.711) and 8/16-bit linear data; maskable codec data
- Independent programmable amplifiers for all analog inputs and outputs
- High-quality voice-data processing for, by way of example:
  - Three-party conference
  - Listening-in
- Two correction filters for transmission and receiving direction
- Adjustable sidetone volume
- Flexible DTMF, tone and ringer generator

With the TE mode, the S_0 transceiver provides the interface to the ISDN network. Layer-1 of the ISDN protocol is thereby completely integrated into the hardware. For the higher protocol layers, a HDLC controller and a module for TIC bus arbitration are available.

The INCA-S (PSB 21483) is a single chip solution for a fully equipped high-end telephone. In addition to the S_0 transceiver, HDLC controller and the TIC bus arbitration, the INCA-S contains a DSP with ROM-based firmware and a 16-bit µC with USB interface (Figure 11.11).

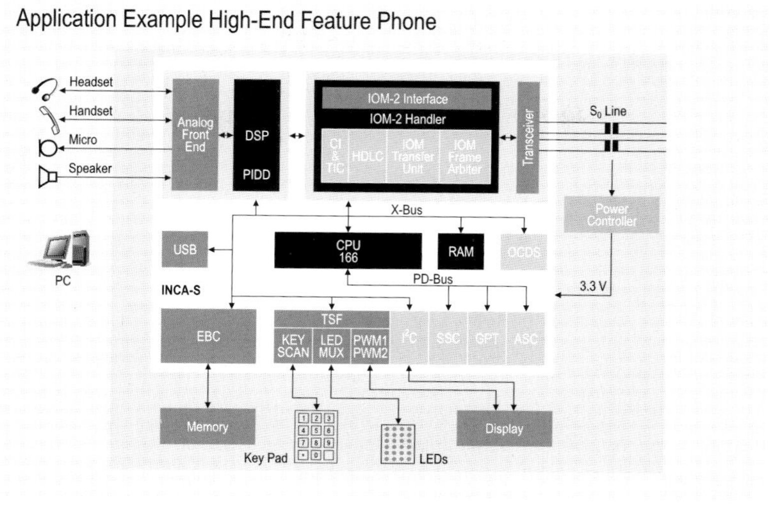

Application Example High-End Feature Phone

Figure 11.11 Block circuit diagram of the INCA-S

The analog frontend is designed for three microphones and three loudspeakers. The user consequently has the option of handset, headset with microphone and handsfree at his disposal. In addition to the telephone-specific modules, such as ringer tone and DTMF generator, the firmware of the DSP also provides a full duplex handsfree algorithm with echo canceller. This allows the call participants to speak and be heard while listening in. Digital filters and a module for suppressing side noises improve the voice quality. A three-party conference circuit can be switched via a summator.

The 16-bit µC is based on the C166 core and is equipped with an 'On Chip Debug Support' module (OCDS). In addition to 2 KByte dual port RAM, 4 KByte XRAM is also provided. Keys and LEDs are connected directly via the terminal-specific functions. A display can be connected via an I^2C interface. A USB V1.1 interface with a data rate of 12 MBit/s provides the option of both remote control of the telephone and data transfer to the PC. The functionality of the module can be extended even further via the IOM-2 interface.

11.3.2 PC plug-in card

A cost-effective implementation for data transfer is provided by the single chip solution, IPAC-X (PSB 21150). The high-grade module has been equipped with special features to visibly reduce the costs for components and circuit board.

This design consists of PITA-2 (PCI Interface for Telephony/Data Applications), IPAC-X (ISDN PC Adapter Interface Cir-

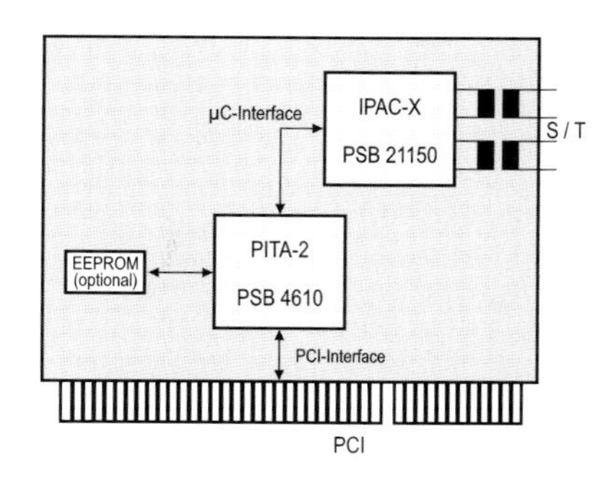

Figure 11.12
PC plug-in card with single chip solution, IPAC-X

cuit) and a line interface (Figure 11.12). IPAC-X includes the S transceiver and three HDLC controllers for the D channel and the two B channels of the BRI (Basic Rate Interface). These channels can be accessed via the parallel interface of the chip. They are buffered via ring buffer FIFOs with variable thresholds (up to 64 bytes for the D channel, up to 2 x 128 bytes for the B channels). It is possible to access the B channels via a serial interface (IOM-2) that allows a codec module like SCOUT, for example, to be connected. The parallel interface of the PITA-2 is used with this design. The IPAC-X is connected to this interface and the access using the software can be carried out as with a normal memory-mapped device. An EEPROM can be optionally connected to the PITA-2 to store manufacturer-specific data like, manufacturer-ID, device ID, subsystem manufacturer-ID and subsystem ID as well as relevant power management information. The subsystem manufacturer-ID and subsystem ID can also be loaded using pin-strapping. This enables the software to distinguish between different cards with the same interface module.

11.3.3 Terminal Adapter (TA) and USB S_0 Adapter

The solution implements ISDN data access plus terminal adapter functionality on the highest possible integration level (see Figure 11.13).

The focus point of the design is the C165UTAH microcontroller, including 16-bit C166 core, four HDLC controllers for B and D channels and 12 MBit/s USB V1.1. Via IOM-2, UTAH provides the connection to the S transceiver module, SBCX-X, and to DuSLIC as dual channel codec for the two POTS (Plain Old Telephone Service) ports. It would also be possible to connect other codecs not compliant with the IOM-2 bus by using the PCM interface instead of the IOM-2 interface.

If the application does not involve a USB port, then the SBCX-X and DuSLIC combination presents a costs-optimized solution. The C165H can, for example, be used as µC.

11.3.4 NT1 and TA combination

A combination of NT1 and TA offered in Germany is allowed in the USA and is frequently used there. This combination establishes a connection to the U interface and POTS telephones can also be connected (Figure 11.14).

This design implements a terminal adapter with an RS-232 port. It consists of the C165H microcontroller, the IEC-Q V 5.3 with its line interface and DuSLIC as dual channel voice codec for POTS services. The 'ISDN Echo Canceller' (IEC-Q V 5.3) includes the U transceiver and hybrid for two to four wire conversion.

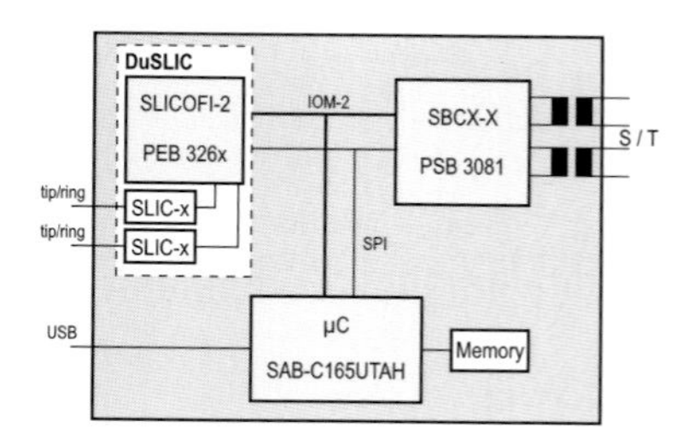

Figure 11.13 TA and USB combination

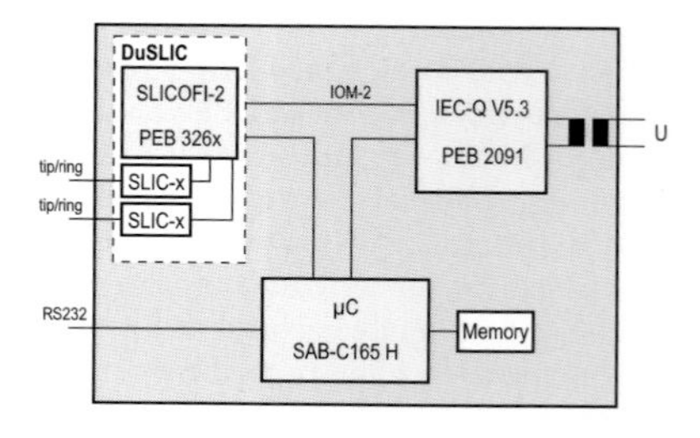

Figure 11.14 NT1 and TA combination

The line interface consists of a transformer. A metal loop termination can be optionally integrated into the interface. The 16-bit microcontroller, C165H, is used to process the D channel and B channel protocols. Because the C165H also contains four bi-directional HDLC controllers, no external HDLC controllers are required to access the 2B+D channels.

11.3.5 High-end telephone with USB-S_0 adapter and TA function

For a design that unifies, for example, the functionality of telephone, USB data transmission and terminal adapter (TA) for an analog fax machine in one device, the telephone module, INCA-S in combination with the DuSLIC provides an inexpensive variant.

As already explained in section 11.3.1, the INCA-S is the most efficient telephone module that has already integrated the USB data adapter function. The DuSLIC is connected via the IOM-2 interface, which provides telephone calls via two analog lines (Figure 11.15).

11.4 Reference designs for ISDN

Anyone who wants to operate on the telecommunications market as a system vendor must be able to develop and provide competitive devices within short periods. Infineon Technologies is the global market leader in the ISDN-ICs area and has decades of experience as a pioneer in this and other fields. This know-how is implemented in development systems and products, with the help of which the development of designs is accelerated and made significantly easier.

11.4.1 Complete solutions facilitate accelerated marketing

Telecommunications is a global growth market. The system manufacturer operating in this market is faced with many hurdles:

- With what communications modules can the desired system be realized?
- Is the storage of hardware functionality in software possible or is active hard-

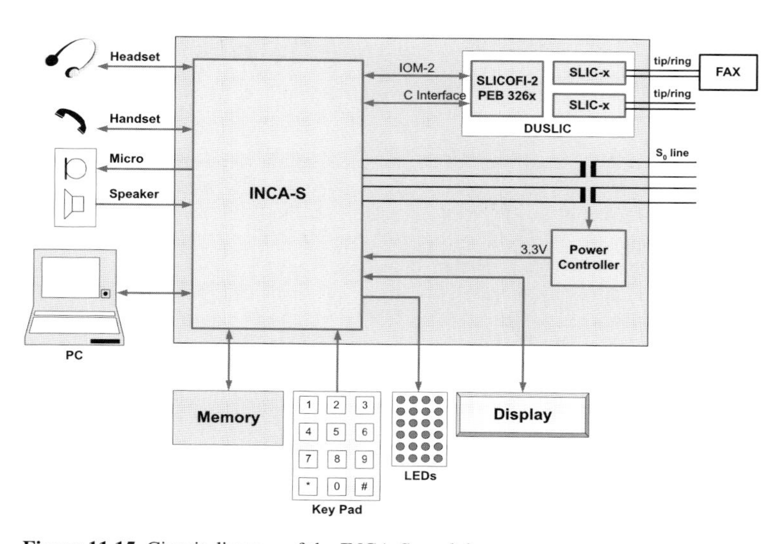

Figure 11.15 Circuit diagram of the INCA-S module

ware with microcontrollers necessary for performance reasons?

- What computing capacity does a specific telecommunications protocol require? Where can support for country-specific variants be found?
- How does one attain an internationally approved system?
- Can an optimized design be developed and quickly launched on the market without major expense?

A competitive positioning on the telecommunications market requires cost-effective products with state-of-the-art technology. To attain this, the system concept stands full to the fore, because alongside the development of hardware and software, the approval of the system is a critical sales argument.

Reference Board Package

The systems described as "Reference Board Package" contain hardware, comprehensive software and documentation for typical applications. Valuable information for some products is supplied by design proposals, circuit plans, application notes, EMC and approval tests.

11.4.2 Hardware

An "ISDN Reference Board" is a PC card or an external module at state-of-the-art technology. With the emergence of the first engineering samples of an ISDN module, an "ISDN Reference Board" is now also available. The circuits and semiconductors are used the world over and experience gained is therefore consistently incorporated into the reference boards. Activation is made via standard interfaces, in the PC area, for example, via PCI bus or USB. The ISDN reference board enables problem-free addressing of the international telecommunications market.

11.4.3 Software

In addition to evaluation software that enables rapid access to the chip features, D and B channel protocol stacks are also available in the reference package for the ISDN area. The D channel protocol stack covers the USA (NI-1, 5ESS, DMS100), Japan (INS) and Europe (DSS-1), i.e. it can be used worldwide. Supplementary services, such as, "Completion of calls to busy subscribers" for example, are also integrated. The B channel protocol stack provides HDLC, transp., X.75, V.110, V.120, T.90NL and ISO 8208.

The data connection can be made with PPP (Point to Point Protocol) or MLPPP (Multi Link PPP). In addition to the protocol stacks, example applications are also supplied, such as "RVS-COM Lite" (RVS data technology) or "Teles Power Pack" (TELES), for example.

11.4.4 ISDN access

The IPAC reference board is an optimized passive ISDN PC card with S interface. This essentially consists of the highly integrated ISDN module, IPAC-X (ISDN PC Adapter Circuit PSB 21150) and a PC interface with PCI controller. The ISDN chip, PSB 21150, integrates 3 x HDLC for the B and D channels and the S transceiver.

For decoupling the data stream, a FIFO with a depth of 128 bytes is available for each B channel, and for the D channel, a FIFO of 64 bytes is available for each direction. The completion and reading of the FIFOs can take place "interrupt" or "DMA" controlled. In addition to freely programmable I/O pins, a serial PCM and an IOM-2 interface enable the communication with further modules. The IPAC-X supports both the TE and the LT-S, LT-T or NT mode. It can therefore be used multi-purpose, including e.g. PBX applications. The connection of the S interface is the first solution worldwide that only requires one single transformer and two capacitors instead of the conventional '2 transformers plus one choke' solution, which contributes to a significant costs advantage.

An intelligent network termination (NT+2ab) for ISDN with terminal adapter functionality is provided by the designs 'SIPB 8191-8/-16 (for 2B1Q)', and 'SIPB 8090-16 (for 4B3T)'. The NT+2ab provides the option of connecting ISDN telephones and two analog telephones. Additionally, internal switching between the telephones is possible, i.e. NT+2ab also acts as a PBX. With the analog lines (a/b), a programmable CODEC (2 channel DSP SICOFI-2TE PSB 2132) ensures A or µ-law coding and the necessary impedance adjustment. A highly integrated ISDN chip (INTC-Q PEB 8191 and NTC-T PEB 8090) is used as transceiver for the U interface and S interface.

Depending on the model, the NT+2ab is controlled via an 8-bit (C513A) or 16-bit (C161RI) microprocessor. The software, IOS (ISDN Operating Software), successfully tested on a global scale, implements the D channel protocols for Europe (DSS1+Supplementary Services), the USA (5ESS, NI-1, DMS100) and Japan (INS).

11.4.5 ISDN telephone

A reference board for a telephone is available as a further ISDN device. The 'SCOUT Telephone Board' (SIPB 21385) with acoustic box provides a high-end ISDN telephone. CLIP and COLP are also implemented as ISDN features, as are the supplementary services 'Call Parking', 'Call Waiting' and 'Three-Party Conference', to name just a few.

As an ISDN module, the 'SCOUT Telephone Board' contains the 'SCOUT-SX' (PSB 21383, P-MQFP-44), which, alongside the standard ISDN functions, contains an audio codec.

Programming of the module is carried out by the 16-bit µC C161OR. 1 MByte Flash and 32 KByte S-RAM are provided as memory. The power supply is sourced from the S_0 line via a DC/ DC converter. A telephone handset and the 8 Ω acoustic box can be directly connected as analog interfaces. The telephone can also be operated via the 20 key comprehensive input field and an LCD display, as with the COM interface of a PC.

The accompanying reference software "IOS" contains all the elements of layer1 to layer3, which are necessary for setting up an ISDN voice connection. A demo version without source code is already provided on the flash memory. Source codes of IOS can be acquired as starting point for some designs.

The program, UCIF (User Control Interface), is available as Windows 95/98/ NT4/2000 compatible software, using which all relevant settings can be made. These settings are, for example, the selection of the B channel (B1 or B2), download of codec parameters, and also the connection set-up to subscribers in the ISDN network.

An expansion module for full duplex handsfree operation is provided with the SIPB 21387. This module is automatically recognized by the IOS and UCIF software and integrated into the signal path. The answering machine module, SAM-EC PSB 4860, which has implemented an adaptive echo canceller algorithm, among others, provides the highest performance in this respect. The SAM-EC module is described in more detail in section 11.8.

New interfaces and new chip technologies (e.g. 3.3 V) will be consistently implemented in the future in new reference designs.

11.5 Quality analysis in the telephone network

Intelligent analysis systems are indispensable in proving the quality of a telephone service. Oskar Vierling GmbH + Co. KG, a specialist with long years of experience in measuring and test technology in telecommunications, provides such a system on the basis of experimental connections.

Providers of telephone services must be able to prove the quality of their network. On one hand, regular reports to the regulatory authorities are prescribed, on the other hand, verifiable quality is especially important for the customer.

"Quality of Service (QoS)" is the standard term used for the totality of all aspects that are important from the customer's point of view. The customer will, for example, ask: "Will I get through immediately when I call, or is it constantly busy?" "How quickly will my data be transmitted?" "Is the bill correct?" Network operators have an interest in examining their performance where these points are concerned and in rapidly detecting weak points in their network.

11.5.1 TIQUS for every telephone network

Vierling has developed the quality analysis system TIQUS for this very purpose. It consists of numerous automatic test units that are connected to the network just like normal subscribers. They carry out thousands of experimental connections around the clock and in doing so they test the network form the users perspective. With the continuous measurement, important statistical conclusions about the quality of the network can be reached, including, for example, the transmittance probability. The system also detects any faults that may be present, such as faulty charging units, for example, and consequently enables rapid counter measures.

The TIQUS system (Figure 11.16) determines the quality characteristics for fixed networks with analog and digital (ISDN) accesses, for mobile networks and for packet-switched X.25 networks. With its modular structure, it can be individually adjusted to the requirements of every network operator.

As a "Partner in Quality of Service (PiQoS)", Vierling therefore also provides measurements with TIQUS as a service and supplies each network operator with the quality characteristics of their network – completely, neutrally and in accordance with standards.

Figure 11.16
The fully automatic test telephoner: TIQUS test unit

11.5.2 With call test: The test connection

A test (experimental) connection is at first nothing more than a normal telephone call. A TIQUS test unit calls another unit, i.e. a connection is set up, a tone or data signal is transmitted and the connection is subsequently terminated. In addition, however, the sequence of each test connection is precisely examined. Delays with the connection set-up, whether or not the called test unit was reached, impairments of the transmitted voice, signal delays, echoes, the transmitted rate information – all essential data is recorded by the test unit. Instead of the "call", a precisely defined test signal is sent in order to test the transmission quality, that is then, the volume, room noises, voice quality or bit errors.

A control center is provided for controlling the test units. It is determined here, which parameters will be measured in which time frames. Using intelligent software, a time and space oriented optimized list of test connections is generated. The data contained in this list is distributed via ISDN switched connections to the test units, which subsequently carry out the test connections. After all measurements are made, the results data is retrieved again and saved and evaluated in the control center.

11.5.3 Infineon ISDN access technology

Each test unit combines different functionalities. The communication with the control center, the saving of the configuration and results data, the controlling of the test connection procedure, the generation of the test signals, the measuring of the different parameters and, of course, not forgetting, the connection to the network. The TIQUS system is modularly structured (Figure 11.17), and with a range of intelligent interface modules, it enables connection to the most diverse terminal subscriber interfaces in telecommunications networks. There are interfaces for the classic analog connection, for ISDN basic and primary multiplex connection, for X.25 network accesses and for mobile communications networks (GSM 900/GSM 1800).

For example, the following ISDN interfaces are implemented with Infineon modules:

- ISDN basic connection in the S_0 configuration with the ISAC-S module.

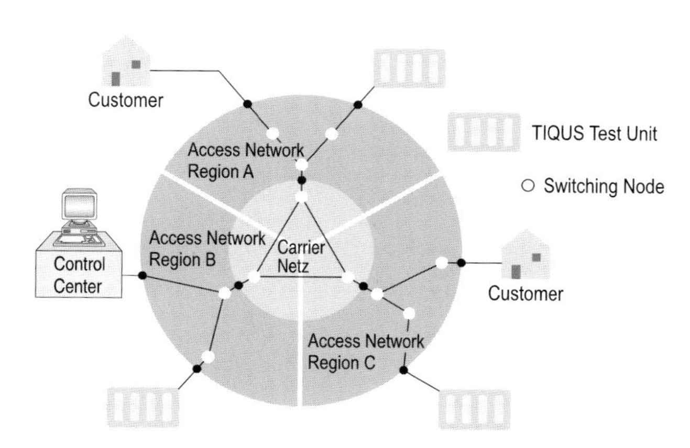

Figure 11.17
TIQUS test units are connected to the telephone network just like normal subscribers

- ISDN basic connection in the $U_{K0\text{-}4B3T}$ configuration with the IEC-T and ICC modules.
- ISDN primary multiplex connection S_{2M} with the FALC module.

The switching unit module, EPIC-S is also used for coupling the user information channels to the measuring DSPs

As part of the determining of the quality characteristics, the switching parameters (layer 1 to layer 3 of the corresponding protocols) and the transmission and data technical parameters of the user information channels are measured using the corresponding communications controller and DSPs.

The measuring and signaling-technical software functions are as interface independent and uniform as much as is feasibly possible.

The ISDN modules that are used provide excellent support for the TIQUS system concept. The completeness of the connection technologies, the module-spanning architectures and, of course, the product support, have all been decisive criteria at Vierling for many years, criteria on which Infineon can base its family of modules.

11.6 Flexible chip concept reduces costs with PBXs

The digitalization of the public network (ISDN) generates new solutions and perspectives for private branch exchanges (PBX). Inexpensive PBXs can be designed with just a small amount of highly integrated ICs.

Small to medium-sized systems with less than 30 subscribers currently account for more than 40 percent of connections. The market volume in this respect is currently 63 million connections. The highest growth rate is forecast at the moment for small to medium-sized systems.

11.6.1 Cost-effective system solutions

Major investments in application-specific software actually obstruct costs reduction with PBX systems, however, they remain inevitable. Low costs semiconductor components are accompanied by the growing integration and functionality of the semiconductor chip. Infineon Technologies is now establishing a further milestone with “Systems on Silicon”, an extremely flexible and costs-optimized system solution for both medium-sized and large digital PBXs.

11.6.2 Trend towards size reduction

A private branch exchange (PBX) consists of:

- Numerous subscriber circuits (layer 1 – controller) for connecting analog and/or digital lines
- Numerous controllers for exchange, signaling, internal communication, sequence interruption, etc.
- A voice and sound processing unit (based on a DSP)
- At least one microprocessor
- A memory
- A power supply

Until now a substantial number of ICs was required for this. These modules were therefore integrated even more and their number was reduced, where possible with simultaneous enhancement of their functionality. This produces several advantages:

- Lower production costs
- Lower space requirement
- Lower power consumption and consequently reduced power supply

11.6.3 ICs tailor-made for digital PBX

Development engineers profit with the new generation of highly integrated ISDN circuits from the abovementioned advantages. At the same time, the investments

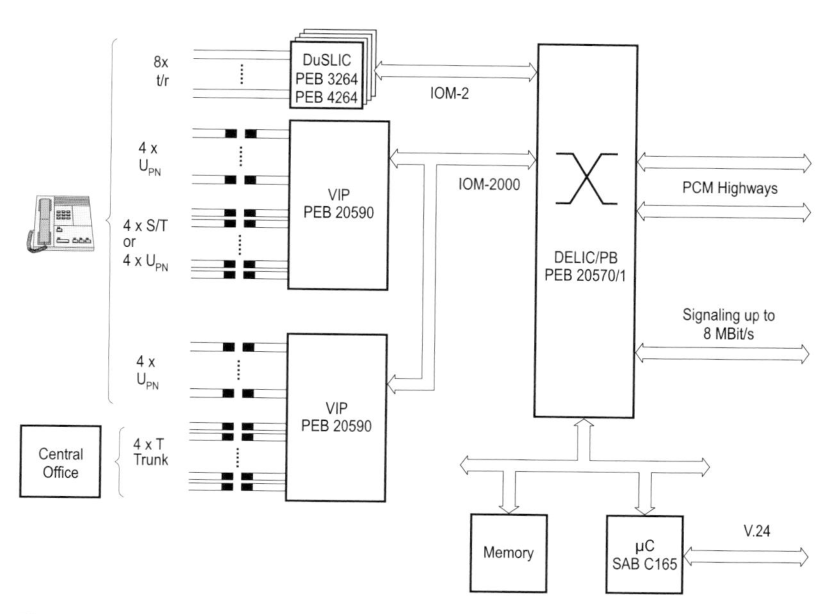

Figure 11.18 Operation of a PBX with just a small number of highly integrated components

already made by customers in software for microprocessors and DSPs with this concept do not lose their potential.

Figure 11.18 shows a small PBX based on the Delic/Vip chipset with 12 digital ($4 \times S_0$ and $8 \times U_{PN}$), 8 analog subscriber lines and 4 digital trunk lines.

The VIP (Versatile Interface Port, PEB 20950) is a layer-1 transceiver which can handle 8 digital interfaces. Up to four of these can be configured as S/T or U_{PN} interfaces, the rest are U_{PN} interfaces. This chip is connected via an IOM-2000 bus to the DELIC chip (PEB 20570/1). Up to three VIPs can be connected to the DELIC, giving up to 24 digital lines (up to 12 of these can then be configured as S/T interfaces). Furthermore, the DELIC has IOM-2 ports, allowing connection of standard layer-1 transceivers like the DuSLIC or SICOFI for analog lines and U_{k0} transceivers like IEC-Q (PEB 2091) for digital lines. Using these IOM-2 lines, connection of up to 16 digital or 32 analog lines to the DELIC is possible.

The DSP of the DELIC is used as a layer-1 controller (the state machines for the digital interfaces when using the VIP are controlled by the DELIC DSP), for switching and for signalling control. If necessary, special PBX software, such as tone generation, conferencing etc. can be integrated (DELIC-PB only). The microcontroller is used to control the DELIC chip via a mailbox interface. (Mailbox: Memory space shared by DELIC and µC). The DSP code is also downloaded in the internal RAM of the DELIC via this interface. For signaling purposes in a linecard application, a link with data rates of up to 8 Mbit/s is available. The solution with VIP-8 (PEB 20591) provides up to 24 S/T interfaces.

Bigger PBXs with 1,000 or more subscribers can be implemented with linecards consisting of Delic/VIP solutions and switching matrix modules.

11.6.4 PCM switching solutions

This section shows solutions for PCM switching matrices. These solutions are frequently used on group switches and switching networks. With the SWITI chips, it is possible to design virtually unlimited PCM switches. Figures 11.19 to 11.21 show a few examples, ranging from a non-blocking PCM switch for 512 channels to a non-polluting switch for 2048 PCM channels. The solutions shown may also be used with different data rates.

With the Memory Time Switch series (MTS), it is possible to implement various PCM switching matrices. The smallest solution is shown in Figure 11.19. With this solution, MTSI (PEF 20450) is used for implementing a blocking-free switch, which connects 512 input time slots with 512 output time slots. The MTSI needs an external oscillator or crystal of 16.384 MHz or 32.768 MHz, only the data rate of the PCM highways is 2048 kbit/s. This gives 8 lines with 32 timeslots each for input and output.

To allow smaller boards in high-capacity switches, two more members of the MTSI series are available. Figure 11.20 shows the MTSI-L (PEF 20470) as a non-blocking switch with a capacity of 1024 incoming and outgoing PCM channels. Using the MTSI-L, the input and output data rates can also be varied from 2048 kbit/s up to 16384 kbit/s.

Finally, the MTSI-XL (PEF 24470) can handle up to 2048 incoming and 2048 outgoing time slots. The external oscillator or crystal frequency can be 16.384 MHz or 32.768 MHz, the data rate of the inputs and outputs can be from 2048 kbit/s to 16384 kbit/s. A sample configuration is shown in Figure 11.21.

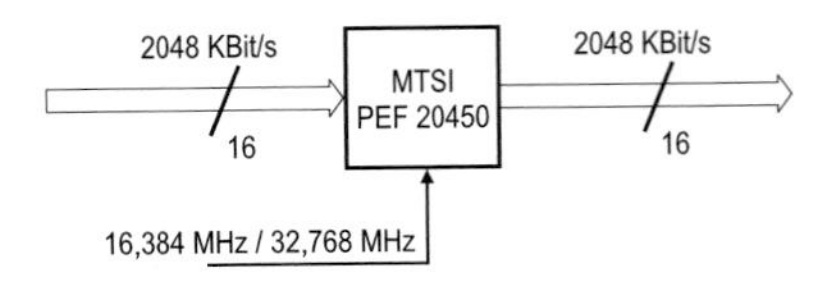

Figure 11.19
Non-blocking PCM switch for 512 time slots

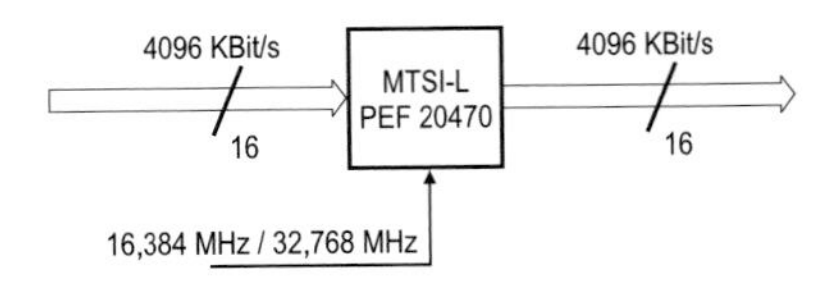

Figure 11.20
Non-blocking PCM switch for 1024 time slots

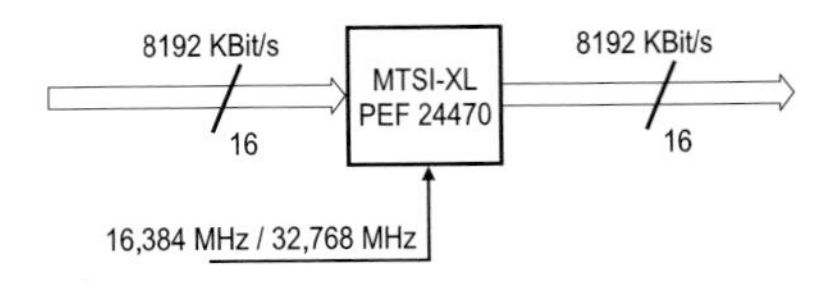

Figure 11.21
Non-blocking PCM switch for 2048 time slots

Figure 11.22 shows a conferencing solution with MTSI and DELIC-PB. Any input channel up to a total number of 64 can be switched in 21 independent conferences simultaneously, if only conferences

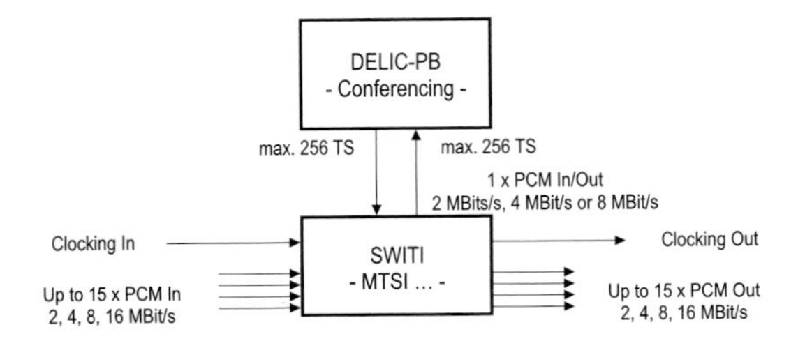

Figure 11.22
PCM switch with explanation of conferencing functionality

with three participants are needed. Any conference combination from 3 subscribers in 21 conferences up to 64 subscribers in only one conference is possible. The input and output channels can be attenuated very precisely in 64-kbit steps (16-bit word).

The conferencing functionality is provided by an integrated DSP chip which also handles conference overflows and noise suppression using different thresholds. The DELIC-PB can be programmed very easily with a PC-based configurator.

11.6.5 Using SWITI to connect to H.100/H.110 buses

The SWITI chip family supports the H.100/H.110 bus, which is the key to high performance Computer Telephony (CT) and Voice-over-IP applications (especially in PCI and Compact PCI systems), one of the most important future technologies in telecommunications. The MTSI chips are normally PCM switches; the HTSI chips can be connected to the H.100/H.110 bus. Both MTSI and HTSI (in the XL versions) can process up to 2048 connections with data rates of up to 16384 kbit/s.

Figure 11.23 shows an application that uses the SWITI as a backplane switch. The linecards are connected to the PCM highways of the MTSI (or HTSI). If no H.100/H.110 bus is used in the system, the MTSI or the HTSI can be used in M mode (thus disabling the H.100 support). The linecards may use the DELIC or the ELIC/EPIC to connect to the PCM highways.

Figure 11.24 illustrates a basic concept for a 'Computer Telephony (CT)' application. Modules such as voice compression, voice recognition, fax server cards, base transceiver stations (BTS), linecards etc. are connected to the H.100/H.110 bus using the HTSI. This system can be designed using the Compact PCI bus standard. The connectors used with this bus include a complete H.110 (H100) compatible bus on the backplane.

HTSI can be used in a Voice-over IP application to connect a conventional PBX to the H-bus (as shown in Figure 11.25). A Vocoder card, which is also connected using HTSI with the H-bus, carries out the voice compression and decompression. A network interface card connects the whole system to the LAN or WAN.

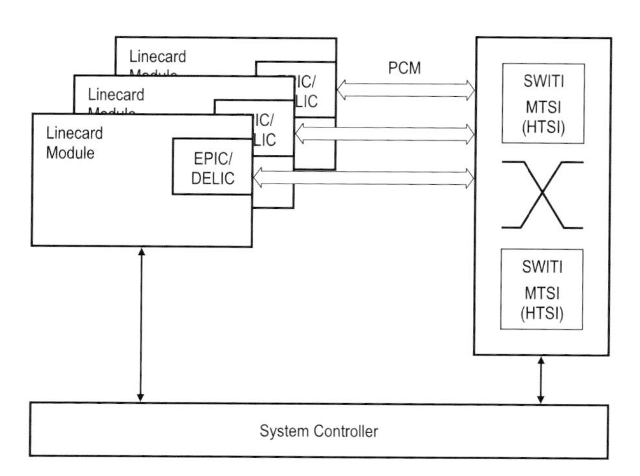

Figure 11.23 Using SWITI as PCM backplane switch

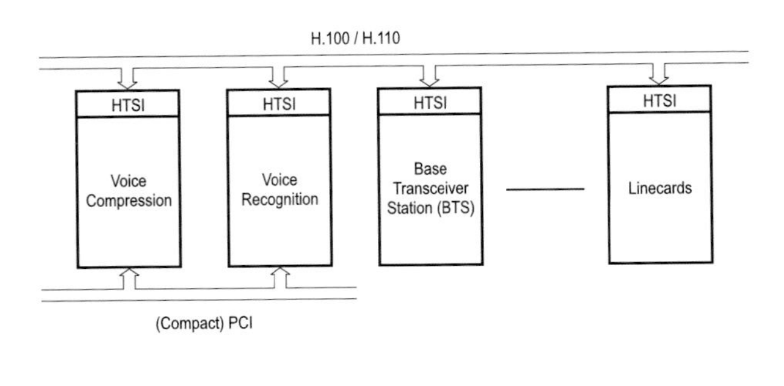

Figure 11.24
Using SWITI in a CT application

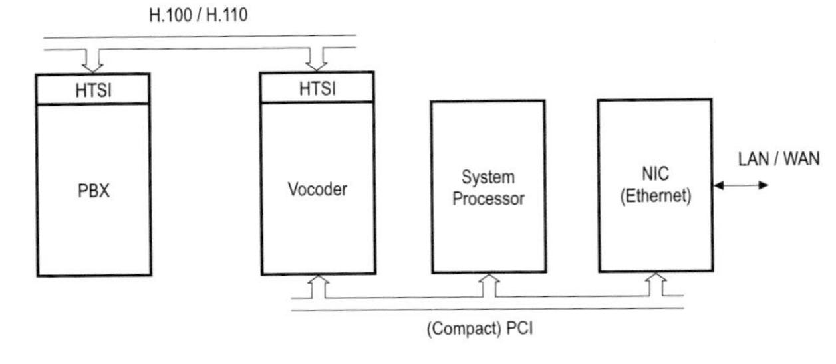

Figure 11.25
Using SWITI in a Voice-over IP (VoIP) application

11.7 Next architecture generation of mobile terminal equipment – GOLDen-future for GSM

Even though the growth of the GSM market had to weather a substantial slump in 2001, the market potential is once again present here and now today. Many factors contribute to the incomparable success of the "Global System for Mobile Communications (GSM)". Included among the most important are the simple operability and convenience that small, light and inexpensive GSM mobile phones provide. In conjunction with the offer of innovative subscriber packages, GSM has developed a strong end consumer basis.

The GSM standard has been constantly supplemented in order to cultivate this growth. Among other factors, the quality of the voice service, the capacity of the networks and the performance capacity of mobile data transmission have all been improved. In the meantime data services (GPRS) with high data throughput rates on the basis of line-switched connections have become available.

In order to meet and exceed the multiple requirements of the GSM mobile terminals, Infineon Technologies has developed the high-performance, high-integration 'E-GOLD Architecture' with ultra-low power consumption for cellular dual or triple band telephones. E-GOLD is the platform for "leading edge" terminal products that provide voice, fax, rapid data transmission and additional "comfort" features.

11.7.1 E-GOLD – Expanding the GOLD Standard

Since the introduction of the GOLD chipset at the beginning of 1993, Infineon Technologies has established itself as the leading provider of standard GSM solutions and has set standards ever since. The GOLD chipset was the first to achieve full GSM type approval.

The E-GOLD platform represents the latest generation of GSM solutions. Building on the abundance of experience that has been gained with the products of the previous platforms, it advances the system integration significant steps forward, and in doing so, guarantees the customer investment security in developments and short "time-to-market" cycles. These advantages will also be maintained with future versions of the GOLD architecture that use the wide-ranging Infineon Technologies know-how and Infineon's leading CMOS technologies in targeting system solutions with the highest performance, lowest power consumption and highest costs optimization.

The architecture is designed for simple and efficient integration with multiband radio solutions, for example, with the single-chip HF circuit, SMARTi (Siemens Multi-Advanced Radio Transceiver IC). The E-GOLD chipset simply includes two baseband ICs: The single-chip E-GOLD microcontroller and DSP and its companion, E-GAIM (GSM Analog Interfacing Module), which implements AD and DA convertors for the radio and voice band interfaces. With the integrated DSP and the static 16-bit CPU core, E-GOLD provides full-rate, half-rate and enhanced full-rate voice coding and supports dual and triple band HF implementations. It additionally supports modern multi-slot data transmission and adaptive multi-rate voice coding (AMR) and packet data (GPRS).

The up-to-date power management system guarantees ultra-low power dissipation and opens up new dimensions for the operating time of mobile phones. The highlights include:

- Higher degree of integration and optimized HF interface for a minimized number of external components
- DSP firmware with numerous features, e.g. 'Triplerate Vocoder', 'High Speed Data (HSCSD)', 'OnChip PRAM' for customer-specific features
- Ultra-low power dissipation
- Small BGA (Ball Grid Array) package

A more detailed integration is attained with E-GOLD, which for the first time combines the analog and digital baseband functions on one single chip.

Special emphasis should be made here on:

- 0.18-μm-, 18-volt semiconductor technology
- DSP core OAK+ with 78 MHz clock frequency
- Multi-function DSP firmware including adaptive multirate (AMR) vocoder
- Complete two-chip GSM system solution with SMARTi.

11.7.2 Application support

Infineon Technologies provides extensive application support that ranges from documentation and tools through to a form-factor reference design that can significantly reduce the development time and resources of the customer and thereby shortens the "time-to-market" period. The E-GOLD evaluation board (Figure 11.26), which provides a complete, tested reference environment with easy access to all system interfaces, is one of the most important of these tools. With a new generation of on-chip debug and monitor features, it makes the parallel development of customer hardware and software significantly easier.

Figure 11.26 E-GOLD evaluation board

Infineon additionally provides layer-1 driver and GSM protocol stack support. Standard development tools for the C166 microcontroller support the software development. E-GOLD/E-GOLD+ provide tested and optimized DSP firmware developed by Infineon Technologies that meets and exceeds ETSI and GSM network requirements.

Recommended partners for hardware and software development can also help to accelerate the product development – Infineon has a partnership with Debis Systemhaus, a subsidiary of DaimlerChrysler Services AG, for supplying a complete GSM software stack, which has been integrated into the E-GOLD platform. A co-operation is also underway with ATL Research A/S, a well-recognized GSM system design house.

11.7.3 The future has already begun

Based on the award-winning, new high-performance Cores TriCore™ DSP, Infineon Technologies will establish an open development platform for future customer-specific solutions and cellular standards, such as, Universal Mobile Telecommunications System (UMTS), for example.

The new architecture is optimized for simple and efficient integration with Infineon multiband radio solutions, for example, with the single-chip HF circuit, SMARTi

With this unique silicon platform, customers will be afforded a high degree of flexibility and performance. The path forward to mobile multimedia devices and services of the future will be consequently opened up.

11.7.4 GSM module

We provide an all-inclusive solution for the dual band 900/1800 GSM terminal device. It consists of the integrated hardware in an extremely small form-factor single-side, sub-200 component count with extremely low BOM, populated PCB and the software protocol stack layer 1 to 3 as shown in Figure 11.27. The semiconductor content is based on the most recent additions to our proven and successful product families to include the monolithic baseband processing chip, E-

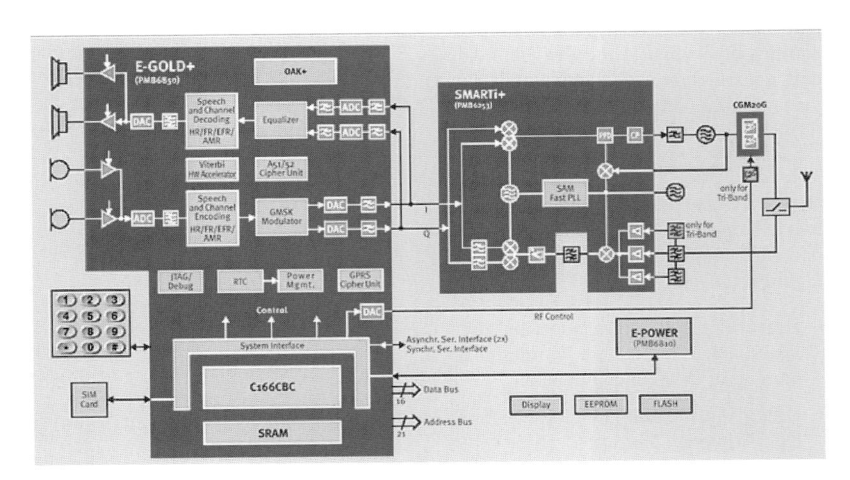

Figure 11.27 Transceiver SMARTi+ and the multi-function PA module (all-in-one-housing)

Gold+, fully integrated power management system, E-Power, single-chip RF transceiver, SMARTi+, and the all-in-one-housing PA module.

11.8 Digital answering machines

Answering machines can be arranged into three market segments. These are differentiated according to how the announcements or messages are recorded:

- Analog machines: Messages and announcements are recorded analog on a magnetic tape.
- Digital machines: Messages and announcements are digitally coded and saved.
- Hybrid machines: The announcement is saved digitally; messages are saved analog.

With the recording of telephone signals, less demand is made on quality than in the HiFi range. The range between 300 and about 3400 Hz is sufficient. The benefits of semiconductor technology are:

- Semiconductors are not subject to mechanical wear and tear.
- The quality of the recording does not become impaired with time.
- The necessary signal processing and memory modules have become much more cost-effective as technology advances and mass production becomes more efficient.

Conversion to digital values generally takes place with a sampling frequency of 8000 Hz with a resolution of 8 bit. However the accruing data volume is still substantial: 64000 bits must be saved every second.

11.8.1 DSP reduced data

One solution with digital answering machines is provided by the Digital Signal Processor (DSP). Using suitable signal processing algorithms, the chip compresses digital data. Data volumes and memory requirements are consequently reduced within specific limits without audible quality loss.

Table 11.1
Quality dependent on compression factor

Compression	Quality
<3	excellent
3–10	good
10–22	sufficient
>22	inadequate

However the compression factor cannot be increased without limit. The comprehensibility of the recording suffers first, particularly with a high degree of interference noise. And even if the voice signal is not significantly disturbed, the recording may seem 'synthetic' with a further increase of the compression. A good voice compression algorithm (coder) shows its real value with a justifiable compromise between compression factor and recording quality (Table 11.1).

'Excellent' means that no difference can be heard between the original and the recording. 'Inadequate' quality is comparable with that of a used magntic tape. Because of the price competition for analog recording, practical compression factors seen from today's perspective are not below five, nor are they above 20 because the quality would no longer be acceptable.

11.8.2 Single-channel codec is sufficient

Although the voice compression presents the most important task for digital answering machines, depending on the application, some additional tasks still remain to be completed. The most important requirements can be illustrated using two applications, that of a standard an-

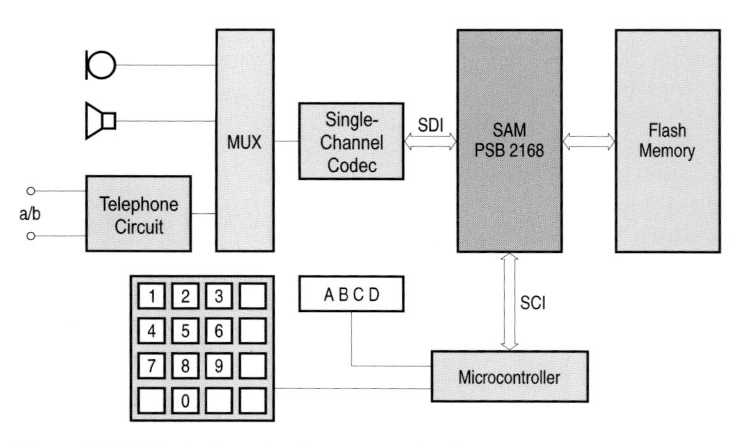

Figure 11.28 Single-channel codec as connection between digital and analog part

swering machine and that of a DECT base station. Microphone, loudspeaker and telephone line are connected to a standard answering machine (Figure 11.28) as signal sources and signal sinks. Amplifiers are required for the microphone and loudspeaker; the line connection requires a hybrid (telephone circuit). As recording is made either from the line or from the microphone, a single A/D converter suffices. The answering machine must never play back two different messages over the loudspeaker and the telephone line at the same time, so that again a single D/A converter suffices. Both converters together are, for the most part, referred to as single-channel codec. Using an analog switching network, the respective necessary connections are set up to the signal sources and signal sinks.

The digitalized signals must be compressed with the recording by a coder and with the play-back, decompressed by a decoder. Furthermore, for the remote enquiry, dial tones must be detected by a DTMF detector. DTMF (Dual Tone Multi Frequency) is a coding procedure with which a paired combination presents several frequencies as one digit. The end of a recording is logged by a CPT detector (CPT: Call in Progress Tone) on the basis of the busy tones. It is useful to detect the signaling tones generated by modems and fax machines with a CNG detector (CalliNG tones) so that these calls can be rejected in good time. If the answering machine also has to forward messages, it must be able to generate dial tones (DTMF generator). A realtime clock (RTC) is now also included as standard with standard answering machines. Memory space for storing the voice messages must also be provided. An option for storage for pre-programmed phrases for voice synthesis is required in case the answering machine does not have a display.

With a 'comfort features' DECT base station, additional tasks must be completed. For handsfree operation in the analog area, a second codec with amplifiers is required. The handsfree operation itself is implemented by a corresponding signal processing algorithm. It should additionally be possible to detect and decode information sent from the exchange to the called subscriber. This procedure has been in use for some time in the USA as 'caller identification' and is become increasingly more available in Europe and South-East Asia. A CID decoder is additionally required for this procedure. In comparison with standard answering machines, a DECT system can include different connections of signal sources and signal sinks, which requires a digital switching network. Administration tasks (e.g. subscriber directory) and user interface (display, keypad/keyboard) also require considerable added expense in comparison with standard answering machines.

11.8.3 SAM provides costs optimization

A costs-optimized solution for all applications would provide the ideal situation. The necessary compromise lies between several somewhat varying solutions that are each designed for a respective application and a single solution that is suitable for all applications. Individual solutions are frequently an unfeasible option because of the high development costs that are incurred for each device. If the costs advantages of mass production are, however, to be exploited, features must be paid for, which many users may never even use.

In order to attain optimized system costs with the optimum distribution of the system tasks among the appropriate components, the SAM (Sophisticated Answering Machine) chipset has been developed with the following components:

- PSB 4851: Double codec with amplifiers and analog switching network
- PSB 2168: DSP for standard digital answering machines
- PSB 4860: Like PSB 2168; additionally with handsfree comfort feature (full duplex with acoustic echo cancellation)

Almost all applications with an integrated digital answering machine are consequently optimally covered with just three modules. Standard modules can be used for all other tasks (standard codec, memory, microcontroller).

Analog amplifiers, programmable independent of each other

The SAM module, PSB 4851, (Figure 11.29) is specially tailor-made for the requirements of analog 'comfort' telephones with digital answering machines. It contains two A/D converters and two D/A converters that serve two independent channels in the digital area. Three signal sources and three signal sinks can even be connected at the analog side via amplifiers that are independently programmable from one another.

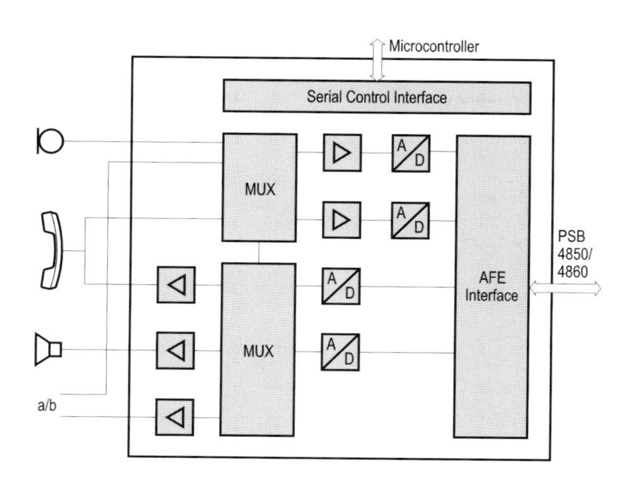

Figure 11.29
Three signal sources and signal sinks can be connected to the SAM analog frontend, PSB 4851

The receiver inset, the receiver microphone, the loudspeaker and the handsfree microphone can be connected without any further active elements. Under normal circumstances, a standard telephone circuit is required for the connection to the a/b line. The assignment of the individual amplifiers to the converters can also be programmed. As a special feature, the converters can be completely switched off. In this case the analog signal sources and signal sinks are directly connected with one another via an integrated switching network. The power consumption of the module has therefore been reduced so much that a supply is possible from the mains. A telephone that uses this feature is therefore still functional within certain limits with a power failure and can be used as an emergency telephone.

The PSB 4851 has its own interface to the PSB 4860 module, which has been developed for this purpose and a serial standard interface for programming with a microcontroller.

Digital module, upward compatible

The modules, PSB 2168 and PSB 4860 (Figure 11.30), are upward compatible.

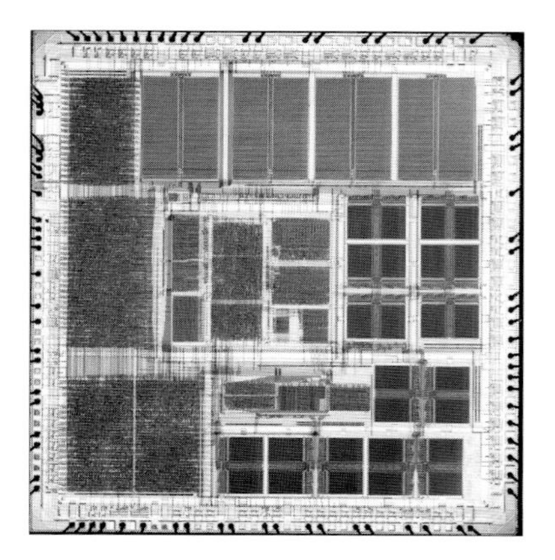

Figure 11.30
Upward compatible digital module, PSB 4860

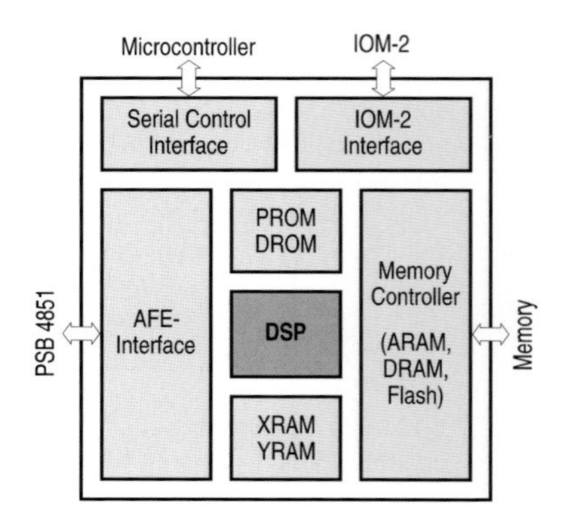

Figure 11.31
The DSP is the heart of all three digital modules

This affects the pin assignment and the programming. They differ only in the number of interfaces and functional blocks. Figure 11.31 shows a block circuit diagram that applies equally for all three modules. The program memory (PROM), the memory for constants (DROM) and the working memory (XRAM, YRAM) are grouped around a DSP core. The memories are scalable, which means they can vary in size depending on the module. The outward connection is set up by interface modules that are independent of one another.

Almost all of these modules can be programmed by the microcontroller within wide-ranging limits. The modules can consequently be integrated in a number of different systems without additional external expense.

Flexibility through modularity

Each of the three digital modules is a single signal processing module that is independent from the others, which can all be implemented using the corresponding DSP program components. Although these modules are implemented with software alone, the user can treat them as if they were hardware components. For the user a module does not correspond with a

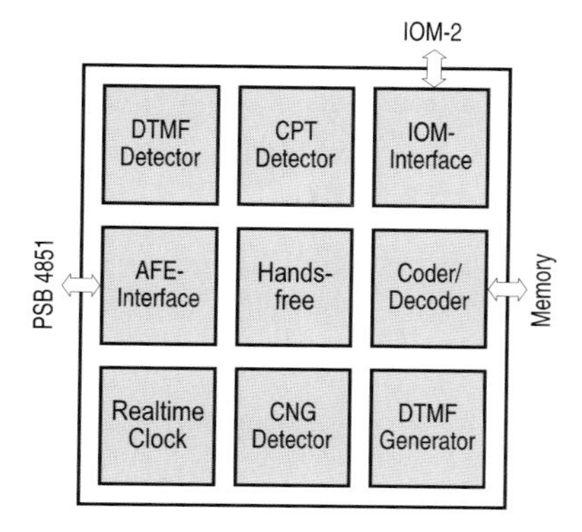

Figure 11.32
Digital module as building block set system

functional block diagram, but rather a building block set that contains different modules (Figure 11.32).

Each module in the building block set can be equipped with several signal inputs and several signal outputs. Using the building block set principle, each input of a module can be connected with each output of another module independently of one another. The options included in this concept are illustrated in the following examples with the same two modules:

DTMF selection

Two of three inputs of an interface module remain open. The third is connected with an output of the DTMF generator. Via the appropriate frequency and amplitude settings of the DTMF generator, the microcontroller can generate each DTMF tone and send to the exchange with the appropriate volume. At the same time, a listening-in of the dialing process is possible with reduced volume via the second output of the DTMF generator.

Ringer tone

If an incoming call is signaled, then a DTMF generator output can be connected with an input of the interface to the loudspeaker (PSB 4851, channel 2). The microcontroller here generally only switches on one of the two frequency generators of the DTMF generator at the appropriate intervals.

Call waiting tone

A call is already in progress here while a second call comes in (example: ISDN). As with the ringer tone, via a second, until now open input of an interface module, any tone can be cut into the call.

This concept covers almost any application case provided the modules in the building block set are sufficient. In contrast to the usual fixed-wire solution ("record until busy tone or more than three seconds silence is detected, then delete the last three seconds"), application cases that have not been planned for can consequently also be covered. The individual modules can be extensively parameterized in the process. The coder presents an example in this respect. It contains, among other elements, three compression factors. With factor 6.5, yet another very good voice quality is attained; with factor 19 with sufficient quality a long recording time can be attained (20 minutes with 4 Mbit memory). In line with video recording technology, these modes are described as HQ (High Quality), SP (Standard Play) and LP (Long Play) Diverse options again result in combination with the modular concept:

- The recording of the announcement is made in HQ; recording of messages in LP or SP.
- The recording type can be switched over by the user – for longer periods of absence, LP; otherwise HQ.
- The recording type can be automatically switched over – only when the memory is running out is a switch made from HQ to SP or LP.
- Recompression – automatic or manual later compression in SP or LP of a message recorded with HQ. Additionally, in accordance with the modular concept, only the decoder must be con-

nected with the coder, whereby the decoder decompresses with HQ and the coder simultaneously compresses with SP or LP. The manufacturer also has room to play around here, e.g. a listened-to message could be automatically recompressed to save memory space.

11.8.4 Development made easy

To keep the development costs for the overall system low, a series of development aids are also provided parallel to the modules.

Of particular importance here are:

- reference design for 'comfort' telephones with digital answering machines and
- modular control software in "C" (source code); suitable for typical microcontrollers.

With a functioning system as the basis to start from, these tools enable the terminal equipment manufacturer to test their own developments and improvements and to quickly work out their own design with the required features.

Strong basis for future developments

A wide range of applications from standard answering machine to ISDN-DECT base station can be covered with just a few variants. At the same time these modules serve as a supportive basis for future developments.

The modular concept simplifies the variation of the modules. Both the previous hardware and the already available software modules can continue to be used here. Beneficial for the user: the modules are compatible with one another and can be transferred without being changed.

A number of SAM modules, for example, have consequently been integrated into the product series of the ISDN telephone module, INCA. While modules like the DTMF and ring tone generator have been taken over without change, with INCA, the handsfree algorithm was improved on the basis of the higher DSP performance capacity.

The principle of the acoustic echo canceller and the level balance, which is required for handsfree, is however the same with both implementations.

11.9 Handsfree algorithms

The operation of a telephone has changed very little since the invention of the phone system. The user still has to place a microphone close to his mouth and a loudspeaker at the ear. When implementing handsfree telephone systems, powerful signal processors with high sophisticated signal processing algorithms are necessary.

Using Infineon Technologies products, high-end full duplex handsfree systems can be developed, the performance of which will be practically unsurpassable and the price of which cannot be beaten. This flexibility has its price – more than 200 parameters must be optimized.

11.9.1 Handsfree systems

Handsfree systems essentially consist of three signal processing algorithms:

- Suppression of acoustic feedback (echo) due to alternating attenuation of a subscriber (echo suppressor, level balance)
- Active compensation of acoustic feedback with adaptable echo filtering (echo compensation)
- Post filtering for better audibility (e.g. noise reduction)

As a result, handsfree systems can be divided into two classes. 'Where' the system is to be integrated is not decisively important here. It can therefore be installed in a car as well as used for video conferencing, office phone systems or somewhere completely different as long as it is a handsfree communication system.

11.9.2 Full duplex systems

Systems with full duplex algorithms (FD) consist of at least one echo canceller. This reduces the loudspeaker signal reflected from the walls and surroundings of the transmission path. Usually an automatically adjustable switched loss algorithm is added to the echo canceller to improve the quality of the system at the beginning of a phone call. After adaptation of the echo filter, the switched loss algorithm is faded out, because full duplex conversations can only take place without additional switched attenuation.

11.9.3 Half duplex systems

Systems with half duplex algorithms (HD) are less highly developed handsfree systems. They enable the calling parties to use the line sequentially only. Simultaneous talking is not possible. The comfortable "stronger wins" algorithm, which always switches the stronger party, still needs a lot of discipline in the conversation. In this case, the weaker signal is attenuated to reduce the speaker's echo.

Infineon Technologies has a wide-ranging product portfolio of devices with handsfree implementations:

SCOUT-SX	(PSB 21383):	HD
SCOUT-PX	(PSB 21393):	HD
SAM-EC	(PSB 4860):	VD
ACE	(PSB 2170):	VD
ACE-R	(PSB 2171):	VD
INCA-S	(PSB 21483):	VD
INCA-P	(PSB 21493):	VD

Because of their flexibility, these devices are perfectly suited for many different applications. They can be optimized to the acoustical behavior of an office phone as well as to the acoustically very difficult surroundings of a car.

11.9.4 Echo cancellation (full duplex systems)

The question, often arises as to why we need echo cancellation to get a full duplex conversation. And there is a very easy answer to this: We don't need echo cancellation for a full duplex conversation. Look at a handset and you will not find any echo cancellation, but it will work in full duplex mode. So lie the handset on the table and you can talk handsfree. The only problem is the loudspeaker volume and the microphone gain. When you start to increase the volume and gain we have mentioned, you will end up in a very unpleasant system. The far end caller will hear a very strong side tone and echo of his voice. When we use such systems on both sides, a 'whistle' is generated. To avoid this, we need to cancel the echo and provide a defined terminal coupling loss.

At first observation, a simple solution would be to use the signal driving the loudspeaker, and subtract this at an appropriate amplitude from the microphone output signal. However this approach assumes the microphone detects exactly the same signal as the output of the loudspeaker. Fine if your environment happens to resemble an echo-free chamber, but problematic elsewhere. Instead all the echoes with delays, phase shifts and reverberations have to be considered.

When we think of a model in signal theory that could reproduce phase shifts, delays and reverberation, we end up in an impulse response $h(k)$ (Figure 11.33). We have to assume that the acoustic environment will act like a linear operator on a signal source. Fortunately the world of physics is good to us and the effects of the room actually are linear operations.

Nonlinear effects, such as distortion in the loudspeaker or resonant vibrations of the telephone housing have to be avoided by the acoustic design of the telephone set. Changes within the room must also be taken into account. The impulse response will change when we open the

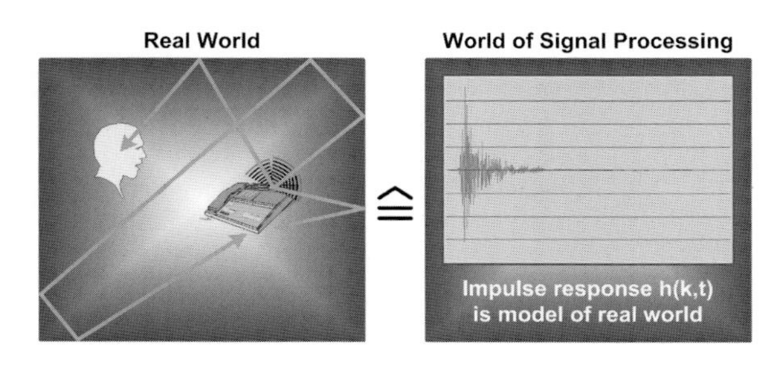

Figure 11.33
Model of the acoustic environment

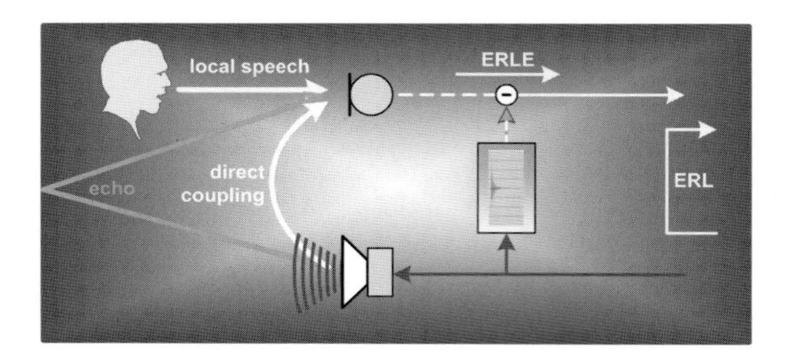

Figure 11.34
Principle of echo cancellation

door, flap a folder over the phone or walk around. So we have to use a time variant impulse response *h(k,t)*. When we assume a filter containing the required impulse response *h(k,t)*, we can feed the filter with the loudspeaker signal and subtract the output of that filter from the microphone output signal. The echo and direct coupling will be cancelled by the system, so that only local speech will pass (Figure 11.34).

To build an echo canceller, you need an algorithm that adapts the filter coefficients of the filter that provides the impulse response *h(k,t)* to the real acoustical world. Due to the fact that the resolution of the filter and the equality of adapted impulse response and the real world are never perfect, an echo canceller can only reduce the echo. The amount that the echo is reduced by is called Echo Return Loss Enhancement (ERLE). Even more important is the Echo Return Loss (ERL), which is the attenuation between the loudspeaker signal (input signal of echo canceller) and the output from the echo canceller. The coupling between loudspeaker and microphone is taken into account in the process. As a result, the microphone gain and loudspeaker volume will also have an influence, when we assume ERLE is a fixed value.

Basically there are two different types of implementations. The fullband mode with a maximum loop delay of 5ms and the subband mode which is much faster in adaptaion, but causes up to 35ms delay. A standard office phone can create a loop delay of up to 5ms. The subband mode is used with GSM applications and video phones.

Fullband mode

The implementation of fullband mode is based on a normalized least-mean square algorithm (NLMS) also called stochastic gradient algorithm. An FIR (finite impulse response) filter is used as echo filter. The filter coefficients are adjusted in such a way that the mean square error rate of the filter is minimized. A correction pro-

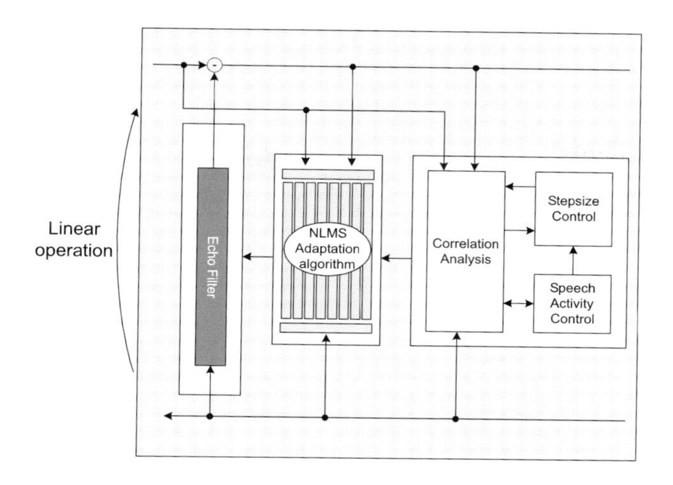

Figure 11.35
Echo canceller, fullband mode

portional to the negative of the gradient will increase the quality of the filter step by step. The number of filter taps can be a maximum of 768 taps. The number of filter taps has an influence on the echo canceller tail length directly, which is calculated as "number of filter taps / 8000". When an echo caused by reverberation of the room stays longer than the echo canceller tail length, it will not be cancelled. For normal offices, an echo canceller 'tail length' of 50 ms is sufficient.

The correlation analysis block in Figure 11.35 is necessary to ensure proper adaptation. It decides if there is enough far end speech activity to start adaptation and if there is local activity to stop adaptation immediately. In the case of adaptation during 'double talk', the algorithm would adapt its impulse response to a mixture of echo and local speech, which is not allowed.

Noise reduction

A noise reduction feature is available which is placed into the transmission path between the echo canceller and the echo suppressor. This spectral noise reduction attenuates in the frequency range, in which no speech activity is recognized. The maximum noise reduction can be adjusted. It is recommended to keep the maximum noise reduction below 15 dB. At higher values, the transmitted voice signal sounds robotic.

Subband mode

As well as in fullband mode, an NLMS algorithm is used in subband mode. There are 8 frequencies of which the lowest (subband 0) is only available as real part of the complex signal. All other subbands are divided into real and imaginary parts (Figure 11.36). As a result, 15 FIR filters are used and each subband filter is adapted separately.

Due to the signal processing in subbands, the noise reduction also works in the subbands. The noise reduction programming is the same as in the fullband mode. The DSP resource requirements in the subband mode are less than in the fullband mode, because the subbands are adapted sequentially. Nevertheless, the adaptation is carried much faster. The detection of significant changes in room acoustics is detected by a room change detector. This detector consists of a small echo canceller with a few filter taps in subband 1 only, which is used to detect faulty adaptations of the echo filters. An example will show the benefit of this: A phone call comes in

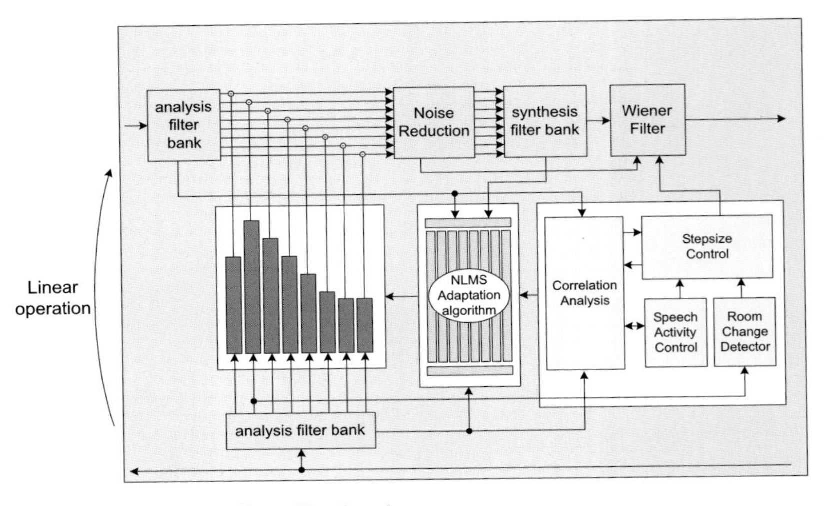

Figure 11.36 Echo canceller, subband mode

and the echo filters are adapted to their optimum. The step size is reduced to a minimum to keep the good adaptation. So if, when the local user flaps a folder cover over the telephone, the adaptation no longer matches the acoustics of the room, but the step size is small, then changes to the echo filters are also small. The room change detector does not have a variable step size, so the small filter adapts to the new situation very quickly. The difference between the actual echo filter and the room change filter exceeds a threshold, so that the step size for adaptation in all subbands is changed to the maximum value.

11.9.5 ITU-T Recommendations

The ITU (International Telecommunication Union) updated most of the handsfree recommendations in 2000. Acoustic engineers can use these recommendations to adjust parameters (Table 11.2). The recommendations advice with regard to how to implement good speech quality in a handsfree system. Nevertheless, a lot of experience is required when constructing a high quality telephone.

11.10 DSL architectures

Since their inception, Digital Subscriber Line (DSL) technologies and products have conquered the market with their wide-ranging diversity. In the process, they have not just opened up new possibilities and opportunities, but they have also created a somewhat unmanageable and confusing market. This section provides an overview of the technology that makes it possible to transmit information over copper wire loops and the evolution of the various DSL technologies.

Table 11.3 provides a summary of the different standards, data rates, transmission techniques and maximum loop lengths.

11.10.1 Basic DSL concepts

The PSTN and supporting local access networks were designed with guidelines that limit transmissions to a 3,400 Hz analog voice channel. For example, telephones, dial modems, fax modems, and private line modems limited their transmissions over the local access phone lines to the frequency spectrum that exists be-

Table 11.2 ITU-T recommendations for handsfree terminal equipment

Transmission characteristics of handsfree terminals		
P.340	05/2000	Transmission characteristics and speech quality parameters of handsfree terminals
P.342	05/2000	Transmission characteristics for the telephone band (300-3400 Hz), digital listening-in and handsfree telephony terminals
Reference attenuations		
P.76		Determination of the reference attenuations; fundamental principle
P.79	09/1999	Calculation of reference attenuations for telephone sets
Talker Echo		
G.131	08/1996	Control of talker echo
Test Signals and Methods		
P.501	05/2000	Test signals for use in volume measurement
P.502	05/2000	Objective test methods for speech communication systems using complex test signals
P.581	05/2000	Use of Head And Torso Simulator (HATS) for handsfree terminal testing
P.832	05/2000	Subjective performance evaluation of handsfree terminals

tween 0 Hz and 3,400 Hz. The highest achievable data rate using the 3,400 Hz frequency spectrum is less than 56 kbit/s. So how does DSL technology achieve data rates in the millions of bits per second over the same copper loops? The answer is simple – eliminate the 3,400 Hz boundary. DSL uses a much broader range of frequencies than the voice channel. An implementation of this kind requires transmission of information over a wide range of frequencies from one end of the copper wire loop to another supplementary device, which receives the broadband signal at the far end of the copper wire loop.

However, the following effects must be minimized here:

1. Attenuation – The power dissipation of a transmitted signal as it is sent over the copper wire line. In-house wiring also contributes to attenuation.
2. Bridged tap – These are un-terminated extensions of the loop that cause additional loop loss with loss peaks in the frequency range of a quarter of the wavelength of the extension length.
3. Crosstalk – The interference between two wires in the same bundle, caused by the electrical energy carried by each.

Attenuation and resulting distance limitations

The transmission of an electric signal can be compared to driving a car. The faster you go, the more energy you burn over a given distance and the sooner you have to refuel. With electrical signals transmitted over a copper wire line, the use of higher frequencies to support higher-speed services also results in shorter loop reach. This is because high-frequency signals transmitted over metallic loops attenuate energy faster than the lower-frequency signals.

Table 11.3 Key data for different DSL technologies

DSL Transceiver Reference Table		DMT ADSL	CAP RADSL	CAP S/HDSL	2B1Q S/HDSL	2B1Q IDSL	CAP SDSL	G.shdsl	ReachDSL
Symmetric Applications (bps)	128 Kbps	×	×	×	×	×	×	×	×
	384 Kbps	×	×	×	×		×	×	×
	512 Kbps	×	×	×	×		×	×	×
	768 Kbps		×	×	×		×	×	×
	1 Mbps		×	×	×		×	×	×
	T1 1.544 Mbps			×	×		×	×	
	E1 2.048 Mbps			×	×		×	×	
Assymetric Downstream		×	×						
Optional Analog POTS		×	×	×					×
Rate Selectable		×	×	Future	Future		×	×	×
Auto-Rate Adaption Option		×	×				×	×	×
Echo Cancelled		*		×	×	×	×	×	
FDM		*							
Typical Loop Reach (24 AWG)		18 kft (1.5 Mbps)	18 kft (1.5 Mbps)	14 kft (HDSL)	10 kft	26 kft	29 kft (128 Kbps)	14.5 kft (1.5 Mbps)	18,000 kft (512 Kbps)**
		6 kft (7 Mbps)	6 kft (7 Mbps)	12 kft (SDSL)			21 kft (768 Kbps)		
Typical Loop Reach (.5 mm)		5.5 km (1.5 Mbps)	5.5 km (1.5 Mbps)	4.3 km (HDSL)	3.0 km	8.0 km	8.9 km (128 Kbps)	4.4 km (1.5 Mbps)	
		1.8 km (7 Mbps)	1.8 km (7 Mbps)	3.6 km (SDSL)			6.4 km (768 Kbps)		

* Certain vendor implementations only ** No loop lenght limit at 128 Kbps when loop has existing telephony service

Bridged Taps

Bridged taps are un-terminated extensions of the loop that cause additional loop loss with loss peaks in the frequency range of a quarter of the wavelength of the extension length. Since wavelength and frequency have an inverse relationship, short bridged taps have the greatest impact on wide-band services, while long bridged taps have a greater impact on narrow-band services. Most loops contain at least one bridged tap, and the effect of multiple taps is cumulative. Building-internal wiring contains additional bridged taps. The additional loss created is greatest on short bridged taps. Consequently, technologies that operate at lower frequencies are affected to a lesser degree.

The effects of crosstalk

The electrical energy transmitted over the copper wire line as a modulated signal also radiates energy onto adjacent copper wire loops that are located in the same cable bundle. This cross coupling of electromagnetic energy is known as "crosstalk". In the telephone network, multiple insulated copper wire pairs are bundled together into one cable referred to as a "bonded cable". Adjacent systems within a bonded cable that transmit or receive information in the same frequency range can create significant crosstalk interference. This is because crosstalk-induced signals combine with the signals that were originally intended for transmission over the copper wire loop. The result is a waveform with a different shape to the one originally transmitted.

11.10.2 Using the surroundings with asymmetry

Maximizing loop reach with various line codes resulted in an extensive study of the characteristics of the line operation itself. This study revealed that we could transmit a signal a greater distance from the CO (Central Office) to a remote home or office than could be achieved in the opposite direction. This was due to the effects of crosstalk, which are more dominant on the telephone company side of the copper wire loops than on the remote subscriber side. This phenomenon is due to the fact that more copper wires, each of which introduces a crosstalk component, are combined in large bundles as they get closer to entering the CO. Conversely, as we traverse the loop from the CO out to the end service user, the loops tend to branch off to different connections, resulting in fewer copper wire loops. Therefore, less aggregated crosstalk is introduced by the transmitters at the far end copper wire bundles. We can also profit from the characteristics of the telephone line operation in that low frequencies are used with the transmission towards the exchange. As the lower frequencies are attenuated less than high frequencies, this procedure ensures that the received signal is as strong as possible when it arrives in the heavily distorted environment of the exchange, where the crosstalk is at its worst.

It can be stated in summarizing that a signal of higher frequency can be transmitted more reliably from the CO to a remote location than in the opposite direction. Devices that have been constructed to support this concept (of a service of higher frequency from the CO to the service user and of a service of lower frequency in the opposite direction) are referred to as ADSL equipment (Asymmetric Digital Subscriber Line – Figure 11.37)

The following requirements have been identified as being essential for increasing ADSL cost efficiency:

- Creation of first-class services and enhancement of the service offering, e.g. with higher data rates, video streaming, voice services
- Increasing of the customer base by extending the loop reach
- Reduction of system, deployment and maintenance costs and operating overheads

In mid-2002, ADSL2 was approved by ITU-T Recommendations G.992.3 (G.dmt.bis) and G.992.4 (G.lite.bis) to address exactly these issues, without (and this is a key point) adding costly implementation expenses for hardware or software. Providing compliance with the ADSL2 standard 'on top' of the existing and well-established ADSL standards

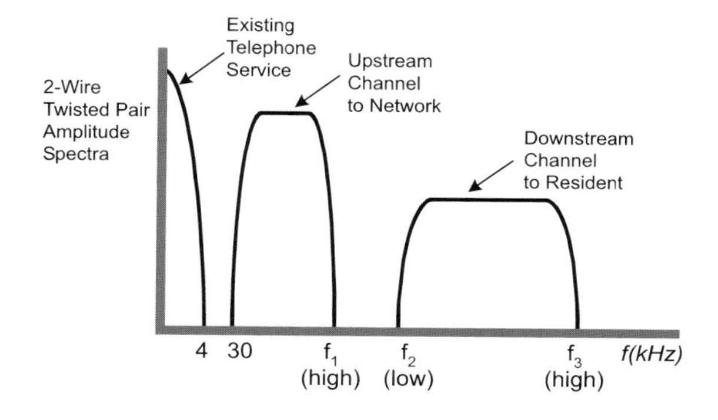

Figure 11.37 Frequency range of ADSL services

will therefore improve the cost position for ADSL services and enable providers to offer new and better services to a higher number of customers.

For emerging markets with no ATM legacy infrastructure, ADSL2 paves the way to a purely IP-based infrastructure by eliminating the need for an ATM termination and an ATM encapsulation from the former ATM-centered ADSL.

By the first quarter of 2003, leading CO (Central Office) and CPE (Customer Premise Equipment) silicon manufacturers like Infineon Technologies had developed solutions that support the ADSL2 standard. System manufacturers incorporate these chipsets to provide their DSLAM, DLC and ONU equipment sold on established and emerging markets with totally new functionality. Finally, the end-customer now benefits from both new and better services and more attractive prices.

Benefits of ADSL2 for users, providers and carriers

The following technical improvements have been integrated into the standard to satisfy the requirements mentioned above:

- Increased data rate and improved performance, especially for long loops
- Power saving modes
- Diagnostic mode
- Seamless Rate Adaption (SRA)
- Channel bonding, IMA (Inversed Multiplexing for ATM)
- Channelized Voice and Voice over DSL (CVoDSL, VoDSL)
- Support of packet-based services

The technical details and how manufacturers and users will benefit from the new features is described in the following. It should be noted that most of the improvements above require ADSL2 compliance on both the CO and CPE side. If the 'handshake' procedure shows that one side does not support ADSL2, the established G.dmt/lite will serve as 'fall-back' (substitute).

ADSL2 provides significant rate/reach improvements for long loops and with the presence of narrowband disturbers, i.e. in case of low signal-to-noise ratio (SNR). This is achieved mainly by one-bit constellations, and mandatory Trellis coding, which was only planned as optional in the ADSL standard. In addition, receiver-determined tone ordering leads to improved robustness against Radio Frequency Interference (RFI), because the receiver can spread out the non-stationary noise caused by RFI to get better gain from the Viterbi decoder.

In the original ADSL standard, a fixed number of overhead bits per frame took away 32kbps data rate, which is a major share of the payload rate on long loops with low upstream rates. To reduce this significant overhead, ADSL2 allows programming of the overhead bits per frame from 4 to 32 kbit/s. A frame structure with optimized use of the 'Reed Salomon (RS)' coding gain is an additional approach to improve performance for long lines. Another area of improvement is the initialization state machine, including transmit power reduction within the scope of the management control to reduce near-end echo and crosstalk, improved robustness against bridged taps with receiver-determined pilot tone, subcarrier blackout to allow RFI measurement during initialization and showtime, optimized training, and short initialization sequence for fast recovery.

The main purpose of all these improvements is to increase the robustness and performance of the long loops and thus increase the customer base for ADSL services. Figure 11.38 shows the rate and reach that can be expected from ADSL2. The main result is an increased coverage area for the service of about 6%, which gives a significant advantage, since more customers can be reached. It should be noted that the improvements for long loops shown here reflect 'real' loop con-

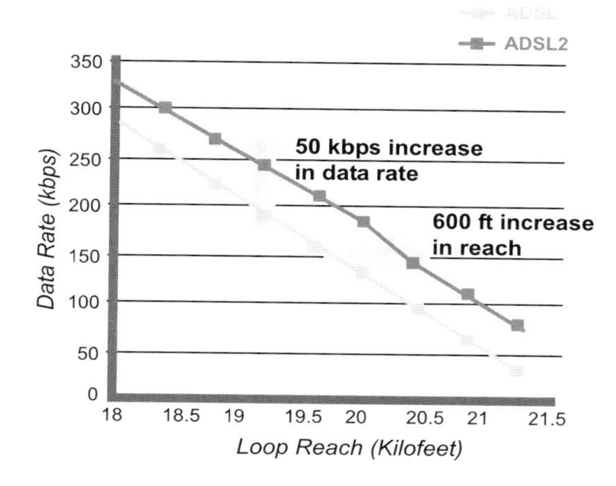

Figure 11.38
ADSL2 rate compared with line range of longer loops

ditions, not theoretical limits in a lab environment. The theoretical maximum value for the ADSL2 downstream data rate is 15.3 Mbit/s, but will never apply under real loop conditions. Not considering manufacturer-specific and proprietary approaches, a more significant increase of upstream and downstream data rates can only be expected by extending the ADSL bandwidth (as planned for the future ADSL+ standard).

Power saving modes

Besides the 12 dB power cutback in ADSL standards and various flexible power cutback and sleep modes offered on the market, there has been no standardized way to reduce the power budget of ADSL systems until now, especially in cases where the CPE modem is in service, but not transmitting data. This is the most common operational mode during a 24-hours always-on connection. Power and thermal overhead is a tremendous problem, especially for 'Digital Loop Carrier (DLC)' systems and remote units with their stringent heat dissipation requirements. ADSL2 addresses this issue by standardizing two low-power modes: L2 power mode is entered in case the CO modem detects low traffic on the active ADSL link data rate. In L2 mode, the CO limits the transmit power, while the ADSL connection still stays active and can be brought back into full blown L0 power mode very quickly, where required. This entire mechanism is not transparent to the user. For CO systems, tremendous power saving effects can be expected from using L2 mode. Infineon's GEMINAX chipset, for example, operates at about half the power dissipation in L2 mode compared to L0 mode at maximum transmit power. In addition, an L3 power mode is standardized as a sleep mode in case the user is offline. Entering L0 mode (Loop Diagnostic Mode) is possible after a short 3-second re-initialization. As ADSL has evolved into a mass market technology with immense deployment potential in various countries, loop qualification, service monitoring and bundle management becomes very important for cost-effective provisioning and maintenance of the ADSL service. The ADSL2 standard specifies new line diagnosis procedures, including measurements of the line noise, loop attenuation and signal noise ratio (SNR). These can be made from both ends of the link, CPE and CO side. CPE results are sent to the CO in a fixed data format, and the collected data can be used for interpretation to quantify quality, noise conditions and disturbance factors of the line under test. The tests can be run during or after installation as well as parallel to an active ADSL connection.

Interpretation of the data can either be performed on the ADSL transceiver level, linecard level or Network Management System (NMS) level. However, the interpretation itself is not part of ADSL. The Loop Diagnostic Mode of ADSL2 accurately incorporates Aware's and Infineon's DrDSL double-ended measurement and handshake procedures. In addition to ADSL2, DrDSL includes interpretation algorithms that can either be implemented within the ADSL datapump or, depending on feature set and system manufacturer requirements, on NMS level.

SRA – Seamless Rate Adaption

Another aspect of bundle management and troubleshooting in an ADSL mass deployment scenario is the increasing crosstalk between wire pairs in a bonded cable, if an increasing number of wires are carrying broadband services. Crosstalk leads to service deterioration or even loss of an active ADSL connection. In addition to bundle crosstalk, external disturbance factors such as AM radio stations can cause interferences.

To avoid these problems, providers often limit the number of wire pairs carrying ADSL in a bundle to a maximum, whereby they calculate an overhead that cannot be precisely specified and monitored. This approach leads to a 'waste' of bandwidth and can lead to bottlenecks and installation problems in densely populated areas. Furthermore, temporary disturbance factors cannot be easily identified and taken into consideration.

ADSL2 addresses this topic by providing 'Seamless Rate Adaption (SRA)' as an optional feature, enabling the ADSL2 system to change the data rate dynamically during operation. For this purpose, the modulation and framing layers have been decoupled in the standard, i.e. data rate parameters can be changed without modifying parameters in the framing layer, which would cause the modems to lose frame synchronization. SRA can be used to optimize bundle management and avoid problems with the ADSL service with temporary or stationary disturbance factors and increasing numbers of subscribers.

Channel Bonding

ADSL2 supports the ATM Forum's 'Inverse Multiplexing for ATM (IMA)' standard, i.e. chipsets can bind two or more copper pairs in a single ADSL link and thus increase the data rate. With this feature, providers can develop their service offering towards high quality services without having to change to other standards or equipment.

Voice transmission via DSL

Although voice and video transmission via DSL has recently lost some momentum because of the capital expenditure situation for providers and Competitive Local Exchange Carriers (CLECs), future systems must be enabled to support service bundles of voice and data services.

Channelized Voice over DSL (CVoDSL) is a method for assigning dedicated ADSL bandwidth to 64kbps TDM voice channels on the physical layer, i.e. avoiding additional expenditure for voice packetization into ATM or IP as happens with VoDSL. Especially considering investment protection for existing TDM switches and costs for ATM Application Layer 2 (AAL2) gateways, this is an interesting and cost-efficient option for combining voice and data transmission.

To support CVoDSL, an ADSL datapump delivers the PCM voice data via a dedicated PCM interface to the backplane.

Since voice and data traffic have different requirements with regard to Quality of Service (QoS, i.e. latency versus error rate), ADSL2 offers flexible framing, including support for up to 4 frame bearers and 4 latency paths. It is therefore possible to split the ADSL bandwidth into different channels with different link characteristics for various traffic types.

Packet-based services

The established G.dmt/lite standards define ATM as default technology for Layer 2 transmission, even in purely IP/Ethernet-based infrastructures in emerging markets. For this reason, all ADSL modems available today for COs and CPEs carry ATM, even if this is no longer included in the DSLAM, and ATM is not used for the uplink.

ADSL2 addresses this issue by defining a Packet Mode Transmission Transconvergence Layer (PTM-TC) in addition to ADSL's existing STM and ATM TSP-TC functions. Details for PTM-TC, such as mandatory, minimum, maximum and reserved data rate, improved configuration capability for latency time, Bit Error Rate (BER), are also provided. Especially for emerging markets without ATM functionality legacy that will have a purely IP/Ethernet-based infrastructure, ADSL2 offers a concrete roadmap to also skip ATM on the downlink, once ADSL2-capable CO and CPE modems are available. Due to the widespread base of ATM-based legacy CPE modems, ATM will still be visible on the ADSL subscriber line for some time, but ADSL2 will help in driving IP/Ethernet technology in both directions.

Various improvements

Besides the key improvements explained in the above sections, there are some minor points that nevertheless offer certain advantages when compared with established ADSL:

- Improved initialization ensures interoperability between products of different chipset manufacturers
- Fast startup mode reduces initialization time to less than 3 seconds
- In the 'All Digital Mode', the voice band can be used for adding up to 265 kbit/s upstream ADSL bandwidth.

ADSL+ and semiconductor modules of the next generation

ADSL2 will not be the last step in improving the ADSL standard with regard to performance and user-friendliness of the service.

In early 2003, the release of ADSL+ was introduced as an improvement to ADSL2. In addition to the already standardized ADSL2 features, ADSL+ provides significantly higher data rates. On loops shorter than 1 km, up to 25 Mbit/s downstream data rate can be expected. This is achieved by extending the ADSL spectrum from 1.1 MHz to 2.2 MHz. The second significant improvement is a standardized double upstream mode, which is already defined as optional in today's ADSL Annex B and is offered by many chipsets. For infrastructures with a large amounts of short loops, ADSL2 and ADSL+ provide significant business opportunities by offering new and high-grade types of broadband service bundles, including high-speed Internet, voice and video transmission.

Chipset manufacturers will bundle their solutions for standardization upgrades to ADSL2 and ADSL+ with next generation ADSL silicon, mainly addressing the second major requirement besides reach and performance, which is lower power and space requirements. As a consequence, ADSL chipset and system generation is picking up speed, enabling telephone companies and system manufacturers to continue improving their cost position and offer diverse features on the broadband market.

11.10.3 VDSL delivers video data and higher bandwidths

VDSL, or Very High Speed DSL is the latest variant of DSL. VDSL systems are still being developed, so the final capabilities have not yet been firmly established, but proposed standards call for downstream bandwidths of up to 52 Mbit/s and symmetric bandwidths of up to 26 Mbit/s.

The high speeds provided by VDSL will bring opportunities for service providers to offer the next generation of DSL services, with video being seen as the principal area of application. At 52 Mbit/s, a VDSL line can provide the customer with multiple channels with full quality MPEG-2 video streams and can even provide one or more channels of full quality 'High Definition Television (HDTV)'. Some service providers have already begun trial deployments of VDSL systems providing these services, with the VDSL endpoint appearing in the customer's house as a cable TV-like set-top box, with an Ethernet or other data interface for connecting to PCs for simultaneous data transmission services.

The basic prerequisite of DSL being a local loop technology in which compatible devices are located on either end of a single copper wire loop ensures that new DSL technologies will continue to emerge again and again over the years. The service provider must therefore ensure that with the selection of a specific DSL technology or a specific network model for today's service provision that they do not limit the options for adopting new technologies in the future.

12 Customer-specific integrated circuits

Introduction

In contrast to standard modules, customer-specific integrated circuits are tailor-made for the requirements of a customer and a specific application. Customer-specific circuits are also designated the abbreviation, ASIC (**A**pplication **S**pecific **I**ntegrated **C**ircuit).

With customer-specific integrated circuits, the difference is made between 'Full-custom' and 'Semi-custom' IC (IC standing for '**I**ntegrated **C**ircuit'. Full-custom ICs are ICs developed at the transistor level by the semiconductor manufacturer. Semi-custom ICs are based on gate arrays or cell ICs. Semi-custom ICs can be developed by the user with support from the semiconductor manufacturer (see Figure 12.1).

Full-custom ICs require longer development periods and have smaller chip areas than comparable semi-custom ICs. As development times and chip dimensions significantly determine the price of the device full-custom ICs are preferred in costs-critical applications, particularly with large unit volumes. One major advantage of the semi-custom ICs is the faster realisation. The faster a new product is launched on the market, the greater the chances of success. This advantage often compensates for the disadvantage of a higher price.

The developers will also have to ask themselves, 'Do standard, semi-custom or full-custom ICs provide an economically viable system solution?' The most important factors to take into account in this respect are system costs, development time, system reliability and flexibility.

12.1 Semi-custom IC

Semi-custom ICs are based on gate arrays or cell ICs. Semi-custom ICs can be developed by the user with support from the semiconductor manufacturer.

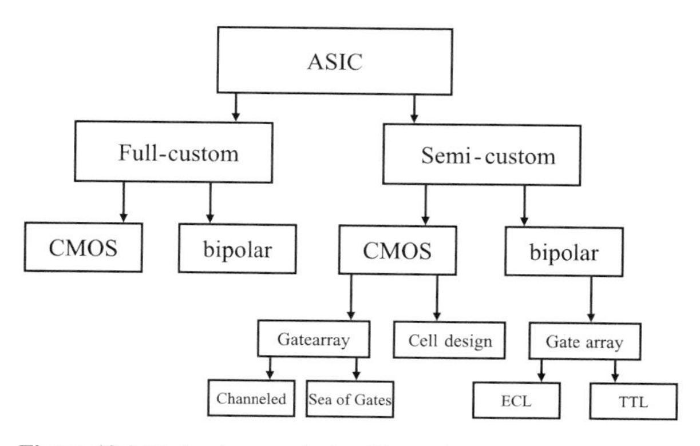

Figure 12.1 Technology variants with semi-custom

12.1.1 Gate arrays

The structure of a gate array consists of what are known as 'core cells' (gates) and the channeled gate array technology between them. Gates consist of between 2 and 8 transistors. With CMOS gate arrays, another procedure provides the option of completely leaving out the channeled gate array technology and of using the inter-cell wiring between the gate array cells (logic cells). This 'sea-of-gates' technology considerably reduces the necessary silicon surface (cost-determining).

The developer wires the gates (via a workstation or PC) to one another in such a way that the required circuit function is created. At the same time, the libraries of the CAD systems contain wiring specifications concerning higher organized functions, such as logic gates, counters, multiplexers etc. During production, the wafers with the unwired gate arrays (master) are cost-effectively prefabricated. The intra and inter-cell wiring 'personalizes' the master.

The 'personal' circuit is developed from the universal master.

12.1.2 Cell design

The methods of a prescribed pattern are no longer used with a cell design. Arrangement and cell width can be selected individually. The option also exists of retrieving and integrating macros stored in the cell library, such as DRAMs, SRAMs, CPU cores, DA/AD converter etc.

Figure 12.2 shows various semi-custom module types.

12.1.3 Gate array or cell design?

This decision is, provided implementation is possible, essentially determined by two factors:

- by the time for the sample creation taking into account potential redesigns
- by the total of the costs for development and series production

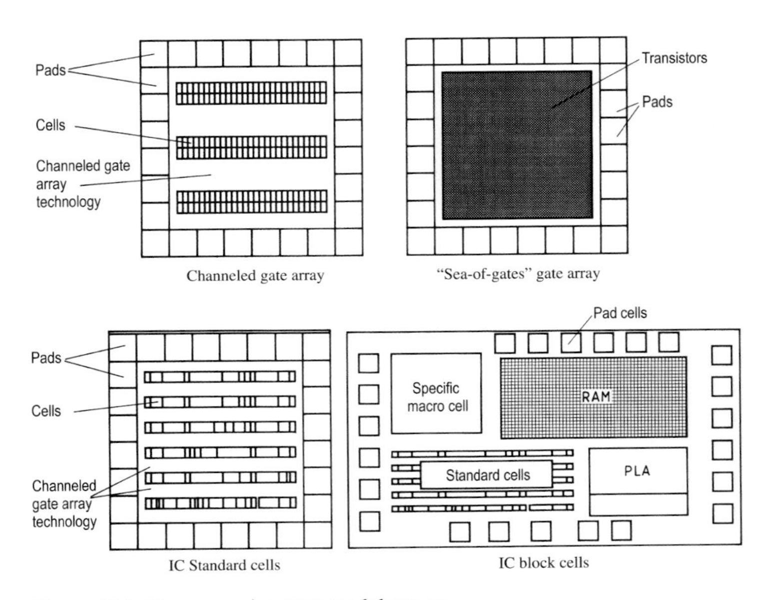

Figure 12.2 Diverse semi-custom module types

Time factor

Samples for gate arrays can be produced significantly quicker than for cell ICs. With gate arrays, on the prefabricated master wafers, only the metalization layers have to be applied, whereby with cell ICs, a complete process sequence is necessary. Sample deliveries after the layout release take 4 to 7 weeks with gate arrays; with cell ICs, 12 to 16 weeks.

Costs factor

With a comparable project, the development of a gate array costs less than half that of a cell IC. The costs with volume production behave differently. With small quantities (approx. 1,000 to 10,000), the costs with the gate array are the same or less (inexpensive pre-fabrication of the master). With medium-sized quantities (10,000 to 100,000 units), the smaller chip surface of the cell IC plays a significant role, with the result that here the cell IC can be more economical in price.

12.2 Technologies

Apart from the logic, the selected technology significantly determines the properties of the IC. The assignment of the bipolar and CMOS technologies offered on the market to the gate arrays and cell ICs is shown in Figure 12.1.

12.2.1 Bipolar semi-custom ICs

Bipolar (gate) arrays are particularly well suited for use with high frequencies and short gate delays in the order of magnitude of 30 to 100 ps.

Typical areas of application:

Data technology

Large computer CPUs

Periphery processors

Memory controls

Table 12.1 Overview of the SH100G family

Master name	SH 100 G1	SH 100 G2/G4*	SH 100 G3
Gate functions (max.)	3300	1000	10000
I/O pads (max.)	100	66	144
Power pads (max.)	32	20	64
Integrated linear arrays on chip: No. of lin. arrays: No. of lin. components	 two 2800	 two 2800	 four 5600
Packages available	C-PGA 88C-PGA 144C-MQFP 152P-TQFP 100 PQ2	C-PGA 64C-PGA 88P-TQFP 64 PQ2P-TQFP 80 PQ2P-TQFP 100 PQ2	P-MQFP 208 PQ2
Packages planned	P-TQFP 144 PQ2		
Power dissipation (typ.)	2.5 W	0.9 W	7.0 W
Available	yes	yes	yes

* only for –3.3 V

Information technology

Broadband communication

Digital transmission systems for different transmission media (fiber optics, coax cable, directional radio)

Message switching/transmission

Medical technology

Digital signal processing in sonography and tomography

Metrology

Logic analyzers

IC testers

The maximum number of gates of the bipolar gate arrays of the SH100G family provided by Infineon is approx. 10,000 gates (see Table 12.1).

12.2.2 CMOS Semi-custom ICs

With CMOS semi-custom ICs, both gate arrays and cell ICs are offered on the market. Products in 0.8 μm and 0.5 μm technology (and smaller) are today's state of the art. The smaller the structure width, the higher the possible integration density and so higher also the possible working frequency. The trend extends to structures of < 0.1 μm.

Each semi-custom design can only be developed with fault-free CAD tools. The user interface plays an important role in this respect. Flexibility during the circuit development must be especially emphasized with the cell design. Software generators can generate RAM, ROM, PLA and computer structures in the individual architecture. Structures with analog functions and microprocessor cores can also be integrated. Hierarchical design technologies enable sub-circuits to be generated, simulated and tested.

12.2.3 Bipolar gate arrays

Gate delays of some 30 ps are attained for the bipolar technology, B6HF, currently produced by Infineon. Cell libraries contain 'macros', which are pre-defined logical circuit functions in CML or ECL circuit technology (Figure 12.3). The functions of these macros are tested using test chips. They range from simple NOR and NAND gates to complex functions, such

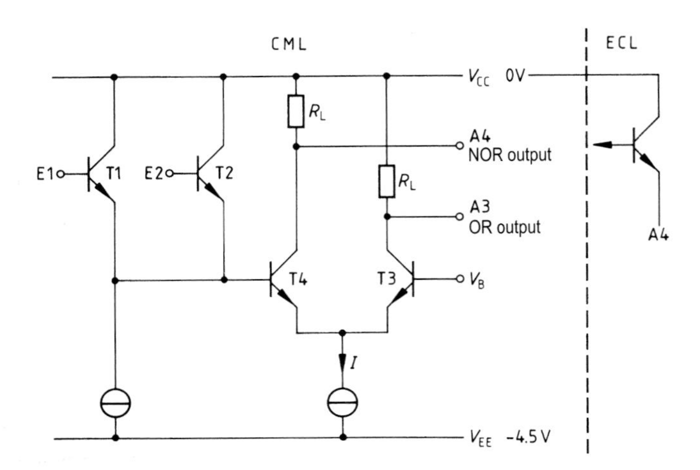

Figure 12.3 Circuit technology differences between ECL and CML

as counter and multiplexer. Over 100 different macro cells are available for the developer in the library provided by Infineon.

With the SH100G family, Infineon provides two circuit technologies for the structure of the logic cells. The better alternative can therefore be selected for the respective application case:

Current Mode Logic (CML)

Especially for use in

- metrology
- transmission technology and
- medical technology,

high signal symmetry, low tolerances and low power dissipation are required. These requirements are best met with "Three Level Series Gating" and the use of CML circuit technology.

Emitter Coupled Logic (ECL)

With applications that require high drive capacities (e.g. in bus systems), the use of ECL circuit technology is especially beneficial. With this technology, the necessary drive capacity is attained using an additional emitter follower transistor.

Combination of ECL/CML

The combination of ECL and CML circuit technology on a chip is possible in the SH100G family and it consequently provides the developer with maximum flexibility.

Use of differential logic

Differential logic can be used for particularly fast circuit parts. This enables the processing of the highest frequencies, with the SH100G family, up to 6 GHz, whereby the maximum input and output frequencies depend on the package that is used.

Speed-power programming

The faster a circuit needs to work, the greater the currents that are required. However, this also increases the power dissipation. On one hand Infineon provides an efficient speed-power product, and on the other hand, it enables the additional option of programming each individual cell of a circuit, which depending on speed and load, eliminates power dissipation. This optimization option is available at all times until the end of the design phase.

The features are enhanced by the series gating principle. More complex cells can be structured with a lot of simple OR, NOR, AND and NAND gates, which however has an adverse effect on speed and power dissipation. With a multiple usage of the cross-current I (up to 3 differential amplifier levels), the space requirement of the logical links and the power dissipation can be minimized and considerable delay reductions can be attained.

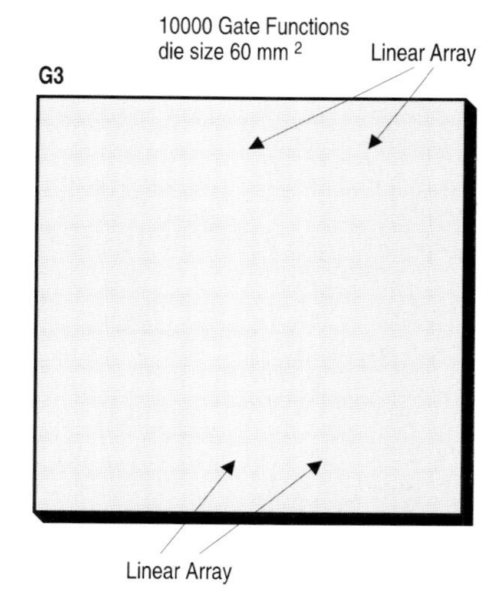

Figure 12.4
The SHG100G master is a linear array with 10,000 gate functions on 60 mm^2

12.2.4 Bipolar transistor arrays (linear arrays)

The cell structure of transistor arrays is already prefabricated, that means that a CAD cell library support is no longer required. As a special feature on the chips of the SH100G family, Infineon provides 2 and 4 linear arrays respectively (Figure 12.4).

Each linear array has

- 600 Transistors
- 800 Resistors
- 20 Capacitors
- 1 Bias driver

This linear array is suitable for the structure of analog and digital circuits in the frequency range up to 6 GHz. New circuits can be developed by the developer or they can use the 'hard macros' provided by Infineon, such as PLL, VCO MUX etc., which have already been tested in other circuits.

12.3 Package variants

A wide range of packages is available for semi-custom ICs. The correct package must be selected according to the number of pins, power consumption and mounting requirements.

Packages available from Infineon are:

- C-PGA package, up to 144 pins
- P-TQFP-(**T**hin **Q**uad **F**lat **P**ackage) package, up to 208 pins
- P-SSOP package, up to 32 pins
- The BGA (**B**all **G**rid **A**rray) package is provided for especially high pin numbers with small dimensions.

12.4 Customer–IC manufacturer cooperation

Cooperation can take place at various levels in accordance with the customer requirements. This can be demonstrated most clearly by the development procedure for gate arrays (Figure 12.5).

Output product for each project is the individual circuit plan (logic plan) of the customer. The activities then follow from the bottom up:

- Implementation of the circuit on the basis of the cell library of the semiconductor manufacturer
- Integration of the circuit in the CAD system
- Pre-placement of critical circuit parts
- Logic simulation of the circuit
- Check routines

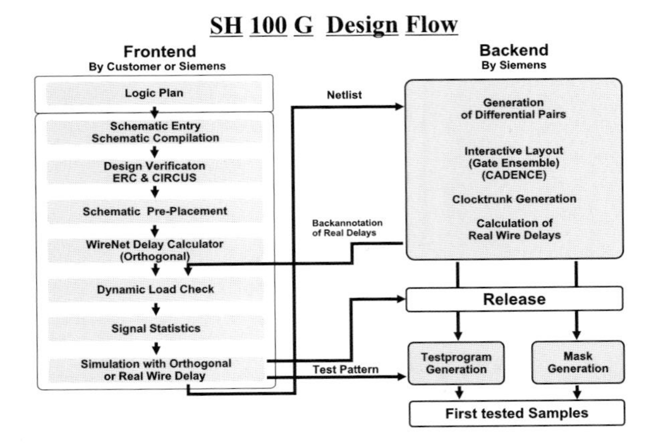

Figure 12.5
Development procedure plan for SH100G

- Interactive placement
- Logic simulation with real delays (back annotation)
- Generation of test bit samples
- Sample release for production

Production of the sample then follows. It consists of:

- Production of the mask set
- Metalization of the master
- Testing and assembly of the chip
- Final testing
- Delivery to the customer

Which of these jobs will be carried out by whom is determined individually by the customer. There are two proven and tested options in this respect:

Handover of the individual circuit plan (interface logic plan)

The customers use a manufacturer-provided design center. The manufacturer carries out all other work for the customer. Infineon provides design centers at its head offices in Munich and at its own company-run branch offices and regional sales offices.

Handover of the simulated network list (simulated network list interface)

Customers that have their own corresponding CAD systems and the necessary experience receive the cell libraries from the semiconductor manufacturer on a suitable data carrier. After the circuit integration and simulation, the customer delivers the results to the manufacturer on a suitable data carrier. The manufacturer creates the layout and gives the actual delays created by the layout back to the customer for back annotation. Following release by the customer, the manufacturer produces the first sample.

13 Electromagnetic Compatibility – EMC

13.1 Fundamentals

Electromagnetic compatibility (EMC) is defined as the ability of an electrical installation to function satisfactorily in its electromagnetic environment without having an impermissible effect on this environment, which includes other electrical installations.

For some time now, the consideration given to the electromagnetic compatibility of electronic devices and modules with individual parts and components has been widening, and increasing in depth. In this chapter, we are particularly interested in integrated semiconductor circuits (ICs), because they are often sources (emitters) or sinks (receiver) of interference in electronic devices. The increasing requirement for enhanced EMC characterizations is due to the continual spread of electronics, the application of system solutions which use microelectronics and semiconductor electronics in all areas of industry, households and traffic. This makes it ever more necessary, right at the early stages of electronic developments, to make use of estimates of the subsequent EMC behavior of the product.

The generic term electromagnetic compatibility covers both radiated and line-borne (conducted) interference transmissions (emissions) and the susceptibility to influence (interference immunity). Here, the frequency range considered extends from 0-400 GHz. Figure 13.1 gives an overview of the relevant topics in this connection.

13.1.1 EMC phenomena

Electromagnetic interference has its origin in natural occurrences and in technical processes. Examples of natural interference are atmospheric discharges during lightning strikes (LEMP, lightning electromagnetic impulse), and electrostatic discharges (ESD). The latter are of

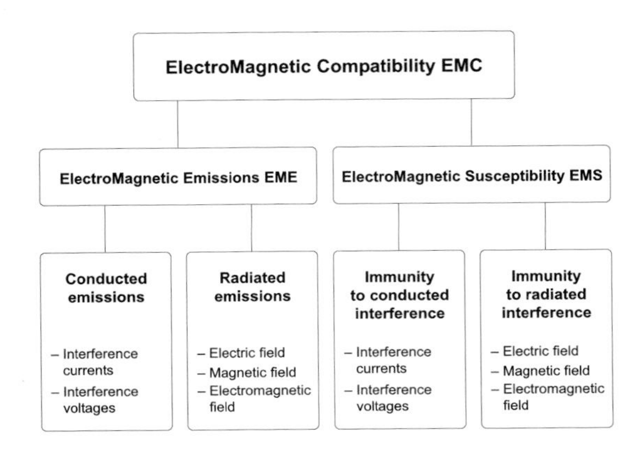

Figure 13.1
The different aspects of electromagnetic compatibility

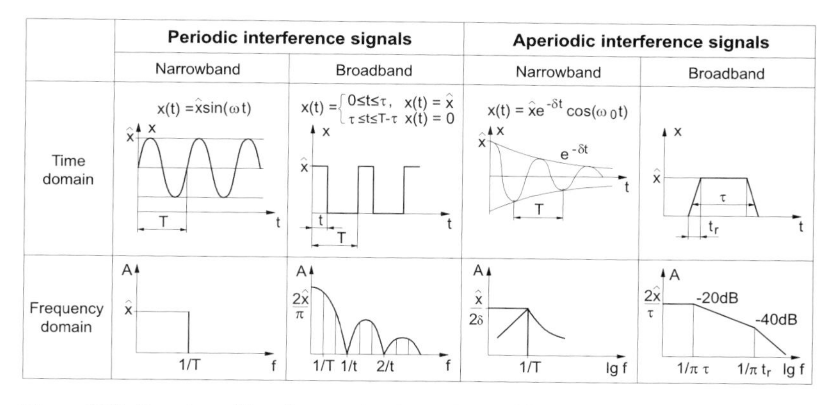

Figure 13.2 Overview of interference signals, as time and frequency domain representations

special importance for semiconductors, in particular. In technological systems, it is mainly circuit-dependent switching operations, involving high-speed changes to current and voltage, which lead to electromagnetic interference. This can arise on a periodic or random basis. In such cases, propagation of the interference can be either conducted, as currents or voltages, or radiated by an alternating electromagnetic field.

Conducted forms of interference, in which a signed interference current flows backwards and forwards with the same amplitude on the connecting wires, is then described as symmetric or differential mode (dm) interference. If the interference current circuit is completed via the ground reference, and if the interference propagates on the connecting wires as a signed current in one direction, it is described as asymmetric, or common mode (cm) interference.

Electromagnetic coupling between a source and a sink can result from:

- galvanic coupling, the most commonly-occurring form of coupling and the reason for symmetric interference
- capacitive coupling, which results from parasitic stray capacitances and an alternating electric field
- inductive coupling, which is caused by the alternating magnetic field around conductors in which a current is flowing, and
- electromagnetic wave coupling, which arises conductively between wires in cable harnesses or conducting tracks, or is established radiatively if the gap r between an interference source and an interference sink is greater than 0.1 times the wavelength λ.

Interference signals

Interference signals, which arise periodically or with random timing in the time domain, can be represented by superimposed sinewave and cosine wave signals of differing frequencies and amplitudes. Figure 13.2 shows typical examples of interference variables in the time domain, and their frequency representations.

Semiconductor switches, logic ICs, µCs mostly produce interference with a wide frequency spectrum, generated by the internal working and clock frequencies. The use of Fourier transformations enables these signal forms, which are periodic in time, to be analyzed in the frequency domain. The approximation equations for calculating the amplitudes and the corner frequencies for a trapezoidal

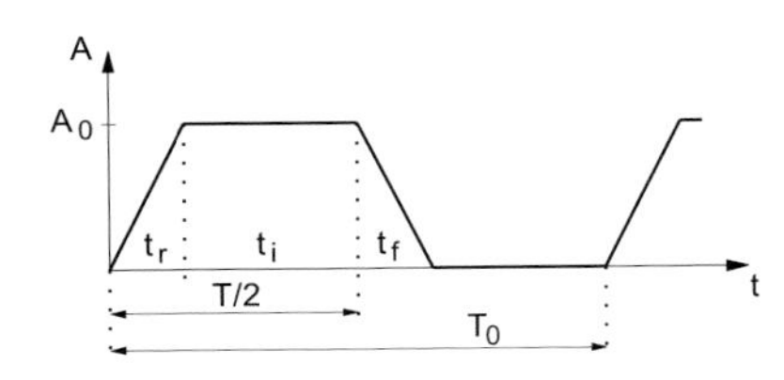

Figure 13.3
Trapezoidal function for a switching operation

signal such as that in Figure 13.3, and the resulting envelope representation in Figure 13.4, are shown below. From the fundamental frequency of the signal up to the first corner frequency f_{g1}, the graph of the spectrum is parallel to the frequency axis. After the first corner frequency, the amplitude reduces by 20 dB/decade up to the second corner frequency f_{g2}, from where the spectrum falls off by 40 dB/decade. The following symbols are used:

A_0 Amplitude of the source signal
A_n n^{th} amplitude
T_0 Period of the fundamental frequency
t_i Pulse width
t_s Switching time ($t_r = t_f$)
n Multiple of the fundamental frequency
n_{g1} Multiple for 1st range corner
n_{g2} Multiple for 2nd range corner
f_0 Fundamental frequency
f_{g1} Corner frequency 1
f_{g2} Corner frequency 2

1st range: f_0 to f_{g1}, flat, frequency-independent spectrum

$$A_n \approx \frac{2 \cdot A_0 \cdot t_i}{T_0} \quad (1)$$

2nd range: f_{g1} to f_{g2}, 20 dB amplitude drop per decade

$$f_{g1} = \frac{1}{\pi \cdot t_i} \quad (2) \qquad A_n \approx A_0 \cdot \frac{2}{\pi} \cdot \frac{1}{n} \quad (3)$$

3rd range: f_{g2} to ∞, 40 dB amplitude drop per decade

$$f_{g2} = \frac{1}{\pi \cdot t_s} \quad (4)$$

$$A_n \approx A_0 \cdot \frac{2}{\pi^2} \cdot \frac{1}{n^2} \cdot \frac{T_o}{t_s} \quad (5)$$

Electromagnetic emission measurements

Electromagnetic emission measurements are carried out in the frequency domain, using a test receiver or a spectrum analyzer. These devices are capable of evaluating different interference characteristics by measurements on their peak, quasi-peak, mean or effective values. In doing this, the measured value is always expressed in terms of the effective value of a sinusoidal voltage which, when tuned to

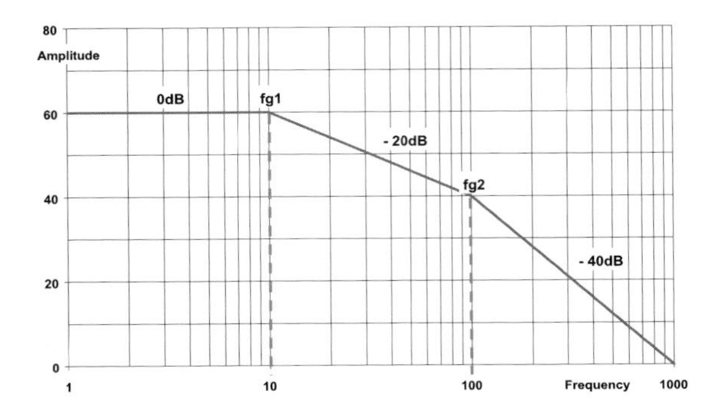

Figure 13.4
Envelope of the spectral graph

Table 13.1 Measurement bandwidths

Bandwidth (B_W)	200 Hz	9 kHz	120 kHz
Frequency range	0.01-0.15 MHz	0.15-30 MHz	30-1000 MHz

the midband, gives the same deflection. The measured value is indicated as a level, to maintain a manageable indication range for showing physical magnitudes from μV up to V. This level is the logarithmic ratio of the magnitude of a signal to a reference value. In telecommunications engineering the reference value has been defined as the effective power or interference power. The values correspond to 10 times the logarithm to the base ten, and are stated in decibels (dB).

Power level:

$$p = 10 \cdot \log_{10} \frac{P_X}{P_0}, \quad \text{dB}$$

In communication technology, the value defined for the reference magnitude P_0 is 1 mW. In order to indicate this reference value, an "m" is added (dBm, dB(m)). The voltages or field strengths measured when making emission measurements are stated in dB (μV) or dB (μV/m), and can be derived from the equation for the power level. The reference value is 1 μV and corresponds to 0 dB (μV). It follows that an indicated value for the power of 0 dBm corresponds to a voltage of 107 dBμV when measured using a 50-Ω system.

Voltage level:

$$p = 20 \cdot \log \frac{U_X}{U_0}, \quad \text{dB } (\mu\text{V})$$

The operation of test receivers is frequency-sensitive, with selectivity curves in accordance with CISPR16. Selection is made using different filter bandwidths depending on the frequency range, as shown in Table 13.1.

If only one spectral line falls within the receiver's reception channel, then the value indicated will be independent of the measurement bandwidth B_W and the characteristics of the indicator. This type of interference is referred to as narrowband interference. If, because the fundamental frequency is lower, several spectral lines fall within the reception channel, the indication will depend on the bandwidth B_W and the waveform analysis. An integrating measurement procedure then no longer permits a unique amplitude-frequency assignment. This type of interference is referred to as broadband interference. This behavior can be detected when broadband signals are being measured from the step jumps in amplitude indications in the area of the switchover between mesasurement bandwidths, at 0.15 MHz and 30 MHz.

13.1.2 EMC: norms and regulations

There are numerous norms and test regulations for the purpose of guaranteeing electromagnetic compatibility. They contain and distinguish between measurement procedures, measuring instruments and test arrangements for the various interference phenomena, lay down test conditions and define limiting values within which, when adhered to, the operation of electrical devices can be assumed to be mutually interference-free.

Standardization work is carried out at the international, European and national levels. The work is divided up between the ISO (International Organization for Standardization) and the IEC (International Electrotechnical Commission) with its CISPR subcommittee (Comité International Spécial des Perturbations Radioélectriques). At the European level there is again a division of IEC responsibilities between CEN (Comité Européen

Table 13.2 EMC measurement methods for ICs

61967	**Integrated Circuits – Measuring Electromagnetic Emission**
IEC 61967-1	General and Definition
IEC 61967-2[1]	TEM-cell method
IEC 61967-3[1]	Surface scan method
IEC 61967-4	1 Ω / 150 Ω direct coupling method
IEC 61967-5	Workbench Faraday cage method
IEC 61967-6	Magnetic probe method
IEC 62132	**Integrated Circuits – Measuring Electromagnetic Interference Immunity**
IEC 62132-1[1]	Fundamentals and Definitions
IEC 62132-2[1]	TEM-cell method
IEC 62132-3[1]	Bulk current injection method
IEC 62132-4[1]	Direct power injection method
IEC 62132-5[1]	Workbench Faraday cage method

[1] Standardisation not yet completed

de Normalisation [European Committee on Standardization]) and CENELEC (European Committee for Electrical Standardization) and in addition the ETSI (European Telecommunications Standard Institute).

Integrated semiconductor circuits (ICs) are a new area for EMC standardization, and represent a special case. The EMC phenomena which arise are similar to those for devices and components, but as individual components ICs can seldom be uniquely assigned to one application. Today, the IEC has two sets of norms, which specify interference emission and interference immunity measurement methods. For electromagnetic emission there is the IEC 61967 set of norms, and for interference immunity IEC 62132. Table 13.2 gives an overview of these standards. The IC measurement methods are intended to help in the earliest stages of the development of a device, if possible even when components are being selected, to make use of EMC considerations about the emission and absorption characteristics of an IC, so that they are taken into account in designing devices and in their usage-specific requirements.

13.1.3 EMC measurement methods for integrated circuits (ICs)

Emission measurement methods

IEC 61967 makes a claim to be a generally applicable standard, for any type of integrated circuit (IC), for characterizing its

Table 13.3 Ovcerview of IC emission measurement methods

	TEM-cell	**Scan**	**1 Ω / 15 Ω**	**WBFC**	**Mag. probe**
Frequency	0.15-1000 MHz	1-1000 MHz	0.15-1000 MHz	0.15-1000 MHz	0.15-1000 MHz
Phenomenon	E/H field	E/H field	RF current RF voltage	Field components RF voltage	H field RF current

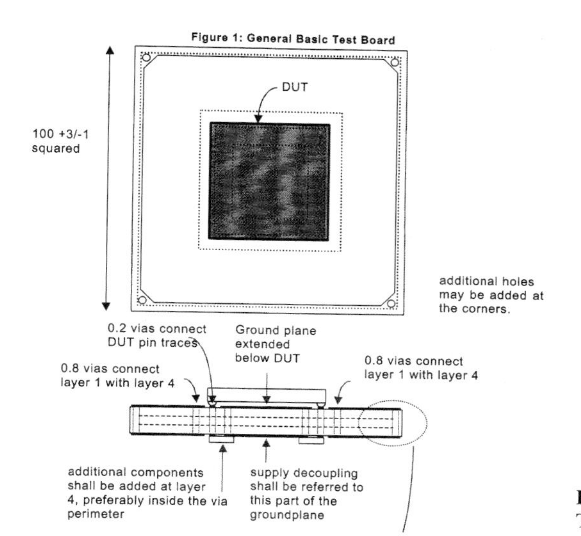

Figure 13.5
TEM-cell test board

electromagnetic interference emission in the frequency range from 150 kHz to 1 GHz (Table 13.3). It contains 5 methods for recording EMC phenomena.

The "TEM-cell method" – Measurement of radiated emissions IEC 61967-2 – is a method for determining the electromagnetic emission an IC radiates into its environment exclusively from its internal structure and the leadframe. The TEM-cell, a coaxially tapered waveguide comprising a flat inner conductor (the septum) and an external conductor (the screen), is used for this purpose as a defined receiving antenna with external screening. This method of measurement requires the development of a special test board for the IC which is to be tested (Figure 13.5). The IC, for which the emissions are to be measured, is mounted on the bottom-side of this test board. Except for the IC connection pads, this side is everywhere metallized, and serves as the GND plane (GND: ground, earth; the negative reference voltage – generally 0 V). Mounted on the top-side are the peripheral components required for the operation of the IC, and here too are the signal and ground conductors. Additional layers, required for the wiring and electrical supply, take the form of intermediate layers. The test setup is sketched in Figure 13.6. To carry out the test, the test board is placed over the opening provided in the TEM cell in such a way that the IC which is to be tested projects into the interior of the TEM cell.

The groundplane on the bottom-side of the test board is connected to the screen of the TEM cell by spring contacts, and closes off the screen of the TEM cell with an RF-tight seal. The emissions radiated by a microcontroller with a system frequency of 40 MHz, and received at the septum, are shown in Figure 13.7. In order to take into account the different positions of the internal antenna structures, the test board is measured in two positions, with orientations of 0° und 90°.

A TEM cell is particularly suitable for characterizing the electromagnetic emissions from integrated circuits with high operating frequencies and/or from the large structures and leadframes which favor emissions. The measured results relate to the entire IC. Selective measurements are not possible.

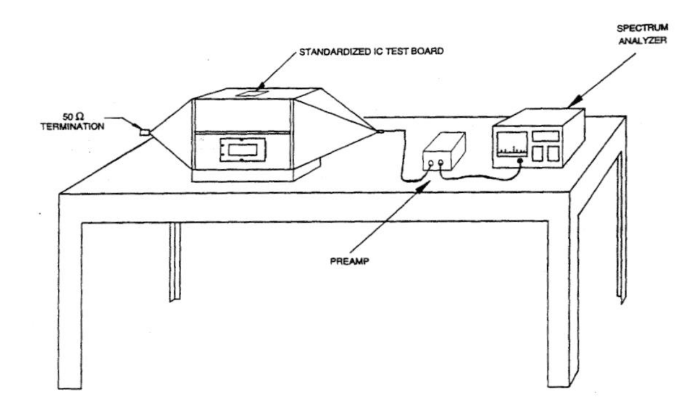

Figure 13.6 Test setup for a TEM cell

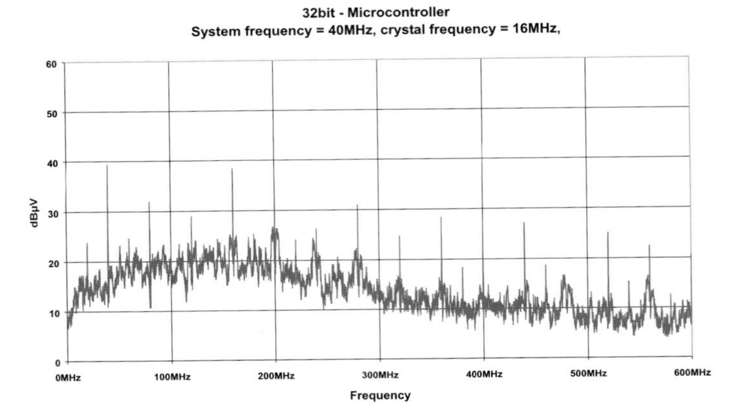

Figure 13.7 µC emissions in position 1 (0°)

The "Surface scan method" – measurement of radiated emissions IEC 61967-3 (TR) – is a measurement method for scanning the surface of ICs for radiation fields, using near-field probes for the E- and H-fields. In order to measure the electrical or magnetic emission field strengths, a rod or loop antenna is moved in the x-y plane over the surface of the IC to be investigated, at a defined distance from it and with a defined orientation, by means of a software-controlled positioning drive. This records the field strength profile at a set frequency. The resolution of the measurement is dependent on the dimensions of the probe and the step size for the positioning drive. During the measurement, the IC is located on a test board, which is fixed non-conductively to a reference GND surface. Figure 13.8 shows a sketch of the arrangement and Figure 13.9 a measured field distribution. The method is suitable for localizing known interference frequencies, but across the entire frequency range it is very time-consuming.

The "1 Ω / 150 Ω direct coupling method" – Measurement of conducted emissions IEC 61967-4 – contains two methods which characterize the conducted electromagnetic interference emissions from ICs in the form of an RF current, using a 1 Ω current shunt, and in the form of an RF voltage using a 150 Ω network on

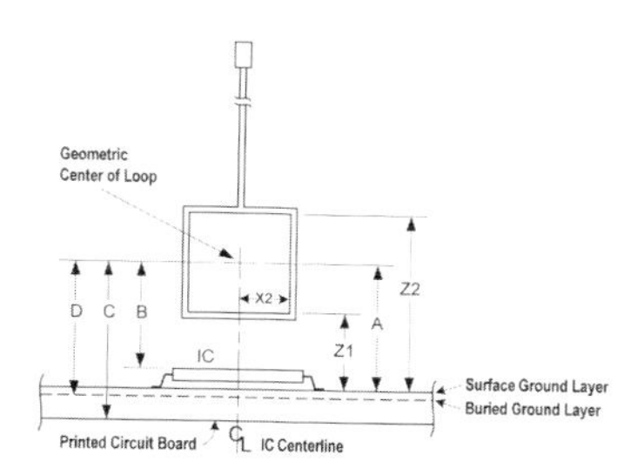

Figure 13.8
Principle of the surface scan method

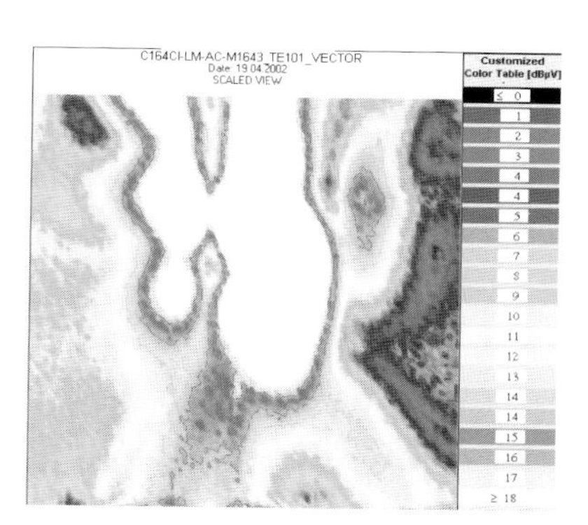

Figure 13.9
Field distribution across the IC surface

the pin. In considering the emissions, it can be assumed that all the interference current circuits in the IC are closed circuits. The way this is effected depends for each IC type on the reference potential, and is either via the common ground reference or via the IC power supply. This offers the possibility of recording the total interference current by a total current measurement in the GND or power supply path, with the help of the 1 Ω current probe. But it is also possible to determine the interference current components for individual pins, which may be of particular interest for an application. The 1 Ω current probes used for the RF current measurement is shown schematically in Figure 13.10. The voltage drop across the 1 Ω precision resistor is analysed by the test receiver. The result characterizes the interference emissions of the IC. The 49 Ω resistor is used for impedance matching to the measurement device. The DC blocking capacitor at the output of the current probe protects the input on the test receiver from excessive DC voltages.

The RF voltage measurement using the 150 Ω network such as that shown in Figure 13.11 is intended for the pins of an IC which has connecting leads, on the PCB

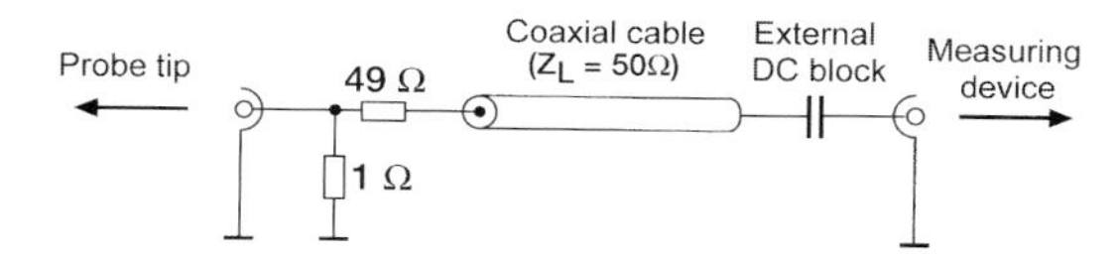

Figure 13.10
Circuit diagram for a 1 Ω current probe

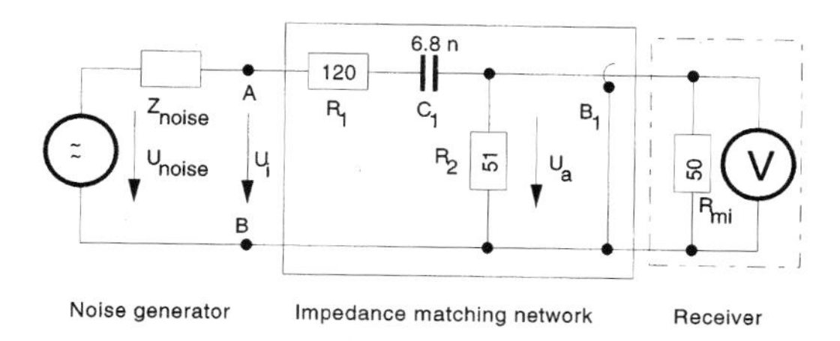

Figure 13.11
Circuit diagram for a 150 Ω voltage probe

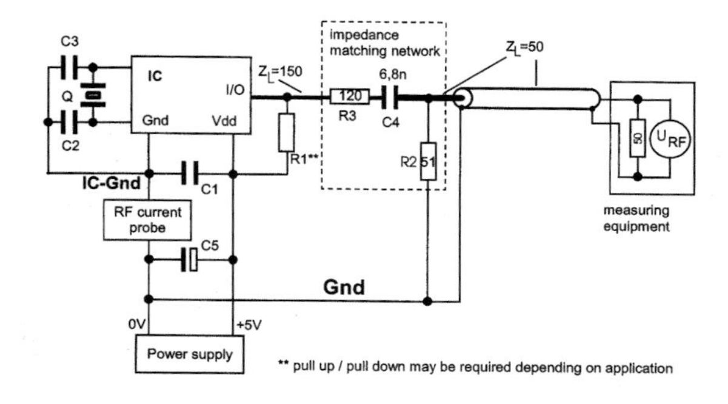

Figure 13.12 Test circuit for the 1 Ω / 150 Ω method

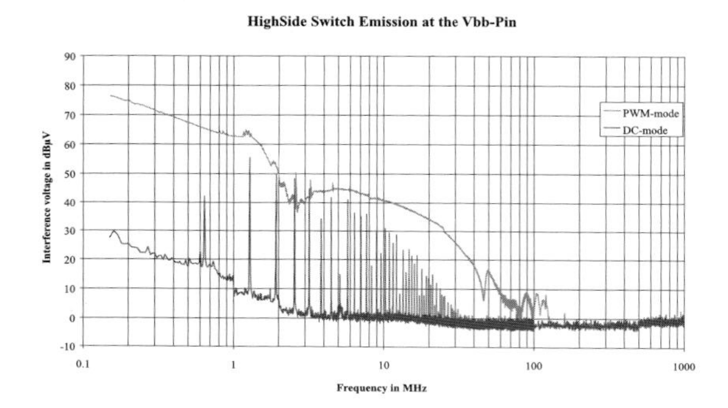

Figure 13.13 Emission spectra

or in the cable harness, which are more than 10 cm long. These pins are loaded with a typical antenna impedance of 150 Ω (IEC 61000-4-6). Figure 13.12 shows a sketch of a test circuit using a 1 Ω current probe and a 150 Ω network.

The narrowband and broadband emissions from a power IC, measured in accordance with the 150 Ω method, are shown in Figure 13.13. Because of its almost constant transmission characteristic, the 1 Ω / 150 Ω method of measurement can be used to characterize the emission behavior of ICs with high resolution across the entire frequency range. This method can be used to determine overall and pin-selective results.

The "Workbench Faraday cage method" – Measurement of conducted emissions IEC 61967-5 – characterizes the emissions from ICs, IC groups and application-related circuit boards. Like the 1 Ω / 150 Ω methods of measurement, it assumes that up to 1 GHz ($\lambda/2$ = 17 cm) the emissions from the connected cables predominate relative to the direct radiation from the IC. The 150 Ω common mode impedance of free cables over the ground (IEC 61000-4-6) is simulated for

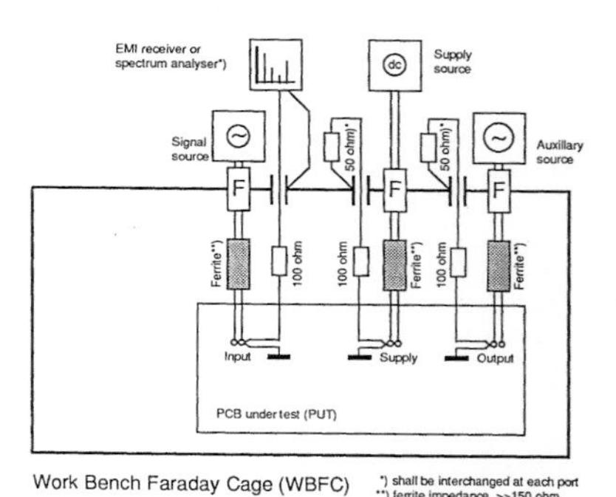

Figure 13.14 WBFC test setup

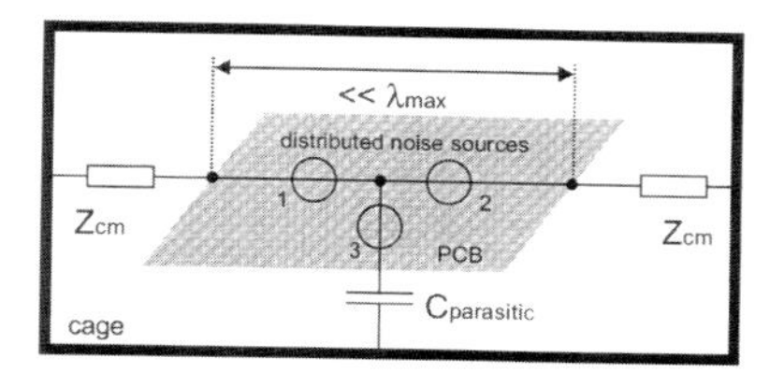

Figure 13.15
Principle of the common mode measurement

the measurement by a 150 Ω network to the reference ground (Faraday cage). This network comprises a 100 Ω resistor and the 50 Ω internal resistor of the test receiver or a 50 Ω terminating resistor. The wall of the Faraday cage also forms the reference ground. The leads for the peripheral power supplies, the outputs and inputs, are fed into the interior of the box via lead-in filters, and are fitted with common mode chokes which should have an impedance 280 μH at 150 kHz (Z = 263 Ω).

To make the measurements, the network is connected to two common mode points on the signal, power supply or output lines on the test board. The principle of the arrangement is shown in Figure 13.14. The recorded interference emission consists of the voltage drop caused by the RF current to GND together with the current, caused by rapid voltage changes (*dv/dt*), which flows via the coupling capacitors to the screen wall of the box. The interference current circuits are closed via the 150 Ω network to the common mode point of the interference source. The lumped-element model in Figure 13.15 should clarify the principle.

The "Magnetic probe method" – Measurement of conducted emissions IEC 61967-6 – is an RF current measurement method using a magnetic field probe, which measures the magnetic field strength at a defined distance above a connecting lead of an IC when this takes takes the form of a strip line. This characterizes RF currents which cause electromagnetic emissions via conducting paths on PCBs, power supply and GND surfaces. In principle it is possible, as can be seen from the circuit in Figure 13.16, to record RF currents at various points in the circuit. The test setup with the arrangement of the magnetic probe on the board under test is shown in Figure 13.17. By applying a calibration factor and using

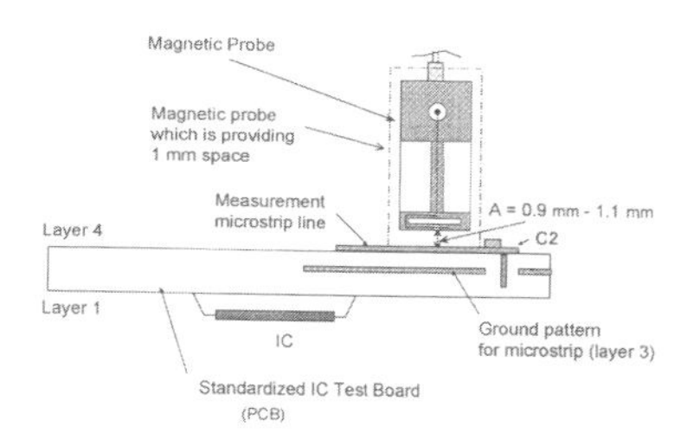

Figure 13.16 Magnetic probe method setup

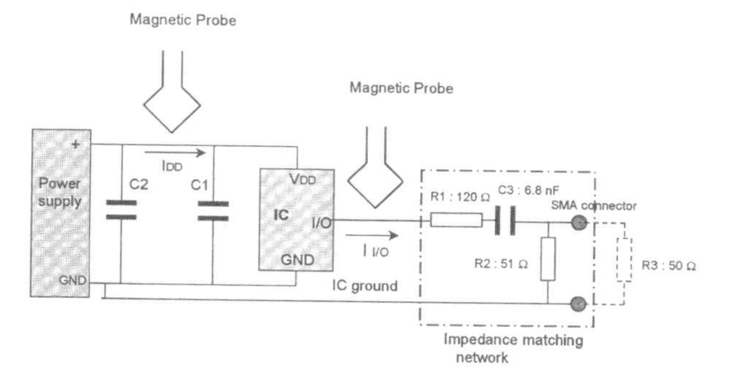

Figure 13.17
Test circuit with magnetic probe

Table 13.4 Overview of the EMC methods for measuring the interference immunity of ICs

	TEM cell	BCI	DPI	WBFC
Frequency	0.15-1000 MHz	1-400 MHz	0.15-1000 MHz	0.15-1000 MHz
Phenomenon	E/H field	RF current	RF current RF voltage	Field components RF voltage

the physical relationships, the magnetic field strength determined using the probe can be converted into an equivalent RF current.

The magnetic probe method is an emission measurement method which should preferably be used for ICs with operating frequencies in the upper MHz range. By comparison with the TEM cell, however, it offers the advantage that it can also be performed for selected pins. The results are comparable with the 1 Ω method.

Interference immunity measurement methods for ICs

IEC 62132 makes a claim to be a generally applicable standard, for any type of integrated circuit (IC), for characterizing its electromagnetic interference in the frequency range from 150 kHz to 1 GHz. Table 13.4 shows an overview of the various measurement methods and the EMC phenomena they take into account.

The "TEM-cell method" – Measurement of radiated immunity IEC 62132-2 – is a method for determining the interference immunity of ICs, where the reception of RF energy takes place exclusively through its own internal structure and the leadframe. The method of measurement works on the same principle as the emission measurement method of the same name. Instead of a test receiver, an RF generator with an amplifier connected to its output is connected to the TEM cell to effect a controlled injection of RF power. The same test board can be used for the interference immunity test as for the emission measurements. The maximum field strength that the IC can be subject to without malfunction is logged by the measurement system.

The "Bulk Current Injection method" – Measurement of conducted immunity IEC 62132-3 – is a method for determining the interference immunity of ICs when an RF current is injected into the connecting leads to those IC pins which, in the application, are connected to peripheral wiring. The basis for this measurement methods, which was drafted on the lines of the component measurement method ISO 11452-4, is the assumption that particularly the connecting wires to electronic components form antennas,

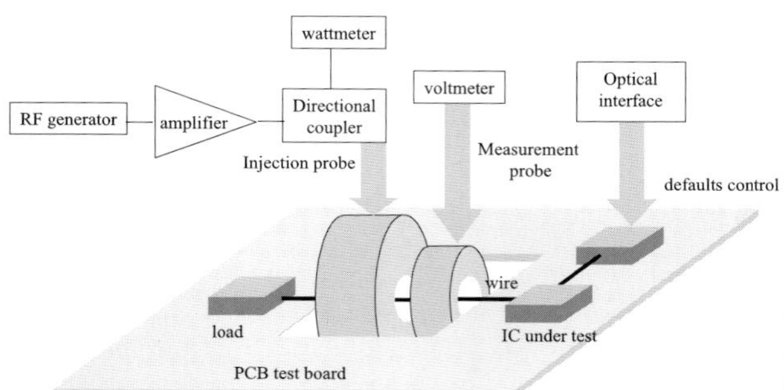

Figure 13.18
Setup for the BCI IC measurement method

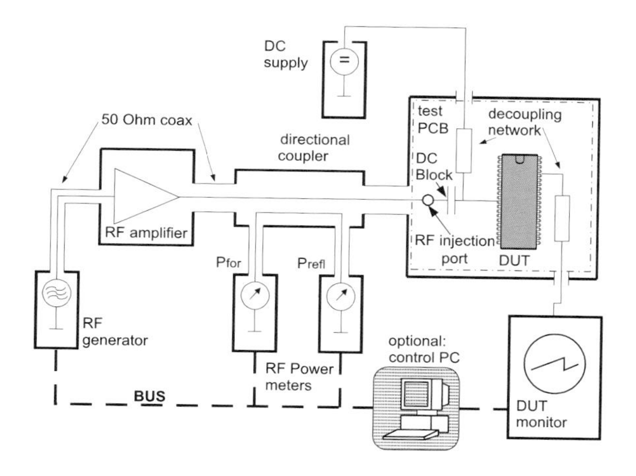

Figure 13.19
Test setup for the DPI method

which absorb RF energy from the electromagnetic environment and conduct it via these connecting leads to the IC which acts as the interference sink, in the form of RF currents. The RF current is injected into the connecting lead of the IC which is to be tested by means of a clip-on current probe. Using a second clip-on current probe, the injected RF current is measured. Figure 13.18 shows a sketch of the principle of the test arrangement.

BCI measurements require a special test board, with an opening in its middle to accommodate the current injection and measurement probes. The device under test (DUT) is placed in its application circuit on one side of this BCI test board. The peripheral loads, sources etc. are located on the opposite side, and are decoupled by means of RF filters. The IC under test is connected to its peripherals via leads. The connecting lead for the pin to be tested is fed through the BCI jaws. All the other leads are routed around the jaws. On the IC side, the RF current circuit is completed directly through the IC, and on the peripheral side via a feedthrough capacitor or a direct galvanic connection to the reference ground. Using the BCI method it is possible to inject RF energy into selected pins, and even into several pins at the same time. The interference immunity of the IC is characterized by reference to the measured RF current and the injected RF power. This is to ensure that high impedance inputs or outputs in integrated circuits are not overtested.

The “Direct Power Injection method” (DPI) – Measurement of conducted immunity IEC 62132-4 – is a method for determining the interference immunity of ICs with which RF power is selectively injected into a pin on the IC. This method starts from the assumption that RF energy from the electromagnetic environment is absorbed mainly by connected wires or antenna structures on the PCB, and is then conducted to the pin of the IC. The test criterion for the interference immunity of the IC is the level of the forward power applied to the pin via a coupling capacitor. The test setup is shown in Figure 13.19. DPI tests require a special test board with at least two layers, which must satisfy the same criteria for the RF injection points as those which apply for the RF tapping points in the case of the 150 Ω emission measurement methods. Peripheral loads or sources are decoupled from RF by a high impedance ($Z \geq 400\ \Omega$). The arrangement of the injection paths for symmetrical injection (differential mode) and asymmetric injec-

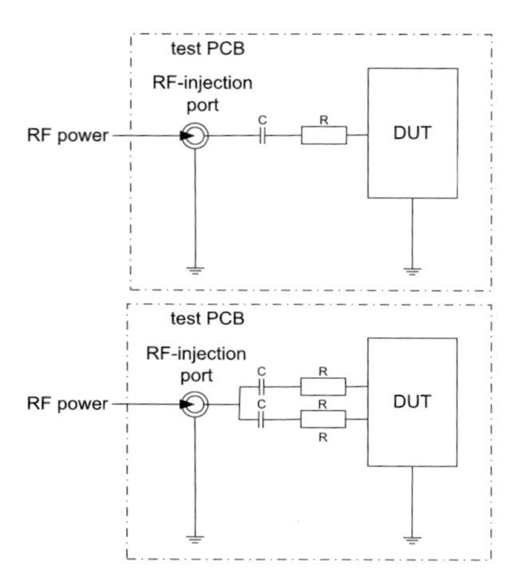

Figure 13.20 DPI injection points

tion (common mode) is shown in Figure 13.20.

The DPI method can be used equally well to make selective measurements for a pin or to inject RF into several pins at the same time. Depending on the pins which are to be tested, anything from 50 mW to 5 W of RF power is injected, as appropriate for the pin class (internal, protected, external).

The "Workbench Faraday cage method" – Measurement of conducted immunity IEC 62132-5 – is a method for determining the interference immunity of ICs in which RF power is applied to those common mode points of the IC test boards or application boards at which the RF voltage is tapped for the emission measurement method of the same name. The injection network consists of a 100 Ω resistor which is connected through a hole in the wall of the Faraday cage to an RF generator via an amplifier ($R_I = 50\ \Omega$). The interference current circuit is completed via a second 100 Ω resistor, which is connected to the Faraday cage through a 50 Ω dummy load. The wall of the Faraday cage also forms the reference ground. The leads for the peripheral power supplies, the outputs and inputs, are fed into the interior of the box via lead-in filters, and are fitted with common mode chokes which should have an impedance 280 µH at 150 kHz (Z = 263 Ω). The test setup for a WBFC interference immunity test is shown in Figure 13.21.

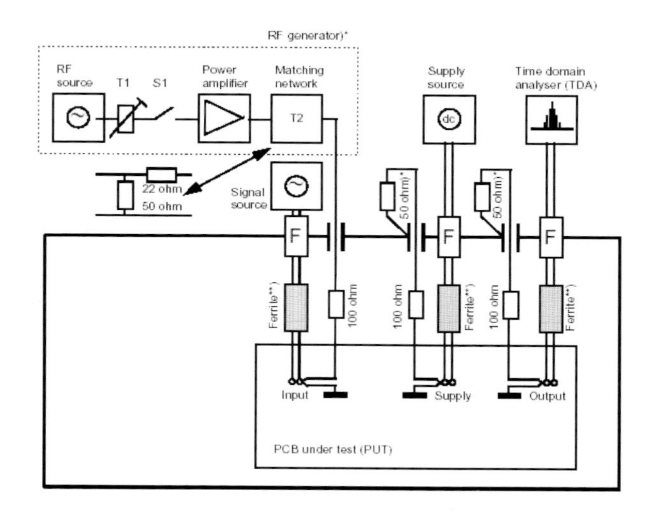

Figure 13.21 Setup for the WBFC interference immunity test

13.1.4 Models for determining the ESD robustness of components

Apart from frequency-dependent EMC phenomena, which arise mainly during the operation of components, the functionality of individual components can also be impaired by electrostatic effects. The current from an ESD (= *E*lectro*S*tatic *D*ischarge) can give rise to damage in the component. The limits of the ESD robustness of components with integrated protective structures is checked by means of internationally standardized models. The best known of these are the Human Body Model (HBM), the Machine Model (MM) and the Charged Device Model (CDM), or its socketed variant the Socketed Device Model (SDM). The models differ in that the HBM and the MM simulate the discharge of a charged object (man/machine) into a component, whereas CDM or SDM represent the rapid discharge of a charged component.

Human Body Model (HBM) and Machine Model (MM)

The best known and most used model for determining the ESD resistance of electronic components is the Human Body Model. This simulates the discharge of a charged person into a (grounded) component (Figure 13.22). The person is specified as having a capacitance of $C = 100$ pF. The resistance $R = 1500\ \Omega$ represents the contact resistance between the person and the component.

In principle, an ESD pulse ought to be applied between each pin and every other pin of the component which is being tested (pin-to-pin test). For devices with numerous pins, however, this is too expensive. For this reason, the standards specify tests in so-called pin combinations. This involves stressing the various pins (I/O pins and power supply pins) against the individual power supplies. In addition, a so-called I/O test is also performed, in which each I/O pin is stressed against all the other (shorted) I/O pins. The first document dealing with HBM tests was MIL- STD883D, method 3015 which, although it is constantly being cited, is no longer being updated. Infineon follows the requirements of the JEDEC Standard: Electrostatic Discharge (ESD) Sensitivity Testing, Human Body Model (HBM) JESD22-A114-B which defines, as well as the pin combinations, the pulse count and also the pulse form of the tester (Figure 13.22).

Unlike the HBM, which simulates the discharge of a charged person, the so-called Machine Model (MM) simulates the discharge of a device or machine. In the equivalent circuit shown in Figure 13.22 it is only the values of the charge capacitance and the discharge resistance which change ($C = 200$ pF and $R = 0\ \Omega$). Because of the low discharge resistance, the pulse form (Figure 13.23) is determined only by the parasitic elements in the arrangement. The reproducibility and comparability of the results are therefore

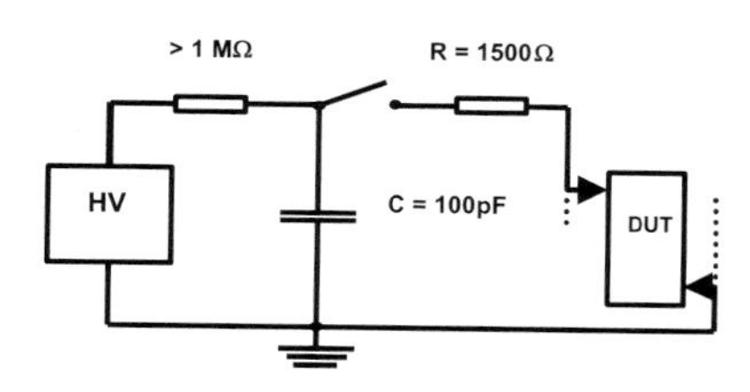

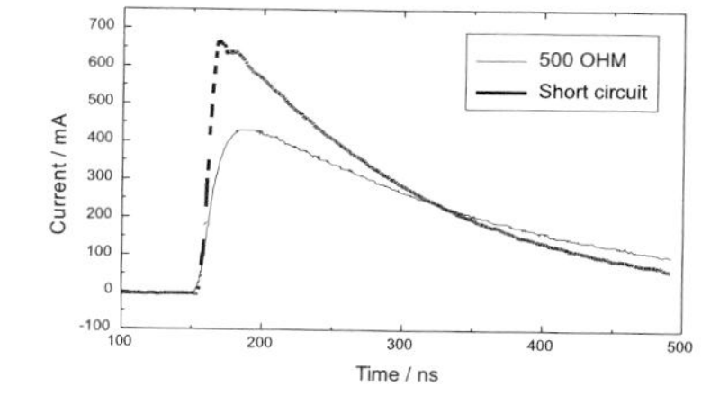

Figure 13.22
Equivalent circuit and discharge current graph for an HBM tester for a charge voltage of 1 kV (DUT = Device Under Test)

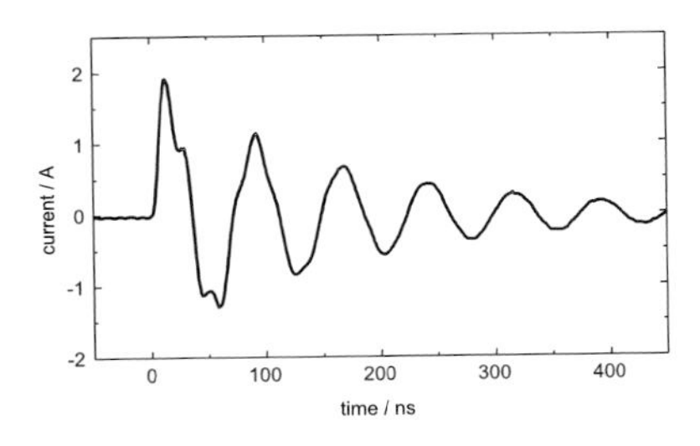

Figure 13.23
Discharge current in the MM test at 100 V

significantly lower in the case of the MM than with the HBM.

The procedure followed in the test – pin combinations, electrical assessment after stressing – is virtually identical to that for the HBM, and is laid down in international standards. Infineon applies the JEDEC Standard: Electrostatic Discharge (ESD) Sensitivity Testing, Machine Model (MM) JESD22-A115-A. The pulse forms for the HBM and MM have comparable rise times, and do not differ greatly in their pulse widths. Both models address the same failure mode (thermal overload of the ESD protection structure) in the device. As the MM therefore provides no additional information, and the reproducibility is significantly better for the HBM than for the MM, Infineon generally only determines the ESD sensitivity in accordance with the HBM.

Charged Device Model (CDM) and Socketed Device Model (SDM)

The Charged Device Model (CDM) specifies another form of discharge, and causes different failure mechanisms. It simulates a device which becomes charged in the automated manufacturing process, which is then discharged to a highly conducting object (ground). In the case of the CDM, this very rapid discharge does normally not produce a thermal overload of the protection structure, as with the HBM or MM stresses, but instead an oxide breakdown. Figure 13.24 shows a typical discharge pulse form for a CDM test. The rising edge is very much faster (approx. 300 ps) and the pulse width is significantly less (approx. 0.5 ns) than in the case of the HBM and MM. This means that, although the current intensities are as high as several amperes even when the charging voltages are low, with the CDM less energy is dissipated in the device than with the HBM and MM.

While it is being stressed the device is not in a socket, but lies on a metal plate with its pins facing upwards. The device is charged up via the ground pin through a high ohmic resistance, or by electrostatic induction from the charged base plate. One at a time, the pins on the charged device are then discharged using an earthed needle. It can be seen from Figure 13.24 that, with the metal plate on which it is lying, the chip forms a capacitor. The magnitude of this capacitance depends on the

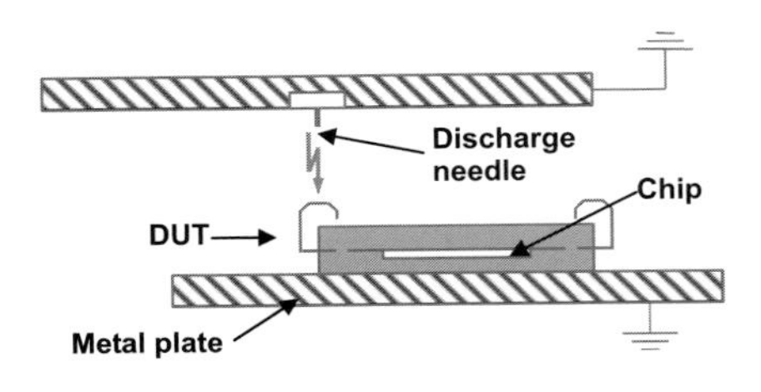

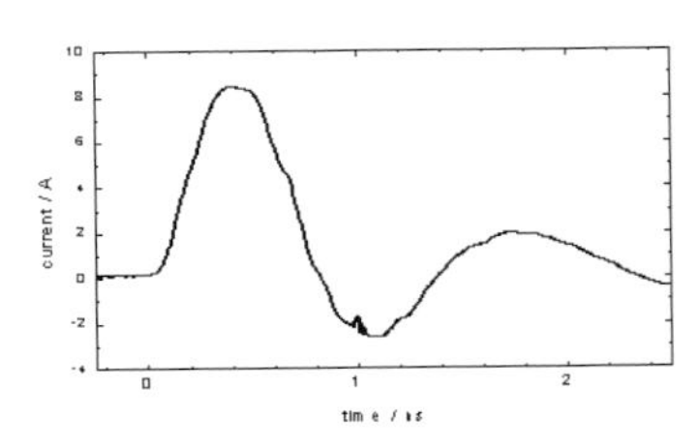

Figure 13.24
Typical arrangement for CDM tests, and the current waveform (PLCC-44 package, 500 V charge)

geometry of the package. This determines how much charge can be stored on the chip, and hence the magnitude of the current during the discharge. Because the discharge begins with a spark even before any actual contact between the device pin and the discharge needle, the reproducibility of the CDM test is not very high. However, it does describe the real situation during the handling of an individual device very well. The standards for the CDM are not yet as well developed as for the HBM. Apart from the methods used for charging up the devices, the present standards also differ in respect of other parameters, such as the peak current or methods of calibration. Infineon applies the JEDEC norm JESD22-C101-A.

Significantly higher reproducibility is shown by the Socketed Device Model (SDM), because the device is plugged into a socket, and the discharge takes place through a relay. However, not only is the device charged up, but also the socket and the entire test setup. Because of this greater capacitance, with the SDM a greater charge is stored for the same charging voltage, and the discharge current is thereby increased. The SDM test is therefore more critical than the CDM test and, because of the larger capacitance, it is independent of the package. Standards in the field of SDM testing are still being worked on (e.g. at the IEC); currently there is only a Technical Report from the American ESD Association (ESD TR 08-00: Socket Device Model (SDM) Tester). Because of the advantages cited, at Infineon CDM weak devices are identified using SDM, and a CDM test is only performed if the SDM shows a low ESD resistance.

13.2 Electromagnetic compatibility of automotive power ICs

Automotive power ICs encompass a wide range of IC applications and functions for automobile electronics. They all have as their primary function the provision, conversion or driving of electrical power, from a few milliwatts (mW) up to several kilowatts (kW). The operating range of these ICs is adapted to the vehicle electrical system voltages of 12 V, 24 V and 42 V. The spectrum ranges from the simple MOSFET (DMOS) through low-side, high-side and bridge switches with integral protective and diagnostic functions, linear and switched-mode power supply regulation ICs, communication ICs, up to highly integrated ASICs for safety applications such as ABS and airbags. The objective of IC development is to reduce interference as far as possible, and make them inherently resistant to interference from the electromagnetic environment, so that the EMC requirements of the system are provided for in the IC itself, avoiding as far as possible measures affecting external circuits. In characterizing the EMC properties of automotive power ICs, use is made of the 1 Ω/150 Ω methods in accordance with IEC61967-4, and the DPI methods in accordance with IEC 62132-4.

13.2.1 Power switch ICs

Power switches (low side, high side, bridge) are used in electronic control units both for DC switching of power sources and also increasingly to realize applications with freely adjustable power flows by means of pulse-width modulation (PWM). In operation, power switches can generate both narrowband and also broadband electromagnetic emissions, which propagate via the supply or output lines. In DC mode, the harmonics of the integral charge pump produce a narrowband noise spectrum, which in PWM mode is overlaid with an additional broadband spectrum, generated by the load current switching at PWM frequency. Some of these interference signals can extend far into the MHz region, and have unwanted effects on radiocommunications and radio reception.

Interference emissions from charge pumps in power switch ICs

Charge pumps are used in integrated circuits to generate internal voltages which are higher than the supply voltage fed in from outside. They are sources of narrowband interference. Figure 13.25 is a sketch of the block diagram of a charge pump.

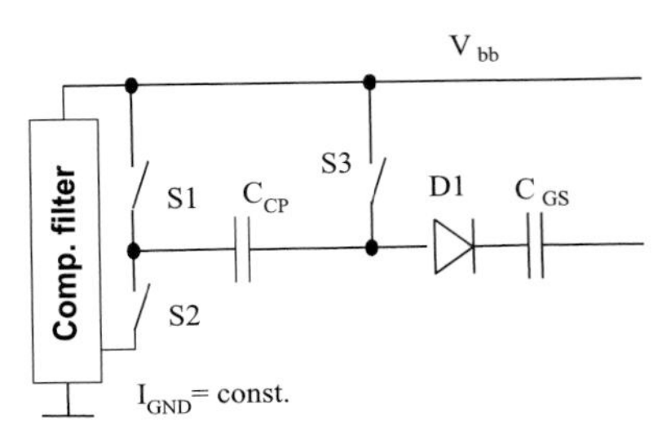

Figure 13.26
Principle of a charge pump with an internal filter

In the charging phase for the pump capacitor C_{CP}, the switches S2 and S3 are closed. After S2 and S3 open, S1 closes and the charge stored in C_{CP} is discharged via the diode in the gate capacitor C_{GS}. Because of the limited areas in the IC, it is only cost-effective to integrate pF capacitances. For this reason, the switches must work at switching frequencies in the MHz range in order to charge up the gate capacitor (nF) of the power transistor in a few microseconds. The periodic unregulated current flow out of the on-board power supply due to the change and discharge cycles, and the problems of cross currents through S1 and S2, cause the unwanted high-frequency currents on the power supply and ground leads of the IC. The typical narrowband spectrum of a charge pump, with the harmonics of its 700 kHz switching frequency, is shown in Figure 13.27. A simple internal measure within the IC for reducing the interference emissions is to move the operating frequency into a frequency range where there are no radio bands. However, regional requirements and special radio ranges strongly restrict this option. Other measures are to control S1 and S2 to avoid cross currents, and to reduce the pump power at the expense of increased switching power losses. More costly in terms of switch technology are new power supply concepts, which keep the interference currents away from the power supply lines. In this case, the charge

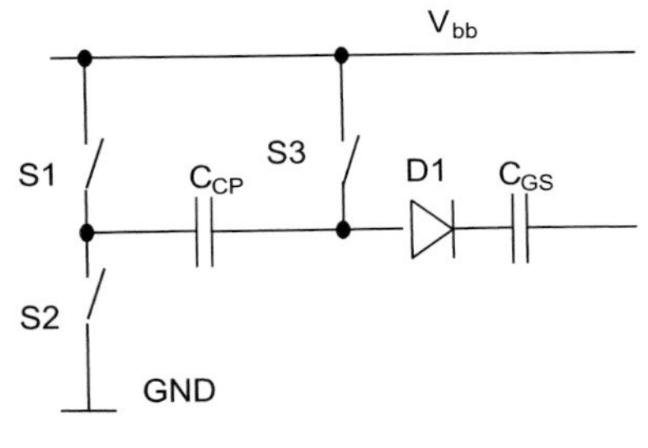

Figure 13.25
Block diagram of a charge pump

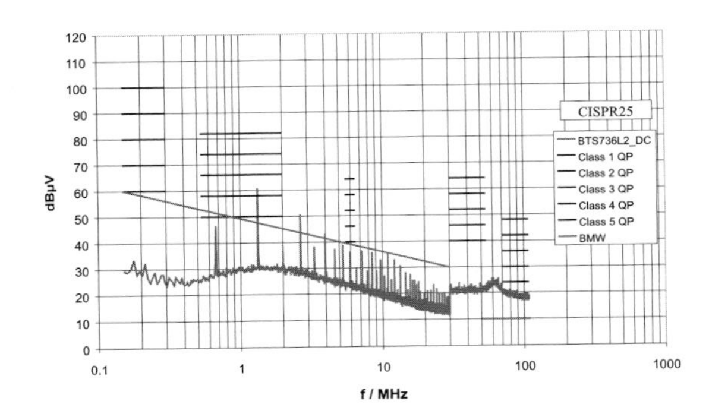

Figure 13.27
Interference emissions from a charge pump

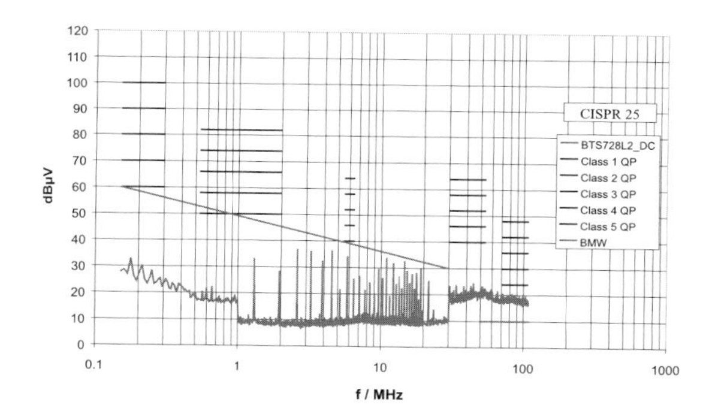

Figure 13.28 Interference emissions from a charge pump with a filter

pump is connected as in Figure 13.26, via a filter circuit with a compensating current, which permits a reduction in emissions of between 20 and 30 dB (Figure 13.28). Other concepts provide for a jitter in the charge pump frequency over a frequency range Δf, in order to distribute the energy content of the interference across several frequencies.

External EMC measures to reduce the narrowband interference emissions on power supply leads can be applied with the help of filter capacitors or RC filters. Figure 13.29 shows a recommended circuit.

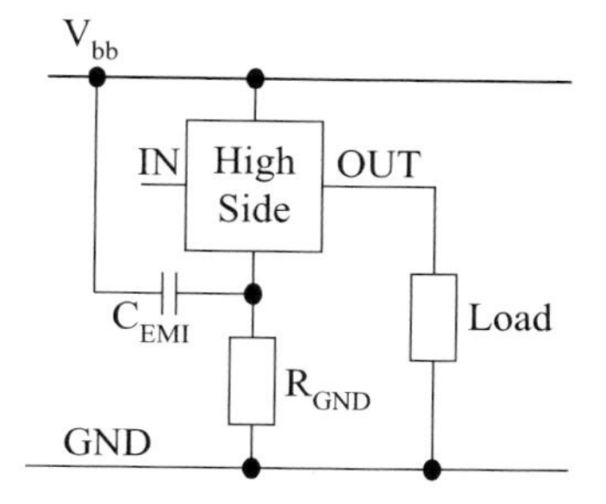

Figure 13.29 External circuitry for suppressing charge pump interference

Between the power supply and GND pins should be placed a ceramic capacitor of about 10 nF, combined with a resistance of about 150 Ω from the GND pin to reference ground potential. The current-limiting effect of the GND resistance simultaneously raises the immunity of the IC to interference from ISO pulses. If a GND resistance cannot be implemented, the interference suppression capacitance must be inncreased to a value between 1 µF and 10 µF in order to achieve the same filtering effect (Figure 13.30).

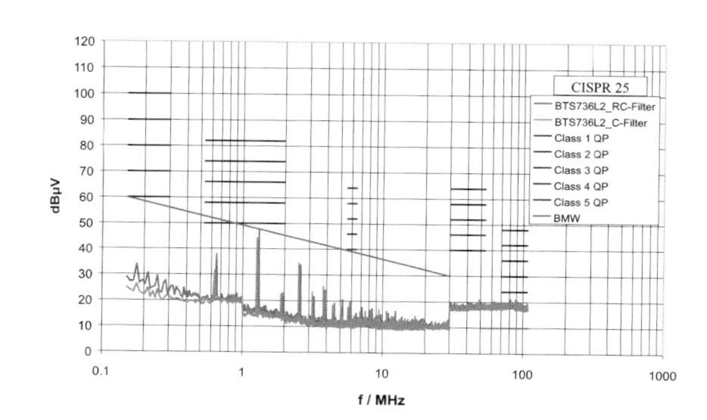

Figure 13.30 BTS736L2 in DC mode with a filter

Interference emissions from power switch ICs in PWM mode

In PWM applications, power switches generate broadband interference emissions if the switching frequency is less than the measurement bandwidth of the test receiver. These interference emissions propagate from the switch via the output and power input connections on the IC, into the connected cable harness. Particularly in the lower and middle frequency range, these interference emissions are very high-energy. The characteristics of such interference sources are deteremined by the current and voltage signal forms, with their amplitude, switching times and period length. In an application, the load current and the switching frequency are generally prescribed. The only remaining control variable in the IC is the switching time or an EMC-optimized form of the switching transition. Longer switching times do reduce the emissions, but there is an increase in the power loss which occurs in the region of the switching transition. Another IC design option, for reducing the emissions caused by PWM switching, is to avoid high-frequency components in the spectrum of the switching process. This is achieved by selective shaping of the current or voltage curve over the switching process. The switching process which is ideal from the emissions point of view provides the desired signal at the maximum switching speed and with low losses, without exceeding the limiting values

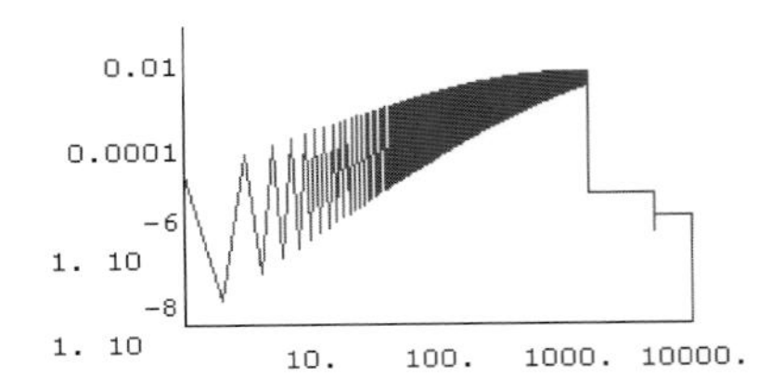

Figure 13.31
Ideal radio noise voltage spectrum

for the interference emissions in doing so. Figure 13.31 shows the ideal spectrum, and the ideal switching process calculated from it, for a 60 W PWM application on a 12 V power supply. Here, above 150 kHz all the frequency components have been limited to the limiting values prescribed in CISPR 25. The theoretical limits on emission minimization lie, as can be seen from Figure 13.32, in the region of a 10 µs switching time, and correspondingly at PWM frequencies of up to a maximum of 1 kHz. In the static-dynamic transition region a rounding of the edges then occurs. Higher switching frequencies, with steeper switching edges require external circuit-dependent filtering components, in order to meet the prescribed limit values. One output signal form which has been realized for a power IC is shown in Figure 13.34, and the reduction in emissions achieved in Figure 13.33. By comparison with the same IC with no edge shaping, it has been possible to re-

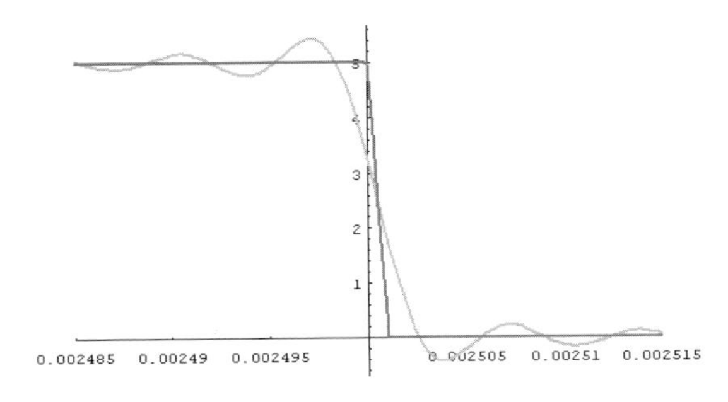

Figure 13.32
Real and ideal switching edge in an operating point

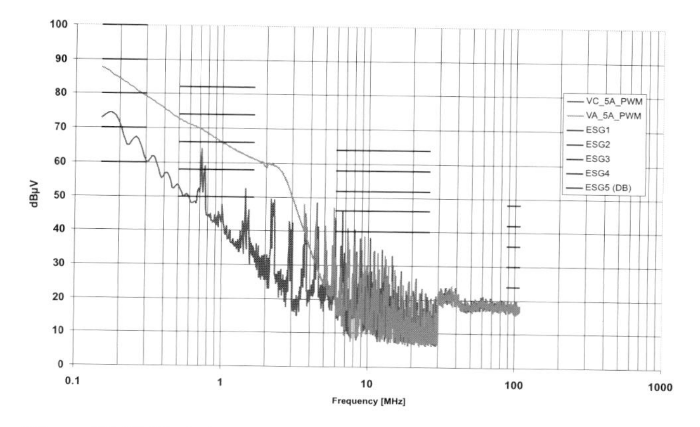

Figure 13.33 Emission spectrum with/without edge shaping

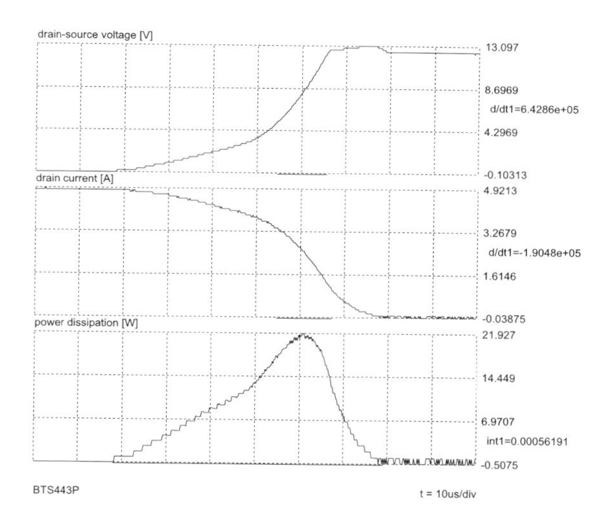

Figure 13.34 Edge shaping

duce the emissions due to the switching in the lower frequency region by about 15 dB, without significantly increasing the total power loss.

Depending on the load current and switching time, external filtering measures for reducing the interference emissions on the supply leads, caused by PWM switching processes, can be reduced below the limit values by energy storage capacitors from the reference ground to the supply leads, LC or π filters. The narrowband emissions, such as for example those from the charge pump, are suppressed at the same time by these measures.

13.2.2 Interference emissions from DC-DC converters

The interference emissions from DC-DC converters are, like those from power switches, generated by the rapid switching of currents and voltages. However, in

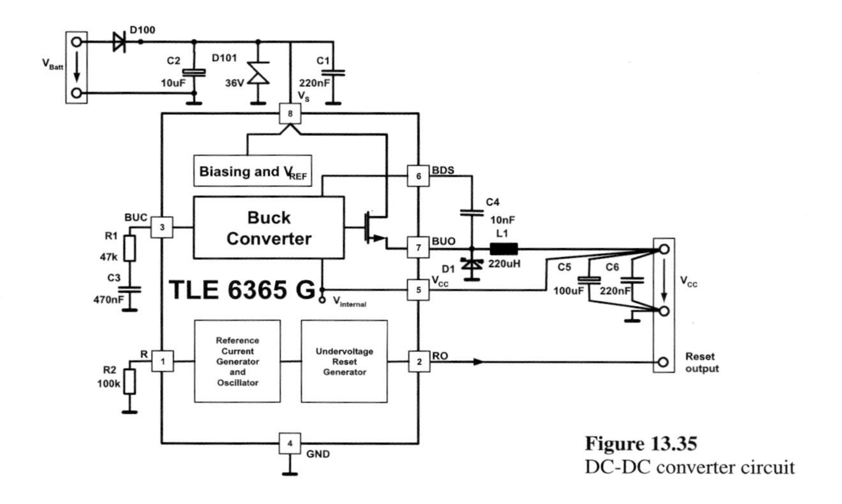

Figure 13.35
DC-DC converter circuit

contrast to switches, DC-DC converters work with significantly higher switching frequencies in the power stage, in order to keep the passive energy stores such as smoothing chokes, input and output capacitors, as small as possible, and to reduce costs and volumes. Switching frequencies between 100 kHz and 500 kHz are here the preferred operating frequencies. That is to say, DC-DC converters work with high switching frequencies in order to keep the switching power loss low. For the interference emissions, which in the case of DC-DC converters propagate via the supply and output connections, this implies narrowband interference emissions with high energy content, extending far into the 100 MHz region. The DC-DC converter datasheets contain suggested circuits (Figure 13.35) for the design of such applications. The application layout is of particular importance for its EMC behavior. The recommendations are for the smallest possible stray inductances in the power circuit, minimum areas with AC voltages (*dv/dt*), a GND area in the second layer which will act as a screen, and optimal filter connections to GND. Additional measures, which can be included in the layout, but which will then have an effect on the power loss and efficiency of the system, are gate dropping resistors, additional gate-drain capacitors and an RC snubber over the freewheeling diode. Figure 13.36 shows such an optimized application layout.

Without additional external filtering measures, the emissions from the DC-DC converter still lie above the limit values (Figure 13.37). Only the inclusion in the layout of a filter, comprising a choke and additional capacitor, which together with the link capacitor forms a π filter, allows

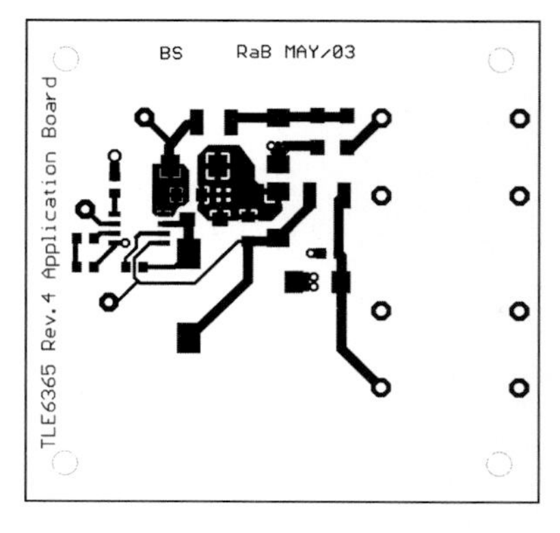

Figure 13.36 Optimized application layout

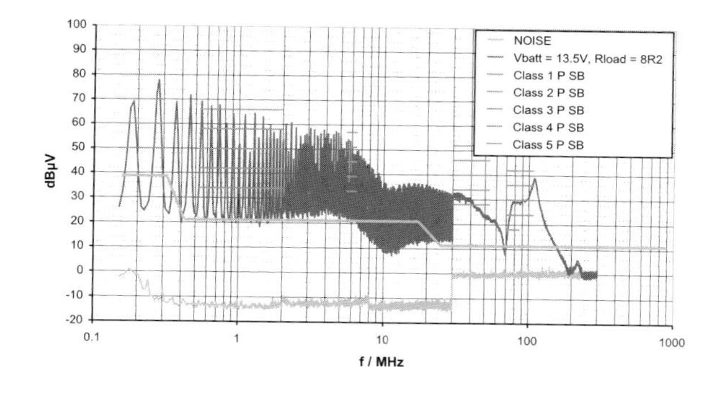

Figure 13.37
DC-DC converter without filter

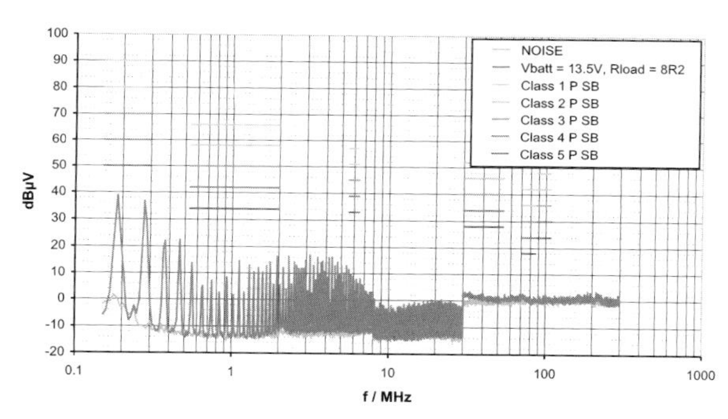

Figure 13.38
DC-DC converter with filter

the emissions to be reliably reduced below the limit values for CISPR 25 (Figure 13.38).

13.2.3 Interference emissions from communication ICs – CAN

The interference emissions from CAN transceivers results from the high frequency data transmission on the communication lines which are connected. There is a high-speed (HS) CAN transceiver (1 MBaud) and a low-speed (LS) CAN transceiver (120 kBaud). With both CAN variants, data is transmitted on two lines. From its idle state at 2.5 V, the HS CAN data signal is set to 5 V by switching on the CAN High transistor, and to 0 V by the CAN Low transistor. The change in signal level on each line is thus only 2.5 V. The principle is identical for the low-speed CAN, only here CAN Low is switched from 5 V to 0 V and CAN High from 0 V to 5 V. Theoretically, if the output signal forms of the two transistors are identical, the currents flowing in the lines and the radiated fields should cancel each other out. In practice, the IC design is subject to the switching possibilities of the semiconductor technology concerned. Figure 13.39 shows the bus signals and the total signal for an LS CAN. The residual asymmetry leads to common mode interference emissions, as shown in Figure 13.40. The emissions from such ICs can be determined with the help of the 150 Ω measurement method using common mode tapping.

External measures for limiting interference emissions at the system level are severely restricted for CAN transceivers. The high data transmission rates mean that it is only possible, for functional reasons and because of the specified maximum capacitive bus load, to make limited use of low-cost measures in the form of

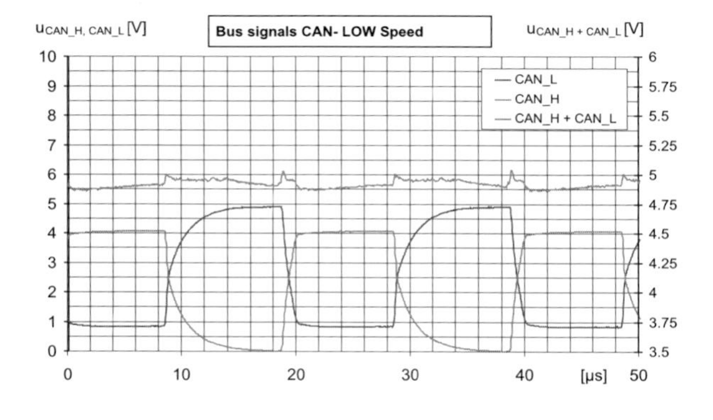

Figure 13.39
Bus signals for the LS CAN

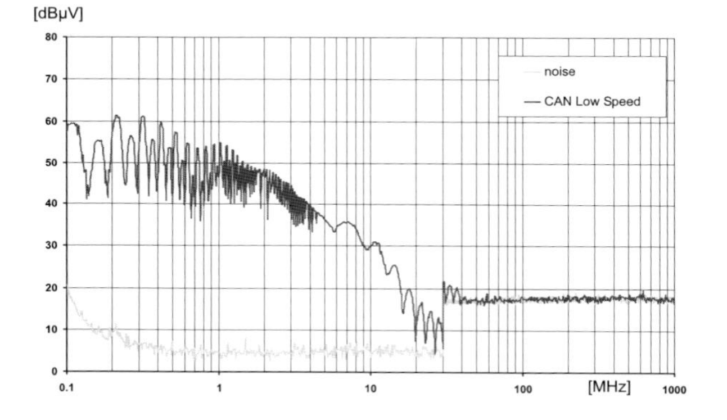

Figure 13.40
Emissions from the LS CAN

capacitor circuits, in the range from 150 pF to 330 pF, for the low-speed CAN. An additional EMC measure which can be used is common mode chokes (51 µH), which prevent the propagation of common mode interference in both directions. Further possibilities are twisted or screened leads, but these are rarely used in vehicles on cost reasons.

13.2.4 Interference immunity of automotive power switch ICs

Automotive power switches in high-side or low-side applications are often operated directly from the vehicle electrical supply. This means that they are exposed to the ISO pulses, which produce on the connections overvoltages which are above the operating level specified for the IC (Figure 13.41). Depending on the design concept, the integrated protection circuits which disconnect the IC and limit the current or voltage, then come into operation.

RF interference emissions reach the IC by the same path, via the cable harness, which acts as an antenna. In order to characterize the RF interference immunity, the DPI method is used during normal operation to inject RF power directly at the IC pin (Figure 13.42). Internal pins, which have short connecting leads on the board, are subject to 50 mW (17 dBm) of RF forward power, and those external pins which in the application have long connecting leads, via which they leave the board, are subject to up to 5 W (37 dBm) of RF forward power.

To raise the internal interference immunity of an IC there is a range of possibilities, which depend on the pulse or RF energy of the interference phenomenon, the effectiveness of which in affected by the circuit concept, the layout and the semiconductor technology used. The effects of ISO pulses on an IC can be divided into three categories:

- The *dynamic effect*, caused by the *dv/dt* for the pulse, can be reduced internally within the IC by raising the internal operating currents and reducing the internal capacitive coupling paths. Externally, capacitor circuits are possible.
- The *effect of the pulse energy* leads, if the limits are exceeded, to the destruction of the semiconductor structure. There are possibilities for IC-internal intervention, in the development of a protection concept against overvoltage, overcurrent and overtemperature. Externally, protective diodes or overvoltage protectors can be used.
- If the diode voltage is exceeded, then depending on the technology the *negative voltage* on the supply can lead to charge carrier injection. A remedy is offered by protective structures which drain off such charge carriers, or paths which can be selectively switched to a low ohmic state, but they must then bear the entire pulse current. External reverse-connect protection diodes can protect against such events.

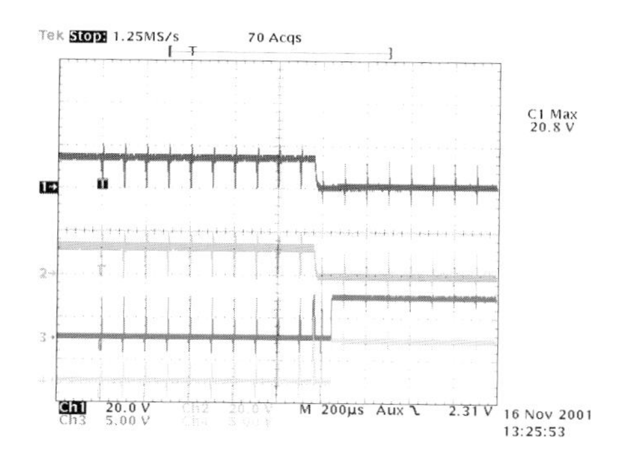

Figure 13.41
ISO pulse injection

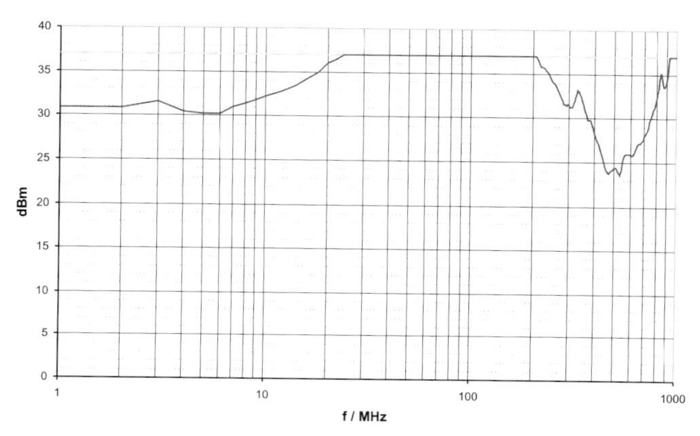

Figure 13.42 Interference immunity of power ICs

RF interference in the IC is reduced by using internal filter structures to divert the RF energy directly to the reference ground or to reflect it directly. Depending on the concept, high RF currents may flow, or high RF voltages exist. The operation of such filter structures can be active or passive. Raising the internal operating currents also raises the interference immunity. External measures are filter capacitors, and in some cases chokes.

13.2.5 Interference immunity of communication ICs – CAN

Investigations into the electromagnetic interference immunity of CAN transceivers is concentrated, as in the case of interference emissions, on the CAN High and CAN Low communication connections. The RF energy reaches the pins directly, via the data lines, which act as antennas. This effect is reproduced in RF interference immunity tests using the DPI method in accordance with IEC62132-4. In these, up to 4 W (36 dBm) of RF forward power is superimposed on the data signal as common mode interference, and the IC is monitored for transmit / receive failures (Figure 13.43). A typical interference immunity limit graph for a high-speed CAN transceiver is shown in Figure 13.44.

In the lower frequency range, around 1 MHz, there are signal errors. Because it is so close to the data transmission range, it is hardly possible to implement filtering measures in this region. As the frequency increases, the interference voltage (VRMS) present at the pins falls, as can be seen from the second graph in Figure 13.44. This gives information about the impedance characteristics of the IC in this frequency range. To achieve high common mode interference immunity, special attention should be paid to the symmetry of the input impedances. External filtering measures correspond to the options and restrictions described for the emissions.

As line drivers, CAN transceivers are often positioned directly and without protection on the plug connectors of control devices (management units) and are thus directly exposed to electrostatic discharges. Typical transceivers achieve an ESD resistance of 4 kV to 8 kV on CAN High / CAN Low when subject to HBM ESD tests in accordance with IEC 1000-4-6, at 150 pF and 330 Ω.

13.2.6 EMC measures in application circuits – external components

The EMC of a device or system is not only a matter of the IC as a source or absorber of interference emissions, it must always be considered in the context of the system and the requirements of the application and its electromagnetic environment. Depending on the application, EMC design measures cover layout recommendations and suggestions for filtering with external capacitors and inductances. In this connection, it should always be noted that every component has

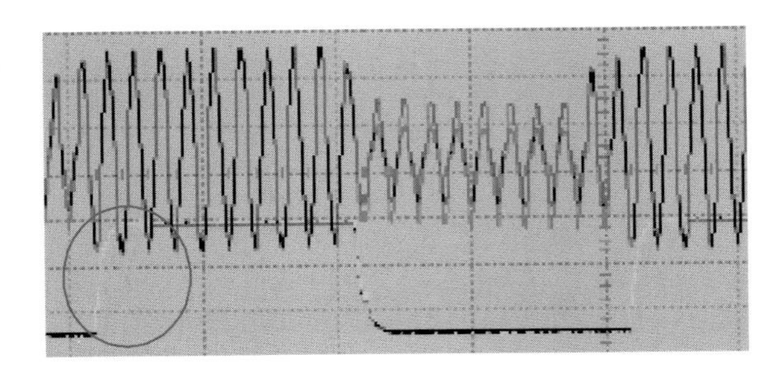

Figure 13.43
Fault pattern and superimposed RF

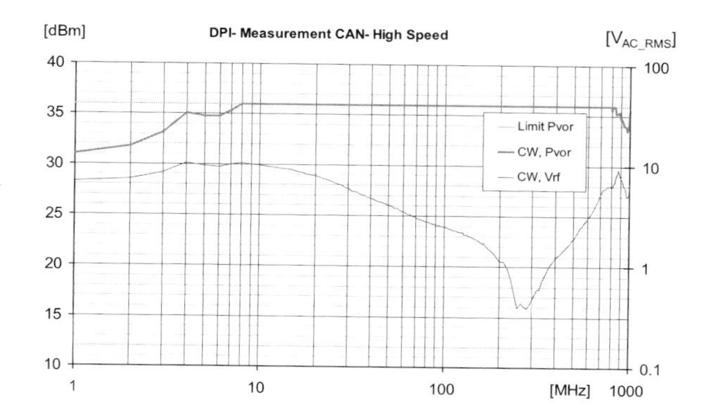

Figure 13.44
CAN interference immunity limit graph

parasitic effects which, in general, become more important as the frequencies rise. That is to say, both the nature of the filter components and also their positioning in the layout have an effect on the EMC behavior of the system. For example, depending on their type (electrolytic, MKP, ceramic) and internal structure, capacitors only behave capacitively up to a certain frequency within the frequency range. The frequency limit

$$f_g = \frac{1}{2\pi\sqrt{L \cdot C}}$$

is determined by the value of the capacitance, its internal inductance and the external connection. Above the frequency limit, the impedance graph changes from capacitive to inductive. The ohmic resistance of the connections and the leakage resistance must also be considered. The same applies for filter inductances. Parasitic capacitances across their windings limit the inductive impedance graph of chokes in an analogous way to capacitors. The equivalent circuit includes the ohmic resistance of the windings, the inductance and, in parallel with these, the parasitic winding capacitances. The magnitude of these parasitic elements is determined by the design. Load and smoothing chokes can have parasitic capacitances between 10 pF and 1 nF. In the case of rod-core radio interference chokes, which are used to attenuate push-pull interference, subdivided multi-chamber windings reduce the parasitic capacitance to values around 2 pF. Current-compensated chokes for attenuating common mode interference have similar values.

RF interference immunity problems on the power supply or output lines of ICs can generally be remedied by using a low-inductance ceramic capacitor, of the order of magnitude of 10 nF to 100 nF, placed between the IC pin and GND. Capacitors of this type have a positive effect on both the interference emissions and also on the interference immunity of the ICs and, depending on the requirements, should be an option considered for the application layout.

13.3 Electromagnetic compatibility of microcontrollers

13.3.1 Automotive microcontroller systems and technological trends

The applications for microcontrollers in automobile electronics range from simple control functions, e.g. window lifts, lights, through safety-related functions, e.g. airbags, up to complex regulatory functions, e.g. fuel injection, ignition timing control or braking activities. The wide-ranging requirements for computing

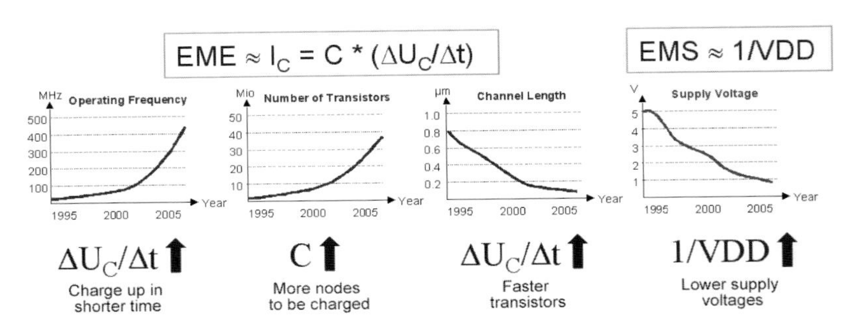

Figure 13.45 Effects of technological trends on the EMC

power and peripheral logic functions such as timers, serial and parallel data interfaces, CAN controllers and fast interrupt processing, call for very complex designs. Recently, far more than 10 million transistors have been incorporated into a high-end microcontroller using 32-bit architecture. The current demand is increasing correspondingly, which is not only noticeable in the required capacity of the voltage regulator, but also in the increasing potential for interference emissions from the power supply system due to the growing switching currents. As a result, the requirements for an optimized concept for blocking RF are steadily growing. Unfortunately, increasing IC performance works against good EMC. Figure 13.45 shows the effect of the significant parameters from the technology roadmap on interference emissions and immunity.

As a result of continuous further miniaturization of transistors, the same or an increased area can be expected to accommodate an increasing number of transistors. Today, there are around 20 million transistors on a silicon chip with an area of less than 100 mm^2. Overall therefore, it can be expected that there will be a steady increase in the switching current. This means increasing interference emissions.

At the same time, the interference immunity of the complex circuits is declining, because the interference caused by the switching currents (Simultaneous Switching Noise, SSN) is of the same order of magnitude as the useful signals. The so called signal-to-noise ratio (S/N) is becoming ever smaller. This makes it necessary, not only from the point of view of reducing the emissions, but also to improve the interference immunity, to reduce the SSN back at the interference sources. Apart from the measures to improve the EMC at the board level, measures to improve the EMC at IC level are also becoming ever more important. The field of safety-related applications in particular, such as automobile electronics, is therefore today unimaginable without EMC-optimized circuit design.

The more electronic systems are deployed in automobiles, the higher will the requirements to be met by each system become, in terms of minimal interference emissions (EME = Electromagnetic Emission) and minimal sensitivity to interference (EMS = Electromagnetic Susceptibility), or conversely maximal interference immunity. For the propagation of interference, a distinction is made between conducted and radiated RF interference. Although the characteristic lead lengths of a microcontroller chip mean that below 1 GHz it hardly radiates any energy directly, the interference generated by it can excite geometrically larger structures acting as antennas on the PCB (Printed Circuit Board). Here it is principally the voltage supply systems and leads in the IC package and on the PCB, with lengths l > $\lambda/20$ (corresponding to 15 cm at 100 MHz, 15 mm at 1 GHz), which act as transmission antennas.

It is in the nature of microcontrollers that their connection paths are mostly purely local on the board, and there are no direct links to wiring which leaves the circuit board. The power supply system is one exception. It is connected to the vehicle power system via the voltage regulator IC. Because the voltage regulator does not block high frequencies, RF interference which is generated by the microcontroller and injected into its power supply system due to non-optimal blocking, is passed on via the regulator into the vehicle electrical system from where it can cause interference at other sites or in the environment of other electronic systems, for example, the automobile radio.

It can therefore be basically assumed that all RF interference which is not neutralized as close as possible to its source on the silicon will necessitate blocking measures further from that source, such as on the board or right at the end-device, with a disproportionately higher cost. Although a blocking concept cannot be integrating on-chip without affecting its area, it is always more efficient than numerous "emergency measures" in the external system. In the final analysis, the design of a microcontroller-controlled electronic system will require measures for EMC optimization both on the chip and also on the circuit board.

Before an electronic vehicle system receives its EMC approval, quite a lot of tests must be carried out on the devices, system and vehicle, and an assessment made of:

- interference emissions at IC level
- interference immunity at IC level
- interference emissions at system level
- interference immunity at system level
- interference emissions at vehicle level
- interference immunity at vehicle level

13.3.2 EMC-optimized circuit board design

Modern microcontroller architectures use synchronous clocks. This means that the active edge of the system clock pulse, used by the logic for switching, should arrive at all parts of the circuit as near as possible simultaneously. A consequence of this is that there are very high and rapid current peaks, i.e. a high value for *di/dt*, of the order of magnitude of 100 mA/ns. In order to generate an approximation to a squarewave clock pulse (i.e. a trapezoidal-shaped clock pulse with the shortest possible signal edges), high recharging currents are required. This high current demand must be provided directly at all transistors which switch simultaneously. The damage actually generated by radiation from an electronic system is thus greater the more RF energy must be supplied from the external system. That is to say, the high-frequency components of the switching current are dissipated into the environment via the conducting tracks and structures which act as antennas. For this reason, if one can succeed in providing the charging requirements for the switching processes on the chip as close as possible to the sources of the emissions, the energy radiated out via the board structures will be less than if the charge replenishment must be delivered from a distant voltage regulator (interference path = IC power supply system – IC

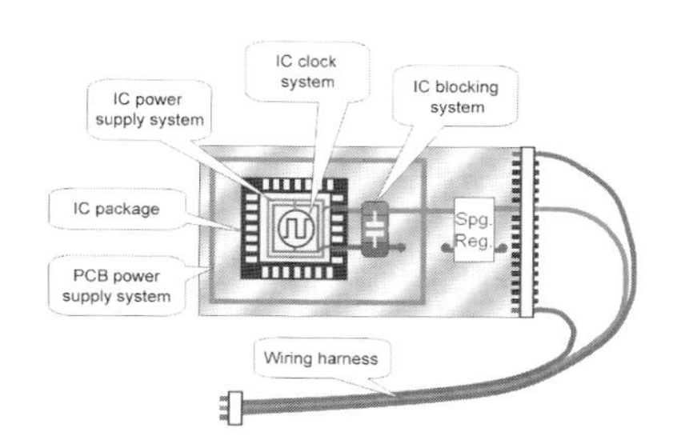

Figure 13.46
Interference paths in a clocked electronic system

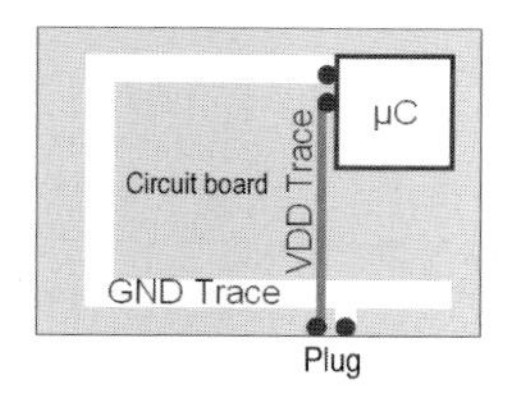

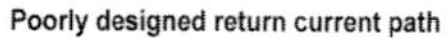

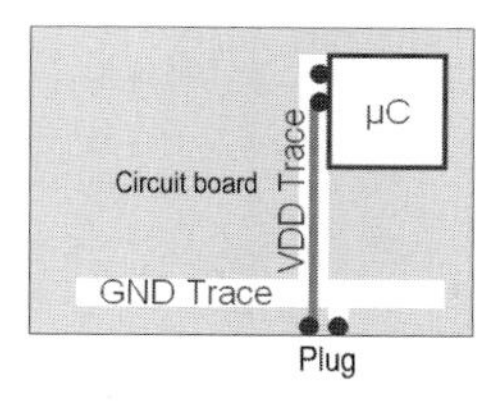

Figure 13.47
Voltage supply on two-layer PCBs

package – PCB power supply structure, see Figure 13.46).

For this reason, numerous blocking capacitors are located in electronic systems where they can provide a portion of the required switching current locally. They are recharged between two active clock pulse edges. Unfortunately discrete capacitors on the circuit board have the disadvantage that they exhibit a non-negligible contact inductance of about 2 nH (ESL = Equivalent Series Inductance) and a contact resistance of approximately 30 mΩ (ESR = Equivalent Series Resistance), which more than cancels out the capacitive behavior of the capacitor in the higher frequency ranges. In contrast, integrated capacitors, arranged close to the on-chip power supply tracks, have a negligibly small contact inductance. Hence, the more such capacitors can be directly integrated on the chip, the more effectively can the emission of RF energy be suppressed.

The problems of electromagnetic emissions can also be overcome by suitable circuit and design measures on the PCB. These include inductive filters (e.g. choke coils and ferrite beads) as RF low-pass filters, ground planes with no through-holes, routing high-frequency signal lines as strip lines, plug-in capacitors and screening sheets (tuner boxes). However, these additional elements represent a significant cost factor, particularly where large numbers are involved.

If, for cost reasons, a board with only two layers is used, one surface should if possible be used throughout as the ground (GND, VSS). If this is not possible, the positive supply potential (= VDD) and VSS should be wired closely alongside each other or one above the other, to minimize the return current conduction loop, and with it the common mode radiation emissions (Figure 13.47).

Multi-layer PCBs should if possible have two complete layers for the supply voltage (VDD and VSS). Any further supply voltages which are present can be constructed as islands under the modules which are to be supplied. In the interest of

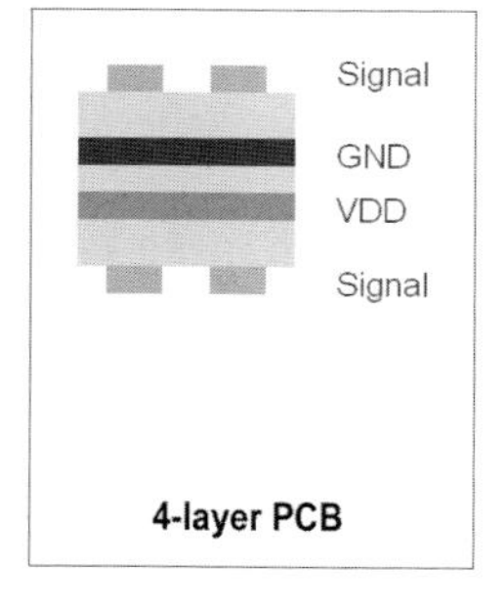

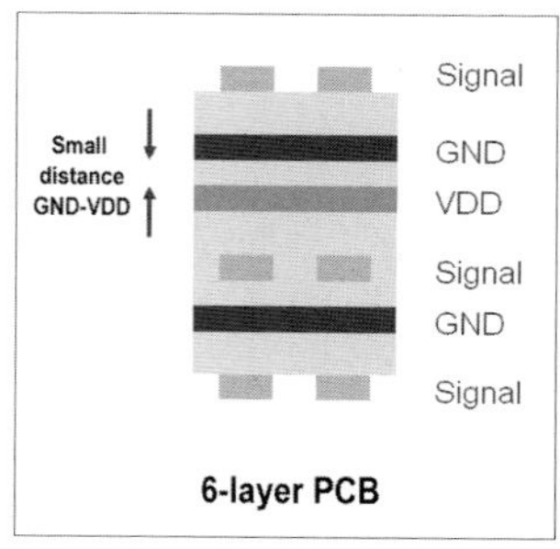

Figure 13.48
Layer arrangement in multi-layer PCBs

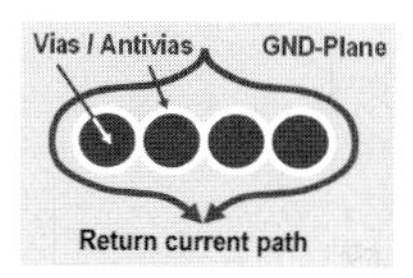

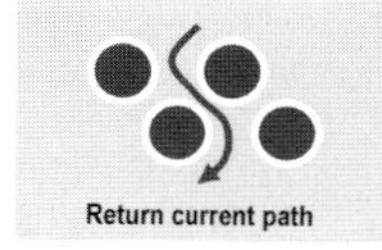

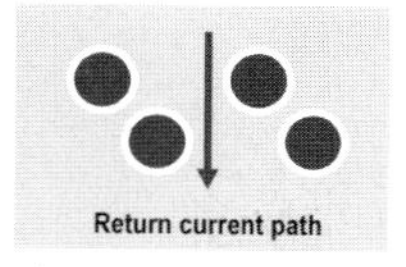

Figure 13.49
Optimal return current path for RF signals

minimal RF radiation emissions, it is desirable to achieve the maximum possible coupling capacitance between the power supply layers. This can be achieved by means of internal layers lying directly one above the other (Figure 13.48).

In a 6-layer PCB, for example, a second layer could be arranged as the GND area to provide a defined impedance for high-frequency data signals. Here again, as in the case of the power supply, attention should be paid to providing uninterrupted return current paths which are as short as possible. In particular, perforation of the GND plane by adjacent via contacts should be avoided (Figure 13.49).

Ultimately, great importance still attaches to the distribution of the several ICs on the PCB. Starting at the voltage regulator, dedicated supply areas should be defined for fast logic, slow logic and sensitive analog circuits. The power supplies should come together in the form of a star at the voltage regulator, in order to minimize the mutual interference effects of the circuits (Figure 13.50).

When using capacitors for blocking, care should be taken that the charge current they provide can be supplied to the consumer via the shortest possible path. When doing so, the power supply current from VSS or VDD, as applicable, should physically flow through the capacitance to the consumer. To clarify this, Figure 13.51 shows a poor RF connection and a good one for a blocking capacitor on the microcontroller power supply pins. The upper two diagrams show the arrangement of the components if only one side of the board is fitted with components, in the two lower diagrams the microcontroller and blocking capacitor are on opposite surfaces of a 4-layer board.

When connecting an integrated oscillator with a ceramic resonator or a quarz crystal with a terminal base capacitance, care should be taken to provide a local VSS island, starting at a VSS pin on the microcontroller. This is the only way to avoid the return current path spreading across the GND surface.

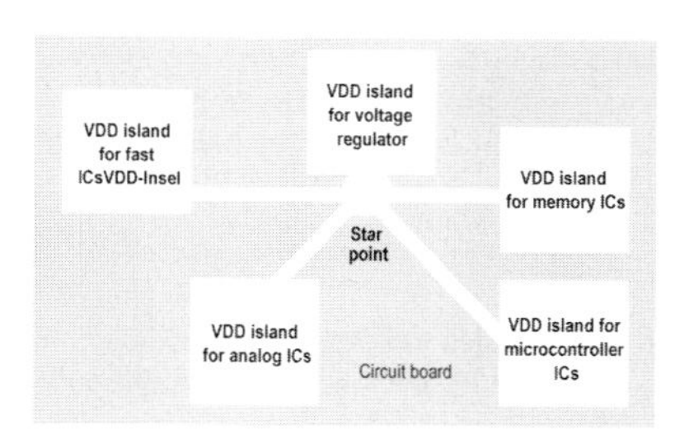

Figure 13.50
Star arrangement of the power supply for circuit types

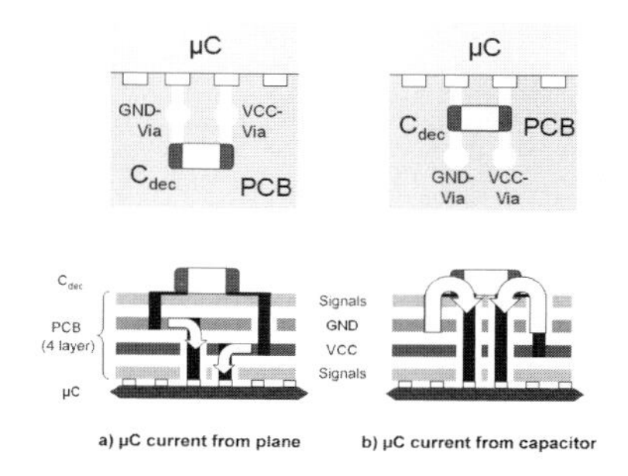

Figure 13.51
Poor (a) and good (b) layout of the blocking capacitors

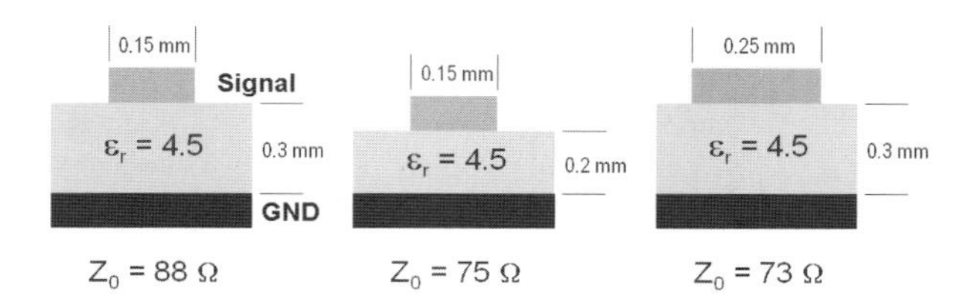

Figure 13.52 Characteristic signal impedances for microstrip signals

Apart from the blocking concept, great importance attaches to the chanelling of high-frequency data signals in relation to RF radiation and signal integrity. The radiation or bridging of RF, for example into the power supply system, can be effectively prevented by GND screening of the lead concerned. When this is done, a micro stripline is created in the case of single-layer screening, and a complete stripline in the case of two-layer screening. Both variants can be arranged to be impedance controlled, which enables them to be matched to the driver impedance. Figure 13.52 shows some numerical examples.

The matching of the line impedance to the driver impedance just mentioned serves both to improve the signal integrity and to reduce interference emissions, because mismatches lead to reflection of the signal and thus overshoots at switching edges, or even to short term oscillation (ringing) in the signal. Further details of these effects will be found in section 13.4.

EMC modeling and simulation

In principle, simulations can be performed in the time domain or in the frequency domain. In the time domain what is of interest is the signal integrity, i.e the speed, edge steepnesses, under-/overshoots and ringing of the signals. The simulation results help to identify overdimensioned drivers, excessively long leads and impedance mismatches. The IC manufacturers provide so-called IBIS models (= I/O Buffer Information Specification) for digital switching circuits, which describe the switching behavior of all the drivers present in the IC. The PCB structures are specified either by transmission line models (TLMs) or by 2D or 3D elements, as appropriate, for which there are various forms of geometric representation. The more irregularly a PCB

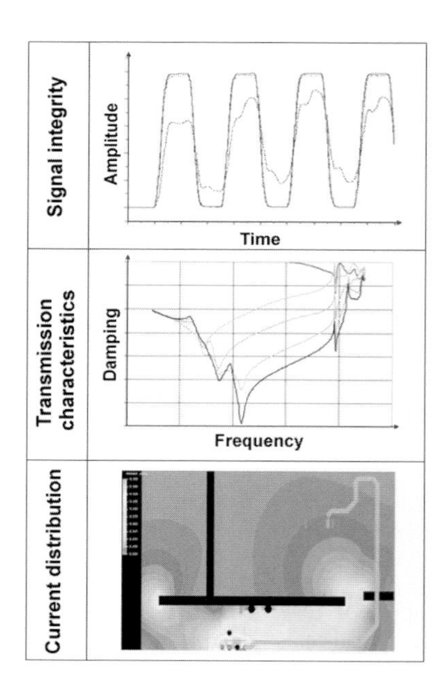

Figure 13.53
Signal integrity in the time domain (above), transmission coefficient S21 in the frequency domain (center), local current distribution on a PCB (below)

is structured, the more finely its environment must be modelled. This is the reason for the high data amount and computing costs. The resulting complex RLCG matrices are solved by fieldsolver programs. In this process, the calculations can be conductor-oriented or field-oriented. Figure 13.53 shows some examples of analyses in the time and frequency domains.

13.3.3 Measurement of interference emissions from microcontrollers

In assessing the interference emissions from microcontrollers, measurements made in accordance with the international norm IEC 61967 serve as the reference. This includes five different measurement methods, with particular use being made of the measurements of conducted emissions (IEC 61967-4, previously VDE 767.13) and radiated emissions (IEC 61967-2, previously SAE 1752/3). The various measurement methods were proposed by national EMC committees, and continue to co-exist in the international norm. A comprehensive description of these standardized measurement procedures is given in section 13.1.3.

The two measurement methods can be combined on a single test board (Figure 13.54). A unified test board for TEM cell measurements requires each edge to have a length of four inches, and the IC to be measured must be located on the screened underside of the board, and the peripherals (memory, oscillator, I/O circuitry, ports) on the upper side. In addition, this board can be operated in a near-field scanner, in order to measure the field distribution as a function of position. This can be used to identify critical supply pins.

Typical emission spectra from microcontrollers show the harmonics of the system clock rate with particularly large amplitude. For the purposes of comparing several development steps, for example, use is made of the envelopes. Figure 13.55 shows a typical emission spectrum for a microcontroller with a 40 MHz clock frequency.

Interference sources and noise propagation paths

A microcontroller consists of numerous functional modules (Figure 13.56). There is always a CPU (= Central Processing Unit), which is responsible for data transfers and calculations. Further modules, the so-called peripheral modules or simply peripherals, include memory (RAM, ROM, E^2PROM, Flash), timers (e.g. capture/compare timer), serial and parallel data interfaces (ASC, SSC, SPI, parallel ports), special modems (CAN, J1850), interrupt controller, and a memory interface (EBU = External Bus Unit). To give the maximum computing power, the data buses must be chosen to be as wide as possible and the processing clock rate as high as possible. A high clock rate does not leave much latitude for internal data

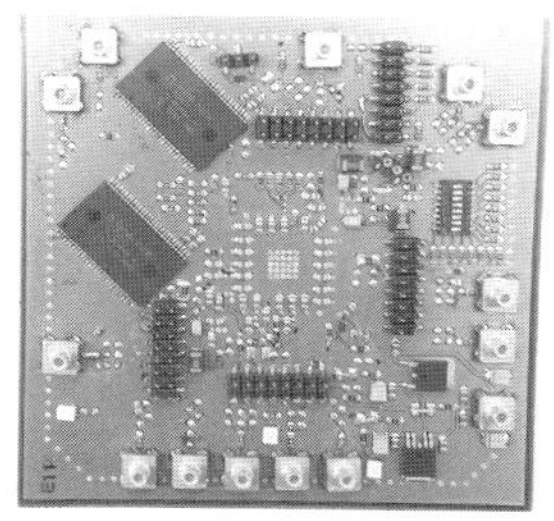

Figure 13.54
Microcontroller emissions test board (left: upper side; right: underside with test specimen)

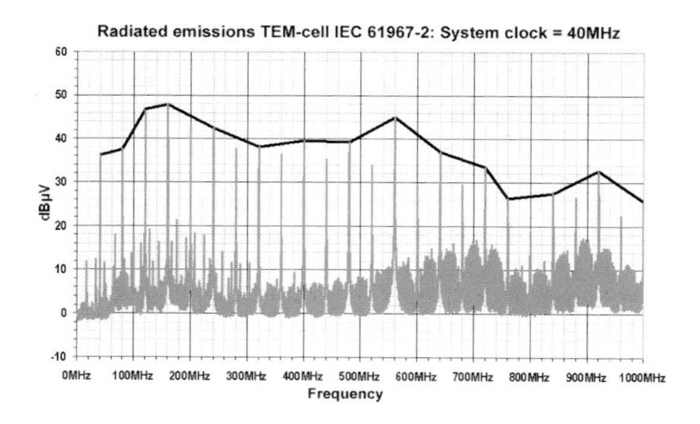

Figure 13.55
Radiated emissions from a microcontroller in a TEM cell in accordance with IEC 61967-2

processing, so that the quiescent phase of the cycle must be as long as possible. This results in very fast switching edges which, moreover, must occur synchronously throughout the system.

Since all the transistors draw their switching current from the power supply system, this latter forms the main path by which about 60% of the RF noise is coupled into the surrounding system. Further coupling occurs through the I/O pins (about 30%), the remainder (about 10%) is directly emitted radiation. That portion of the high-frequency interference which has not been eliminated by design measures on the silicon itself (e.g. integrated blocking capacitors) gets into the surrounding system via the power supply pads. In the first instance, this system is the board on which the microcontroller is mounted. Discrete blocking capacitors mounted as close as possible to the power supply pins help to eliminate some of the RF interference. Due to their non-optimal contact values (ESL = Equivalent Serial Inductance, approx. 2 nH; ESR = Equivalent Serial Resistance, approx. 30 mΩ) they are only capable of effective blocking at lower frequencies, up to some 10 MHz, higher frequency interference components are hardly attenuated. For this reason it is necessary to counter the RF interference back at its sources and along the coupling paths, directly on the silicon chip.

Synchronous clock pulse systems are the central source of high-frequency narrowband interference. These are overlaid by derived clock pulses and aperiodic signals, which contribute the broadband component in the interference spectrum (in this connection cf. Figure 13.55). Today's complex and high-performance digital circuits use synchronous timing pulses for their complete logic, i.e. the clock pulse signal arrives at all points in the circuit at the same time. The many hundreds of thousands of transistors switching simultaneously draw an enormous peak

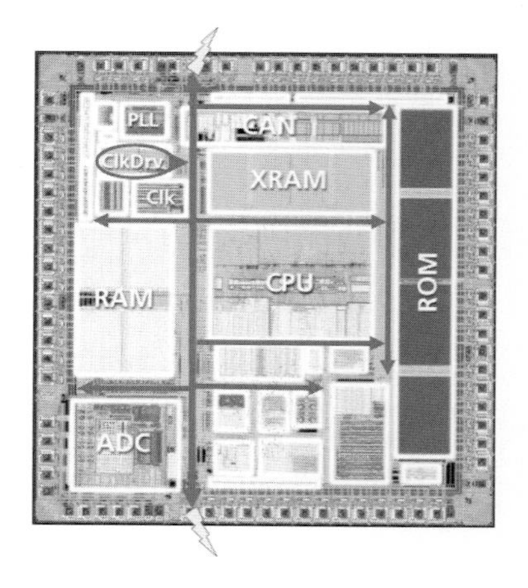

Figure 13.56
Floorplan of a microcontroller with a central clock pulse system

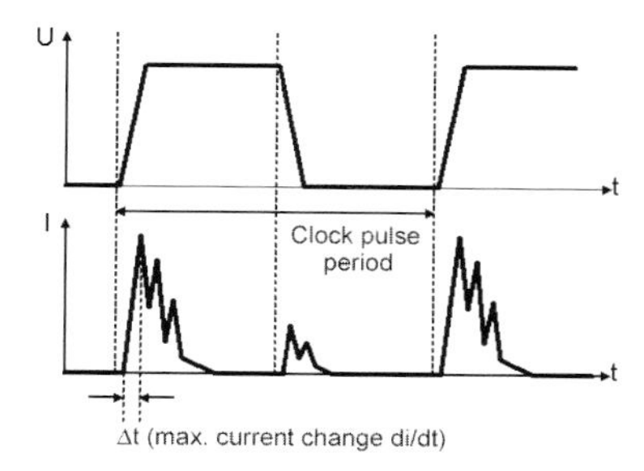

Figure 13.57
Voltage and current curves for the clock signal

current, of the order of several amperes, from the power supply system. The power supply system will ideally be arranged to be low-impedance, to avoid significant voltage drops, so that by comparison with the supply voltage it is the supply current which makes by far the greatest contribution to the generation of RF interference. Depending on the arrangement of the internal logic, these current peaks will occur once or twice per clock pulse period. They have very steep edges, and decay as the switching signal edges propagate through the various stages of the logic circuits (Figure 13.57).

The switching currents required by the switching logic on the silicon chip must normally be supplied from an external voltage source. However, because of the steep switching edges required (large *di/dt*) and the parasitic resistances and inductances in the supply system, this is not possible without generating "simultaneous switching noise" (SSN). The RF components of this are fed on within the electronic system, or are radiated out into the environment. For this reason, so-called blocking capacitors are inserted in the supply system close to the current sinks, the integrated circuits. This means that each microcontroller ideally has a capacitor as close as possible to its power supply contacts. The effect of this capacitor is advantageous in two respects:

1. It makes a portion of the switching current available locally.
2. To a certain extent, it prevents the propagation of the high-frequency interference towards the voltage regulator or the battery.

Circuit measures for avoiding interference emissions

The closer they are located to the interference sources, the more effective are measures for reducing the interference emissions. Measures for EMC improvement can thus be: transistors and logic gates with less abrupt switching edges, but also the availability of additional charge stores (capacitors) in the immediate neighborhood of the switching logic. Slower switching edges are only possible if the performance of the microcontroller permits this. Basically, the transistors and standard gates incorporated should not generate any unnecessarily fast switching edges. With the circuit synthesis which is nowadays in widespread use, the selection should be managed by a software tool.

In order to reduce the switching current peaks, most semiconductor manufacturers integrate capacitors on the chip. In combination with a low-ohmic series resistance, this produces an RC low pass filter. This both provides the RF part of the switching current locally, and also successfully prevents it from leaving the chip. Although a series inductance would be preferable because of its low series resistance, it would require an unjustifiably large surface area for values of a few 100 nH. Systematic investigations into microcontrollers show that the planned realization of such RC implementations produces a significant reduction in interference emissions. The left-hand side of Figure 13.58 shows two emitted radiation measurements for a 16-bit microcontroller, the upper one when there are no special EMC design measures and the lower one with a good on-chip blocking concept. The interference emissions have

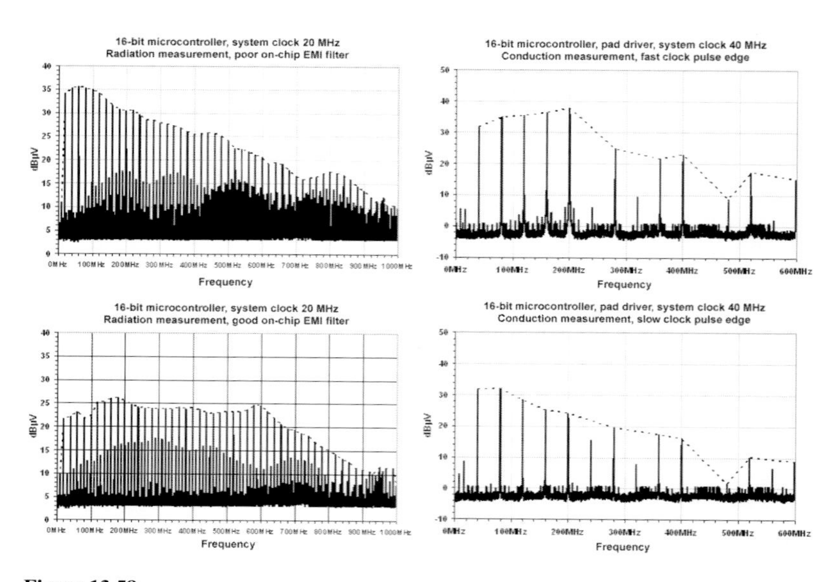

Figure 13.58
Reduction in interference emissions by on-chip circuit measures: left – without (above) or with (below) blocking capacitors, right – fast (above) or slow (below) switching signal edges from the pad driver

been reduced in the critical frequency range around 100 MHz (FM radio band) by about 10 dB.

The interference generated by the switching pad drivers also contributes to the total potential interference from the microcontroller. The signal lines on the circuit board connected to the I/O pins are, in a similar way to the power supply system, very effective antennas, with lengths which can easily be in the region of a few centimeters. Design measures should also be taken to reduce the RF interference in the case of the I/O pad driver stages. Common methods are to reduce the drive power of the output driver or to slow down the switching edges (slew rate control). As the sizes of driver transistors are arranged to satisfy the device specification under worst-case conditions there are certain reserves, particularly for slower system clocks or where there is a smaller capacitive load, which permit the output signal edges to be made slower. The right-hand side of Figure 13.58 shows two emitted radiation measurements for a 16-bit microcontroller, the upper one with steep switching edges from the system clock driver pad, and the lower with switching edges which have been slowed down. The interference emissions have been reduced in the critical frequency range around 100 MHz (FM radio band) by about 6 dB.

A further possible circuit measure to improve the EMC behavior is to use a frequency-modulated system clock. Such oscillators are known under the generic term “spread spectrum oscillator”. Even a limited level of modulation, in the region of 0.5 - 1%, effects a reduction in the narrowband amplitudes of the clock pulse harmonics. At the same time, somewhat higher emissions occur in the sidebands. Although the energy content of the emission spectrum is not reduced in total, its

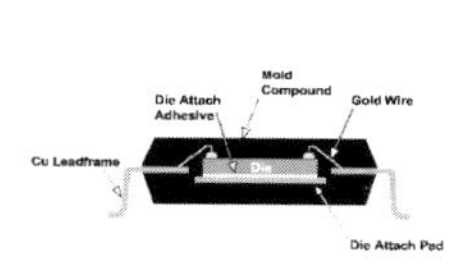

Leadframe Package

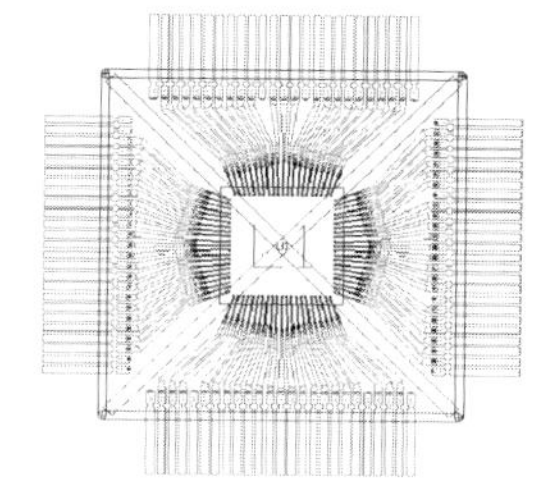

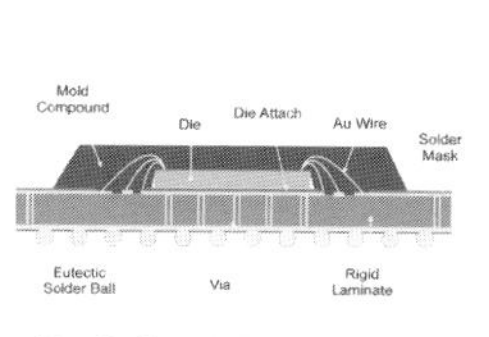

Ball Grid Package

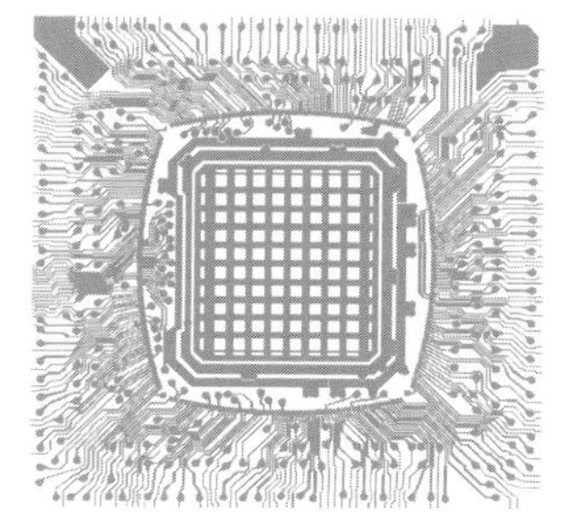

Figure 13.59
Package cross-section and internal plan view, for a leadframe (above) and a ball grid array (below)

more favorable distribution does ultimately result in smaller maximum values for the interference emissions.

The interference effect of a microcontroller can also be decreased by appropriate modification of the package and pinning. Widespread use is made of so-called leadframe packages (Figure 13.59, above), in which the power supply and signal connection leads are routed to the terminal pins on the package via “spider's legs”. The familiar representatives of this genus are DIL, PLCC, M-QFP and T-QFP. The greater the distance between the silicon pad and the package pin, the greater too is the contact inductance. This is made up of the inductance of the leadframe and the inductance of the bondwire, which creates the electrical link between the leadframe and the silicon chip. When designing the pinning of the microcontroller, care should therefore be taken to arrange the power supply pins in each case in pairs, as nearly as possible in the geometric center of the package edge, to keep the inductance of the connections small and to permit optimal blocking.

Whereas the signal routing is prescribed in the case of a leadframe package, with a ball grid array (BGA, Figure 13.59, below) there is a possibility of optimizing critical signals and supply systems with respect to EMC criteria by the design of the interposer. For these packages, the pins consist of solderable balls on the underside which, when heated briefly, form connections with the conducting tracks on the circuit board. The silicon pads and the tracks of the interposer are connected either by bondwires or, in the case of face-down assemby, by direct contact via so-called bumps.

EMC modelling and simulation

Whereas EMC simulations have been used for a long time now in designing PCBs, EMC modelling and simulation for microcontrollers is still in its infancy. The main reason for this is the complexity of these devices, which has the consequence that transistor-level simulations would require unjustifiably large volumes of data and long simulation times. However, one possible approach is to generate

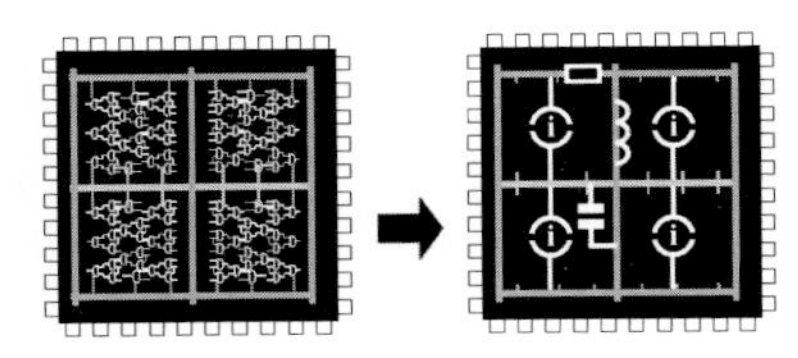

Figure 13.60
Modeling a complex IC by equivalent current sources and RLC extraction

equivalent current sources, which reproduce the switching current behavior of the microcontroller for the entire chip or parts of it.

For system-specific EMC simulations, a development of the IBIS model has been devised. However, this describes only the behaviour of the I/O pins. The so-called ICEM standard (Integrated Circuits Electromagnetic Emission Model) specifies equivalent current sources at the IC power supply contacts. As it is mainly the potential for interference at the microcontroller's supply pins which needs to be assessed, the simulation model must also include data for the package and the PCB in the close vicinity of the IC (e.g. blocking concept). On the chip side, module-specific equivalent current sources and the most accurate possible RLC model of the voltage supply system are required (Figure 13.60).

The quality of the models is heavily dependent on the correct calculation of the dynamic switching current and the accurate extraction of the RLC parasitic effects. If it is required only to make relative statements, for example about the EMC improvements to be expected from a redesign, it is possible within certain limits to forgo quantitative accuracy in the models.

13.3.4 Interference immunity of microcontrollers

Although interference immunity has until now been of little importance for microcontrollers, because at 5 V or 3.3 V the supply voltages are very high, interference immunity must be expected to decrease at a voltage of 1.5 V and below. The assessment of interference immunity by measurement techniques is very expensive, because circuit elements which are at risk of failing must first be identified and a customized monitoring program must be written for making the pass/fail measurements. At the international level, efforts are being made to standardize interference immunity measurements for ICs, starting from the familiar PCB measurement methods. These are mainly concerned with high-frequency fields or conducted RF currents. The measurement methods are brought together in the norm IEC 62132 (see section 13.1.4).

Work is in progress on test arrangements for assessing the pulse interference immunity of ICs, and the ESD resistance of microcontrollers is also an important topic. As microcontrollers normally have, apart from their power supply system, no interface pins which have a direct plug connection to an external wiring system, their requirements for pulse resistance are not as high as for other ICs or complete systems.

The greatest influence on the interference immunity of microcontrollers is that from the variation in the power supply system caused by high switching currents (SSN). These disturbances are caused either by the microcontroller itself or by neighboring devices, in particular power semiconductors, and can lead to a significant reduction in the signal-to-noise ratio (SNR).

13.4 EMC objectives for wire-line communications

The term wire-line communication systems covers a wide range of devices and technologies. These include elements of the telecommunication networks, LAN

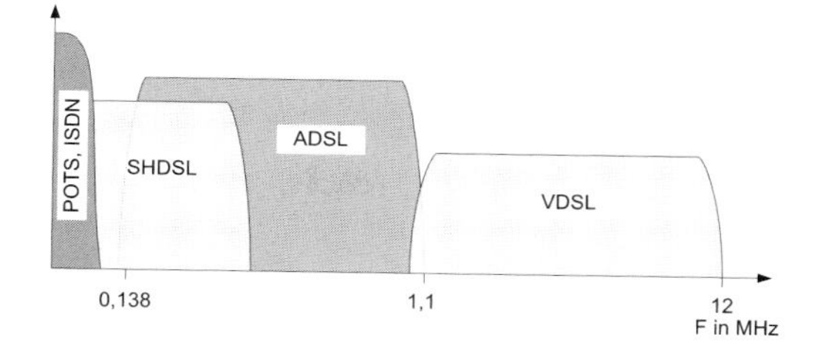

Figure 13.61
Bandwidth of the broadband services in wire-based communications

switching centers and LAN routers (Local Area Networks), plus devices in private households, for example telephones, PCs or network termination devices.

However, all the EMC requirements to be met by communication systems relate to the overall system, and not to its components. From this it follows that the EMC tests are generally carried out at the interfaces to the external world:

- Signal lines (requirements in terms of RF voltage and current level)
- Power supply lines (requirements in terms of RF voltage)
- Packages (requirements in terms of the electric and magnetic field strengths)

Data transmission rates are becoming higher and higher. As a result the analog bandwidth for the services is increasing. Today, ADSL or VDSL uses the same frequencies as the radio services, which should be protected by the EMC regulations. The spectrum used by the various wire-line communication services is shown in Figure 13.61.

Fulfilling the EMC requirements in this frequency range is not easy. It is not possible simply to filter the interface, because this would also corrupt the signal which is being transmitted. Figure 13.62 shows the typical signal spectrum of an ADSL line. The common mode voltage was measured using an EMI receiver, with the help of a standard coupling network for telecommunication lines.

The ADSL signal spectrum is present between 30 kHz and 1.1 MHz. The diagram actually shows two plots. The first plot shows the measured peak values (Pk), the second the mean values (Avg). This is required by the standard, which specifies a separate limit value for each measurement.

But DSL services are symmetrical, with reference to ground and should not appear in a common mode measurement! The reason it does is because of the slight

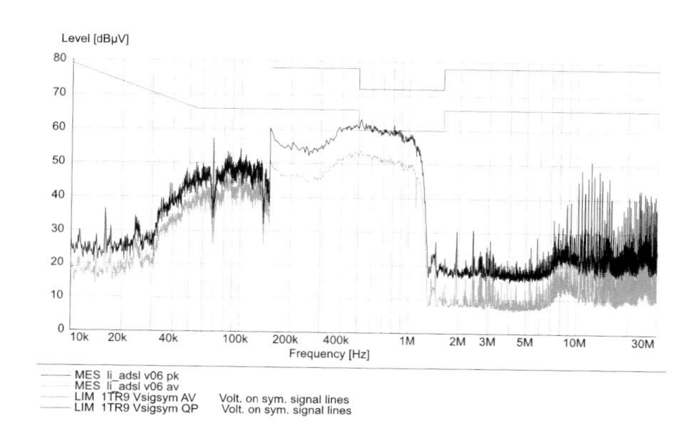

Figure 13.62
Signal spectrum in an ADSL line

asymmetry of the test circuit, which produces this signal. There is also narrowband interference in the spectrum, the origin of which is to be found in the voltage converter on the board or in the clock pulse sources (harmonics). These signals absolutely must not be injected onto the ADSL line by the system.

Quite clearly, this mixture of signal and interference demands a certain amount of thought to determine the best EMC strategy. Sooner or later it will become apparent that the best way to ovecome this problem is by improving the circuits within the system, that is at circuit board or component level.

A few ideas will now be put forward, as to how signals can be managed within the system.

13.4.1 System, components and fundamentals

A telecommunication device consists in general of the following components:

- Power supply
- Digital front-end (DFE): e.g. a digital signal processor (DSP) and microcontroller
- Analog front-end (AFE): an analog-to-digital converter (ADC) and digital-to-analog converter (DAC)
- Amplifier: transmission amplifier (line drivers) – or receive amplifier

Figure 13.63 is a sketch of a very simple but typical communication system.

Every communication system must be specified so that it conforms to certain EMC requirements. These EMC requirements must be met in the design cycle. The important points are:

- Specification of the EMC levels for new products
- System design: subdivision of the system into smaller blocks, i.e. modules, if EMC cannot be handled at system level.
- Module design: simulate or emulate the EMC of the design before designing the overall system, or use an early prototype for testing the EMC.
- Ensure adherence to the regulatory compilance in all the target markets.
- Use of EMC attributes in quality assurance (QA).
  EMC is affected by changes to the module, including the replacement of components or cables, sometimes indeed even by the elimination of software faults.
- Sales of the system exclusively in the markets for which it was planned.
- Installation of fault reports for EMC problems and feedback for product improvements which are already available or under development.

13.4.2 Design of high-speed PCBs – signal integrity (SI)

Figure 13.63 shows: the digital signal processor (DSP) performs the majority of the signal processing in the system, and in

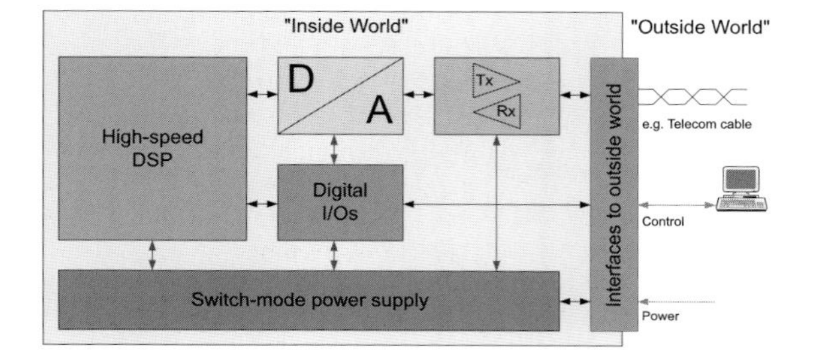

Figure 13.63
A typical system for wire-based communication

general the clock cycles lie within 10 ns or even less. But does 10 ns already represent high-speed?

The term "high-speed" is context-dependent, but in the case of PCB design it can be relatively clearly defined for a signal, with the following symbolism: rise time t_r, decay time t_f, and pulse time during which the signal is 'high' t_p. The frequency distribution of such a signal can easily be calculated using the Fourier transformation, see also section 13.1.1.

$$\lambda = \frac{c \cdot \pi \cdot t_{fr}}{\sqrt{\varepsilon_r}},$$

$$f_{highspeed} = \frac{1}{\pi \cdot t_{fr}},$$

where $t_{fr} = \min(t_f, t_r)$, c = speed of light in vacuum, ε = rel. permittivity of PCB.

If the length of the track l_{track} is not much smaller than the wavelength λ (for example only one $\lambda/20$) on a PCB, then we speak of a high-speed signal. In this connection, it should be remarked that the spectra of digital signals with fast rise and fall times include very high frequencies even at low clock frequencies! This means that a signal with a rise time of 500 ps is a high-speed signal if the track conductor length on the printed circuit board (PCB) is larger than 11.75 mm. The good news is that the signal is no longer high-speed if one is in a position to connect the devices using a track conductor which is shorter than 11.75 mm. However, the reverse side of this coin is that it is sometimes impossible to keep the track conductor lengths within this limit. In this case it is particularly important to consider the EMC and signal integrity.

According to an extract from the web site for Signal Consulting Inc. (http://signal integrity.com/integrty.htm), signal integrity (SI): "is a field of study half-way between digital design and analog circuit theory. It's about ringing, crosstalk, ground bounce, and power supply noise. It's all about how to build really fast digital hardware that really works. It's about practical, real-world solutions to high-speed design problems."

"Maximize the performance, and minimize the cost, of interconnection technology used in high-speed digital designs."

It follows that SI is concerned with a small but important area of EMC, in the "inner world" of system design (see Figure 13.63). SI is in some senses one level of abstraction above electrodynamics. With the increasing speed of integrated circuits, engineers must take more and more account of EMC. Digital design engineers are in general not particularly familiar with analog or high-frequency technology, and there is no requirement for them to concern themselves with the fundamentals of electrodynamics.

For SI, the focus is on:

- Crosstalk
- Ringing
- Ground Bounce
- Power supply noise

We indicate below how to produce a good SI design, and hence produce the basis for good EMC.

Crosstalk

Crosstalk is the electromagnetic coupling of signal conductors, whether in a semiconductor circuit, on a PCB or in a bundle of telephone lines.

Crosstalk is an unwanted but unavoidable phenomenon in electrical engineering. On the other hand, crosstalk can be kept within certain tolerances, so that it is possible to prevent any reduction in system performance.

PCB conductors or telephone lines are so-called distributed elements, and cannot be specified by formulae comprising lumped elements such as resistors, capacitors, coils and transformers. Nevertheless, the common approach in engineering is generally to approximate distributed elements by a chain of lumped elements.

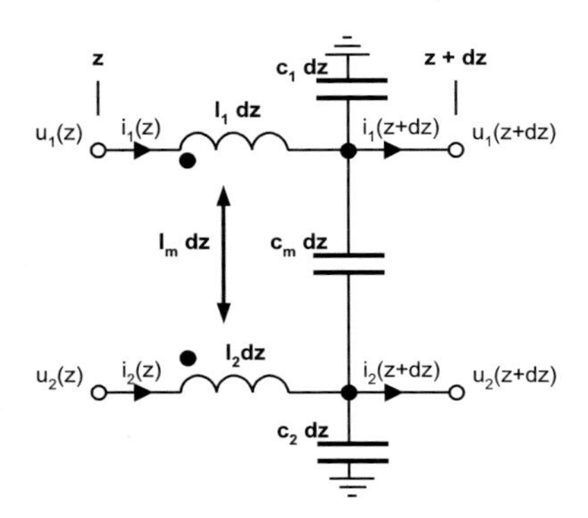

Figure 13.64 A simple loss-free crosstalk model with two tracks (track length dz)

This is possible because a track with a length l_{track} and a chain of n tracks each of length dz = l_{track} / n is always identical. If the length dz of the track element is short enough, then the signal is not considered as a high-speed signal, as has already been mentioned above. The element dz may then be approximated by lumped elements such as resistors capacitors, inductors and transformers.

This model is used in Figure 13.64, which shows two loss-free conductors, one is modelled by c_1-l_1 and the other by c_2-l_2. Crosstalk between the conductors is represented by c_m (capacitive crosstalk) and l_m (inductive crosstalk).

The solution of the partial differential equations (PDEs) for this model defines a wave which moves forward and backward: $u_i(z, t) = A_i u_i\ (z + v_i t) + B_i u_i\ (z - v_i t)$ for i = 1, 2. This can be effected by resolving the PDEs for the specified model, by diagonalizing the matrix (linear algebra). To simplify the explanation, the types of crosstalk are shown separately in the following diagrams.

Inductive crosstalk:

To clarify forward and backward crosstalk, Figure 13.65 shows the conductor, subdivided into n unit elements. Each unit element consists of a transformer and a delay element τ = dz/v. This discrete model simplifies the explanation, and is subdivided into forward and backward crosstalk.

Forward crosstalk increases the amplitude (towards larger z) and is inverted with respect to the aggressor (conductor). Assume a measurement is made on the victim (conductor) at $z = 1 \cdot$ dz. This measurement shows a small negative pulse at the time $t = 1 \cdot \tau$. But a measurement on the victim at $z = 2 \cdot$ dz shows a pulse with twice the amplitude at time $t = 2 \cdot \tau$. This arises because of the superposition of the pulse transformed in "unit element 2" and because of the propaga-

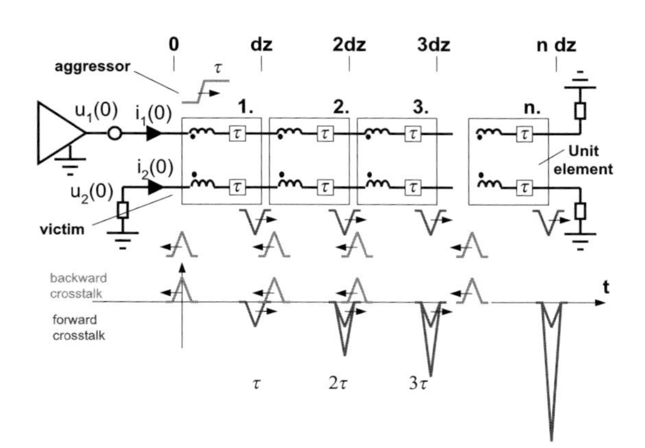

Figure 13.65
Inductive crosstalk

tion of the pulse from $z = 1 \cdot dz$ to $n \cdot dz$ on the victim.

Backward crosstalk increases the width (towards smaller z) but not the amplitude, and is not inverted with respect to the aggressor. Assume a measurement is again made at various points along the conductor. At $z = (n - 1) \cdot dz$, this measurement shows a single small positive pulse at time $t = (n - 1) \cdot \tau$, moving backward towards $z = 0$. At $z = (n - 2) \cdot dz$, two pulses occur; one at time $t = (n - 2) \cdot \tau$, and another one at time $t = n \cdot \tau$. This additional pulse arises from $z = (n - 1) \cdot dz$, which is moving backward and appears after a delay of τ at $(n - 2) \cdot dz$. At $z = 0$, n pulses are measured, and each pulse has a delay of $2 \cdot \tau$ due to the delays on the aggressor conductor and on the victim conductor.

As $dz \leftarrow 0$, the pulses can longer be separated, and hence they superpose one another to form one long pulse with a low amplitude.

Figure 13.66 shows the backward crosstalk qualitatively for several points, z, against the time variable t.

Capacitive crosstalk:

The model in Figure 13.67 is very similar to that for inductive crosstalk, but the inductive transformer has been replaced by a capacitive transformer (capacitive coupling).

Forward crosstalk increases the amplitude (towards larger z) and is not inverted with respect to the aggressor. Assume a measurement is made on the victim (conductor) at $z = 1 \cdot dz$. This measurement shows a small positive pulse at the time $t = 1 \cdot \tau$. But for a measurement on the victim at $z = 2 \cdot dz$, a pulse with twice the amplitude has developed at time $t = 2 \cdot \tau$. This arises from the superposition of the pulse transformed in "unit element 2" and because of the propagation of the pulse from $z = 1 \cdot dz$ to $n \cdot dz$ on the victim.

Backward crosstalk increases the width (towards smaller z) but not the amplitude, and is not inverted with respect to the aggressor. Assume a measurement is again made at various points along the track. At $z = (n - 1) \cdot dz$, this measurement shows a single small positive pulse at time $t = (n - 1) \cdot \tau$, moving backward towards $z = 0$. At $z = (n - 2) \cdot dz$, two pulses occur; one at time $t = (n - 2) \cdot \tau$, and another one at time $t = n \cdot \tau$. This additional pulse arises from $z = (n - 1) \cdot dz$, which is moving backward and appears after a delay of τ at $(n - 2) \cdot dz$. At $z = 0$, n pulses are measured, and each pulse has a delay of $2 \cdot \tau$ due to the delays on the aggressor conductor and on the victim conductor.

As $dz \leftarrow 0$, the pulses can no longer be separated, and hence they superpose one another to form one long pulse with a low amplitude.

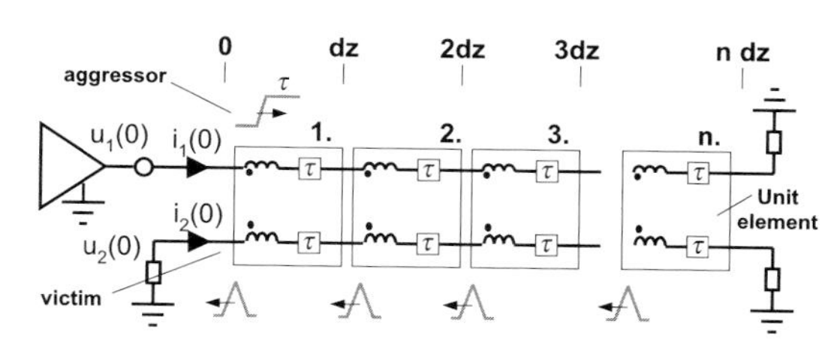

For dz ← 0, we will have qualitatively the following backward crosstalk:

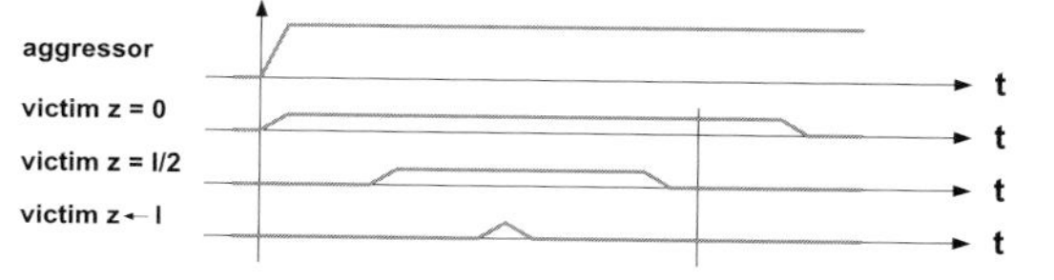

Figure 13.66
Inductive backward crosstalk

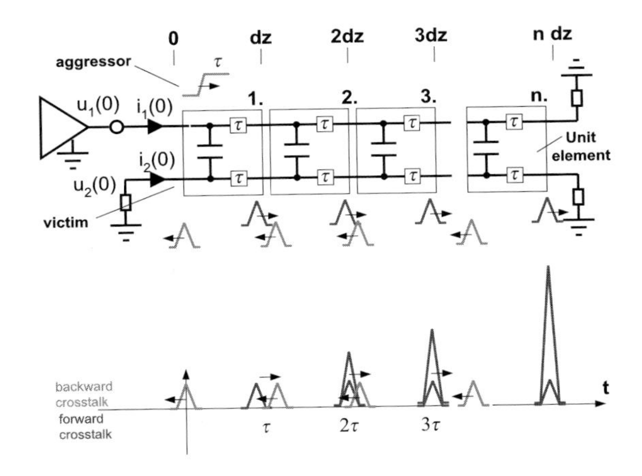

Figure 13.67
Capacitive crosstalk

Superposition of the inductive and capacitive crosstalk:

If the capacitive and inductive crosstalk are well balanced, as is the case for track conductors between the power supply or ground layers (i.e. a stripline), then the forward crosstalk is very small (almost 0!). For external signal layers, i.e. a signal track conductor above the ground layer (i.e. a microstrip), the capacitive coupling is smaller than the inductive, and the forward crosstalk is inverted compared to the aggressor. The reason for the lower capacitive coupling is that the majority of the electrical field lines pass through the air instead of through the dielectric of the PCB.

Backward crosstalk is always uninverted by comparison with the aggressor, and is independent of whether microstrip or stripline geometry is used. For the sake of completeness, the difference between a microstrip and a stripline is shown in Figure 13.68.

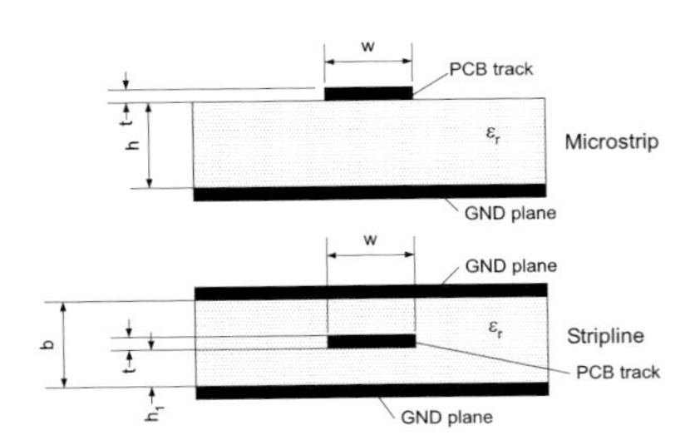

Figure 13.68
A microstrip compared to a stripline

From the preceding exposition, the following rules can be formulated for reducing crosstalk:

- Reduce the speed of the signal edges (i.e. increase the rise/fall time, reduce $\delta/\delta t$)
- Appropriate layout (track conductor geometry)
  - Track conductors as short as possible
  - Increase the gap between the track conductors
  - Do not run sensitive conductors parallel to conductors with fast rise/fall times
  - In the case of differential conductors, the gap must be as small as possible, and the conducting tracks must run parallel to each other. The reason for this is that with differential lines the aim is to subject each conducting track to the same interference, and thus to eliminate the interference signal.
- Add ground layers so as to avoid ground-slots. Only use ground-slots if

it is certain that no high-speed track conductors cross the ground-slot!

- Ground-slots under a high-speed track conductor prevent the signal current from returning by the shortest path, i.e. l_m increases overall!
- The external signal layers are worse than internal layers.
  - Striplines have advantages over microstrips
  - With striplines, c_m and l_m are balanced, i.e. there is very little forward crosstalk.
  - With microstrips, the electric field runs mainly through the air instead of through the dielectric, so that the coupling is less capacitive than inductive.

Crosstalk increases if the waves are reflected at the ends of tracks. The next section explains how reflection can be controlled.

Ringing

Ringing is an unwanted oscillation which frequently occurs with digital high-speed signals. Figure 13.69 shows an example of this.

Ringing arises due to different source and sink impedances. Because of these, part of the energy transferred (or possibly all of it) is reflected. The ringing amplitude depends on the reflection coefficient, but on the other hand the ringing frequency depends on the length of the track conductor (Figure 13.70).

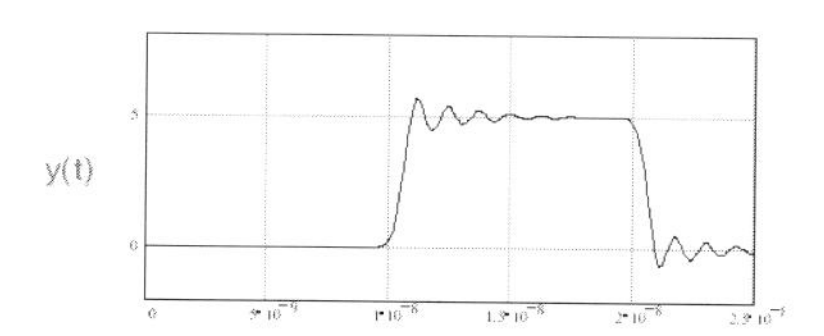

Figure 13.69
Example of ringing with a digital signal

Ringing frequency

$$f_{\text{ring}} = \frac{c}{2 \cdot \sqrt{\varepsilon_r} \cdot l_{\text{track}}}$$

reflection at source

$$\rho_{\text{S}} = \frac{Z_{\text{S}} - Z_0}{Z_{\text{S}} + Z_0}$$

reflection at load

$$\rho_{\text{L}} = \frac{Z_{\text{L}} - Z_0}{Z_{\text{L}} + Z_0}$$

The bandwidth required for digital high-speed designs is large, as a result of which it is impossible to adjust the reflection coefficient down to zero. However, there are other good options for reducing ringing so far that the system requirements are satisfied (Figure 13.71).

The following rules are recommended for calculating appropriate resistors and capacitors:

- $R_{\text{S}} = |Z_0| - |Z_{\text{S}}|$ usual $(|Z_0| > |Z_{\text{S}}|)$
- $R_{\text{L}} = |Z_0|$
- $|pC_L| >> |Z_0^{-1}|$ @ $p = j \cdot 2\pi \cdot f_{\text{fr}}$
- $C_{\text{L}} \sim 10 \cdot \dfrac{1}{2\pi \cdot f_{\text{fr}} \cdot |Z_0|} = 5 \cdot \dfrac{t_{\text{fr}}}{|Z_0|}$

with $t_{\text{fr}} = \min(t_{\text{f}}, t_{\text{r}})$, i.e. critical edge

Using this information about termination, the following prescriptions can be made for reducing ringing:

- Look for the critical (i.e. the high-speed) signals in the design.
- Keep the track conductors for high-speed signals as short as possible (i.e. make the track conductors shorter than $\lambda/20$ for the fastest signal edge)
- Use impedance matching on the track conductors.
- Use a continuous ground layer (no slots!)

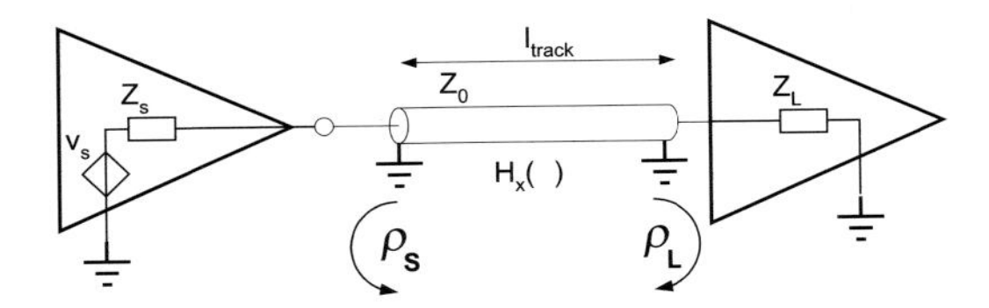

Figure 13.70
Reflection model

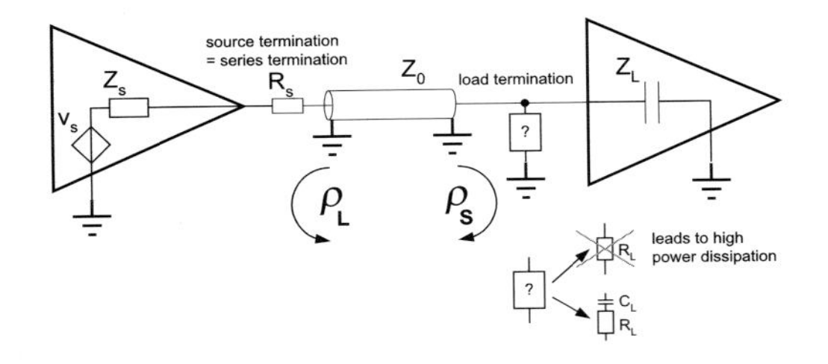

Figure 13.71
Recommended termination resistor for digital signals

- Keep the series termination resistor as close as possible to the source (i.e. to the driver).
- Keep load termination resistor as close as possible to the sink (load).
- Implement load termination as an alternating current impedance, i.e. no static current should flow (because this would increase the power loss!)

Crosstalk and ringing have a direct effect on the signals themselves. But attention should also be paid in the design to other aspects, such as provision of proper ground areas and the correct power supply for the system.

Ground bounce

Ground bounce is an unwanted voltage drop on ground connection conductors. This applies both for the PCB level and also for the chip level (Figure 13.72). For digital high-speed designs, *di/dt* is generally very high because of the very short *dt values* (switchover time from high to low, and vice versa). The inductance of the ground connection conductor is the most critical factor. Because at high speed, the inductance dominates the resistance component of the impedance. The mean current intensity is generally not very high.

To reduce ground bounce, the following rules must be observed:

- Reduce the signal edge speeds (i.e. increase the rise/fall times).
- Use ground connection conductors with low inductance. Preference should be given to ground layer(s) instead of a ground fingers and ground-slots, because current flows back along the path of least impedance.
- Use numerous local decoupling capacitors in the immediate neighborhood of the power supply and ground pins.
- Use several capacitors in parallel, with different nominal values (e.g. 10 pF, 100 nF and 1 nF) to handle the different switching frequencies in the design properly. Note: in real applications, a capacitor with a larger nominal value has a smaller capacitance at higher frequencies or may even be more inductive than capacitive.
- Chip designers are recommended to implement a sense line for internal reference; because this signal conducts little ground current, no additional ground bounce occurs. Around the chip die, numerous distributed ground pins should be used.

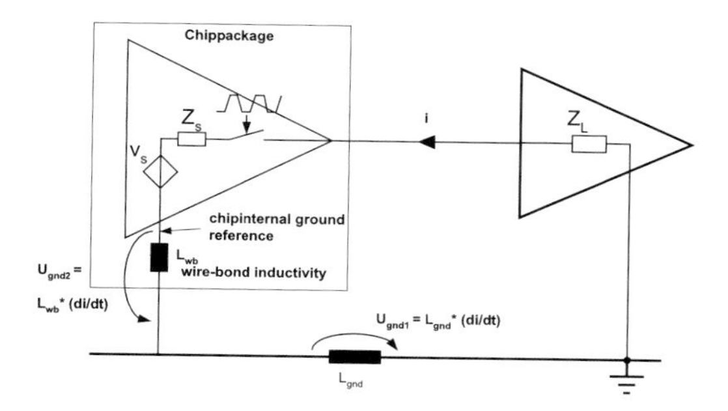

Figure 13.72
Ground bounce model

Power supply noise

Power supply noise can have several causes: it may occur as a result of switching in several current paths in the design, or it can also be injected from a DC/DC converter operating in switch mode. Such DC/DC converters always contain paths with high currents and high *di/dt*. The main problem lies in the fact that power supply noise degrades the analog performance of analog circuits (for example in A/D and D/A converters, operational amplifiers) because of a limited power supply rejection ratio (PSRR). I.e. noise from the power supply causes noise at the (analog) output of the integrated circuit, which is lowered by the PSSR factor.

This type of noise can be reduced as follows:

- Use of power supply conduction tracks with low impedance (e.g. by using wider tracks or even better using a ground plane), to minimize the voltage drop (power supply layers)
- Use of beads or resistors in the power supply path to reduce the noise. Note their working range (I_{max}, f_{20dB}, R_{DC}, ...)! An energy store in the form of a capacitor should of course be positioned close to the current sink.
- Use of decoupling capacitors in the immediate neighborhood of the power supply and ground pins.
- Attention must be paid to the parasitic effects of capacitors, for example the equivalent series resistance (ESR), the equivalent series inductance (ESL). These effects can have a powerful influence on the proper operation of the decoupling capacitor.
- Careful choice of the power supply, i.e. choice of the working frequency for the DC/DC converter, so that:
  - the required noise performance and noise margin for the integrated circuit is satisfied (note: the PSRR decreases with increasing frequency), or
  - the DC/DC converter uses a switching frequency which is outside the frequency band used for the communication device, or
  - a linearly regulated voltage source is used, if this is possible.

Many of the EMC or SI recommendations for PCB design apply also for design at chip level. This applies not only to the pure "silicon design", it should permit good PCB practice (e.g. proper pinning, timing). It is not sufficient to develop a chip which is appropriate in terms of its function and EMC, rather should a system solution be developed which is appropriate in its function and EMC. In order to achieve this objective, the chip designer, system developer and customer must work closely together. The following conclusions can be drawn about this cooperative work:

- Keep high-speed track conductors as short as possible.

- Use power supply and ground layers; avoid slots in these layers.
- Use adjacent power supply and ground layers, because they then act as a distributed capacitor.
- Decoupling capacitors close to the power supply/ground pins.
- Check the data sheets for the capacitors; their parasitic effects (ESR, ESL, resonant frequency) may not be neglected in high-speed designs.
- Impedance matching to minimize reflections.
- Proper termination resistances on high-speed track conductors.
- Ensure that each signal has the shortest possible current return path. The intensity of the RF return current is greatest underneath the signal conducting track (path of least impedance means lowest inductance at high frequencies).
- Keep current loops which have a high *di/dt* as short as possible. This reduces crosstalk as well as EMC, EMI and EMR.
- Keep digital high-speed track conductors away from sensitive analog conductors.
- Parallelism of conducting tracks:
  - avoid parallel conducting tracks where crosstalk must be minimized
  - use parallel conducting tracks which are close together for differential signals.
- Increase the gap between any conducting tracks where a mutual influence is unwanted.
- Route the digital signals which have the highest speeds on the inner layers.
- Ensure that the communications between all the engineers (e.g. chip designers, system developers and layout engineers) involved in the system development are excellent.

13.5 ESD protection measures during handling

Electrostatic discharges (ESD) can cause severe damage to semiconductor devices. For this reason, ESD protection concepts are implemented in most semiconductor components nowadays (Figure 13.73). The ESD protection structures integrated into the chip define a current path over which the ESD current can flow without causing damage. They have the additional function of reducing the current and voltage which affect the sensitive circuits within the actual chip.

However, protective components of this type can definitely have an effect on the functionality of the device. It is mainly in respect of such requirements as minimum leakage current, small chip area or high frequencies, at which parasitic elements begin to have a disruptive effect, that the ESD protection structures can have negative consequences. For this reason, when it comes to ESD protection on the chip it is generally necessary to compromise between ESD robustness and the performance of the device. Consequently, the ESD protection integrated into the device is generally not sufficient to permit handling without endangering it.

In addition, the ESD stresses during the manufacture and processing of the device must be reduced. This is effected by the introduction of special measures, and the use of ESD protective materials. These must meet the following requirements:

- Prevention of high charging of the objects which come into contact with the device
- Reduction of the uncontrolled discharge current during contact with the pins of a charged device, by means of a high electrical resistance

For this reason, the manufacture and processing of ESD-sensitive devices takes place in so-called EPAs (ESD Protected Area, Figure 13.74), in which the objects comprising the equipment are made

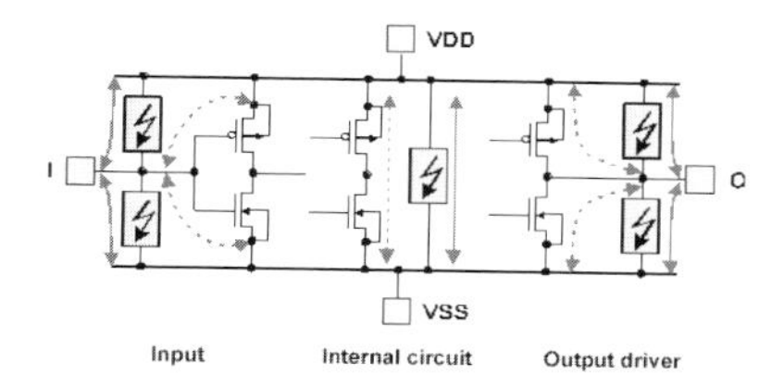

Figure 13.73
Typical ESD proctection concept

mainly of ESD protective materials. In these protected areas, there is only a minimal residual risk of damage to the devices from ESD.

If the ESD-sensitive devices or components are taken out of the EPA, to minimize the risk of damage they must be transported in packing with adequate ESD protection.

Help is provided in the selection of the correct materials and installed equipment for an EPA, together with suitable packaging materials, by the standards EN 61340-5-1 (or IEC 61340-5-1, as appropriate) and ANSI ESD S20.20. These standards not only define the requirements for ESD protection measures, but also provide aids for the design of an ESD protection concept.

13.5.1 Measures for protecting against charged objects (man/machine)

Objects which become charged during the manufacturing process and then discharge into the device or board represent a major danger to the latter. The first of these which should certainly be mentioned is people. To prevent uncontrolled discharges from charged personnel, they must wear wrist straps, or alternatively have conductive shoes on a conducting floor. Investigations have shown that, if correctly grounded, a person does not charge to more than 100 V, and hence represents no danger to the devices.

An ESD risk can also arise in the manufacturing machinery. Conductive parts of a machine, which are not grounded, can become charged during manufacture and

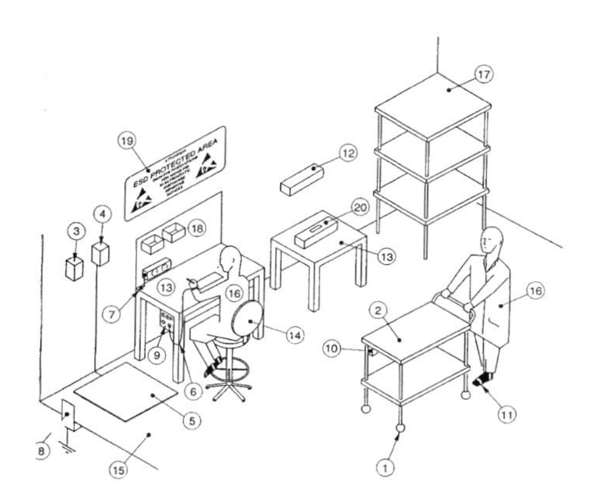

1 = Rollers which can be grounded
3 = Wristband tester
5 = Footplate for footwear tester
7 = Grounding cable
9 = Ground connection point
11 = Shoe grounding band/footwear
13 = Worksurface
15 = Floor
17 = Rack with grounded surfaces
19 = EPA sign

2 = Surface which can be grounded
4 = Footwear tester
6 = Grounding wristband
8 = Ground
10 = Ground connection point for trolley
12 = Ionizer
14 = Seat which can be earthed
16 = Clothing
18 = Containers which can be grounded
20 = Device

Figure 13.74
Example of an ideal EPA conforming to IEC 61340-5-1

discharge into the component or module. A regular check is therefore essential on the grounding of all (moveable) parts of the machines.

13.5.2 Protective measures against charged devices

Apart from charged persons or machine parts, charged devices or boards also represent a danger, if they discharge into very conductive objects (CDM risk). A risk of this sort can be countered either by reducing the charging or by avoiding any (hard) discharge. For this reason, all the installed equipment such as benches, chairs, racks etc, should, as far as possible, be made of ESD protective materials, and be grounded. Charged objects in manufacturing can transfer their charge to nearby devices or boards by induction. If due to process related reasons ESD protective materials can not be used, either the charge must be neutralized by the use of ionizers, or any hard discharges must be avoided. A more detailed analysis of the individual processing steps is thus absolutely necessary for this purpose.

14 Packages

An increasing challenge for system development is the growing importance of packaging and interconnect technology.

14.1 From physics to innovation – the growing importance of package development

The major success of GSM ("Global System for Mobile Communication") is recognized throughout the world as a success for joint European efforts in the systems area. A system such as GSM has been made possible, above all, by outstanding developments in microelectronics.

Apart from developments in communications technology, fundamental physical research, material research, semiconductor technology, the design of integrated circuits (IC's) and package development and design have also been required. Figure 14.1 shows the chain of innovation, from the physics to a system and its applications. Ever greater importance is being assumed by package design and package technology. The individual links in this chain of innovation interact strongly with each other. In today's world, so-called "T-shaped" individuals are in demand as researchers and developers, i.e. researchers and developers who have a broad knowledge of physics, materials research, semiconductor technology, packaging technology, up to the system itself (from transistor physics through to communication systems). At the same time, however, the demand is also for a specialist knowledge in the technical area concerned.

The package, including the two interfaces – chip/package (1st level interconnect) and package/board (2nd level interconnect) – is increasingly limiting the performance of a semiconductor chip (e.g. for

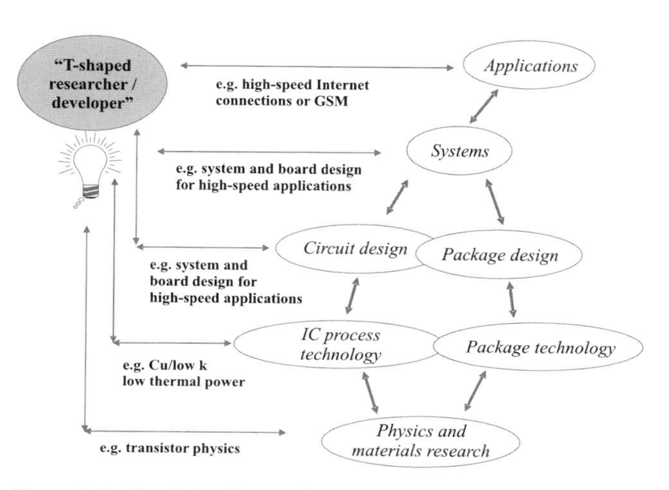

Figure 14.1 The chain of innovation from physics to applications

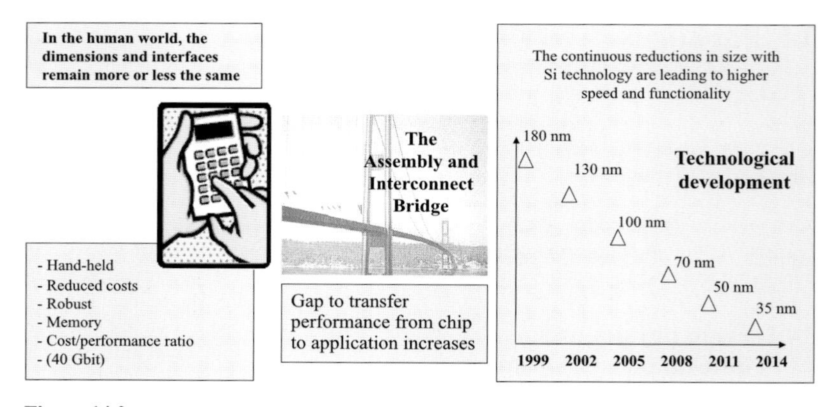

Figure 14.2
Package development and interconnect technology is the bridge between the semiconductor chip and the application

high-frequency applications or due to inadequate heat dissipation), and hence also of the system. On the one hand the dimensions in chip technology are following Moore's Law (i.e. the number of transistors per chip will double approximately every 18 months) and becoming ever smaller, and on the other hand the dimensions of the human world remain unchanged (see Figure 14.2). The challenges for package development and interconnect technology, which provide the bridge between the chip and the human interface, are thus getting ever greater. The market opportunities are thus greater for products which have not only innovative chip technologies but also forward-looking packaging and processing technology.

14.2 Packages for semiconductor chips – an overview

In packaging technology, a distinction is made between the 1st level interconnect (the chip/package interface) and the 2nd level interconnect (the package/board interface). Even today, wire bonding is still the most used chip/package interconnect technology (Figure 14.3a). With increasing frequencies and numbers of connections on the chip, the flip-chip technology is becoming increasingly important (Figure 14.3b).

For the package/board (2nd level interconnect) there are four main types of interconnection technology, which are summarized in Figure 14.4 with typical values of their pitch (the standardized spacing of the contacts, between centers). A distinction is made here between surface mounted devices (SMD; or surface mount technology: SMT) and devices with pin connections (through hole technology: THT). Nowadays, increasing use is being made of SMD devices.

The growing product variety, the ever shorter product life cycles and the rapid succession of technology changes, make it imperative to concentrate on the strategically important package families and interconnection technologies.

Figure 14.3
A semiconductor chip in a modern BGA (Ball Grid Array) package, bonded a) using wire bonding and b) by flip-chip bonding

THD through hole DIP, TO 220	SMD gull wing SO & QFP	SMD solder balls BGA	SMD leadless VQFN, TSLP
Pitch: 2.54mm	DSO-Pitch: 1.27mm SSOP-Pitch: 0.65...0.5mm QFP-Pitch. 0.8...0.65...0.5...0.4mm	Pitch: 1.5...1.27...1.0mm CSP BGA: 1.0...0.75...0.5mm	Pitch: 0.8, 0.65, 0.5 mm

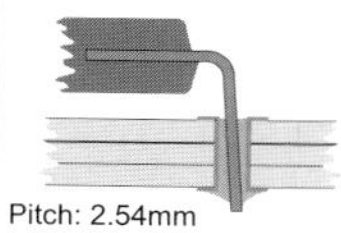

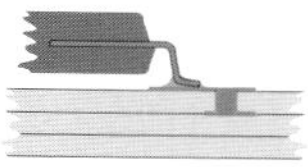

Figure 14.4 Overview of the package/board interconnection technologies

Infineon Technologies AG is conscious of the importance of the subject of packaging and the interface to the customer. For this reason, the focus of development effort continues to be on a consistent expansion of the standard package families. These are shown in Figures 14.5 and 14.6; the proportion of through hole technology (THT, plug in) packages is continuing to decline. However, Infineon also focuses on new innovative solutions in the development of packages and interconnection technology, such as for example leadless packages (TSLP), wafer-level packaging (WLP) or system-in-package (SiP) solutions, to satisfy the market trends.

Now even more than before, the semiconductor manufacturers and their customers, as the users, must collaborate more intensively on the interface – the package – to find the right technological development path, leading to the most cost-effective and hence most competitive system solution.

In the case of consumer and industrial electronics, the customer requirements are increasingly defined by the attributes: small, light, low cost but nevertheless of high perfomance. This means that one of the critical factors in the development of new electronic products is the IC package (Figure 14.7). For future applications,

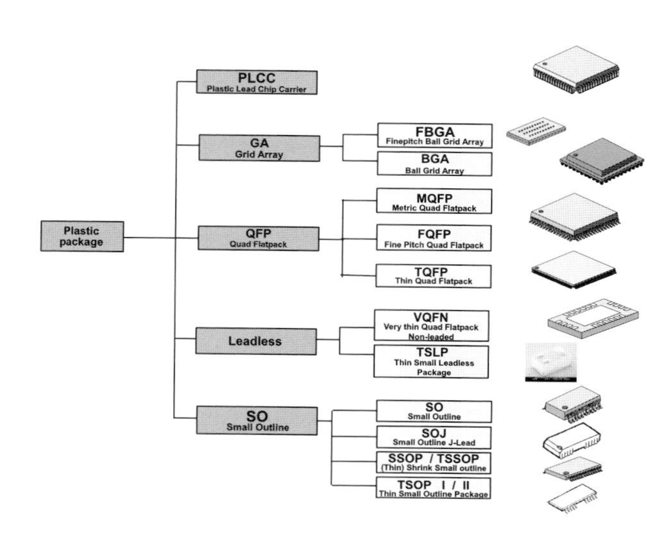

Figure 14.5 Family tree for SMD packages

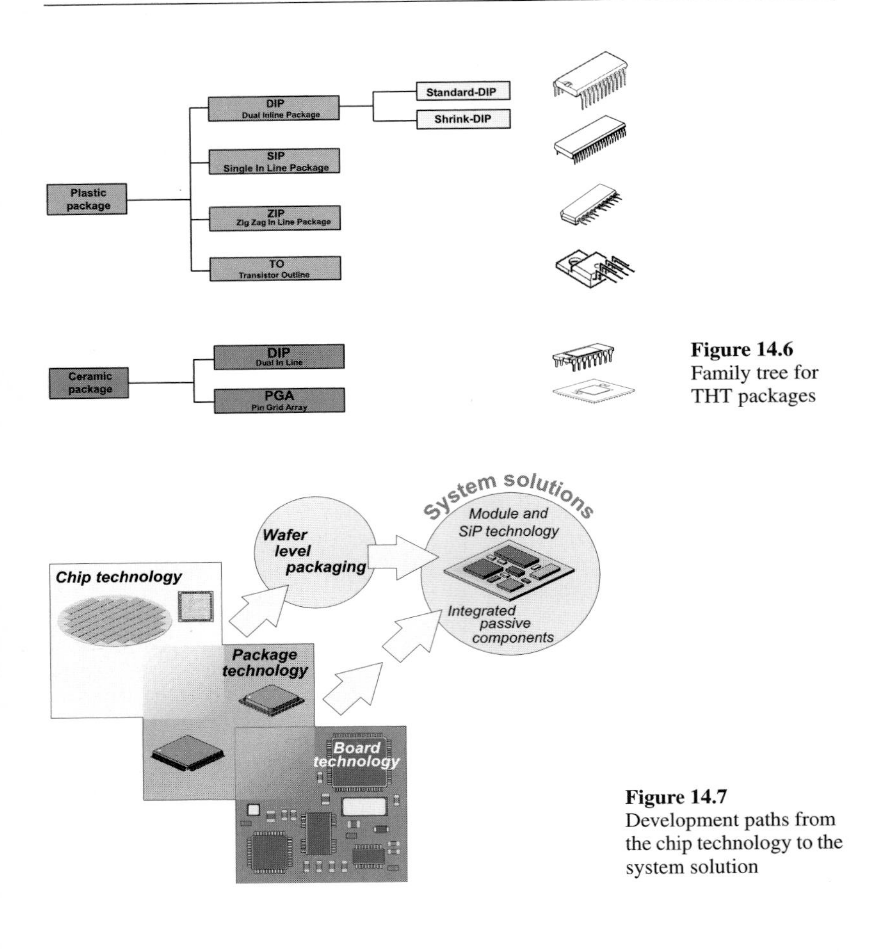

Figure 14.6
Family tree for THT packages

Figure 14.7
Development paths from the chip technology to the system solution

there is increasing discussion of wafer scale assembly (wafer level packaging: WLP).

14.3 The driving forces behind package development

Figure 14.8 shows what today's customers demand of microelectronic products. The customer increasingly expects high frequencies and high data transfer rates combined with good thermal performance. On the one hand this means adequate heat dissipation at rising frequencies and with large powerful chips, and on the other hand the ability to function reliably under extreme environmental conditions, for example close to the engine in the case of automobile electronics. Depending on the field of application, what is demanded is either the lowest power consumption, for example in the case of battery powered mobile products, or the highest performance, such as with various electrical switches. For almost all of these applications, the customer expects the smallest possible components, which must meet the necessary reliability criteria. In future the trend towards forward integration, i.e. ever higher functionality

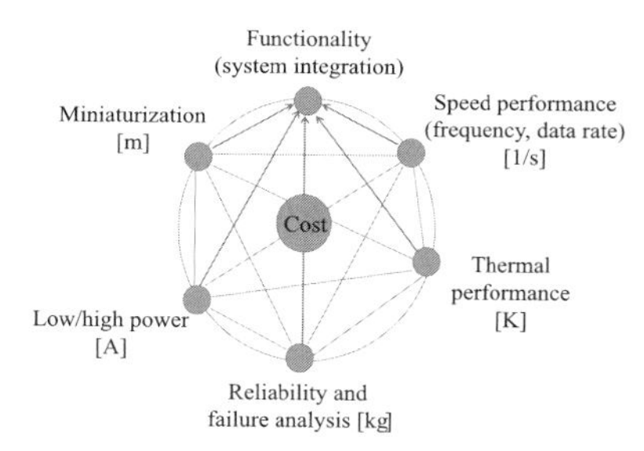

Figure 14.8
Driving forces for package development

and system integration, will continue. Ever more of the functions, which previously were on a semiconductor board, will be integrated into the package and the chip (see also Figure 14.14).

Due to the steady trend towards further miniaturization, and the steadily rising frequencies, ever more physical effects are coming into play in package development. The customer's expectations reflect the basic units of classical physics: the high frequencies involve 1/s, and so involve the basic unit of the second (s), the thermal performance involves the basic unit of the Kelvin (K), the miniaturization the basic unit of the meter (m), the power involves the Ampere (A) via the unit of the Joule, and finally the reliability – which is mainly about the consequences of forces and stresses – means that the unit of kilograms (kg) must also be added, via the agency of the Newton ($N = kg\ m/s^2$). Of course, all the customer's wishes must be met at the lowest cost and with the shortest time to market.

The most important technological drivers for new package developments with major market penetration are memory products, microprocessors with their peripherals, ICs for communications technology, and ASICs. For example, the ultra-flat packaging technology was developed primarily for memory products, in particular DRAMs, and then transferred across to other application areas.

14.4 Package development around the world

14.4.1 Standardization

Standardization plays an important role in package development. The collaboration between the main national associations of interested parties with responsibility for the standardization of packages, JEITA (Japan), JEDEC (USA) and IEC (worldwide), has become closer as a result of the mutual agreements within the general standardization trends. For processability on the board, i.e. for usage by the customer, standardized dimensions are essential. This applies to the metric dimensions for fine pitch spacings, short connecting pins, reduced clearance heights and package dimensions.

14.4.2 Worldwide trends: memory packages

In line with the general trend, package development for DRAMs has proceeded in the direction of reduced volumes – smaller installed height, body thickness and weight. Currently, for the latest memory products, ultra-flat housings are being launched on the market, with an installed height of a mere 1 mm, development work has begun on packages with a thickness of 0.7 mm and 0.8 mm. For the present-day generations of DRAMs, such as 256 Mbits, the P-TSOP package (where P indicates a plastic package) remains dominant. The latest development trends indicate that in future, particularly because of the rising frequencies and the associated parasitic effects, there will be a growing proportion of CSPs (chip scale packages) for 256 Mbit DRAMs, and the following generations of 512 Mbit and 1 Gbit. In the area of DRAM memory, increasing use will be made of 3D chip stacks, i.e. several chips stacked in one package (Figure 14.9). Throughout the world, WLP is also being considered by a wide variety of memory manufacturers.

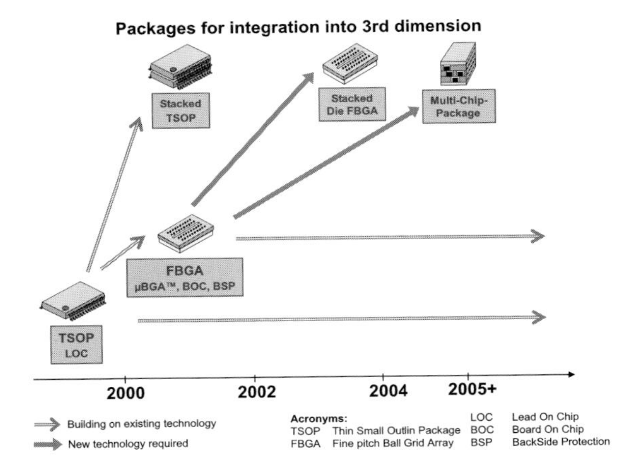

Figure 14.9
The trend in DRAM memory packages

14.4.3 Worldwide trends: IC packages

The trend towards layouts with large pin counts will continue in future, forced on by the developments in ASICs, gate arrays, microprocessors and microcontrollers. In this context, ever greater market shares are being forecast for P-BGA packages, on which the maximum pin count will increase even further in the next few years. BGA-based packages with pin counts of over 1000 are nowadays no longer anything special.

The continuing changes to individual package families, towards larger numbers of connections, smaller gaps between the connections (lead pitch) and thinner packages will be determined by the semiconductor circuits which are defining the market (Figure 14.10).

Both in the case of plug-in forms (THT) and also in the case of SMD packages, plastic packages take the major share. Of these, P-SO, P-LCC, P-QFP, SOD and SOT have the largest marked share. In the case of ceramic constructions, C-LCC, C-Flat-Pack and C-PGA are dominant. According to the forecasts from market research companies, IC package structures will develop ever more towards the SMD technology. In 2002, more than 90% of all ICs around the world were already integrated in SMD packages, and in 2007 the share of THT packages will be down to just 6%. As a result of the technological change-over in the field of flat module production, from plug-in to SMD technology, there will be a rising demand for surface-mountable packages. For SMD packages, a growth rate of 11% p.a. is forecast.

There are also changes for the SMD package families: P-SO and P-QFP package families are diversifying into subfamilies, with finer lead pitches, thinner packages, more diverse package geometries and larger pin areas. As an alternative there is also the BGA family. A low-profile fine-pitch BGA is, for example, characterized by an average spacing of the solder balls (the ball pitch) of 0.8/0.65/0.5 mm and a package height of 1.2 - 1.7 mm; a TFBGA (Thin Profile FBGA) is characterized by a package height of 1.0 - 1.2 mm. The innovative BGA packages offer a better ability to incorporate the high performance of modern ICs into flat modules. The FBGA package family is particularly interesting for semiconductor chips with numerous connections.

Beside the two classical package families, SMD and THT, there are also for so-called 'bare dice', i.e. unpackaged chips, so-called high-density-interconnect (HDI)

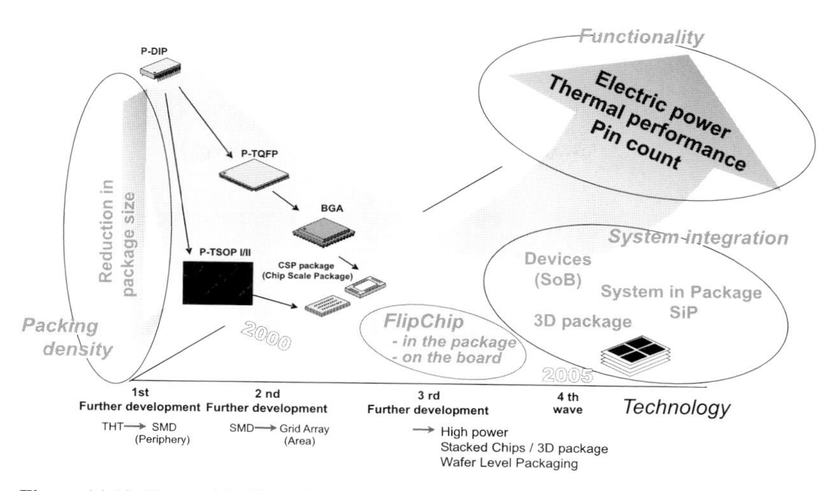

Figure 14.10 The worldwide package development

connection technologies, such as tape automated bonding (TAB), chip on board (CoB) and flip chip (FC), which by comparison with the standard die/wire bonding technology permit even higher packing densities on a flat module. Figure 14.11 shows the area occupied by a semiconductor chip in various packages when mounted on a board. Apart from the standard packaging technology, i.e. die/wire bonding on a metal leadframe (e.g. the QFP package in Figure 14.11) followed by encapsulation in molding compound (transfer molding process), other technologies permit further space savings on the board. Thus, using the CoB and flip-chip technology it is possible to double or treble the packing density on the circuit board. The least area requirement is achieved by mounting the chip directly on the board, using the flip-chip technology (FCoB = flip chip on board).

In the case of the flip-chip technology, a distinction is made between flip chip in package (FCiP, see Figure 14.3) und flip chip on board (FCoB, see Figure 14.11). Whereas FCoB is already being used in special products, there is an increasing trend today towards using flip-chip technology to mount semiconductor chips in P-BGA or P-FBGA packages (FCiP). Market research forecasts have been saying for years that the areas for the use of FCoB should be increasing. However, these forecasts have been too optimistic. The introduction of the CSP packages, that is packages with a footprint which is by definition only 1.2 times larger than the area of the chip, has made FCiP significantly more attractive. The FCiP variant gives the user the option of a fully checked device which can be processed

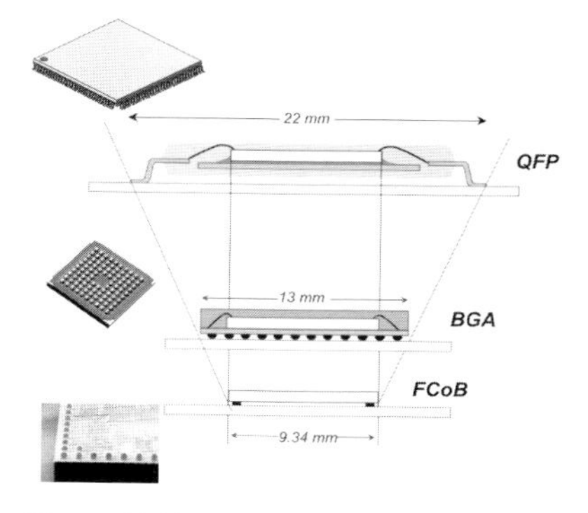

Figure 14.11
Area on a board required by a semiconductor chip in various packages

using standard SMT processes. The electrical characteristics of the IC – with significantly shorter connecting leads because the bondwires have been replaced – approach those of a flip chip. An additional advantage for the user is that it is repairable. The FCoB variant customarily requires an "underfill", which makes it impossible to repair the FCoB component by unsoldering and replacing it. Although direct connection methods minimize the discrepancies between the chip and circuit board structures on a flat module, the mass application of bare dice is inhibited by the problems of KGD (Known Good Die), the difficulties of controlling the underfill process, the curing and the clean room conditions. Because of these disadvantages, today's entry barriers for mass application will remain.

Both the semiconductor manufacturers and also the users have an interest in the continuation of existing processing techniques. The user of standard packaging technology profits from a decade of experience, which minimizes the risk of continuing developments in the direction of ultra-flat packages. Even products such as portable TVs, CD players, camcorders and memory cards, but also palmtops, personal digital assistants and notebooks, in which extremely small dimensions are important, today still contain a high proportion of P-TSOP, Thin P-QFP packages and CSP.

Discrete semiconductors: SMD package development

More than 20 years ago, the first SMD design – the SOT-23 – was incorporated into the production line at Siemens. This design became the worldwide basis for the SMD technology, and the model for all subsequent designs, which today are manufactured on the most modern equipment in their billions.

In the 80s, the production processes, testing technology and insertion technologies were modified appropriately for this design format, so that today cost-effective production and further development are possible, particularly in conjunction with the SMD packaging technology (Figure 14.12).

The users' demands for higher electrical performance and smaller dimensions have been satisfied with the standard SOT-223 construction or the mini-packages SCD-80, SOT-3x3. It has been possible to further optimize these packages for ever higher working frequencies, up to over 50 GHz (e.g. using SOT-343). Reducing parasitic effects, variations in the connections, reducing the size of the

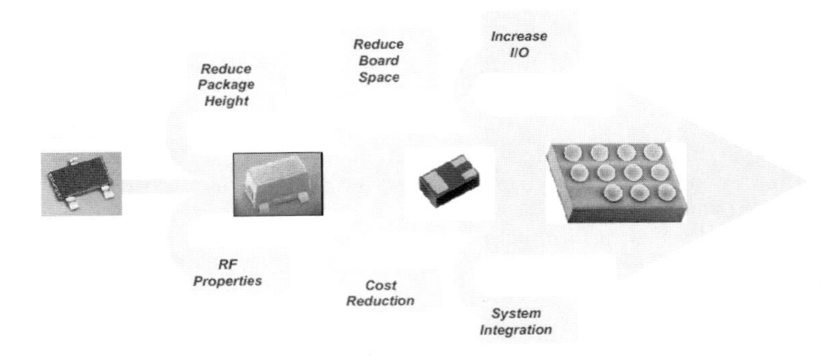

Figure 14.12 Trends for discrete semiconductors

package and improving the electrical efficiency are further development tasks.

In the field of discrete components, the increasing trend towards the optimization of not only the costs but also the performance have led over recent years to ever more attention being paid to ultraflat (flatlead) packages and pin-free (leadless) packages. Here, Infineon Technologies has developed the TSLP (Thin Small Leadless Package), which will be further developed in future for very high frequencies and larger numbers of contacts. In the field of discrete semiconductors, wafer-level packaging is also being discussed. Wafer level packaging should also be considered for chips with integral passive components for future system solutions.

14.4.4 Worldwide trends: passive modules

Originally it were passive elements – starting with the chip resistors and ceramic capacitors – which were the first components in SMT. In Europe, the 0805 (2.0 mm × 1.2 mm) and 0603 (1.6 mm × 0.8 mm) packages are the state of the art. A start has been made on the processing of the 0402 (1.0 mm × 0.5 mm) package which, just like the fine-pitch package, represents a challenge for the processing methods. In Japan, the 0402 package and now also the 0201 package (0.5 mm × 0.25 mm) are in widespread use.

Applications in wireless communications are driving today's trend of forward integration for passive components. In this case – as also with chip technology (see also Figure 14.14) – whenever possible the components are taken from the board and integrated into the package or indeed onto the silicon, i.e. into the BEOL (backend of line). Infineon is also following this trend towards forward integration, as a means of achieving significantly higher overall system integration combined with miniaturization.

14.5 Usability for the customer: fine pitch and alternatives

A critical determinant of the time at which the packages based on high pincount leadframes (e.g. the QFP package), or constructions such as the BGA package family, are launched on the market is the users having a secure grasp of how to process the new construction format. This is dictated by the ever finer lead pitch combined with an increasing pin count. Of particular importance here is the matter of the reliability of the package/board connection.

The processing of the fine-pitch package is a challenge for circuit board manufacturing: the component insertion machines must place components more precisely, and are equipped with Visio systems. For the printing of soldering paste, there are higher demands on the soldering paste and the screen printing machinery. This makes the devices more expensive, and the costs of production rise.

The main problem in processing finepitch packages with a pitch of 0.5 mm consists in ensuring that the filigree contact leads do not get bent during handling processes, both at the semiconductor manufacturer and at the subsequent processor. Ever more demanding requirements are being placed on the coplanarity of the connecting leads. Even a bent contact will cause an electrical failure. Warping of the circuit board at the installation location, and its surface quality, also play a part.

The BGA packages are attracting ever more interest. The ability to affix the connecting leads flush across the entire underside of the package permits more connections to be accommodated on a smaller area. It is an advantage that the pitch of the solder balls, 1.5 mm, 1.27 mm and 1 mm, is not critical to the processing. Apart from this, it is possible to do the processing with the available machinery. Developments for this new package format are proceeding rapidly. For example, the package thicknesses are constantly being reduced, so that today thicknesses of 1.2 mm no longer represent any particular challenge. One variant is the CSP, already mentioned, before, which is only insignificantly larger than the chip and ultraflat. LFBGA packages can be made in CSP form. Some of the data: ball diameter from 0.25 mm - 0.45 mm, package thickness 1.2 mm - 1.7 mm, ball pitch 0.3 mm - 0.8 mm. With an FBGA it is possible, using a ball pitch of 0.3 mm, to achieve a packing density similar to that of a flip chip.

By comparison with plug-in assembly, the SMD technique for surface-mountable components leads to higher thermo-mechanical loadings on the component during reflow soldering. As a result of their take-up of moisture, large-area chips in plastic packages are particularly sensitive to the 'popcorn' effect which occurs during reflow soldering. This can lead to cracks in the package and the chip. For this reason, large P-LCC, P-QFP and P-DSO packages are generally supplied in dry-packs. The subject of the packaging of the finished components is an interface between the semiconductor manufacturer and the processor/customer which is becoming increasingly important.

14.6 IC Packaging road map – where is the journey taking us?

Figure 14.13 shows an overview of an analysis of the flip-chip technology, the FBGA package technology and the SMT technology. In the field of device packaging, the cost factor plays a special part.

The following are the directions of future development, the consequences of which are an increasing diversification of the processing technologies and package types, and which will make the packaging more application-specific, and hence more customer-specific:

- High pin-count packaging as a consequence of the increasing functional complexity of the chips, in particular for ASICs.

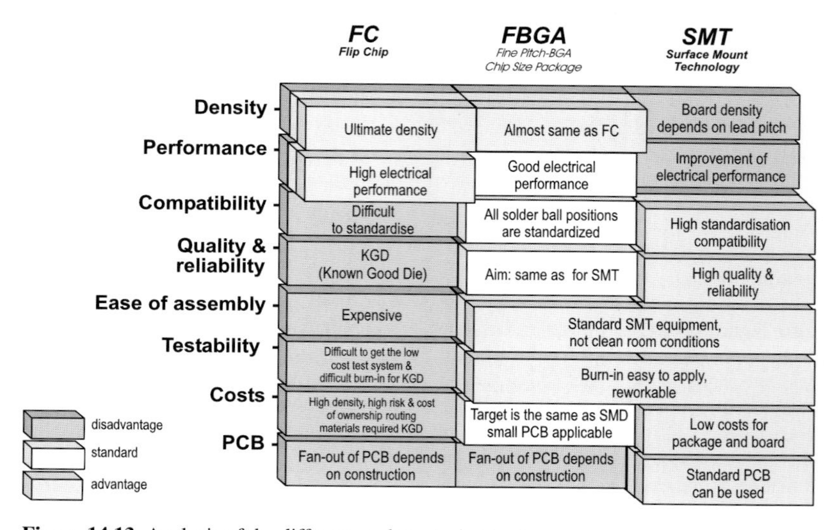

Figure 14.13 Analysis of the different package technologies

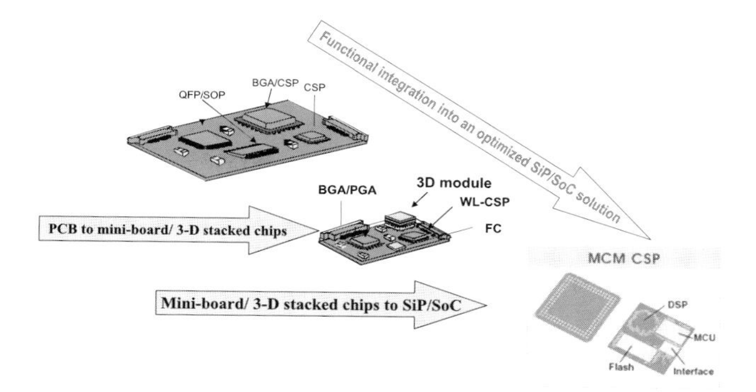

Figure 14.14 The trend towards forward integration

- System packaging, by which an optimal integrated solution is applied to the entire complex, consisting of the chip, package and flat module (increasing trend towards forward integration).
- Further merging of the front-end technology (chip technology including BEOL) and back-end technology (packaging technology).
- Wafer level packaging, which in future should combine front-end and back-end technologies in an optimal way.

As already discussed in section 14.4.3, new challenges are being posed, as a result of the Known Good Die problem and the possibilities of repair for chips mounted directly on the board by means of flip-chip technology. In order to solve these for the high pin-count problem, increasing use is being made nowadays of packages in the BGA package family. The trend here is ever more towards LFBGA with a flip-chip-in-package (FCIP) solution (see Figure 14.3). The LFBGA technology now permits the positive characteristics of the flip-chip technology to be combined with SMT-compatible processing. As P-FBGA packages provide the characteristics which are today being demanded for many application areas, in terms of performance and miniaturization, over the next three to five years they will replace some of the P-QFP, SO and P-BGA construction formats. P-LFBGA packages are accompanied by increasing demands on circuit board technology. Until now, standard PCB types (2- or 4-layer) with a track conductor width or spacing of 150 µm have been adequate for P-SO, P-QFP and P-BGA packages. With P-FBGA packages with a ball pitch of 0.8 mm or 0.5 mm, the signal artwork can now only be realized using multilayer PCBs in fine-line or built-up technology

The increasing trend towards forward integration (Figure 14.14), and with it miniaturization, call increasingly for the integration of several IC chips and memory components into common packages. This includes also the integration of passive components (trend towards system-in-package solutions). The consequence is an increasing merging of packaging and chip technology. In order to satisfy future trends, many semiconductor manufacturers have already brought together flip-chip technology, i.e. UBM (under bump metallization) and the application of the flip-chip balls, and chip technology in one semiconductor plant.

In order to continue to meet the increasing challenges, it would seem today that wafer-level packaging technologies will assume ever greater importance. These involve a close merging of chip and package technologies. Around the world, research and development is being carried out in this direction.

14.7 Materials aspects

In February 2003, the EU agreed directives on the recycling of electrical and electronic equipment. In these, various materials are declared to be banned substances. The two directives (WEEE, RoHS) are the main driving force for industry to give attention to the replacement of lead in soldered connections. Allowing for a few exceptions, by mid-2006 appropriate changes must be made to all products.

14.7.1 Lead-free, halogen-free packages

The biggest effect of the directives will be on the soldered surfaces. For SMT packages, the SnPb (tin-lead) surface will be replaced by a surface of pure Sn (tin). For the BGA package technology, SnAgCu will be used as the replacement material. In addition, the change in the solder paste used by the customer will have effects on the circuit board soldering process. A large scale change will be made from SnPb paste to SnAgCu paste (a tin-silver-copper alloy). This raises the theoretical melting point of the paste from 179°C to about 217°C. As a result, the semiconductor device will be subject to substantially higher loadings during soldering.

The new soldering materials call for additions to numerous standards and norms. Experts from Infineon are keeping an eye on this process, and providing supporting data. Although there is a legal deadline for Europe, there is no global changeover scenario, so allowance must be made for a transition phase. During this period, both lead-free and lead-containing products will be being processed. On operational and logistic grounds, this phase should be as short as possible. The prerequisites for this are that the new soldering processes fulfill the requirements, and the provision of a compatible lead-free technology.

On top of the matter of "lead-free" products, bromine-containing flame-retardants will also be eliminated from plastics. Flame retardants satisfy the demands for a material to be difficult to set alight. The bromine-containing materials are included in the discussion from the environmental protection point of view. Replacement materials will be used in parallel with the introduction of lead-free materials.

The package of a semiconductor device will be designated as "green" if it contains no bromine-based flame-retardants and no lead. This status presumes that they can be processed if a lead-free soldering paste is used.

14.7.2 Constituents in devices and materials

In the most diverse electronic end products, use is made of a large number of material and consumables together with process chemicals, the choice and appropriate use of which affect the quality, safety and environmental compatibility of their applications throughout their service life.

Optimizing these characteristics requires that an exchange of information and a dialog are maintained along the value-creation chain – from raw material extraction, through the manufacture of preproducts, components, the application itself, down to their use and ultimately their recycling or disposal.

To achieve this, constituents must be provided in a form that the customer can use. The solution to this need is based on making up special product families from typical representatives. These product families are, in the nature of their constituents and materials, and in their percentage composition, all the same within a certain range of variation. The resulting instructions are designated as umbrella specifications (Figure 14.15).

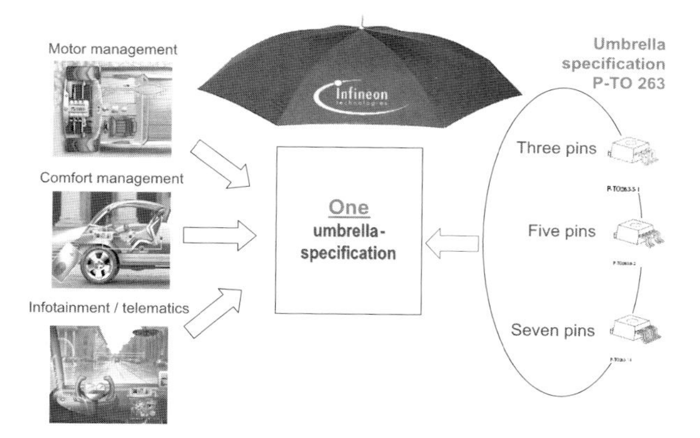

Figure 14.15 Different applications – one umbrella specification

In order to make this approach applicable to all components, the criteria by which families are formed for the individual product areas must be defined in accordance with customer needs.

In the context of the increased requirements for environmental friendliness, plastic materials which are normal components of products may only contain the following heavy metals in especially low concentrations:

- proportion by weight of cadmium: less than 5 ppm, i.e. 0.0005 percent
- proportion of total weight which is cadmium, mercury, lead and chromium (6-valent): maximum 100 ppm.

This will assist in the objective of either avoiding environmental pollution, possibly more than required by the applicable regulations, or reducing it to a minimum.

14.7.3 Soft errors due to radioactive impurities in the package material

The molding compounds for plastic packages contain filler materials which normally have minimal natural impurities of such radioactive elements as uranium and thorium. These emit alpha particles. The lead used for the flip-chip solder balls is also such an alpha particle emitter. The problems of alpha particles in packaging materials was already known long from DRAM technology, and for them appropriate precautionary measures were taken.

As a result of the trend towards ever smaller transistors with low voltage differences (MOS transistors are operated at 1.2 or 1.5 Volts), alpha particles from the radioactive materials can now also affect the state of an SRAM, i.e. the setting of a bit can be changed from the “1” state to the “0” state by an incident alpha particle. For this reason, appropriate precautionary measures must be taken for SRAM with technology structures 0.18 μm and smaller. For example, error correction can be built into the basic circuit, or molding compounds can be used with filler materials which contain no alpha-radiating elements. The problem of alpha radiation from the lead content of the flip-chip solder balls will be significantly reduced by the changeover to lead-free materials, such as SnAg or SnAgCu. Recently there have also been initial investigations into whether logic circuits could also react to alpha radiation.

15 Quality

The quality of our products is an essential element in determining our commercial success – for Infineon this is therefore an exceptionally important competition factor. The high quality demands that the customer makes on our products are characterized by the high requirements and expectations for the cost-effectiveness of the product processing and, of course, by their high reliability factor – especially when the components are used in safety or environment-relevant applications.

Quality is generally provided in accordance with DIN 55350, which states that, "Quality is the totality of features and characteristics of a product or a service that bear on its ability to satisfy given needs". Quality is therefore the measure by which products and services meet the demands placed upon them. From the user's point of view, quality is defined by usage efficiency under conditions that comply with regulations.

15.1 Elements that determine quality

Quality is composed of different elements. The most important elements that contribute to product quality are:

- Product characteristics and features, including function and specific characteristic values, specified in the data sheet and the product description.
- Manufacturability: Ability to produce a product on a robust and stable process.
- Number of the electrically and/or mechanically defective components in a shipment. Applicable here as defective is each part that does not comply with the values or characteristics specified in the data sheet.
- Delivery Quality e.g. delivery in time
- Processability in the customer's application (board assembly – especially soldering and cleaning processes).
- Reliability: Stability of the component characteristics relevant for the application and their degradation, which can impair the component function.

Quality management as the basis for optimum quality

Optimum quality of products and services, seen from our internal viewpoint and from that of our customers, is not attained as a matter of coincidence. In our experience, it is the direct result of the perfect controlling of all business processes that are run in the company, culminating in "*Business Excellence*".

We view quality management as being the inter-dependent activities of management and every employee in our company aimed at achieving the targeted goal of business excellence. Our quality policy therefore incorporates a proactive quality planning, a continuous improvement program (CIP), preventive quality control and quality control to detect and analyze individual production faults. The fundamental concepts of a modern quality management as the essential basis for business excellence were first developed in the 1970s. The self-evident view that quality deficiencies would not be prevented for all intents and purposes with just reactive quality assurance in the classical sense alone (through output control) became an increasingly commonly held view.

The basic causes of product quality deficiencies can be found primarily in problems with their development and insufficient control of the production process. As a consequence, the emphasis is on *proactive procedures in quality management* – optimum planning of quality measures and their consistent implementation.

This lead to the development in the 1980s of the demanding and promising concept of *"Total Quality Management (TQM)"*, the motto of which is impressively simple and clear: Do everything right from the beginning until the successful conclusion – a perfect zero failure strategy! This proactive basic tenet for quality consciousness, enriched by today's immediate, matter-of-course commitment to consistent business process management has in the meantime also reinforced its presence with the annual bestowal of awards (USA: Malcolm Baldridge Award, EU: European Quality Award, etc.). Infineon has also already qualified twice for the European Quality Award (EQA) as a finalist in the top class.

Modern proactive quality management as an essential part of a successful business process management stands out from the old style quality assurance as a result of its orientation towards personal commitment. Only through our own actions is the immense potential released; potential that optimally satisfies the quality requirements of all interested parties; those of our customers, suppliers, shareholders, the environment, etc. – and, of course, our own requirements as well.

15.2 Quality measures in business processes

By way of example, quality measures integrated in the development, qualification and production process of products are dealt with in more detail in the following.

Quality measures for assuring product characteristics and features as well as delivery quality

The product development process, well structured and described in the Product Development Handbook is based on the specifications.. The essential measures for guaranteeing product characteristics and features are controlled in the design reviews to be carried out at each milestone of the development process, which are supported by the corresponding checklists in the development handbooks.

Depending on the progress of the development, the following factors are controlled: comparison of block diagram with customer specification, comparison of layout with block diagram, comparison simulated electrical characteristics with target values from customer specifications.

After samples have been produced, these are tested in operation within the scope of the product qualification on suitable test systems in the specified voltage and temperature ranges. Important here is a test severity level as high as possible, which means that the test program must be able to cover all specified functional and electrical product characteristics.

After ramping up the volume to mass production, single fails continue to occur, even with the best production process control with the state of the art methods like SPC, These fails require final tests as quality measures.

Quality measures for assuring processability and reliability

Processability and reliability of semiconductor components are evaluated according to the following criteria:

- Manufacturability
- Compatibility with further processing (e.g.board assembly)
- The number of components that fail because of defects (early failure rate)

cumulative failures vs.time

F [ppm]

wear-out failures

useful life

defect related failures

t

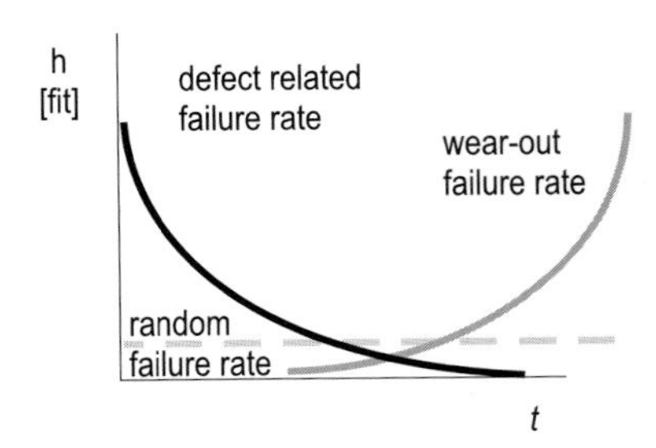

Figure 15.1 Comparison of the histograms for the cumulative failures and the failure rate

- Performance of the components under application conditions over the required operating time (time until start of the wear-out period > required operating time)
- Robustness with regard to external stressors (ESD, EMI, voltage spikes, etc.)
- Figure 15.1 shows a graphic comparison of the terms “Useful life” and “Failure rate”.

In examining the different criteria, special test concepts are used which are summarized in Table 15.1.

Within the scope of the product qualification as the last phase of the product development process before ramp-up, the influence and effect of the different physical failure mechanisms on the function and useful life of the component are determined.

Table 15.1 Test concepts for different quality criteria

Criterion	Risk	Test concept
Processability for the customer	Problems with the board assembly	Stress simulation with assembly procedures (soldering stress, cleaning measures)
Defect-related failures	Early failures	Test of a representative product or of a monitor product of a technology
Product lifespan	Influence of mechanisms that depend on the construction of individual elements, as well as operation and environment stresses	Test structure examination; use of acceleration models; simulations carried out
Electrical resistance	Increased sensitivity with regard to external electrical charges and fields	ESD test, latch up, EMI, radiation sensitivity

15.3 Processability for the customer

The assembly of components on printed circuit boards at the customer or in other modules is simulated during the product qualification by the manufacturer using the corresponding component stress. These conditions for soldering and stress in the customers production chain, which are shown in detail in our packages handbook, comply with internationally applied standards.

Components also see the same pretreatment for other internal reliability tests to make relevant reliability assessment for the application at the customer.

Defect-related failures

Defects that occur during the production process can still remain undetected with the final testing of the product, and they can then cause a failure after only a short period in application (early failure). These are known as statistical failures that are derived directly from the defect density level of the respective production process.

Because of the high quality level of today's production processes and the resulting low defect densities, large volumes of components are required to demonstrate the early failure rate. The testing is carried out using relatively large random samples accumulated over long monitoring periods.

As a test vehicle the most complex product as representative of the production technology is selected. Operation with the worst case use conditions is selected for testing

The cumulative failure distribution and the progression of the failure rate can normally be shown by a simple exponential distribution.

Figure 15.2 shows the "model" for describing the defect-oriented cumulative failures and immediate and mean failure rates *F(t)/t*. The model curve refers to 1000 dpm failures after 10^5 hours. Parameter b was determined instantaneous with 0.7 average for logic components.

Reliability behavior during product life

It is checked with these tests to see if the component can fail during the required lifetime because of wear-out effects

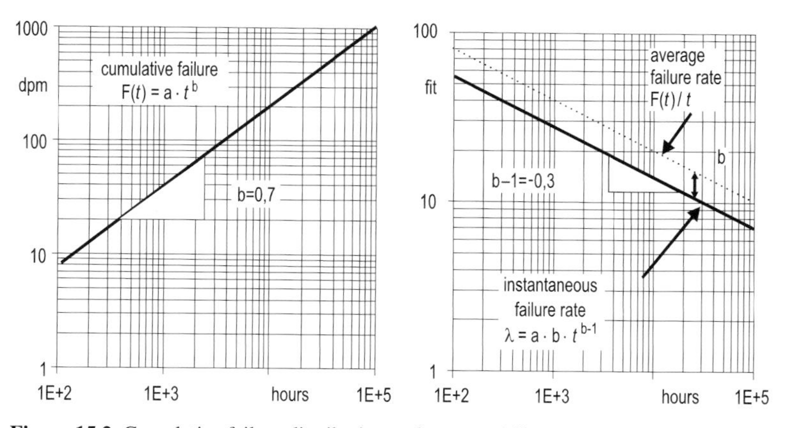

Figure 15.2 Cumulative failure distribution and average failure rates

Table 15.2 Reliability tests schema

stress/load	tested units	test conditions	remark
Tests for expected life (specific to mechanisms dependent on construction and stress)			
operational static/dynamic: – temperature – electr. field strength – current density	individual integrated elements (test devices) sensitive to the evaluated mechanism	conditions specific to the mechanism relevant for element and load	beyond ratings of product for high test acceleration
environmental	individual package types (plastic)	max ratings for package, e.g.:	
– temperature – humidity – temperature change		150°C 85°C, 85% r.h.el.bias –40 / +150°C	
Tests for product function (usually at max. ratings or at manufacturer defined conditions)			
electrical operation	product	dependent on product function: e.g. static / dynamic / cycled power / write / erase cycling	for coverage of expected life
Tests for early life reliability (usually at max. ratings or at manufacturer defined conditions)			
early life reliability	product representative for the product family (same technology)	dependent on product function: e.g. static / dynamic / cycled power	for coverage of expected life
Tests for processability (simulation of OEM processes) frequently applied as preconditioning before environmental tests			
soldering heat	individual package types (plastic)		
	through hole packages	soldering heat profile	
humidity at storage and transport	SMD packages	preconditioning *) + soldering heat profile	*) as simulation of humidity absorption
Tests for functional robustness against external electrical loads			
electrostatic discharge (ESD)	product	– human body discharge (HBM) – charged device discharge (CDM)	
latch-up	product		
electromagnetic interference (EMI)	product		

(caused by material properties and the construction of the individual elements). Evaluation is made using principally 'known' relations between physical mechanisms and applied stress conditions.

For the quantitative development of this "model" as shown in Table 15.2, simplified test structures are utilized, which can be specifically loaded with high stresses. Assessments can be made much sooner with this test concept (see the "Acceleration models" section).

The results of such lifetime tests, which must be provided as soon as possible (at the early phase of the development of a new production technology), are implemented immediately in obligatory design rules to provide a long life time for all products to be produced on the new technology. Significant reliability factors for the evaluation with regard to the component behavior during the operating time that ultimately determine the lifetime are shown in Table 15.3.

Electrical robustness

This aspect includes the behavior of the components with excessive electrical stress caused by electrostatic charges/fields (ESD) and conditions that occur during the application (EMI, latch-up).

These component characteristics are determined for the most part by the circuit design of the products. Simulations and tests in accordance with internationally recognized procedures in the early phase of a new production technology also result in obligatory design rules to predict the behavior of our new products to be developed.

Acceleration model

In order to be able to map the results of reliability tests that were carried out with increased stress conditions onto application conditions, we use "acceleration models" as shown in Figure 15.3.

Table 15.3 Reliability factors for determining the life time of components

Product element	Mechanisms	Stress factors
Stability of the electrical parameters with active switching elements	– Charge trapping (hot carrier effects)	E, T
	– Drift ionic impurities	E, T
Time-dependent breakthrough of dielectric layers (thin oxides)	– Charge trapping	E, T
Stability of conductor paths and contacts	– Inter-diffusion of different metals with formation of gaps in the metal	T
	– Electromigration	J, T
Product robustness:		
– Thermo-mechanical instability of product elements	– Crack formation	ΔT
	– Mechanical material fatigue at interfaces	ΔT – Number of cycles
– Density	– Corrosion	Relative humidity
	– Stressmigration	

T: Temperature, ΔT: Temperature interval, E: Electric field strength, J: Current density

Temperature: (Arrhenius Model)	$\mathrm{AF}(T) = \exp\frac{\Delta E}{k}\left(\frac{1}{T_{\mathrm{use}}} \times \frac{1}{T_{\mathrm{stress}}}\right)$
Temperature and Bias: (Eyring Model)	$\mathrm{AF}(V, T) = \mathrm{AF}(T)\exp \mathrm{B}(V_{\mathrm{stress}} - V_{\mathrm{use}})$ B depends on mechanism (element/technology), default B=1
Temperature and Hunidity: Model for corrosion failures in plastic packages (Peck-Model)	$\mathrm{AF}(\mathrm{rh}, T) = \left(\frac{\mathrm{rh}_{\mathrm{stress}}}{\mathrm{rh}_{\mathrm{use}}}\right)^{\mathrm{n}} \times \mathrm{AF}(T)$ n=3, ΔE=0,9 eV
Temperature Cycling: Model for mechanical fatique failures of soldered/welded contacts (Coffin-Manson-Model)	$\mathrm{AF}(\Delta T) = \left(\frac{\Delta T_{\mathrm{stress}}}{\Delta T_{\mathrm{use}}}\right)^{\mathrm{C}} \times \frac{\mathrm{f}_{\mathrm{c\ stress}}}{\mathrm{f}_{\mathrm{c\ use}}}$ C depends on mechanism (element/technology), default C=2

ΔE apparent thermal activation
k Boltzmann constant = 8,617 · 10^{-5} ev/K
T chip temperature (K)
V bias voltage (V)
rh relative humidity (%)
ΔT temperature (°C or K)
$\mathrm{f_c}$ number of cycles per unit time (h^{-1})

Figure 15.3 Acceleration models that can be applied to products

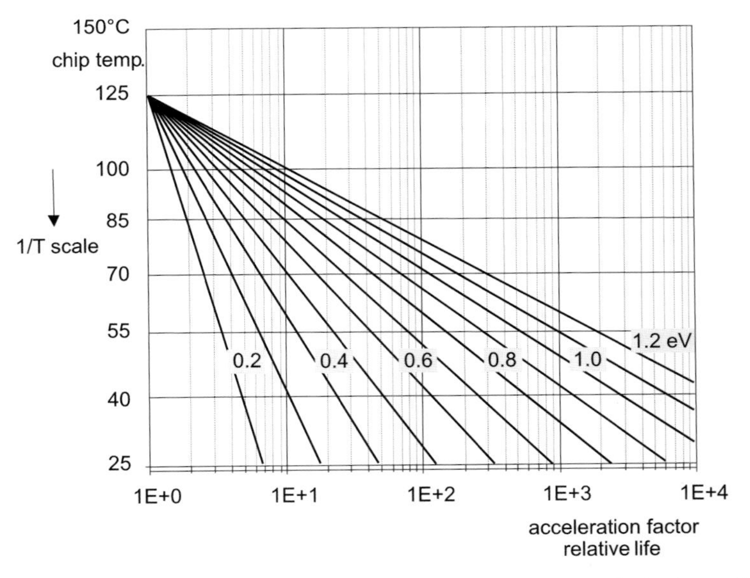

Figure 15.4 Test acceleration/relative life dependent on temperature (Arrhenius model)

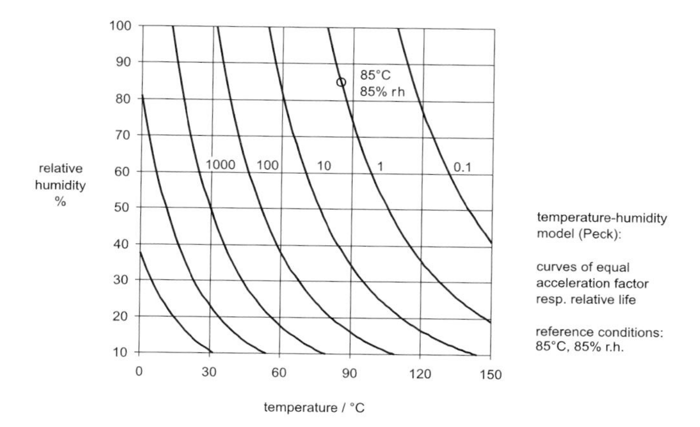

Figure 15.5
Test acceleration/relative life dependent on temperature and humidity (Peck model)

Four models are listed in Figure 15.3:

Thermal stress (Arrhenius model)

0.5 ... 0.7 eV are typically accepted as average apparent activation energies.

Thermal and electrical stress (Eyring model)

In this model, the operating voltage is an additional stress activator. It is used to describe oxide breakthrough in MOS structures. With this model, the empiric factor *B* takes the density of the oxide layer into account.

Thermal and humidity stress (Peck model)

This model is based on the statistical evaluation of a comprehensive number of experiments of varying origin. The value for rated operation is accepted here as operating voltage. The reference point of 85°C at 85% relative air humidity is a typical test condition.

Cyclic thermal stress (Coffin-Manson model)

This model refers to the failure mechanism that is activated by the material fatigue of soldered and welded contacts. It is seen as a feasible model for components in plastic encapsulation.

Figure 15.3 shows a combination of the most important acceleration models that are incorporated as part of reliability evaluations. The models do not refer for the most part to individual mechanisms, but rather take the consequences of a mixture of different effects into account.

The models enable the determining of acceleration factors that subsequently enable the conversion to the operating time using the equation:

$$T_{\mathrm{use}} = \mathrm{AF} \cdot t_{\mathrm{stress}}$$

Figures 15.4 and 15.5 show how enhanced test conditions have an accelerated effect on potential mechanisms. The attainable acceleration depends on the values of the model parameters, which for the most part reproduce a mixture of varying mechanisms (with or without dominance of a single mechanism).

16 Glossary

This glossary contains terms that are in part explained in the book, simply mentioned and/or crop up frequently in the specialist literature. The glossary focuses on terms from semiconductor technology, optoelectronics, production engineering, the associated electromechanics, circuit technology and the basic principles of electronics and semiconductors. Suitable attention is also paid to the relevant standardization bodies as well as interest groupings.

N.B.: Most terms in semiconductor technology originate from English. If this glossary contains a translation from the German, the definition can be found under the German term. Terms that occur mostly only in English (such as chip, interpreter or assembler) are explained directly. Otherwise English nouns and adjectives have only been written in capitals when explaining acronyms (apart from articles and prepositions), and if they are generally always written in capitals in practice.

3GPP Third Generation Partnership Project

A/D converter
A converter that converts analog signals into digital code, see also D/A converter.

Accumulator

1. A charge store that can store and deliver electric charge. In contrast to the primary cell, the accumulator can be recharged many times. In principle, every capacitor is an accumulator. In practice, however, its charge is generally not sufficient, which is why electrochemical accumulators tend to be used.
2. A register for logic and arithmetic operations, normally to count elements or for summation.

ACI Advanced Chip Interconnect.

ACL Advanced CMOS Logic
A new CMOS technology for logic circuits.

ACTFEL Alternating Current Thin Film Electroluminescence.

Active area
In electronics, the electrically or optically effective area of a component.

Active component
Unlike a passive component, one that can add useful electrical power to a signal, see also amplification. Active components are often semiconductors, but also occasionally (still) vacuum tubes.

Active filter
A filter circuit with an integrated amplifier to improve the filter properties, (cf. Butterworth filter, Bessel filter, Tschebyscheff filter).

Adapter
A cross-system adapter that balances differences in signal levels, synchronization, mechanics or similar due to varying systems or standards, e.g. a plug-in card for a computer to adapt interfaces to each other or a coupling for a non-system connector.

Adaptive control
A control strategy whereby the control algorithms allow for adaptation to the prevailing conditions.

Adder

1. A CPU component that adds two numbers together.
2. A circuit that adds the amplitudes of two input signals together.

Add-in
A hardware enhancement that is subsequently built into a computer, see also add-on.

Add-on

1. A hardware enhancement that is connected to a computer, see also add-in.
2. An additional program that enhances the functionalities of an application program.

Address conversion
The conversion of one address into another, e.g. a virtual address into a physical address or a relative address into an absolute address.

Address space
The total position area in the main memory that a computer can address.

Addressing
Activating a certain cell or group of cells, particularly in storage technology.

Adjustment correction

ADSL **A**symmetrical **D**igital **S**ubscriber **L**ine

Adsorption
A physical process whereby gases or fixed particles adhere to a fixed surface.

AIM **A**valanche **I**nduced **M**igration
Permanent charge carrier displacement in semiconductors triggered by an avalanche effect. It is used amongst other things to program non-volatile memories, see also ROM.

ALE **A**ddress **L**atch **E**nable
A control signal for semiconductor memories. Also: retention buffer stand-by for address acceptance.

Allocation
(In particular) semiconductor manufacturers' response to delivery bottlenecks. With allocation, only selected customers receive part of the semiconductors they have ordered. Other customers have to wait.

Alternating current
An electrical current that periodically changes its direction, see also direct current.

ALU **A**rithmetic and **L**ogic **U**nit
Part of an electronic computer.

AMC **A**nalog **M**icro**c**ontroller
A digital microcontroller with additional analog circuit technology.

Analog circuit
An electrical functional block that provides and/or processes analog signals, see also logic circuit, storage circuit.

Angled cut
A technique used with optical waveguide plug connectors to keep the optical junction non-reflective.

Anion
A negatively charged ion, see also cation.

Anode
A positively charged electrode.

ANSI **A**merican **N**ational **S**tandards **I**nstitute, New York

Antiparallel
A type of circuit in which two diodes are connected in parallel in opposite all-pass directions.

APD **A**valanche **P**hoto **D**iode
A photo diode in which the photocurrent generated releases secondary charge carriers, leading to an avalanche effect. This avalanche effect only occurs with a high field strength that can only be built up if the boundary layer is very homogeneous, a major technological challenge. Large-area APDs are thus rare and expensive.

AR coating
Anti-reflective coating, see coating.

ARIB **A**ssociation **R**adio **I**ndustries and **B**usinesses

AROM

1. **A**lterable **ROM**. See also EEPROM, EAROM.
2. **A**ssociative **ROM**. Contents-addressable read-only memory.

Array

1. A field or structured arrangement of similar elements.
2. In programming, a list of data values (also: "variables fields") with the same

data type. Arrays are fundamental data structures.

Arrhenius Law
The observation that the failure rate amongst components increases exponentially with the temperature.

Arrhenius model
A model used for testing components that accelerates the failure rate through increased temperature, see also Eyring model, Peck model.

ASBC **A**dvanced **S**tandard **B**uried **C**ollector
An improved epitaxy double diffusion procedure for manufacturing bipolar ICs.

ASCII **A**merican **S**tandard **C**ode for **I**nformation **I**nterchange
A worldwide code for computer representation of character sets. The basic set has 128 characters (7-bit code). Extremely common with small computers. Adopted by CCITT as code no. 5. The ANSI table offers greater scope with 256 characters.

ASE **A**mplified **S**pontaneous **E**mission
Spontaneous light emission in fiber amplifiers (a considerable noise factor).

ASER **A**ccelerated **S**oft **E**rror **R**ate
An enforced much higher and/or faster SER caused by radioactive irradiation for assessing a semiconductor module's susceptibility to errors.

ASIC **A**pplication **S**pecific **I**ntegrated **C**ircuit
See also gate array.

ASIC library
A collection of wiring regulations concerning higher-organized functions such as logic gates, counters, multiplexers, etc.

ASM **A**pplication **S**pecific **M**emory

Assemble
In programming, converting a program written in an assembler language into the relevant machine code.

Assembler language
A (machine-oriented) programming language that uses abbreviations or mnemonic codes; each of these codes corresponds to a certain instruction from the machine code, see also assemble. The advantages of assembler language are faster sequencing and direct access to the system hardware. It is however not as easy and intuitive to use as higher programming languages.

Associative storage
Also: contents-related storage. A storage method that does not address data elements via a fixed address or position in the main memory, but by evaluating its contents.

ATM **A**synchronous **T**ransfer **M**ode

ATM-F ATM Forum

Attenuation loss, damping
The loss of signal strength in a circuit or transmission line, generally expressed in dB.

Auger effect
An atomic effect whereby a high-energy state is left by donating an electron without radiation occurring.

Automatic device robot
A device that executes sequences automatically after receiving the order to do so from a trigger.

Automatic testing machine
A device that performs tests on specimens automatically using prescribed parameters and routines.

Automation
Equipping a technical system with automatic devices.

Avalanche effect
If very high voltages drop in the space charge zone with PN junctions, the space charge zone loses its insulating property. The high voltage accelerates the charge carriers in the space charge zone so much that new charge carriers are generated in an avalanche-like fashion through impact ionization. The avalanche effect is a high-field effect in the semiconductor.

AVLSI **A**nalog **V**ery **L**arge **S**cale **I**ntegration

Back annotation
Customer release for custom circuits.

Back end

1. In semiconductor manufacturing, the last stage of production involving final inspection and testing and placement in the package.
2. In a client/server application, the part of a program that runs on the server.
3. In programming, the part of a compiler that converts the source code into machine code, see also compiler, front end, interpreter, source code.

Back lash
Post-pulse oscillation, hysteresis, lost motion.

Back reflection return

Back scattering

Backplane
To a certain extent, the road network of a computer, control or telecommunications system. It includes wiring ducts, power supply leads and is often located on the rear of a switchgear cabinet, hence the name.

Backward wave
A reflected wave in a traveling-wave tube that moves against the electron direction of flow.

Bandpass filter bandwidth filter

Bandwidth
The interval between the upper and lower limiting frequencies of an analog signal transmission system. It is expressed in Hz and depends on the definition of the limiting frequency. The 3-dB definition is very common, i.e. halving the signal level. With digital systems, the bandwidth is the maximum data transfer capacity and is expressed in bits per second.

Bandwidth filter
A filter that only permits radiation and/or signal transmission within a certain frequency range (frequency band) without any appreciable attenuation. See also blocking filter.

BARITT Barrier **I**njected **T**ransit **T**ime diode
Similar to the tunnel diode and the IMPATT diode, the semiconductor diode has negative differential resistance in parts. It is used predominantly in microwave circuits.

Barkhausen effect
The simultaneous and sometimes audible fold-over of molecular structures (Weiss' domains) that occurs when certain materials undergo magnetic reversal. Discovered in 1917 by H. Barkhausen.

Barnett effect
The effect observed in some ferromagnetic materials of spontaneous magnetization through simple rotation without an external magnetic field.

Barrel shifter
A device in a microprocessor or microcontroller that can perform several shift operations in one instruction cycle, resulting in a considerable increase in speed.

Base
The very thin semiconductor layer in a bipolar transistor with one boundary layer to the collector and one to the emitter.

Base circuit
A transistor circuit whose base is common to the input and output circuit. The base circuit has low input resistance.

Basic Beginner's **A**ll-Purpose **S**ymbolic **I**nstruction **C**ode
A higher programming language developed in the mid-60s by John Kemeny and Thomas Kurtz that is very easy to learn.

Batch see lot

Bathtub curve
It describes the statistical distribution of component failures throughout the service life of the entire population. At the beginning of the service life, a certain proportion accumulate early failures. After this phase, the failure rate falls but then increases again towards the end of the service life. The gradient is similar to the longitudinal profile of a bath tub.

BCD Binary **C**oded **D**ecimal
Decimal digits binary-coded with four bits each (tetrads, nibbles), see also EBCDIC. This representation prevents

rounding errors when converting into binary code, see also EBCDIC.

BCT Ballistic **C**ollection **T**ransistor

BDI Base **D**iffusion **I**solation
A dielectric isolation technology for bipolar ICs.

Bellcore Bell **C**ommunications **R**esearch (now Telcordia Technologies)

Bending coupler
An optical waveguide coupler with two adjacent fibers bent to such a degree that light emerges from the core at the bend, see also melting coupler.

BER

1. **B**asic **E**ncoding **R**ules.
2. **B**it **E**rror **R**ate, see also BERT. The relative number of received corrupted bits. An important characteristic quantity for the reliability of a data connection.

BERT BER Test(er)

Beta version
The preliminary as yet unreleased version of a hardware or software product. It is generally tested by a representative pilot customer (beta site) under realistic operating conditions.

BFL Buffered **FET L**ogic
An IC family in GaAs D-MESFET technology.

BGA Ball **G**rid **A**rray
Ball matrix; an SMT device for ICs.

BH Buried **H**etero (laser)
A structure for semiconductor lasers that is not on the surface of the substrate but "buried" to a certain extent.

Bias
Bias voltage, biasing current, magnetic bias, biasing. Required to specifically condition a circuit, a component or a material.

Biased pn junction
At the junction of a p-type and a n-type semiconductor layer the electrons of the n-type layer are attracted by the holes in the p-type layer. Without a potential applied to this pn junction a domain with reduced number of holes and free electrons is created (depletion layer). The pn junction is called forward biased if a positive potential is applied to it. Reverse biased relates to a negative potential. The positive potential increases the free electrons and holes, reduces or even removes the depletion layer and may lead to a current flow, the negative potential widens the depletion layer reducing the current flow towards 0. Therefore the pn junction behaves like a current valve, see also diode, rectifier.

BICFET Bipolar **I**nversion **C**hannel **F**ield **E**ffect **T**ransistor

BiCMOS Bipolar **CMOS**
An IC technology that combines bipolar transistors and CMOS FETs on one chip, see also BiMOS.

Bidirectional in two directions

1. With remote data transmission, the possibility of using both directions either simultaneously or time-staggered, see also duplex.
2. The capability of printers to print in both directions of movement with the print head.
3. The property of switches to guide and switch currents in two directions.

BiFET Bipolar **FET**
An IC technology for analog applications with junction FETs and bipolar transistors on a single chip, see also BiCMOS.

Big-endian
A storage method in which the MSB of a number appears first, as opposed to the little-endian method where the LSB is first. Motorola processes use the big-endian format, Intel processors the little-endian format.

BiMOS Bipolar **CMOS**
An IC technology that combines bipolar transistors and MOS FETs on a single chip, see also BiCMOS.

Binary dual
The existence of exactly two components. Nearly all digital computers work on a binary basis, see also ternary.

Binary file
A file that consists of an 8-bit data sequence or executable code.

BIOS **B**asic **I**nput **O**utput **S**ystem
A basic operating system for executing the basic functions of a computer. The BIOS must generally be activated before the actual operating system can be loaded.

Bipolar
A property of semiconductor components that use both p-doped as well as n-doped semiconductors. The simplest structure is the PN diode, see also unipolar.

Bistable
A property of a system or component that can have two possible states such as "on" and "off".

Bit **B**asic **I**ndisoluble Information Uni**t**
Individual binary digits. The smallest indisoluble unit of information that recognizes two states. (The plural, when specifying a quantity, is also "bit", and designates the amount of information; only with a specific, non-numerical plural is it "bits", for instance the last two bits of a data word).

Bit slice processor
A processor chip for microprocessors developed for special purposes based on customer requirements. They perform the same tasks as other processors but work with smaller information units, typically 2 or 4 bits. To process larger data items, individual bit slice processors are combined to form processor units.

Bit width
The number of bytes that a processor can process with an instruction. It ranges from a half-byte (nibble) up to 32 bytes or more.

BJT **B**ipolar **J**unction **T**ransistor
The traditional transistor as developed in 1947 by Bardeen, Brittan & Shockley.

Blackout
A state in which the energy level falls to zero; a total failure of the power supply. Bouts of unconsciousness are also known as blackouts, see also brownout.

Bluetooth
A technology for the wireless transfer (via radio) of voice and data.

BNC

1. **B**ayonet **N**ut **C**oupling.
2. **B**ayonet **N**eill **C**oncelman.
3. **B**ritish **N**aval **C**onnector.
4. **B**ritish **N**ational **C**onnector.
   Various different designations for the same coaxial connector system with one bayonet coupling. It was developed several decades ago and is often used in measuring equipment.

Board electronics board.
See also PCB, Electronic printed circuit board.

Bode diagram
A special way of showing the frequency response of the amplitude and/or phase of a circuit in a circular diagram.

BOL

1. **B**eginning **o**f **L**ife, see also bathtub curve.
2. **B**ehörden **Onl**ine [Public Authorities Online]. A joint venture between Viag Interkom and the IZB (Bavarian Information Technology Center) set up in 1998 to connect all the ministries, subordinate public authorities and district offices in Bavaria to the voice network of Viag Interkom.

BOM **B**ill **o**f **M**aterial

Bond pad
A relatively large metal surface on a semiconductor die which provides the electrical contact with a package or test probes.

Bond wire
A connection wire for semiconductor chips, mostly made of gold.

Bonder
A machine used for (mostly automatic) bonding.

Bonding
A wiring and connection technique for ICs and other semiconductors, see also bond wire.

Boolean algebra
A system of logic operations developed by George Boole and particularly well suited for use in digital technology.

Boot
To initialize a computer. Comes from "boot strap loader", initial program loader. The process involves writing a defined value in different registers and storing certain memory contents in certain places.

Booth algorithm
An algorithm that permits simple multiplication and division in a few steps. The Booth algorithm is often used in microcontroller and DSP technology.

Bootstrap loader
An operating system routine that ensures that data can be transferred from the data carrier to the working memory.

BORSCHT **B**attery Feed – **O**vervoltage Protection – **R**inging – **S**ignaling – **C**oding – **H**ybrid – **T**esting
These features represent the basic functions of a subscriber circuit in telephony.

Breadboard
In electronics, a test board with holes for the provisional insertion of components. Sometimes wooden boards are actually used, but breadboards are normally made of pertinax or another plastic.

Breakdown
In semiconductors, overcoming a voltage or current barrier and the subsequent surge in the injection current.

Breakpoint
The point in a sequential program at which an error condition stops the program proceding. The breakpoint is an important tool for debugging.

Brightness
The subjective impression of the amount of light, e.g. corresponding to the objective photometric size of the LED.

Brownout
A state in which the power supply is considerably reduced for a certain period of time, see also blackout.

BRS **B**uried **R**idge **S**tructure
(for laser diodes).

BST **B**arium **S**trontium **T**itanate
A material with a high dielectricity constant for the manufacture of small DRAM memory capacitors.

BTDL **B**asic **T**ransient **D**iode **L**ogic

BTRS **B**uried **T**win **R**idge **S**ubstrate

Bug
A fault in the software or hardware. With hardware, a bug is a wiring or linkage fault that can be temporarily removed by a patch. With software, a bug is a fault in the code or a logical error that leads to misoperations or incorrect result output.

Bulk noise shot noise

Buried layer
A buried semiconductor layer, i.e. a layer that is covered by other layers.

Burn-in
A hardness test during which electronic components are operated under pressure for several hours or days with increased voltage and temperature so that early failures occur before delivery to the customer. This measure increases the quality of the remaining components, see also bathtub curve.

Burrus diode
An IRED with the radiation being, so to speak, extracted in a downward direction through a "hole" in the substrate (etched in after production); its PN junction is located on the underside. This produces particularly good conditions for coupling it up to fiber optic cables. As the internal construction of Burrus diodes makes them particularly fast, their use is preferred for fiber optic transmission.

Burst
The intermittent transmission of a signal.

Bus
A wiring system that facilitates access to a data-processing system by means of insertion. The slots are like "bus stops" at which data streams (like the "passengers") can get on and off. Most bus systems are parallel, but there are also serial systems, see USB.

Bus width
The number of transmission lines on a parallel bus.

Byte
From "**By** Eigh**t**" or "**Bi**nary **T**erm". A logically associated group of 8 bits that permits 256 combinations, also: octet. See also nibble.

BZT **B**undesamt für **Z**ulassungen in der **T**elekommunikation
Germany's Central Bureau of Approval in Telecommunications

C
A programming language developed in 1972 by Dennis Ritchie at Bell Laboratories. The name comes from its immediate predecessor, language B.

C++
An object-oriented variant of programming language C. It was developed in the early 80s by Bjarne Stroustrup at Bell Laboratories.

C3L **C**omplementary **C**onstant **C**urrent **L**ogic
A DTL circuit with Schottky diodes for high-scale integration.

Cable packaging
A form of back-end production in which the pieces of cable are cut to the required length and equipped with the requisite connectors.

CACA **C**omputer-**A**ided **C**ircuit **A**nalysis

Cache
French for "hiding place". In computer technology, it is the buffer facility intended to prevent waiting times. The precautionary buffering of data likely to be required in the near future from the mass storage improves the data throughput and/or reduces the effective data access time, see also write-behind cache.

CAD **C**omputer **A**ided (Assisted) **D**esign

CAE **C**omputer **A**ided **E**ngineering

Calibrate
To bring into line with a reference value, also: gauge.

CAMP **C**omputer **A**ided **M**ask **P**reparation
(In IC technology).

CAN **C**ontrol **A**rea **N**etwork
A field bus system that originated in the automotive sector and was developed by Robert Bosch.

Capacitance electrical capacity

Capacitor
A component with a specific electrical capacity.

Capacity capacitance

1. Generally: content or capability.
2. The ability to store electric charge, expressed in farad. The ratio of charge (expressed in coulombs or ampere-seconds) to voltage applied.

CAPI **C**ommon **A**pplication **P**rogramming **I**nterface
A standardized interface for programming ISDN cards.

Carbon resistor
A type of resistor that has a spiral carbon layer on a cylindrical ceramic body. The carbon resistor is cheap to manufacture but has high intrinsic noise as well as high process tolerances and can not withstand very high operating temperatures and mechanical loads.

Carrier
This can refer both to a carrier frequency but also to a mechanical carrier that holds a component. The operator of a telecommunications network is also known as a carrier.

Carry bit
A bit position of an adder circuit that signals that an addition has resulted in a carry-over.

Carry flag

CAS **C**olumn **A**ddress **S**trobe
The release of the column address. A control signal for memory chips, see CE.

Cascade
In technology, the series connection of units, also known as concatenation connection. For example, amplifiers are cascaded to achieve the highest possible

overall amplification. The Darlington circuit is also an example of a cascade.

Cat's eye
The graphical representation of the gradient of a cyclical switching process over time. As long as the zero crossings are broken down well in terms of time, the switching process is easy to detect. As the frequency increases, dispersion causes faults to occur at some point. When transferring data, this means an increase in the error rate, see also BER. The graphical representation changes increasingly from discretely representable periodic squares into increasingly flatter and intertwining "cat's eyes".

Cathode
A negatively charged electrode.

Cation
A positively charged ion, see also anion.

Cavity
A local indentation or surface change relevant for semiconductor structures. Cavities that have a lower reflectivity than their environment are also created in CDs and CD-ROMs.

1. A resonance chamber in laser technology.
2. A discrete, extremely small space in semiconductor material.

CB **C**omplementary **B**ipolar
A semiconductor structure (for transistors).

CBEMA **C**omputer and **B**usiness **E**quipment **M**anufacturers **A**ssociation
A US association of hardware suppliers and manufacturers that promotes the standardization of information-processing and associated equipment.

CBIC **C**ell-**B**ased **IC**
A standard-cell IC.

CBR **C**AS **b**efore **R**AS
A DRAM refresh cycle.

CC
1. **C**hip **C**ard
2. **C**hip **C**arrier
3. **C**ontinuity **C**heck

CCC **C**eramic **C**hip **C**arrier

CCD **C**harge **C**oupled **D**evice
Also: charge transfer device or bucket-brigade device. An MOS structure in which charges are gradually transported and then evaluated.

CCFL **C**apacitor **C**oupled **FET** **L**ogic
An IC technology with GaAs MESFETs.

CCIR **C**omité **C**onsultatif **I**nternational des **R**adiocommunications
International Radio Consultative Committee

CCITT **C**omité **C**onsultatif **I**nternational **T**élégraphique et **T**éléphonique
International Telegraph and Telephone Consultative Committee

CCL **C**omposite **C**ell **L**ogic
A design concept for ASICs that uses a cell library.

CCMD **C**hip **C**arrier **M**ounting **D**evice

CCPD **C**harge **C**oupled **P**hoto**d**iode Array

CDA **C**ustomer **D**efinable **A**rray

CDI **C**ollector **D**iffusion **I**solation

CD-ROM **C**ompact **D**isc **R**ead-**O**nly **M**emory
It has a capacity of around 650 megabytes and records in the same way as a CD.

CE
1. **C**hip **E**nable. A signal that facilitates access to an electronic component.
2. **C**olumn **E**nable. A control signal for memory chips (formerly CAS).

CECC **C**enelec **E**lectronic **C**omponents **C**ommittee
Based in Brussels, see also ECQAC.

Cell design
A development technology for custom semiconductor circuits using prescribed macros (standard cells).

CEN **C**omité **E**uropéen de **N**ormalisation **Elec**trotechnique
The European Committee for Electrotechnical Standardization.

CENELEC **C**omité **E**uropéen de **N**ormalisation **Elec**trotechnique

The European Committee for Electrotechnical Standardization.

CEPT The European **C**onference of Postal and Telecommunications Administrations

Cerdip **Cer**amic **D**ual **i**n Line **P**ackage

Cermet
An artificial word derived from **Cer**amic **Met**al, i.e. a mixture of ceramic and metal that can be used as a resistance material.

Cerpack
An artificial word derived from **Cer**amic **Pack**age, i.e. a ceramic package for electronic components.

Certification
The awarding of a particular qualification or skill, see also qualification.

CFT **C**hirp **F**ourier **T**ransform
Fourier analysis of the chirp behavior of laser diodes.

Channeled Gate Array Technique
A technique for joining basic gate cells (gates) using the wiring ducts between them.

Characteristic curve
The parameter gradient of a component under the influence of a changed input parameter.

Chip card
A plastic card with an integrated semiconductor chip. The chip card can therefore store and process information. The telephone card is a well-known example.

Chip die, semiconductor component

Chip set
A collection of different chips that act as a single unit and perform a common task.

Chirp

1. Shifting the central wavelength of a laser by changing the injection current, see also CFT.
2. The unwanted frequency modulation of a radar signal.

CIC **C**ustomized **I**ntegrated **C**ircuit
See also ASIC.

CICC **C**ontactless **IC** **C**ard

CID **C**harge **I**njection **D**evice
or Detector. A semiconductor structure in which charge carriers are injected to a certain extent, in contrast to CCD.

CIM **C**omputer **I**ntegrated **M**anufacturing
The merging of planning and production with the aid of a computer.

CISC **C**omplex **I**nstruction **S**et **C**omputer
The counterpart to RISC.

Cladding sheath

Clamping circuit
A (diode) circuit for "clamping" an amplitude, see also clipping.

Clamping diode
Also freewheeling diode or suppressor diode. The clamping diode also serves to suppress the interference from commutation.

Class A, AB, B, C
Operating mode of output tubes. In class A, a transistor or a tube guides the entire output signal since the closed-circuit current corresponds at least to the maximum amplitude. In class B, one circuit respectively takes over the positive and negative half-wave, see also push/pull. Class AB is between the two. In class C, the output stage only conducts from a certain threshold value. The efficiency of the output stage increases from A to C.

C-LCC **C**eramic **L**eaded **C**hip **C**arrier
A ceramic carrier for semiconductor chips with bond wires, see also PLCC.

Clean room technology
This is extremely important for the yield in semiconductor manufacturing since even the smallest dirt and dust particles are large in comparison to the semiconductor structures on microchips and can very easily render them unusable. There are various classes of clean room technology, depending on the purity required.

Clear
Free, empty, deleted, set to zero, brighten, plain writing.

Cleave
e.g. the controlled breaking of optical fibers or semiconductor wafers by giving the material a rupture joint and then subjecting it to a mechanical load.

Climatic chamber
A room with adjustable temperature and air humidity ratios for simulating possible environmental conditions, see also burn-in.

Clipping
A procedure for limiting the amplitude of an oscillation. Peaks above the limit are cut off. With audio signals, this may result in audible distortions, see also clamping.

CLIW **C**onfigurable **L**ong **I**nstruction **W**ord
A processor architecture especially for DSPs.

Clone chained copy
Due to the linkage, subsequent changes to the original also have a direct effect on the copy.

CLSI **C**ustom **L**arge **S**cale **I**ntegration
A customized semiconductor IC with high-scale integration.

CMBH **C**apped **M**esa **B**uried **H**etero-structure.

CMD **C**onductivity **M**odulated **D**evice
A module with conductivity modulation for power electronic circuits (for example IGBT).

CMFS **C**eramic **M**ultilayer **F**unctional **S**ubstrate
The basis for a technology to integrate passive components.

CML **C**urrent **M**ode **L**ogic
A current switch technology for bipolar ICs, e.g. ECL and E2CL. It switches quickly because the semiconductor switches remain in a non-saturated state.

CMMU **C**ache **M**emory **M**anagement **U**nit

CMOS **C**omplementary **MOS**
An MOS technology that works with complementary logic elements. This circuit technology considerably reduces the power requirements as one of the (complementary) logic elements of a number in series is not conductive in the idle state. Power is required essentially due to dynamic reloading of the switching capacities when switching the transistors.

CMP **C**hemical **M**echanical **Polishing**
Leveling: a surface finishing procedure for semiconductor wafers.

CMRR **C**ommon **M**ode **R**ejection **R**atio
The suppression of co-phasal (interference) signals with differential amplifiers.

Coating
Applying a protective layer to semiconductor dice or their package. With opto-electronic components, often a low-reflective additional layer is applied.

CoB **C**hip **on** **B**oard
A hybrid circuit technology which assembles unhoused chips on one PCB.

COD **C**atastrophic **O**ptical **D**amage
Optical destruction, e.g. with semiconductor lasers.

CODEC

1. **Co**der/**Dec**oder. A key module for digital telephony, which converts voice signals from analog into digital and vice versa.
2. Also: **Co**mpressor/**De**-**C**ompressor, i.e. a circuit that compresses and decompresses signals.

Coding
The representation of concepts using agreed characters.

Coercive field strength
Also: coercive force. This is the field strength to be applied, which returns the flow density and/or the magnetization to zero, see also remanence.

COG **C**hip **on** **G**lass
A semiconductor on a glass substrate.

Coherent
In optics, a property of electromagnetic waves with a constant phase relationship. Depending on the type, laser diodes have strong coherence, in contrast to LEDs.

Coincidence circuit
A detector circuit that only responds

when two events happen at the same time.

Cold soldered joint
An incorrectly soldered connection that can spontaneously develop (and/or through heating, vibration, etc.) a high contact resistance. It can cause considerable malfunctions and is often difficult to detect.

Collector
The part of a bipolar transistor into which the dopants drain (collector current).

Collector follower
A transistor circuit in which the output signal is generated by current flow in the collector resistor, see also emitter follower.

Color coding
The marking of resistors and other electrical components with different color combinations which contain the encrypted value of the component.

COMFET **Co**nductivity **M**odulated **FET**
See also IGBT.

Communication
An exchange, most commonly applied in the context of information.

Communications technology
A collective term for all technologies that serve communication in the sense of information exchange. They include above all telecommunications, radio and television engineering as well as computer, storage and printing technology.

Commutation load reversal
A process in drive and power electronics which can cause unwanted effects such as overshoots.

Comparator
A circuit that compares two input voltages and indicates which is the higher of the two.

Compatible consistent, compliant

Compile
To translate a program language into executable machine code.

Compiler
A program that translates a program written in a higher programming language into machine code. RISC compilers also arrange the machine code instructions so as to ensure optimum supply to the RISC pipeline.

Computer
Originally made from mechanical then electromechanical and, since the 50s, predominantly electronic logic elements. Most computers are digital and most have binary switches, see also ALU, analog computer, CPU, PC.

Concatenation
Cascade, cascading, concatenation connection.

Condensation soldering
Also known as vapor phase soldering. This involves soldering under a constant temperature, which means that the complete PCB is immersed in the vapor of a boiling soldering fluid.

Conducting track
A wiring plan of a circuit etched out of the metal film on a PCB to connect the components on the PCB.

Conductor
A material with high electrical conductivity made possible by the large number of free charge carriers with good mobility. See also electron, semiconductor, insulator.

Contact force
The mechanical force with which an electrical contact is established and/or maintained.

Contact resistance
The resistance to the electric current generated at the junction of two electrical conductors. Contamination, inhomogeneities and irregularities in the geometry of the contact surfaces can cause the contact resistance to be considerably higher than would be expected with the material properties.

Controller
A control module or device.

Cooling fin heat fin
A construction made of heat-conducting metal (generally aluminum) that increases the surface area for the purpose of heat removal.

COS Calculator **on** Substrate

Coupling
In electrical engineering, the possibility of two adjacent current-carrying lines (in particular coils) exchanging electrical energy through induction.

C-PGA Ceramic **P**in **G**rid **A**rray

Cpi **C**lock Cycles **p**er Average **I**nstruction

CPLD **C**omplex **P**rogrammable **L**ogic **D**evice

CPU
Central **P**rocessing **U**nit see also MPU

CRC **C**yclic(al) **R**edundancy **C**hecking
A procedure used for data transfer error checking. The sender transmits a checksum that has been calculated in advance that is compared by the recipient with the data received, see also parity.

CRD **C**apacitor **R**esistor **D**iode
A network (e.g. rectifier layout) consisting of capacitor, resistor and diode.

Creepage current
A parasitic current on the surface of an insulator particularly as a result of high voltage, see also leakage current.

Critical frequency
The frequency that may not be fallen short of or exceeded in order to retain a certain performance. Most systems have an upper and lower critical frequency, see also bandwidth.

Critical load
The load that a component can just about withstand without incurring any damage.

Cross assembler
A translation program for converting machine-oriented programs into a form that can also be executed on other computers.

Cross compiler
A compilation program for converting machine code for use with a different computer.

Crossover
The current transfer in push/pull circuits.

Crosstalk

Crystal
A uniform material structure that develops under certain conditions when transferring from liquid into the solid aggregation state.

CS **C**hip **S**elect

CSIC **C**ustomer **S**pecific **IC**
See also ASIC.

CSP

1. **C**hanneled **S**ubstrate **P**lanar. A basic structure for optosemiconductors.
2. **C**hip **S**cale **P**ackage. A package that is adjusted to the chip dimensions.
3. **C**hip **S**ize **P**ackaging. A minimum packaging technology for ICs.

CTD **C**harge **T**ransfer **D**evice
A semiconductor component that transfers charges, see also CCD.

CTL **C**apacitive **T**hreshold **L**ogic
MOSFET activation from the charge voltage of discrete capacitors.

CTR **C**urrent **T**ransfer **R**atio
A measure for the current transfer ratio of the output current to the input current for optocouplers with phototransistor outputs.

Current amplification current gain
The ability of a component to control a larger current with a small current, see transistor.

CVD **C**hemical **V**apor **D**eposition
A procedure to precipitate thin films during IC manufacture using chemical vapor deposition, for (opto)semiconductors. There are various procedures, for example from the vapor phase, plasma, etc., see also MCVD, MOCVD, PCVD, OVPO, PVD.

CW **C**ontinuous **W**ave
For lasers, or continuous transmission (uninterrupted periodic signal).

D/A converter
A converter that converts digitally coded signals into an analog (generally the original) signal, see also A/D converter.

DA **D**epletion **A**pproximation
A semiconductor boundary layer that is almost free of charge carriers.

DAM **D**irect **A**ccess **M**emory

Dark current
The (reverse) current in a photovoltaic cell in the absence of light.

Darlington
A concatenation of transistors whereby the emitter of the first level supplies the base current for the following level. This results in an extremely high level of current amplification.

Data carrier
A medium for the permanent storage of data. These include diskettes, exchangeable discs, CD-ROMs, hard drives, magnetic tapes, etc.

Data compression
A procedure to reduce the scope of a data collection without losing the actual information. It makes use of the redundancy that nearly always exists. This facilitates more efficient storage (also: packing) or transfer.

Data format
The defined structure of a data record that can vary from program to program. The exchange of data between different programs is only possible if they understand the corresponding data format, see also data structure.

Data integrity
The proper state of data.

Data protection
Special legally regulated protection against the misuse of personal information, such as, for example, the unauthorized passing on of personal data to third parties.

Data security
Protecting data against unwanted access by unauthorized parties or also against physical damage.

Data structure
The internal layout of a data record, see also data format.

Data transmission
The transfer of data between two data-processing units (e.g. computers).

Data type
A data category, such as constant, variable, string, etc.

DC

1. **D**ark **C**urrent.
2. **D**irect **C**urrent.
3. **D**uty **C**ycle.

DCFL **D**irect **C**oupled **F**ET **L**ogic
In GaAs E-MESFET technology and/or with D-MESFETs or HEMT.

DCPBH **D**ouble **C**hannel **P**lanar **B**uried **H**eterostructure
For the manufacture of semiconductor lasers.

DCTL **D**irect **C**oupled **T**ransistor **L**ogic
i.e. without using other coupling components.

DDR **D**ouble **D**ata **R**ate
A technique to increase the data throughput of semiconductor RAMs achieved by data being read or written both on the negative and rising edge of the clock signal.

De facto standard
A technical standard that has come into being almost naturally, generally because it has proven to be expedient or/and was championed by sufficiently influential lobbying. De facto standards often lack the legitimation of an official standardization organization, e.g. ANSI or ISO, see also de jure standard.

De jure standard
A standard that has been specified or approved during the course of a formal process by an institute for standardization, see also de facto standard.

Deadlock
A situation that arises if two programs or devices are waiting for the response of the other before it can continue, or if a device is waiting for a condition that it

must yet establish. Since a deadlock can never be fully ruled out, even with the most careful programming, the device should have a hardware mechanism that recognizes deadlocks reliably, see also watchdog.

Debugger
A test aid for checking programs (in microcontroller technology, etc.). It is generally a program used to locate faults in another program, see also bug.

Decay
The decrease in signal amplitude over time.

Decay decline

Decimal system
An internationally standardized number system with a base of ten.

Declaration
In particular determining a data type for variables.

Decoder
A device for decoding binary signals.

Decrement
Reducing a numerical value by the value 1, see also increment.

DECT **D**igital **E**nhanced **C**ordless **T**elecommunications

Dedicated hardware
Hardware designed for a specific, very narrowly defined purpose.

Default
Basic setting, standard setting, unaltered.

Degradation
All radiation-emitting semiconductors suffer declining output power over their operating lifetime that has not yet been completely explained.

DEK **D**oppel-**E**uropa**k**arte, double europecard
A PCB format for industrial computers and similar.

Delay line

Demodulation
The (most faithful possible) recovery of the original signal from a modulated signal, see modulation.

Density
Generally the degree of concentration of a material. Also the packaging density of data on a data carrier.

Depletion

Depletion layer
See also barrier layer.

Depletion-type FET
A self-conducting FET in which only a negative control voltage blocks the channel, see also enhancement-type FET.

Derating
Adjusting an operating parameter to extreme operating conditions, for example, limiting the power consumption once a particular ambient temperature is reached.

Detectivity
The detectability limit of a detector, i.e. the signal strength that can still be perceived. The reciprocal of the NEP.

Detector
A device for discovering, revealing, exposing, often in the form of an electronic circuit for tracing and recording signals or other physical phenomena.

Development system
A package of software and hardware components required to develop application programs.

Development tool
An aid for designing circuits or software.

DFB **D**istributed **F**eed**b**ack
A technique to increase coherence and reduce line width for laser diodes.

DFM **D**esign **f**or **M**anufacturability

DFSM **D**ispersion **F**lattened **S**ingle **M**ode

DFT

1. **D**esign **f**or **T**estability.
2. **D**iscrete **F**ourier **T**ransform. A mathematical application used in digital signal processing, see also FFT and IFDT.

DG **D**iode **G**ate

DH **D**ouble **H**eterostructure
In (opto)semiconductors.

DHL **D**ouble **H**eterostructure **L**aser

DIAC **Di**ode **A**lternating **C**urrent (Switch)
With a four-layer diode or thyristor, see also TRIAC.

Die (plural: dice)
In semiconductor technology, a separate part on a finished wafer that contains all the functions of the finished semiconductor product. It is often called a semiconductor chip, particularly once it has been detached from the wafer (by sawing or breaking).

Die bonder
A device to bond dice, see also bonding.

Die shrink
The scaling of semiconductor chips. Instead of scaling the existing layout of a module, developments enable the semiconductor components to be accommodated on a smaller area. One proven method is to reduce the individual structures, which is by no means trivial from a technological point of view.

Dielectric punch-through
The permanent destruction of an MOS transistor that occurs when the gate substrate voltage is so high that conductive channels are generated in the oxide. ESD in particular triggers dielectric punch-through very easily.

DIFET **D**ielectrically **I**solated **FET**
A BiFET variant for monolithically integrated operational amplifiers.

Differential amplifier
An amplifier whose output amplitude corresponds to (the increased) difference between the two input signals, see also CMRR.

Differential detector
A detector that responds to signal changes.

Differentiator
An analog circuit, generally with operational amplifier, whose output signal corresponds to the differential of the input signal.

Diffusion
The spontaneous spread of molecules and/or charge carriers within the semiconductor material, see also drift, barrier layer.

Digital
From "digitus" meaning thumb. It describes the representation of results using countable values (natural numbers). Interim values are therefore not possible. In data processing, nearly all information is represented digitally. For computers, this is virtually only possible in binary (two-valued) form since the logic used is nearly always binary for technical reasons, see also bit, byte. The counterpart to digital is analog.

Digitize
Recoding an analog signal into a digital (usually binary) one. Digitization generally requires a sampling procedure that periodically samples and buffers the analog signal values. In the broader sense, the digital conversion and storage of characters is also digitization, see also sampling rate, sample and hold.

Digitizer
A device capable of digitizing.

DIL **D**ual **I**n-**l**ine
A component mounting technology for ICs with two parallel rows of pins.

DIMM **D**ual **I**n-line **M**emory **M**odule
Memory module with contact rows on both sides.

DIMOS **D**ouble Diffused **I**on Implanted **MOS**
A variant of DMOS technology.

DIN **D**eutsches **I**nstitut für **N**ormung e.V., Berlin, the German Institute for Standardization

Diode
A component that blocks the flow of electrons in one direction (from anode to cathode) and lets it through in the other. There are semiconductor diodes and vacuum diodes, but the latter have become almost meaningless, see also barrier layer.

DIP **D**ual **I**n-line **P**ackage
A package in DIL technology.

Direct current
An electrical current whose polarity does not change over time, see also alternating current.

Direct voltage
An electrical voltage with unchanging polarity and a constant or only slightly fluctuating amplitude. Alternating voltages and disturbances can influence the direct voltage, see also alternating voltage.

Disable obstruct, block

Disassembler
A program that converts (translates) machine code back into the assembler language.

Disk operating system
The most well known is MS-DOS, which was developed by Microsoft and became the operating system of the IBM PC in 1981. The disk operating system is fully installed on a data carrier from which it can be imported.

DKE Deutsche **E**lektrotechnische **K**ommission
The German Commission for Electrical, Electronic & Information Technologies of DIN and VDE, responsible for standardization in these fields.

DL **D**iode **L**ogic

DMA **D**irect **M**emory **A**ccess
The DMA module takes control of the system bus and transfers memory contents within the memory or to and/or from the peripheral equipment much more efficiently than the processor could.

D-MESFET **D**epletion **M**ode **M**etal **S**emiconductor **FET**
See also E-MESFET.

DMF **D**ielectric **M**ultilayer **F**ilter

DMOS

1. **D**iffusion **M**etal **O**xide **S**emiconductor.
2. **D**ouble Diffused **MOS**. A procedure involving the double diffusion of doping atoms for MOS manufacture.

Donor atom see majority charge carrier

Dopant majority charge carrier

Doping
The specific admixing of foreign atoms in a semiconductor material. A distinction is made between donor atoms that bring in free electrons and acceptor atoms that deprive the semiconductor atoms of electrons and cause "holes". Both types increase the conductivity of the semiconductor, see also dopant.

Downtime outage
The absolute or percentage time span during which a device or system is not operational, see also failure.

Downward compatible
Compatibility with products of an earlier generation, see also upward compatible.

DPL **D**iode **P**umped Solid State **L**aser

DPPM **D**efective **P**arts **p**er **M**illion

Draft blueprint

Drain
It designates the power outlet end of a conductive channel in an FET, for example.

DRAM **D**ynamic **RAM**
Dynamic random access memory. A memory cell consists in principle of one very small capacitor integrated in the semiconductor that controls an MOS transistor. The charge state of the capacitor determines whether the transistor conducts or not (corresponds to logical ONE or ZERO). Spontaneous discharge of the capacitor results in the information being lost. The memory contents must therefore be regularly "refreshed", hence the name "dynamic". The advantage of this technology is above all the small per-bit space requirements. It is therefore the most cost-effective RAM technology. The disadvantage is the requisite refreshing logic as well as the longer access time.

DRC **D**esign **R**ule **C**heck
A CAD term for checking compliance with design rules, see also ERC.

Drift
In semiconductors, the tendency of charge carriers to migrate.

Driver
A circuit that switches or controls at least one subsequent circuit, see also fan out.

Droop
Also: flow curve. Decreasing power or a drop in power.

DRTL **D**iode **R**esistor **T**ransistor **L**ogic
A logic circuit that uses resistors, diodes and transistors, see also TTL, DTL.

Dry joint cold soldered joint

DSL Digital Subscriber Line

DSP **D**igital **S**ignal **P**rocessor
An IC that performs rapid data manipulation in audio, communications and image-processing technology as well as data reocrding, storage and control. The DSP facilitates operations that are not possible with analog technology or that require great lengths to do so.

DSW **D**irect **S**tep on **W**afer
A procedural step in semiconductor manufacturing.

DTL **D**iode **T**ransistor **L**ogic
A logic whereby diodes are responsible for the logic operation and transistors for the (inverting) level amplification.

DTZL **D**iode **T**ransistor with **Z**ener Diode **L**ogic
A DTL variant with a high interference voltage ratio through the use of Zener diodes.

Ductility workability

Dummy display pack, mock-up

Duplex alternate communication
The ability to transmit information via a communication channel in both directions. Full duplex means that information can be transmitted in both directions simultaneously. With a half-duplex system, the information is transmitted alternately in one direction only respectively, see also simplex.

DVB **D**igital **V**ideo **B**roadcasting

DWV **D**ielectric **W**ithstanding **V**oltage

Dyadic in pairs
For example, a system with two microprocessors. Dyadic Boolean operations are operations, e.g. AND or OR, whose results depend on both values.

E/O **E**lectrical to **O**ptical

E^{2}PROM EEPROM

EACEM **E**uropean **A**ssociation of **C**onsumer **E**lectronics **M**anufacturers, Brussels

EAM **E**lectro**a**bsorption **M**odulator
A modulator that modulates a light beam by means of varying absorption. The absorption is changed by applying a varying electrical factor.

EAP **E**lectroabsorption **A**valanche **P**hotodiode

EAPLA **E**lectrically **E**rasable **P**rogrammable **L**ogic **A**rray

EAROM **E**lectrically **A**lterable **ROM**
Unlike the EEPROM, when it is deleted only part of the data content is deleted (rather like with flash memory). Only certain types permit the deletion of individual cells.

EBCDIC **E**xtended **B**inary **C**oded **D**ecimal **I**nterchange **C**ode
An extension of the BCD code to 8 bits.

EBIC **E**lectron **B**eam **I**nduced **C**urrent

ECD **E**mitter **C**ollector **D**otted
A chip technology.

ECIL **E**mitter **C**oupled **I**njection **L**ogic

ECL **E**mitter **C**oupled **L**ogic
With applications that require extensive driver capabilities (e.g. in bus systems), the use of ECL circuit technology is particularly cost-effective. This technology achieves the requisite driving power with an additional emitter follower transistor.

ECMA **E**uropean **C**omputer **M**anufactures **A**ssociation

ECTF **E**nterprise **C**omputer **T**elephony **F**orum

ECTL **E**mitter **C**oupled **T**ransistor **L**ogic

EDA **E**lectronic **D**esign **A**utomation
The automation of PCB and chip development, see also CAE, CAD.

EDO **E**xtended **D**ata **O**ut
A RAM output technique whereby the output signal is available for a longer period of time at the data output, which favors secure readout. The EDO mode is essentially an FPM variant.

EECA **E**uropean **E**lectronic **C**omponent **M**anufacturers **A**ssociation

EEPROM **El**ectrically **E**rasable **Pro**grammable **ROM**
Also: E^{2}PROM. Deleting deletes all data contents.

EFL **E**mitter **F**ollower **L**ogic

EFQM **E**uropean **F**oundation for **Q**uality **M**anagement

EHP **E**lectron **H**ole **P**air
A pair of charge carriers consisting of one electron and the (positively charged) hole left behind in a semiconductor layer, see also boundary layer, majority charge carrier, recombination.

EI **E**lectron Impact **I**onization.

EIA **E**lectronic **I**ndustries **A**ssociation of America, Washington DC.

EIAJ **E**lectronic **I**ndustries **A**ssociation of **J**apan

EICTA **E**uropean **I**nformation and **C**ommunications **T**echnology **I**ndustry **A**ssociation

EIL **E**lectron **I**njection **L**aser.

Elasticity
The reciprocal of electrical capacity.

ELD **E**lectro**l**uminescent **D**iode

Electroluminescence cold light
The emission of light under the influence of through current, which can occur with some materials under certain conditions, see also ELD.

Electromigration
A self-diffusion process in semiconductors with a high current density as a result of the intensive interaction of the electrons with the lattice. Material is transported in the conductive tracks in the direction of the flow of electrons, which is dependent on temperature and current density.

Electron trap
A small semiconductor structure in which electrons can be trapped by quantum mechanical effects. This gives rise to a memory component, see also ETOM.

Electronic
The property of physical processes based on electronics.

Electronic printed circuit board
A plastic board with electronic components and lines and circuits that have been fitted using a special procedure. See also PCB; board.

Electronics
The practical application of the physical effects of electrons in semiconductors, gas and vacuum.

Electrosmog
The "contamination" of the environment with magnetic and electrical fields by current-carrying conductors. The health consequences of electrosmog are disputed.

ELED **E**dge Emitting **LED**

ELPC **E**lectro**l**uminescent **P**hoto**c**onductive

ELSI **E**xtremely **L**arge **S**cale **I**ntegration
For electronic components, see also ULSI, VLSI.

EMBH **E**tched **M**esa **B**uried **H**eterostructure

EMC **E**lectro**m**agnetic **C**ompatibility
This designates the capability of an electrical circuit to cut itself off from external electromagnetic radiation and not itself emit any disturbing electromagnetic radiation, see also shielding, EMS.

E-MESFET **E**nhancement-Mode **Me**tal **S**emiconductor **FET**
Self-locking, see also D-MESFET.

EMI **E**lectro**m**agnetic **I**nterference, see also EMC

Emission
The release of radiation or charge carriers into the environment.

Emitter
The layer of a bipolar transistor from which the majority charge carriers are emitted into the base when a control voltage is applied.

Emitter follower
A transistor circuit in which the output signal is generated by current flow in the emitter resistor, see also collector follower.

EMR **E**lectro**m**agnetic **R**adiation

EMS **E**lectro**m**agnetic **S**usceptibility
See also EMC.

Emulation
The reproduction of a system. It is often used as a test aid in computer technology. In contrast to simulation, which takes place totally separately from the system to be investigated, emulation takes place on the system itself.

Emulator
A technical device used for emulation.

Enable standby, deallocation, release.

Enhancement-type FET
A self-locking FET in which current only flows once a control voltage has been applied, see also depletion-type FET.

EOL **E**nd **o**f **L**ife
See bathtub curve.

EOLM **E**lectro-**O**ptic **L**ight **M**odulation

EOS **E**lectrical **O**ver**s**tress

Epitaxy
A semiconductor layer with a homogeneous crystal structure and varying doping. From the Greek "epi" (on) and "taxis" (arrangement).

EPLD **E**rasable
(also: **E**lectrically) **P**rogrammable **L**ogic **D**evice. See also PLD.

EPROM **E**rasable **P**rogrammable **ROM**
Using UV light. Deleting deletes all data contents. The deletion of individual cells is not possible.

EQA **E**uropean **Q**uality **A**ward

Equivalent circuit alternate circuit
A simplified hypothetical circuit that has essentially the same properties as the relevant physical circuit.

ERA **E**lectrically **R**econfigurable **A**rray

ERC **E**lectrical **R**ule **C**heck
To check compliance with electrical design rules in the CAE, see also DRC.

ES **E**uropean **S**tandard
See also CEN, CENELEC.

ESD

1. **E**lectro**s**tatic **D**ischarge. The sudden draining of electrostatic charge. Even with small charges, it poses a considerable risk to small semiconductor structures, in particular MOS structures. It is therefore essential to take precautions when dealing with unprotected semiconductors.
2. **E**lectrostatic **S**ensitive **D**evice.

ESDS **E**lectrostatic **D**ischarge **S**ensitive **D**evices

ESIA **E**uropean **S**emiconductor **I**ndustry **A**ssociation

ETOM **E**lectron **T**rapping **O**ptical **M**emory

ETSI **E**uropean **T**elecommunication **S**tandards **I**nstitute

eV **E**lectron **V**olt
The energy released/stored by an electron when passing through a potential jump of 1V. It corresponds to $0.16022.10^{-18}$ Nm.

EVÖ **E**lektrotechnischer **V**erein **Ö**sterreichs, Wien
The Austrian Electrotechnical Society, based in Vienna.

Extinction
Enforced thyristor switch-off.

1. The mutual cancellation of two oscillations by interference.
2. The suppression of commutating current spikes.

Extrinsic
Not intrinsic, see also intrinsic.

Eye safety
If a continuous optical wave power of more than 1 mW reaches the retina of the human eye, it can cause (with visible

radiation) not only glare, but also permanent damage. Since laser diodes emit this type of power, protective measures must be taken. These depend on the respective laser safety class, which in turn depends on the power emitted and the wave length (with invisible radiation, there is no protective pupil reflex).

Eyring model
A model used for testing components that accelerates the failure rate through increased temperature and electrical load, see also Arrhenius model, Peck model.

Fab
Abbreviation for fabrication facility. All fabrication facilities for semiconductor circuits.

Failover
A mode of operation in which a backup device always runs in the background as a "hot" reserve.

Failure
The temporary or permanent loss of a major function. There is a failure if an originally fault-free component or device transfers to a state in which it no longer fulfils the requirements. A failure can be the result of an internal hardware or software fault, but it can also have external causes (power failure or similar), see also reliability.

Failure frequency
The number of failures within a specified period of time. The failure frequency is one of the benchmarks for the reliability of a device or component, see also reliability.

Failure quota
The proportion of a total volume of components that fails during burn-in.

Failure rate
The proportion of components that becomes inoperative within a specified period of time. The failure rate is generally expressed in "fit" (failures in time). 1 fit = 1 failure/10^9 component hours.

Fall time decay time, release time
The length of time between triggering a switch-off process and reaching the switched-off state. It depends on both the properties of the switch and the signal level definition. Generally it is 10% of the initial value.

Fallback
An emergency solution or emergency operation in the event that not all functions are available, e.g. the possibility of switching over to manual operation if an automatic device fails or maintaining disturbed communication by reducing the transmission speed.

FAMOS **F**loating Gate **A**valanche **MOS**
A field effect transistor with an electrically floating gate and avalanche injection. The basic component of flash memories.

Fan out
The power rating of a driver output in terms of the number of inputs that can be connected. It therefore depends on the condition of these inputs and is not defined in absolute terms.

FC **F**lip **C**hip
A monolithic IC without connecting wires that is mounted on a hybrid circuit using a special interconnection technology.

FCHI **F**lip **C**hip **H**ybrid **I**ntegration

FCT **F**ast **C**osine **T**ransform, see also FFT.

FDDI **F**iber **D**istributed **D**ata **I**nterface
A data interface for distributed connection to fiber optic networks as per ANSI X3T9.5 in ring topology with 100 MBit/s, see also CDDI.

Feature size
The minimum width of a semiconductor structure that is reduced each year. Examples: chips from the 300 mm wafer line of Infineon – 110 nm, for GaAs chips – 0.5 and 0.8 µm. The standard for logic chips nowadays is 0.18 µm, as of 2004 Infineon will produce 90-nm memories.

Feedback
A basic requirement for closed control loops. The acknowledgment from the output signal and the resulting possible comparison with the target value permits

correction of any error signals using actors.

FEFET **F**erro**e**lectric **FET**
An FET with a ferroelectric insulating layer between channel and gate.

FEL **F**ree **E**lectron **L**aser

FEM **F**inite **E**lement **M**ethod
An algorithmic analysis method whereby the complex object to be investigated is virtually dismantled into any number of discrete parts. These individual parts are then investigated separately. The results are combined successively in accordance with certain rules. The method has extensive computational requirements, but this is no longer a problem with modern computers.

FET **F**ield **E**ffect **T**ransistor
In contrast to bipolar transistors, this transistor does not work more or less with conductive junction layers. Instead, the control voltage applied at a high-ohmic control electrode (gate) generates an electrical field in the conductive channel. This field effect influences the conductivity of the channel, whose connections are known as source and drain, see also MOSFET, IGFET, MESFET, MODFET, JFET.

FFC **F**lexible **F**lat **C**onductor

FFT **F**ast **F**ourier **T**ransform
A mathematical operation described by the French mathematician Fourier that enables a periodic oscillation to be broken down into individual harmonic oscillations. Such operations can be performed by DSPs in real time, see also DFT.

Fiberoptics
The theory and technology of light propagation in optical fibers through total reflection, see also optical waveguide.

FIFO **F**irst **i**n, **F**irst **o**ut
Whatever goes in first arrives at the front of the queue and is the first to be processed. A potential batch-processing principle, see also FILO.

File
A complete, designated collection of information, e.g. a program, a data record used by a program or a document created by the user. The file is the fundamental unit of storage that enables a computer to differentiate between individual sets of information.

File protection
Securing files against unwanted access or damage.

Film resistor
A passive component that derives its electrical resistance from a conductive film. This film is applied to the resistor body using special manufacturing techniques.

FILO **F**irst **in**, **L**ast **out**
A primitive batch principle: whatever goes in first arrives at the end of the queue and is the last to be processed. A potential batch-processing principle, see also FIFO.

Filter

1. A device with selective admission or rejection characteristics. Mechanical filters only allow certain materials through (or reject them), electrical and optical filters only certain wavelengths.
2. An application add-on program that can convert external file formats so that they can be read in and further processed by the application.

Fine pitch
In component mounting technology, typically less than 0.1 mm.

Firmware
A program that is permanently linked to a piece of hardware, e.g. the BIOS.

Fit **F**ailures **in** **T**ime
1 fit = 1 failure/10^9 hours, see also failure rate.

Fixed-point number
A decimal number with a fixed number of decimal places, i.e. the opposite of floating-point number.

Flag
A variable that indicates whether a certain state exists, see also semaphore.

Flake
A fragment of a semiconductor chip.

Flash
An electrically alterable ROM, see also EAROM. A voltage pulse ("flash") deletes a delimitable data area of the ROM that can then be rewritten, see also FAMOS.

FLC Ferroelectric **L**iquid **C**rystal
Also FELIX.

Flip-flop bistable trigger circuit
A basic component of digital technology, also for static memories with random access, see also SRAM.

Floating at zero potential

Floating point floating decimal point

Floating-point number
The internal representation of binary numbers in the processor that allows different numbers of decimal places. Changing the exponent migrates (floats) the decimal point to the corresponding position, see also fixed-point number.

FM Frequency **M**odulation
A modulation procedure whereby the frequency of a carrier signal is varied. The frequency change is directly related to the instantaneous value of the useful signal, see also AM, PM, modulation.

Foundry
The outsourcing of wafer production without design. A subcontracted company.

Four-pole model
A model with four poles to represent a transistor. The four-pole model describes all the circuit attributes of the transistor using the relationships of the instantaneous values for input/output current/voltage at the four poles.

FP

1. **F**abry-**P**erot. An optical resonance effect recorded by Fabry and Perot and used in laser diode technology, see also FPI.
2. **F**ixed **P**art. A DECT measurement technique, see also PP.

FPC Flexible **P**rinted **C**ircuit
See also RPC.

FPGA Field **P**rogrammable **G**ate **A**rray
A logic module that can be programmed by the user, see also FPLA.

FPLA Field (or Fuse) **P**rogrammable **L**ogic **A**rray
See also FPGA.

FPM Fast **P**age **M**ode
A page-by-page method of accessing RAM resulting in a rapid flow of data. The enforced sequential polling is opposed to the principle of random access but hardly has any effect in practice since most data is stored sequentially.

FPU Floating-**P**oint **U**nit
A CPU area for computation with floating-point numbers.

FQFP Fine Pitch QFP

FRAM Ferro-Electrical **RAM**

FREDFET

Free **R**unning **E**xtinction **D**iode **FET**.

Fast **R**ecovery **E**pitaxial **D**iodes **FET**.

Freewheeling diode
Also: suppressor diode. The freewheeling diode offsets voltage spikes outside the supply voltages and derives commutation currents.

Front end

1. In semiconductor manufacturing, the preliminary production stages or prefabrication.
2. A computer or a processing unit that produces and manipulates data before it is received by another processor.
3. In communications technology, a computer between the transmission lines and a main computer (host) that relieves the host of administrative tasks in terms of data transfer, see also back end.

FSA Fabless **S**emiconductor **A**ssociation, New York

Full-custom IC
An integrated circuit developed individually for the customer. Full-custom ICs are

normally developed by the semiconductor manufacturers.

Fuzzy logic
A technology that causes control flows based on approximate rules, i.e. dependent on probabilities.

GaAs
The chemical symbol for **Ga**llium **Ar**senide. A semiconductor material with highly mobile charge carriers that is very well suited to applications in the HF field and to Hall sensors.

Gain amplitude factor

GAL **G**eneric **A**rray **L**ogic

Gate

1. In electronics, this generally denotes a control electrode that, when voltage is applied, either sets a current channel into a conductive state or blocks it.
2. The basic structure of the gate array consisting of 2 to 8 transistors.
3. An electronic logic element that creates logic operations e.g. NAND, AND, OR gates, see also Boolean algebra.

Gate array
Also known as ASIC or logic array. A special type of IC that initially only represents a non-specific accumulation of logic gates. It consists of basic cells (gates) and wiring channels between them (channeled gate array technology). The gates consist of 2 to 8 transistors. Only towards the end of the manufacturing process is a level added that links the gates for a specific function. One disadvantage is that large parts of the chip remain unused.

GBL **G**iga**b**it **L**ogic
Fast logic for processing more than 1 GBit/s.

Ge
The chemical symbol for **Ge**rmanium, Element No. 32 in the Periodic Table. Ge was the material used for the first transistor. It was then mostly later displaced by silicon (Si), which has more suitable properties for most applications.

GEMFET **G**ain **E**nhanced **MOSFET**
A component of CMD technology.

Gigabit chip
An integrated circuit with approximately 1 billion memory components ($2^{30} \approx 10^9$)

GIMOS **G**ate **I**njection **M**etal **O**xide **S**emiconductor
An MOS structure in which the gate absorbs and retains charge via a type of injection, thus not suitable for non-volatile memories.

GLSI **G**iant **L**arge **S**cale **I**ntegration
With more than 100 million transistor functions, see also ULSI, ELSI.

GMCF **G**lobal **M**obile **C**ommerce **F**orum

GMOSTS **G**ated **M**etal **O**xide **S**ilicon **T**ransistor **S**witch

GMR **G**iant **M**agneto **R**esistor
A resistor that is sensitive to magnetic fields and comprises numerous extremely thin board layers made of a magnetic material.

GPS **G**lobal **P**ositioning **S**ystem

Graded-index fiber
An optical waveguide with a predominantly parabolic curve of the refractive index profile. This curve reduces the modal dispersion, see also optical waveguide.

Group delay time group transmission time, envelope delay
The period of time during which simultaneously emitted signal parts arrive at the receiving end. Ideally it should be zero, in practice, however, dispersion occurs, which in turn leads to phase shifts and loss of bandwidth.

Grove's Law
The cynical observation of Andrew Grove, a founder of Intel, that the telecommunication bandwidth doubles every hundred years. He is alluding here (in contrast to Moore's Law) to the repressive technological effect of the (then still) stringent regulations regarding telecommunications markets.

GSM **G**lobal **S**ystem for **M**obile **C**ommunication

GTO **G**ate **T**urn-**o**ff thyristor

Guard ring arcing ring
A ring structure comprising well and/or substrate contacts in a semiconductor. A guard ring reduces the resistance between the base and the emitter, which in turn makes self-triggering more difficult, see also latch up.

Gunn diode
A GaAs or InP diode in which the Gunn effect takes place, i.e. above a certain field strength the charge carriers group themselves into domains. Thus when a current flows, an oscillating drift develops, which can be used for HF oscillators with an extremely high frequency (over 10 GHz).

Gyrator

1. An electronic circuit that rotates a signal by 180° in phase.
2. Also: a circuit that provides a current proportional to the input voltage (current source).

H parameter
Also: hybrid parameter. A set of transistor parameters that combines the input impedance and output admittance parameters, see also transistor parameter, y parameter, z parameter.

Hall effect
Changing the direction of current flow using a transverse magnetic field as the result of the Lorentz force on moving electrons. This effect generates an electrical voltage that is perpendicular to the direction of current flow and the magnetic field (the so-called Hall voltage). In certain materials, this voltage is high enough to be used in metrology or sensor technology.

Hall sensor
A semiconductor based on the Hall effect, see Hall effect.

Hard error permanent fault
See also soft error.

Hardware fault
A fault that is based on the hardware.

Hardware trap
A stop condition when testing (debugging) microcontroller systems.

Hardware
The material part of a computer. It denotes all the equipment but without the data and programs, the latter being known in turn as software, for the purpose of distinction.

Hard-wired
Permanently wired processing units on a microcontroller.

Harmonic distortion
A particular form of distortion that results from the non-linear behavior of many electronic components. According to the Fourier analysis, entire multiples of the basic oscillation can arise and/or, the other way round, any non-harmonic periodic oscillation can be broken down into several harmonic oscillations.

Harvard architecture
A computer architecture which stores the data and programs in different memory areas, see also von Neumann architecture.

HBT **H**eterojunction **B**ipolar **T**ransistor
The base of this transistor is being changed such that there is a junction between different materials. Due to different effects, the transistors are faster. A famous proponent is the SiGe HBT.

HDGA **H**igh **D**efinition **G**ate **A**rray

HDL **H**ardware **D**escription **L**anguage
A programming language to describe (integrated) circuits; it supports the concurrence of electrical processes in particular. See also Verilog, VHDL.

HDLC **H**igh **L**evel **D**ata **L**ink **C**ontrol
An ISO-standardized bit-oriented, synchronous protocol for the transfer of information in layer 2 of the ISO/OSI model. The transfer involves frames with a variable data quantity and a particular organizational structure.

Heat sink heat slug

Heat slug
A component that absorbs the heat generated by an electrical component and dis-

tributes it to prevent overheating. Heat slugs are generally made of metal and equipped with cooling fins.

HEMT **H**igh **E**lectron **M**obility **T**ransistor
An extremely fast depletion layer FET in a heterostructure. The FET channel consists of a quantum well, which produces a two-dimensional cloud of extremely mobile electrons, see also HJBT.

Hertz
A unit of frequency. 1 Hz = 1 oscillation per second. Named after Heinrich Hertz, a pioneer of broadcasting technology.

Heterojunction
Also heterostructure. A boundary layer between two semiconductor materials, e.g. AlAs/GaAs.

Hexadecimal system
A method for representing numbers on the basis of 16. Each digit in this system can be represented with 4 bits. This makes the hexadecimal system suitable for data processing. For numerical representation, the digits 0-9 and the letters A to F are used with the additional "H".

HEXFET **Hex**agonal Cell **FET**

HF **H**igh **F**requency
A frequency range that is not clearly defined whose lower end represents the radio frequencies according to most definitions.

HFET **H**eterostructure FET

HIC

1. **H**igh Voltage **I**ntegrated **C**ircuit.
2. **H**ybrid **I**ntegrated **C**ircuit, see also hybrid.

Hidden refresh
With DRAMs, it aims to prevent the enforced pause caused by a refresh in normal operation to impact on the system throughput. As with the CBR refresh, the counter address is used as the refresh address. What is different, however, is that with the read cycle for example, the output remains active whilst a refresh cycle is launched.

HIFET **H**eterojunction **I**on-Implanted **F**ield **E**ffect **T**ransistor
See also HJBT.

HIGFET **H**eterostructure **I**nsulated **G**ate **F**ield **E**ffect **T**ransistor
An IGFET whose properties correspond both to the MESFET as well as the MODFET.

High performance
A significant criterion, although not clearly defined, for the speed of computer systems and their individual components, see also benchmark, Whetstone

High-pass filter
A filter that only slightly attenuates frequencies above its critical frequency, see also low-pass filter.

HJBT **H**etero**j**unction **B**ipolar **T**ransistor
The bipolar counterpart of the HEMT.

H-LPBGA **H**igh Temperature **L**ow **Pro**file **BGA**

HMOS **H**igh Density **M**etal **O**xide **S**emiconductor
Used for FETs.

HNIL **H**igh **N**oise **I**mmunity **L**ogic

Hole
A vacant electron position in the valence structure of semiconductor crystals. The hole is a moveable gap that behaves like a free charge carrier.

Hot carrier diode
Another name for Schottky diode.

Hot electrons
Free electrons with a kinetic (thermal) energy significantly above kT (k is the Boltzmann constant, T is the temperature in K).

Hot spot
A certain point in a semiconductor chip that becomes very hot due to a concentration of internal power dissipation. Semiconductor developers always face the challenge of aniticpating these hot spots and countering their occurrence by taking appropriate measures regarding chip layout. There are numerous simulation options available in this context, see also FEM.

HTL High **T**hreshold **L**ogic

HTML Hyper**t**ext **M**arkup **L**anguage
A markup language for text pages on the Internet that supports cross-references (links) in addition to formatting.

HVIC High **V**oltage **IC**

Hybrid
In the broader sense, the merging of two technologies.

I²L Integrated **I**njection **L**ogic
For bipolar high-scale integration. See also IIL.

IC Integrated **C**ircuit
The first IC was developed by Jack Kirby in 1958 in the TI laboratory. Robert Noyce and Gordon Moore (two of the founders of Intel) invented a planar IC independently in 1959 at Fairchild.

ICC Internal **C**urrent **C**onfinement
A geometrically enforced current flow within a semiconductor structure.

ICE In-**C**ircuit **E**mulator
i.e. an equivalent circuit for the purpose of function simulation, see also emulation, ICT.

ICT In-**C**ircuit **T**esting
Testing using the target circuit, see also ICE.

IEA International **E**lectrical **A**ssociation

IEC International **E**lectrotechnical **C**ommission
The international committee responsible for drawing up electrotechnical standards.

IEEE Institute of **E**lectrical and **E**lectronic **E**ngineers, Inc., New York (Often pronounced "I triple e"). See also IRE. Similar to the VDE in Germany. It works on legislation and standardization and proposes standards.

IEEE-488
The standard for a parallel interface, mainly for the computer connection of measuring equipment, based on a development by Hewlett Packard. Also: General Purpose Interface Bus (GPIB) or IEC bus (with other plug connectors).

IEGT Injection **E**nhanced **G**ate **T**ransistor
A power switch with a lower saturation voltage than GTO.

IFL Integrated **F**use **L**ogic
A flexible customized gate array concept.

IFU Instruction **F**etch **U**nit

IGBT Insulated **G**ate **B**ipolar **T**ransistor
A MOSFET developed by Siemens with conductivity modulation. It combines the advantages of MOSFETs and bipolar transistors. It has a particularly low on-state resistance and is therefore well suited for use as a power switch.

IGCT Integrated **G**ate **C**ommutated **T**ransistor

IGFET Insulated **G**ate **FET**
A field effect transistor with a very high ohmic (up to one billion GW) insulation layer separating the gate and source drain channel. This insulation is often made of metal oxide, see MOSFET. The extremely thin layer is very sensitive to ESD so that components in this technology (also ICs) must be protected against static charge and handled with particular care.

IGT Insulated **G**ate **T**ransistor
A power electronics component in CMD technology, see also IGBT.

IIL
Integrated **I**njection **L**ogic. A bipolar logic module family. See also: I²L.

ILD Injection **L**aser **D**iode
A laser diode with injection of charge carriers to pump the active area.

Illuminance illumination, luminance
The luminous flux per unit area, expressed in Lux (= lumens/m²). The human eye perceives illuminance (with a given wavelength) subjectively as brightness, see also photometry.

IMOS Ion-implanted **MOS**
The manufacture of MOS structures with a self-adjusting gate using ion implantation.

IMPATT Impact **A**valanche **T**ransit **T**ime

A semiconductor diode in which an avalanche effect occurs in places so that the current drops as the forward voltage rises. These diodes thus exhibit, in a similar way to the tunnel diode and the BARITT diode, a negative differential resistance and are used predominantly in microwave circuits, see also TRAPATT.

Impedance
The alternating current resistance of a component or circuit.

Implementation
The real, functioning realization of an idea or invention.

In phase
When two signals with the same frequency tally also in their phase relation.

Increment
To increase a numerical value by the value 1, see also decrement.

Indexed sequential
Data search or access via the data record key. This is used with databases and similar programs involving data being input and stored generally unsorted.

Inductance
The inductive resistance of a circuit.

Induction
1. A method for logical reasoning that is based on a finite number of legalities and that determines a general conclusion, see also deduction.
2. The generation of an electrical voltage by a change in the magnetic field strength.

Inductivity
A measure for self-induction, expressed in H (Henry).

Infant mortality
Early failure, see bathtub curve.

Information technology
The science concerned with the nature and processing of information.

Infrared
Also: ultrared. Optical radiation with a wavelength between about 780 nm and 400 µm perceived by humans at most as thermal radiation but invisible to them.

Initialize
To put into the initial state, see also boot.

Input entry

Input impedance
The alternating current resistance to the electric current at the input of a circuit, see also output impedance.

Installation
1. Setting up the hardware components of a computer system.
2. Fitting additional hardware to the computer.
3. Copying software on to the hard drive of a computer and simultaneously adapting it to system operation.

Instruction set command set
The total number of machine instructions that a processor understands and can execute, and/or that are supported by a programming language, see also assembler, microcode.

Insulator
A material with very low electrical conductivity.

Integrated injection logic
A type of logic circuit with only NPN and PNP transistors. These circuits switch relatively quickly and require little energy and space. Also known as merged transistor logic.

Integrator
An analog circuit, generally with operational amplifier, whose output signal corresponds to the integral of the input signal.

Interleaving
Also: interlacing. Skip, jump. It designates a nesting procedure that initially addresses not the adjacent point or address, but the next but one or two, etc. The adjacent point is only addressed in a new cycle. In storage technology, for instance, the procedure results in a higher throughput; in screen technology, it produces less flickering, see also line skip.

Intermittent
Interrupted, at irregular intervals. Intermittent faults, for example, are non-

reproducible faults that are difficult to isolate due to their unpredictable nature.

Interpreter translator
In data processing, a program that converts the instructions of a programming language into machine code and executes them immediately.

Interrupt discontinuation, abort
With microcontrollers, a signal that stops the intended program run and initiates a sub-routine prepared for this scenario. After executing the routine, the program continues as scheduled, see also trap.

Intrinsic i-type
Also: immanent. It designates semiconductor material that is either lightly doped or completely pure and thus has only a few free charge carriers. Intrinsic semiconductor material therefore has a high electrical resistance.

Inversion layer
An oxide layer in which electrons with inverted polarity in relation to the charge carriers in the substrate material form an electron layer.

Inverter
A digital circuit that outputs the negated value of the input signal.

IOC **I**ntegrated **O**ptical **C**ircuit

IOLC **I**ntegrated **O**ptical **L**ogic **C**ircuit

IOM **I**SDN-**O**riented **M**odular interface
A standard interface defined by Siemens that enables the ISDN user to devise the best configuration for his application.

Ion
An atomic nucleus without any associated electrons, and therefore positively charged.

Ion implantation
A process whereby high energy is used to implant ions in a semiconductor substrate, see also doping.

IP **I**nstruction **P**ointer

IPC **I**nstitute for Interconnecting and **P**ackaging Electronic **C**ircuits

IPG **I**n-**P**lan **G**ate
A CMOS alternative.

IPLSI **I**ntelligent **P**ower **L**arge **S**cale **I**ntegration

IrDA **I**nfra**r**ed **D**ata **A**ssociation
An industry organization set up in 1993 for providers of computers, components and telecommunications equipment. It lays down standards for infrared communication between computers and peripheral equipment.

IRED **I**nfra**r**ed **E**mitting **D**iode
A light emitting diode or luminescent diode. It consists essentially of one PN junction where the excess charge carriers injected during all-pass operation recombine in the n- or p-layer and emit a long-wave photon (luminescence). There are also diodes that emit visible radiation (LED).

IRLD **I**nfra**r**ed **L**aser **D**iode

ISDN **I**ntegrated **S**ervices **D**igital **N**etwork
The public telephone network with digital transmission technology and thus the possibility of (simple) implementation of services. The basic channel offers a transmission rate of 64 Kbit per second. All services for voice, text and data transfer are offered in a single communications network with a single network access.

ISFET **I**on-**S**ensitive **FET**
An FET with a special insulating layer.

ISL **I**ntegrated **S**chottky **L**ogic
A logic family for customized ICs.

ISO **I**nternational **S**tandardization **O**rganization, Geneva
See also OSI.

ITG **I**nforma**t**ionstechnische **G**esellschaft
The Information Technology Society of the VDE.

ITO **I**ndium **T**in **O**xide
A transparent material for optoelectrical components, e.g. LEP.

ITU **I**nternational **T**elecommunication **U**nion
The worldwide umbrella organization for telecommunications carriers and the industry as a whole.

ITU-T
Formerly CCITT, the **T**elecommunication **S**tandardization **S**ector.

JCCD Junction **C**harge **C**oupled **D**evice
A special type of CCD that uses non-conductive PN junctions instead of MOSCs as the capacitor.

JEDEC Joint **E**lectronic **D**evice **E**ngineering **C**ouncil
The US standardization organization for electronic components.

JFET Junction **FET**
A barrier-layer FET in which the space charge zone of a blocked PN junction constricts the current channel and thus influences the current flow. It was proposed by W. Shockley in 1952 and manufactured for the first time in 1953 by G. C. Dacey and I. M. Ross. The precursor to the IGFET.

JIT Just in **T**ime
A byword for lean production.

Jitter
In particular the irregularly fluctuating zero crossing point of a periodic signal. Excessive jitter leads to synchronization problems.

JMOS Joint Gate **MOS**
A CMOS variant in which a PMOS and NMOS transistor share a control electrode (gate).

JOERS Joint **O**pto **E**lectronic **R**esearch **S**cheme

Josephson effect
This occurs when two superconducting materials separated by an insulator are brought close together. The electric current can jump or tunnel through the insulation gap.

Joy's Law
The observation of Bill Joy, a founder of Sun Microsystems, that the computing power of processors increases exponentially with time. RISC processors have complied with this legality for more than ten years; with CISC processors, development is somewhat slower. Also attributed to Joy: "The smartest people in every field are never in your own company".

Jumper
A wire strap to activate a particular function.

Junction butt joint, connecting point
In semiconductors, the boundary layer between p- and n-conducting material.

KEMA
NV tot **K**euring van **E**lektrische **M**aterialen, **A**rnhem. The official body in the Netherlands for testing electrical materials.

Kirchhoff rules
Two basic rules of electricity: firstly that the sum of all voltages is zero with a complete loop of a closed circuit, and secondly, that the algebraic sum of the currents at any point in the individual branches of a circuit is likewise zero.

Kitting
Putting together a delivery comprising individual components that together produce a meaningful component kit which can generally be used to manufacture a complete product.

LAE Large **A**rea **E**lectronics

Latch hold latch

Latch up
An unwanted effect in semiconductors caused by the irreversible through-connection of parasitic thyristor structures in CMOS circuits. Latch up can lead to the destruction of the component and can only be eliminated by removing the supply voltage.

Layer model see OSI

Layer

1. In semiconductor technology, a thin, active or passive area that in turn separates or connects other layers. An IC generally consists of very many layers.
2. In electronics or optoelectronics, a thin area that has an electrical or optical separation and/or protective function, for example.

3. In communications technology, a logically functional, hierarchically assigned area.
4. A display type of graphics programs that show associated elements from drawings on different virtual levels that can be hidden or highlighted where necessary.

LBIC **L**aser **B**eam **I**nduced **C**urrent

LCA **L**ogic **C**ell **A**rray

LCC

1. **L**ateral **C**harge **C**ontrol.
2. **L**eaded **C**hip **C**arrier.
3. **L**eadless **C**hip **C**arrier.
4. **L**iquid **C**rystal **C**ell.
5. **L**ogic **C**ontrol **C**ell.

LCCC **L**eadless **C**eramic **C**hip **C**arrier

LCD **L**iquid **C**rystal **D**isplay
This uses liquid crystals with a polar molecular structure that are encased as a thin layer between two transparent electrodes. If an electrical field is applied to the electrodes, the molecules align themselves with the field and form crystalline arrangements that polarize the penetrating light. A polarization filter arranged in lamellar form above the electrodes blocks the polarized light. A cell (pixel) that is selectively activated in an electrode grid and that contains liquid crystals generates black spots.

LCDTL **L**ow **C**urrent **DTL**

LCI **L**ateral **C**urrent **I**njection
In optosemiconductors.

LCOS **L**iquid **C**rystal **on** **S**ilicon
A technology for microdisplays.

LD **L**aser **D**iode
A semiconductor structure made from material from groups III-V of the Periodic Table of the Elements. It behaves like an ELED until the threshold current is reached. A large injection current produces an amplification effect: the LED becomes the laser.

LDD **L**ightly **D**oped **D**rain

LDR **L**ight **D**ependent **R**esistor

Leadframe chip carrier
A right-angled metal frame with pin wiring and linking elements in between for package mounting. The wiring is punched during mounting. Because of its appearance, it is sometimes also called "spider".

Leading edge
The first section of an electrical pulse, see also trailing edge.

Leakage current
The current flow through an insulator as a result of material defects or impurities, see also creepage current.

Lean production
A collective term and/or byword for production with maximum efficiency, see also JIT.

LED **L**ight **E**mitting **D**iode
Luminescent diode. It consists essentially of one PN junction where the excess charge carriers injected during all-pass operation recombine in the n- or p-layer and emit a long-wave photon (luminescence). There are also diodes that emit infrared radiation (IRED).

LEOMA **L**aser and **E**lectro-**O**ptics **M**anufacturers' **A**ssociation
A US association, formerly LAA.

LEOS **L**asers and **E**lectro-**O**ptics **S**ociety
An IEEE society.

LEP **L**ight **E**mitting **P**olymers
Polymers (plastics) whose injected charge carriers (excitons) emit light by recombining.

LFBGA **L**ow **F**ine **P**itch **BGA**

LGA **L**and **G**rid **A**rray

License
A usage authorization, for example for a book or software. The license is generally acquired by buying a copy.

LIF **L**aser **I**nduced **F**luorescence

Light
Electromagnetic radiation perceivable by the human eye with a wavelength of around 380 to 700 nm. Outside this range, it can only be perceived in exceptional cases. The human eye is most sen-

sitive to a wavelength of around 550 nm, which is perceived as green.

Light barrier optical obstacle
In practice, a light beam that irradiates a photodetector. Objects or people interrupting the light beam interrupt the photocurrent, which produces a signal.

Limiter
A circuit that ensures that a certain amplitude is not exceeded. Limiting can be via clipping, but also as a result of less abrupt continuous limiting procedures.

Line width pattern width

1. In microelectronics, the dimension of the smallest active structure, also called feature size.
2. The spectral width of a monochrome light source, e.g. a laser.

Link connection
A data channel in communication technology and/or a link to related information in data technology.

Little-endian see big-endian

LIW **L**ong **I**nstruction **W**ord

LLCC **L**eadless **C**hip **C**arrier
See also LC, LCCC.

LNA **L**ow **N**oise **A**mplifier

Load balancing load sharing

Load
A process whereby data stored on the data carrier is transported to the working memory so that it can be read or processed by the computer.

LOC

1. **L**arge **O**ptical **C**avity. A semiconductor laser structure with a wide optical resonator.
2. **L**ead **on** **C**hip. An assembly technique for semiconductor chips whereby the leadframe is mounted on the silicon chip and not the other way round.
3. **L**ine **of** **C**ode. A program line or an instruction line of a source program.

LOCMOS **L**ocally **O**xidized **CMOS**
An oxide-insulated CMOS technology whereby the individual structures are electrically insulated by local oxidation of the silicon.

LOCOS **Loc**al **O**xide of **S**ilicon
An SOI procedure in which small, insulated substrate structures are produced by specific path etching.

Logic
A scientific method for drawing or making clear, non-contradictory conclusions or decisions.

Logic algebra see Boolean algebra

Logic analyzer
An intelligent device that can be used to check the correct function of logic circuits and processors to a large extent automatically and in a machine-oriented fashion. This includes monitoring the bus signals during the program run, stopping execution if a given memory location is being read or written, and tracing a certain number of instructions if execution has stopped for some reason, see also break-point, debugger.

Logic circuit
An electrical functional block that is based on the binary representation of information and interlinks this information, see also analog circuit, storage circuit.

Logical operation
A logic operation based on Boolean algebra.

Logistics materials management
For example, the system used during product manufacture for transporting and assembling the individual components.

Loop cycle

Lorentz force
A force that is exerted on a charge when the latter traverses a magnetic field. The Lorentz force is perpendicular both to the direction of current flow and to the direction of the magnetic field.

Loss attenuation

Lot
A quantity of components that have passed through the manufacturing process together. Changes in the manufactur-

ing process affect all parts of a batch. Lot also called batch.

Low-cost
Implies for hardware or software products and/or generally with processes, that little expense is required. Low-cost can be an indication of economic efficiency.

Low-pass filter
A filter that only lets frequencies up to a certain level through. Above this level it becomes a reject filter, see also high-pass filter.

LPLD **L**ow **P**ower **L**aser **D**iode
Generally not harmful to the eye.

LQ **L**imiting **Q**uality
This specifies the defective lot fraction as a percentage of a delivered quantity, whereby it is still accepted with 10 percent probability of acceptance.

LQFP **L**ow **P**rofile **Q**FP

LRU **L**east **R**ecently **U**sed
A storage strategy of a cache to speed up data access. The least recently used bytes are replaced with information that is currently required.

LSB **L**east **S**ignificant **B**it
The binary digit with the lowest value in a number, see also LSD, MSB, MSD.

LSD **L**east **S**ignificant **D**igit
The digit with the lowest value in a number, see also LSB.

LSI **L**arge **S**cale **I**ntegration
At least 100 to 1000 transistor functions or other elementary circuits on an IC chip, see also ELSI, GLSI, MSI, VLSI, ULSI.

LSL **L**ow **S**peed **L**ogic
Slow but immune to interference.

LSTTL **L**ow Power **S**chottky **TTL**
A fast TTL family with integrated Schottky diodes.

LTT **L**ight **T**riggered **T**hyristor

Luminance illuminance

LVC **L**ow **V**oltage **C**MOS
A CMOS with a 3.3 V supply voltage.

LVTTL **L**ow **V**oltage **TTL**
An interface for semiconductor modules with a 3.3 V supply voltage.

Machine code
Instruction coding that can be executed by a processor directly, but that is barely understood by humans. The machine code consists of a sequence of logical ones and zeros. It is only indirectly associated with the underlying programming code. The latter must first be compiled or assembled, whereby it is converted into machine code based on certain rules. Machine code is also known as object code or machine language.

Machine instruction
A machine-oriented instruction that can be executed without translation by the processor.

Macro
A small, very application-specific program.

Macroassembler
A programming language that is still very close to the machine code but that uses symbols humans can understand.

Main board motherboard
The electronic board in a device, e.g. a computer, which houses the most important electronics. The main board is often supplemented by additional boards that are either inserted or interconnected via a bus interface and/or other junctions.

Main memory
Also: working memory or RAM.

Majority charge carrier
Charge carriers that originate from (specifically added) foreign atoms and that change the conductivity of a semiconductor material. An intrinsic semiconductor material (see also intrinsic) has a low and balanced concentration of free charge carriers; these are either free electrons or positively charged “holes”, i.e. vacant electron positions. The conductivity is low. If dopants (donor atoms, foreign atoms) with an excess or lack of electrons are added, the semiconductor material becomes p- or n-conducting. The concentration of minority charge carriers is auto-

matically lower since they can recombine with the abundant majority charge carriers, see also minority charge carrier.

Markup language
Such as, for example, HTML Hypertext Markup Language, see also SGML, HTML.

MASFET **M**etal **A**lumina **S**ilicon **FET**
An FET with an aluminum oxide layer between gate and channel. The basic component of EPROM modules.

Master
1. In circuit technology, a leading circuit (for example a master oscillator) on which other circuits depend, see also slave.
2. A preassembled silicon wafer for gate array manufacture.
3. Initiator. A component that can automatically trigger actions such as data transfer via the PC bus, see also slave.

Matching oil
An oil with an optical density roughly identical to that of glass, which is embedded between the adjacent end faces of optical fibers to prevent optical transition to air, which would cause reflections.

Maximum rating
See critical load and derating.

MBE
1. **M**olecular **B**eam **E**pitaxy. A layering procedure using molecular beams that vapor deposit on the semiconductor material, oxide or metal in a thin, precisely measured layer.
2. **M**ulti-**B**oard **E**mulator.

MC **M**emory **C**ontroller
Integrated in the chip set of a PC. It relieves the CPU of DRAM-specific communication.

MCBF **M**ean **C**ycles **b**etween **F**ailures. See also MTBF.

MCM **M**ulti-**C**hip Module

MCP **M**ulti-**C**hip **P**ackage

MCT **MOS** **C**ontrolled **T**hyristor

MCU **M**icro**c**ontroller **U**nit

MCVD **M**odified **C**hemical **V**apor **D**eposition
A coating technology for (opto)semiconductors; a modified vapor deposition procedure, see also CVD.

MDA
1. **M**onolithic **D**iode **A**rray. A diode array on a single semiconductor die.
2. **M**onochrome **D**isplay **A**dapter. A graphics card without color support. The precursor to the Hercules graphics card launched by IBM in 1981.

MDD **M**edium **D**oped **D**rain

MDM **M**etal **D**ielectric **M**ultilayer
A structure with several conducting layers that are separated by a dielectric.

MEA **M**olded **E**lectronic **A**ssembly

Mechatronics
A core technology that combines sensor technology, actuator technology and signal processing.

Melting coupler
An optical waveguide coupler in which two fiber cores are bonded so closely together through melting that light exchange is possible, see also bending coupler.

Memory
1. In general, a medium that can store information.
2. The repository of a computer. In semiconductor technology, a binary data memory. When reading out almost all types of data memory, the content is retained. Some memory types do however lose the stored data when the supply voltage is removed (volatile memories). Some memories can only be filled with information once, see also ROM, others on multiple occasions, some of these even on a random basis, see also RAM.

MEMS **M**icro **E**lectro **M**echanical **S**ystem

Merged transistor logic integrated injection logic

Mesa structure
A rising mesa-like semiconductor struc-

ture whose ambient has been etched away.

MESC **M**odular **E**quipment **S**tandards **C**ommittee
A SEMI committee concerned with standardizing tool interfaces in semiconductor fabrication facilities. There have been very few attempts at standardization in this field although the production sites represent a huge investment volume.

MESFET **Me**tal **S**emiconductor **FET**

Metallization
The application of a metal layer to produce/increase conductivity or as protection. A common procedure in IC manufacture.

MFLOPS **M**illion **Fl**oating Point **O**perations **p**er **S**econd

MIC

1. **M**edia **I**nterface **C**onnector. An interface for FDDI to the transfer medium (optical waveguide connection standard), defined for the PMD standard. The MIC is a pluggable duplex connection that can be mechanically coded to prevent inadvertent confusion.
2. **M**icrowave **I**ntegrated **C**ircuit. Generally a hybrid or multi-chip circuit, see also MMIC.

Microchip
A small wafer with an integrated circuit.

Microcomputer
A collective term for computers whose CPU is a microprocessor.

Microcontroller
A control module based on a microprocessor that facilitates an operable computer system. Generally all essential system components are integrated in the microcontroller.

Microelectronics
A generic term for the technology of integrating electronic circuits, see also microchip.

Micrologic
A set of electronic logic circuits or instructions that are stored in binary form and define and control the switching and transfer processes within a microprocessor.

Microprocessor
The programmable module based on the mainframe computers. The microprocessor includes the arithmetical logical unit (ALU) as the computational part, instruction decoding and sequence control. An increasing number of peripheral functions are being integrated on a silicon chip, see also microcontroller.

Microprogram
The internal microcode for a processor that controls the elementary processes at the lowest processor level. The machine code must be based on it. Some systems (essentially minicomputers and mainframe computers) permit modifications to the microcode, even with installed processors.

Microwave
A radio wave with a frequency of around 1 GHz. It is used for point-to-point connections and radar purposes as well as to heat substances (microwave oven).

Miller effect
With transistors, the feedback of the collector emitter voltage via the collector base capacity to the base connection. With FETs, the analog effect between drain source and gate.

Miller integrator
An amplifier circuit in which the voltage via the feedback capacitor corresponds (approximately) to the integral of the input signal, named after the Miller effect.

MIL-Spec
As per military specification.

MIM **M**etal **I**nsulator **M**etal
A semiconductor layer sequence. A capacitive structure consisting of two conducting layers with an insulating interlayer.

Minority charge carrier
See also majority charge carrier.

MIPS **M**illion **I**nstructions **p**er **S**econd
This is a combination of the most common instructions from a statistical point

of view. The lack of standardization limits the significance. MIPS are only suitable for assessing computers of a family or comparable architecture.

MIRS **M**icromachined **I**ntegrated **R**elay **S**ystem
(An ESPRIT project).

MIS **M**etal Conductor **I**nsulator **S**emiconductor
A semiconductor layer sequence/structure comprising a conducting metal layer and semiconductor layer separated by an insulator, see also FET.

MISFET **M**etal Conductor **I**nsulator **S**emiconductor **FET**
A generic term, see also MOSFET, MESFET, MNS-FET.

MISS **M**etal Conducting **I**nsulator **S**emiconductor **S**witch

ML **M**ono**l**ayer
A semiconductor layer with the strength of an atom, as is possible nowadays in industrially manufactured ICs.

MMIC

1. **Millim**eter Wave **I**ntegrated **C**ircuit.
2. **M**onolithic **M**icrowave **IC**.

MMU

1. **M**ass **M**emory **U**nit.
2. **M**emory **M**anagement **U**nit.

Mnemonic
A memory aid, e.g. a word similar to an acronym, or a rhyme for complicated facts or longer sets of information. Most programming languages and operating systems use a mnemonic vocabulary. For example: "Ctrl" for Control, "RTR" for **R**o**t**ate **R**ight or "JMP" for **Jump**.

MNOS FET **M**etal **N**itride **O**xide **S**emiconductor **FET**
An FET in MNOS technology with a double insulation layer between gate and channel.

MNOS **M**etal **N**itride **O**xide **S**emiconductor
A semiconductor structure with a double insulation layer made of silicon nitride and silicon oxide, in which charges can be stored, see also SNOS/ SONOS.

MNS FET **M**etal **Ni**tride **S**emiconductor **FET**
An FET with a nitride layer between gate and channel.

MNS

1. **M**etal **N**itride **S**emiconductor. Like MNOS, but without the oxide layer.
2. **M**ulti-**N**umbering **S**cheme. Several numbers per telephone subscriber for different services.

Mobility
In semiconductor technology, the measure by which free charge carriers can move around in the crystal, and thus a measure for the conductivity of the semiconductor material.

MOCVD **M**etal **O**rganic **C**hemical **V**apor **D**eposition
CVD which also produces pure metal during the process, see also MOVPE.

Mode dispersion
The distribution of the signal propagation time in optical waveguides due to the different propagation modes of simultaneously injected photons, which then arrive time-staggered at the end of the fiber. Mode dispersion leads to a reduction in bandwidth as the optical waveguide length increases.

Mode hopping
A characteristic of some laser diodes whereby the oscillation mode changes spontaneously in discrete steps, e.g. due to thermal expansion or contraction of the resonance chamber.

Modem
An artificial word derived from **Mo**dulator-**Dem**odulator. A device that converts digital data into analog signals so that they can be transferred via the analog telephone network. It also includes the inverse function, as well as a program to support the communication process.

MODFET **Mo**dulation **D**oped **FET**
Extremely fast, see also HEMT, HIGFET, MESFET, TEGFET, SDHT.

Modulation
Changing or regulating the characteristics of a carrier signal that oscillates in a cer-

tain amplitude (level) and frequency (clock) so that these changes correspond to a useful signal based on a set rule. The carrier can be transmitted and with it the useful signal. Demodulation enables this signal to be regained at the end of the transmission. There are many modulation procedures, see also AM, FM, PM, PPM, PCM.

Monocrystal
A crystal with a homogeneous structure, ideally without imperfections.

Monolith
A stone or crystal that is cast in one piece to a certain extent.

Moore's Law
An observation made in 1964 by Gordon Moore (one of the founders of Intel) that the number of transistors on IC chips doubles every year. This trend continued unbroken until the end of the century, even though one and a half years has been required since the late 70s. It was only recently that Moore observed that the investment sums for a fabrication facility double every chip generation, see also Grove's Law, IC, Joy's Law.

MOS GTO Gate **T**urn **O**ff Thyristor
To switch currents of more than 1000 A and at the same time block voltages of more than 1000 Volts, the thyristor principle is still indispensable even today. A thyristor, however, once activated, can only be deactivated by interrupting the flowing current. The GTO is deactivated by a negative gate current that is around one fifth of the load current. Infineon provides a family of wafer GTOs for a voltage range between 1.8 and 4.5 kV and currents up to 3000 A.

MOS Metal **O**xide **S**emiconductor
A semiconductor structure that produces an FET structure, which is extremely well suited to high-scale integration. The semiconductor structures are separated by metal oxide layers. Depending on the polarity of the minority charge carriers, a distinction is made between NMOS and PMOS. There are also complementary structures, see also CMOS.

MOSC MOS Capacitor
The capacitor between a metal and a semiconductor with an oxide layer as a dielectric.

MOSFET
A combination of the acronyms **MOS** and **FET**, i.e. a field effect transistor based on MOS technology. It is therefore a special form of IGFET. The very high-ohmic insulation layer between the gate and source drain channel consists here of metal oxide.

MOST MOS Transistor
See also MOSFET.

MOV Metal **O**xide **V**aristor
A varistor with a symmetrical characteristic curve consisting of a metal oxide junction (mostly consisting of zinc).

MOVPE Metal **O**rganic **V**apor **P**hase **E**pitaxy
Semiconductor layer growth from the vapor of metal organic materials. The process is practically equivalent to MOCVD, but with the emphasis on the epitaxy element.

MPU Micro**p**rocessing **U**nit
The central processing unit on a microprocessor chip.

MQFP Metric **Q**uad **F**lat **P**ack

MQW Multiple **Q**uantum **W**ell
See also SQW. A semiconductor structure for laser diodes.

MRAM Magnetic **RAM**
A storage technology with two magnetic layers separated by a tunnel barrier.

MSB Most **S**ignificant **B**it
The binary digit with the highest value in a number, see also LSB, LSD, MSD.

MSD Most **S**ignificant **D**igit
The digit with the highest value in a number, see also LSB, LSD, MSB.

MSI Medium **S**cale **I**ntegration
With around 10 to 100 elementary circuits on an IC chip, see also LSI, VLSI.

MSM Metal **S**emiconductor **M**etal
A layer sequence for a special photo diode consisting of a semiconductor layer with two boundary layers to metals.

MTBF **M**ean **T**ime **b**etween **F**ailures
Therefore also: mean time to failure. A statistical variable for determining the reliability of technical equipment, see also MTTF and MCBF.

MTL **M**erged **T**ransistor **L**ogic
A bipolar logic IC with multi-collector NPN transistors and lateral PNP transistors, also I^2L.

MTNS **M**etal **T**hick **N**itride **S**emiconductor

MTOS

1. **M**etal/**T**unnel **O**xide/**S**emiconductor. A tunnel element that consists of a tunnel oxide layer with a boundary layer to metal and a boundary layer to a semiconductor.
2. **M**etal **T**hick **O**xide **S**emiconductor. A special MOS structure with a thick oxide layer.

MTTF **M**ean **T**ime **t**o **F**ailure
See also MTBF.

MTTR **M**ean **T**ime **t**o **R**epair
(also: **R**ecover).

Multilayer
Both semiconductor structures as well as PCBs often consist of many functional or interconnection layers. The multiple layers permit the circuit complexity.

Multiplex
Multiplex procedures or circuits interleave individual signals. There are space division, time division and frequency division multiplexes that are separated by a demultiplexer.

Multiprocessor system
In this system, two or more processors work closely together to a greater or lesser degree (fixed or loose coupling). The synchronization of this cooperation, generally by means of semaphor mechanisms, is of particular importance here. The purpose of multiprocessor systems is to achieve better system performance through division of labor and parallel working.

Multitasking
The simultaneous execution of more than one program on a computer.

Multithreading
The parallel processing of several program runs in one single program. The tasks can also be held simultaneously in the memory and processed.

Multiuser operation
The possibility of accessing resources such as data or peripheral equipment within a network of several work stations.

Multivibrator trigger circuit
A circuit that can have two circuit states. These can be stable, see flip-flop or monostable (also: monoflop), i.e. the circuit returns to the stable state after a while. An external trigger is always required for a state change. There are however also oscillating multivibrators, which change their state periodically without external assistance, see also oscillator.

MVL **M**ulti-**V**alued **L**ogic
Not binary but at least ternary.

NA **N**umerical **A**perture
The acceptance angle of an optical waveguide. The NA is calculated from the refractive indices of core and sheath.

Nanoscience
The science and technology of researching and manufacturing material structures smaller than 1 μm.

N-channel transistor
An N-channel FET that is controlled with a positive gate source voltage and blocked with a positive drain source voltage.

NEP **N**oise **E**quivalent **P**ower
The output power that a signal generates at the output of an amplifier or detector, which is the same as the intrinsic noise.

Nesting program nesting, interleaving

Network
An interconnection structure for transport or communication purposes. There are different topologies: star, ring or mesh. An exchange, change of direction or other type of signal flow change takes place at the node points.

NF **N**oise **F**igure

NI **N**egative **I**ntrinsic
The junction between n-doped and intrinsic semiconductor zone, see also PI.

Nibble
A data word of 4 bits. This produces 16 combination options. Also: half-byte, quadbit or tetrad.

NMOS **N**-Channel **M**etal **O**xide **S**emiconductor
An MOS technology with n-channel transistors on a p-doped substrate. The channel refers to the inverted polarity of the p-doped zone between source and drain with an activated transistor.

Noise
Deviation from the ideal signal gradient due to uncontrollable interference voltages and currents. These are generally the result of thermal effects in the conducting medium, scattering and reflection effects, as well as external interference sources, see also bulk noise, NF, NEP, RIN.

Non-volatile
A characteristic of memories, which means that they do not lose the stored information even when the supply voltage is switched off, in particular magnetic memories, but also flash memories, see also ROM.

NPIN **N**egative **P**ositive **I**ntrinsic **N**egative
A layer sequence for transistors with an intrinsic layer between the p-zone and one of the two n-zones.

NPN **N**egative **P**ositive **N**egative
A layer sequence for bipolar transistors, see also PNP.

NRE **N**on-**R**ecurring **E**ngineering
One-off development efforts.

NTC **N**egative **T**emperature **C**oefficient
This means that modifying a semiconductor (or other) parameter produces the opposite sign to the causal temperature change, e.g. a decrease (-) in electrical resistance with a rise (+) in temperature (thermistor). The correlation can be proportional, but in most cases it is non-linear, see also PTC. An NTC resistor is also known as an NTC thermistor.

NTC resistor see NTC

OBIC **O**ptical **B**eam **I**nduced **C**urrent
This is used to check electronic components, see also EBIC.

OC **O**pen **C**ollector
A type of circuit that allows the collectors of several bipolar transistors to be integrated.

Octal
A method for representing numbers based on 8, see also decimal, binary, hexadecimal.

OD **O**pen **D**rain
The FET counterpart to the open collector circuit, see also OC.

OE **O**utput **E**nable

OEIC **O**pto **E**lectronic **I**ntegrated **C**ircuit

OFDM **O**rthogonal **F**requently **D**ivision **M**ultiplexing **F**orum

OIC **O**ptical **I**ntegrated **C**ircuit

OIF **O**ptical **I**nternetworking **F**orum

OIG **O**ptically **I**solated **G**ate

OPA **O**ptoelectronic **P**ulse **A**mplifier

OPAMP
Also: opamp. **Op**erational **Amp**lifier.

Opcode **Op**eration **Code**
The instruction part of a computer word.

Operating point set point
The point on a characteristic curve that is set for the application-compatible operation of a component.

Operating system
The basic program of a computer that enables it to perform all the necessary basic operations (input/output, storage, administration, etc.).

Operational amplifier
Instrumentational amplifier. An amplifier with very good linear characteristics, high bandwidth, low noise and low drift. It was originally developed for computing operations in analog computers, hence also the name operational amplifier.

Optical fiber see optical waveguide

Optical waveguide
It consists of a thin transparent core made of optic or plastic fiber, which is surrounded by a sheath with a low refractive index. Incident light within the so-called acceptance angle in the core is totally reflected at the boundary layer to the sheath. It then propagates in the core area. If the core has a sufficiently short diameter in relation to the wavelength of the transported light, only a single propagation mode is possible (single-mode fiber). Since this makes the differences in propagation time very small, mono-mode optical waveguides are very wideband. The thin core, however, means that the injection of light is less efficient, see also graded-index fiber.

Optics
The science of visible electromagnetic radiation. Optics is about studying the generation, propagation and interaction of light.

Optocoupler
A module that optically transfers electrical signals internally. As a result, there is electrical isolation between the transmitter and receiver.

Optoelectronics
A sub-area of electrical engineering concerned with electronic components that emit, convert, transmit and modulate electromagnetic radiation in the infrared, visible and ultraviolet range of the electromagnetic spectrum.

OROM Optical **R**ead **O**nly **M**emory
For example, a CD.

Oscillator
An electronic circuit that generates a periodically changing output signal. In practice, almost all oscillators consist of a positive-feedback amplifier with a frequency-determining element, see also quartz, VCO.

Oscilloscope
Literally: oscillation viewer. An oscilloscope enables signal gradients to be viewed on a screen. It has an adjustable amplifier, a time base with a trigger circuit for synchronization with the time reference of the signal and often also via a memory, in particular for the representation of one-off signals.

OSEK
German abbreviation for **O**ffene **S**ysteme und deren Schnittstellen für die **E**lektronik im **K**raftfahrzeug: open systems and the corresponding interfaces for automotive electronics. An OS standard that resulted from initiatives by French and German motor vehicle manufacturers.

OSI Open **S**ystems **I**nterconnect(ion)
Communication between open systems according to the ISO reference model, also seven-layer model. A protocol structure in seven layers based on ISO standard 7498, Basic Reference Model OSI, also layer model, that permits the continuous and open exchange of information. The junctions between the layers are standardized so that access can be controlled. The individual devices and conventions are inevitably subordinate to the model.

OTP One **T**ime **P**rogrammable
A feature of certain ROMs, see also PROM.

Output
The circuit connection that supplies the signal.

Outsourcing
Transferring corporate tasks to external, usually specialized contractors. Tasks such as data input, PCB production and assembly as well as programming are often outsourced. It is beneficial above all because it allows the client to concentrate on his core competencies.

Overhead base load

Overshoot overswing

OVPO Outside **V**apor **P**hase **O**xidation
A coating technology from Corning using vapor deposition, for (opto)semiconductors.

Packaging density component density

1. The number of memory cells in terms of the length or surface area of a storage medium. One measure for the packaging density is bit per inch or square inch.

2. The number of integrated components on a certain area such as, for example, a board.

Page mode RAM
A RAM access concept that supports access to consecutive memory cells with a reduced cycle time. This method is advantageous for example for video, since image information is usually stored sequentially, see also FPM.

PAL

1. **P**rogrammable **A**rray **L**ogic. An IC with programmable AND matrix and fixed OR matrix, see also PGA, PLA.
2. **P**hase **A**lternating **L**ine. A technique for color stabilization in television. Developed by Dr. Walter Bruch at Telefunken, this standard offsets spontaneous phase shifts by rotating the phase by 180° each time the television line changes, see also NTSC, SECAM.

Parity
Equality or homogeneity. In remote data transmission, a convention that is specified between transmitter and receiver prior to transmission. In the most straightforward case, it involves transmitting a redundant parity bit. There are however also more complex methods, see also CRC.

Passivation
Attaching a scratch-resistant layer to finished semiconductor dice. It protects the module against mechanical damage and other harmful influences from impurity. The passivation layer is the last coating step in semiconductor manufacturing, see also coating.

Passive component
In contrast to an active component, a passive component only stores, consumes or transfers useful electrical power. Examples of such components are resistors, capacitors and coils.

Patch
On a PCB, provisional rewiring or additional wiring to eliminate a design fault. With software, the correction of a program fault, see also bug and debugger.

PBGA **P**lastic **BGA**

PC

1. **P**arity **C**heck.
2. **P**eak **C**lipping. Restricting current or voltage peaks.
3. **P**enta**c**onta. An ITT switching system with crossbar technology.
4. **P**ersonal **C**ommunicator.
5. **P**ersonal **C**omputer. A term coined by IBM. The first PC appeared in 1981.
6. **P**hoto**c**onductor.
7. **P**hysical **C**ontact.
8. **P**ocket **C**alculator.
9. **P**rinted **C**ircuit. See also PCB.
10. **P**rogram **C**ounter.
11. **P**rogrammable **C**ontroller.
12. **P**rotocol **C**onverter.
13. **P**ulsating **C**urrent..

PCB **P**rinted **C**ircuit **B**oard
Often also simply called an electronics board.

PCD **P**lasma **C**oupled **D**evice
A device or component in which electrical coupling uses plasma.

P-channel transistor
A P-channel FET that is controlled with a negative gate source voltage and blocks with a negative drain source voltage.

PCMCIA **P**ersonal **C**omputer **M**emory **C**ard **I**nternational **A**ssociation
An association of manufacturers and dealers concerned with the maintenance and further development of a general standard for peripheral equipment on the basis of PC cards with a corresponding slot for portable PCs and similar. The PCMCIA standard of the same name was adopted in 1990 as Version 1.

PCT **P**hoton **C**oupled **T**ransistor

PCVD **P**lasma **A**ctivated **C**hemical **V**apor **D**eposition
A coating technology for optical waveguides by means of vapor deposition from the plasma phase.

PD

1. **P**hoto**d**iode.
2. **P**ublic **D**omain.

PDA **P**hoto**d**iode **A**rray
A multiple arrangement of photodiodes (for example in matrix or linear format).

PDB **P**lanar **D**oped **B**arrier
A planar-doped semiconductor area with a barrier function.

Peak clipping
The elimination of unwanted peak voltages or currents, generally by means of a special equalizing circuit, see also clipping.

Peak-to-peak
The total value for the amplitude hub of an oscillation between the positive and negative maximum.

Peck model
A model used for testing components that accelerates the failure rate through increased temperature and air humidity, see also Arrhenius model, Eyring model.

PEEL **P**rogrammable **E**lectrically **E**rasable **L**ogic

Peeling
The specific removal of semiconductor layers.

Peltier effect
Also: thermoelectric effect. This can be observed when a direct current flows through a boundary between two metals. The boundary is either deprived of or supplied with heat depending on the direction of current flow. Peltier elements are used to cool critical components and/or to stabilize wavelengths or frequency. The inverse effect also occurs (Seebeck effect), see also thermocouple.

PGA

1. **P**in **G**rid **A**rray. A pin pattern for assembling chips on boards with a large number of connections.
2. **P**rogrammable **G**ate **A**rray. An IC with a programmable AND and NAND matrix, see also PAL, PLA.

Phase control
Switching the alternating voltage when a particular phase angle is reached. This enables the amplitude of the resulting current to be regulated almost loss-free.

Phonons
The lattice vibrations that occur in a semiconductor crystal.

Photo diode
A diode that carries current under incident light, i.e. whose reverse current increases under incident light. Incidence from a sufficiently high-energy photon into a PN structure generates an electron hole pair. If voltage is applied, the electrical field causes migration of the charge carriers, the photocurrent.

Photo resistor
A component whose ohmic resistance decreases under the influence of light. Most photo transistors are made of cadmium sulphide (CdS).

Photolithography
A manufacturing technique for ICs. A reduced image of the IC structure, the photo mask, is used to expose a semiconductor wafer coated with photo-resist. The light penetrating through the mask changes the wafer structure. Non-exposed photo-resist is then washed off. Etching then creates the required circuit pattern on the wafer.

Photometry
The measuring of visible light. Definition of this unit took account of the spectral gradient of perception sensitivity in the human eye.

Photon
An energy quantum, see also light

Phototransistor
A transistor whose collector current increases under incident light. Instead of an injection current, the incidence from a sufficiently high-energy photon into the base layer generates an electron hole pair, see also EHP.

PI **P**ositive **I**ntrinsic
The junction between p-doped and intrinsic semiconductor zone, see also NI.

PIC **P**ower **I**ntegrated **C**ircuit

Piezoelectric effect
A physical effect in some crystals that develop an electrical voltage when they are exposed to mechanical strain. Vice

versa, they respond with mechanical strain to the application of an electrical voltage.

Piezo-resistive effect
The change in the electrical resistance of a metal or semiconductor under the influence of mechanical stress. With semiconductors, this is caused by the band structure and with metals it is the Fermi surface that is deformed by the mechanical stress. This then changes the conductivity. Silicon has a very distinctive piezo-resistive effect and is therefore excellent for pressure sensors.

Pigtail
In optical waveguide technology, a fiber that is directly connected to a component and generally coiled.

PIL Picosecond **I**njection **L**aser
An injection laser that generates extremely short light pulses.

Pin grid see PGA

PIN Positive **I**ntrinsic **N**egative
A layer sequence for fast photodiodes. The positively and negatively doped zones are separated in this structure by a very lightly doped (intrinsic) semiconductor layer. Due to the low concentration of free charge carriers, the latter has a high resistance and, due to the broad space charge zone, a low capacity. Incidence from a sufficiently high-energy photon generates an electron hole pair. If voltage is applied, the electrical field generates a migration of the charge carriers, the photocurrent.

PINFET Positive **I**ntrinsic **N**egative **F**ield **E**ffect **T**ransistor

Pinning
The pin assignment of an IC or other electronic module.

Pipeline
In data processing, a continuous shifting of instructions or data.

Pipelining
The simultaneous processing of instructions for accelerated program run. With a full pipeline, an instruction is completed after each time interval.

Pit groove

PLA Programmable **L**ogic **A**rray
A logic circuit with programmable AND and OR matrix that consists of a large number of linkable circuits, see also PAL, PGA.

Planar flat
With planar semiconductors, the chemical elements are diffused to control the electrical conductivity into (and below) the surface of a silicon wafer, which remains flat throughout the entire process, in contrast for example to mesa structures, whereby a type of stage is formed.

P-LCC Plastic **L**eaded **C**hip **C**arrier
With bond wires, see also CLCC.

PLD
1. **P**igtailed **L**aser **D**iode.
2. **P**rogrammable **L**ogic **D**evice. See also EPLD.
3. **P**ulsed **L**aser **D**iode.

PLL Phase **L**ocked **L**oop
Serves the frequency stabilization or synchronization of an oscillator.

PM
1. **P**hase **M**odulation. Changing the phase of an oscillation in dependence of a useful signal.
2. **P**hoto**m**ultiplication. See also PMT.
3. **P**hysical **M**edium. The bottom layer of the OSI layer model.
4. **P**reventive **M**aintenance.
5. **P**ulse **M**odulation. Modulation whereby a pulse is used to carry the useful signal.

PMOS see MOS

PN diode see diode and/or barrier layer

PN junction see barrier layer

PNIP Positive **N**egative **I**ntrinsic **P**ositive
A layer sequence for transistors with an intrinsic layer between the n-zone and one of the p-zones.

PNP Positive **N**egative **P**ositive
A layer sequence for bipolar transistors, see also NPN.

POH **P**ower-**o**n-**H**ours

Port

1. In electronics, an I/O junction.
2. Transmission with simultaneous reformatting/translation for adapting to different conditions, e.g. adapting machine code to different computer architectures.

Portability
The ability to load files into different programs (also: import/export).

Position sensor
A sensor that provides an output signal whose evaluation permits conclusions to be drawn about the position of an object. There are inductive, capacitative, resistive, optical and ultrasonic sensors of this genus.

Potential-free
Also: floating. A point is potential-free vis-à-vis another point if it has no electrical voltage vis-à-vis the latter due to its configuration, and thus no equalizing current flows as a result.

Potentiometer
A resistive voltage divider with a linear or rotary regulator.

Power gain
The power amplification factor of an amplifier circuit.

Power supply
A power pack that supplies electrical equipment with direct current. The power supply generally takes alternating current from a mains network and converts it into direct current. Accumulators and batteries can also supply power.

Power-down
Disable, reduced power, echo mode.

P-PGA **P**lastic **P**in **G**rid **A**rray
See also PGA, CPGA.

ppm **P**arts **p**er **M**illion
A common measure for the frequency of certain parts in a population (1 million) of other parts.

P-QFP **P**lastic **Q**uad **F**lat **P**ackage

Precharge time
The time required by a storage device for data recharge.

Prefetching
A fetching strategy whereby the memory contents (probably required soon) are retrieved from the mass storage in the cache as a precaution.

PRO Electron
The European type designation system for semiconductors. A European organization under the umbrella of the EECA, which coordinates and registers product type designation amongst European semiconductor manufacturers.

Process computer
An especially rapid-response computer that processes data in real time on the basis of incoming signals.

Processor
An integrated circuit for processing computer programs, see also microprocessor, MPU, CPU.

PROFET®
SIPMOS transistors from the smart family with integrated protective functions and feedback to the control logic.

Program nesting
Calling up subprograms from a subprogram.

Programming language
A system of characters and rules, similar to language, to generate instruction sequences for a computer. There are so-called higher programming languages that are approaching natural human language and machine-oriented languages that are difficult for humans to understand, see also machine code, assemble.

PROM **P**rogrammable **R**ead **O**nly **M**emory
A ROM that can be programmed (generally only once) by the user, see also EEPROM and EPROM.

Protocol
A set of regulations and agreements for a particular purpose or occasion. In remote data transmission, the protocol covers all control and behavioral regulations for the

intended processes. Also: signaling, handshake.

Prototype
Forerunner, first product implementation.

Proximity switch
A switch that responds when a solid body approaches it.

PRTN **P**iano della **R**egolomentazione **T**elefonica **N**azionale
The Italian Regulatory Authority for National Telecommunications.

PTC **P**ositive **T**emperature **C**oefficient
This means that modifying a semiconductor (or other) parameter has the same sign as the causal temperature change, e.g. an increase (+) in electrical resistance with a rise (+) in temperature. The correlation can be proportional, but in most cases it is non-linear, see also NTC.

Pulse duty factor
The ratio between the in and out interval of a periodic signal.

Punch-through penetration, reach through
An effect seen in MOS transistors that occurs with a short channel: the space charge zones can make contact even with a low drain-source voltage.

PVC

1. **P**ermanent **V**irtual **C**ircuit (also: **C**onnection) in the ATM. From the point of view of the subscriber, it is a fixed connection.
2. **P**oly**v**inyl**c**hloride. A plastic that can be manufactured relatively cheaply and is very well suited to use in insulators and/or cable covering.

PVD

1. **P**hoto**v**oltaic **D**iode. A diode that generates an electrical voltage using light.
1. **P**hysical **V**apor **D**eposition.

Pyroelectric detector
A sensor that emits a voltage when it encounters long-wave radiation. The pyroelectric detector can be used for temperature telemetry and to detect living things.

Pyroelectric effect
A charge movement due to the incidence of long-wave infrared radiation (from around 2 μm) and/or thermal radiation.

PZT **P**ie**z**oelectric **T**ransducer
See also piezoelectric effect.

QA **Q**uality **A**ssurance

QCCN **Q**uad **C**hip **C**arrier **N**on Leaded

QE **Q**uantum **E**fficiency

Quadrant detector
A photodetector whose circular surface is divided into quadrants. The position of the incident light beam can be calculated from the ratio of the four output signals.

Qualification
The result of a test in which the relevant characteristics of a component are recorded and evaluated in full. Qualification is necessary in order to ascertain with certainty that the component is suitable for its intended use, see also certification.

Quality
According to DIN 5530 "the totality of characteristics of a product or activity that bear on its ability to satisfy given requirements".

Quality assurance
All measures to ensure a specified product quality, also for services. The measures begin during product design and cover all steps, even beyond delivery to the customer.

Quantum dot
A metallic or semiconducting nanoparticle that holds a single charge carrier. It is suitable for use as a micro memory cell, see also electron trap.

Quantum efficiency
Also: quantum yield. The number of charge carriers released by a photon as a result of the photoelectric effect. The quantum efficiency depends on the respective photon energy, i.e. the wavelength.

Quantum well
A microscopically small semiconductor heterostructure in which charge carriers

are trapped to a certain extent or can only move on one particular level, see also SQW/MQW.

Quartz
Silicon dioxide in a crystalline structure. Thin quartz plates have a very distinctive, discrete and stable self-resonant frequency. They are therefore excellently suited to oscillator frequency stabilization. If they are also temperature-stabilized, extremely stable frequency standards result, see also OCXO.

QW **Q**uantum **W**ell

Race condition
A uncontrollable state in which data passes through a logic circuit faster than the controlling clock signal.

RADAR **Ra**dio **D**etection **a**nd **R**anging

RAM **R**andom **A**ccess **M**emory
Write/read memory with random access. The only disadvantage of the currently available types is the volatility (loss of data during removal of the supply voltage), see also memory, ROM, DRAM.

Ramp-down run down, phase out (production).

Ramp-up start up (production)

Random arbitrary, stochastic, optional

RAS **R**ow **A**ddress **S**trobe
Row address release. A control signal for memory chips, see also RE.

RBSOA **R**eversed **B**iased **SOA**
Secure operation of a diode in the reverse direction (i.e. sufficiently well below the breakdown voltage).

RCEEA **R**adio **C**ommunication and **E**lectronic **E**ngineering **A**ssociation, UK

RCTL **R**esistance **C**apacitance **T**ransistor **L**ogic
An RTL variant with C as the coupling element to increase speed.

RE **R**ow **E**nable
A control signal for memory chips (formerly RAS).

Read only
Refers for example to file attributes, access rights or protective devices.

Real time
A term from open and closed loop control engineering. As opposed to batch processing that is not time-critical. Real-time systems however must be able to control processes and respond to signals quickly enough for the process to run properly.

Recombination
In semiconductor materials, electrons can leave the valence band of an atom through thermal influences, but also through the force of attraction of external atoms, and move to the conduction band. This makes them charge carriers that sooner or later come across a "hole" and so return to the valence band. This process is referred to as recombination. It is quantified by the expression "charge carrier service life", see also majority charge carrier.

Rectifier
Also power converter. A component that allows current to pass in only one direction, for example a diode. With a bridge circuit, both alternating current half-waves can be rectified into one direct current. The oscillating direct current generated in this way with dual power frequency must then be filtered for most applications.

Redundant superfluous, excessive, superabundant.
The purpose of redundant devices is generally to prevent loss of function or to at least guarantee emergency operation in the event of faults, or safely detect and, if applicable, correct errors. Ideally, redundancy should be on as small a scale as possible or as large a scale as required. It makes a system more reliable.

Reed relay
An encapsulated gas-tight relay, often filled with protective gas. The contacts contain ferromagnetic material and thus move towards each other when an external magnetic field is applied.

Reference model see OSI

Reflectometer
A measuring device that records the temporal gradient of the signal reflection and thus permits conclusions to be drawn

about the position of imperfections that attenuate or reflect to an above-average extent. Most reflectometers emit a strong, short electrical or optical pulse, see also OTDR.

Reflow soldering
A soldering procedure whereby a solder already attached to the PCB is refused so that the PCB and components are connected to each other. The solder is generally applied as a soldering paste.

Refresh
A periodic mechanism for data retention with DRAM memory cells, see also retention.

Refresh rate
The frequency at which the entire contents of a screen are regenerated. The refresh rate should be high enough to guarantee a constant and flicker-free picture, see also flicker.

Reject
A product that is rejected due to non-compliance with the quality criteria.

Rejection rate scrap rate
The proportion of faulty, unusable products.

Rejection scrapping

Relay switching point
An electronic relay is an electric circuit that changes the output current greatly as a result of a small change to its supply (control) voltage. This type of relay functions more or less without inertia and without mechanical wear and tear. The combination of metal surface micromechanics and anisotropic silicon etching facilitates the manufacture of small-size microrelays on a single silicon substrate.

Reliability
The ability of a technical device to meet set requirements during a certain period of time, see also MTBF, MTTF, MTTR, failure.

Reluctance magnetic resistance

Remanence retentivity
A measure for the level of magnetic flux density remaining following full magnetization in a closed circuit, see also coercitive field strength.

Reset
A procedure whereby an electronic circuit is returned to its initial or idle state.

Resistance
(as a physical variable).

Resistor
(as a physical object).

Retention
The time after which the contents of a DRAM cell is still sensed correctly despite drained charges, therefore also called refresh time.

RHET **R**esonance Tunneling **H**ot **E**lectron **T**ransistor

RIBE **R**eactive **I**on **B**eam **E**tching
An etching procedure that involves an ion beam that triggers chemical reactions, see also RIE.

RIE **R**eactive **I**on **E**tching
See also RIBE.

RIN **R**elative **I**ntensity **N**oise

RISC **R**educed **I**nstruction **S**et **C**omputer
This type of computer is not flexible when it comes to programming, but is very fast, see also CISC.

Rise time
The length of time between triggering a switch-on process and reaching the switched-on state. It depends on both the properties of the switch as well as the signal level definition. Generally it is 90% of the final value.

RMOS **R**efractory **MOS**
An FET technology in which the gate is made of a difficultly fusible metal (molybdenum, tungsten).

ROM **R**ead **O**nly **M**emory
The memory contents are written in mechanically, magnetically, electrically or optically during the manufacturing stage.

ROR **R**AS **O**nly **R**efresh
A series refresh strategy for DRAM cells.

RPC **R**igid **P**rinted **C**ircuit
See also FPC.

RTD Resonant **T**unneling **D**iode
With storage effect, see also RTT, TSRAM.

RTL Resistor **T**ransistor **L**ogic
With links via resistors, inversions via transistors.

RTT

1. **R**esonant **T**unnel **T**ransistor. A fast transistor type with a resonant tunnel effect, see also RTD.
2. **R**oad **T**ransport **T**elematics.

S Siemens
The unit for electrical conductivity, i.e. the inverse value of electrical resistance (named after Werner von Siemens).

SAGMOS Self **A**ligning **G**ate **MOS**

SAM-APD Separate **A**bsorption and **M**ultiplication Region **APD**

SAMOS Stacked Gate **A**valanche Injection **MOS**
See also SIMOS.

Sample and hold
A sampling procedure whereby the instantaneous value of a variable (e.g. voltage) is determined and then stored in an analog retention buffer.

Sample specimen

Sampling
In signal processing, the periodic sampling of the signal value, generally for the purpose of digitization, see also sampling rate.

Sampling rate
The frequency at which samples of a physical variable, such as sound, for example, are taken.

Saturation
The completely conductive state of semiconductor components, and/or the full magnetization of a ferromagnetic material. An increase in the control voltage, the injection current and/or the magnetic field strength no longer results in any increase or amplification.

SBD Schottky **B**arrier **D**iode

SBR Selectively **B**uried **R**idge
A semiconductor structure for index-guided laser diodes.

Scale of integration
The packaging density on an IC.

SCCD Surface **C**harge **C**oupled **D**evice
A charge-coupled semiconductor component (a "bucket-brigade device") with charge transfer at the surface, see also CCD.

SCEW
The propagation time differences on a chip (indicated by clock signals at both ends).

SCFL Source **C**oupled **F**ET **L**ogic

SCH Separate **C**onfinement **H**eterostructure
A semiconductor heterostructure with a separate resonance chamber for optoelectronic components (lasers).

Schmitt trigger
A bistable multivibrator that changes its circuit state when an input level is exceeded. The special feature of the circuit is its hysteresis: only once the input level falls short of an adjustable value does the circuit switch back.

Schottky diode
Also: hot carrier diode. A semiconductor diode with a junction formed from a semiconductor layer and a metal layer. It has extremely short switching times and is therefore used in fast logic circuits.

SCL Source **C**oupled **L**ogic
The FET counterpart to ECL.

SCLA Semi**c**onductor **L**aser **A**mplifier
See also TWSLA.

SCLC Space **C**harge **L**imited **C**urrent

SCOS Smart **C**ard **O**perating **S**ystem

SCR Silicon **C**ontrolled **R**ectifier
A thyristor, a semiconductor component with four layers in PNPN structure and three junctions. A thyristor has two stable states and is therefore suited to use as an electronic switch. Area of application: power electronics.

Scrambling

1. Address encryption in storage technology. This means that the address is recalculated so that a defined cell in the chip is addressed. This is necessary because the cells in the chip do not form an ideal matrix.
2. In addition to "address scrambling", "data scrambling" is also used. Depending on the word line or bitline addressed, the actual signal is stored in inverted or unchanged form. For cell field analysis, the actual physical information of the cell is required.

SCSOA **S**hort **C**ircuit **S**afe **O**perating **A**rea
Designates in power electronics the parameter field of a module in which it still functions safely under short circuit conditions without being damaged.

SCT **S**urface **C**harge **T**ransistor

SDFL **S**chottky **D**iode **FET** **L**ogic
An IC family with GaAs MESFETs.

SDHT **S**electively **D**oped **H**eterojunction **T**ransistor
An extremely fast switching FET based on GaAs.

SDRAM **S**ynchronous **DRAM**
A storage technology for semiconductor RAMs whereby the information is stored and read in a synchronous time reference. It is therefore no longer necessary to wait for data to arrive before the next address is created. This technology considerably increases data throughput.

Sea-of-gates technology
An ASIC technology developed by Siemens AG. The gates are connected not via the wiring ducts, but directly via the gate array cells (logic cells). This technology reduces the silicon area requirements.

SECAP **S**emiconductor **E**quipment **C**onsortium for **A**dvanced **P**ackaging

SEED **S**elf **E**lectro Optic **E**ffect **D**evice

SEL **S**urface **E**mitting **L**aser

SELD **S**urface **E**mitting **L**aser **D**iode

SEM **S**ilicon **E**lectron **M**ultiplication

Semaphore signal
In process technology and data processing, a flag that is set under certain conditions. The program-led process polls the semaphor under certain circumstances and evaluates the state. Traffic lights are one example of a semaphore.

SEMATECH **S**emiconductor **Ma**nufacturing **Tech**nology

SEMI **S**emiconductor **E**quipment and **M**aterials **I**nternational

Semiconductor
A crystalline material that in its pure form has a conductivity between a conductor and an insulator. This conductivity increases in the presence of certain foreign matter that disturbs the crystal structure. The specific addition of certain foreign matter can influence the properties of the semiconductor in a very particular and reproducible way. In the broader sense, "semiconductor" also denotes electronic components that are manufactured on the basis of semiconductor materials, see also diode, MOS, donor atom, barrier layer, transistor.

Semicustom IC
These are developed on the basis of prefabricated gate arrays or standard cells. Semicustom ICs can be developed by the user himself with the support of the semiconductor manufacturer.

SEMKO **S**venska **E**lektriska **M**aterielkontrollanstalten
The Swedish Electrical Material Inspectorate, Stockholm.

Sensing technology
The technology that converts physical variables such as pressure, temperature, etc. into electrically analyzable signals like voltage, frequency and pulse width.

Sensor a component of sensing technology

SEP **S**towarzyszenie **E**lektryków **P**olskich
The Association of Polish Electrical Engineers, Warsaw.

SER **S**oft **E**rror **R**ate

Sexadecimal see hexadecimal

SFR Special **F**unction **R**egister

SGML Standard **G**eneralized **M**arkup **L**anguage
A standard for information administration purposes based on ISO standard 8879. SGML describes a procedure for providing platform- and application-independent documents with retention of formatting, indexing and linked information and it also includes a grammar-like schema for determining a general document structure, see also HTML.

SGRAM Synchronous **G**raphics **RAM**
A specially enhanced variant of SDRAM modules that is used for graphics cards. With the aid of an internal command pipeline, access sequences can be buffered on the chip, thereby achieving larger access bandwidths. If a memory address is specified, SGRAMs can also change the column address internally within a system clock, which produces very fast sequential access.

SGT Surrounding **G**ate **T**ransistor

Sheath modes
Modes that propagate in the optical waveguide sheath.

Sheath
The outer, optically active layer of an optical waveguide.

Shielding screening
A material barrier that prevents electrical, electromagnetic and/or magnetic fields from penetrating or escaping. For electrical and electromagnetic shielding, generally a cage, a foil or another jacket made of conductive material, usually metal, is sufficient. For shielding against magnetic fields, a jacket made of high-permeability material (e.g. μ metal) is required, see also EMC.

Shift register
A data memory that is able to shift its contents by a certain number of positions. This operation is imperative for digital multiplication, see also barrel shifter.

Short circuit
A low-impedance connection between two points with different potential. A short circuit results in practice in a strong equalizing current that has interfering, often even destructive consequences.

Shot noise bulk noise, granular noise
The noise level increases with the signal amplitude due to the quantum nature of the current, see noise.

Si Silicon
Element number 14 in the Periodic Table. It is a semiconductor material with relatively low intrinsic conduction, which can be increased specifically by doping it with external atoms. Further cost-effective attributes made the material almost ideal for manufacturing electronic components through to memory chips and microprocessors.

SIA Semiconductor **I**ndustry **A**ssociation, USA

SIC Semiconductor **I**ntegrated **C**ircuit

SiC Silicon **C**arbide
A semiconductor material that still functions even at very high temperatures (up to more than 400°C) due to its high energy gap.

SID Slewing **I**nduced **D**istortion

SIEGET® Siemens Grounded Emitter Transistor
An HF transistor with a critical frequency of 25 GHz.

SIL Single **in L**ine Package

Silicon Valley
A region of California south of the Bay of San Francisco, also known as Santa Clara Valley, between Palo Alto and San Jose. Silicon Valley is the traditional hub for researching, developing and manufacturing electronics and computers. Many leading semiconductor companies were set up here and still have their headquarters in SV.

SilVer Silicon **Ver**ified
A draft of an electronic circuit whose serviceability has been confirmed by implementation in a silicon chip.

SIMM Single **I**n-Line **M**emory **M**odule

SIMOS Stacked Gate **I**njection **MOS**
See also SAMOS.

Single-mode fiber
A type of optical waveguide with only one mode in the fiber, i.e. the modal dispersion is zero. These fibers have a considerably higher bandwidth than graded-index fibers, see also optical waveguide.

SIP **S**ingle **i**nline **P**ackage

SIPMOS
An MOS transistor product family from Siemens for switching high currents and voltages. It is directly addressable using TTL/CMOS levels.

SIPP **S**ingle **I**nline **P**in **P**ackage

SIRET® Siemens Ring Emitter Transistor
SIRET permits both high switching speeds in the voltage range between 500 and 1200 V and thus also the construction of power converters in the power range of several hundred kilowatts with frequencies over 10 KHz, as required for low-noise drives.

SIT **S**tatic **I**nduction **T**ransistor
Another name for FET.

Skew
A (generally unwanted) spatial or temporal offset.

Skin effect
Current displacement, occasionally also called the Kelvin effect: inhomogeneous current distribution in current-carrying conductors. As the frequency increases, the alternating current is displaced to the edge of the conductor by high-frequency interaction. The effective resistance thereby increases. The skin effect can be reduced by a mesh of current-carrying flexible leads.

SLA **S**emiconductor **L**aser **A**mplifier

SLD **S**uper**l**uminescent **D**iode
See also LED. A light-emitting diode that achieves great light efficiency.

SLED **S**urface **e**mitting **LED**
In contrast to an ELED, this light emitting diode emits in a large solid angle from the semiconductor surface. It facilitates efficient injection with optical waveguides with a high NA. The optical power fluctuates less with the operating temperature than is the case with the ELED.

Slew rate rate of rise
A measure for the pulse response of a circuit.

Slip-fit mounting insertion mounting
A method of assembly whereby the components are inserted in holes on the printed circuit board.

Slope
The first derivation of a rising curve. The slope denotes the increase of a variable as a result of the increase in an underlying variable, e.g. the increase in the light power of a laser diode depending on the increase in the injection current.

SLSI **S**uper **LSI**
A chip technology with more than 100,000 transistor functions.

SMD **S**urface **M**ounted **D**evice
An electronic component without any connecting wires that is placed directly on the surface of an electronics board, where it is contacted in the solder bath.

SMH **S**ociété Suisse de **M**icroélectronique et d'**H**orlogerie
The Swiss Corporation for Microelectronics and Watchmaking Industries.

SMIF **S**tandard **M**echanical **I**nter**f**ace
A production standard proposed by Hewlett Packard for electrical engineering in ultra-clean rooms.

SMPGA **S**urface **M**ount **P**in **G**rid **A**rray

SMPS **S**witched **M**ode **P**ower **S**upply
A pulsed power supply that chops the rectified, screened mains voltage with semiconductor switches. Since the semiconductors either through connect or block conductively, there is little switching and on-state loss. This is responsible for the characteristically high efficiency of a pulsed power supply by comparison with analog methods with in-phase voltage control.

SMT **S**urface **M**ounting **T**echnology
See also SMD.

SNOS **S**ilicon **N**itride **O**xide **S**emiconductor
A semiconductor structure for EPROM cells, see also MNOS/SONOS.

SNR **S**ignal to **N**oise **R**atio

SO **S**mall **O**utline Package

SOA

1. **S**emiconductor **O**ptical **A**mplifier. See also EFA, EDFA.
2. **S**afe **O**perating **A**rea. Defining the SCSOA is particularly critical.

Socket
A plug connection mounted on the board for electrical or electronic components.

SOD

Silicon-**o**n-**D**iamond. Similar to SOS.

Small **O**utline **D**iode.

Software
The immaterial part of a computer. It designates data and programs, in contrast to the hardware.

SoG **S**ea-**o**f-**G**ates

SOIC

1. **S**ecret **O**bject **I**dentification **C**ode. For example, for access authorization or immobilization, see also POIC.
2. **S**mall **O**utline **I**ntegrated **C**ircuit.

SOJ package Small Outline J-leaded
The "J" stands for the pin form.

Solar cell
A photoelectrical component that generates electrical power when exposed to light irradiation.

Solder bridge
An unintentional electrical connection through soldering flux.

Soldering tag
A metal tongue that is permanently attached to an electrical component and solders a connecting wire to the component.

Solid state relay
A relay on a semiconductor basis, i.e. practically wear-free. Solid state relays are used above all as replacements for mechanical miniature relays. As the detector, they have a photodiode array with a control circuit plus two MOSFETs.

SONOS **S**ilicon **O**xide **N**itride **O**xide **S**emiconductor.
See also SNOS/MNOS.

SOP **S**mall **O**utline **P**ackage

SOS **S**ilicon **on** **S**apphire
A CMOS variant that uses a sapphire substrate instead of silicon.

SOT **S**mall **O**utline **T**ransistor

Source
e.g. the current-supplying end of the conducting channel in an FET.

Source code
The source program for a computer as it was noted by the programmer (e.g. in a higher programming language). The source code is not executable and must first be translated into machine code using a compiler program, see also machine code, assemble.

Source file
The file from which data is read (during a copy run, for example), see also target file.

SP **S**tack **P**ointer

Space charge zone see barrier layer

SPICE **S**imulation **P**rogram for **I**n-**C**ircuit **E**mulation

Spike
An uncontrolled voltage peak that is triggered by switching processes and similar. Spikes can cause considerable interference.

Splicing
This involves fusing optical fibers together using a controlled voltaic arc. The decisive factor here is that the optical cores are accurately aligned, which is a particular challenge with mono-mode fibers.

SPT **S**iemens **P**ower **T**echnology
A special power semiconductor technology from Siemens that enables bipolar,

C-MOS and D-MOS structures to be implemented on one chip.

SQUID **S**uperconducting **Qu**antum **I**nterference **D**etector
A superconducting sensor that records minute changes in the magnetic field. The SQUID loop is based on YBCO.

SRAM **S**tatic **RAM**
A memory cell consisting of several transistors that are switched as two feedback inverters (bistable flip-flops). Once achieved, a logical state is only terminated by specific switchover or removal of the supply voltage. Hence the name "static", which should not belie the fact that this type of memory has much shorter access times than the DRAM technology. It is however also more complex and therefore more costly.

SRD **S**uper **R**adiant **D**iode

SS **S**olid **S**tate
A solid body, semiconductor.

SSD **S**olid **S**tate **D**isk
Mass storage based on a semiconductor.

SSI **S**mall **S**cale **I**ntegration
With a maximum of ten elementary circuits on one IC, see also MSI, LSI.

SSL

1. **S**olid **S**tate **L**ogic, i.e. with semiconductors.
2. **S**olid **S**tate **L**aser.

SSOP **S**hrink **S**mall **O**utline **P**ackage

SSPC **S**olid **S**tate **P**ower **C**ontroller

SSR

1. **S**olid **S**tate **R**elay.
2. **S**ub-**M**ode **S**uppression Rate. For lasers.

Stack pointer
A register in a microprocessor that manages the addresses of the stack.

Stack push-down storage
This storage technique reserves subprogram and interrupt return addresses. Parameters can also be transferred between the subprogram and the main program in the stack. The stack is managed via the stack pointer. It includes an address that indicates the next free position in the stack. With each read (POP) or write (PUSH) process into the stack, the stack pointer is increased or reduced by one address.

Standby
An operating mode in which the functions of a system can be activated at short notice.

STI **S**hallow **T**rench **I**solation

Storage capacity
The capacity of a module or medium to store data, expressed in bits or bytes.

Storage circuit
An electrical functional block that retains information and can return it unchanged, see also logic circuit, analog circuit.

STR **S**elf**t**ime **R**efresh
A special refresh cycle for DRAMs.

Submicron technology
The technology that has made it possible to manufacture semiconductor line widths of less than 1 µm. This further increased the scale of integration.

Substrate
An inactive base material that is used as a carrier in semiconductor manufacturing, see also wafer.

Surface mounting
A PCB assembly method whereby conducting tracks and components are located on one side. The components specially developed for this technology are not, as was the case in the past, soldered into pre-bored holes but attached directly to the surface of the PCB. The PCB is consequently compact and vibration-resistant, see also SMD.

Surge
An intermittent voltage or current increase that can impair the function of a circuit or, in emergencies, results in damage, see also spike.

TAB **T**ape **A**utomated **B**onding
A (now obsolete) automatic procedure for bonding electronic modules on film.

Tag
A special identification that signals a state, see also flag, semaphore.

Tailing
A phenomenon that occurs after deactivating a semiconductor switch, caused by the final drain time of the charge carriers.

TAZ **T**ransient **A**bsorption **Z**ener Diode

TEGFET **T**wo-Dimensional **E**lectron **G**as **FET**
A very fast FET structure with an undoped GaAs layer. It contains a heterojunction that keeps the electrons in this layer, giving them great mobility, see also FET.

Telecommunication communication over long distances

TFBGA **T**hin **F**ine Pitch **BGA**

TFEL **T**hin **F**ilm **E**lectro**l**uminescent Display

TF-FET **T**hin **F**ilm **FET**

TFT **T**hin **F**ilm **T**ransistor
See also TFTLC.

TFTLC **T**hin **F**ilm **T**ransistor **L**iquid **C**rystal
The basic component of active flat screens.

TGG **T**erbium **G**allium **G**arnet
A special semiconductor material.

THC **T**hrough **H**ole **C**ontact
The plated-through hole on electronics boards.

Thermal schock
The sudden cooling or heating of a component. This often reveals hidden weaknesses.

Thermistor
A passive component whose electrical resistance decreases with rising temperature along a material-dependent characteristic curve. It is also known as an NTC resistor.

Thermocouple
When two dissimilar polarized metallic junctions in a circuit differ in temperature, a thermoelectric voltage builds up in this circuit (Seebeck effect). The resulting current flow then eventually cancels out the temperature difference by heating the colder point, see Peltier effect. The thermoelectric voltage depends on the material combination and can exceed 50 µV/K.

Thick film
A technology used to manufacture integrated circuits whereby several films from special pastes with different properties are applied to a ceramic substrate with the aid of templates using so-called photo screen printing, see also thin film.

Thin film
A technology to manufacture integrated circuits, similar to thick film technology, but the layers are made from metals and metal oxides by vapor deposition on the substrate (under vacuum), see also MBE.

Threshold current
The minimum current required to activate a function, for example the laser effect with laser diodes.

Threshold value threshold
The minimum value to achieve a certain state (conducting, radiating, etc.) with an electronic component. A direct consequence of the non-linear behavior of electronic elements.

Thyristor see SCR

TJS **T**ransverse **J**unction **S**tripe
Active in a laser diode structure.

Touch-sensitive switch touch contact
A switch with a conductive surface that can therefore be activated by light contact.

TQFP **T**hin **Q**uad **F**lat **P**ack
A very thin package for electronics circuits that can also be easily mounted on tapes.

TQM **T**otal **Q**uality **M**anagement
Operating sequence control that is consistently oriented towards achieving and complying with certain quality criteria.

Trailing edge
The last section of an electrical pulse, see also leading edge.

Transceiver
An artificial word derived from **Trans**-mitter/Re**ceiver**. A module that can both transmit and receive signals.

Transformer
An arrangement of inductively coupled coils that permits the transfer of electrical energy with a reduction or increase in the alternating current if there is electrical isolation.

Transient suppressor clipping

Transistor
A combination of "transformer" and "resistor", also "transfer resistor". It is quite simply the basic component of modern electronics, invented in 1947 by Shockley, Bardeen and Brittain. It is so named because the transistor behaves like a controllable resistor. There are unipolar and bipolar transistors. Bipolar transistors consist of two boundary layers that are separated by a very thin base. When a suitably polarized basic emitter voltage is applied, the potential jump is reduced so that free charge carriers can penetrate from the emitter into the base and can migrate further to the collector (if they do not recombine in the base layer). This current is generally greater than the basic emitter current. The ratio is known as the current amplification of the transistor. A unipolar transistor consists of one semiconductor channel whose conductivity is influenced by a control voltage at a gate, see also FET.

Transistor logic
A method used to reproduce logic functions by interconnecting/linking several transistors, often in combination with diodes and resistors. The electrical signals symbolize the logical state, see also DTL, RTL.

Transistor parameter
A parameter based on the four-pole model of the transistor.

Trap
In processor technology, a capture condition that generally results in an interrupt.

TRAPATT Trapped **P**lasma **A**valanche **T**riggered **T**ransit
A semiconductor structure for fast microwave diodes. It uses the avalanche effect in a trapped plasma, one consequence of which is that it has a negative differential resistance, see also IMPATT.

Trench groove
A structure often used in semiconductors to drain or prevent current flow. The trench simultaneously creates space for other structures so that it can be used, for example, to increase capacity in DRAMs.

TRIAC Triode **A**lternating **C**urrent (switch)
See also DIAC.

TriCore
A 32-bit microcontroller architecture from Infineon designed for constantly increasing microcontroller performance requirements.

Trigger release
A signal that starts up a particular process.

TRILITHIC
A trilithic integrated circuit as opposed to a monolithic circuit cast in one piece. This semiconductor concept developed by Siemens is able to switch high electrical powers quickly and therefore with low loss.

TRIOS® TRansparent **Io**n **S**creen
A weakly conducting layer on Infineon phototransistor chips that prevents ions precipitating under the influence of high field strengths, which could impair or even damage the coupler function, see also passivation.

TSL Tunable **S**emiconductor **L**aser

TSOJ Thin **S**mall **O**utline **J**-Lead

TSOP Thin **S**mall **O**utline **P**ackage

TSRAM Tunneling-based **SRAM**
A static RAM that uses bistable tunnel diodes as the basic component.

TSSOP Thin **S**hrink **S**mall **O**utline **P**ackage

TTG Tunable **T**win **G**uide
A structure for tunable laser diodes.

TTL Transistor-to-**T**ransistor **L**ogic
Formerly a very widespread technique for integrated logic circuits using bipolar transistors as the logic element.

Tunnel diode
Also known as Esaki diode, after the (Japanese) engineer Esaki. A diode in whose very heavily-doped and thus very thin boundary layer the tunnel effect occurs. This quantum mechanical effect results in charge carriers traversing (tunneling through) the barrier layer even with a low forward voltage (well below the potential jump). An increasing voltage results not only in a type of saturation effect, but the current then also drops at some point until the normal diode characteristic curve is reached. The differential resistance of the diode is negative in this area, which makes it suitable for example for attenuation reduction and/or amplification in HF circuits or for oscillators, see also IMPATT diode and BARITT diode.

TWA Traveling **W**ave Semiconductor Optical **A**mplifier
See also TWSLA.

TWSLA Traveling **W**ave **S**emiconductor **L**aser **A**mplifier
See also SCLA, TWA.

UART Universal **A**synchronous **R**eceiver/**T**ransmitter
An interface module for asynchronous peripheral equipment performing serial data communication operations, see also USART, V.24.

UJT Uni**j**unction **T**ransistor

ULA Uncommitted **L**ogic **A**rray
A semicustom logic circuit that can be programmed using interconnection masks, see also PLA.

ULSI Ultra **L**arge **S**cale **I**ntegration
With more than a million transistor functions, see also VLSI.

Ultrared see infrared

Ultraviolet
The optical wavelength range between 30 and 380 nm. Ultraviolet is invisible to the human eye and is harmful to health in large doses and over long periods and as the wavelength decreases.

UMTS Universal **M**obile **T**elephone **S**ystem

Unijunction transistor
A special transistor with only one barrier layer. This can suddenly become very conductive if the threshold voltage (which is based proportionally on the collector voltage) is reached. It is occasionally used as the logic element in trigger circuits.

Unipolar
An attribute of semiconductor structures in which there is only one doping semiconductor material type, i.e. either p-doped or n-doped material. The MOSFETis a well-known example, see also bipolar.

Update
A process that brings data up-to-date.

Upward compatible
Compatible with products of a later generation. Compliance with standards and conventions makes it easier to achieve upward compatibility, see also downward compatible.

UQFP Ultra thin **QFP**

USART Universal **S**ynchronous **A**synchronous **R**eceiver **T**ransmitter
A universal interface module for both asynchronous as well as synchronous communication between peripheral equipment, see also UART.

USB Universal **S**erial **B**us
A peripheral PC bus for the straightforward connection of a number of peripheral devices even during operation. It replaces the various PC connections such as RS-232 and Centronics.

Vapor phase soldering see condensation soldering

Varistor
A passive semiconductor component whose electrical resistance depends on the voltage applied, see also VDR. A varistor is very good for clipping voltage peaks, see also spike.

VCD Variable **C**apacitance **D**iode
Also: Varicap. It is based on a non-linear effect in the boundary layer: as the reverse voltage increases, the barrier layer becomes wider, which causes its capacitance to drop.

VDE Verband **d**er **E**lektrotechnik, Elektronik und Informationstechnik e.V., Frankfurt am Main
The German Association for Electrical, Electronic and Information Technologies, Frankfurt am Main.

VDI Verein **D**euscher **I**ngenieure e.V., Düsseldorf
The German Association of Engineers, Düsseldorf.

VDL Visible **D**iode **L**aser
A diode laser with emission in the visible (red) range.

VDMOS Vertical **D**ouble **D**iffused **MOS**

VDR Voltage **D**ependent **R**esistor

Verilog
A hardware description language (used predominantly in America), see also HDL, VHDL.

VFET Vertical **F**ield **E**ffect **T**ransistor
See also VDMOS, VMOS. The VFET is particularly well suited to use as a fast power switch.

VHDL VHSIC **HDL**
A language that can be used to describe chips (chip design). A hardware descriptive language (used predominantly in Europe), see also Verilog, HDL, VHSIC.

VHSIC Very **H**igh **S**peed **I**ntegrated **C**ircuit

Vibration table
A mobile support that can be used to generate shocks in order to investigate the corresponding behavior of equipment and components.

Vibration test
Testing for hardness by exposing the sample to defined repeated shock.

VLD Visible **L**aser **D**iode
A laser diode with visible (generally red) emission.

VLSI Very **L**arge **S**cale **I**ntegration
With (sometimes far) more than 1000 elementary circuits on one IC chip, see also LSI, MSI.

VMOS Vertical (also: V-Groove) **MOS**
A fast MOSFET structure with a V-shaped, short channel, see also VFET.

Volatile memory floating memory
See also memory.

Von Neumann architecture
A computer architecture that is now used for most modern computers. It is characterized amongst other things by the fact that data and programs can be stored in the same homogeneous memory, in contrast to the Harvard architecture.

VQFP Very thin **QFP**

VRAM Video **RAM**
A module for storing image information on a graphics card. It is now no longer used and has been replaced by SGRAM.

VTL Variable **T**hreshold **L**ogic

W Write
Write enable. A control signal for memory chips (formerly WE).

Wafer
A disc of semiconductor material on which a number of identical dies (chips) are created using wafer stepping.

Wafer stepping
A consequence of (often numerous) individual work steps whereby semiconductor structures are assembled or modified on a wafer.

WAP Wireless **A**pplication **P**rotocol

WARC World **A**dministrative **R**adio **C**onference
It assigns and manages radio frequencies within the scope of international consultation.

Watchdog
In control engineering, it designates a supervisory circuit that triggers an alarm or outputs reports when predefined error conditions occur (in this way, the watchdog timer monitors compliance with time limits, see also deadlock).

WE Write **E**nable
A control signal for memory chips, see W.

Whetstone
A synthetic program for computer comparisons. It was developed in Whetstone, England, and is based on ALGOL but is mostly used in FORTRAN. There is a version with single (32-bit) precision and a version with double (64-bit) precision.

Wire-wound resistor
A resistor constructed with a wound wire.

WPB Write **p**er **B**it
A write cycle for RAMs.

Write-behind cache
A data access strategy in the computer whereby the data intended for mass storage stays initially in the cache memory, after experience has shown that recently archived data is often accessed at least once a short while later. The procedure therefore improves general system performance since relatively slow read and write access to the mass storage (hard disk) is required less frequently.

WSI Wafer **S**cale **I**ntegration
The manufacture of circuits in wafer size.

WSTS World **S**emiconductor **T**rade **S**tatistics

XDSL
The abbreviation for **X D**igital **S**ubscriber **L**ine; a generic term for different technical concepts for broadband digital data transfer using conventional twisted copper wire loops. Depending on the configuration, the X may be replaced by another letter. ADSL and HDSL have become more important.

X-radiation
Extremely short-wave electromagnetic radiation (wavelength less than 30 nm) discovered by Wilhelm Conrad Röntgen. "X-ray" = unknown ray, although it has not been an unknown quantity for a long time.

Y parameter
A set of admittance transistor parameters (for short-circuit operation).

Yield
In semiconductor technology, the proportion of chips in a production lot found to be good in accordance with predefined criteria.

Zener diode
A diode in which the Zener effect occurs. It is operated in the reverse direction. Only a low reverse current flows until the Zener voltage is reached. Above the Zener breakdown, the current increases in an avalanche-like fashion as the reverse voltage increases. For this reason, Zener diodes are suitable for voltage limiting in electrical circuits.

Zener effect
The occurrence of a current in a semiconductor material due to free charge carriers, which have moved from the valence band into the conduction band due to a high voltage (the Zener voltage).

Zero crossing
The point in time at which the amplitude of a periodic oscillation equals zero. It is beneficial to apply an alternating voltage exactly at the zero crossing so that there is no line surge, and/or switch off an alternating current exactly at the zero crossing to avoid induction currents.

ZIP
1. **Z**inc **I**mpurity **P**hotodetector.

2. **Z**igzag-**i**n-Line **P**ackage. A packaging technique with offset pin rows that permits a high packaging density using the push-through technique due to the vertically built-in chips.

ZVEI Zentral**v**erband **E**lektrotechnik- und Elektronik**i**ndustrie (ZVEI) e.V., Frankfurt am Main
The German Electrical and Electronic Manufacturers' Association, Frankfurt am Main.

ZZF Zentralamt für **Z**ulassungen im **F**ernmeldewesen, Saarbrücken
Germany's Central Bureau of Approval in Telecommunications, Saarbrücken. Has been renamed BZT.